Foundations of Classical Mechanics

Classical mechanics is an old subject. It is being taught differently in modern times using numerical examples and exploiting the power of computers that are readily available to modern students. The demand of the twenty-first century is challenging. The previous century saw revolutionary developments, and the present seems poised for further exciting insights and illuminations. Students need to quickly grasp the foundations and move on to an advanced level of maturity in science and engineering. Freshman students who have just finished high school need a short and rapid, but also thorough and comprehensive, exposure to major advances, not only since Copernicus, Kepler, Galileo, and Newton, but also since Maxwell, Einstein, and Schrodinger. Through a careful selection of topics, this book endeavors to induct freshman students into this excitement.

This text does not merely teach the laws of physics; it attempts to show how they were unraveled. Thus, it brings out how empirical data led to Kepler's laws, to Galileo's law of inertia, how Newton's insights led to the principle of causality and determinism. It illustrates how symmetry considerations lead to conservation laws, and further, how the laws of nature can be extracted from these connections. The intimacy between mathematics and physics is revealed throughout the book, with an emphasis on beauty, elegance, and rigor. The role of mathematics in the study of nature is further highlighted in the discussions on fractals and Madelbrot sets.

In its formal structure the text discusses essential topics of classical mechanics such as the laws of Newtonian mechanics, conservation laws, symmetry principles, Euler–Lagrange equation, wave motion, superposition principle, and Fourier analysis. It covers a substantive introduction to fluid mechanics, electrodynamics, special theory of relativity, and to the general theory of relativity. A chapter on chaos explains the concept of exploring laws of nature using Fibonacci sequence, Lyapunov exponent, and fractal dimensions. A large number of solved and unsolved exercises are included in the book for a better understanding of concepts.

P. C. Deshmukh is a Professor in the Department of Physics at the Indian Institute of Technology Tirupati, India, and concurrently an Adjunct Faculty at the Indian Institute of Technology Madras, India. He has taught courses on classical mechanics, atomic and molecular physics, quantum mechanics, solid state theory, foundations of theoretical physics, and theory of atomic collisions at undergraduate and graduate levels. Deshmukh's current research includes studies in ultrafast photoabsorption processes in free/confined atoms, molecules, and ions using relativistic quantum many-electron formalism. He also works on the applications of the Lambert W function in pure and applied physics.

Foundations of Classical Mechanics

P. C. Deshmukh

CAMBRIDGE
UNIVERSITY PRESS

University Printing House, Cambridge CB2 8BS, United Kingdom

One Liberty Plaza, 20th Floor, New York, NY 10006, USA

477 Williamstown Road, Port Melbourne, vic 3207, Australia

314 to 321, 3rd Floor, Plot No.3, Splendor Forum, Jasola District Centre, New Delhi 110025, India

79 Anson Road, #06–04/06, Singapore 079906

Cambridge University Press is part of the University of Cambridge.

It furthers the University's mission by disseminating knowledge in the pursuit of education, learning and research at the highest international levels of excellence.

www.cambridge.org
Information on this title: www.cambridge.org/9781108480567

First published 2019

Printed in India by Rajkamal Electric Press

A catalogue record for this publication is available from the British Library

ISBN 978-1-108-48056-7 Hardback

ISBN 978-1-108-72775-4 Paperback

To my teachers, with much gratitude

and

to my students, with best wishes

CONTENTS

FIGURES

TABLES

FOREWORD

I am a theoretical astrophysicist, and my professional work requires a foundation in classical mechanics and fluid dynamics, and it then draws on statistical mechanics, relativity, and quantum mechanics. How is a student to see the underlying connections between these vast subjects? Professor Pranawa Deshmukh's book '*Foundations of Classical Mechanics*' (FoCM) provides an excellent exposition to the underlying unity of physics, and is a valuable resource for students and professionals who specialize in any area of physics.

It was 2011 and I was Chair of the Department of Physics and Astronomy at the University of Western Ontario (UWO). An important global trend in education is internationalization: the effort to increase the international mobility of students and faculty, and increase partnerships in research and teaching. The present-day leading universities in the world are the ones that long ago figured out the benefits of academic mobility and exchanges. As part of our internationalization efforts at UWO, I was keen on building ties with the renowned Indian Institute of Technology (IIT) campuses in India. I invited Professor Deshmukh of the IIT Madras to come to UWO to spend a term in residence and also teach one course. That course turned out to be Classical Mechanics. How enlightening to find out that Professor Deshmukh had already developed extensive lecture materials in this area, including a full videographed lecture course 'Special Topics in Classical Mechanics' that is available on YouTube. Not surprising then, the course he taught at UWO was a tour-de-force of classical mechanics that also, notably, included topics that are considered to be 'modern', including chaos theory and relativity. Many of our students appreciated, highly, the inclusion of special relativity and, for example, to be able to finally understand a resolution to the mind-bending Twin Paradox. We were lucky to have Professor Deshmukh bring his diverse expertise to Canada, and UWO in particular, and it established personal and research links between individuals, and more generally between UWO and the IITs, which continue today.

The FoCM book is a further extension of the broad approach that Professor Deshmukh has brought to his teaching. This is not just another book on Classical Mechanics, due to both its approach and extensive content. The chapters are written in a conversational style, with everyday examples, historical anecdotes, short biographical sketches, and pedagogical features included. The mathematical rigor of the book is also very high, with equations introduced and justified as needed. Students and professionals who want a resource that synthesizes the unity of classical and modern physics will want to have FoCM on their shelves. The classical principles of conservation laws are introduced through the beautiful ideas of symmetry that have found resonance in modern physics. FoCM also introduces the Lagrangian and Hamiltonian approaches and shows their connection with relativity and quantum theory, respectively. The book covers Fourier methods and the equations of fluid mechanics, plus Maxwell's equations, topics that are not always present in classical mechanics books. Chaos theory is a modern incarnation of classical physics and finds significant coverage. The inclusion of special relativity and even the basic tenets of general relativity makes the book a very special resource. There is a simplified derivation of the precession of the perihelion of Mercury's orbit as deduced from general relativity.

After reading this book I am convinced that classical physics is a very 'modern' subject indeed. Over the years, the offering of classical mechanics courses has been reduced at many physics departments, sometimes to just a single term. Physics departments around the world would be well advised to ensure a two-term series on the subject that makes the crucial connections between classical and modern physics as expounded in FoCM.

Shantanu Basu, Professor, Department of Physics and Astronomy,
University of Western Ontario, London, Canada

PREFACE

It gives me great pleasure in presenting *Foundations of Classical Mechanics* (*FoCM*) to the undergraduate students of the sciences and engineering.

A compelling urge to comprehend our surroundings triggered inquisitiveness in man even in ancient times. Rational thought leading to credible science is, however, only about two thousand years old. Science must be considered young, bearing in mind the age of the homo sapiens on the planet Earth. Over two millennia, extricating science from the dogmas and superstitious beliefs of the earlier periods has been an inspiring struggle; a battle lamentably not over as yet. Progress in science has nonetheless been rapid in the last few hundred years, especially since the fifteenth and the sixteenth century.

To be considered *classical*, just as for the arts or literature, a scientific formalism needs to have weathered the tests of acceptability by the learned over centuries, not just decades. The work of Nicolas Copernicus (1473–1543), Tycho Brahe (1546–1601), and Johannes Kepler (1571–1630) in the fifteenth, sixteenth, and seventeenth centuries was quickly followed, in the seventeenth through the nineteenth centuries, by a galaxy of stalwarts, many of whom excelled in both experimental and theoretical sciences, and some among them were proficient also in engineering and technology. They developed a large number of specialized fields, including of course physics and mathematics, and also chemistry and life sciences. Among the brilliant contributions of numerous researchers who belong to this period, the influence of the works of Galileo Galilei (1564–1642), Isaac Newton (1642–1727), Leonhard Euler (1707–1783), Joseph Lagrange (1736–1813), Charles Coulomb (1736–1813), Thomas Young (1773–1829), Joseph Fourier (1768–1830), Carl Friedrich Gauss (1777–1855), Andre-Marie Ampere (1775–1836), Michael Faraday (1791–1867), William Hamilton (1805–1865), and James Clerk Maxwell (1831–1879) resulted in a lasting impact which has withstood the test of time for a few hundred years already. Their contributions are recognized as *classical physics*, or *classical mechanics*. The theory of relativity, developed later by Albert Einstein (1879–1955), is a natural fallout of Maxwell's formalism of the laws of electrodynamics, and is largely considered to be an intrinsic part of *classical* physics. The theory of chaos is also an integral part of it, though it is a little bit younger.

The quantum phenomena, discovered in the early parts of the twentieth century, exposed the limitations of the classical physics. The quantum theory cannot be developed as any kind of modification of the classical theory. It needs new tenets. To that extent, one may have to concede that classical physics has failed. This would be, however, a very unfair appraisal. It does no justice to the wide-scale adequacy of classical physics to account for a large number of physical phenomena; let alone to its beauty, elegance and ingenuity. Niels Bohr established the 'correspondence principle', according to which classical physics and quantum theory give essentially the same results, albeit only in a certain limiting situation. Besides, the dynamical variables of classical physics continue to be used in the quantum theory, even if as *mathematical operators* whose analysis and interpretation requires a new mathematical structure, developed by Planck, Louis de Broglie, Schrödinger, Heisenberg, Bohr, Einstein, Pauli, Dirac, Fermi, Bose, Feynman, and others. Notwithstanding the fact that classical physics is superseded by the quantum theory, it *continues* to be regarded as *classical*, outlasting the times when it was developed. *Many of the very same physical quantities of classical physics continue to be of primary interest in the quantum theory*. Classical physics remains commandingly relevant even for the interpretation of the quantum theory. The importance of classical mechanics is therefore stupendous.

The laws of classical mechanics are introduced early on to students, even in the high-school, where they are taught the three laws of Newton. Many kids can recite all the three laws in a single breath. The simple composition of these laws, however, hides how sophisticated and overpowering they really are. The subtleties are rarely ever touched upon in high-school physics. The first course in physics for students of science and engineering after high-school education is when Newton's laws must therefore be *re-learned* to appreciate their depth and scope. The central idea in Newton's formulation of classical mechanics is inspired by the principles of causality and determinism, which merit careful analysis. Besides, students need an exposure to the alternative, equivalent, formulation of classical mechanics based on the Hamilton–Lagrange principle of variation. The theory of fractals and chaos which patently involves classical determinism is also an intrinsic part of classical mechanics; it strongly depends on the all-important initial conditions of a dynamical system. The main topics covered in this book aim at providing essential insights into the general field of classical mechanics, and include Newtonian and Lagrangian formulations, with their applications to the dynamics of particles, and their aggregates including rigid bodies and fluids. An introduction to the theory of chaos is included, as well as an essential summary of electrodynamics, plus an introduction to the special and the general theory of relativity.

FoCM is the outcome of various courses I have offered at the Indian Institute of Technology Madras (IITM) for over three decades. During this period, I also had the opportunity to offer similar courses at the IIT Mandi. Over the National Knowledge Network, this course was offered also at the IIT Hyderabad. During the past four years, I have offered courses at the same level at the IIT Tirupati, and also at the Indian Institute of Science Education and Research Tirupati (IISER-T). Besides, during my sabbatical, I taught a course at the University of Western Ontario (UWO), London, where Professor Shantanu Basu had invited me to teach Classical Mechanics. I am indebted to IITM, IITMi, IITH, UWO, IITT, and IISERT for the teaching opportunities I had at these institutions. My understanding of the subject has been greatly enhanced by the discussions with many students, and colleagues, at the institutions where I taught, especially with Dr C. Vijayan and Dr G. Aravind at IITM, and Dr S.R.Valluri, Dr Shantanu Basu, and Dr Ken Roberts at the UWO. The gaps in my understanding of this unfathomable subject are of course entirely a result of my own limitations. The colleagues and students whom I benefited from are too numerous to be listed here; and even if I could, it would be impossible to thank them appropriately. Nevertheless, I will like to specially acknowledge Dr Koteswara Rao Bommisetti, Dr Srijanani Anurag Prasad, and Dr Girish Kumar Rajan for their support. Besides, I am privileged to acknowledge the National Programme for Technology Enhanced Learning (NPTEL) of the Ministry of Human Resource Development, Government of India, which videographed three of my lecture courses, including one on 'Special/Select Topics in Classical Mechanics: STiCM' which strongly overlaps with the contents of this book. The various courses that I offered at the four different IITs, the IISER-T, the UWO, and the STiCM (NPTEL) video-lecture courses have all contributed to the development of *FoCM*.

The topics included in *FoCM* are deliberately chosen to speed up young minds into the foundation principles and prepare them for important applications in engineering and technology. Possibly in the next twenty, or fifty, or a hundred years, mankind may make contact with extra-terrestrial life, understand dark matter, figure out what dark energy is, and may comprehend comprehensively the big bang and the very-early universe. The 'multiverse', if there were any, and possibly new physics beyond the 'standard' model, will also be explored, and possibly discovered. Students of science must therefore rapidly prepare themselves for major breakthroughs that seem to be just around the corner in the coming decades. *FoCM* aspires to provide a fruitful initiation in this endeavour. Engineering students also need a strong background in the foundations of the laws of physics. The GPS system would not function without accounting for the special theory of relativity, and also the general theory of relativity. GPS navigation, designing and tracking trajectories of ships, airplanes, rockets, missiles, and satellites, would not be possible without an understanding, and nifty adaptation, of the principles of classical mechanics. Emerging technologies of driverless cars, drone technology, robotic surgery, quantum teleportation, etc., would require the strategic manipulation of classical dynamics, interfacing it with devices which run on quantum theory and the theory of relativity. *FoCM* is therefore designed for both students of science and engineering, who together will innovate tomorrow's technologies.

A selection of an assorted set of equations from various chapters in *FoCM* appears on the cover of this book. This choice is a tribute to Dirac, who said "a physical law must possess a mathematical beauty", and to Einstein, who said "... an equation is for eternity". Of the top ten most-popular (and most-beautiful) equations in physics that are well known to internet users, two belong to the field of thermodynamics, and three are from the quantum theory. The remaining five, *and also some more which are of comparable importance and beauty*, are introduced and discussed in *FoCM*.

FoCM can possibly be covered in two undergraduate semesters. Contents from other books may of course be added to supplement *FoCM*. Among other books at similar level, those by J. R. Taylor (*Classical Mechanics*), David Morin (*Introduction to Classical Mechanics*), and by S. T. Thornton and J. B. Marion (*Classical Dynamics of Particles and Systems*) are my favourites. Along with these books, I hope that *FoCM* will offer a far-reaching range of teaching and learning option.

It has been a pleasure working with the Cambridge University Press staff, specially Mr Gauravjeet Singh Reen, Ms Taranpreet Kaur, Mr Aniruddha De, and Mr Gunjan Hajela on the production of this book. At every stage, their handling reflected their admirable expertise and generous helpfulness. The compilation of problems that are included in *FoCM* has been possible due to superb assistance from Dr G. Aarthi, Dr Ankur Mandal, Mr Sourav Banerjee, Mr Soumyajit Saha, Mr Uday Kumar, Dr S. Sunil Kumar, Mr Krishnam Raju, Mr Aakash Yadav, Mr Aditya Kumar Choudhary, Mr Pranav P. Manangath, Ms Gnaneswari Chitikela, Mr Naresh Chockalingam S, Ms Abiya R., Mr Harsh Bharatbhai Narola, Mr Parth Rajauria, Mr Kaushal Jaikumar Pillay, Mr Mark Robert Baker, and Ms Gayatri Srinivasan. A number of books and internet sources were consulted to compile the problem sets.

I am grateful to Dr G. Aarthi, Mr Uday Kumar, Mr Sourav Banerjee, Mr Pranav P. Manangath, Dr Ankur Mandal, Ms Abiya R., Mr Kaushal Jaikumar Pillay, Mr Mark Robert Baker, Mr Harsh Bharatbhai Narola, and Mr Krishnam Raju, who have helped me in plentiful ways which have enriched the contents of this book. They provided absolutely indispensable support for the development of *FoCM*. Dr Aarthi Ganesan, Mr Pranav Sharma, Mr Pranav P. Manangath, Mr Mark Robert Baker, and Dr Jobin Jose also helped with the proof correction of various chapters. I am indebted, of course, to a large number of authors of various books and articles from whom I have learned various topics. It is impossible to acknowledge all the sources. A few select references have nevertheless been mentioned in the book, where it seemed to me that readers should be directed to them. I am especially grateful to Dr Shantanu Basu for writing the 'Foreword' for this book. Dr Balkrishna R. Tambe helped me in various ways, directly and indirectly, during the progress of this book. *FoCM*, of course, could not have been written without the forbearance of my family (Sudha, Wiwek, and his wife Pradnya). They all supported me in many a challenging times. The joy in playing with my grand daughter Aditi energized me the most during the final stages of *FoCM*, when numerous hectic commitments were rather demanding. I dedicate this book to the young and promising students of science and engineering. May they see beyond.

Introduction

We believe that physics would make it possible for us to comprehend the nature of the physical universe from our observations and investigations. We expect the methods of physics to equip us with a capability to take cognizance of a physical reality, and describe it in terms of crisp, succinct *laws*. Physics endeavors to recognize consistent patterns in natural events, and formulate them in a precise, unambiguous, and verifiable manner. When you look around, you would notice that there is mostly *irregularity* in much of what we experience in our daily life. When dust is raised by blowing winds, the dust particles seem to move in random directions, chaotic; the tree leaves flutter unpredictably. If a dense forest were to catch fire, it would be hard to tell which the next tree would be that the fire would gut. On the other hand, if you hold an object and let go of it, it always falls down, accelerating toward the ground at 9.8 m/s^2 (with only minor alterations over the Earth's surface). This would be so no matter what the mass or the size of the falling object is—let alone what its color or smell—in fact, just no matter which object was dropped. The *free fall* occurs at an acceleration which is even quantitatively *predictable*, of course only if the object fell under gravity alone, undisturbed by any other interaction, including friction with air-molecules or their buoyant effects. Likewise, if you have two electric charges, Q_1 and Q_2 at a distance r, they would repel or attract each other, depending only on the nature of the charges being respectively like or unlike. No matter what the individual values Q_1 and Q_2 of the two charges are or who performed the experiment, where it was performed, or for that matter, at what distance the charges were initially held from each other, the magnitude of the force between the charges would always be proportional to $\frac{Q_1Q_2}{r^2}$.

If you look at the sky on a clear night, you would see lovely stars, even galaxies, and supernovae, sometimes amid recognizable patterns and shapes that we call constellations. Further, if you measure the Doppler shifts of the spectral lines present in the light from the stars, you could compile some systematic information and discover that the universe is actually expanding. By further carrying out some clever calculations, you can possibly discover at least some of the *physical laws* which describe the expansion of the universe. In fact, you can even determine the age of the universe from the big bang by studying such observable phenomena. On the other hand, in some *local* part of the cosmos, we may as well observe galaxies not flying away from each other; they may even be on a collision path toward each other. Detailed observations and calculations have now established that the Andromeda galaxy would actually collide with ours, the 'Aakaash Ganga', commonly called the 'Milky Way', in about four billion years. Notwithstanding such irregular events prompted by rather special situations, *most* of the galaxies, nonetheless, fly *away* from each other.

As Eugene P. Wigner described in his Nobel Prize Lecture (1963), what we call as a *law of nature*, is a statement on the *regularities* in nature. Physicists arrive at the laws of nature after systematically separating the infrequent departures from the same, like the collision of the galaxies. Systematic data analysis from extensive studies shows that when not affected by buoyant effects and wind currents, even feathers fall just as do stones, toward the earth at an acceleration of $\sim 9.8\ m/s^2$, and that the universe is actually expanding; in fact, the galaxies seem to be not merely moving *away* from each other, but are actually *accelerating away* from each other. It is then the *regularities in natural phenomena* that physicists revere as the laws of nature.

It is, however, necessary, every once so often, to review and reassess, what we believe the laws of nature are. This is of course mandatory when a physical event is detected in a reliable experiment that does *not* conform to one of the hitherto-known-laws. We must then improvise the formulation of that law. We must even be prepared to completely abandon its basis if a mere modification to it might turn out to be insufficient. Our understanding of the universe thus progresses incrementally, requiring improvisations every now and then. The process of discovering the laws of nature, modifying, or even completely replacing them by new enunciations as and when necessary, is the very signature of the scientific pursuit.

The backdrop canvas of the scientific inquiry is huge. The range of values which the fundamental physical quantities—*mass* (M), *length* (L) *and time* (T)—take in the universe is mind-boggling. It is well beyond the day-to-day human experience: from extremely tiny to absolutely humongous. On a cognizable mass scale, the photons and the gluons are virtually massless. In the tiny world of the elementary particles alone, we already have a range of mass values covering more than 10^{10} orders of magnitudes. The mass of an electron is of the order of $\sim 10^{-30}$ kg, a one-rupee coin has a mass of $\sim 10^{-3}$ kg, the Sun has a mass of $\sim 10^{30}$ kg, and the mass of the universe (including 'dark' matter) is estimated to be $\sim 10^{52}$ kg. On the length scale, we have again an enormously vast range. The tiniest of elementary particles have a size of the order of an *am* (an 'attometer' being 10^{-18} m), the Sun has a diameter of $\sim 10^9$ m, and the radius of our, cognizable, universe is $\sim 4\times 10^{26}$ m, and viable speculations on multiverses are strongly building even as I write. The range of the order of magnitude numbers for mass and size is staggering, from the exceedingly tiny to the enormously gigantic. Physicists thus deal with objects from the very tiny to the very big.

It seems therefore almost arrogant that physics aims at understanding the dynamics of an object in such a multifarious, intricate, and enormous universe in terms of only a few parameters, which would not merely describe the 'state of a physical system' but also account for its dynamical evolution in terms of an appropriate equation of motion. What is breath-taking is that advances in physics have already provided a huge amount of insight and understanding about the universe, *and* its evolution, in terms of just a few parameters and a very few 'laws'. What is especially remarkable is that these advances have been made merely in the last few hundred years, despite the much longer life-time (about 200,000 years) of the human species on our planet.

The first known conception of what is referred to as the atomistic approach to analyze the material world is perhaps found in the works of Maharshi Kanad (~600 BCE), an Indian sage who developed the *Vaishesikha* philosophy. This scheme seeks to explore the cognizable universe in terms of nine elements: air, earth, fire, mind, self, sky, space, time, and water. The Vaishesikha scheme has of course now been superseded by the periodic table of elements, and further by the elementary particles of the standard model of physics, which would perhaps be upgraded someday to include what is now considered to be 'dark matter' (and may be even the 'dark energy'?) What has come to be regarded as the scientific method is also based on a somewhat similar principle in the sense that it begins with a set of *ansatz*, which is an initial assumption or an axiom that we make. It is a basic tenet from which we develop our physical model. The ansatz is, in some sense, an intelligent guess at an underlying fundamental physical principle that is intuitively recognized and verifiable, even if not derivable from anything else. It needs to be robust enough to stand scrutiny by physical observations, and must not lead to logical inconsistencies. A scientific theory is then built in a logically consistent formalism based on the ansatz.

Advances in science have been phenomenal. It is now possible to fathom distances from as tiny as 10^{-35} m to very vast that even light would take billions of years to traverse. Even as physicists are toiling to determine what may have happened in the first 10^{-43} s since the beginning of the universe, they estimate that the universe itself is over 10^{+17} s (~13.8 billion years) old. Major events during the period over which the universe has evolved are depicted in Fig. I.1 [1]. The physical events and the dynamics in the ponderable evolution of the universe are absolutely stupefying. The universe is made up of just a few elementary particles. The atoms were once thought to be the smallest ingredients of matter. In the early part of the twentieth century, following the experiments such as those conducted by J. J. Thomson and E. Rutherford, it became known that the atom itself has an internal structure consisting of the central nuclei and the electrons. Even tinier ingredients exist, namely the quarks and the leptons (and their anti particles), which together with the so-called gauge bosons and the Higgs boson, constitute the fundamental particles as we know now. The quarks participate in the making up of the protons and the neutrons, which along with the electrons are ingredients of atoms, which further partake in the composition of the molecules and clusters which constitute condensed matter that make up both 'beings' and 'things' – living and inanimate. From grains of sand to mountains, from planets and stars to the millions of galaxies in the universe, and from life-cells to flowers and trees, from microbes to animals and humans, every *thing* and every *being* is made up of elementary particles. By and large, physics does not directly endeavor to address just what constitutes *life* and *consciousness*, *at least not quite as yet*. Nevertheless, attempts to address even these mind-boggling mysteries have also been undertaken already by physicists. Broadly speaking, physicists seek to discover '*what*' the laws of nature are, and '*how*' they influence the physical events; rather than '*why*' the laws are what they are. Toward this study, physics aims at discovering the *minimal* set of laws of nature in terms of which all else that happens in the universe can be described. It should now be clear why the most general enunciations of the fundamental principles are worthy of being called the 'fundamental universal laws of nature'.

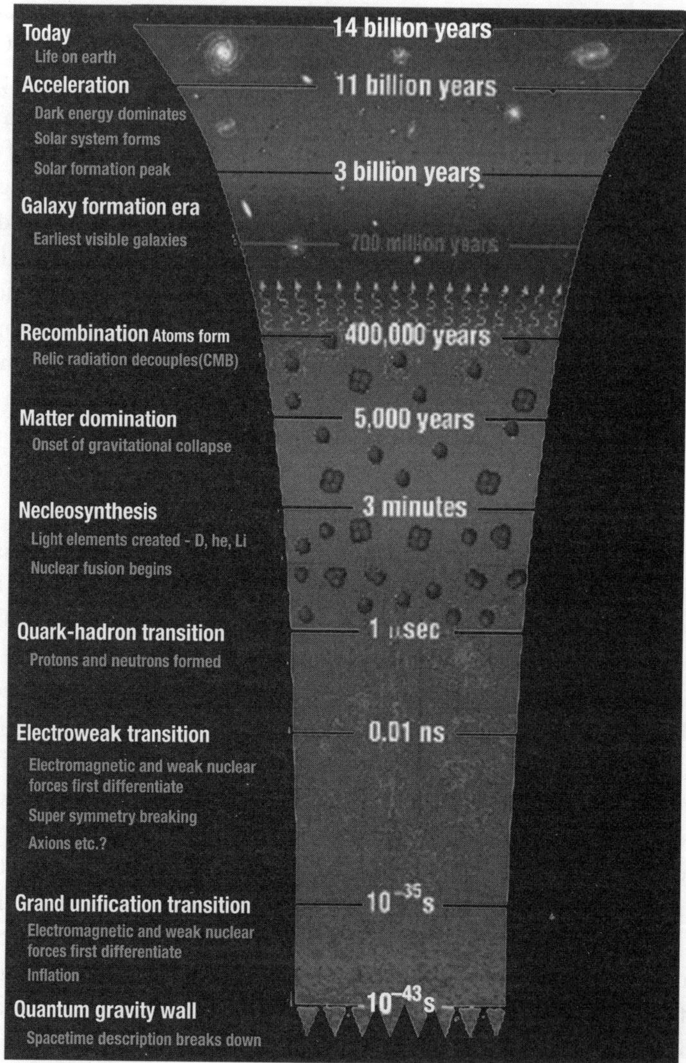

Fig. I.1 Major events in the evolution of the universe [1]. Even as much is known now, much also remains to be known. Some of this ignorance is referred to as 'dark matter' and 'dark energy', whose effects have been sensed, but it poses itself as a huge challenge to physics. This is only a small part of the reason why getting into a career in physics is a very exciting proposition today. Mankind is in an exciting time-slot in which exciting discoveries are just round the corner.

The laws of nature that account for the effects of relativity and quantum physics are relatively recent (considering the age of man), just over a hundred years old. Until about the seventeenth century, even 'gravity' was a total mystery. In earlier times, the Greeks *explained away* why an object falls when let to do so by stating that the earth is the natural abode of things, and objects fall just as horses return to their stables. Earth was earlier believed to be flat, and other planets in our solar system were considered to be wandering stars. Ignorance led to fear and superstition. Shadows cast in eclipses influenced life-styles and contaminated human thought, forming prejudices that are regrettably hard to break, even as science has scooped out much of the truth from the shadows of ignorance and fear.

A few early deductions known to some ancient Indian and Greek astronomers have of course turned out to be correct. For example, Aryabhatta, in the fifth century CE, had inferred and proclaimed that Earth has a spherical shape, not flat. Brahmagupta, in the seventh century, had estimated to an extremely high degree of accuracy the circumference of Earth. His estimate of ~36,210 km is remarkably close to the value obtained using modern technology, which is 40,075 km (~24,900 miles). The Kerala astronomers [2], in the fifteenth century CE, *even before Copernicus*, were quite aware of the Sun's central position in our solar system, even if this knowledge is often referred to as the *Copernican* revolution. Nevertheless, many other thinkers continued to maintain incorrect, often absurd, ideas about the universe, including about our solar system. It was in the backdrop of such prejudices that Nicolus Copernicus (1473–1543) advanced his heliocentric frame of reference to describe our solar system. It was indeed courageous of him to do so, for a system that did not have Earth at the center of the universe was in sharp contradiction with the belief promoted at that time by the church. Copernicus' book could be published only well after he died, so that he would not be persecuted by the church. When it was finally published, the church prohibited it [3]. Copernican view was upheld by Galileo Galilee (1564–1642; Fig. I.2), who was then predictably prosecuted by the Catholic Church for going against its doctrine. The trial of Galileo is one of the most famous ones [4]. It is against the dogmas of his times that Galileo arrived at robust scientific conclusions based on systematic measurements. Apart from upholding the Copernican viewpoint, Galileo carried out ingenious experiments which laid the very foundations of classical mechanics. Galileo is thus fittingly regarded as the *father of experimental physics*.

Nicolaus Copernicus (1473–1543)

Galileo Galilee (1564–1642)

Fig. I.2 The scientific methods are old, but often the most significant beginnings are referenced from the works in the fifteenth and the sixteenth centuries. The names of Copernicus and Galileo are among the tallest in this period.

Classical mechanics refers to the developments in science mostly during the fifteenth to the nineteenth century through the *classic* works of Nicolaus Copernicus (1473–1543), Tycho Brahe (1546–1601), Johannes Kepler (1571–1630), Galileo Galilei (1564–1642), Isaac Newton (1643–1727), Leonhard Euler (1707–1783), Joseph-Louis Lagrange (1736–1813),

William Rowan Hamilton (1805–1865), Charles Augustine de Coulomb (1736–1806), Andre-Marie Ampere (1775–1836), Michael Faraday (1791–1867), Heinrich Lenz (1804–1865), James Clerk Maxwell (1831–1879), etc.

Several results of the scientific studies based on classical mechanics were subsequently found, nevertheless, to be only approximately correct, no matter how good the approximation is. The redress required revolutionary ideas which led to the development of the quantum theory in the twentieth century. Classical mechanics, however, provides the essential nucleus, and continues to be adequate in a large number of situations. Even the special theory of relativity developed by Albert Einstein (1889–1955) connects intimately to the classical electromagnetic theory and thus integrates seamlessly into the philosophy of classical mechanics.

This book aims at laying down the foundations of classical mechanics. Methodologies developed to account for the analysis of motion and dynamics of simple physical systems like point-particles are dealt with in Chapters 1 through 6. Extensions of these formalisms to study the dynamics of composite rigid bodies are discussed in Chapters 7 and 8. Chapter 9 provides an introduction to the theory of chaos and fractals, which judiciously belongs to the increasingly important domain of classical mechanics which is very sensitive to the initial conditions that describe the physical system under investigation. Chapters 10 and 11 provide the framework for studying fluids. The formalism in these two chapters is applicable to study the state of *matter* that flows, and it also provides the framework to describe imperative properties of the electromagnetic *field*. A condensed summary of the laws of electromagnetism is provided in Chapter 12 and 13. Finally, in Chapters 13 and 14, the foundations of the *special* and the *general* theories of relativity are laid down. The role of symmetry in natural laws is exhibited in several chapters, permeating through the whole book. The astute reader will hopefully find this insightful. This approach links prefatory concepts in the first course in Physics to contemporary research in frontiers areas, even if the intermediate links are challenging. As the students advance through the contents of this book, it is earnestly hoped that they will enjoy the romance in physics and beauty in its simplicity. Also, very importantly, the students will, I hope, get adept at the necessary rigor in the formulation of the physical laws. Mathematics lends itself as a constructive collaborator in the development of the physical laws. No wonder it is then that Newton developed the law of gravity and differential calculus together. In this book too, then, mathematical methods are introduced as and where required, sometimes as an important technique, but more often as an inseparable collaborator in the very nature of the laws of physics.

The mind-boggling, counter-intuitive, quantum theory is, by the very connotation of *classical mechanics*, outside the scope of this book. *Classical mechanics* provides the *prerequisite* foundation on which the student must stand firmly before she/he may leap into the quantum world. An attempt is therefore made in this book to comprehensively elucidate the path that led to the *classical*, non-quantum, laws of nature. I hope that the student-readers will rediscover the educative excitement felt by physicists in the last few hundred years in learning how simple enunciations of the classical laws account for a really vast range of phenomena in the physical universe. This text-book aims at providing a robust under-structure to students of physics and engineering, to prepare them to appreciate advance discoveries in science, and also developments in new technologies, whose frontiers are pushed by the hour even as these pages are written.

References

[1] *The Universe in a Nutshell.* 2015. http://cjuniverseinanutshell.blogspot.com/2015/04/a-brief-history-of-relativity-cont.html. Accessed on 26 April 2019.

[2] Ramasubramanian, K. M. D. S., M. D. Srinivas, and M. S. Sriram. 1994. 'Modification of the earlier Indian planetary theory by the Kerala astronomers (c. 1500 CE) and the implied heliocentric picture of planetary motion.' *Current Science* 66(10): 784–790.

[3] *Vatican Bans Copernicus' Book.* https://physicstoday.scitation.org/do/10.1063/PT.5.030911/full/. Accessed on 26 April 2019.

[4] Carlo Maccagni. 2009. *Galileo Galilei: A New Vision of the Universe.* http://www.ncert.nic.in/publication/journals/pdf_files/school_science/sc_June_2009.pdf. Accessed on 26 April 2019.

CHAPTER 1

Laws of Mechanics and Symmetry Principles

Nature seems to take advantage of the simple mathematical representations of the symmetry laws. When one pauses to consider the elegance and the beautiful perfection of the mathematical reasoning involved and contrast it with the complex and far-reaching physical consequences, a deep sense of respect for the power of the symmetry laws never fails to develop.

—*Chen Ning Yang*

1.1 THE FUNDAMENTAL QUESTION IN MECHANICS

What is it, exactly, that characterizes the *mechanical state of a system*? A physical object of course has several attributes; but which of these properties *characterize* its mechanical state? For instance, could we associate properties such as the object's mass, energy, volume, temperature, color, shape, etc., with its *characteristic mechanical state*? All of these surely are important attributes of an object. However, when we talk about the 'mechanical state of a system' in the context of the *fundamental laws of mechanics*, we must remember that the laws of nature would not change only if the object's mass or shape or color were to change. Otherwise, one would end up in a mess, with different laws for objects having different shapes, sizes, colors and smell. In fact, even the object's angular momentum, energy, etc., are not appropriate to *characterize* its mechanical state, because these properties are *derivable* easily from its more fundamental mechanical properties. We look for the *fewest* number of independent physical properties of the system which are of consequence toward describing its 'motion'. These properties, for a point-sized object are, *for each of its degree of freedom*, (i) *position*, represented by a *coordinate*, commonly denoted by the letter q and (ii) *momentum*, commonly denoted by p. Equivalently, the mechanical state of that object can also be described by its (i) position q and (ii) velocity, $v = \frac{dq}{dt}$, usually denoted by putting a dot on q, i.e., $v = \dot{q}$. The mechanical state of a system is then described by the *pair* (q, p), or, equivalently by $(q, \dot{q})$, for *each* of its degree of freedom. However, I should give you a heads up here that, in Chapter 6, we will *refine*[1] our notions of q and p.

Having described the characteristic mechanical state of the system by (q, p) or $(q, \dot{q})$, we now need the *physical law* which governs the temporal evolution of the state (q, p). Such a law would give us the equation of motion for $\dot{p} = \frac{dp}{dt}$, which is identified as the 'force' in Newtonian mechanics. The equation of motion is a rigorous mathematical relationship involving the position, velocity and acceleration of an object. If the equation of motion is applicable to *every* object, regardless of its mass, color, etc., it is called a *law of nature.* The reason Newton's equation is revered as a law of nature is that it applies to apples and coconuts falling from trees, and also to Mars and Jupiter orbiting around the Sun. Newton's equation is a law because it explains the motion of *every* physical object.

The equation of motion embodies the law of physics. Its solution would provide complete knowledge about $(q(t), p(t))$ of the object at any arbitrary time t if the corresponding values $(q(t_0), p(t_0))$ are known at a reference time t_0. Equivalently, in the alternative description of the object in terms of position and velocity, the values $(q(t), v(t))$ of these parameters for each degree of freedom at any arbitrary time t can be obtained by solving the equation of motion from the values $(q(t_0), v(t_0))$ at the reference time t_0. *The equation of motion thus represents the law of nature according to which the mechanical state of the system evolves with time.* The classical equation of motion is either a second order differential equation of motion, or equivalently two first order differential equations. The second order differential equation is the familiar *Newton's equation* $F = \dot{p}$, or alternatively, it is the *Lagrange's equation*, which we shall meet in Chapter 6. The two first order differential equations which provide an alternative and equivalent scheme are the *Hamilton's equations*, which also are discussed in Chapter 6.

Solution to the equation of motion, using advance knowledge about the state of the system at the reference time t_0, often called initial time by setting $t_0 = 0$, predicts the *state of the object* at an arbitrary later (or even earlier) time. The classical equation of motion is symmetric with respect to time reversal, i.e., time t going to $-t$. Hence, it not only enables us to determine the state of the system at a later time, but also determines what the mechanical state would have been in the past, at time $t < t_0$. Classical mechanics is thus a precise *deterministic* theory. The predictive capacity of the equation of motion hinges on the simultaneous accurate knowledge of $(q(t_0), p(t_0))$, or equivalently, of $(q(t_0), v(t_0))$. *Determinism* is thus an important cornerstone of classical mechanics. This scheme, however, breaks down when we reconcile with the impracticality of obtaining precise values of (q, p) at some particular instant of time by a measurement. Consideration of this issue would take us beyond classical mechanics, into the realm of quantum theory.

1 In Chapter 6, a more comprehensive definition of momentum, more appropriately called the 'generalized momentum', will be given. It will continue to be denoted by p. It will turn out that the usual 'mass times velocity' with which we are familiar from high school is only a special case of the 'generalized momentum'. The corresponding coordinate will be referred to as the 'generalized coordinate', also denoted by q.

1.2 THE THREE LAWS OF NEWTONIAN MECHANICS

Galileo Galilei (1564–1642) was the first to recognize, on the basis of careful experiments that a force is *not* required to *keep* an object in *sustained* motion. This conclusion was counter-intuitive in his time. It is common experience to require application of an external force on an object to keep it in uniform motion, since an object set in motion comes to a halt *due to friction*. A force is therefore usually necessary to keep an object in motion. Galileo dropped objects from the top of the mast of a ship and observed that the object would fall at the bottom of the mast, *regardless of whether the ship was anchored at the shore, or it was in a state of uniform motion in the sea* (Fig. 1.1). The motion of the ship of course had to be *uniform* for this conclusion to be drawn, as is easy to understand. Surely, the inertia of the object dropped from the top of the mast would not allow it to hit the base of the mast, if the ship was either accelerated forward, or backward, or if the ship had turned, *during* the object's fall. Such observations led Galileo to the 'law of inertia'. This law is fundamental; it lays down the very foundation of the principles of mechanics.

Fig. 1.1 The point at which a ball dropped from the top of the mast of a ship hits the base is the same irrespective of whether the ship is at rest, or is sailing at a constant velocity. This is, however, not so if the ship undergoes any kind of acceleration during the object's fall.

Galileo's law of inertia basically states that in the absence of interactions (i.e., absence of external forces), motion of an object is self-sustaining. It is determined by the initial conditions alone. If the object is at rest initially, it will remain so; and if it is already moving at a constant momentum at the initial time, it would continue to do just that. *This principle, essentially, enables the identification of an 'inertial frame of reference': it is one in which an object that is not interacting with any force has its motion governed entirely by initial conditions.* The motion so observed is self-sustaining—*whether at rest or at a constant momentum.* Uniform motion at constant velocity (zero or non-zero) seeks no cause; that such was its initial condition is reason enough. Essentially, the first law of mechanics is a statement of sustenance of the state of motion determined only by the initial conditions. Thus, tomato ketchup held *inside* a glass bottle flows *out* of the bottle's mouth when you suddenly bring the bottle to a halt after rapidly moving it. The ketchup flows out of the bottle because it merely *keeps moving* by its inertia, since it is not rigidly bound to the bottle which is brought to halt. Inertia is, therefore, a property: a physical object has to sustain its state of motion, and resist changes in the same (including the state of rest, which is simply 'no motion'). The more the quantity of matter in a body, the more is its mass, and greater its inertia. It is the tendency of the object to defy changes in its state of motion. *Equilibrium* is the state of the body in which its momentum is invariant with time. The inertial frame of reference therefore is identified by the condition that an object's equilibrium in that frame is independent of time, when it is not interacting with any external force. When in equilibrium, the object's motion is completely determined by its inertia alone, i.e., by its inertial, initial, conditions alone.

The states of 'rest' and of 'uniform motion' are, both, completely 'equivalent'. This is because an object at uniform motion in one frame of reference (say, frame A) can be at rest in another frame of reference (say, frame B), *if* B is so moving at a uniform velocity with respect to A as to exactly compensate for the object's motion in the frame A. The constancy of the velocity of the second frame with respect to the first is an essential element in this consideration. Both such reference frames provide an interchangeable description of 'equilibrium', and each of the two frames of reference is referred to as an inertial frame of reference. This fact is encapsulated in the 'Galilean principle of relativity': the laws of physics remain invariant under a transformation from one inertial frame of reference to another. The condition for this principle to hold of course is that the second frame moves at a constant velocity with respect to the first. This principle is even retained in the STR (special theory of relativity). The STR brings under its range of applications also the electromagnetic phenomena, in addition to the mechanical (including gravitational) interactions. We shall introduce *classical* electromagnetic phenomena and the theory of relativity in Chapters 12, 13 and 14.

The First Law of Mechanics – The Law of Inertia: The *law of inertia*, and its implication that the *laws of mechanics are the same in all inertial frames of reference*, follow from experiments carefully carried out by Galileo. Experiments provide an essential pillar on which the scientific method is raised. A systematic compilation of experimental observations helped Galileo discover the fundamental platform on which the theory of mechanics is developed. *All* theories of mechanics—Newtonian, Lagrangian, Hamiltonian, and even Schrodinger's and Einstein's—use it. Galileo's law of inertia is incorporated in Newtonian mechanics as the first law of mechanics.

Isaac Newton (1643–1727), born just about a year after Galileo died, went on to address the following question: *when a departure from the equilibrium of an object occurs*, exactly what is the *cause* that is responsible for the *change* in its equilibrium? In answering this question, Newton integrated physics and mathematics. Carl Sagan wrote of Newton thus: "*He was not immune to the superstitions of his day... in 1663... he purchased a book on astrology, ... he read it till he came to an illustration which he could not understand, because he was ignorant of trigonometry. So he purchased a book on trigonometry but soon found himself unable to follow the geometrical arguments, so he found a copy of Euclid's Elements of Geometry, and began to read. Two years later, he invented differential calculus*" [1]. This, is magnificently outstanding, isn't it? The methods of differential and integral calculus did not even exist in Newton's time. The idea of a function possessing analytical properties, being continuous and derivable, did not exist. Newton conceived, and developed, these ideas into the mathematical discipline of 'differential and integral calculus'.

Fig. 1.2 Isaac Newton (1643–1727) synthesized mathematics, astronomy, and physics and laid the foundation for what is now called classical mechanics, so called because his theory has withstood the test of times for a long period.

It should be remarked, however, that Gottfried Leibniz (1646–1716) also developed differential calculus around the same time; but we shall refrain from getting engaged in the often hot debate on whether it was Newton or Leibniz who was the first to have developed 'calculus'.

Isaac Newton (1643–1727), our hero of classical mechanics, was one of the most brilliant physicists and mathematicians of all times. He made pioneering contributions to the mathematics of calculus, to the laws of classical mechanics, to the theory of optics, to the understanding of dispersion of light, to the understanding of the fundamental theories of astronomy, and to the comprehension of 'gravity'. Subsequently, Newton developed an equation of motion that now bears his name. In doing so, Newton fully (*or at least, almost fully,* notwithstanding small but important relativistic effects) accounted for the trajectories of planets in the solar system. Newton's scheme brought the dynamics of falling apples and coconuts, and also the motion of moons and planets, all under a common roof, thus formulating the (first set of) *universal* laws of nature.

Newton incorporated Galileo's principle of inertia as the cornerstone for his scheme of mechanics. Thus, Galileo's principle of inertia got to be known, rather, as Newton's first law: *An object in a state of rest remains at rest, and an object moving with a constant momentum continues to do so, unless and until acted upon by an external force impressed on that object.* What is implicit in this statement, and often dangerously ignored, is the fact that this *first law* actually provides the very means to identify an inertial frame of reference. An inertial frame of reference is recognized as one in which *equilibrium (whether of rest, or of constant momentum) is self-sustaining* and is determined *completely by initial conditions.* What is intrinsic to this law is the complete *equivalence* of the mechanical state at 'rest' and that 'at constant momentum'. The mechanical state of equilibrium in the inertial frame is determined only by the initial conditions, $(q(t_0), p(t_0))$, or equivalently by the initial position and velocity. This does *not* seek any physical cause; after all, there is nothing that precedes the *initial* condition in a causal, deterministic world.

Newton's Second Law – Linear Response, Causality, and Determinism: What is it, then, that seeks an explanation? What is it that requires a cause? It is *not* the equilibrium itself, but rather a *departure* from equilibrium that seeks a cause. Newton addressed this issue comprehensively. He inferred that the *departure from equilibrium* in an object's state of motion is due to a stimulus by an external agency which interacts with that object. This stimulus is commonly termed as 'force'. It is the central physical factor in Newton's second law: *An object responds to an external force through a change in its momentum at a temporal rate* $\frac{d\vec{p}}{dt}$ *which is exactly equal to the external force,* $\vec{F}$:

$$\vec{F} = \frac{d\vec{p}}{dt}. \tag{1.1a}$$

The mechanical momentum of a point-sized object of mass m moving at a velocity $\vec{v}$ is $\vec{p} = m\vec{v}$. For such an object, Newton's second law takes the following form:

$$\vec{F} = \frac{d\vec{p}}{dt} = \frac{d(m\vec{v})}{dt} = m\frac{d\vec{v}}{dt} = m\frac{d^2\vec{r}}{dt^2} = m\vec{a}. \tag{1.1b}$$

In the above equation, the rate of change of the velocity given by $\vec{a} = \lim_{\delta t \to 0} \frac{\overrightarrow{\delta v}}{\delta t} = \frac{d^2\vec{r}}{dt^2}$ is the acceleration of the object in the chosen inertial frame of reference. The reference frame in which the momentum is changing at the rate $\frac{d\vec{p}}{dt}$ must be chosen *first*, and as stated earlier, it is the *first law* that enables this choice.

On setting $\vec{F} = \vec{0}$ in Eq. 1, we get, on straight forward integration a result that could tempt one to believe that the first law is merely a special case of the second law. This is certainly not the case. The Galileo–Newton first law is a *separate* law, and it *precedes* Newton's second law, both *historically*, and in terms of *conceptual hierarchy.* If one were to suspect that the first law is only a special case of the second, one might as well then ask why the first law should be studied as a *separate* law at all? The apparent paradox is easily resolved on recognizing that the second law is applicable only in an inertial frame of reference, but it is the first law that is needed in the *first* place to even *identify* such a frame of reference.

Equation 1.1 delivers, both, the qualitative and the quantitative content of Newton's second law. One can read this statement as a linear relationship between the *effect* (namely the acceleration $\vec{a}$) and the *cause* (which is the force $\vec{F}$); the proportionality between them being given by the inertia (i.e., mass) m of the object. Newton's second law thus expresses a causality principle stated as a linear response: ***the effect*** $\vec{a}$ ***is linearly proportional to the cause*** $\vec{F}$; i.e., the cause determines the effect. The principle of causality/determinism is expressed here rigorously as a differential equation of motion. The second order differential equation can be solved, and towards that goal one must integrate the differential equation *twice*, requiring *two* initial conditions to be employed as the constants of integration. These two conditions are the initial position $\vec{r}(t_0)$ and the initial momentum $\vec{p}(t_0)$ or, equivalently, the initial velocity $v(t_0)$. It should now be clear that the 'position' and 'velocity' of an object are physically *independent* parameters. Both of these must be known to describe the mechanical state of the system. Without the knowledge of both the initial conditions, it will be impossible to predict the value of (q, p) or (q, v) at an arbitrary time, regardless of the fact that the equation of motion itself is deterministic. Only *one* of the two aforementioned parameters is not sufficient to provide the *two* constants of integration we need to obtain an appropriate solution to the differential equation.

The trajectory of the object in the position–momentum phase space (Fig. 1.3a) given by $(q(t), p(t))$, at each instant t of time, is then the *solution* to the mechanical problem. Equivalently, it is the trajectory described by $(q(t), \dot{q}(t))$ in the position–velocity phase space (Fig. 1.3b). Solving a mechanical problem essentially entails obtaining this trajectory. The form $\vec{F} = \frac{d\vec{p}}{dt}$ of Newton's second law emphasizes that the *change* in the momentum $\vec{p}$ of an object exhibits its departure from equilibrium, at a rate $\frac{d\vec{p}}{dt}$ which is exactly equal to the very cause that effects this change, namely the force. The form $\vec{F} = m\vec{a}$ highlights the linear response measured in terms of the acceleration $\vec{a}$ of the object to a stimulus $\vec{F}$. The second law of Newton therefore encapsulates the principle of causality and determinism of classical mechanics.

The phase space for an object having N independent degrees of freedom will be 2N dimensional. *The 'phase space' is also called 'state space', because a point in it represents the mechanical state of a system.* This enables one to identify the instanteneous mechanical state of a system to be designated by a point in its phase space. The temporal evolution of the state is depicted by a trajectory in the state space. A 2-dimensional space spanned by the two orthogonal (thus independent) degrees of freedom, position and momentum, is termed the 'position–momentum phase space'.

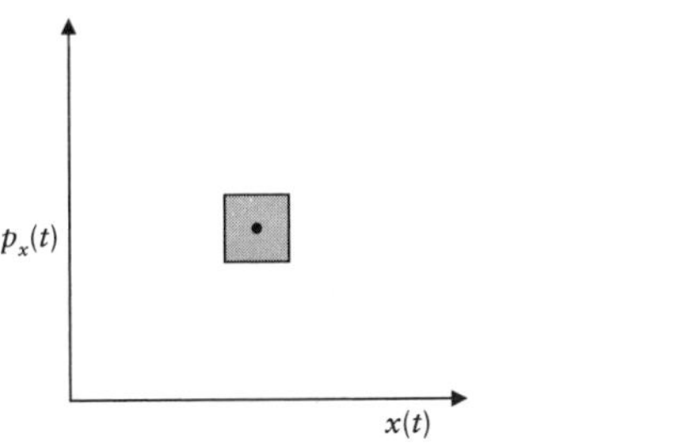

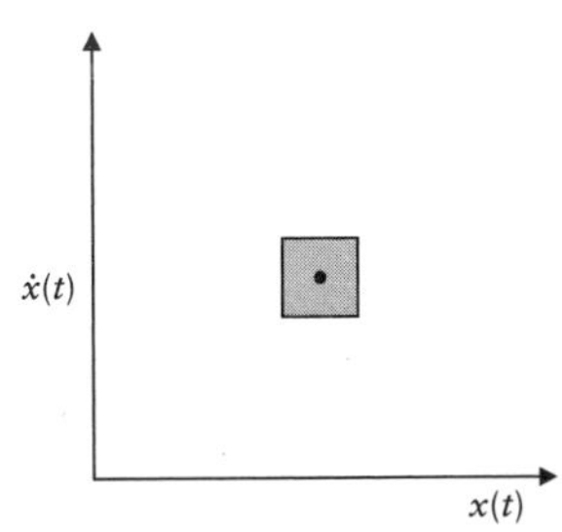

Fig. 1.3 (a) Position–Momentum Phase Space. (b) Position–Velocity Phase Space.

The physical dimensions of a piece of area in this planar space will not be L^2 as would be the case of a plane spanned by the usual x and y axes of a Cartesian frame of reference. Instead, the dimensions of the area of the position–momentum phase space would be ML^2T^{-1}, which are dimensions of angular momentum. If the object could, however, move in two dimensions, such as the xy-plane, then it has 2 degrees of freedom, and we require four parameters, x, p_x, y, p_y to fully describe the mechanical state of the system. These four variables are usually written in a different, but equivalent, order as x, y, p_x, p_y. All coordinates are written first, and all the momenta written next. The phase space of this system is thus 4-dimensional. In general, an object having n degrees of freedom will have a $2n$ dimensional phase space. If there are N particles which can move independently of each other in 3 dimensions, then there are 3N degrees of freedom. The phase space of this N-particle system will then be $3 \times 2 \times N = 6N$ dimensional. The 'tripling' coming from the number of independent degrees of freedom for each particle, and the 'doubling' comes from the fact that q and p corresponding to each degree of freedom must both be specified to describe the system's evolution. The mechanical state of the N-particle system is thus defined by a point in the phase space that is specified by 6N parameters. *The two parameters position and momentum complement each other*. As discussed in Chapter 6, they are referred to as 'canonically conjugate parameters'. An equivalent description of the mechanical state of the system would of course be in terms of its position and velocity, instead of position and momentum. Thus we have the 'position–velocity phase space' analogue of the position–momentum phase space.

Newton's Third Law – Two-body Interactions and Conservation of Linear Momentum: Newton's third law makes an emphatic qualitative, as well as a precise quantitative, statement about the interaction between a pair of objects which exert a mechanical force on each other. This law states that when two bodies interact with each other, the impact/force by one object on the other is opposite (with regard to the direction), and exactly equal (in magnitude), to that by the other on the first. The third law is commonly stated as '*action and reaction are equal and opposite*'. The relationship of the forces between a pair of objects being mutual, one may label either of the two interacting objects as the first, and the other as the second. It is therefore pointless to worry about which of these forces is to be regarded as the 'action', and which as the 'reaction'. This law is best expressed as a succinct mathematical equation:

$$\vec{F}_{12} = -\vec{F}_{21}. \tag{1.2}$$

Like the first two laws, this law can also be tested, and found to hold good accurately within the domain of classical mechanics. It is applicable to all paired mechanical interactions. Applying Newton's second law to Eq. 1.2, we get

$$\frac{d\vec{p}_2}{dt} = \frac{-d\vec{p}_1}{dt}, \quad hence \quad \frac{d(\vec{p}_1 + \vec{p}_2)}{dt} = \vec{0} = \frac{d\vec{P}}{dt}, \tag{1.3}$$

where $\vec{P}$ is the total momentum of the system.

The fact that total time derivative of momentum of the combined system is invariant with time is a conservation principle: *the total momentum of an isolated system is constant*. Newton's third law thus has the principle of conservation of momentum built into it. We underscore, however, at this point that our considerations are restricted to 'mechanical' (including gravitational) forces. When electromagnetic interaction is involved, we must include the momentum carried by electromagnetic field in our analysis, and thus the third law would need to be appropriately expanded. The electromagnetic momentum is discussed in Chapter 13.

When you turn the wheels of your car by operating the steering wheel, the forward force due to friction on the car acquires a component which provides the centripetal acceleration toward the center of a circle along whose arc the car would turn. The force of friction is generated at the contact between the tyres of the *powered* wheel and the road. This is understood in terms of Newton's third law; the tyres of the powered wheels push the road backward, and in turn the road pushes the car forward. The same principle is involved when we walk: our feet push the ground backward, and the friction at the contact between the ground and the feet pushes us forward. Friction does oppose motion, and here it does essentially the same. We must consider horizontal components of the forces which enable movement in walking which results in forward motion. The frictional force always acts in the opposite direction to the horizontal component of the force of your foot on the ground. When you advance your foot to walk, the friction at the contact with ground acts backward and prevents you from slipping. However, as your center of mass shifts forward, you push the ground backward, the ground in turn pushes you forward, in a direction in which you wish to move. The internal biomechanics is extremely complex involving thousands of muscles and ligands which are in play even as we go through these mechanisms uncaring for the complexities. This is a subject of intense research in many advanced laboratories.

The dynamics of all macroscopic motion is addressed in classical mechanics using Newton's laws. It is amazing that a principle *enunciated so simply* that 'action and reaction are equal and opposite' accounts for changes in momenta of *all* the objects around us. It has subtle and deep ramifications. Thus, when you ride a bicycle, the wheel is turned due to the forces transmitted by your legs, through the chain that is locked into (usually) the rear wheel, which pushes the ground backward. The frictional forces by the ground, on the cycle's rear wheel, in turn push the cycle forward. The front wheel is pushed forward, only because it is rigidly bound into the frame of the bicycle. If there were no friction at the front wheel, it would only *skid* forward. The frictional force by the ground on the front wheel being backward, opposes its motion. This results in a torque on the front wheel about its axle at the center, making the wheel turn in the *same* direction as the rear wheel. Motion of all vehicles is accounted for by these principles. The momentum conservation law encapsulated in Newtonian dynamics accounts for the 'strike and rebound' operations on a caroms board, as also the physical phenomenology in all collisions.

Conservation of momentum is a fundamental principle among the laws of nature that govern physical phenomena. In all collisions, momentum is conserved, and so is the *total* energy. The total momentum can get distributed in a variety of ways amid the colliding objects, including their fragments, if the collision would result in the breakup (or coalescence) of one or more of colliding objects. Likewise, total energy would also get distributed, but not necessarily only in the *kinetic energy* of the objects after collision. Some energy may also be used up in changing the internal energy of some, or all, of the constituents after the collision.

Collision is said to be elastic if the total kinetic energy of the system does not change after the collision; otherwise it is inelastic.

Did we derive Newton's laws from anything? We got the *first law* from Galileo's brilliant experiments which led him to conclude that the state of rest is completely equivalent to that of uniform motion. We got the *second law* from Newton's identification of the time-rate of change of momentum with the stimulus (force) that caused the momentum to change. Predictions of trajectories that would emerge from solving the equation of motion $\vec{F}=\frac{d\vec{p}}{dt}$ can be tested. The third law comes from Newton's insight into the equivalence of pairs of mutual forces (interactions) between objects (isolated from other physical systems). One can solve any problem in mechanics using these three laws. The entire formulation of Newtonian mechanics is in terms of the minimal set of assumptions made in these three laws; *no more, and no less*. Within the experimental accuracy that was available for a long time since Newton, no violation of these predictions was ever found, justifying their status as *laws of nature*.

We may nonetheless ask: are Newton's laws truly universal? Could there be any situation that limits the scope of applications of these laws? To determine what the limitations may be, let us examine the premise of these laws. Clearly, the three laws of Newton are based on the assumption that position, and momentum (or velocity), are knowable simultaneously and accurately. However, we did not discuss any mechanism to acquire accurate simultaneous information about these two physical properties. If the process of acquisition of information about the position q were to disturb the momentum p, the very foundation of this scheme collapses. We will merely keep this point at the back of our mind, in anticipation of the compulsions that eventually led to Quantum Theory. Reconciliation with this compulsion leads to the breakdown of the principle of causality and determinism. This is an exciting offshoot, but is beyond the scope of this book. It cannot be studied prematurely. Yet, these exciting developments must be anticipated, even as patiently awaited. For most everyday phenomena, however, the accuracy with which the position and momentum are simultaneously knowable is quite sufficient to carry out fruitful analysis and quantum theory is not required.

Furthermore, even if we were to *assume* that simultaneous and accurate information about the position and momentum is possible at least in principle, we can still ask if a *slight change* in initial condition would produce a trajectory that would be only *marginally* different, or *vastly* so. Sometimes, a minuscule change in the initial condition produces a very different trajectory in the phase space. Temporal evolution of the dynamical system is very sensitive to initial conditions in such cases. We are then led to 'chaos', discussed in Chapter 9. Also, I would alert the reader to the fact that the entire formalism of Newtonian mechanics tacitly employs an intuitive notion about what we mean by *simultaneous* events. It turns out that this is intimately connected with the speed of light. The speed of light is finite, though large, and this places a limit on the validity of Newtonian laws of mechanics, and also on Newton's law of gravity. We shall discuss this issue further in Chapters 13 and 14.

1.3 NEWTON'S THIRD LAW: AN ALTERNATIVE VIEWPOINT

We can of course discuss Newton's third law on the same footing as the first two laws. We would then *not* expect the third law to be deducible from anything deeper. In this section, we present an alternative perspective, inspired by the works of Einstein, Noether and Wigner in the twentieth century. This alternative has for its foundation a very different principle, from which Newton's third 'law' can be actually *deduced*. This alternative principle will thereby illustrate to us a new pathway to discover hitherto unknown laws of nature.

The alternative principle mentioned above is the *principle of translational invariance in homogenous space*. What this means is that if we consider a system of N particles in a medium that is homogenous; a displacement of the *entire* N-particle system through this medium would result in a new configuration that is completely indistinguishable from the previous. The invariance of the environment of an entire N particle system following a translational displacement is a result of *translational symmetry*. A region of space in which translational invariance exists is said to be *homogenous*.

There are three essential considerations that need to be spelled out to underscore the translational invariance of space in which the above mentioned displacement through $\overrightarrow{\delta s}$ of the N-particle system is considered:

(i) We consider each particle of the system to be under the influence only of all the remaining particles. The system is isolated, and there are no external forces acting on any of its particles.

(ii) It is the entire N-particle system that is deemed to have undergone simultaneous and instantaneous displacement through $\overrightarrow{\delta s}$, leaving the inter-particle separation and orientations invariant.

(iii) The medium in which the system undergoes such a translational displacement is essentially homogeneous, spanning the entire system, both before, during, and after, the displacement through $\overrightarrow{\delta s}$.

The implication of the above three considerations is that the displacement under consideration is only a *virtual displacement*. It is only a mental thought process. Real physical displacements would require a certain time *duration* over which the displacements would occur. Virtual displacements can be thought to occur at an *instant*, rather than over a *duration*. The displacement of the N-particle system being only *virtual*, it is redundant to ask which agency has caused it. All the internal forces acting between pairs of objects in the N-particle system remain unchanged, before during, and after, the virtual displacement in the homogeneous space.

Now, the internal forces do no work in this virtual displacement, and we must therefore have the total work done in this displacement to be zero:

$$0 = \delta W = \left\{ \sum_{k=1}^{N} \vec{F}_k \right\} \cdot \overrightarrow{\delta s} = \left\{ \sum_{k=1}^{N} \sum_{i=1, i \neq k}^{N} \vec{F}_{ik} \right\} \cdot \overrightarrow{\delta s}, \tag{1.4}$$

where $\vec{F}_{ik}$ is the force on the kth particle by the ith, and is the total force on the kth particle due to the remaining N-1 particles. The work done (rather, 'not done') considered in Eq. 1.4 is called 'virtual work'. The mathematical techniques involved in the principle of virtual displacement and that of virtual work were developed by Jean le Rond d'Alembert (1717–1783). Equation 1.4 holds if either vector participating in the scalar (dot) product is a null vector, or if the angle between the two is 90^0. Since we have considered the displacement $\overrightarrow{\delta s}$ to be arbitrary (both in magnitude and direction), we must conclude that

$$\vec{0} = \left\{ \sum_{k=1}^{N} \vec{F}_k \right\} = \sum_{k=1}^{N} \frac{d\vec{p}_k}{dt} = \frac{d}{dt} \sum_{k=1}^{N} \vec{p}_k = \frac{d\vec{P}}{dt}. \tag{1.5}$$

In writing the above result, we have implicitly used Newton's first and the second law, but certainly not the third law. Since the time derivative of the total momentum $\vec{P} = \sum_{i=1}^{N} \vec{p}_i$ of the isolated N-particle system vanishes, we conclude that $\vec{P}$ is conserved. We have arrived at this conclusion from the properties of translational symmetry in homogenous space. If we write the above relation for just two particles interacting with each other, we get $\frac{\overrightarrow{dp_2}}{dt} = -\frac{\overrightarrow{dp_1}}{dt}$, which yields a startling result: $\vec{F}_{12} = -\vec{F}_{21}$, since the only forces to recon with are those which the two particles exert on each other. Effectively, we have actually *deduced* Newton's third law from translational invariance in homogeneous space. *This is fascinating since it suggests a path to discover hitherto unknown laws of physics by exploiting the connection between symmetry and conservation laws.*

1.4 CONNECTIONS BETWEEN SYMMETRY AND CONSERVATION LAWS

The third law of mechanics was formulated by Newton, essentially as a fundamental ansatz. It was not derived from any other *more* fundamental principle, but it is of course amenable to tests. In Section 1.3 we found, however, that the third law could be deduced as a conservation principle (conservation of momentum) resulting from translational invariance in homogenous space. The latter path to arrive at a fundamental law provides an illustration of an extremely powerful mechanism that can be employed to discover new laws of nature. Its strength comes from a beautiful theorem, known as Noether's theorem, which connects every symmetry principle with a conservation law.

Until Albert Einstein (1879–1955) developed the special theory of relativity in 1905, it was believed that the conservation principles are results of laws of nature. We shall illustrate this further in Chapter 8, in which we discuss gravitational interaction. As such, in Section 1.2, we have already seen that the conservation of momentum results from Newton's third law of mechanics. Physicists, however, began, since the work on the special theory of relativity (Chapters 13 and 14) by Einstein, to analyze *conservation principles* as a consequence of some underlying *symmetry*, from which the law of physics can be inferred. This line of reasoning finds a rigorous expression in a famous theorem, known after Emmy Noether (1882–1935). She was a brilliant German mathematician-physicist of whom Einstein wrote (in New York Times, 1935): "*In the judgment of the most competent living mathematicians, Fräulein Noether was the most significant creative mathematical genius thus far produced since the higher education of women began.*" Amongst those who made Noether's theorem a powerful tool to explore laws of nature, Eugene Wigner (1902–1995), a towering figure, used originative group theoretical methods to expound symmetry. Wigner's insightful impact on physics is that his explanations of symmetry considerations resulted in a change in the very perception of just what is most fundamental. Physicists began to regard 'symmetry' as the most fundamental entity, whose form would lead us to physical laws [2,3]. Noether's theorem is employed in advanced topics in physics, which require years of learning. It is beyond the scope of this book. We shall, however, have further opportunities in this book to see illustrations of Noether's theorem, especially in the Chapters 6 and 8. The formalism in Chapter 6 will display the connection between symmetry and conservation principle in a beautiful and concise manner, even if only in an abecedarian manner.

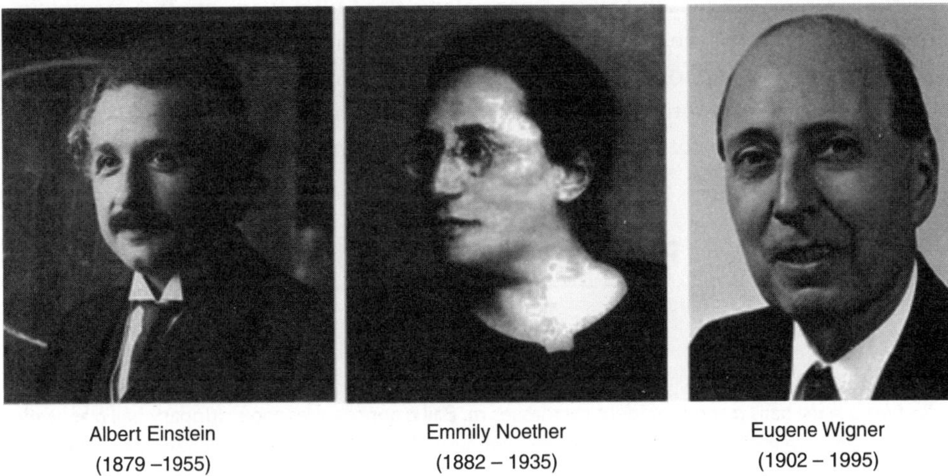

Albert Einstein (1879 –1955) | Emmily Noether (1882 – 1935) | Eugene Wigner (1902 – 1995)

Fig. 1.4 The recognition that symmetry is an extremely important physical attribute seems to have been made first by Albert Einstein, who exploited the symmetry in Maxwell's equations to conclude that space and time are not absolute, and that the geometry of spacetime continuum is non-Euclidean. We shall discuss these points further in Chapters 12 and 13. Noether formulated the connection between symmetry and conservation laws, and Wigner elucidated these principles rigorously using group theory.

The guiding principle in Newtonian mechanics is the principle of causality which governs the linear response of a mechanical system expressed in Newton's second law. It relies on the notion of 'force' whose meaning grows on one's mind from the experience gained from the applications of Newton's laws. The connection between symmetry and conservation laws, on the other hand, becomes more transparent in an *alternative* formulation of mechanics, which employs neither 'force', nor the 'principle of causality'. This alternative formulation stems from the 'principle of variation'. It provides a mathematical framework which produces results that are completely equivalent to those that we get from Newton's formulation of mechanics. We shall discuss the 'principle of variation' in Chapter 6. The principle of variation would lead us to alternative, but equivalent, equations of motion, known as the Lagrange's equation and Hamilton's equations.

In Chapter 3, we shall discuss some deeper, exciting, aspects of Newton's laws. Specifically, we shall discuss how to use them in a *real* frame of reference that an observer finds herself/himself. In a practical scenario, we do not find ourselves in an ideal inertial frame of reference. The Earth is our frame of reference, and there is nothing absolute about it. Our place in the universe is by no means unique. The Earth is rotating about its own axis, which is what gives us progression through the day and the night. Furthermore, we revolve around the Sun completing one revolution in ~365 days. The whole solar system is accelerated toward the solar apex, which is seen in the direction of the beautiful star Vega, in the background of the constellation Hercules. *Aakaash Ganga*—our galaxy, the Milky Way—itself is moving in the universe. The dynamics of the universe is thus incredibly humbling. The observations that we carry out on Earth are therefore not in the inertial frame of reference, which is what provides the essential premise of Newtonian mechanics. Newton's laws need clever adaptation in the observer's frame of reference. We shall take up this question in Chapter 3.

Problems with Solutions

P1.1:

Consider two billiard balls *a* and *b* of identical masses *m*. Ball *a* approaches the stationary ball *b* at a velocity $v\breve{e}_x$ along the positive *x* direction. After an elastic collision, the balls *a* and *b* move away at angles ξ and δ, respectively, with X-axis as shown in the figure. Determine the final speeds of the balls and determine the angle between the lines along which they escape the collision region.

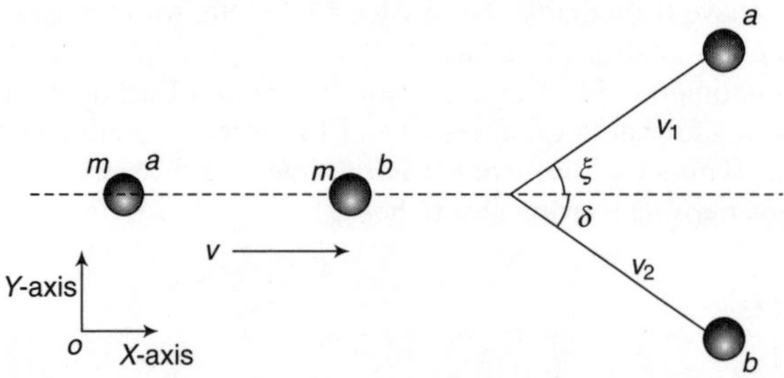

Solution:

Let v_1 and v_2 be the final speeds of the balls *a* and *b*, respectively.

Conservation of momentum along the *X* and *Y* axes:

$$mv = mv_1 \cos \xi + mv_2 \cos \delta \quad \text{(i)}$$

and $mv_1 \sin \xi = mv_2 \sin \delta$ (ii)

Eliminating δ: $v^2 - 2vv_1 \cos \xi + v_1^2 = v_2^2$ (iii)

Conservation of kinetic energy in elastic collision:

$$v^2 = v_1^2 + v_2^2 \quad \text{(iv)}$$

From Eq. (iii) and Eq. (iv), we have: $v_1 = v \cos \xi$ and $v_2 = v \sin \xi$ (v)

From Eq. (ii) and Eq. (v); $\cos\xi = \sin\delta = \cos\left(\frac{\pi}{2} - \delta\right)$ where $\xi = \frac{\pi}{2} - \delta \Rightarrow \delta = \frac{\pi}{2} - \xi$

Thus the balls *a* and *b* bounce off at right angles with respect to each other after collision.

P1.2:

Frogs can stretch their muscles and leap as much as over 10 meters, more than 20 times their length. A frog at point *A* is trying to jump from the point *A* to *B* as shown in the figure. The horizontal distance between the frog and the step is *L*, and the point *B* is at a height *h*. The angle that the line *AB* makes with the horizontal line is β. If the frog jumps out an angle α with the horizontal with an initial speed v_0, determine the angle α in terms of the angle β such that the frog expends minimum energy to reach point *B*.

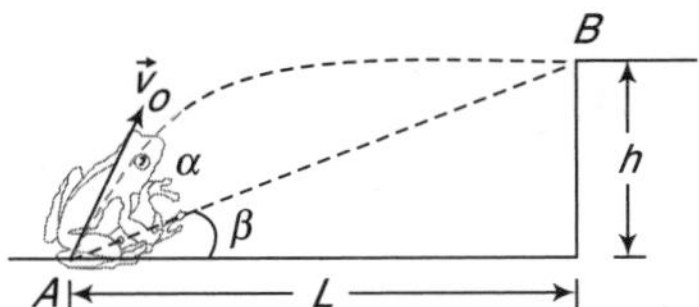

Solution:

To expend minimum energy, the frog must have the minimum kinetic energy at the launch of its jump. The speed at which it must launch itself at the angle α must therefore be minimum.

$$v_x = v_o \cos \alpha \text{ and } v_y = v_o \sin \alpha - gt. \quad \text{(i)}$$

Hence: $x = v_o t \cos \alpha$ and $y = v_o t \sin \alpha - \frac{gt^2}{2}$ (ii)

After the frog lands at point B the distance it covers along the X-axis is *L* and along the Y-axis is *h*. From Eq. (ii), we have, $t = \frac{L}{v_o \cos\alpha}$ and $h = v_o t \sin\alpha - \frac{1}{2}gt^2$.

Substituting the value of *t* in the previous expression for *h*, we get

$$h = L\tan\alpha - \frac{gL^2}{2v_o^2\cos^2\alpha}. \quad \text{(iii)}$$

From Eq. (iii), $v_o^2 = \frac{gL^2}{2(L\sin\alpha\cos\alpha - h\cos^2\alpha)}$; and from the figure, $\frac{h}{L} = \tan\beta$.

Therefore, $v_o^2 = \dfrac{gL\cos\beta}{\sin 2\alpha\cos\beta - \sin\beta - \sin\beta\cos 2\alpha} = \dfrac{gL\cos\beta}{\sin(2\alpha-\beta)-\sin\beta}$.

For the launch speed to be minimum, sin $(2\alpha - \beta)$ should be maximum, and hence,

$2\alpha - \beta = 90°$ and $\alpha = 45° + \dfrac{\beta}{2}$. Thus the minimum launch-speed is $v_0 = \sqrt{\dfrac{gL\cos\beta}{1-\sin\beta}}$.

P1.3:

A rocket of mass M moves in a free space at a velocity $\vec{v}_i = 16\,\text{kms}^{-1}\breve{u}$ at some instant of time. At that instant, the rocket ejects fuel of mass Δm at a relative velocity $v_e = 4\ \text{kms}^{-1}$ to attain a final velocity of 18 kms^{-1}. Determine the ratio $\dfrac{M_i}{M_f}$ of the initial mass of the rocket to its final mass after the ejection of the fuel.

Solution:

At time t the momentum of the rocket is given to be $\vec{P}_i = M\vec{v}_i$.

Conservation of linear momentum:

$Mv_i = (M - \Delta m)(v_i + \Delta v) + \Delta m(v_i - v_e) \simeq Mv_i + M\Delta v - (\Delta m)v_i + (\Delta m)v_i - (\Delta m)v_e$.

$M\Delta v \simeq (\Delta m)v_e$, i.e., $dv \simeq \left(\dfrac{dM}{M}\right)v_e$. Integrating: $\int_{v_i}^{v_f} dv = -v_e\int_{M_i}^{M_f}\dfrac{dM}{M}$.

Hence: $v_f - v_i = v_e \ln\dfrac{M_i}{M_f}$. Therefore, $2 = 4\ln\dfrac{M_i}{M_f}$. Hence: $\dfrac{M_i}{M_f} = e^{0.5} \approx 1.65$.

P1.4:

(a) A conveyer belt transporting coal moves at a velocity $\vec{v} = v\breve{e}_x$ with $v > 0$. The conveyer belt is loaded with coal continuously. The mass of the 'belt + coal' system therefore increases continuously as a function of time. The coal added increases the mass at a rate $\dfrac{dm}{dt} = \lim\limits_{\delta t\to 0}\dfrac{\delta m}{\delta t}$, where Δm is the extra mass of the coal that lands on the belt in a time-interval Δt. The coal that is added to the belt descends on it at an initial velocity $\vec{u}$.

(b) How much *extra power* must be supplied to the belt to keep it moving at a constant velocity, as a larger mass is to be driven on addition of coal to the belt?

(c) As the mass on the belt increases, its kinetic energy will increase. The kinetic energy increases at the rate at which the extra mass of coal accumulates on the belt. If the coal is dropped on the belt at zero velocity ($\vec{u} = \vec{0}$), determine if the extra power delivered to the belt (to keep it moving at constant velocity) goes into increase in the rate at which kinetic energy increases.

Solution:

An electric motor powers the conveyer belt to keep it moving at a constant speed. This power is required to overcome friction that the belt experiences as it runs on the rollers. The mass of the belt is M. Coal of mass δm lands on the belt in time δt at a uniform rate $\dfrac{dm}{dt} = \lim\limits_{\delta t\to 0}\dfrac{\delta m}{\delta t}$. The total momentum of the 'belt + coal' at time *t*, just an instant before the mass Δm drops on the belt is the sum of the momentum of the belt, and the momentum of the coal: $\vec{p}_{\text{initial}}(t) = M\vec{v} + (\delta m)\vec{u}$.

At time $t + \delta t$, the momentum is: $\vec{p}_{\text{after time }\Delta t} = M\vec{v} + \vec{v}\delta m$, since (i) the belt continues to travel at the ***same*** velocity as the initial one, and (ii) the added mass settles down on the belt and moves at the same velocity, $\vec{v}$, as that of the belt.

Therefore, the ***change*** in momentum: $(\vec{p}_{\text{after time }\delta t} = \vec{p}_{\text{initial}}) = \delta\vec{p} = (\vec{v} - \vec{u})\delta m.$

The ***rate*** of change of momentum: $\dfrac{\vec{p}_{\text{final}} - \vec{p}_{\text{final}}}{\delta t} = \dfrac{\delta\vec{p}}{\delta t} = (\vec{v} - \vec{u})\dfrac{\delta m}{\delta t}$.

Thus, an ***extra*** force $\vec{F} = (\vec{v} - \vec{u})\dfrac{dm}{dt}$ is required to keep the conveyer belt moving at a constant velocity. Note that if $(\vec{u} = \vec{v})$, the extra force needed is zero, so it is best to dump coal on the belt in the direction in which the belt is moving, and at the velocity of the belt. This extra force would be delivered by the electric motor which drives the belt, and hence the belt would consume more electrical power to drive the belt when coal is accumulating on it.

(b) It is given in the statement of the problem that coal lands on the belt at velocity $\vec{u} = \vec{0}$. Its kinetic energy would thus be zero as it lands on the belt. The extra source power delivered to the belt drives it along with the added coal, whose velocity increases from zero to $\vec{v}$, which is the velocity of the belt. Therefore, the ***extra power*** required to drive the belt at the constant velocity is just the rate at which the extra work has to be done:

$P_{\text{extra}} = \vec{F}\cdot\left(\dfrac{dx}{dt}\vec{e}_x\right) = (\vec{v} - \vec{u})\dfrac{dm}{dt}\cdot\vec{v} = \dfrac{dm}{dt}(\vec{v}\cdot\vec{v}) = \left(\dfrac{dm}{dt}\right)(v^2)$, since $\vec{u} = \vec{0}$. Now, the rate at which the kinetic energy of the coal increases is $R_{KE} = \dfrac{1}{2}\left(\dfrac{dm}{dt}\right)(v^2)$. Since $\dfrac{P_{\text{extra}}}{R_{KE}} = 2$, half of the extra power delivered to the belt goes into the increase of the kinetic energy of the coal.

P1.5:

In the above problem, an extra power $\left(\dfrac{dm}{dt}\right)(v^2)$ is delivered to the conveyer belt to keep it going at a constant velocity, but the kinetic energy of the coal increases at only half that rate. Where is the other half of the power delivered to the belt lost?

Solution:

After the coal falls on the belt at zero velocity, it needs to be ***accelerated*** to pick up speed, and then continue to move with the belt. Coal would first slide backward, opposite to the direction of motion of the belt, due to its inertia along the horizontal motion. Kinetic friction between the coal and the belt halts the backward motion of the coal. Thereafter, coal move ***along with*** the belt. The frictional force exerted by the belt on coal is $\delta F_f = \mu g \delta m$, where μ is the coefficient of kinetic friction between the belt and coal. The acceleration which coal will receive from the frictional force is $a_f = \dfrac{\delta F_f}{\delta m} = \mu g$. The distance d over which the frictional force would act, is the distance through which coal would move till it catches up with the conveyer belt's speed, which we know is *v*. From the kinematic equation, this distance is $d = \dfrac{v^2}{2a_f} = \dfrac{v^2}{2\mu g}$. The work done by the frictional force is $\delta W_f = (\delta F_f)d = (\mu g \delta m) \times$

$\frac{v^2}{2\mu g}$. Hence, the rate at which the power will have to be supplied to overcome the frictional loss is $\frac{dW_f}{dt} = \frac{1}{2}v^2\left(\frac{dm}{dt}\right)$. This explains why only half the extra power supplied to the belt goes into the rate of increase of kinetic energy of the added coal. The other half is expended in overcoming friction on the belt.

P1.6:

Five steel balls of equal mass *m* are hung as shown in the figure (Newton's cradle).

At the beginning of the experiment, the balls are at rest, barely touching each other. The ball 1 is now pulled to the left side, and released to hit the ball 2. Describe the mechanism and consequence of the impact if the collision is perfectly elastic.

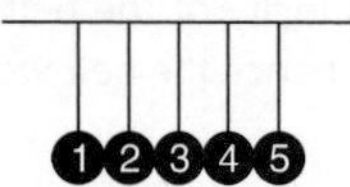

Solution:

As the ball 1 strikes the ball 2, the ball 2 gets momentarily compressed, since it does not have free space available on its right to get displaced. Since the balls are made of a material such as steel, the entire kinetic energy of the ball 1 gets stored as the potential energy in the compressed ball 2. The compression results in generating elastic potential energy in the ball 2, which then tries to recover to its state of equilibrium. In doing so, it passes the impulse to its neighboring ball, ball number 3, and the impulse thus travels to the right. When the ball 5 gets this impulse from the ball 4, it flies off to its right, and then coming back under gravity, hitting the ball 4, now from the right. The impulse then travels from right to left, the same wave as it had traveled from left to right. This device is called 'Newton's cradle', although John Wallis, Christopher Wren, and also Christian Huygens seem to have set up this experiment rather than Newton himself. It was Marius J. Morin who named the device as Newton's cradle. If the cradle has only two balls, the expressions of the conservation of energy and the conservation of momentum are sufficient to determine the final outcome. If, however, there are more than two balls suspended, and two (or more) balls are pulled aside and let go to strike the remaining balls, the dynamics is quite complex. A detailed solution can be found in the paper by Stefan Hutzler et al., ***Rocking Newton's cradle*** Am. J. Phys. 72:12, page 1508 (December 2004). The net outcome is that if two balls are pulled aside and let go, two balls at the opposite end are pushed away, rather than only one with twice the velocity. This is because ***both*** the conservation of momentum and energy must be satisfied. If the cradle has 5 balls and ***three*** of these are pulled aside and let go, then the oscillations continue with the central ball oscillating ***continuously***. In a 3-balls cradle, the ball in the middle stays practically static, merely transferring the energy and momentum to its neighbor through the compression/expansion of the steel material of the ball. The mechanism to transmit a shock across a medium without affecting intermediate links has applications in surgery by urologists. They blast off kidney stones by ultrasonic shock waves that pass through the skin, ***without*** harming it.

Additional Problems

P1.7 A ball b_1 of mass m_1 and velocity v_1 makes a head-on collision with another ball b_2 of mass m_2 at rest. After collision, the ball b_1 rebounds straight back with $\left(\frac{9}{16}\right)^{\text{th}}$ of its initial kinetic energy. Find the ratio of the mass of the two balls, assuming that the collision is elastic.

P1.8 A ball b_1 of mass m is initially moving in the positive X-direction with a speed $v_{1,i}$ and collides elastically with another ball b_2 of mass $2m$, which is initially at rest. After the collision, the ball b_1 moves with an unknown speed $v_{1,f}$, and at an unknown angle $\theta_{1,f}$ with respect to the positive x-direction. After the collision, the ball b_2 moves with an unknown speed $v_{2,f}$ at an angle $\theta_{2,f} = 45°$ with respect to positive X-direction. Find $\theta_{1,f}$.

P1.9(a) In a 1-dimensional problem, a mass m_1 initially moving in the direction of the unit vector $\hat{e}$ at a speed u_1 collides head-on, elastically, with another mass m_2 approaching it from the opposite direction at a speed u_2. Determine the final velocities $\vec{v}_1$ and $\vec{v}_2$, respectively of the two masses m_1 and m_2.

(b) What will be the final velocities of the two masses if $u_2 = 0$? Also, determine the final velocities of the two masses if (i) $m_1 = m_2$, (ii) $m_1 < m_2$, and (iii) $m_1 > m_2$?

(c) A heavy truck of mass 15000 kg moving at a speed of $u_1 = 70 \text{ kms}^{-1}$ hits a car of mass 1500 kg in a head-on collision. Determine the final velocities of both the vehicles if (i) the car is initially at rest and (ii) if it is moving in a direction ***opposite*** to that of the truck at a speed of 70 kms^{-1}.

P1.10 Consider a particle moving along a curve as shown in Fig. (a). The position of the particle at point A is specified with reference to the origin *O*. The distance between the points *O* and A is the arc length *l* in Fig. (a).

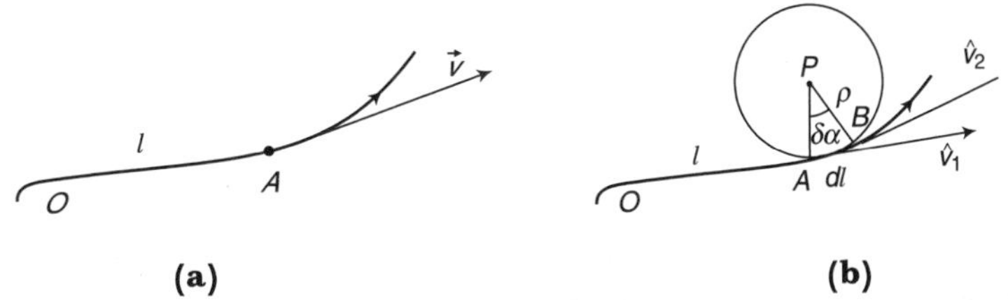

(a) **(b)**

The particle's velocity is along the direction $\hat{v}_1$ when at point A, and along $\hat{v}_2$ when it is at point B, as shown in Fig. (b). The points A and B make an arc of a circle, with center at P and radius ρ. The particle's movement through *dl* is therefore along an arc length of the circle. Determine the components of the net acceleration of the particle if it makes an infinitesimal movement *dl* from the point A to a point B.

P1.11(a) Determine the angle at which a road should be banked if a car cruising at 108 km/hr is to go without sliding on a frictionless curved road having a turning radius of 500 meters? (b) If there is no banking, would the car turn when the steering wheel is turned?

P1.12 Consider a particle moving along a circle of radius ρ as shown in the figure. It traverses from the point *O* along the arc of a circle. Its instantaneous speed at any point on the arc is proportional to $l^{3/2}$, where l is the arc length traversed till that instant. The net acceleration of the particle is indicated by $\vec{a}$ in the figure. It includes the centripetal acceleration $\vec{a}_c$ which is responsible for making it go along the circle. Determine the angle α between the net acceleration $\vec{a}$ and the direction $\hat{v}$ of the instantaneous velocity at any instant of its motion.

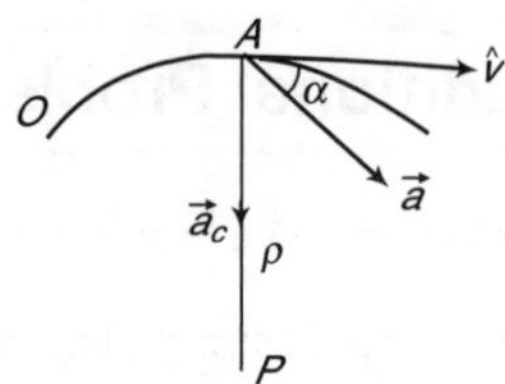

P1.13 Two point-sized positive charges Q_1 and Q_2 are placed at a distance 'd' apart in free space having permittivity ε_0. The two charges are moved through a simultaneous instantaneous displacement δs through a surface of separation into a medium having permittivity ε. In the second medium also, they are to be held at exactly the same distance 'd' apart, as before. Compare the energy of the pair of charges in the new medium, with that in vacuum. Comment on the possibility of obtaining Newton's third law from this situation.

P1.14 A block of mass 2.5 kg slides on an inclined plane that makes an angle of 30° with the horizontal. The coefficient of friction between the block and the surface is $\frac{1}{\sqrt{2}}$.

(a) What force should be applied on the block so that it moves down without any acceleration?

(b) What force should be applied on the block so that it moves up without any acceleration?

P1.15 The ballistic pendulum is used to measure the speed of a bullet. It consists of a wooden plank hung from a horizontal plane. If a bullet is shot into the wooden plank, the bullet gets lodged into the plank. Thereafter, the plank and the embedded bullet move together. The collision time is very small; i.e., the time taken by the bullet to get embedded in the wooden plank is very short.

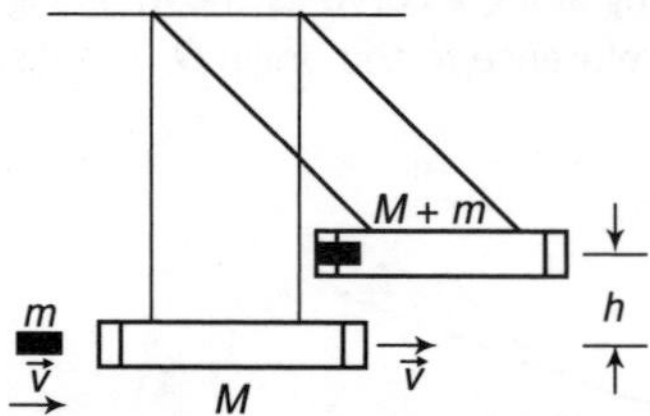

There is no other external horizontal force acting on the system during collision. The bullet gets lodged so fast that during this collision process, the wire with which the wooden plank is suspended remains vertical at, and during, the time of collision. A bullet of mass m traveling from the left (as shown in the figure) with initial velocity $\vec{v}$ striking the wooden plank at rest. The 'plank plus the bullet' gets raised to a maximum height h. (a) Obtain the expression for conservation of momentum for this system (b) If the mass of the bullet and that of the wooden plank is, respectively, 8 g and 2400 g, and the horizontal speed at which the bullet impacts the wooden plank is 1500 ms^{-1}, determine the maximum displacement of the wooden plank.

P1.16 A mass M_1 is placed on an inclined plane having an inclination θ as shown in the figure. The mass M_2 hangs over the side as shown in the figure, attached to M_1 over a pulley. The two masses are connected by a string (whose mass may be ignored). Ignore the mass of the pulley as well.

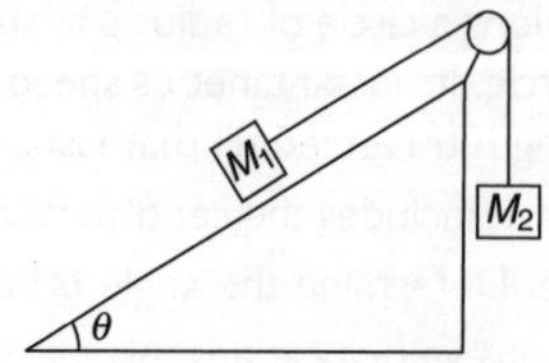

The coefficient of kinetic friction between M_1 and the plane on which it slides is μ. Now, M_1 is released from rest. Assume that $M_2 >> M_1$. (a) Write down all the forces on the two masses. (b) What are the accelerations of the two masses? (c) What is the tension in the string? (d) What is the range of M_2 for which the system does not accelerate?

P1.17 Consider collisions in which various atomic targets (a) hydrogen atom (b) beryllium atom and (c) gold atoms are bombarded by neutrons. Determine the ratio $\frac{\Delta k}{k}$ where k is the kinetic energy of the incident projectiles (i.e., neutrons) and Δk is the ***difference*** in the kinetic energy of the projectile ***before*** and ***after*** the collision in each of the three cases.

P1.18 An Atwood's machine is often used as a demonstration experiment to illustrate Newton's laws. It also has applications in machines such as the operations of the garage doors and elevators. The machine consists of two masses m_1 and m_2, usually unequal ($m_2 > m_1$), attached to a rope of negligible mass which runs over a pulley of mass M and radius R as shown in the figure. Determine the acceleration of the masses m_1 and m_2, and that of the pulley (on account of its rotary motion). The pulley may be assumed to be frictionless for the purpose of this problem.

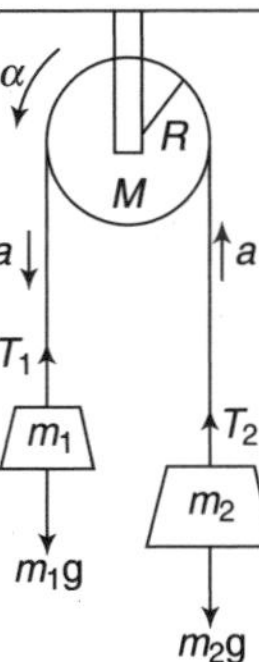

P1.19 Two loads are hung between two walls as shown using ropes. An unknown mass is suspended from the point **B** as shown in the figure. The ropes **AB**, **BC** and **CD** have negligible masses. The angles these ropes make with the horizontal are as shown in the figure. Determine (a) the forces in each of the three ropes and (b) the unknown mass **M** if the system is in equilibrium.

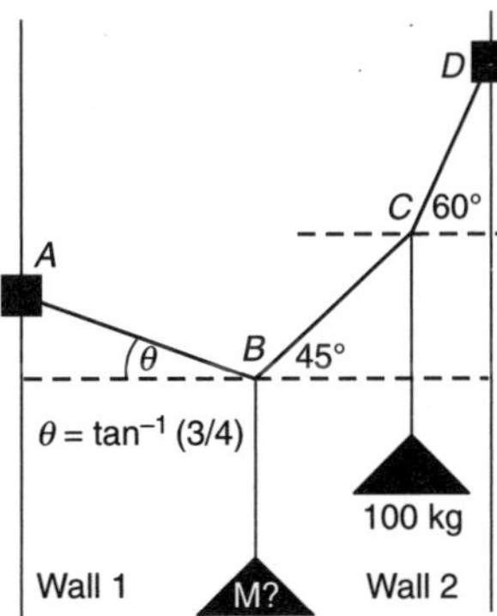

P1.20 A load is hung between two walls from ropes from the point **B** as shown in the figure. The walls are separated by 0.8 m, and the equilibrium length of the spring is 0.5 m. The spring constant of the spring is $k = 200$ N/m. What length of the rope **BC** will hold the weight hung from the point **B** in equilibrium, if the weight suspended is 20 kg?

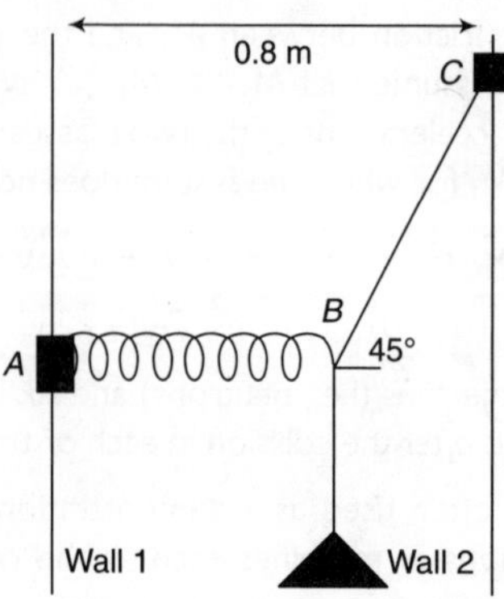

References

[1] Sagan, Carl. 1980. *Cosmos*. New York: Random House

[2] Deshmukh, P. C., and J. Libby. 2010. 'Symmetry Principles and Conservation Laws in Atomic and Subatomic Physics–1.' *Resonance* 15(9): 832–842.

[3] Deshmukh, P. C., and J. Libby. 2010. 'Symmetry Principles and Conservation Laws in Atomic and Subatomic Physics–2.' *Resonance* 15(10): 926–940.

CHAPTER 2

Mathematical Preliminaries

The miracle of the appropriateness of the language of mathematics for the formulation of the laws of physics is a wonderful gift which we neither understand nor deserve.

—*Eugene Wigner*

Some important mathematical methods that will be useful to appreciate seamless applications in the select topics in classical mechanics dealt with in this book are briefly reviewed in this chapter.

Fig. 2.1 **Eugene Paul 'E. P.' Wigner 1902–1995** The 1963 Nobel Prize in Physics was divided. Half of it was awarded to Eugene Paul Wigner *"for his contributions to the theory of the atomic nucleus and the elementary particles, particularly through the discovery and application of fundamental symmetry principles."* (The other half was awarded jointly to Maria Goeppert Mayer and J. Hans D. Jensen *"for their discoveries concerning nuclear shell structure."*)

2.1 MATHEMATICS,THE NATURAL ALLY OF PHYSICS

In the above quote, Wigner (Fig. 2.1) puts forth a far-reaching idea. Indeed, what makes mathematics splendidly well-suited to describe natural phenomena is a thrilling gift, no matter what its reasons are. Galileo Galilei called mathematics as the very language of physics. The relationship between mathematics and physics is very intertwined. No matter what its complexities, *all* of natural sciences, and also many other disciplines such as economics, sociology, etc., benefit from mathematics. Intricate analysis in entrepreneurship, marketing, human resource management, even the analysis of critical thinking and human values benefits from mathematics. Arithmetic logic and geometry have been employed for over two thousand years. Algebra has been used for well over a thousand years. In all civilizations—Indian, Greek, Arabic, Chinese and European—mathematics has developed, independently in the beginning, and later influenced by each other. In this chapter, I provide an elementary introduction to some mathematical methods used in this book. Admittedly, the breadth of these topics will be brief. Students who are already well familiar with these methods may skip these topics and move on to subsequent chapters. A quick review is, however, recommended to consolidate the mathematical machinery used in succeeding chapters. Mathematical methods are not merely tools to study physics, but are seamlessly integrated in the very description of the physical laws of the universe.

Fig. 2.2 French mathematician, Laplace: *"The ingenious method of expressing every possible number using a set of ten symbols (each symbol having a place value and an absolute value) emerged in India. The idea seems so simple nowadays that its significance and profound importance is no longer appreciated. Its simplicity lies in the way it facilitated calculation and placed arithmetic foremost amongst useful inventions."*

One of the greatest impetus to mathematics came from the use of the 'zero' (Fig. 2.2) in the decimal system, invented in India. It provides a *different* value to a number depending on its *absolute value* and its *place value*. There also are other interesting values which numbers acquire, by manipulating their positions. For example, one can arrange numbers in a pyramidal structure, beginning with '1' at the top in a block, and then again in two such blocks displaced half-way beneath the first block (Fig. 2.3). You can then sum the numbers '1' and '1' in the second row, and put the resulting number '2' in a block below them, centered in the middle, so that the two numbers ('1' each) which are summed, appear right above '2'. In the third row, the sum '2' may now be sandwiched between the '1' and '1' (as seen in the figure), and we may then extend the previous steps repeatedly, thus generating a triangular structure of numbers, (Fig. 2.3). It is called the Varahamihira triangle, and *also* as Pascal's triangle. Varahamihira (the sixth century mathematician, astronomer, from Ujjain) recognized the patterns in the binomial coefficients and arranged them in this triangular structure. *Addition of two numbers in the rows produces the number below it in this endless triangle*. Fascinating arithmetic patterns emerge as you examine the various diagonals of the Varahamihira's triangle, about which the interested reader may learn from other sources.

$$\begin{array}{ccccccccccc} & & & & & 1 & & & & & \\ & & & & 1 & & 1 & & & & \\ & & & 1 & & 2 & & 1 & & & \\ & & 1 & & 3 & & 3 & & 1 & & \\ & 1 & & 4 & & 6 & & 4 & & 1 & \\ 1 & & 5 & & 10 & & 10 & & 5 & & 1 \end{array} \quad \begin{array}{ccccccccccc} & & & & & \binom{0}{0} & & & & & \\ & & & & \binom{1}{0} & & \binom{1}{1} & & & & \\ & & & \binom{2}{0} & & \binom{2}{1} & & \binom{2}{2} & & & \\ & & \binom{3}{0} & & \binom{3}{1} & & \binom{3}{2} & & \binom{3}{3} & & \\ & \binom{4}{0} & & \binom{4}{1} & & \binom{4}{2} & & \binom{4}{3} & & \binom{4}{4} & \\ \binom{5}{0} & & \binom{5}{1} & & \binom{5}{2} & & \binom{5}{3} & & \binom{5}{4} & & \binom{5}{5} \end{array}$$

Fig. 2.3 Varahamihira's triangle is also called the Pascal triangle. Varahamihira studied the binomial expansions of $(a + b)^n$ and organized the coefficients in a simple triangular pattern, known as the binomial coefficients, ${}^nC_p = \frac{n!}{(n-p)!p!} = \begin{pmatrix} n \\ p \end{pmatrix}$, commonly read as '*n choose p*'.

As stated in the 'Preface', a selection of an assorted set of equations from various chapters in this book appear on its cover. This choice is a tribute to Dirac, who famously said "a physical law must possess a mathematical beauty", and to Einstein who said, "....an equation is for eternity." In this chapter, we shall equip ourselves with some basic, but powerful, mathematical tools. This book does not attempt a comprehensive review of mathematical methods routinely employed in physics. That would be a huge task even for a complete book specifically on mathematical physics. Only a few salient features of some of the mathematical methods will be visited. While the choice of topics their coverage are both restricted, I trust that it will be adequate to equip the student-readers with the tools that will be needed in the rest of the book.

Vector and matrix algebra are sometimes studied as separate topics in mathematics. Vectors are, however, often represented by matrices, and transformed using matrix transformations. Vectors and matrices often appear together in the analysis of physical problems. In this section, vector algebra and matrix algebra will appear interwoven together, in a form that will be most useful for applications in physics. A few important aspects of the matrix algebra are included at the end of this chapter, as an appendix.

Physics is about the natural laws of the physical world, the description of its state, and that of its evolution. As discussed in Chapter 1, the rationale behind designating a mechanical system by its physical properties $(q, \dot{q})$ or (q, p) is that if you know this pair at a certain instant of time, then you can determine the pair at any later (or even prior) time. In the 3-dimensional Euclidean world our senses commonly perceive, the position of a point object is described by its position vector. In some elementary books, one often defines a scalar as a quantity that is defined by magnitude alone, having no directional attribute. A vector is defined in these texts as a quantity that requires *both* its magnitude and its direction. These definitions are unfortunately misleading. If we consider the space-rate at which the height h of a mountain at a point P above the mean sea level is changing, we have to examine, $\frac{dh}{ds} = \lim_{\delta s \to 0} \frac{\delta h}{\delta s}$, where δs is an infinitesimal step size away from that point (Fig. 2.4a). This ratio is of course a *scalar*; it is just the limiting value of one scalar divided by another. However, its value indubitably depends on the *direction* in which the step away from the said point is taken. In general, the value of $\frac{dh}{ds}$ is not the same when the step is taken in *different directions*, such as east and west. The derivative $\frac{dh}{ds}$ is actually called the *directional derivative*. It is a scalar, but it has a directional attribute.

We considered the height of a mountain above the mean sea level in this example, but it would be true for every scalar field, such as the rate $\frac{dT}{ds}$ at which the temperature $T(\vec{r})$ changes from point to point, in a region of space in which various sources and sinks of heat are present. One may have an ice box in a room, a hot iron box at another place, a glass of water on the table and perhaps also hot cup of coffee too. The temperature gradient along which the heat would flow will have complex patterns in this region of space; it will be determined by the directional derivative $\frac{dT}{ds}$ in different directions at each point. This rate would be different *toward* a heat source, compared to the direction *away* from it. It is a direction-dependent scalar ratio. It is therefore not appropriate to define a scalar merely as a quantity that is devoid of any directional attribute. We shall shortly introduce a satisfactory definition of the 'scalar'.

Having found that the definition of a scalar as a quantity that is described by magnitude alone is not satisfactory, we also now have to reconcile with the fact that the definition of a vector as a quantity that has both magnitude and direction is not acceptable. A clockwise (or anticlockwise) rotation of an object about an axis through a finite angle, such as 90°, has both the magnitude (90°, in this illustration), and direction (clockwise, or anticlockwise). Despite these two properties, the rotation through a finite angle is not a vector. Consider, for example, the rotation of an object, such as a blackboard duster, about two orthogonal axes through its center (Fig. 2.4b). Let us attempt to represent the first of these rotations by a vector $\vec{A}$, and the second by a vector $\vec{B}$. The two operations of rotations do not, however, commute since the result of the two operations depends on the order in which they are performed. We, however, expect vector addition to commute. We could assign magnitude and direction to the finite rotations of the duster, but non-commutation of these operations makes them unsuitable to be accepted as vectors. Thus, we cannot regard every quantity that has a magnitude and direction as a vector. Definitions of a scalar and a vector must be free from such ambiguities.

Fig. 2.4a The slope of the mountain at any point on it is not the same in every direction. It is a *scalar*, called the *directional* derivative.

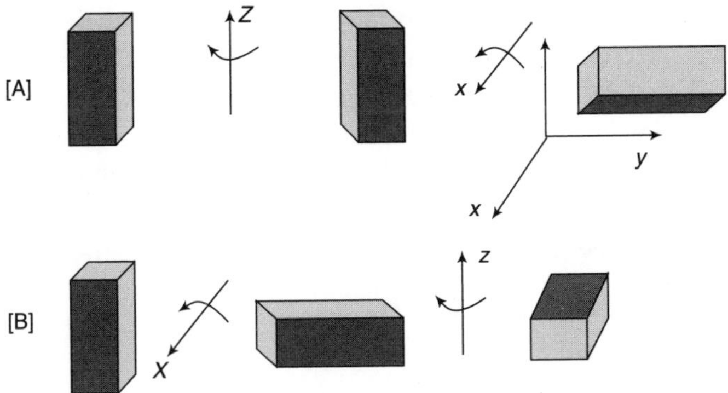

Fig. 2.4b In [A] and [B], rotations of a blackboard duster are carried out about the Z and the X-axis, but in opposite order. Finite rotations do NOT commute, unlike the law of vector addition, even if the rotation has direction, and also magnitude.

To obtain an acceptable definition of a vector, we must study how a vector transforms under the rotation of a coordinate frame of reference in which it is observed. Let us consider the Cartesian representation of a vector (Fig. 2.5) in two alternative basis sets, $(\hat{e}_x, \hat{e}_y, \hat{e}_z)$ and $(\hat{e}'_x, \hat{e}'_y, \hat{e}'_z)$, both orthonormal:

Thus, in the frame $(\hat{e}_x, \hat{e}_y, \hat{e}_z)$, $\vec{V} = V_x\,\hat{e}_x + V_y\,\hat{e}_y + V_z\,\hat{e}_z$. (2.1a)

In the coordinate frame $(\hat{e}'_x, \hat{e}'_y, \hat{e}'_z)$, which is rotated with respect to the previous, the same vector is written in terms of its components as

$$\vec{V} = V'_x\,\hat{e}'_x + V'_y\,\hat{e}'_y + V'_z\,\hat{e}'_z. \tag{2.1b}$$

This can be easily generalized to N > 3 dimensions.

The vector $\vec{V}$ has a simple decomposition in the basis $(\hat{e}_x, \hat{e}_y, \hat{e}_z)$ and also in $(\hat{e}'_x, \hat{e}'_y, \hat{e}'_z)$. Its components along the alternate base vectors are of course different. Both the sets of components (V_x, V_y, V_z) and (V'_x, V'_y, V'_z), contain complete information about the vector they represent, and hence they must be easily obtainable from each other. We may use a different notation, such as $\begin{bmatrix} V_x \\ V_y \\ V_z \end{bmatrix}$, to write the components of the vector. This form is a column matrix. It consists of 1 column and 3 rows. One can also use a slightly different notation, obtained by spreading out the elements of the previous matrix in a row, as $\begin{bmatrix} V_x & V_y & V_z \end{bmatrix}$. It is a row matrix. It consists of 3 columns and 1 row. In the basis $(\hat{e}'_x, \hat{e}'_y, \hat{e}'_z)$, the column matrix representation of the same vector is $\begin{bmatrix} V'_x \\ V'_y \\ V'_z \end{bmatrix}$, and the row matrix representation is $\begin{bmatrix} V'_x & V'_y & V'_z \end{bmatrix}$.

Matrices[1] provide us with an extremely powerful tool to organize information about the components of a vector. Matrices can represent a vector, as we have seen. They also help us provide the very *definition* of a vector. This is obviously important, since we have seen from our discussion on Fig. 2.4b that defining a vector as a physical quantity by its magnitude and direction is unsatisfactory. Using matrices, we shall arrive at a satisfactory and precise definition of a vector. The geometrical connections between the components (Eq. 2.1) of the vector can be easily obtained from Fig. 2.5, and written as

$$V_x' = V_x(\hat{e}_x' \cdot \hat{e}_x) + V_y(\hat{e}_x' \cdot \hat{e}_y) + V_z(\hat{e}_x' \cdot \hat{e}_z), \tag{2.2a}$$

$$V_y' = V_x(\hat{e}_y' \cdot \hat{e}_x) + V_y(\hat{e}_y' \cdot \hat{e}_y) + V_z(\hat{e}_y' \cdot \hat{e}_z), \tag{2.2b}$$

and $$V_z' = V_x(\hat{e}_z' \cdot \hat{e}_x) + V_y(\hat{e}_z' \cdot \hat{e}_y) + V_z(\hat{e}_z' \cdot \hat{e}_z). \tag{2.2c}$$

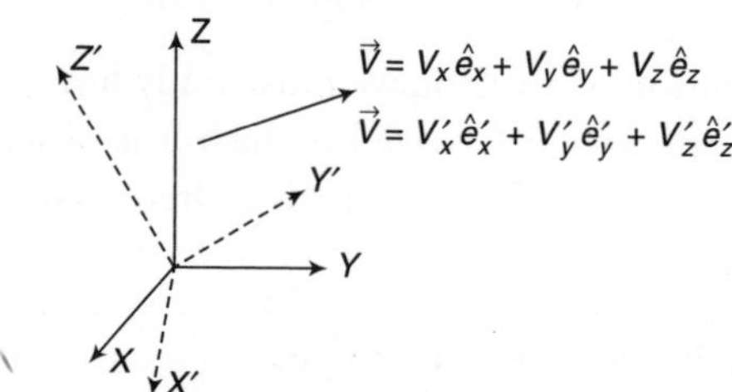

Fig. 2.5 We examine how the same vector $\vec{V}$ is represented in two orthonormal basis sets, one rotated with respect to the other.

The three linear equations, Eq. 2.2a, b, and c, are best written in a compact matrix form:

$$\begin{bmatrix} V_x' \\ V_y' \\ V_z' \end{bmatrix} = \begin{bmatrix} \hat{e}_x' \cdot \hat{e}_x & \hat{e}_x' \cdot \hat{e}_y & \hat{e}_x' \cdot \hat{e}_z \\ \hat{e}_y' \cdot \hat{e}_x & \hat{e}_y' \cdot \hat{e}_y & \hat{e}_y' \cdot \hat{e}_z \\ \hat{e}_z' \cdot \hat{e}_x & \hat{e}_z' \cdot \hat{e}_y & \hat{e}_z' \cdot \hat{e}_z \end{bmatrix} \begin{bmatrix} V_x \\ V_y \\ V_z \end{bmatrix}. \tag{2.3}$$

We see that the elements of the transformation matrix are essentially cosine functions of the angles between the pairs of the two orthonormal basis sets, $(\hat{e}_x, \hat{e}_y, \hat{e}_z)$ and $(\hat{e}_x', \hat{e}_y', \hat{e}_z')$. The cosine-functions which go into the above matrix of transformation *uniquely* tell us how the *components* of a vector transform, when one rotates the coordinate system to which the components are referenced. Unlike the high-school definition of a vector which landed us into inconsistency, the cosine-law expressed in Eq. 2.3 provides us with an unambiguous

1 I refer uninitiated readers to the appendix, placed at the end of this chapter, for some salient features of matrix algebra. It includes a brief discussion on matrix multiplication, obtaining the inverse of a matrix, and solving matrix eigenvalue equations. Some of these details will be useful for appreciating the matrix methods used here, and have been placed in the appendix. This arrangement allows the reader to continue with the rest of this chapter without digression. She/he may refer to the appendix if and when necessary.

signature, a defining criterion, and therefore the very definition, of a vector. We can now safely *define* a vector as a quantity whose components transform under the rotation of a coordinate system according to the cosine-law given in Eq. 2.3. Likewise, a scalar is *defined* as a quantity that remains *invariant* under the rotation of the coordinate system. You can thus speak about the temperature at a point P in a region having arbitrary distribution of heat sources with reference to any coordinate system, regardless of how it is oriented or turned. The components of a vector would change under rotation of a coordinate system, but the value of a scalar does not.

In fact, both scalars and vectors are special cases of a larger family of mathematical entities, called tensors. Tensors are extremely important in physics, since all physical objects have a property which is described by a certain number (one or more) of components which *together* constitute a tensor. In an n-dimensional space, a tensor of rank k is specified by n^k parameters, known as the components of the tensor. The rank k can be any of the natural numbers (such as 0, 1, 2, 3, ...). A scalar is a tensor of rank 0, and thus has a single component. In the usual 3-dimensional space R^3, a vector is a tensor of rank 1, and thus it has 3 components, which transform according to the cosine law which we discussed above. Later, we shall also come across pseudo-scalars and pseudo-vectors, which are special cases of *pseudo-tensors*.

Now, you can rotate a coordinate system, or reverse one or two of its axes, or even invert them all. When you do any of these operations, you get a new coordinate system. The original vector can again be written as a linear superposition of the base vectors in the new coordinate system, but the *components* of the vector may transform differently under different operations.

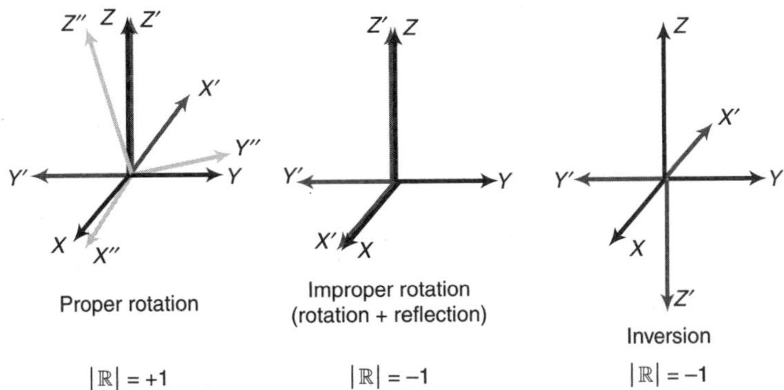

Fig. 2.6 The right-hand screw rule, which gives $\hat{e}_x \times \hat{e}_y = \hat{e}_z$, is preserved under *proper* rotation, but not under *improper* rotation, or inversion. Can you see it from this figure? See text (specially Eq. 2.4 and 2.5) to appreciate the significance of the determinant of the matrix (see Eq. 2.4), $|\mathbb{R}| = \pm 1$.

No matter how the coordinate axes (Fig. 2.6) are rotated *or* flipped, or rotated *and* flipped, the components of a position vector (x', y', z') in the new basis are related to the original components (x, y, z) of that vector in the first system. This relation is always expressible as a matrix equation of the following form:

$$\begin{bmatrix} x' \\ y' \\ z' \end{bmatrix} = \begin{bmatrix} R_{xx} & R_{xy} & R_{xz} \\ R_{yx} & R_{yy} & R_{yz} \\ R_{zx} & R_{zy} & R_{zz} \end{bmatrix} \begin{bmatrix} x \\ y \\ z \end{bmatrix}. \tag{2.4}$$

We shall denote the 3 x 3 matrix in Eq. 2.4 by $\mathbb{R}$. In general, this transformation matrix is an *orthogonal* matrix which preserves the *length* of the vector. It leaves the inner-product (scalar product) of the vector with itself invariant. This property guarantees that the determinant of the transformation matrix is:

$$|\mathbb{R}| = \begin{vmatrix} R_{xx} & R_{xy} & R_{xz} \\ R_{yx} & R_{yy} & R_{yz} \\ R_{zx} & R_{zy} & R_{zz} \end{vmatrix} = \pm 1. \tag{2.5}$$

If the transformation of the coordinate system involves *only* a rotation of the original coordinate system, about *any* arbitrary axis (or rotation plus a displacement of the origin), then the determinant of the matrix of the transformation is $|\mathbb{R}| = +1$. If the transformation involves a flip inversion of one axis, or of *all* the three axes, then the determinant $|\mathbb{R}| = -1$. We note that inversion of *two* coordinate axes amounts to a rotation about the third through 180°, and thus it will have $|\mathbb{R}| = +1$.

The essential point is that rotation is a *different* kind of symmetry operation compared to reflection. All symmetry operations form what is called a *mathematical group*. The group of *all* symmetry transformations which can be represented by orthogonal matrices is called O(3). The determinant of these matrices is $|\mathbb{R}| = \pm 1$. The *special subgroup* for which the determinant of the matrix is +1 is the one corresponding to pure, proper rotations. The group of these transformations is called SO(3). We shall, however, leave the mathematical properties of these groups to an advanced text.

Reflection in a mirror is a symmetry transformation, effected by what is called 'parity'. The image is just like the original object. There is, however, a minor, but significant, difference: the 'left' goes to 'right', and the 'right' to the 'left' (Fig. 2.7, left panel). It is a classic case of improper rotation. Have you ever wondered why the 'top' does not go to the 'bottom' and the 'bottom' to the 'top'? It is only because in the first case you may be thinking about reflection in the YZ plane, whereas in the second, it would be reflection in the ZX plane, like the reflection of trees in a lake. In the latter, the 'top' does go to the 'bottom' and the 'bottom' to the 'top' (Fig. 2.7, right panel), but the 'left' and the 'right' are not interchanged. The transformation matrix would be $\begin{pmatrix} 1 & 0 & 0 \\ 0 & 1 & 0 \\ 0 & 0 & -1 \end{pmatrix}$ instead of $\begin{pmatrix} 1 & 0 & 0 \\ 0 & -1 & 0 \\ 0 & 0 & 1 \end{pmatrix}$, but in both cases its determinant would be −1. The reflection symmetry, or symmetry under parity, has another important difference from the rotation symmetry. The rotation symmetry is *continuous*, in the sense that the rotation of a sphere about its (any) diameter leaves its configuration invariant regardless of the angle of rotation, whether infinitesimal or finite. The reflection symmetry, in contrast, is a *discrete* symmetry; it is a one-step phenomenon. There is no intermediate position between an object and its reflection, but an object that is spherically symmetric can undergo infinite symmetric orientations between any two orientations, however, close.

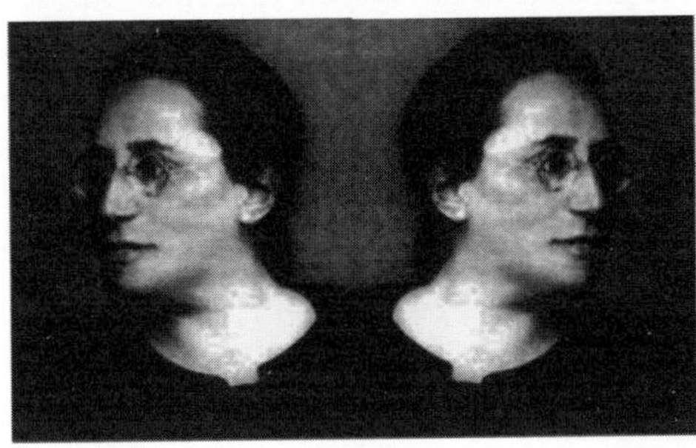

Fig. 2.7 Under reflection symmetry, depending on the plane of reflection, the left and right are interchanged, or the top and the bottom are interchanged under parity transformation. The transformation of matrix in the two cases is different, but in both cases its determinant is −1. The lady in the photograph is Emmy Noether who made crucial contributions to our understanding of the connections between physical laws and symmetry. The picture on the right is the National Geographic award-winning photograph taken by Mr Soumyajit Saha.

The position vector discussed above is the prototype of a class of vectors called the *polar* vectors. Other vectors which transform like the position vectors are velocity, momentum, acceleration, force, electric field, etc. All such vectors which transform like the position vector are called polar vectors. There are other vectors, such as the cross-product of two polar vectors which, however, transform slightly differently. They transform just like the polar vectors under *rotation* of a coordinate system, but *differently* under *reflection* (parity). These vectors are called *pseudo-vectors* (also as *axial vectors*). The angular momentum, $\vec{\ell}$, which is given by the cross product ($\vec{\ell} = \vec{r} \times \vec{p}$) of the position vector and the momentum vector, is a classic example of pseudo-vectors. Fig. 2.8 illustrates this.

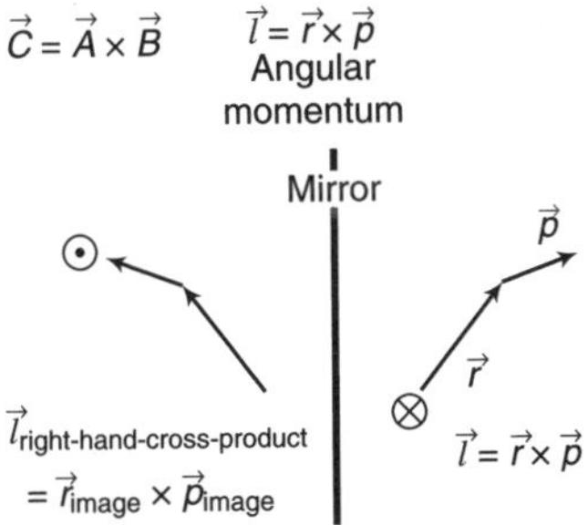

Fig. 2.8 When we use the right hand screw rule to determine the cross-product of two polar vectors, the mirror image of the angular momentum has a direction that is opposite to the cross product of the image of $\vec{r}$ with the image of $\vec{p}$.

opposite to the cross product of the image of $\vec{r}$ with the image of $\vec{p}$.

Under inversion or reflection (i.e. under improper rotations), $\vec{r} \to -\vec{r}, \vec{p} \to -\vec{p}$ but $\vec{\ell} \to \vec{\ell}$, and not $-\vec{\ell}$. The Lorentz force, $\vec{F} = q(\vec{E} + \vec{v} \times \vec{B})$, like any other force, is a polar vector. The term involving the magnetic field in the Lorentz force is a cross product of the polar vector $\vec{v}$ with a pseudo-vector, $\vec{B}$. Thus, the cross-product of two polar vectors is an axial vector, but the cross product of a polar vector with an axial vector is a polar vector. Likewise, the scalar product (dot product) of two polar vectors gives a scalar, but a dot product of a polar vector with an axial vector is a pseudo-scalar. The scalar triple product $\vec{a} \cdot \vec{b} \times \vec{c}$, also called the 'box' product, of three polar vectors is thus a pseudo-scalar. Apart from the angular momentum $\vec{\ell}$ and the magnetic field, $\vec{B}$, torque $(\vec{\tau} = \vec{r} \times \vec{f})$ is another example of a pseudo-vector.

Symmetry under *parity*, along with that under *charge conjugation* and *time reversal*, is one of the cornerstones of what is called the *standard model of physics*. It is referred to as the PCT (or CPT) symmetry.

2.2 EQUATION OF MOTION IN CYLINDRICAL AND SPHERICAL POLAR COORDINATE SYSTEMS

We now move on to other coordinate systems, which are sometimes more elegant than the Cartesian coordinate system. In particular, in this section, we shall discuss the cylindrical polar and the spherical polar coordinate systems. These will be defined with reference to the Cartesian coordinate system. The Cartesian coordinate system is well suited for describing points that are placed at corners of a rectangular parallelepiped. The atoms of sodium and chlorine are so placed in a crystal of common salt (Fig. 2.9). It has the structure of a rectangular room, with a rectangular floor and roof, and all the 'walls' are also rectangular. The position of a particular atom of sodium or chlorine must be specified by its set of *three* coordinates x, y, and z. These stand for the distances in three dimensions from one of its corners, chosen as the origin of a Cartesian coordinate frame of reference.

However, if you want to describe the motion of an object on the surface of a sphere (Fig. 2.9), then we note that the distance of that object from the center of the sphere is always fixed. It is therefore sufficient to describe its position unambiguously by referring only to two variables, one for the 'latitude' and the other for the 'longitude'. Similarly, objects like the DNA or RNA, or the carbon nanotubes, or even monuments like the *Qutub Minar*, have a cylindrical symmetry (Fig. 2.9) which allows one to use yet again only two degrees of freedom instead of three. The two degrees of freedom required in the cylindrical symmetry are, however, different from those of the spherical symmetry. The variables of the cylindrical symmetry are in fact (i) the z-coordinate of the Cartesian system, and (ii) the longitude, same as in the spherical polar coordinate system, since each point on the surface of a cylinder is at a fixed distance from the cylinder's axis.

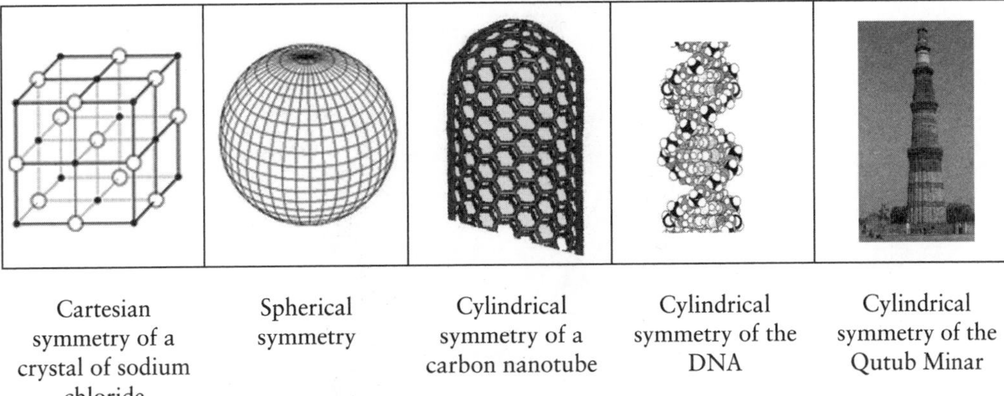

Fig. 2.9 In a Euclidean 3-dimensional space, three degrees of freedom have to be specified to indicate the position of a point. Some symmetries of the object, however, permit us to focus on a smaller number of the degrees of freedom, for example, when one of the three degrees of freedom is constant.

There is nothing fundamentally more 'correct' about using any particular coordinate system, and not another. The mathematical analysis of a problem is, however, often more *elegant* and simple in one coordinate system compared to that in another. It is therefore useful to learn about different commonly used coordinate systems. The reason to celebrate the sun-centered coordinate system of Nicolus Copernicus (1473–1543) to describe our planetary system as a major breakthrough over Ptolemy's earth-centered coordinate system (Fig. 2.10), is not just the fact that the equations of motion become simpler to handle, but also that it reveals the gravitational interaction in an elegant and direct manner. One can always carry out coordinate transformations from one coordinate system to another. If the choice of the coordinate system is not made appropriately, the equations of motion can become complicated. The complexity then gets into the way of gaining insights into the physical parameters of the problem at hand.

There have been important contributions made by some Indian astronomers to the understanding of the heliocentric coordinate frame of reference. In particular, one should mention the works of Aryabhatta (b. 476 CE), Bhaskara I (600 CE), Brahmagupta (591 CE), Vateshwa (880 CE), Manjulacharya (932 CE), Aryabhatta II (950 CE), and Bhaskaracharya II (1114 CE).

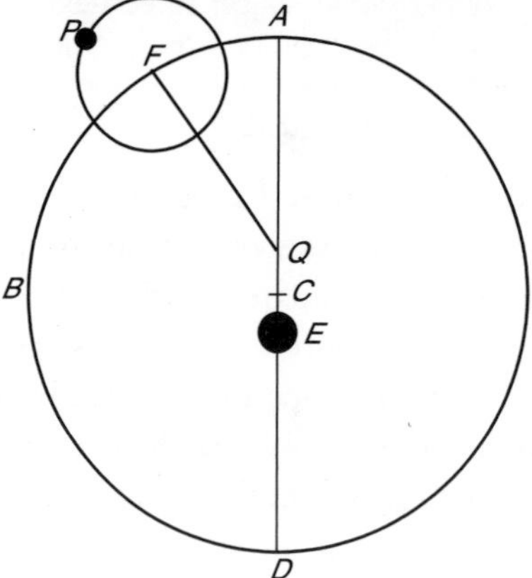

Fig. 2.10 Claudius Ptolemaeus, called Ptolemy (100–170 CE). He worked in the library of Alexandria.

In Ptolemy's coordinate system, the Sun and the planets were considered to move on a small circle (called 'epicycle') whose center would move on a large circle (called 'deferent'). The description of planetary motion becomes extremely complicated.

We have learned that physical quantities are represented by scalars, vectors, and tensors. We now proceed to discuss how the components of a vector are described in the plane polar coordinate system. We begin our consideration with 2-dimensional vectors and then extend it to three-dimensions. On a 2-dimensional surface, we know that the position vector is expressed as a linear superposition of the Cartesian base vectors, $(\hat{e}_x, \hat{e}_y)$:

$$\vec{r} = x\hat{e}_x + y\hat{e}_y. \tag{2.6a}$$

The same vector can also be expressed as a superposition of *any* two linearly independent base vectors, which are easily chosen to be orthogonal to each other. Therefore we can use a pair of base vectors $(\hat{e}_\rho, \hat{e}_\varphi)$ such that $\hat{e}_\rho$ points along the direction of the position vector itself, and $\hat{e}_\varphi$ is orthogonal to it in the direction in which the angle φ, which the position vector $\vec{r}$ makes with the coordinate X-axis, increases in the anticlockwise sense (Fig. 2.11). The angle φ is called the azimuthal angle. The location of the point at a certain distance from the origin, at the tip of the position vector,

$$\vec{r} = \rho\hat{e}_\rho, \tag{2.6b}$$

is uniquely specified by the pair (x, y) of the Cartesian coordinates, and also, equivalently, by the pair (ρ, φ) of the *plane polar coordinates.*

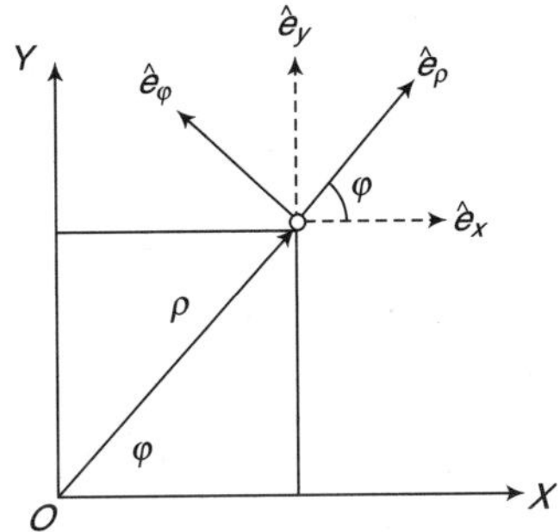

Fig. 2.11 Plane polar coordinates of a point on a 2-dimensional surface. An arbitrary point on this surface can be uniquely represented by the pair (ρ, φ) just as well as by its Cartesian coordinates (x, y).

Note that while the Cartesian coordinates x and y both have the dimensions of length, only the distance ρ of the plane polar coordinates has the dimension of length. The angle φ is of course dimensionless. Thus φ = constant is an infinite line, or, rather, an infinite *half-line*, since it resides strictly in only one quadrant. Its extension in the diagonally opposite quadrant corresponds to the angle $\varphi + \pi$, and not φ.

The Cartesian and the plane polar coordinates are obtainable from each other. The range of values these coordinate have is given by

$$-\infty < x < \infty; -\infty < y < \infty; \text{and } 0 \leq \rho < \infty; 0 \leq \varphi < 2\pi. \tag{2.7a}$$

The relations between these alternative pairs of coordinates is

$$x = \rho\cos\varphi \text{ and } y = \rho\sin\varphi. \tag{2.7b}$$

The inverse relations are $\rho = +\sqrt{x^2 + y^2}$ and $\varphi = \tan^{-1}\left(\dfrac{y}{x}\right)$. (2.7c)

The basis set pair $(\hat{e}_\rho, \hat{e}_\varphi)$ of the plane polar coordinate system is alternative to the basis $(\hat{e}_x, \hat{e}_y)$ of the Cartesian coordinate system. It is, of course, possible to express the basis vectors of one set in terms of the other:

$$\hat{e}_\rho = \cos\varphi\hat{e}_x + \sin\varphi\hat{e}_y; \qquad \hat{e}_\varphi = -\sin\varphi\hat{e}_x + \cos\varphi\hat{e}_y; \tag{2.8a}$$

$$\hat{e}_x = \cos\varphi\hat{e}_\rho - \sin\varphi\hat{e}_\varphi; \qquad \hat{e}_y = \sin\varphi\hat{e}_\rho + \cos\varphi\hat{e}_\varphi. \tag{2.8b}$$

These *linear* relations can be expressed in a compact form as the following matrix equation:

$$\begin{pmatrix} \hat{e}_\rho \\ \hat{e}_\varphi \end{pmatrix} = \begin{pmatrix} \cos\varphi & \sin\varphi \\ -\sin\varphi & \cos\varphi \end{pmatrix} \begin{pmatrix} \hat{e}_x \\ \hat{e}_y \end{pmatrix}. \tag{2.9a}$$

The inverse relations are

$$\begin{pmatrix} \hat{e}_x \\ \hat{e}_y \end{pmatrix} = \begin{pmatrix} \cos\varphi & -\sin\varphi \\ \sin\varphi & \cos\varphi \end{pmatrix} \begin{pmatrix} \hat{e}_\rho \\ \hat{e}_\varphi \end{pmatrix}. \tag{2.9b}$$

The important thing here is to note the fact that the Cartesian base vectors $(\hat{e}_x, \hat{e}_y)$ are constant vectors, regardless of which point in the plane is being described. The plane polar base unit vectors $(\hat{e}_\rho, \hat{e}_\varphi)$, however, need to be judiciously adapted for each point separately in the plane, since they may change from one point to another. The tails of the unit vectors $(\hat{e}_\rho, \hat{e}_\varphi)$ are placed at the point in the plane for which they are being defined. The direction of $\hat{e}_\rho$ is chosen for each point in such a way that it is always directed in the direction of the radial vector away from the origin. The unit vector $\hat{e}_\varphi$ is chosen to be orthogonal to $\hat{e}_\rho$, in the direction in which the azimuthal angle φ increases.

From Fig. 2.11, it is obvious that the unit vectors $(\hat{e}_\rho, \hat{e}_\varphi)$ do not change along a line radially outward from the origin of the coordinate frame. However, if you go along the tangent to a circle centered at the origin, then both $(\hat{e}_\rho, \hat{e}_\varphi)$ change as a function of the azimuthal angle, φ. Thus, if a point '2' is infinitesimally close on the unit circle (Fig. 2.12a) to the point '1', then as one goes from point '1' to '2', the unit vectors change with respect to the azimuthal angle at the following rate:

$$\frac{\partial\hat{e}_\rho}{\partial\varphi} = \lim_{\delta\varphi\to 0}\frac{\hat{e}_\rho(2) - \hat{e}_\rho(1)}{\delta\varphi} = \hat{e}_\varphi \quad \text{and} \quad \lim_{\delta\varphi\to 0}\frac{\hat{e}_\varphi(2) - \hat{e}_\varphi(1)}{\delta\varphi} = \lim_{\delta\varphi\to 0}\frac{\delta\hat{e}_\varphi}{\delta\varphi} = \frac{\partial\hat{e}_\varphi}{\partial\varphi} = -\hat{e}_\rho; \tag{2.10a}$$

$$\text{also,} \quad \frac{\partial\hat{e}_\rho}{\partial\rho} = \vec{0} \quad \text{and} \quad \frac{\partial\hat{e}_\varphi}{\partial\rho} = \vec{0}. \tag{2.10b}$$

One arrives at essentially the same result by differentiating the polar unit vectors (Eq. 2.8a) with respect to φ, and expressing the Cartesian unit vectors in terms of the plane polar system. Having seen that the polar unit vectors change from point to point (on an arc of a circle centered at the origin) in a plane, it is easy to see that if these unit vectors are *tagged* to a point whose motion is being tracked on the plane, the polar unit vectors will change from time to time, as the particle moves as a function of time.

Thus, the time-dependence of the polar unit vectors is given by:

$$\frac{d\hat{e}_\rho}{dt} = \frac{\partial \hat{e}_\rho}{\partial \varphi}\dot{\varphi} = \hat{e}_\varphi \dot{\varphi} \quad \text{and} \quad \frac{d\hat{e}_\varphi}{dt} = \frac{\partial \hat{e}_\varphi}{\partial \varphi}\dot{\varphi} = -\hat{e}_\rho \dot{\varphi}. \tag{2.11}$$

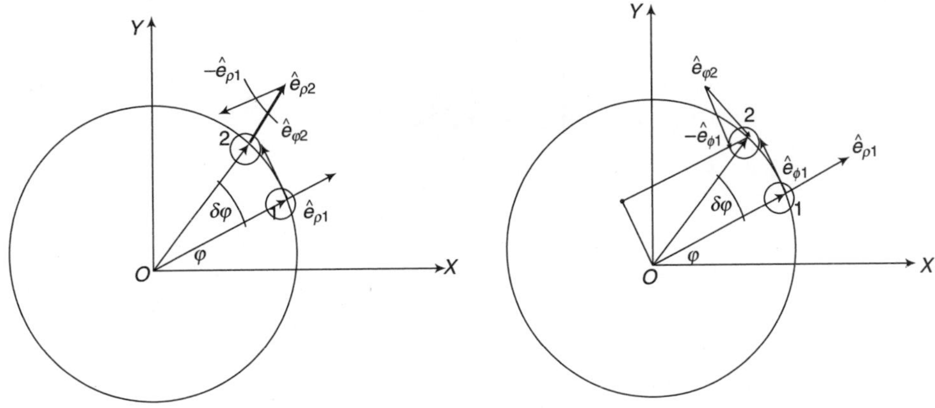

Fig. 2.12a $\frac{\partial \hat{e}_\rho}{\partial \varphi} = \lim_{\delta\varphi \to 0} \frac{\hat{e}_\rho(2) - \hat{e}_\rho(1)}{\delta\varphi} = \hat{e}_\varphi.$ **Fig. 2.12b** $\lim_{\delta\varphi \to 0} \frac{\hat{e}_{\varphi 2} - \hat{e}_{\varphi 1}}{\delta\varphi} = \lim_{\delta\varphi \to 0} \frac{\delta\hat{e}_\varphi}{\delta\varphi} = \frac{\partial \hat{e}_\varphi}{\partial \varphi} = -\hat{e}_\rho.$

The point 2 is considered infinitesimally close to point 1, but the separation is shown exaggerated in this figure to consider how the polar unit vectors change with azimuthal angle.

We can now readily set up the *equation of motion* for a particle using polar coordinates. For this, we find that the expressions for a particle's position vector, velocity and acceleration are respectively given by

$$\vec{r} = \vec{\rho} = \rho\hat{e}_\rho, \tag{2.12a}$$

$$\vec{v} = \dot{\vec{\rho}} = \frac{d\vec{\rho}}{dt} = \frac{d(\rho\hat{e}_\rho)}{dt} = \dot{\rho}\hat{e}_\rho + \rho\frac{d\hat{e}_\rho}{dt} = \dot{\rho}\hat{e}_\rho + \rho\frac{\partial \hat{e}_\rho}{\partial \varphi}\dot{\varphi} = \dot{\rho}\hat{e}_\rho + \rho\dot{\varphi}\hat{e}, \tag{2.12b}$$

$$\text{and } \vec{a} = \frac{d\vec{v}}{dt} = \ddot{\rho}\hat{e}_\rho + \dot{\rho}\frac{d\hat{e}_\rho}{dt} + \dot{\rho}\dot{\varphi}\hat{e}_\varphi + \rho\ddot{\varphi}\hat{e}_\varphi + \rho\dot{\varphi}\frac{d\hat{e}_\varphi}{dt} = (\ddot{\rho} - \rho\dot{\varphi}^2)\hat{e}_\rho + (2\dot{\rho}\dot{\varphi} + \rho\ddot{\varphi})\hat{e}_\varphi. \tag{2.12c}$$

The cylindrical polar coordinate system is a straightforward extension of the plane polar coordinate system to include the third Euclidean dimension. It is shown in Fig. 2.13. Only the Cartesian Z-axis is added through the origin to the plane polar coordinates. Since it is orthogonal to the XY plane containing the (ρ, φ) coordinates, the 3 coordinates $(\rho, \varphi. z)$ provide us a convenient way of describing points in 3-dimensional space, when we wish to exploit cylindrical symmetry. Vectors $z\hat{e}_z$, $\dot{z}\hat{e}_z$ and $\ddot{z}\hat{e}_z$ respectively get added to the expressions for the position vector, velocity and acceleration of a point. In the plane polar coordinate system, the azimuthal angle φ was defined with respect to the X-axis, but here it is defined with respect to the ZX plane.

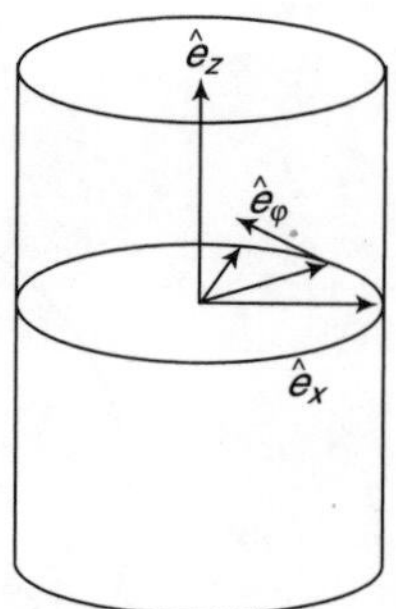

Fig. 2.13 The cylindrical polar coordinate system adds the Z-axis to the plane polar coordinate system. Corresponding terms are then added to the expressions for velocity and acceleration.

Often, the spherical symmetry is more useful than the cylindrical. In fact, this is a common symmetry observed in nature. Several objects in the physical world possess it. In the spherical polar coordinate system, a point in space is denoted by the three coordinates (ρ, θ, φ) shown in Fig. 2.14. The coordinate r is merely the distance of a point from the origin; the angle θ is measured with respect to the Z-axis as shown in Fig. 2.14. The Z-axis is therefore normally termed as the 'polar axis' and the angle θ as the 'polar angle'. The azimuthal angle φ (Eq. 2.13a, b) is defined with respect to the ZX plane, just as in the case of the cylindrical polar coordinates.

Equations 2.13 a,b describe how the Cartesian coordinates are related to the spherical polar coordinates:

$$r = \sqrt{x^2 + y^2 + z^2}; \quad \theta = \tan^{-1}\frac{\sqrt{x^2 + y^2}}{z}; \quad \varphi = \tan^{-1}\frac{y}{x}. \tag{2.13a}$$

The inverse relations are

$$x = r\sin\theta\cos\varphi; \quad y = r\sin\theta\sin\varphi; \quad z = r\cos\theta. \tag{2.13b}$$

The range of the spherical polar coordinates is given in Eq. 2.14.

$$0 \leq r < \infty; \quad 0 \leq \theta \leq \pi; \text{ and } 0 \leq \varphi < 2\pi. \tag{2.14}$$

Similar to Eq. 2.9, the representation of the unit vectors in spherical polar coordinate system (Fig. 2.14) can be expressed in terms of the Cartesian base vectors, and vice versa. Again, as in the case of cylindrical polar coordinate system, *tails* of the basis set unit vectors $\{\hat{e}_r, \hat{e}_\theta, \hat{e}_\varphi\}$ of the spherical polar coordinate system are tagged to each point in space. Their directions are chosen such that $\hat{e}_r$ is in the direction of increasing r, the direction of $\hat{e}_\theta$ is in the direction of the increasing polar angle θ, and finally the direction of $\hat{e}_\varphi$ is in the direction in which the azimuthal angle fi (with respect to the ZX plane) increases.

The linear transformations between unit vectors of the spherical polar coordinates system and the Cartesian coordinates system can be expressed as a simple matrix equation:

$$\begin{bmatrix} \hat{e}_r \\ \hat{e}_\theta \\ \hat{e}_\varphi \end{bmatrix} = \begin{bmatrix} \sin\theta\cos\varphi & \sin\theta\sin\varphi & \cos\theta \\ \cos\theta\cos\varphi & \cos\theta\sin\varphi & -\sin\theta \\ -\sin\varphi & \cos\varphi & 0 \end{bmatrix} \begin{bmatrix} \hat{e}_x \\ \hat{e}_y \\ \hat{e}_z \end{bmatrix}, \tag{2.15a}$$

with the inverse relations given by

$$\begin{bmatrix} \hat{e}_x \\ \hat{e}_y \\ \hat{e}_z \end{bmatrix} = \begin{bmatrix} \sin\theta\cos\varphi & \cos\theta\cos\varphi & -\sin\varphi \\ \sin\theta\sin\varphi & \cos\theta\sin\varphi & \cos\varphi \\ \cos\theta & -\sin\theta & 0 \end{bmatrix} \begin{bmatrix} \hat{e}_r \\ \hat{e}_\theta \\ \hat{e}_\varphi \end{bmatrix}. \tag{2.15b}$$

Similar to the case of plane polar and cylindrical polar coordinate systems, if the unit vectors of the spherical polar coordinate system are tagged to a moving particle in space, then in general they would change from point to point. The spherical polar unit vectors will change with respect to the spherical polar coordinates as can be seen easily by taking partial derivatives of the unit vectors in Eq. 2.15 a with respect to the individual polar coordinates. This would give a result that is a superposition of the Cartesian unit vectors, but then, using the inverse relations (Eq. 2.15b) they can all be expressed in terms of the spherical polar unit vectors. The results are summarized in the nine equations compiled in Eq. 2.16.

$$\left\{\begin{matrix} \dfrac{\partial\hat{e}_r}{\partial r} = \vec{0} \\ \dfrac{\partial\hat{e}_r}{\partial\theta} = \hat{e}_\theta \\ \dfrac{\partial\hat{e}_r}{\partial\varphi} = \sin\theta\hat{e}_\varphi \end{matrix}\right\}, \left\{\begin{matrix} \dfrac{\partial\hat{e}_\theta}{\partial r} = \vec{0} \\ \dfrac{\partial\hat{e}_\theta}{\partial\theta} = -\hat{e}_r \\ \dfrac{\partial\hat{e}_\theta}{\partial\varphi} = \cos\theta\hat{e}_\varphi \end{matrix}\right\}, \left\{\begin{matrix} \dfrac{\partial\hat{e}_\varphi}{\partial r} = \vec{0} \\ \dfrac{\partial\hat{e}_\varphi}{\partial\theta} = \vec{0} \\ \dfrac{\partial\hat{e}_\varphi}{\partial\varphi} = -\cos\theta\hat{e}_\theta - \sin\theta\hat{e}_r \end{matrix}\right\}. \tag{2.16}$$

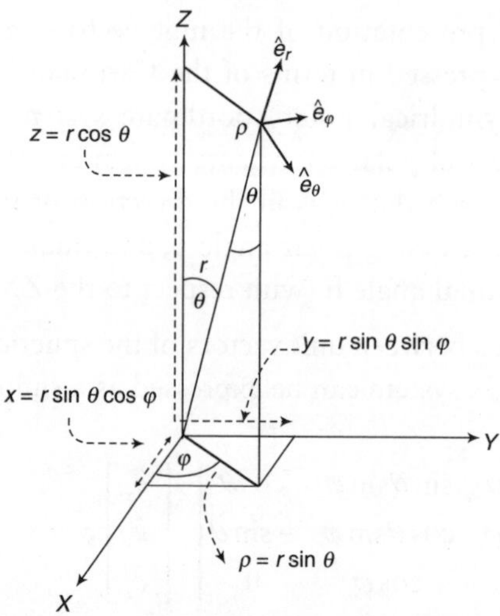

The spherical polar coordinate system showing the relationship between the Cartesian coordinates (x, y, z) and the spherical polar coordinates (r, θ, φ).

Fig. 2.14 A point in the 3-dimensional space is at the intersection of three surfaces:

(i) 'r = constant' sphere,

(ii) 'θ = constant' cone, whose vertex is at the origin, and

(iii) 'φ = constant' plane, through the Z axis. The coordinate ρ of the cylindrical polar coordinate system is also shown and used to display the interrelationships of these coordinates.

The above results can also be obtained using *geometry* to determine the partial derivatives. To do so, consider the Fig. 2.15a, which shows how the unit vector $\hat{e}_r$ changes for points spaced at an infinitesimal distance apart when only the polar angle θ changes. Likewise, Fig. 2.15b shows how the unit vector $\hat{e}_r$ changes for points spaced at an infinitesimal distance apart when only the azimuthal angle φ changes. From these figures, we can determine the partial derivatives of the spherical polar unit vectors with respect to the polar coordinates, and once again obtain Eq. 2.16, using geometry.

The position vector of a point, which in the Cartesian coordinate system is $\vec{r} = x\hat{e}_x + y\hat{e}_y + z\hat{e}_z$, is given in the spherical polar coordinates, simply by $\vec{r} = r\hat{e}_r$.

An infinitesimal volume element in Cartesian coordinate system is $dV = dxdydz$. In the spherical polar coordinate system, one can see from Fig. 2.16 that a similar infinitesimal volume element would be given by $dV = (dr)(rd\theta)(r\sin\theta d\varphi) = r^2\sin\theta\, dr\, d\theta\, d\varphi$. Volume integration would therefore be carried out over the three spherical polar coordinates, over their respective range of integration, as would be appropriate.

We can now set up the equation of motion in spherical polar coordinate system easily. The expressions for the position vector, $\vec{r}$, for an infinitesimal displacement $\overrightarrow{dr}$, for velocity, and for acceleration, are readily obtained and compiled in Eq. 2.17.

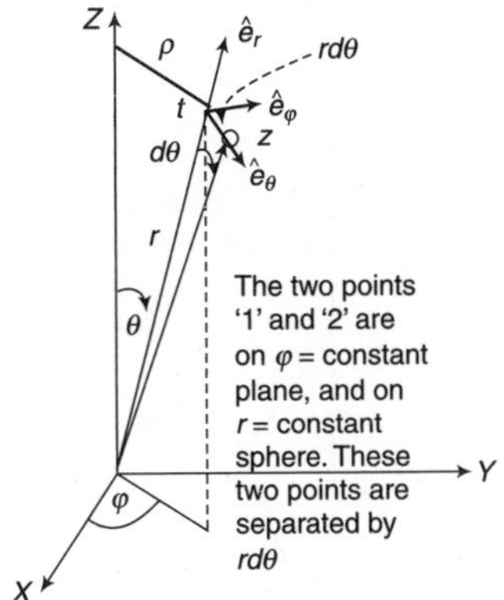

Fig. 2.15a From this figure, we see that $\frac{\partial \hat{e}_r}{\partial \theta} = \lim_{\delta\theta \to 0} \frac{\delta \hat{e}_r}{\delta \theta} = \lim_{\delta\theta \to 0} \frac{\hat{e}_r(2) - \hat{e}_r(1)}{\delta \theta} = \hat{e}_\theta$, in accordance with Eq. 2.16.

Position vector $\vec{r} = r\hat{e}_r$, (2.17a)

Infinitesimal displacement:

$$\overrightarrow{\delta r} = \delta r \hat{e}_r + r \frac{\partial \hat{e}_r}{\partial \theta} \delta\theta + r \frac{\partial \hat{e}_r}{\partial \varphi} \delta\varphi = \delta r \hat{e}_r + r\delta\theta \hat{e}_\theta + r \sin\theta \delta\varphi \hat{e}_\varphi, \tag{2.17b}$$

Velocity: $\vec{v} = \dot{r}\hat{e}_r + r\dot{\theta}\hat{e}_\theta + r\sin\theta\dot{\varphi}\hat{e}_\varphi$. (2.17c)

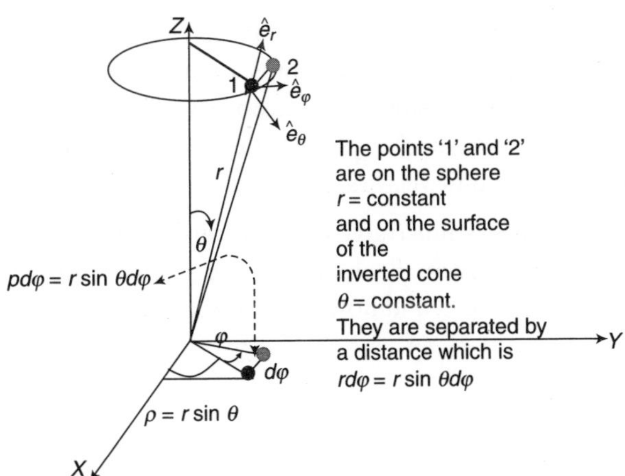

Fig. 2.15b From this figure, we see that $\frac{\partial \hat{e}_r}{\partial \varphi} = \lim_{\delta\theta \to 0} \frac{\delta \hat{e}_r}{\delta \varphi} = \lim_{\delta\theta \to 0} \frac{\hat{e}_r(2) - \hat{e}_r(1)}{\delta \varphi} = \sin\theta \hat{e}_\varphi$, in accordance with Eq. 2.16.

$$\text{Acceleration: } \vec{a} = \frac{d\vec{v}}{dt} = (\ddot{r} - r\dot{\theta}^2 - r\sin^2\theta\dot{\varphi}^2)\hat{e}_r + \tag{2.17d}$$
$$(2\dot{r}\dot{\theta} - r\sin\theta\cos\theta\dot{\varphi}^2 + r\ddot{\theta})\hat{e}_\theta + (2\dot{r}\dot{\varphi}\sin\theta + 2r\dot{\varphi}\dot{\theta}\cos\theta + r\sin\theta\ddot{\varphi})\hat{e}_\varphi$$

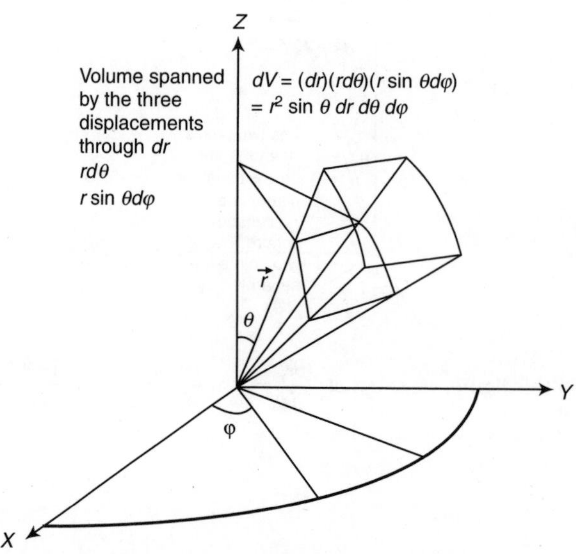

Fig. 2.16 Volume element expressed in spherical polar coordinates.

2.3 CURVILINEAR COORDINATE SYSTEMS

The basis set unit vectors for the Cartesian, Cylindrical Polar, and the Spherical Polar Coordinate systems employed in Section 2.2 are, respectively, $\{\hat{e}_x, \hat{e}_y, \hat{e}_z\}$, $\{\hat{e}_\rho, \hat{e}_\varphi, \hat{e}_z\}$ and $\{\hat{e}_r, \hat{e}_\theta, \hat{e}_\varphi\}$. These three coordinate systems are special, and most common, illustrations of a general class of coordinate systems in the 3-dimensional Euclidean space, known as *curvilinear* coordinate systems. A general curvilinear coordinate system is defined by a set of coordinates (q_1, q_2, q_3) and its basis set is described by unit vectors, $(\hat{e}_1, \hat{e}_2, \hat{e}_3)$.

The following table manifestly illustrates the relationships of various coordinate systems to the Cartesian coordinate system. Each of these coordinate systems is defined by its relationship to the Cartesian system, and how its basis set of unit vectors $(\hat{e}_1, \hat{e}_2, \hat{e}_3)$ are related to the Cartesian unit vectors, $(\hat{e}_x, \hat{e}_y, \hat{e}_z)$.

General Curvilinear Coordinates	Circular Cylindrical Polar Coordinates	Spherical Polar Coordinates
q_1, q_2, q_3	ρ, φ, z	r, θ, φ
$x = x(q_1, q_2, q_3)$ $y = y(q_1, q_2, q_3)$ $z = z(q_1, q_2, q_3)$	$-\infty < x = \rho\cos\varphi < \infty$ $-\infty < y = \rho\sin\varphi < \infty$ $-\infty < z = z < \infty$	$-\infty < x = r\sin\theta\cos\varphi < \infty$ $-\infty < y = r\sin\theta\sin\varphi < \infty$ $-\infty < z = r\cos\theta < \infty$
Unit Basis		dimension of each q_i $\left\{\begin{matrix} = \\ or \\ \neq \end{matrix}\right\}$ L
$\{\hat{e}_1, \hat{e}_2, \hat{e}_3\}$	$\{\hat{e}_\rho, \hat{e}_\varphi, \hat{e}_z\}$	$\{\hat{e}_r, \hat{e}_\theta, \hat{e}_\varphi\}$

Let us for consider the relationship between the Cartesian coordinates and those of the curvilinear coordinates:

$$x = x(q_1, q_2, q_3);\ y = y(q_1, q_2, q_3);\ z = z(q_1, q_2, q_3). \tag{2.18}$$

Using chain rule, we express differential increments in the Cartesian coordinates in terms of differential increments in the curvilinear coordinates as follows:

$$dx = \sum_{i=1}^{3} \frac{\partial x}{\partial q_i} dq_i; \qquad dy = \sum_{i=1}^{3} \frac{\partial y}{\partial q_i} dq_i; \qquad dz = \sum_{i=1}^{3} \frac{\partial z}{\partial q_i} dq_i. \tag{2.19}$$

A fundamental property of geometry of space is inspired by how the distance between two points in this space is measured. In Greek, the word *metron* means a *measure*, so the property that defines how distance between two points is measured is often referred as the 'metric'. In the 3-dimensional Euclidean space, the metric of distance ds between two points placed at an infinitesimal distance from each other is best expressed in terms of its square, given by the scalar product of the position vector $\overrightarrow{dr}$ of the second point (with respect to the first) with itself.

Thus, $ds^2 = |\overrightarrow{dr}|^2 = \overrightarrow{dr} \bullet \overrightarrow{dr} = dx^2 + dy^2 + dz^2,$ (2.20a)

since we have $\overrightarrow{dr} = dx\hat{e}_x + dy\hat{e}_y + dz\hat{e}_z.$ (2.20b)

The distance itself is just the positive square root of Eq. 2.20a. A slight modification in the notation will make it easy for us to generalize these ideas not only to curvilinear coordinates, but also to the geometry of higher-dimensional spaces, including the Non-Euclidean space of spacetime continuum which we shall meet in Chapters 13 and 14. Toward this goal, we set the equivalence $(x, y, z) \equiv (x_1, y_2, z_3)$ and denote the Cartesian coordinates by the same letter, x but subscripted by three different numerals, 1, 2 and 3, instead of the alphabets x, y and z. We thus write the summation on the right hand side of Eq. 2.20a as a double sum,

$$ds^2 = \sum_{i=1}^{3} \sum_{j=1}^{3} g_{ij} dx_i dx_j, \tag{2.21a}$$

where $g_{ij} = \delta_{ij}$, being the Kronecker-δ, (2.21b)

whose value is equal to the number 1 when the subscripts i and j are equal, and zero otherwise.

The orthogonal curvilinear coordinates are readily obtained now from a generalization of Eq. 2.21 by allowing the elements to take various *convenient* values, explained below. Thus, we have

$$ds^2 = \sum_{i=1}^{3}\sum_{j=1}^{3} g_{ij}dq_i dq_j, \tag{2.22}$$

where g_{ij} $\forall i, j = 1, 3$ are elements of the matrix

$$\mathbf{g} = \left[\, g_{ij} \,\right] = \begin{bmatrix} g_{11} & g_{12} & g_{13} \\ g_{21} & g_{22} & g_{23} \\ g_{31} & g_{32} & g_{33} \end{bmatrix}. \tag{2.23}$$

The choice $g_{ij} = \delta_{ij}$ (the Kronecker delta of Eq. 2.21b) gives

$$\mathbf{g} = g_{ij} = \begin{bmatrix} 1 & 0 & 0 \\ 0 & 1 & 0 \\ 0 & 0 & 1 \end{bmatrix}. \tag{2.24}$$

We see that a subset of Eq. 2.22 for two dimensions, which restricts the upper limit in both the double sums on the right hand side to 2 instead of 3, recovers for us the Pythagoras theorem in Euclidean geometry, with $g_{ij} = \delta_{ij}$. The value g_{ij} $\forall i, j$ provides what is called the *signature* of the metric $\mathbf{g}$.

We now *define Orthogonal Curvilinear Coordinates* by making the following choice for the elements of the $\mathbf{g}$ matrix.

$g_{ij} = 0$ for $i \neq j$, (2.25a)

and $g_{ij} = h_i^2$ for $i = j$, (2.25b)

where the three values of h_i, for $i = 1, 2, 3$ are called *scale factors*. On account of the property stated in Eq. 2.25a, such a coordinate system is called orthogonal curvilinear coordinate system. The use of the Kronecker delta as a special case of Eq. 2.25 then defines the Cartesian coordinate system.

We immediately recognize that $ds^2 = \sum_{i=1}^{3} (h_i dq_i)^2$, (2.26a)

and we can therefore write $ds_i = (h_i dq_i)$ $\forall$ $i = 1, 2, 3$. (2.26b)

The differential increment (displacement) vector becomes $\overrightarrow{dr} = \sum_{i=1}^{3} h_i dq_i \hat{e}_i$. (2.26c)

Equation 2.26 explains why the factors h_i in $ds_i = (h_i dq_i)$ $\forall i = 1, 2, 3$ are called 'scale factors'. The increment dq_i is essentially *scaled* by the factor h_i toward generating a *measure* of the distance ds_i. From Eq. 2.26c, it also follows that

$$h_i = \left|\frac{\partial \vec{r}}{\partial q_i}\right| = +\sqrt{\frac{\partial \vec{r}}{\partial q_i} \bullet \frac{\partial \vec{r}}{\partial q_i}} \quad \forall\ i = 1, 2, 3. \tag{2.26d}$$

Replacing the differential increment displacement vector $\overrightarrow{dr}$ of Eq. 2.20b by its new representation as per Eq. 2.26c, we admit the possibility that the dimension of the curvilinear coordinate q_i may, or may not, be L. It is the *product* of the dimension of the curvilinear coordinate q with that of the scale factor h which must have the dimension L. We place a physical quantity in a rectangular bracket to denote its dimension, and recognize that

$$[h_i dq_i] = L \text{ and } [h_i] = L[q_i]^{-1}. \tag{2.27}$$

The spherical polar coordinates system can now immediately be recognized as a special case of the curvilinear coordinate system by establishing the equivalence, $(q_1, q_2, q_3) \leftrightarrow (r, \theta, \varphi)$. Mapping the differential increment vector of the spherical polar coordinate system,

$$\overrightarrow{dr} = dr\hat{e}_r + rd\theta\hat{e}_\theta + r\sin\theta\, d\varphi\hat{e}_\varphi, \tag{2.28a}$$

to the new notation $\overrightarrow{dr} = \sum_{i=1}^{3} h_i dq_i \hat{e}_i,$ (2.28b)

we recognize that the three scale factors of the spherical polar coordinate system are, respectively identified as $(h_1, h_2, h_3) \leftrightarrow (1, r, r \sin \theta)$.

Using now Eq. 2.25, we get the matrix for the g-metric for the spherical coordinate system:

$$\mathbf{g} \text{ matrix} = \left[g_{ij} \right] = \begin{bmatrix} g_{11} & g_{12} & g_{13} \\ g_{21} & g_{22} & g_{23} \\ g_{31} & g_{32} & g_{33} \end{bmatrix} = \begin{bmatrix} 1 & 0 & 0 \\ 0 & r^2 & 0 \\ 0 & 0 & r^2 \sin^2\theta \end{bmatrix}. \tag{2.29}$$

It is now straight forward to verify that the measure of distance using this coordinate system is

$$ds^2 = dr^2 + r^2 (d\theta)^2 + r^2 \sin^2\theta\, (d\varphi)^2 = \sum_{i=1}^{3}\sum_{j=1}^{3} g_{ij} dq_i dq_j, \tag{2.30}$$

in accordance with Eq. 2.22.

In another set of curvilinear coordinates, called the *parabolic coordinate system*, the three degrees of freedom in Euclidean space are described by the parameters, (u, v, φ). The Cartesian coordinates (x, y, z) and the parabolic coordinates (u, v, φ) are specified in terms of each other as follows:

$$x = \sqrt{uv}\cos\varphi,\ y = \sqrt{uv}\sin\varphi,\ z = \frac{1}{2}(u - v); \tag{2.31a}$$

$$u = r + z,\ v = r - z \text{ and } \varphi = \tan^{-1} = \left(\frac{y}{x}\right). \tag{2.31b}$$

Thus,

$$\vec{r} = x\hat{e}_x + y\hat{e}_y + z\hat{e}_z = \sqrt{uv}\cos\varphi\,\hat{e}_x + \sqrt{uv}\sin\varphi\,\hat{e}_y + \frac{1}{2}(u - v)\,\hat{e}_z. \tag{2.31c}$$

Using Eq. 2.26d, the scale factors are

$$h_1 = \left|\frac{\partial \vec{r}}{\partial u}\right| = \frac{1}{2}\left(\frac{u+v}{u}\right)^{1/2}; h_2 = \left|\frac{\partial \vec{r}}{\partial u}\right| = \frac{1}{2}\left(\frac{u+v}{u}\right)^{1/2}; h_3 = \sqrt{uv}. \tag{2.32}$$

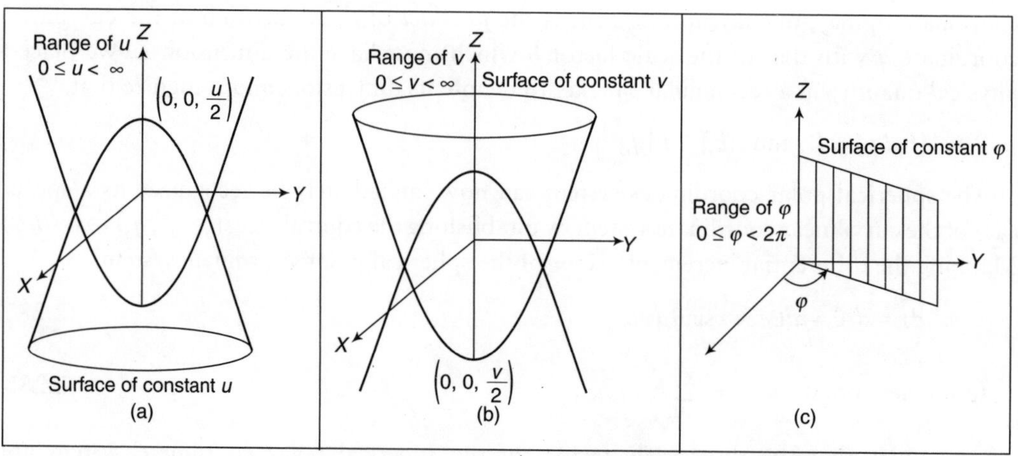

Fig. 2.17 The surfaces of constants u, v, φ in parabolic coordinates are respectively given in Figs. 2.17 (a), 2.17 (b) and 2.17 (c). The range of u, and v, is $0 \to \infty$, and that of φ is $0 \to 2\pi$.

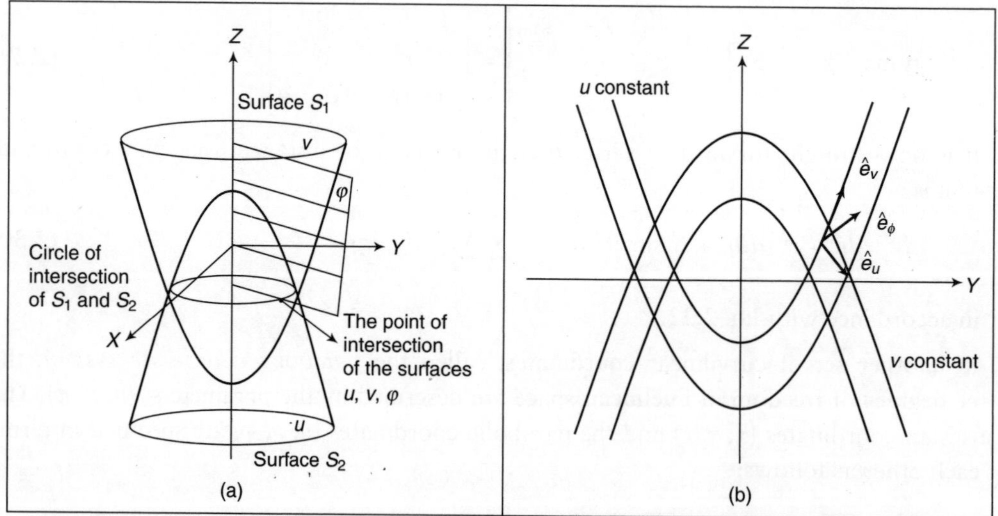

Fig. 2.18 The point of intersection of the three surfaces of constant (u, v, φ), and the unit vectors $(\hat{e}_u, \hat{e}_v, \hat{e}_\varphi)$ of parabolic coordinates are respectively shown in Figs. 2.18(a) and 2.18 (b).

The basis set unit vector, $\{\hat{e}_u, \hat{e}_v, \hat{e}_\varphi\}$ in terms of the Cartesian unit vectors $\{\hat{e}_x, \hat{e}_y, \hat{e}_z\}$ are given by the following relations, expressed in the compact matrix form:

$$\begin{bmatrix} \hat{e}_u \\ \hat{e}_v \\ \hat{e}_\varphi \end{bmatrix} = \begin{bmatrix} \dfrac{v^{1/2}\cos\varphi}{(u+v)^{1/2}} & \dfrac{v^{1/2}\sin\varphi}{(u+v)^{1/2}} & \dfrac{u^{1/2}}{(u+v)^{1/2}} \\ \dfrac{u^{1/2}\cos\varphi}{(u+v)^{1/2}} & \dfrac{u^{1/2}\sin\varphi}{(u+v)^{1/2}} & \dfrac{-v^{1/2}}{(u+v)^{1/2}} \\ -\sin\varphi & \cos\varphi & 0 \end{bmatrix} \begin{bmatrix} \hat{e}_x \\ \hat{e}_y \\ \hat{e}_z \end{bmatrix}. \tag{2.33a}$$

The inverse relations are

$$\begin{bmatrix} \hat{e}_x \\ \hat{e}_y \\ \hat{e}_z \end{bmatrix} = \begin{bmatrix} \dfrac{v^{1/2}\cos\varphi}{(u+v)^{1/2}} & \dfrac{u^{1/2}\cos\varphi}{(u+v)^{1/2}} & -\sin\varphi \\ \dfrac{v^{1/2}\sin\varphi}{(u+v)^{1/2}} & \dfrac{u^{1/2}\sin\varphi}{(u+v)^{1/2}} & \cos\varphi \\ \dfrac{u^{1/2}}{(u+v)^{1/2}} & \dfrac{-v^{1/2}}{(u+v)^{1/2}} & 0 \end{bmatrix} \begin{bmatrix} \hat{e}_u \\ \hat{e}_v \\ \hat{e}_\varphi \end{bmatrix}. \tag{2.33b}$$

The unit vectors of the parabolic coordinates are not fixed. They change from point to point with respect to the coordinates, (u, v, φ).

Partial derivatives of $\hat{e}_u$ with respect to (u, v, φ) are given by:

$$\frac{\partial \hat{e}_u}{\partial u} = -\frac{v^{1/2}}{u^{1/2}}\frac{1}{2(u+v)}\hat{e}_v, \tag{2.34a}$$

$$\frac{\partial \hat{e}_u}{\partial v} = +\frac{u^{1/2}}{v^{1/2}}\frac{1}{2(u+v)}\hat{e}_v, \tag{2.34b}$$

and $$\frac{\partial \hat{e}_u}{\partial \varphi} = \frac{v^{1/2}}{(u+v)^{1/2}}\hat{e}_\varphi. \tag{2.34c}$$

Partial derivatives of $\hat{e}_v$ with respect to (u, v, φ) are given by:

$$\frac{\partial \hat{e}_v}{\partial u} = +\frac{v^{1/2}}{u^{1/2}}\frac{1}{2(u+v)}\hat{e}_u, \tag{2.35a}$$

$$\frac{\partial \hat{e}_v}{\partial v} = -\frac{u^{1/2}}{v^{1/2}}\frac{1}{2(u+v)}\hat{e}_u, \tag{2.35b}$$

and $$\frac{\partial \hat{e}_v}{\partial \varphi} = \frac{u^{1/2}}{(u+v)^{1/2}}\hat{e}_\varphi. \tag{2.35c}$$

Finally, partial derivatives of $\hat{e}_\varphi$ with respect to (u, v, φ) are given by

$$\frac{\partial \hat{e}_\varphi}{\partial u} = 0, \tag{2.36a}$$

$$\frac{\partial \hat{e}_\varphi}{\partial v} = 0, \tag{2.36b}$$

and $$\frac{\partial \hat{e}_\varphi}{\partial \varphi} = -\frac{v^{1/2}}{(u+v)^{1/2}}\hat{e}_u - \frac{u^{1/2}}{(u+v)^{1/2}}\hat{e}_v. \tag{2.36c}$$

The position vector in the parabolic coordinates is easily obtained by carrying out the transformations from any other coordinate system, for example, the Cartesian coordinate system, to the parabolic coordinates.

We have, $\vec{r} = x\hat{e}_x\, y\hat{e}_y + z\hat{e}_z = \sqrt{uv}\,\cos\varphi\hat{e}_x + \sqrt{uv}\,\sin\varphi\hat{e}_y + \frac{1}{2}(u-v)\,\hat{e}_z$. Hence, writing $\hat{e}_x, \hat{e}_y, \hat{e}_z$ in terms of $\hat{e}_u, \hat{e}_v, \hat{e}_\varphi$, we get

$$\vec{r} = \sqrt{uv}\cos\varphi\left[\frac{v^{1/2}\cos\varphi}{(u+v)^{1/2}}\right]\hat{e}_u + e\frac{u^{1/2}\cos\varphi}{(u+v)^{1/2}}\hat{e}_v - \sin\varphi\hat{e}_\varphi\Bigg] +$$

$$\sqrt{uv}\sin\varphi\left[\frac{u^{1/2}\sin\varphi}{(u+v)^{1/2}}\hat{e}_u - \frac{u^{1/2}\sin\varphi}{(u+v)^{1/2}}\hat{e}_v + \cos\varphi\hat{e}_\varphi\right]\frac{1}{2}(u-v)\left[\frac{u^{1/2}}{(u+v)^{1/2}}\hat{e}_u - \frac{v^{1/2}}{(u+v)^{1/2}}\hat{e}_u\right]. \tag{2.37a}$$

Combining terms along each unit vector, we get

$$\vec{r} = \frac{1}{2}\{[u^{1/2}\,(u+v)^{1/2}]\hat{e}_u + [v^{1/2}\,(u+v)^{1/2}]\hat{e}_v\}. \tag{2.37b}$$

We need the expressions for velocity and acceleration to write the equation of motion in the parabolic coordinate. Velocity is given by

$$\vec{v} = \frac{d\vec{r}}{dt} = \frac{1}{2}\frac{d}{dt}\{[u^{1/2}\,(u+v)^{1/2}]\hat{e}_u + [v^{1/2}\,(u+v)^{1/2}]\hat{e}_v\},$$

i.e., $$= \frac{1}{2}\left\{\frac{d}{dt}\right\}[u^{1/2}(u+v)^{1/2}]\hat{e}_u + [u^{1/2}(u+v)^{1/2}]\frac{d\hat{e}_u}{dt} + \frac{d}{dt}[v^{1/2}(u+v)^{1/2}]\hat{e}_v + [v^{1/2}(u+v)^{1/2}]\frac{d\hat{e}_v}{dt}\Bigg\},$$

Finally, using Eqs. 2.34, 2.35 and 2.36 for the time derivatives of the unit vectors, we get

$$\vec{v} = \frac{1}{2\sqrt{u}}(u+v)^{1/2}\dot{u}\,\hat{e}_u + \frac{1}{2\sqrt{u}}(u+v)^{1/2}\dot{v}\,\hat{e}_v + \sqrt{uv}\,\dot{\varphi}\,\hat{e}_\varphi. \tag{2.38}$$

Acceleration is then given by taking the derivative with respect to time of Eq. 2.38:

$$\vec{a} = \frac{d\vec{v}}{dt} = \frac{\sqrt{u+v}}{2\sqrt{u}}\left[\ddot{u} - \frac{2uv\dot{\varphi}^2}{(u+v)} + \frac{\dot{u}\dot{v}}{(u+v)} - \frac{\dot{u}^2 v}{2u(u+v)} - \frac{\dot{v}^2 u}{2v(u+v)}\right]\hat{e}_u \tag{2.39}$$

$$+ \frac{\sqrt{u+v}}{2\sqrt{v}}\left[\ddot{v} - \frac{2uv\dot{\varphi}^2}{(u+v)} + \frac{\dot{u}\dot{v}}{(u+v)} - \frac{\dot{u}^2 v}{2u(u+v)} - \frac{\dot{v}^2 u}{2v(u+v)}\right]\hat{e}_v + \sqrt{uv}\left[\ddot{\varphi} + \frac{\dot{u}\dot{\varphi}}{u} + \frac{\dot{v}\dot{\varphi}}{v}\right]\hat{e}_\varphi$$

2.4 CONTRAVARIANT AND COVARIANT TENSORS

We learnt in Section 2.1 that a scalar is a tensor of rank 0, and that it has a single component. We also learnt that a vector is a tensor of rank 1, and it has 3 components. These components have specific values in a given frame of reference. Tensors acquire only increasing importance in physics, since it turns out that every physical property is expressible as a tensor of appropriate rank. We have mentioned earlier that the number of components of a tensor of rank k defined on an n-dimensional space is n^k. We shall illustrate the tensor notations with respect to a 3-dimensional space, in which the tensor of rank k has 3^k components. The rank of the tensor itself can take values among the natural numbers 0, 1, 2, 3, 4. The components transform according to a *specific law* when you rotate the coordinate system to which the components of the tensor are referenced. The transformation law provides the *signature* property of the tensor, which is defined by the specific nature of the tensor.

We consider a vector whose 3 components in a frame of reference are given by

$$\vec{A} \equiv \{A_1, A_2, A_3\}. \tag{2.40a}$$

In another frame of reference which is rotated with respect to the previous frame, the same vector will be represented by three different components, written with prime on each:

$$\vec{A} \equiv \{A_1', A_2', A_3'\}. \tag{2.40b}$$

The components of the vector in the rotated frame are expressible as a linear superposition of components in the first frame:

$$A_i' = \sum_{j=1}^{3} a_{ij} A_j. \tag{2.41}$$

The coefficients a_{ij} *define* the very law of transformation of the components.

Let us now specifically carry out this analysis to examine how the components of a position vector of a point transform under the rotation of a coordinate frame of reference. Its components in the first frame of reference are given by

$$\vec{r} \equiv (x_1, x_2, x_3) \tag{2.42a}$$

and in the rotated frame of reference, the components of the same vector are given by

$$\vec{r} \equiv (x_1', x_2', x_3'). \tag{2.42b}$$

The first set of components is uniquely determined by the second set, and vice versa. Thus,

$$x_i' = x_i'(x_1, x_2, x_3) \ \forall\ i = 1, 2, 3. \tag{2.43a}$$

and likewise, $x_i = x_i(x_1', x_2', x_3') \ \forall\ i = 1, 2, 3.$ (2.43b)

Hence, using the 'chain rule', we get $\forall\ i = 1, 2, 3,$

$$\delta x_i' = \sum_{j=1}^{3} \frac{\partial x_i'}{\partial x_j} \delta x_j. \tag{2.44a}$$

and likewise, $\delta x_i = \sum_{j=1}^{3} \frac{\partial x_i}{\partial x'_j} \delta x'_j$. (2.44b)

From Eqs. 2.44a and 2.44b, we see that we have two alternatives to choose the coefficients a_{ij} in Eq. 2.41. We may choose either $a_{ij} = \frac{\partial x'_i}{\partial x_j}$, or $a_{ij} = \frac{\partial x_i}{\partial x'_j}$.

If we set, $a_{ij} = \frac{\partial x'_i}{\partial x_j}$, we get

$$A'_i = \sum_{j=1}^{3} \frac{\partial x'_i}{\partial x_j} A_j, \tag{2.45a}$$

and if we set $a_{ij} = \frac{\partial x_i}{\partial x'_j}$, we get

$$A'_i = \sum_{j=1}^{3} \frac{\partial x_i}{\partial x'_j} A_j. \tag{2.45b}$$

The two alternatives, represented by Eqs. 2.45a and 2.45b, provide alternative laws of transformation of the components of the vector from one frame to another which is rotated with respect to the first. Vectors whose components transform according to Eq. 2.45a are called *contravariant* vectors and those whose components transform according to Eq. 2.45b are called *covariant* vectors. We denote the components of the *contravariant* vectors using *superscripts*, and components of a *covariant* vectors are denoted by *subscripts*. Thus, $\vec{A}$ (contravariant)→ (A^1, A^2, A^3) and $\vec{A}$ (covariant)→(A_1, A_2, A_3).

Let us now denote the position vector of a point in Cartesian coordinate system by its components:

$$\vec{r} = x\hat{e}_x + y\hat{e}_y + z\hat{e}_z \equiv (x^1, x^2, x^3). \tag{2.46}$$

Consider now the position vector to be a contravariant vector, with components which transform as per Eq. 2.45a.

For $i = 2$, for example, we get: $A'_2 = \frac{\partial x'_2}{\partial x_1} A_1 + \frac{\partial x'_2}{\partial x_2} A_2 + \frac{\partial x'_2}{\partial x_3} A_3$. (2.47a)

To understand the transformation laws and the use of contravariant and covariant tensors, we consider the components of a position vector in two Cartesian coordinate systems, (i) {X, Y, Z}, and (ii) {X', Y', Z'}, which is rotated from the X-axis through an angle δ.

Thus, we shall have

$$x' = x \; ; \; y' = y\cos\delta - z\sin\delta \; ; \; z' = y\sin\delta + z\cos\delta, \tag{2.47b}$$

which can be written as a matrix equation,

$$\begin{bmatrix} x' \\ y' \\ z' \end{bmatrix}_{\substack{\text{after}\\\text{rotation}\\\text{about X-axis}\\\text{through } \sphericalangle\delta}} = \begin{bmatrix} 1 & 0 & 0 \\ 0 & \cos\delta & -\sin\delta \\ 0 & \sin\delta & \cos\delta \end{bmatrix} \begin{bmatrix} x \\ y \\ z \end{bmatrix}. \tag{2.47c}$$

The above relation merely represents a simple application of the cosine-law of transformation for components of a vector under rotation of a coordinate frame of reference (even if it has sine terms). After all, the sine function and the cosine function are only phase shifted: $\sin\phi = \cos\left(\frac{\pi}{2} - \phi\right)$.

Using the notation in Eq. 2.47a, we write

$$y' = \frac{\partial y'}{\partial x}x + \frac{\partial y'}{\partial y}y + \frac{\partial y'}{\partial z}z = y\cos\delta - z\sin\delta. \tag{2.48a}$$

The inverse transformation is given by

$$\begin{bmatrix} x \\ y \\ z \end{bmatrix} = \begin{bmatrix} 1 & 0 & 0 \\ 0 & \cos\delta & \sin\delta \\ 0 & -\sin\delta & \cos\delta \end{bmatrix} \begin{bmatrix} x' \\ y' \\ z' \end{bmatrix}, \tag{2.48b}$$

i.e., $x = x'$; $y = y'\cos\delta + z'\sin\delta$; $z = -y'\sin\delta + z'\cos\delta$. (2.48c)

We see, from Eqs. 2.47 and 2.48 that

$$\left\{ \begin{matrix} \frac{\partial x'}{\partial x} = 1 = \frac{\partial x}{\partial x'} & ; & \frac{\partial x'}{\partial y} = 0 = \frac{\partial y}{\partial x'} & ; & \frac{\partial x'}{\partial z} = 0 = \frac{\partial z}{\partial x'} \\ \frac{\partial y'}{\partial x} = 0 = \frac{\partial x}{\partial y'} & ; & \frac{\partial y'}{\partial y} = \cos\delta = \frac{\partial y}{\partial y'} & ; & \frac{\partial y'}{\partial z} = -\sin\delta = \frac{\partial z}{\partial y'} \\ \frac{\partial z'}{\partial x} = 0 = \frac{\partial x}{\partial z'} & ; & \frac{\partial z'}{\partial y} = \sin\delta = \frac{\partial y}{\partial z'} & ; & \frac{\partial z'}{\partial z} = \cos\delta = \frac{\partial z}{\partial z'} \end{matrix} \right\}. \tag{2.49}$$

It is clear from Eq. 2.49 that in the Cartesian coordinate system, we have $\forall i, j = 1,2,3$, $\frac{\partial x'_i}{\partial x_j} = \frac{\partial x_j}{\partial x'_i}$, and there is therefore no difference between contravariant and covariant vectors. The differences in the contravariant and the covariant vectors are of importance in curved Riemann space; in the General Theory of Relativity (GTR). We shall need this in Chapters 13 and 14.

Extension of Eq. 2.22 to a 4-dimensional Euclidean space necessitates the upper limit of the double sum to be 4 instead of 3. The g-metric with $g_{ij} = \delta_{ij}\ \forall i, j = 1, 2, 3, 4$ corresponds to this. The squared distance of an infinitesimal step in such a 4-dimensional space will be given by

$$d\xi^2 = +(dw)^2 + (dx)^2 + (dy)^2 + (dz)^2. \tag{2.50}$$

However, we can also choose a *different signature* of the g-metric. Consider, for example, the g-metric given by Eq. 2.51. In this, you will notice that the sign of the matrix element in the last row is opposite to that of the other three elements, whereas in the g-metric corresponding to Eq. 2.50, all the signs of the diagonal elements are the same.

$$[g_{\mu\nu}] = \begin{bmatrix} 1 & 0 & 0 & 0 \\ 0 & -1 & 0 & 0 \\ 0 & 0 & -1 & 0 \\ 0 & 0 & 0 & -1 \end{bmatrix}, \quad \text{or} \quad [g_{\mu\nu}] = \begin{bmatrix} 1 & 0 & 0 & 0 \\ 0 & 1 & 0 & 0 \\ 0 & 0 & 1 & 0 \\ 0 & 0 & 0 & -1 \end{bmatrix}. \tag{2.51}$$

This change in sign of the diagonal elements is a property of the space. The 4-dimensional space with the g-metric given by Eq. 2.51 is Non-Euclidean. The first of the above two signatures correspond to the Non-Euclidean metric,

$$ds^2 = +(dw)^2 - (dx)^2 - (dy)^2 - (dz)^2. \tag{2.52}$$

If we use spherical polar coordinates instead of Cartesian coordinates, the g-metric of the 4-dimensional Non-Euclidean space will have for its signature the following matrix:

$$g_{\mu\nu}(x) = \begin{bmatrix} 1 & 0 & 0 & 0 \\ 0 & -1 & 0 & 0 \\ 0 & 0 & -r^2 & 0 \\ 0 & 0 & 0 & -r^2 \sin^2\theta \end{bmatrix}. \tag{2.53}$$

The signs of the diagonal elements of **g**, $(1,-1,-1,-1)$, contain detailed information about how distance between two points in this space is measured, and hence called the *metric signature*. The Non-Euclidean signature declares that distance is measured in a fundamentally different manner from how it is measured in the 4-dimensional Euclidean space. This signifies a different internal structure of the space. We shall see in Chapter 13 that the signature of the g-metric given in Eq. 2.51 is employed in the '*flat*' Non-Euclidean spacetime continuum in Einstein's Special Theory of Relativity (STR). It turns out, on account of the constancy of the speed of light in all inertial frames of reference, that we shall be led to stunning counter-intuitive consequences with regard to our understanding of a metric. The General Theory of Relativity (GTR), to which we shall present a brief introduction in Chapter 14, will involve a *further* modification of the g-metric. It will require a g-metric that would corresponds to a *curved* 4-dimensional spacetime, which would give

$$ds^2 = \sum_{\mu,\nu} g_{\mu\eta}(x)(dx^\mu)(dx^\nu) = U\,dt^2 - V\,dr^2 - r^2 d\theta^2 - r^2\sin^2\theta d\phi^2, \tag{2.54a}$$

$$\text{i.e.,}\quad [g_{\mu\nu}] = \begin{bmatrix} U & 0 & 0 & 0 \\ 0 & -V & 0 & 0 \\ 0 & 0 & -r^2 & 0 \\ 0 & 0 & 0 & -r^2 \sin^2\theta \end{bmatrix}. \tag{2.54b}$$

In a *flat* space, one may transport (propagate) a vector rigidly from its tail to tip to a different point in space parallel to itself and the vector at the new position would be completely equivalent to the vector's earlier location. In a *curved* space (Chapter 14), on the other hand, this would not remain so.

This chapter has introduced you to various mathematical methods used in classical mechanics. These methods have important applications also in many other areas of physics, including quantum physics and the theory of relativity. We shall use the mathematical framework summarized in this chapter freely throughout this book, as and when required. You will find out how mathematics emerges as the natural medium not merely to *describe* the physical laws, but also that it gets elegantly interwoven in the very *formulation* of the same.

APPENDIX: Some Important Aspects of Matrix Algebra

Matrices help us manipulate information which can be organized in a set of numbers, or functions of numbers, organized in several columns and several rows. In general, these numbers are arranged in an array that may consist of a tabular form consisting of M rows and N columns, giving us the capability to organize $M \times N$ objects, called elements of the matrix. We shall denote a matrix by a fat letter, such as $\mathbb{A}$. For example, a 4 × 5 matrix, having 4 rows and 5 columns, is written as

$$\mathbb{A}_{4\times 5} = \begin{bmatrix} A_{11} & A_{12} & A_{13} & A_{14} & A_{15} \\ A_{21} & A_{22} & A_{23} & A_{24} & A_{25} \\ A_{31} & A_{32} & A_{33} & A_{34} & A_{35} \\ A_{41} & A_{42} & A_{43} & A_{44} & A_{45} \end{bmatrix}. \tag{2A.1}$$

The number of rows and columns can be written explicitly as a subscript, in the order—rows first and columns next—but for simplicity this detail is often suppressed. Each of the $M \times N$ number of elements appears at a specific location in this array. The element that lies at the intersection of the ith row and j^{th} column, (i=1,2,3,......,M and j-1,2,3.....N) is denoted by writing the row index and the column index as subscripts, as A_{ij}, the row index first, and the column index next.

The arrangement of the matrix elements may appear to be abstract to some, and elegant to others, but more importantly it is a smart way to manipulate multiple information when some physical laws govern how this information must be processed. This quality empowers matrices to represent physical quantities such as (i) mass distribution in a uniform or anisotropic manner inside a rigid body, (ii) polarization susceptibility to electrical field inside materials in different directions etc. The use of matrices in physics only grows bigger with advanced topics. In fact, the whole formulation of the quantum theory by Werner Heisenberg is in terms of matrix algebra. The algebra of matrices is a classic example of the *appropriateness of the language of mathematics for the formulation of the laws of physics* that Wigner talked about. You will find several applications in the realm of several domains of physics, including classical mechanics, quantum theory and also in the theory of relativity. In this book, matrices will be used, directly or indirectly, in a few chapters, such as Chapters 7, 13, 14. Matrix algebra has a long history. Some of the most important contributions to matrix algebra have been made by Carl Gauss (1777–1855), Augustin Cauchy (1789–1857), Arthur Cayley (1821–1895), etc. We cannot really review this huge branch of mathematics, but will only summarize some of its important results in order to directly and immediately benefit from those.

For a *square* matrix, i.e., when the number of rows is equal to the number of columns, we can define the *determinant* of the matrix, which is only a number. It is obtained readily from the elements of the matrix written as a determinant.

For example, in the case of the 3×3 matrix,

$$\mathbb{A} = \begin{bmatrix} A_{11} & A_{12} & A_3 \\ A_{21} & A_{22} & A_{23} \\ A_{31} & A_{32} & A_{33} \end{bmatrix}, \tag{2A.2a}$$

its determinant is given by

$$a = \begin{vmatrix} A_{11} & A_{12} & A_3 \\ A_{21} & A_{22} & A_{23} \\ A_{31} & A_{32} & A_{33} \end{vmatrix}. \tag{2A.2b}$$

If the determinant of a matrix is zero, the matrix is said to be *singular*; otherwise, it is called *not-singular* (or non-singular). For a square matrix, an important property is the sum of its diagonal elements, $\sum_{i=1}^{M=N} A_{ii}$. It is called the *trace*, or the *spur*, of the matrix. It is also called the *character* of the matrix. Two matrices, $\mathbb{A}$ and $\mathbb{B}$ are said to be equal if their number of columns, and of rows, is exactly the same. and very importantly, if their *corresponding* elements are equal to each other, i.e., $A_{ij} = B_{ij}$, for every $i, j = 1, 2,...., N$. When two matrices $\mathbb{A}$ and $\mathbb{B}$ have the same number of rows and columns, they can be added (also, of course, subtracted), to give a third matrix $\mathbb{C}$ by carrying out the arithmetic operation on corresponding elements:

$$\mathbb{C} = \mathbb{A} \pm \mathbb{B}, \tag{2A.3a}$$

with Cij $= A_{ij} \pm B_{ij}$, (2A.3b)

for every $i = 1, 2..., M$ and for every $j = 1, 2,..., N$.

We also define matrix multiplication, $\mathbb{C} = \mathbb{A} \times \mathbb{B} = \mathbb{A}\mathbb{B}$, but only when the number of columns of the first matrix $\mathbb{A}$ is equal to the number of rows of the second matrix $\mathbb{B}$. Thus, $\mathbb{A}$ can be an $M \times N$ matrix, and $\mathbb{B}$ can be $N \times L$ matrix. Matrix multiplication is defined by the following prescription, which gives the $(ij)^{\text{th}}$ element of the product matrix in terms of elements of the factor matrices: $\mathbb{C}_{ij} = (\mathbb{A}\mathbb{B})_{ij} = \sum_{k=1}^{N} A_{ik}B_{kj}$. In special situations, matrix multiplication may be *commutative*, i.e., $\mathbb{A}\mathbb{B} = \mathbb{B}\mathbb{A}$; but often it is not so: $\mathbb{A}\mathbb{B} \neq \mathbb{B}\mathbb{A}$. The order in which the matrix multiplication is carried out is thus important, unlike the multiplication of simple arithmetic numbers.

When all the elements of the matrix are zero, we have what is called a *null* matrix. We also have a *unit matrix* if the matrix is a square matrix, all of its diagonal elements are equal to the number 1, and all the off-diagonal elements equal to the number 0. For example, the unit 4×4 unit matrix is

$$\mathbb{I} = \begin{bmatrix} 1 & 0 & 0 & 0 \\ 0 & 1 & 0 & 0 \\ 0 & 0 & 1 & 0 \\ 0 & 0 & 0 & 1 \end{bmatrix}. \tag{2A.4}$$

The unit matrix enables us define the *inverse* of a matrix for a non-singular square matrix. The inverse of the matrix $\mathbb{A}$ is written as $\mathbb{A}^{-1}$, and it has the same dimensions as the original matrix whose inverse it is. The inverse exists if and only if the determinant of the matrix $\mathbb{A}$, $a = |\mathbb{A}| = 0$, is not zero. If the determinant , the matrix is said to be *singular*. When it exists, the inverse of a matrix is unique, and has the property $\mathbb{A}\mathbb{A}^{-1} = \mathbb{A}^{-1}\mathbb{A} = \mathbb{I}$.

The power of matrix algebra is exploited in solving simultaneous linear algebraic equations, in representing quantum operators, in image analysis, in solving large problems in statistics, amongst a large domain of its unparalleled advantages. In this chapter, we shall not use the full scope of matrix algebra, but only employ some simple properties of matrix multiplication and matrix inversion. Hence, we describe below the *recipe* to obtain the inverse of a matrix. Toward this end, we first define a few other matrices, related to any given matrix $\mathbb{A}$, in a simple manner. First, we define the *transpose* $\mathbb{A}^T$ of a matrix. For a rectangular matrix $\mathbb{A}$ of size $M \times N$ (including the special case of the square matrix with $M = N$), the transpose $\mathbb{A}^T$ (also often denoted as $\tilde{\mathbb{A}}$), is an $N \times M$ matrix, whose elements are given by $\mathbb{A}_{ij}^T = \tilde{\mathbb{A}}_{ij} = \mathbb{A}_{ij}$. Essentially, the transpose of the matrix interchanges (i.e. *transpositions*) rows and columns of a given matrix.

We now define, for an $N \times N$ square matrix $\mathbb{A}$, another matrix, called the *cofactor* of $\mathbb{A}$. We shall denote the cofactor of $\mathbb{A}$, as ${}^{C}\mathbb{A}$, by writing the superscript C to the left of the symbol for the matrix. The cofactor ${}^{C}\mathbb{A}$ has the same dimensions $N \times N$ as $\mathbb{A}$, and is defined by the prescription for each of its $(ij)^{\text{th}}$ element,

$$\chi_{ij} = ({}^{C}\mathbb{A})_{ij} = (-1)^{i+j}\, d(\mathbb{M}(\overline{i}, \overline{j})), \tag{2A.5}$$

where $(\mathbb{M}(\overline{i}, \overline{j}))$ is the *determinant* of the $(N-1) \times (N-1)$ *minor* matrix obtained by *removing* the ith row and the jth column of the matrix $\mathbb{A}$.

The inverse of the matrix is now easily obtained as

$$\mathbb{A}^{-1} = \frac{1}{|\mathbb{A}|}\{({}^{C}\mathbb{A})^T\}, \tag{2A.6}$$

where $\{({}^{C}\mathbb{A})^T\}$ is the transpose of the cofactor of the original square matrix $\mathbb{A}$.

We illustrate the process of obtaining the inverse of a square matrix by obtaining the inverse of a 3×3 square matrix,

$$\mathbb{A} = \begin{bmatrix} \frac{2}{3} & -\frac{1}{4} & -\frac{1}{4} \\ -\frac{1}{4} & \frac{2}{3} & -\frac{1}{4} \\ -\frac{1}{4} & -\frac{1}{4} & \frac{2}{3} \end{bmatrix}. \tag{2A.7}$$

As can be readily seen, the determinant of this matrix is $a = |\mathbb{A}| = 121/864$. The inverse of $\mathbb{A}$ therefore exists, since the essential condition $a \neq 0$ is satisfied. The cofactor $({}^{C}\mathbb{A})$ matrix is, therefore,

$$ {}^{C}\mathbb{A} = \begin{bmatrix} (-1)^{1+1}\left(\frac{4}{9}-\frac{1}{16}\right) & (-1)^{1+2}\left(\frac{-2}{12}-\frac{1}{16}\right) & (-1)^{1+3}\left(\frac{1}{16}+\frac{2}{12}\right) \\ (-1)^{2+1}\left(\frac{-2}{12}-\frac{1}{16}\right) & (-1)^{2+2}\left(\frac{4}{9}-\frac{1}{16}\right) & (-1)^{2+3}\left(\frac{-2}{12}-\frac{1}{16}\right) \\ (-1)^{3+1}\left(\frac{1}{16}+\frac{2}{12}\right) & (-1)^{3+2}\left(\frac{-2}{12}-\frac{1}{16}\right) & (-1)^{3+3}\left(\frac{4}{9}-\frac{1}{16}\right) \end{bmatrix}, \tag{2A.8a}$$

$$\text{i.e.,}\quad {}^{C}\mathbb{A} = \begin{bmatrix} \frac{55}{144} & \frac{-11}{48} & \frac{11}{48} \\ \frac{-11}{48} & \frac{55}{144} & \frac{-11}{48} \\ \frac{11}{48} & \frac{-11}{48} & \frac{55}{144} \end{bmatrix}. \tag{2A.8b}$$

The *transpose* of the *cofactor* matrix is

$$({}^{C}\mathbb{A})^{T} = \begin{bmatrix} \frac{55}{144} & \frac{-11}{48} & \frac{11}{48} \\ \frac{-11}{48} & \frac{55}{144} & \frac{-11}{48} \\ \frac{11}{48} & \frac{-11}{48} & \frac{55}{144} \end{bmatrix}. \tag{2A.8c}$$

The inverse $\mathbb{A}^{-1}$ of the matrix $\mathbb{A}$, using Eq. 2A.7, therefore is

$$\mathbb{A}^{-1} = \frac{1}{|\mathbb{A}|}({}^{C}\mathbb{A})^{T} = \frac{864}{121}\begin{bmatrix} \frac{55}{144} & \frac{-11}{48} & \frac{11}{48} \\ \frac{-11}{48} & \frac{55}{144} & \frac{-11}{48} \\ \frac{11}{48} & \frac{-11}{48} & \frac{55}{144} \end{bmatrix}. \tag{2A.9}$$

It is easy to verify that, $\mathbb{A}^{-1}\mathbb{A} = \mathbb{A}\mathbb{A}^{-1} = \mathbb{I}_{3\times 3}$.

In many problems in physics, one is interested in what are called eigenvalues and eigenvectors of a non-singular square matrix. We shall encounter an important application of this in Chapter 7, to obtain the principal axes of the inertia tensor of a rigid body. A non-singular 3×3 square matrix has three linearly-independent eigenvectors which are three 3×1 column vectors $[\mathbb{V}_1]_{3\times1}$, $[\mathbb{V}_2]_{3\times1}$ and $[\mathbb{V}_3]_{3\times1}$ three eigenvalues; α_1, α_2 and α_3. We therefore have three eigenvalue equations:

$$\begin{bmatrix} A_{11} & A_{12} & A_{13} \\ A_{21} & A_{22} & A_{23} \\ A_{31} & A_{32} & A_{33} \end{bmatrix} \begin{bmatrix} V_{11} \\ V_{12} \\ V_{13} \end{bmatrix} = \alpha_1 \begin{bmatrix} V_{11} \\ V_{12} \\ V_{13} \end{bmatrix} = \begin{bmatrix} \alpha_1 V_{11} \\ \alpha_1 V_{12} \\ \alpha_1 V_{13} \end{bmatrix}, \quad (2A.10a)$$

$$\begin{bmatrix} A_{11} & A_{12} & A_{13} \\ A_{21} & A_{22} & A_{23} \\ A_{31} & A_{32} & A_{33} \end{bmatrix} \begin{bmatrix} V_{21} \\ V_{22} \\ V_{23} \end{bmatrix} = \alpha_2 \begin{bmatrix} V_{21} \\ V_{22} \\ V_{23} \end{bmatrix} = \begin{bmatrix} \alpha_2 V_{21} \\ \alpha_2 V_{22} \\ \alpha_2 V_{23} \end{bmatrix}, \quad (2A.10b)$$

and
$$\begin{bmatrix} A_{11} & A_{12} & A_{13} \\ A_{21} & A_{22} & A_{23} \\ A_{31} & A_{32} & A_{33} \end{bmatrix} \begin{bmatrix} V_{31} \\ V_{32} \\ V_{33} \end{bmatrix} = \alpha_3 \begin{bmatrix} V_{31} \\ V_{32} \\ V_{33} \end{bmatrix} = \begin{bmatrix} \alpha_3 V_{31} \\ \alpha_3 V_{32} \\ \alpha_3 V_{33} \end{bmatrix}. \quad (2A.10c)$$

The three equations, Eqs. 2A.10a, b, and c, can also be written by using a variable index i=1,2,3:

$$\begin{bmatrix} A_{11} & A_{12} & A_{13} \\ A_{21} & A_{22} & A_{23} \\ A_{31} & A_{32} & A_{33} \end{bmatrix} \begin{bmatrix} V_{i1} \\ V_{i2} \\ V_{i3} \end{bmatrix} = \alpha_i \begin{bmatrix} V_{i1} \\ V_{i2} \\ V_{i3} \end{bmatrix} = \begin{bmatrix} \alpha_i V_{i1} \\ \alpha_i V_{i2} \\ \alpha_i V_{i3} \end{bmatrix}. \quad (2A.11)$$

The three equations, Eqs. 2A.10a, b, and c, or Eq. 2A.11, can be written together in a compact form as a single matrix equation by stacking the three *one-column* (3×1) eigenvector matrices together in a single *three-column* (3×3) matrix:

$$\begin{bmatrix} A_{11} & A_{12} & A_{13} \\ A_{21} & A_{22} & A_{23} \\ A_{31} & A_{32} & A_{33} \end{bmatrix} \begin{bmatrix} V_{11} & V_{21} & V_{31} \\ V_{12} & V_{22} & V_{32} \\ V_{13} & V_{23} & V_{33} \end{bmatrix} = \begin{bmatrix} \alpha_1 V_{11} & \alpha_2 V_{21} & \alpha_3 V_{31} \\ \alpha_1 V_{12} & \alpha_2 V_{22} & \alpha_3 V_{32} \\ \alpha_1 V_{13} & \alpha_2 V_{23} & \alpha_3 V_{33} \end{bmatrix}. \quad (2A.12a)$$

Factoring the matrix on the right hand side of the above equation, we have

$$\begin{bmatrix} A_{11} & A_{12} & A_{13} \\ A_{21} & A_{22} & A_{23} \\ A_{31} & A_{32} & A_{33} \end{bmatrix} \begin{bmatrix} V_{11} & V_{12} & V_{13} \\ V_{21} & V_{22} & V_{23} \\ V_{31} & V_{32} & V_{33} \end{bmatrix} = \begin{bmatrix} V_{11} & V_{12} & V_{13} \\ V_{21} & V_{22} & V_{23} \\ V_{31} & V_{32} & V_{33} \end{bmatrix} \begin{bmatrix} \alpha_1 & 0 & 0 \\ 0 & \alpha_2 & 0 \\ 0 & 0 & \alpha_3 \end{bmatrix} \quad (2A.12b)$$

In matrix notation, we have

$$\mathbb{A}\mathbb{V} = \mathbb{V}\mathbb{A}_d, \quad (2A.13a)$$

i.e., $$\mathbb{V}^{-1}\mathbb{A}\mathbb{V} = \mathbb{A}_d, \quad (2A.13b)$$

with the diagonalized matrix on the right given by the eigenvalues of $\mathbb{A}$:

$$\mathbb{A}_d = \begin{bmatrix} \alpha_1 & 0 & 0 \\ 0 & \alpha_2 & 0 \\ 0 & 0 & \alpha_3 \end{bmatrix}. \quad (2A.14)$$

The matrix $\mathbb{V}$ is called the transformation matrix. In this case, it transforms the matrix $\mathbb{A}$ to its diagonal form, $\mathbb{A}_d$. Other transformations, such as $\mathbb{S}^{-1}\mathbb{A}\mathbb{S} = \mathbb{A}'$ are also possible where $\mathbb{A}'$ is not diagonal. In general, transformations of the kind $\mathbb{S}^{-1}\mathbb{A}\mathbb{S} = \mathbb{A}'$ are called *similarity* transformations as some properties, for example, the *trace* (or *character*), of $\mathbb{A}$ remain the same in $\mathbb{A}'$. Transformation of a matrix $\mathbb{A}$ effected by a transforming matrix $\mathbb{V}$ for which $\mathbb{V}^{-1} = \mathbb{V}^T = \tilde{\mathbb{V}}$ (inverse is equal to the transpose), are called *orthogonal transformations*. For such a transformation, it follows from Eq. 2A.13 that

$$\tilde{\mathbb{V}}\mathbb{A}\mathbb{V} = \mathbb{A}_d. \tag{2A.15}$$

Equation 2A.15 provides a recipe for diagonalizing a square matrix. To use this recipe, we need the three linearly independent eigenvectors, $[\mathbb{V}_1]_{3\times1}$, $[\mathbb{V}_2]_{3\times1}$ and $[\mathbb{V}_3]_{3\times1}$, stacking which in three columns (as in Eq. 2.12a) we can compose the 3 × 3 transformation matrix $\mathbb{V}_{3\times3}$, which is required to implement the diagonalization recipe (Eq. 2A.15). We therefore proceed to determine the eigenvectors, $\mathbb{V}_1$, $\mathbb{V}_2$, and $\mathbb{V}_3$, and the eigenvalues α_1, α_2, and α_3, (Eq. 2A.10 and Eq. 2A.10), of a non-singular square matrix $\mathbb{A}_{N\times N}$. This is a simple process; all one has to do is to get the roots of a polynomial equation involving powers of a parameter λ, which has an unpretentious form, $P(\lambda) = 0$, called as the *characteristic equation*. It is written as the determinant of a difference matrix to be zero:

$$|\mathbb{A}_{N\times N} - \lambda\mathbb{I}_{N\times N}| = 0, \tag{2A.16}$$

where $\mathbb{I}_{N\times N}$ is *unit matrix* of dimensions N × N and λ is a scalar.

We illustrate the method by taking a specific example of a particular square matrix:

$$\mathbb{A} = \begin{bmatrix} \frac{2}{3} & -\frac{1}{4} & -\frac{1}{4} \\ -\frac{1}{4} & \frac{2}{3} & -\frac{1}{4} \\ -\frac{1}{4} & -\frac{1}{4} & \frac{2}{3} \end{bmatrix} = \frac{1}{12}\begin{bmatrix} 8 & -3 & -3 \\ -3 & 8 & -3 \\ -3 & -3 & 8 \end{bmatrix}. \tag{2A.17}$$

The above matrix is chosen with a specific purpose; it will be used in Chapter 7 where it will appear in the discussion on the inertia tensor for a cubic rigid body. The inertia tensor would represent how mass is distributed inside the cube, in relation to alternate orientations of a Cartesian coordinate frame of reference relative to the cube's sides.

The characteristic equation for the above matrix is

$$\left| \frac{1}{12}\begin{bmatrix} 8 & -3 & -3 \\ -3 & 8 & -3 \\ -3 & -3 & 8 \end{bmatrix} - \lambda\begin{bmatrix} 1 & 0 & 0 \\ 0 & 1 & 0 \\ 0 & 0 & 1 \end{bmatrix} \right| = \frac{1}{12}\begin{vmatrix} 8-\lambda & -3 & -3 \\ -3 & 8-\lambda & -3 \\ -3 & -3 & 8-\lambda \end{vmatrix} = P(\lambda) = 0\,, \tag{2A.18a}$$

where $P(\lambda)$ is a polynomial in λ:

$$P(\lambda) = \frac{1}{12}\{(8-\lambda)[(8-\lambda)(8-\lambda)-9] + 3\,[-3(8-\lambda)-9] - 3\,[9+3(8-\lambda)]\},$$

i.e.,
$$P(\lambda) = \frac{1}{12}\{-\lambda^3 + 24\lambda^2 - 165\lambda + 242\}. \tag{2A.18b}$$

The characteristic equation is satisfied by the roots of the equation $P(\lambda) = 0$, which are

$$\lambda = \frac{2}{12}, \frac{11}{12}, \frac{11}{12}, \tag{2A.19}$$

which are the eigenvalues of $\mathbb{A}$. Two of these eigenvalues are degenerate (i.e., they are the same).

Now, having obtained the eigenvalues of $\mathbb{A}$, we proceed to obtain the eigenvectors $[\mathbb{V}]_{N\times 1}$ of the matrix $\mathbb{A}$ bwy recognizing that $|\mathbb{A}_{N\times N} - \lambda\mathbb{I}_{N\times N}| = 0$ for *each* of the solutions $\lambda_1 = \frac{2}{12}, \lambda_2 = \frac{11}{12}, \lambda_3 = \frac{11}{12}$. Thus, for each of the three eigenvalues, we have $[\mathbb{A}_{3\times 3} - \lambda\mathbb{I}_{3\times 3}]$ $[\mathbb{V}]_{3\times 1} = [\mathbb{O}]_{3\times 1}$, which is a set of 3 simultaneous linear algebraic equations for $[\mathbb{V}] = \begin{bmatrix} X \\ Y \\ Z \end{bmatrix}$.

We shall find that even if two of the eigenvalues are degenerate, linearly independent eigenvectors exist. It may surprise, and possibly also excite, you to know that these modest unassuming methods are used to obtain quantum eigenstates of microscopic particles.

The first eigenvector $\begin{bmatrix} X \\ Y \\ Z \end{bmatrix}$ is obtained by recognizing that pre-multiplying it with the matrix $\frac{1}{12}\begin{vmatrix} 8-\lambda & -3 & -3 \\ -3 & 8-\lambda & -3 \\ -3 & -3 & 8-\lambda \end{vmatrix}$, replacing λ by the first root, $\lambda_1 = \frac{2}{12}$, would give a 3×1 null matrix (as per Eq. 2.18a). Thus,

$$\frac{1}{12}\begin{bmatrix} (8-2) & -3 & -3 \\ -3 & (8-2) & -3 \\ -3 & -3 & (8-2) \end{bmatrix}\begin{bmatrix} X \\ Y \\ Z \end{bmatrix} = \frac{1}{12}\begin{bmatrix} 6 & -3 & -3 \\ -3 & 6 & -3 \\ -3 & -3 & 6 \end{bmatrix}\begin{bmatrix} X \\ Y \\ Z \end{bmatrix} = \begin{bmatrix} 0 \\ 0 \\ 0 \end{bmatrix}. \tag{2A.20}$$

On solving the three simultaneous equations (Eq. 2A.20), we get $X = Y = Z$ which is a set of points along the diagonal of a cube placed with a corner at the origin of a Cartesian coordinate frame of reference, and its three edges of the cube along the Cartesian axes. We can choose $x = 1, y = 1, z = 1$, or any other set proportional to it, such as $x = 2, y = 2, z = 2$. This ambiguity is not surprising, since any vector multiplied by a scalar will also be an eigenvector. The ambiguity can be removed by a process called *normalization* by choosing the *normalization constant* N such that

$$\{N[X \quad Y \quad Z]\}\left\{N\begin{bmatrix} X \\ Y \\ Z \end{bmatrix}\right\} = 1, \tag{2A.21a}$$

i.e., $$N^2(X^2 + Y^2 + Z^2) = 1. \tag{2A.21b}$$

Thus, if we choose $X=Y=Z=1$, the normalization constant would be $N=\frac{1}{\sqrt{3}}$, giving us the normalized vector $\hat{\mathbb{V}}_1=\begin{bmatrix}\frac{1}{\sqrt{3}}\\ \frac{1}{\sqrt{3}}\\ \frac{1}{\sqrt{3}}\end{bmatrix}$. If we had chosen $X=Y=Z=2$, the normalization constant would be $N=\frac{1}{\sqrt{12}}$, giving us the same, unambiguous, normalized eigenvector $\hat{\mathbb{V}}_1$.

For the second and the third root, we have $\lambda_2=\frac{11}{12}$, and $\lambda_3=\frac{11}{12}$, and instead of Eq. 2A.20, we now have :

$$\frac{1}{12}\begin{bmatrix}(8-11) & -3 & -3\\ -3 & (8-11) & -3\\ -3 & -3 & (8-11)\end{bmatrix}\begin{bmatrix}X\\ Y\\ Z\end{bmatrix}=\frac{1}{12}\begin{bmatrix}-3 & -3 & -3\\ -3 & -3 & -3\\ -3 & -3 & -3\end{bmatrix}\begin{bmatrix}X\\ Y\\ Z\end{bmatrix}=\begin{bmatrix}0\\ 0\\ 0\end{bmatrix}. \tag{2A.22}$$

Solving the above set of simultaneous equations, we get $X+Y+Z=0$, which is readily solved, for example, for $X=1$, $Y=-1$, $Z=0$, and also $X=-1$, $Y=-1$, $Z=2$, giving us the remaining two (un-normalized) eigenvectors.

The three un-normalized linearly independent eigenvectors of the matrix are, therefore,

$$\mathbb{V}_1=\begin{bmatrix}X=1\\ Y=1\\ Z=1\end{bmatrix},\ \mathbb{V}_1=\begin{bmatrix}1\\ -1\\ 0\end{bmatrix},\ \mathbb{V}_1=\begin{bmatrix}-1\\ -1\\ 2\end{bmatrix}. \tag{2A.23}$$

We normalize the eigenvectors, so that $\hat{\mathbb{V}}_i^{\dagger}\hat{\mathbb{V}}_i=1$, for i = 1, 2, 3, using the criterion,

$$\hat{\mathbb{V}}_i=\frac{1}{\sqrt{\sum_{j=1}^{3}(\mathbb{V}_{ij})^2}}\mathbb{V}_i. \tag{2A.24}$$

The normalized eigenvectors therefore are:

$$\hat{\mathbb{V}}_1=\begin{bmatrix}\frac{1}{\sqrt{3}}\\ \frac{1}{\sqrt{3}}\\ \frac{1}{\sqrt{3}}\end{bmatrix},\ \hat{\mathbb{V}}_2=\begin{bmatrix}\frac{1}{\sqrt{2}}\\ -\frac{1}{\sqrt{2}}\\ 0\end{bmatrix},\ \hat{\mathbb{V}}_3=\begin{bmatrix}-\frac{1}{\sqrt{6}}\\ -\frac{1}{\sqrt{6}}\\ \frac{2}{\sqrt{6}}\end{bmatrix}. \tag{2A.25}$$

We now construct the 3 × 3 matrix $\tilde{\mathbb{V}}$ of Eq. 2.13 from the three normalized eigenvectors (Eq. 2A.25) of $\mathbb{A}$:

$$\mathbb{V}_{3\times 3} = \left[(\hat{\mathbb{V}}_1)_{3\times 1}\ (\hat{\mathbb{V}}_1)_{3\times 1}\ (\hat{\mathbb{V}}_1)_{3\times 1}\right] = \begin{bmatrix} \frac{1}{\sqrt{3}} & -\frac{1}{\sqrt{2}} & -\frac{1}{\sqrt{6}} \\ \frac{1}{\sqrt{3}} & \frac{1}{\sqrt{2}} & -\frac{1}{\sqrt{6}} \\ \frac{1}{\sqrt{3}} & 0 & \frac{2}{\sqrt{6}} \end{bmatrix}. \tag{2A.26}$$

We can easily verify that the above matrix diagonalizes $\mathbb{A}$ since,

$$\begin{bmatrix} \frac{1}{\sqrt{3}} & \frac{1}{\sqrt{3}} & \frac{1}{\sqrt{3}} \\ -\frac{1}{\sqrt{2}} & \frac{1}{\sqrt{2}} & 0 \\ -\frac{1}{\sqrt{6}} & -\frac{1}{\sqrt{6}} & \frac{2}{\sqrt{6}} \end{bmatrix} \begin{bmatrix} \frac{2}{3} & -\frac{1}{4} & -\frac{1}{4} \\ -\frac{1}{4} & \frac{2}{3} & -\frac{1}{4} \\ -\frac{1}{4} & -\frac{1}{4} & \frac{2}{3} \end{bmatrix} \begin{bmatrix} \frac{1}{\sqrt{3}} & -\frac{1}{\sqrt{2}} & -\frac{1}{\sqrt{6}} \\ \frac{1}{\sqrt{3}} & \frac{1}{\sqrt{2}} & -\frac{1}{\sqrt{6}} \\ \frac{1}{\sqrt{3}} & 0 & -\frac{1}{\sqrt{6}} \end{bmatrix} = \begin{bmatrix} \frac{2}{12} & 0 & 0 \\ 0 & \frac{11}{12} & 0 \\ 0 & 0 & \frac{11}{12} \end{bmatrix},$$

i.e., $\tilde{\mathbb{V}}\mathbb{A}\mathbb{V} = \mathbb{V}^{-1}\mathbb{A}\mathbb{V} = \mathbb{A}_{\text{diagonal}}$. (2A.27)

Problems with Solutions

P2.1:

Sketch the following figures:

(a) (i) $\rho = 5$ (ii) $\varphi = \frac{\pi}{4}$ (iii) $z = 5$ where ρ, φ, z represent cylindrical polar coordinates

(b) (i) $r = 1, \theta = \frac{\pi}{4}$ (ii) $\varphi = \frac{\pi}{3}, z = 5$ (iii) $\theta = \frac{\pi}{4}, \varphi = \frac{\pi}{4}$ where r, φ, θ represent spherical polar coordinates

Solution:

(a) (i) $\rho = 5$
$0 \le \varphi < 2\pi$
$-\infty < z < +\infty$

(ii) $0 < \rho < +\infty$
$\varphi = \frac{\pi}{4}$
$-\infty < z < +\infty$

(iii) $0 < \rho < +\infty$
$0 \le \varphi \le 2\pi$
$z = 5$

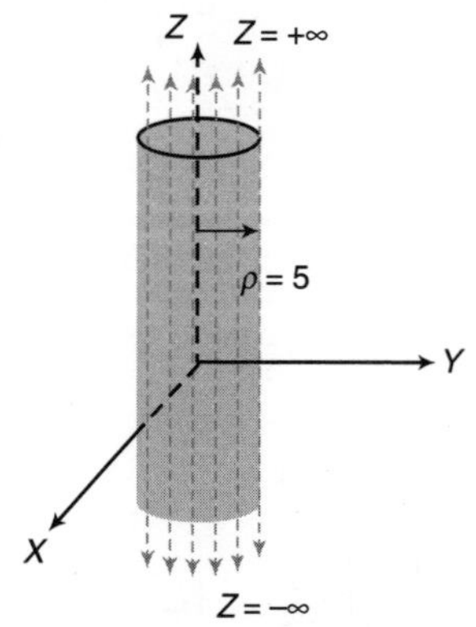

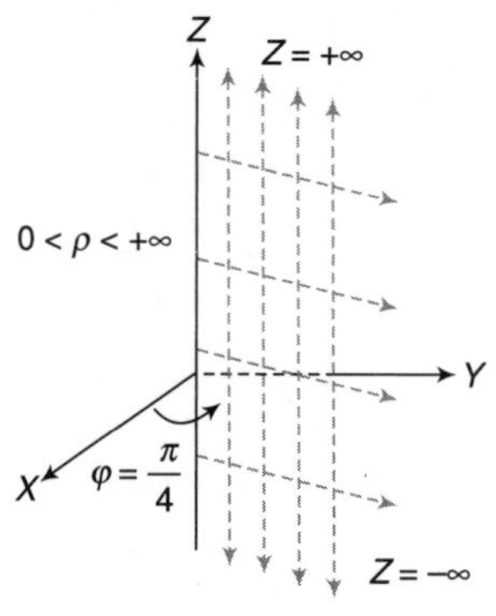

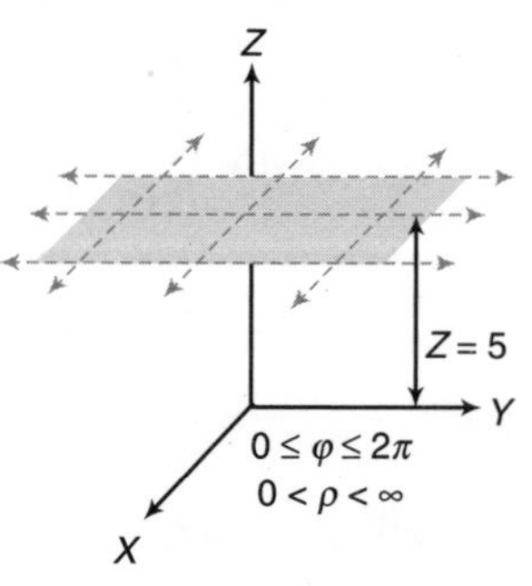

ρ = 5 is the curved surface of an infinitely long cylindwer of radius 5 and axis along the Z-axis. It would stretch all the way for $-\infty < Z < +\infty$.

$\varphi = \dfrac{\pi}{4}$ is the surface of an infinite plane in the first quadrant of the Cartesian coordinate system at an angle $\varphi = \dfrac{\pi}{4}$. The $\varphi = \dfrac{\pi}{4}$ plane is orthogonal to the XY-plane and extends for all values $-\infty < Z < +\infty$, and thus resides both above, and below, the XY-plane.

z = 5 surface is an infinite flat plane parallel to the XY-plane at z = 5. It extends for all values $-\infty < X < +\infty$, and for all values $-\infty < y < +\infty$

(b) (i) $r = 1, \theta = \dfrac{\pi}{4}$ (ii) $\varphi = \dfrac{\pi}{3}, z = 5$

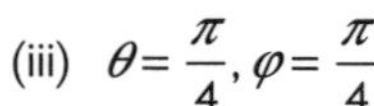

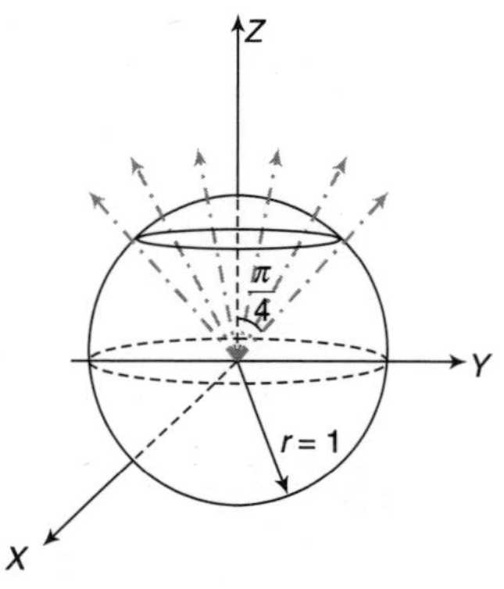

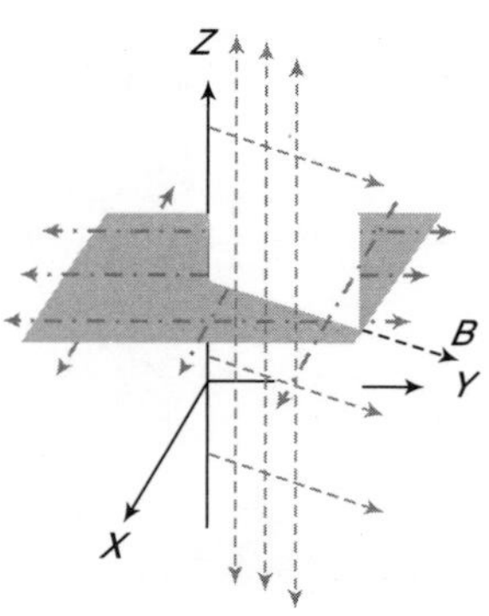

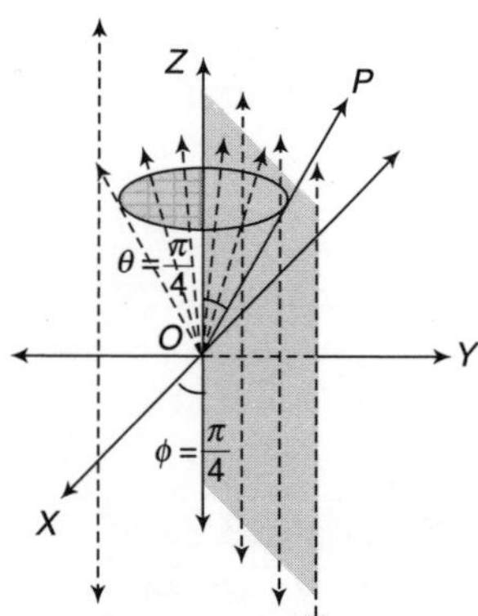

$r = 1$ represents a sphere of unit radius, and $\theta = \frac{\pi}{4}$ represents a cone with 45° semi-vertical angle. Hence, $r = 1, \theta = \frac{\pi}{4}$ represents the intersection of the above mentioned surfaces. The intersection is a circle, parallel to XY-plane, with radius $R = r\sin\frac{\pi}{4} = \frac{1}{\sqrt{2}}$ units, and at height $z = r\cos\frac{\pi}{4} = \frac{1}{\sqrt{2}}$ units.

$\varphi = \frac{\pi}{3}$ is the surface of an infinite plane in the first quadrant of the Cartesian coordinate system at an angle $\varphi = \frac{\pi}{3}$. The $\varphi = \frac{\pi}{3}$ plane is orthogonal to the XY-plane and extends for all values $-\infty < z < +\infty$.

$z = 5$ is an infinite plane parallel to the XY-plane. Hence, the intersection of these two planes represents an infinite straight line (AB) in first quadrant at a height $z = 5$ and making an angle $\varphi = \frac{\pi}{3}$ with X-axis.

$\theta = \frac{\pi}{4}$ represents a cone with 45° semi-vertical angle. $\varphi = \frac{\pi}{4}$ is the surface of an infinite plane in the in the first quadrant of the Cartesian coordinate system at an angle $\varphi = \frac{\pi}{4}$. The intersection of these two surfaces is an infinite straight line (OP)

P2.2:

A particle moves on a plane along the curve $\rho = k(1 + \cos\varphi)$ where (ρ, φ) are plane polar coordinates. Find its acceleration, and determine the magnitude of the acceleration.

Solution:

$\rho = k(1 + \cos\varphi)$. Hence: $\dot{\rho} = -k\dot{\varphi}\sin\varphi$, and $\ddot{\rho} = -k[\dot{\varphi}^2\cos\varphi + \ddot{\varphi}\sin\varphi]$

The velocity is: $\vec{v} = \dot{\rho}\hat{e}_\rho + \rho\dot{\varphi}\hat{e}_\varphi$.

Hence: $\vec{v}^2 = \vec{v}\cdot\vec{v} = (\dot{\rho}\breve{e}_\rho + \rho\dot{\varphi}\breve{e}_\varphi)\cdot(\dot{\rho}\breve{e}_\rho + \rho\dot{\varphi}\breve{e}_\varphi) = 2k^2\dot{\varphi}^2(1+\cos\varphi)$.

Accordingly, $\dot{\varphi} = \left[\frac{v^2}{2k(1+\cos\varphi)}\right]^{1/2} = \frac{v}{\sqrt{2k\rho}}$, and $\ddot{\varphi} = \frac{v}{\sqrt{2k}}\cdot\frac{d}{dt}\left(\rho^{-\frac{1}{2}}\right) = \frac{v}{\sqrt{2k}}\left[-\frac{1}{2}\rho^{-\frac{3}{2}}\dot{\rho}\right]$.

i.e., $\ddot{\varphi} = \frac{\dot{\varphi}^2\sin\varphi}{2(1+\cos\varphi)}$.

Therefore, the acceleration is $\vec{a} = (\ddot{\rho} - \rho\dot{\varphi}^2)\breve{e}_\rho + (\rho\ddot{\varphi} + 2\dot{\rho}\dot{\varphi})\breve{e}_\varphi = \left(-\frac{3v^2}{4k}\right)\breve{e}_\rho - \left(\frac{3}{4}\frac{v^2}{k}\frac{\sin\varphi}{1+\cos\varphi}\right)\breve{e}_\varphi$.

Hence, the magnitude of acceleration is: $|\vec{a}| = \sqrt{a_\rho^2 + a_\varphi^2} = \frac{3v^2}{4k}\left[\frac{2}{1+\cos\varphi}\right]^{\frac{1}{2}}$.

P2.3:

Given: The *Levi–Civita symbol* ε_{ijk} has the following properties:

$$\varepsilon_{ijk} = \left\{ \begin{array}{ll} +1 \text{ if } & (i, j, k) \text{ is cyclic} \\ -1 \text{ if } & (i, j, k) \text{ is non-cyclic} \\ 0 \text{ if } & i = j, \text{ or } j = k, \text{ or } k = i \end{array} \right\} = \breve{e}_i \cdot (\breve{e}_j \times \breve{e}_k)$$

Using the above properties, show that:

(a) $\vec{a} \times \vec{b} = \vec{c} = c_i\hat{e}_i$ if $\varepsilon_{ijk}a_jb_k$

(b) Show that, $\vec{a} \cdot (\vec{b} \times \vec{c}) = \vec{c} \cdot (\vec{a} \times \vec{b}) = \vec{b} \cdot (\vec{c} \times \vec{a})$

Solution:

(a) Suppose we denote the components of some vector $\vec{c}$ as $c_i = \varepsilon_{ijk}a_jb_k$; then we have

$$\vec{a} \times \vec{b} = c_i\breve{e}_i \Rightarrow \left\{ \begin{array}{l} c_1 = a_2b_3 - a_3b_2 = \varepsilon_{123}a_2b_3 + \varepsilon_{132}a_3b_2 \\ c_2 = a_3b_1 - a_1b_3 = \varepsilon_{231}a_3b_1 + \varepsilon_{213}a_1b_3 \\ c_3 = a_1b_2 - a_2b_1 = \varepsilon_{312}a_1b_2 + \varepsilon_{321}a_2b_1 \end{array} \right\}$$

(b) Scalar triple product

$$\vec{a} \cdot (\vec{b} \times \vec{c}) = \vec{a}_i \cdot [\vec{b} \times \vec{c}]_i = a_i\varepsilon_{ijk}b_jc_k = \varepsilon_{ijk}a_ib_jc_k$$

Since $\varepsilon_{ijk} = \varepsilon_{kij} = \varepsilon_{jki}$

$$\vec{a} \cdot (\vec{b} \times \vec{c}) = \varepsilon_{ijk}a_ib_jc_k = \varepsilon_{kij}a_ib_jc_k = c_k\varepsilon_{kij}a_ib_j = c_k[\vec{a} \times \vec{b}]_k = \vec{c} \cdot (\vec{a} \times \vec{b})$$

$$\vec{a} \cdot (\vec{b} \times \vec{c}) = \varepsilon_{ijk}a_ib_jc_k = \varepsilon_{jki}a_ib_jc_k = b_j\varepsilon_{jki}c_ka_i = b_j[\vec{c} \times \vec{a}]_j = \vec{b}\ \mu\ (\vec{c} \times \vec{a})$$

Therefore $\vec{a} \cdot (\vec{b} \times \vec{c}) = \vec{c} \cdot (\vec{a} \times \vec{b}) = \vec{b} \cdot (\vec{c} \times \vec{a})$

P2.4:

(a) Express the operators for the partial derivatives with respect to the Cartesian coorsinates, i.e., $\frac{\partial}{\partial x}, \frac{\partial}{\partial y}, \frac{\partial}{\partial z}$, in the spherical polar coordinates.

(b) Show that the operator $\left(x\frac{\partial}{\partial y} - y\frac{\partial}{\partial x} \right) \equiv \frac{\partial}{\partial \varphi}$.

Solution:

(a) We shall use the relationship between the Cartesian coordinates (*x*, *y*, *z*) and the spherical polar coordinates (r, θ, φ)

$x = r \sin \theta \cos \varphi$; $y = r \sin \theta \sin \varphi$; $z = r \cos \theta$ and $r = \sqrt{x^2 + y^2 + z^2}$; $\cos\theta = \frac{z}{r}$; $\tan\varphi = \frac{y}{x}$

Hence: $\frac{\partial r}{\partial x} = \sin\theta\cos\varphi$; $\frac{\partial \theta}{\partial x} = \frac{1}{r}\cos\theta\cos\varphi$; $\frac{\partial \varphi}{\partial x} = -\frac{\sin\varphi}{r\sin\theta}$

$$\frac{\partial r}{\partial y} = \sin\theta \sin\varphi;\quad \frac{\partial \theta}{\partial y} = \frac{1}{r}\sin\theta\cos\varphi;\quad \frac{\partial \varphi}{\partial y} = \frac{\cos\varphi}{r\sin\theta}$$

$$\frac{\partial r}{\partial z} = \cos\theta;\quad \frac{\partial \theta}{\partial z} = \frac{1}{r}\sin\theta;\quad \frac{\partial \varphi}{\partial z} = 0$$

Using the chain rule:

$$\frac{\partial}{\partial x} = \frac{\partial}{\partial r}\frac{\partial r}{\partial y} + \frac{\partial}{\partial \theta}\frac{\partial \theta}{\partial y} + \frac{\partial}{\partial \varphi}\frac{\partial \varphi}{\partial y} \Rightarrow \frac{\partial}{\partial x} = \sin\theta\cos\varphi\frac{\partial}{\partial r} + \frac{1}{r}\cos\theta\cos\varphi\frac{\partial}{\partial \theta} - \frac{\sin\varphi}{r\sin\theta}\frac{\partial}{\partial \varphi}$$

$$\frac{\partial}{\partial y} = \frac{\partial}{\partial r}\frac{\partial r}{\partial y} + \frac{\partial}{\partial \theta}\frac{\partial \theta}{\partial y} + \frac{\partial}{\partial \varphi}\frac{\partial \varphi}{\partial y} \Rightarrow \frac{\partial}{\partial y} = \sin\theta\sin\varphi\frac{\partial}{\partial r} + \frac{1}{r}\cos\theta\sin\varphi\frac{\partial}{\partial \theta} - \frac{\cos\varphi}{r\sin\theta}\frac{\partial}{\partial \varphi}$$

$$\frac{\partial}{\partial z} = \frac{\partial}{\partial r}\frac{\partial r}{\partial z} + \frac{\partial}{\partial \theta}\frac{\partial \theta}{\partial z} + \frac{\partial}{\partial \varphi}\frac{\partial \varphi}{\partial z} \Rightarrow \frac{\partial}{\partial z} = \cos\theta\frac{\partial}{\partial r} - \frac{1}{r}\sin\theta\frac{\partial}{\partial \theta}$$

(b) Using the above results:

$$x\frac{\partial}{\partial y} = (r\sin\theta\cos\varphi)\left(\sin\theta\sin\varphi\frac{\partial}{\partial r} + \frac{1}{r}\cos\theta\sin\varphi\frac{\partial}{\partial \theta} - \frac{\cos\varphi}{r\sin\theta}\frac{\partial}{\partial \varphi}\right)$$

and $$y\frac{\partial}{\partial x} = (r\sin\theta\sin\varphi)\left(\sin\theta\cos\varphi\frac{\partial}{\partial r} + \frac{1}{r}\cos\theta\cos\varphi\frac{\partial}{\partial \theta} - \frac{\sin\varphi}{r\sin\theta}\frac{\partial}{\partial \varphi}\right).$$

Finally, taking the difference: $\left(x\frac{\partial}{\partial y} - y\frac{\partial}{\partial x}\right) \equiv \frac{\partial}{\partial \varphi}$.

[Note: By the way, on multiplying both sides of the above equation by $-i\hbar$ (where $i = \sqrt{-1}$ and $\hbar = \frac{h}{2\pi}$, h being the Planck's constant), we get: $-i\hbar\left(x\frac{\partial}{\partial y} - y\frac{\partial}{\partial x}\right) = -i\hbar\frac{\partial}{\partial \varphi} = L_z$. This is the quantum operator for the z-component of the angular momentum].

P2.5:

Determine the vector $\vec{A} = (y - z)\hat{e}_x + x\,\hat{e}_y$ at the point P(–3, 2, 4) and express it in (a) the spherical polar and (b) the cylindrical polar coordinates.

Solution:

At the point P, $x = -3$; $y = 2$; $z = 4$. Hence, $\vec{A}(at\ P) = -(2\hat{e}_x + 3\hat{e}_y)$.

From Eq. (2.15a):

$$\begin{bmatrix} A_r \\ A_\theta \\ A_\varphi \end{bmatrix} = \begin{bmatrix} \sin\theta\cos\varphi & \sin\theta\sin\varphi & \cos\theta \\ \cos\theta\cos\varphi & \cos\theta\sin\varphi & -\sin\theta \\ -\sin\varphi & \cos\varphi & 0 \end{bmatrix}\begin{bmatrix} A_x \\ A_y \\ A_z \end{bmatrix} = \begin{bmatrix} \sin\theta\cos\varphi & \sin\theta\sin\varphi & \cos\theta \\ \cos\theta\cos\varphi & \cos\theta\sin\varphi & -\sin\theta \\ -\sin\varphi & \cos\varphi & 0 \end{bmatrix}\begin{bmatrix} y - z \\ x \\ 0 \end{bmatrix}$$

Hence $A_r = (y - z)\sin\theta\cos\varphi + x\sin\theta\sin\varphi;\ A_\theta = (y - z)\cos\theta\cos\varphi + x\cos\theta\sin\varphi$

and $A_\varphi = -(y - z)\sin\varphi + x\cos\varphi.$

The right hand side of the above equations has both Cartesian and spherical polar coordinates. We must therefore transform all the Cartesian coordinates on the right hand side to appropriate spherical polar equivalents, for which we use Eq. 2.13b.

Hence $A_r = (r\sin\theta\sin\varphi - r\cos\theta)\sin\theta\cos\varphi + r\sin\theta\cos\varphi\sin\theta\sin\varphi,$

$$A_\theta = (r\sin\theta\sin\varphi - r\cos\theta)\cos\theta\cos\varphi + r\sin\theta\cos\varphi\cos\theta\sin\varphi,$$

and $A_\varphi = -(r\sin\theta\sin\varphi - r\cos\theta)\sin\varphi + r\sin\theta\cos\varphi\cos\varphi.$

Accordingly, we get:

$$\vec{A} = r\,(2\sin^2\theta\cos\varphi\sin\varphi - \cos\theta\sin\theta\cos\varphi]\,\hat{e}_r +$$

$$r[2\sin\theta\cos\theta\cos\varphi\sin\varphi - \cos^2\theta\cos\varphi]\,\hat{e}_\theta +.$$

$$r[\sin\theta(\cos^2\varphi - \sin^2\varphi) + \cos\theta\sin\varphi]\hat{e}_\varphi$$

At the point $P(-3, 2, 4)$,

$$r = \sqrt{x^2 + y^2 + z^2} = \sqrt{29}\,,\ \theta = \tan^{-1}\frac{\sqrt{x^2 + y^2}}{z} = \tan^{-1}\frac{\sqrt{13}}{4} \text{ and } \varphi = \tan^{-1}\frac{y}{x} = \tan^{-1}\frac{2}{-3};$$

and . $\sin\theta = \sqrt{\dfrac{13}{29}}, \cos\theta = \dfrac{4}{\sqrt{29}}, \sin\varphi = \dfrac{2}{\sqrt{13}}, \cos\varphi = \dfrac{-3}{\sqrt{13}}$

Hence $\vec{A} = 0\hat{e}_r + 0\hat{e}_\theta + \sqrt{13}\breve{e}_\varphi = \sqrt{13}\breve{e}_\varphi$

(b) Similarly, in the case of the cylindrical polar coordinate system:

Hence: $A_\rho = (y - z)\cos\varphi + x\sin\varphi.$

$$A_\varphi = -(y - z)\sin\varphi + x\cos\varphi.$$

$$A_z = 0$$

Therefore, $\vec{A} = (2\rho\cos\varphi\sin\varphi - z\cos\varphi)\,\hat{e}_\rho + (-\rho\sin^2\varphi + z\sin\varphi + \rho\cos^2\varphi)\hat{e}_\varphi.$

At $P(-3, 2, 4)$, $\rho = \sqrt{x^2 + y^2} = \sqrt{13}\,;\ \varphi = \tan^{-1}\left(\dfrac{2}{-3}\right)$ and $z = 4$,

and $\cos\varphi = \dfrac{-3}{\sqrt{13}}, \sin\varphi = \dfrac{2}{\sqrt{13}}$.

Therefore, $\vec{A} = 0\hat{e}_\rho + \sqrt{13}\breve{e}_\varphi = \sqrt{13}\breve{e}_\varphi.$

P2.6:

(a) Consider two tensors, $T_\gamma^{\alpha\beta}$ and R_κ^η. Prove that $G_{\gamma\kappa}^{\alpha\beta\eta} = T_\gamma^{\alpha\beta} R_\kappa^\eta$ is also a tensor.

(b) Find the rank of the tensor which is the *inner* product of $T_\gamma^{\alpha\beta}$ and R_κ^η.

[Note: A contravariant vector, and a covariant vector, is represented as T^{α} and T_{α} respectively. Details in Section 2.4 may be referred to. Tensors of rank 2 are represented using double indices, such as $T^{\alpha\beta}$ and $T_{\alpha\beta}$. A mixed tensor of rank 2 is expressed as T^{α}_{β}. The transformation of a mixed tensor from one coordinate system to another coordinate system is expressed as:

$$\bar{T}^{\alpha}_{\beta} = \sum_{\delta=1}^{N}\sum_{\gamma=1}^{N} \frac{\partial \bar{x}^{\alpha}}{\partial x^{\gamma}}\frac{\partial x^{\delta}}{\partial \bar{x}^{\beta}} T^{\gamma}_{\delta} = \frac{\partial \bar{x}^{\alpha}}{\partial x^{\gamma}}\frac{\partial x^{\delta}}{\partial \bar{x}^{\beta}} T^{\gamma}_{\delta}$$

[Summation convention: index appearing twice is summed over.]

Solution:

(a) Since $T^{\alpha\beta}_{\gamma}$ and R^{η}_{κ} are two tensors, we have:

$$\bar{T}^{\mu\nu}_{\lambda} = \frac{\partial \bar{x}^{\mu}}{\partial x^{\alpha}}\frac{\partial \bar{x}^{\nu}}{\partial x^{\beta}}\frac{\partial x^{\gamma}}{\partial \bar{x}^{\lambda}} T^{\alpha\beta}_{\gamma}$$

$$\bar{R}^{\sigma}_{\xi} = \frac{\partial \bar{x}^{\sigma}}{\partial x^{\eta}}\frac{\partial x^{\kappa}}{\partial \bar{x}^{\xi}} R^{\eta}_{\kappa}$$

Multiplication of these two tensors yields $G^{\alpha\beta\eta}_{\gamma\kappa}$:

$$\bar{T}^{\mu\nu}_{\lambda}\bar{R}^{\sigma}_{\xi} = \frac{\partial \bar{x}^{\mu}}{\partial x^{\alpha}}\frac{\partial \bar{x}^{\nu}}{\partial x^{\beta}}\frac{\partial x^{\gamma}}{\partial \bar{x}^{\lambda}}\frac{\partial \bar{x}^{\sigma}}{\partial x^{\eta}}\frac{\partial x^{\kappa}}{\partial \bar{x}^{\xi}} T^{\alpha\beta}_{\gamma}R^{\eta}_{\kappa}.$$

$T^{\alpha\beta}_{\gamma} R^{\eta}_{\kappa}$ is a tensor of rank 5. Here, α, β and η are contravariant indices, whereas γ, κ are covariant indices. This is an example of outer product of tensors.

(b) First, we carry out contraction of one contravariant and one covariant index. The outer multiplication of two tensors, followed by a contraction, gives an ***inner product***. [Note: The process of lowering the order of tensor by equating a covariant index with a contravariant index and performing the summation is called the ***contraction*** of a tensor.]

Consider $\eta = \gamma$. Thus, we consider $G^{\alpha\beta\gamma}_{\gamma\kappa}$.

From the above expression in (a), putting $\sigma = \lambda$, we get:

$$\bar{T}^{\mu\nu}_{\lambda}\bar{R}^{\lambda}_{\xi} = \frac{\partial \bar{x}^{\mu}}{\partial x^{\alpha}}\frac{\partial \bar{x}^{\nu}}{\partial x^{\beta}}\frac{\partial x^{\gamma}}{\partial \bar{x}^{\lambda}}\frac{\partial \bar{x}^{\lambda}}{\partial x^{\eta}}\frac{\partial x^{\kappa}}{\partial \bar{x}^{\xi}} T^{\alpha\beta}_{\gamma}R^{\eta}_{\kappa}, \quad \text{i.e.,} \quad \bar{T}^{\mu\nu}_{\lambda}\bar{R}^{\lambda}_{\xi} = \delta^{\gamma}_{\eta}\frac{\partial \bar{x}^{\mu}}{\partial x^{\alpha}}\frac{\partial \bar{x}^{\nu}}{\partial x^{\beta}}\frac{\partial x^{\kappa}}{\partial \bar{x}^{\xi}} T^{\alpha\beta}_{\gamma}R^{\eta}_{\kappa},$$

i.e., $\bar{T}^{\mu\nu}_{\lambda}\bar{R}^{\lambda}_{\xi} = \frac{\partial \bar{x}^{\mu}}{\partial x^{\alpha}}\frac{\partial \bar{x}^{\nu}}{\partial x^{\beta}}\frac{\partial x^{\kappa}}{\partial \bar{x}^{\xi}} T^{\alpha\beta}_{\gamma}R^{\gamma}_{\kappa}$. The last expression shows $G^{\alpha\beta\gamma}_{\gamma\kappa} = T^{\alpha\beta}_{\gamma} R^{\gamma}_{\kappa}$ is a tensor of rank 3.

P2.7:

Determine the inverse of the matrix $\mathbb{A} = \begin{bmatrix} \cos\theta & -\sin\theta & 0 \\ \sin\theta & \cos\theta & 0 \\ 0 & 0 & 1 \end{bmatrix}$. Also check $\mathbb{A}\mathbb{A}^{-1} = \mathbb{A}^{-1}\mathbb{A} = \mathbb{I}$

***Solution*:**

Using Eq. 2A.5, the cofactor matrix is: $({}^{C}\mathbb{A}) = \begin{bmatrix} (-1)^{1+1}\cos\theta & (-1)^{1+2}\sin\theta & (-1)^{1+3}0 \\ (-1)^{2+1}(-\sin\theta) & (-1)^{2+2}\cos\theta & (-1)^{2+3}0 \\ (-1)^{3+1}0 & (-1)^{3+2}0 & (-1)^{3+3}1 \end{bmatrix}$

The transpose of the cofactor matrix is the adjoint of $\mathbb{A}$:

$$({}^{C}\mathbb{A})^T = \begin{bmatrix} \cos\theta & -\sin\theta & 0 \\ \sin\theta & \cos\theta & 0 \\ 0 & 0 & 1 \end{bmatrix}^T = \begin{bmatrix} \cos\theta & \sin\theta & 0 \\ -\sin\theta & \cos\theta & 0 \\ 0 & 0 & 1 \end{bmatrix}$$

The determinant, $|\mathbb{A}| = \cos\alpha \cdot \cos\alpha + \sin\alpha \cdot \sin\alpha + 0 = 1$, therefore $\mathbb{A}^{-1}$ exists.

Using Eq. 2A.6, $\mathbb{A}^{-1} = \dfrac{1}{|\mathbb{A}|}({}^{C}\mathbb{A})^T \Rightarrow (1)\begin{bmatrix} \cos\theta & \sin\theta & 0 \\ -\sin\theta & \cos\theta & 0 \\ 0 & 0 & 1 \end{bmatrix}$

Check: $\mathbb{A}\mathbb{A}^{-1} = 1_{2\times 2} = \mathbb{A}^{-1}\mathbb{A}$:

$$\begin{bmatrix} \cos\theta & -\sin\theta & 0 \\ \sin\theta & \cos\theta & 0 \\ 0 & 0 & 1 \end{bmatrix}\begin{bmatrix} \cos\theta & \sin\theta & 0 \\ -\sin\theta & \cos\theta & 0 \\ 0 & 0 & 1 \end{bmatrix} = \begin{bmatrix} 1 & 0 & 0 \\ 0 & 1 & 0 \\ 0 & 0 & 1 \end{bmatrix} = \begin{bmatrix} \cos\theta & \sin\theta & 0 \\ -\sin\theta & \cos\theta & 0 \\ 0 & 0 & 1 \end{bmatrix}\begin{bmatrix} \cos\theta & -\sin\theta & 0 \\ \sin\theta & \cos\theta & 0 \\ 0 & 0 & 1 \end{bmatrix}.$$

Additional Problems

P2.8 Find the angle between two vectors $\vec{A}$ and $\vec{B}$ if:

(a) $|\vec{A}| = |\vec{B}| = |\vec{A} + \vec{B}|$ (b) $(\vec{A} + \vec{B}) \times \vec{A} = \vec{B} \times (\vec{B} \times \vec{A})$ and $|\vec{A}|^2 = |\vec{B}|^2$

P2.9 Express the components of $\vec{A} = 2y\hat{e}_x - z\hat{e}_y - x\hat{e}_z$ in cylindrical polar, and in spherical polar, coordinates.

P2.10 Determine the velocity and acceleration of a particle moving at constant ρ and at the azimuthal time-dependent angle $\varphi = \omega t + \dfrac{\alpha t^2}{2}$. Here, ω is the angular speed, and α is the magnitude of the angular acceleration. Express your answer in (a) plane polar coordinate system and (b) Cartesian coordinate system.

P2.11 Determine the volume of a four dimensional unit sphere:

$x_1 = r\sin\theta_2 \sin\theta_1 \cos\varphi$, $x_2 = r\sin\theta_2 \sin\theta_1 \sin\varphi$, $x_3 = r\sin\theta_2 \cos\theta_1$, and $x_4 = r\cos\theta_2$;

with $0 \le \varphi \le 2\pi$; $0 \le \theta_1 \le \pi$; $0 \le \theta_2 \le \pi$,

P2.12 Let $\vec{Q} = 3\hat{e}_x + 2\hat{e}_y$; Switch to the parabolic coordinate system, and:

(a) write $\vec{Q}$ in terms of the basis vectors at the point ($u = 2$. $v = 1$).

(b) write $\vec{Q}$ in terms of the basis vectors at point ($u = 3$, $v = 3$).

P2.13 Express the vector $\vec{B} = -\frac{5}{r}\breve{e}_r + r\cos\theta\hat{e}_\theta + \hat{e}_\varphi$ in cylindrical polar and Cartesian coordinate systems.

P2.14 A particle moves in such a way that in the spherical polar coordinate system, its motion is described by the properties that $\dot{\theta}$ is constant, and $r = r_0 e^{\varepsilon t}$. Both r_0 and ε are constants. (a) Determine the condition on ε such that the particle's radial acceleration becomes zero. (b) When the radial acceleration is zero, is the radial velocity constant? [Caution. Write down the vector expressions for velocity and acceleration before you attempt a quick answer to this question].

P2.15 Let $\vec{P} = 2\hat{e}_r + 5\hat{e}_\theta - 3\hat{e}_\varphi$ and $\vec{Q} = 3\hat{e}_r + 5\hat{e}_\varphi$;

Determine the cross-product $(\vec{P}\times\vec{Q})$, and the component of $(\vec{P}\times\vec{Q})$ along $\hat{e}_z$ at $\left(1, \frac{\pi}{3}, \frac{5\pi}{4}\right)$.

P2.16 A covariant tensor has components $2x + 5z$, x^3y, y^2z in rectangular coordinates. Find the covariant and contravariant components in cylindrical coordinates.

P2.17 Diagonalize the following matrix $\mathbb{A} = \begin{bmatrix} 1 & 0 & 0 \\ 1 & 2 & 0 \\ -1 & 1 & -1 \end{bmatrix}$.

P2.18 (a) Determine the eigenvalues and the eigenvectors of the matrix $\mathbb{A} = \begin{bmatrix} 1 & 0 & 1 \\ 1 & 0 & 1 \\ 3 & 1 & 1 \end{bmatrix}$.

(b) Diagonalize the above matrix.

P2.19 Prove that the sum of the eigenvalues of the matrix $\begin{pmatrix} 3 & 4 \\ x & 1 \end{pmatrix}$ for real and negative values of x is greater than zero.

P2.20 If a matrix is given as $S = \begin{bmatrix} 1 & 1 & 1 \\ 1 & a^2 & a \\ 1 & a & a^2 \end{bmatrix}$, where $a = e^{2i\pi/3}$, prove that, $S^{-1} = \frac{1}{3}\bar{S}$ where $\bar{S}$ is the complex conjugate matrix of S.

CHAPTER 3

Real Effects of Pseudo-forces: Description of Motion in Accelerated Frame of Reference

The creative principle resides in mathematics. In a certain sense, therefore, I hold true that pure thought can grasp reality, as the ancients dreamed.

—*Albert Einstein*

3.1 GALILEO–NEWTON LAW OF INERTIA—AGAIN

One of my favorite experiments when I have fastened the seatbelt in an airplane ready to take-off, is to drop an object, say a pen (since they do not give a chocolate anymore.), from one hand into the other. At the beginning of the run, when the plane is at rest, the pen drops 'vertically' down, straight into the palm of the other hand beneath it. As the plane speeds up on the runway, the pen drops *backward*, missing the palm, hitting my chest. The line along which the pen falls would generate for me a perception of the 'vertical'. The line along which the pen falls is clearly different when the plane is at rest, and when it is accelerating on the runway. It could also change my perception of gravity, as I would think of it to be along the space curve of the falling pen's path. The perception of gravity which an observer in an inertial frame of reference has, is therefore clearly different from that of another in an accelerated frame of reference. To understand acceleration in different frames, we must therefore examine the cause that changes the equilibrium of an object.

We all have an intuitive idea of what equilibrium is. It is, however, important to underscore that objects not merely at rest but also those which are moving at a constant momentum, are essentially in a state of equilibrium. The latter form of equilibrium is almost counter-intuitive. It is contrary to common experience, because motion is resisted by friction. In order to maintain constant momentum, application of force is then necessary to overcome friction. Galileo recognized the essence of equilibrium by carrying out innovatory experiments, such as rolling objects on inclined planes, and dropping objects from the top of the mast of a ship (discussed in Chapter 1).

Galileo Galilei was born in 1564 to a distinguished musician, an expert lutenist, Vincenzo Galilei. Galileo not only grew into an expert lutenist himself, but also became a man of exceptional scholarship and contributed pioneering works in astronomy, physics, mathematics, philosophy and also engineering. Even as he is known as the 'father' of experimental physics (and thus the 'grandfather' of engineering, and the 'great-grandfather' of technology), he made seminal contributions to theoretical physics by recognizing what constitutes an inertial frame of reference. The inertial frame of reference provides absolutely the *ground zero* to understand fundamental physical interactions in nature. Galileo was also one of the first ones to recognize the incredible relationship between physics and mathematics, describing the latter as the very language of physics. He said: "*Philosophy is written in this grand book, the universe, which stands continually open to our gaze. But the book cannot be understood unless one first learns to comprehend the language and read the letters in which it is composed. It is written in the language of mathematics, and its characters are triangles, circles, and other geometric figures without which it is humanly impossible to understand a single word of it.*" Galileo's greatest works include (a) the fascinating experiments he performed to recognize that objects of different masses accelerate synchronously, (b) that projectiles follow parabolic trajectories from a combination of uniform horizontal motion and constant vertical acceleration, and (c) most importantly, how to recognize if an object is in a state of equilibrium.

The Galileo–Newton *law of inertia* is *central to all theoretical formulations of the laws of physics*. The identification of a fundamental interaction (whether gravitational, electromagnetic or nuclear-weak, or nuclear-strong) hinges on its effect on the object it influences, resulting in a *departure* from equilibrium of that object *in an inertial frame of reference*. The equivalence of the two alternate expressions of equilibrium, as a state of rest, or of uniform motion (i.e., at constant velocity/momentum), comes from the fact that *the very same physical interaction* accounts for any *departure* from equilibrium in alternate frames of reference moving at a constant velocity with respect to each other. This is illustrated below by considering the motion of an object P in two such frames.

We consider, in Fig. 3.1, the motion of an object P. We shall consider its motion in three different frames of reference, U, U' and U''. We consider these three frames of reference to have the same position in space at $t = 0$, with their respective X, Y, Z axes parallel to each other. These conditions only make the description of motion of the object P simple. However, the conclusions that we shall draw would not depend on any such restriction about their relative positions, or orientations, at $t = 0$.

We begin with the *ansatz* that the frame U is an inertial frame, and the frame U' moves with respect to it a constant velocity $\vec{u}_c$ (Fig. 3.1). The third frame of reference we consider, the frame U'' (Fig. 3.2), has a velocity $\vec{u}_i$ at $t = 0$ with respect to the frame U. It also has a constant acceleration $\vec{f}$ with respect to the frame U. What qualifies the frame U to be recognized as an 'inertial frame' is that all motion at constant momentum—*whether none* (i.e., at rest), *or, at a constant velocity*—in that frame is self-sustaining, determined solely by its initial (at $t = 0$) conditions. There is neither any need, nor the possibility, of any cause that would explain motion at the initial condition, since nothing can precede $t = 0$ in a causal analysis.

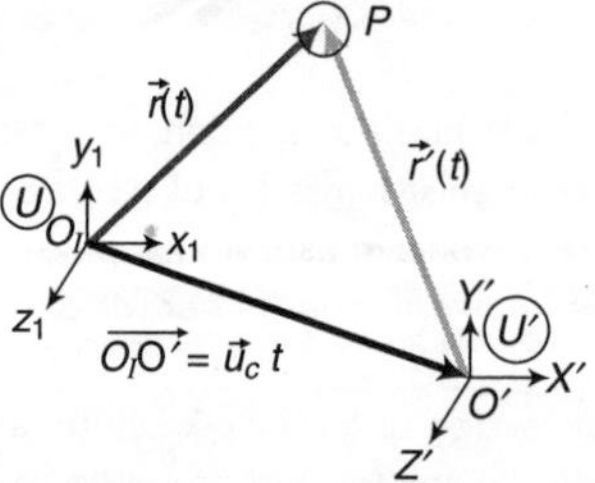

Fig. 3.1 Motion of the object *P* studied in an inertial frame of reference, *U*, and in another frame *U′* which is moving with respect to *U* at a *constant velocity* $\vec{u}_c$.

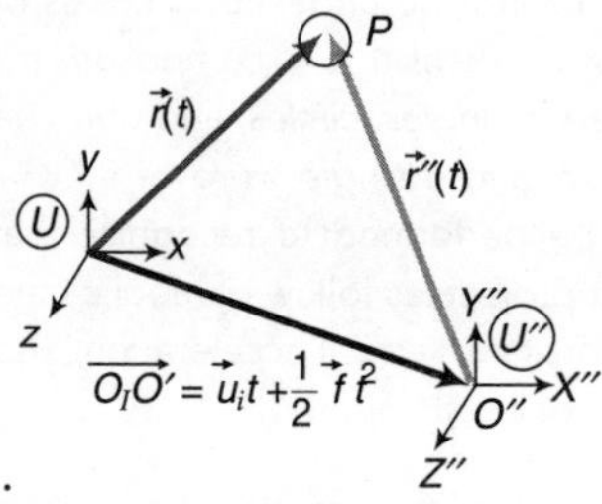

Fig. 3.2 Motion of the object *P* studied in the inertial frame *U*, and in the frame *U″* which is moving with respect to *U* at a *constant acceleration* $\vec{f}$.

Let us denote the position of an object *P* in frame *U* by its position vector $\vec{r}(t)$ at time *t*, and by $\vec{r}'(t)$ in frame *U′*. In frame *U″*, we shall denote it by $\vec{r}''(t)$. We see from Fig. 3.1 that

$$\vec{r}(t) = \overrightarrow{O_IO} + \vec{r}'(t), \tag{3.1}$$

i.e., $\vec{r}(t) = \vec{u}_c t + \vec{r}'(t)$,

and from Fig. 3.2 we see that

$$\vec{r}(t) = \overrightarrow{O_IO} + \vec{r}''(t), \tag{3.2}$$

i.e., $\vec{r}(t) = \vec{u}_i t + \dfrac{1}{2}\vec{f}t^2\,\vec{r}''(t)$.

The velocity of the object *P* in the frame *U* is $\vec{v}(t) = \dfrac{d}{dt}\vec{r}(t)$, and in the frames *U′* and *U″* the velocity of that object is respectively given by the two relations,

$$\vec{v}'(t) = \frac{d}{dt}\vec{r}'(t) = \vec{v} - \vec{u}_c, \tag{3.3a}$$

and $$\vec{v}''(t) = \frac{d}{dt}\vec{r}''(t) = \vec{v} - \vec{u}_i - \vec{f}t. \tag{3.3b}$$

The acceleration of the object P can now be determined easily in the three frames of reference, U, U' and U'', by taking the *time-derivative* of the velocity vectors in the respective frames.

Toward a far-reaching scrutiny of the time-derivative of the position, and of velocity, it is important to note that the perception of *time* itself is essentially the same in the three frames of reference, U, U' and U''. This *ansatz* requires a revolutionary modification when one reconciles it with the fact that it is the speed of light, and not *time* itself or the Euclidean distance between two points, that is the same in all the inertial frames. The analysis of relative motion, by observers in different frames of reference, with *time-interval* and *space-interval* considered to be *absolute* in the frames of reference, is called *Galilean* (or Newton-Galileo) *relativity*. We shall discuss later (in Chapters 13 and 14) that accurate description of simultaneity requires reconciliation with the fact that it is the *speed of light* (and not the *time-interval*, nor the *space-interval*) that is *absolute* in all inertial frames of reference. The *ansatz* of Galilean relativity, on the other hand, is that time-intervals and space-intervals are both *absolute*. This seems intuitively obvious, but only because the speed of light is extremely high, even if not infinite. The crucial factor that governs the dynamics discussed in the present chapter comes from the fact that it is the *derivative* with respect to time (of the position and the velocity vectors), and not *time* itself, that has imperative consequences in different frames of reference, well within the framework of Galilean Relativity.

3.2 MOTION OF AN OBJECT IN A LINEARLY ACCELERATED FRAME OF REFERENCE

We have seen that motion at constant momentum in an inertial frame of reference seeks no cause, since it is completely determined by the initial conditions. In an inertial frame, what *does* need an explanation is therefore the *change* in the momentum, and not the momentum $\vec{p}$ itself. In order to study this change, we must examine the rate of change of momentum with respect to time. For a point mass m, the momentum is simply mass times velocity, we must therefore examine the *rate of change of velocity*, i.e., the acceleration of the object under observation.

Let us determine the acceleration of an object P as would be measured in the three frames of reference, U, U' and U'' (Fig. 3.1 and Fig. 3.2). From Eq. 3.3, we find that the acceleration $\vec{a}'$ in the frame U' is the same as the acceleration $\vec{a}$ in the frame U, but in the frame U'', the acceleration is different; it is $\vec{a}''=\vec{a}-\vec{f}$. This essentially means that if a force $\vec{F}=m\vec{a}$ accounts for the change in equilibrium (i.e., change in momentum) in the inertial frame U, then the very same force also accounts for the change in equilibrium in the frame U', since $\vec{F}=m\vec{a}'$ as well. This is because $\vec{a}'$ is exactly equal to $\vec{a}$. The *dynamics* of changing equilibrium in both the reference frames, U and U', is exactly the same. Validity of Newton's laws in *either* of these frames guarantees that they would be valid in the other frame. The laws of mechanics hold good in the frame U', as much as in the inertial frame U. We therefore regard the frame U' *also* as an inertial frame. Every frame of reference which moves with respect to an inertial frame at a constant velocity is therefore also an inertial frame. It is of no importance to debate which of these frames is fundamental; all frames at constant velocity with respect to each other are completely equivalent.

The situation is, however, different in the frame U'', since $\vec{a}''=\vec{a}-\vec{f}$. This leads us to the relation

$$m\vec{a}''=m\vec{a}-m\vec{f}. \tag{3.4}$$

Mathematically, we can surely the left hand side of the above equation denote by $\vec{F}_{U''}$, a quantity which has the physical dimensions for force, such that

$$\vec{F}_{U''}=m\vec{a}'', \tag{3.5}$$

i.e., $$\vec{F}_{U''}=\vec{F}-m\vec{f}. \tag{3.6}$$

$\vec{F}_{U''}$ obviously is *different* from the physical interaction $\vec{F}$. Nonetheless, $\vec{F}_{U''}$ plays essentially the same (Newtonian) role in accounting for the acceleration $\vec{a}''$ measured in the frame U'', as $\vec{F}_U$ plays in the frame U, to account for the acceleration $\vec{a}$ measured in the frame U. The difference $m\vec{f}$ between $\vec{F}_{U''}$ and $\vec{F}$ involves only the fact that that the frame U'' is accelerated with respect to U. Accordingly, $\vec{F}_{U''}$ is known as 'pseudo-force'. It contains the term $\vec{F}$, the real physical interaction that was identified in the inertial frame U, but also the pseudo-force $m\vec{f}$. The term $m\vec{f}$ is a mathematical artefact whose inclusion enables us to write Eq. 3.5 in a form conventionally used to express the principle of causality in Newton's second law. This form makes it possible for us to interpret the acceleration $\vec{a}''$ in U'' as a linear response to the (*pseudo-*) force $\vec{F}_{U''}$. Even as the acceleration $\vec{a}''$ would be real (i.e., measurable and perceptible) in the frame U'', its cause $\vec{F}_{U''}$ is unreal; only the part $\vec{F}$ of $\vec{F}_{U''}$ represents the real physical interaction. The acceleration $\vec{a}''$ in the frame U'' is thus a *real effect of the pseudo-force* $\vec{F}_{U''}$. Happily, for example, if you are keen on loosing your weight ($\vec{W}=m\vec{g}$), all you need to do is to weigh yourself in an elevator accelerating downward (Fig. 3.3).

$\downarrow\vec{f}\downarrow$

Fig. 3.3 The effective weight measured in an accelerating elevator would not be $\vec{W}=m\vec{g}$, but $\vec{W}''=m\vec{g}-m\vec{f}$.

Thus $\vec{W}''<\vec{W}$ when $\vec{f}$ is parallel to $\vec{g}$ (i.e., the elevator is accelerating downward).

Unfortunately, if $\vec{f}$ is anti parallel to $\vec{g}$ (elevator accelerating upward), $\vec{W}''>\vec{W}$. An underweight person may, however, actually desire this.

The difference between the relation $\vec{F}=m\vec{a}$ (valid in the inertial frame U) and the relation $\vec{F}_{U''}=m\vec{a}''$ (which is valid in the frame U'') is of crucial importance to understand *fundamental* interactions in nature. The four fundamental interactions commonly recognized are: (1) gravitational, (2) electromagnetic, (3) nuclear-weak, and (4) nuclear-strong. The number of fundamental interactions is actually *less* than the four listed above. Sheldon Lee Glashow, Abdus Salam and Steven Weinberg developed the theory that unified the nuclear-weak and the electromagnetic interaction. For this work, they were jointly awarded the 1979 Nobel Prize in Physics. Attempts toward further integration of the fundamental interactions in a unified theory are still in progress. These studies represent major frontiers of theoretical physics. The term $\vec{F}$ which would go into the principle of causality in Newton's second law in the inertial frame of reference alludes to the sum total of all the fundamental *physical* interactions that cause the object *P*'s acceleration in the inertial frame *U*. The term $\vec{F}_{U''}$ which is used in the *imitation* of Newton's second law in the frame U'' however includes not just the physical interactions, but also the mathematical artefact $m\vec{f}$. The pseudo-force $m\vec{f}$ does not represent any fundamental interaction. It has only resulted from the acceleration of the frame U'' with respect to *U*. We note that in the present case, the acceleration of U'' with respect to *U* is linear, i.e., along a straight line.

The difference between the weight $\vec{W}$, and the effective weight $\vec{W}''$, in a vertically accelerating elevator has interesting consequences. The effective weight of a person in an elevator which is in a state of free fall (i.e., $\vec{f} = \vec{g}$) would be zero. This is often referred to as 'weightlessness'. It does not, however, refer to gravity itself being switched off, even if such a state is sometimes loosely called zero-gravity. The gravitational interaction cannot be switched off. The state of weightlessness is only an experience of an observer who is in *free fall* under gravity. In such a frame, real physical interactions alone do not account for the acceleration measured. The observations can be accounted for in the accelerated frame using an *imitation* of Newton's second law by including a mathematical construct using the pseudo-force.

An object in a state of free fall would thus experience weightlessness. A cat that jumps (or falls) from a wall is in a state of free fall. It has to overcome only the inertia of its limbs, and not gravity. This enables the cat to swirl itself sharply to land safely on all four. The agility with which a cat survives the fall has earned it the reputation of having nine lives. Likewise, pole-vaulters and high-jumpers flex their bodies, exploiting their agility, and clear heights often allowing their center of mass to pass *below* the bar even as their bodies , bend over, and clear greater hight. Isn't it tempting to imagine that you could actually lift an elephant in a freely falling room? A dramatic consequence of 'weightlessness' can be illustrated by considering the shape of a liquid in a sealed glass cell, in a satellite orbiting the Earth. The satellite is continually *free falling* towards the Earth. The liquid inside the cell, and the cell itself, are both in the state of free fall. If the cell was placed on a desk in a laboratory on Earth, the liquid would of course settle down in the container, *on account of gravity*. The liquid's open surface would then be horizontal (except for the meniscus). However, when freely falling along *with* the cell in a satellite, gravity causes *both* the cell and the liquid inside it to 'fall' *together*. It would then be the competition between (a) the weak adhesive forces between the molecules of the liquid and those of the container, and (b) the cohesive forces within the liquid molecules, that would determine the shape of the liquid. If the cohesive forces are stronger, the liquid would coalesce into globules that float and fall freely along the cell in the satellite. NASA carried out various such experiments in the Skylab which you will find interesting to read [1].

3.3 MOTION OF AN OBJECT IN A ROTATING FRAME OF REFERENCE

In general, the acceleration of a frame of reference moving with respect to an inertial frame of reference may include a linear component, considered above, and also a rotating component (Fig. 3.4), which we now discuss.

Fig. 3.4 The 'Superman Ultimate Flight' roller coaster rides in Atlanta's 'Six Flags over Georgia' provides thrilling experience to its riders who are in an accelerated frame of reference. The riders undergo both linear and rotational acceleration with respect to Earth.

We consider a frame of reference U_R (Fig. 3.5) which rotates with respect to the inertial frame U at a constant angular velocity $\vec{\omega}$ about an axis OC. The axis OC is oriented along the direction of a unit vector $\hat{n}$. Now, a few new and different kinds of pseudo-forces will have to be mathematically introduced to finally express the motion of an object P in the frame U_R to employ Newtonian principle of causality. We already know all too well that Earth is rotating about its South–North axis, and thus the observations of motion of each object we carry out on Earth must be correctly analyzed using the methods of this section. It will turn out that this has major ramifications in today's high-technology space-age. Aeroplane and sea-ship navigation, working of the GPS system, functioning of mobile telephones, calibration of the atomic clocks, etc., are affected by the pseudo-forces that must be accounted for in the earth-centric frame of reference.

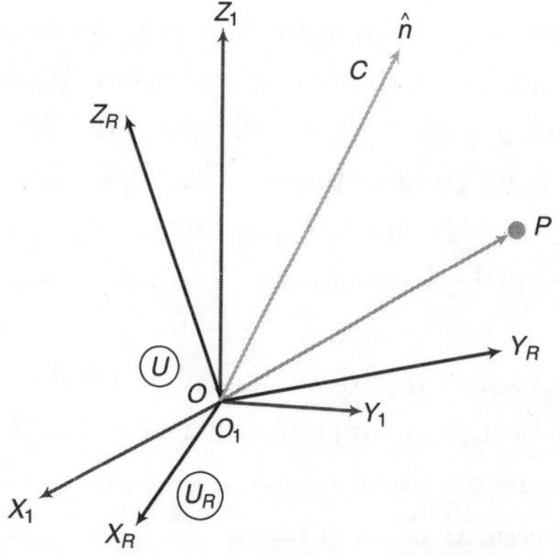

Fig. 3.5 The frame U_R rotates about the axis OC, oriented along the direction of a unit vector $\hat{n}$, at a constant angular velocity $\vec{\omega}$. A point P that appears to be static in the rotating frame is rotating in the inertial frame U, and vice versa.

Centuries ago, man believed that the Earth was static, and that it was the center of the universe. It was natural for him to believe, what he saw, that every other astronomical object revolved around it. Aristotle (384 to 322 BCE), for example, theorized that the heavens rotate around the Earth. Ptolemy (second century CE) also believed the same. That the sky seems to turn around from evening to the next morning, and then again to the following evening, is on account of Earth's rotation about its own axis was recognized by early Indian astronomers [2,3], notably Aryabhata around 500 CE. He wrote (in Sanskrit):

अनुलोमगतिर्नौस्थः पश्यत्यचलं विलोमगं यद्वत् ।
अचलानि भानि तद्वत् समपश्चिमगानि लङ्कायाम् ।। ९ ।।

Meaning: Just as a man in a moving boat sees the stationary objects (on either side of the river) as moving backward, so are the stationary stars seen by the people at Lanka (i.e., reference coordinate on the Equator) as moving exactly toward the west.

We consider an object P which appears to be at rest in the rotating frame U_R. As such, you are also an observer on Earth that is rotating about the axis which runs between the *South Pole and the North Pole*. The object P could be anything around you that is at rest with respect to you. To an observer in an inertial frame of reference, this object would appear to traverse on the rim of a circle that would constitute the base of a cone, with its vertex at the origin O and axis along $\hat{n}$ (Fig. 3.6). We shall denote the angle between the instantaneous position vector $\vec{r}(t)$ of the point P, and the axis of the cone, as ξ.

The velocity $\vec{v}$ of the object P in the inertial frame U_I is obviously different from the velocity $\vec{v}_R$ in the rotating frame U_R. The acceleration $\vec{a}_I$ of the object P in the inertial frame U_I and the acceleration $\vec{a}_R$ in the rotating frame U_R would be different. Given the fact that the acceleration in the inertial frame U_I is accounted for by *physical* stimuli (forces) that act on P, it is obvious that *unreal*, *pseudo-forces*, must be introduced to account for the acceleration, $\vec{a}_R \neq \vec{a}_I$. Introducing pseudo-forces would enable us cast the equation of motion for P in the rotating frame in the Newtonian causality form: $\vec{F}_R$ (pseudo-force in U_R) $= m\vec{a}_R$.

As can be seen from Fig. 3.6, the change in the position vector of the object P in the inertial frame U_I over a time interval δt is given by $\overrightarrow{\delta r} = \vec{r}(t+\delta t) - \vec{r}(t)$, even as the object P would appear to be totally unmoved in the rotating frame U_R.

Clearly, we have, in the frame U_I,

$$\overrightarrow{\delta r} = \vec{r}(t+\delta t) - \vec{r}(t) = |\overrightarrow{\delta r}|\,\hat{u}, \tag{3.7}$$

wherein we have factored the displacement vector $\overrightarrow{\delta r}$ in its magnitude $|\overrightarrow{\delta r}|$ and direction, which is given by

$$\hat{u} = \frac{\hat{n}\times\hat{r}}{|\hat{n}\times\hat{r}|}, \tag{3.8}$$

$\hat{r}$ being the *instantaneous* unit vector in the direction of the position vector $\vec{r}(t)$. Note that the object P appears (in the inertial frame U_I) to traverse on the rim of the circle (whose plane is orthogonal to $\hat{n}$). The axis OC is oriented along $\hat{n}$, and intersects the plane of this circle at the point Q. The circle subtends an angle ξ at the origin. The radius of this circle is therefore $r \sin \xi$.

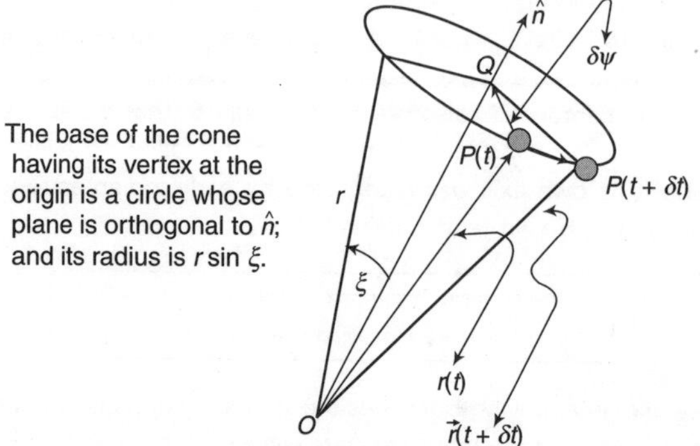

Fig. 3.6 In time δt, the object P which appears to be static in the rotating frame U_R would appear to have moved in the inertial frame U to the point $P(t + \delta t)$. Its position vector $\vec{r}(t)$ with respect to the origin O in the inertial frame would appear to have changed to $\vec{r}(t + \delta t)$ over the same time-interval.

We see that the displacement vector $\overrightarrow{\delta r}$ subtends an angle $\delta\psi$ at the point Q.

Hence, $$\overrightarrow{\delta r} = (r \sin \xi \delta\psi)\left(\frac{\hat{n} \times \hat{r}}{|\hat{n} \times \hat{r}|}\right) = (r\delta\psi)\hat{n} \times \hat{r} = \delta\psi \hat{n} \times \vec{r}, \tag{3.9}$$

since $\vec{r} = r\hat{r}$.

It immediately follows from Eq. 3.9 that the object P which is stationary in the rotating frame U_R, has a velocity $\vec{v}_I$ in the inertial frame U, given by:

$$\left[\frac{d}{dt}\right]_I \vec{r} = \vec{v}_I = \lim_{\delta t \to 0} \frac{\overrightarrow{\delta r}}{\delta t} = \lim_{\delta t \to 0} \frac{\delta\psi}{\delta t} \hat{n} \times \vec{r} = \vec{\omega} \times \vec{r}. \tag{3.10}$$

Note that in Eq. 3.10, we have identified the angular velocity vector. $\left[\frac{d}{dt}\right]_I$ is the differential operator which gives the rate of change with respect to time in the inertial frame of reference U_I. This result is valid for every object that is at rest in a rotating frame, no matter what its position vector. We therefore get an operator *identity* for the time-derivative operator in the inertial frame of reference U_I:

$$\left[\frac{d}{dt}\right]_I \equiv \vec{\omega}\times \tag{3.11}$$

The above *operator identity* implies that the effect of taking the time-derivative in the inertial frame of reference is completely equivalent to obtaining the cross product with the angular velocity vector. Thus, if the operator in Eq. 3.11 operates on the position vector $\vec{r}$, we get Eq. 3.10.

Now, if the object P happens to already have a *non-zero* velocity $\vec{v}_R = \left[\frac{d}{dt}\right]_R \vec{r}$ in the rotating frame of reference U_R, then this velocity must be added to the right hand side of Eq. 3.10 to get its velocity in the inertial frame of reference, i.e.,

$$\left[\frac{d}{dt}\right]_I \vec{r} = \vec{\omega}\times\vec{r} + \left[\frac{d}{dt}\right]_R \vec{r}. \tag{3.12}$$

The operator equivalence in Eq. 3.11 must therefore be modified to the following:

$$\left[\frac{d}{dt}\right]_I \equiv [\vec{\omega}\times\ ] + \left[\frac{d}{dt}\right]_R . \tag{3.13}$$

We are now ready to determine the relation between the acceleration $\vec{a}_I$ of an object in the inertial frame U_I, with the acceleration $\vec{a}_R$ in the rotating frame U_R. All we have to do now is to apply the operator in Eq. 3.13 *twice* on the position vector:

$$\left[\frac{d}{dt}\right]_I\left[\frac{d}{dt}\right]_I \vec{r}(t) = \left\{[\vec{\omega}\times] + \left[\frac{d}{dt}\right]_R\right\}\left\{[\vec{\omega}\times] + \left[\frac{d}{dt}\right]_R\right\}\vec{r}(t). \tag{3.14}$$

Therefore,

$$\vec{a}_I = \left[\frac{d^2}{dt^2}\right]_I \vec{r}(t) = \left\{[\vec{\omega}\times] + \left[\frac{d}{dt}\right]_R\right\}\left\{[\vec{\omega}\times\vec{r}(t)] + \left[\frac{d\vec{r}(t)}{dt}\right]_R\right\}, \tag{3.15a}$$

i.e., $$\vec{a}_I = [\vec{\omega}\times\ ]\left\{[\vec{\omega}\times\vec{r}(t)] + \left[\frac{d\vec{r}(t)}{dt}\right]_R\right\} + \left[\frac{d}{dt}\right]_R\left\{[\vec{\omega}\times\vec{r}(t)] + \left[\frac{d\vec{r}(t)}{dt}\right]_R\right\}, \tag{3.15b}$$

i.e., $$\vec{a}_I = [\vec{\omega}\times\{\vec{\omega}\times\vec{r}(t)\}] + \left[\vec{\omega}\times\frac{d\vec{r}(t)}{dt}\right]_R + \left[\frac{d}{dt}\right]_R[\vec{\omega}\times\vec{r}(t)] + \left[\frac{d}{dt}\right]_R\left[\frac{d\vec{r}(t)}{dt}\right]_R \tag{3.15c}$$

$$\vec{a}_I = \left[\{\vec{\omega}.\vec{r}(t)\}\,\vec{\omega} - \omega^2\vec{r}(t)\right] + \left[\vec{\omega}\times\frac{d\vec{r}(t)}{dt}\right]_R$$

$$+\left[\frac{d\vec{\omega}}{dt}\right]_R\times\vec{r}(t) + \vec{\omega}\times\left[\frac{d\vec{r}(t)}{dt}\right]_R + \left[\frac{d^2\vec{r}(t)}{dt^2}\right]_R . \tag{3.15d}$$

Rearranging the terms, we get

$$\vec{a}_I = \left[\{\vec{\omega}\cdot\vec{r}(t)\}\,\vec{\omega} - \omega^2\vec{r}(t)\right] + \left[2\vec{\omega}\times\frac{d\vec{r}(t)}{dt}\right]_R + \left\{\left[\frac{d\vec{\omega}}{dt}\right]_R\times\vec{r}(t)\right\} + \left[\frac{d^2\vec{r}(t)}{dt^2}\right]_R . \tag{3.15e}$$

The last term in Eq. 3.15e, $\left[\frac{d^2\vec{r}(t)}{dt^2}\right]_R$, is the acceleration $\vec{a}_R$ of the object P which an observer in the rotating frame U_R would see. Multiplying Eq. 3.15 by the mass of the object, we get from the left hand side the mass times acceleration as identified in Newton's second law. The causality principle in Newtonian mechanics identifies it as the physical force (interaction) $\vec{F}_I$ which causes the departure from equilibrium of P in the inertial frame of reference. Thus,

$$\vec{F}_I = m\left[\frac{d^2\vec{r}(t)}{dt^2}\right]_I = m\vec{a}_I, \tag{3.16a}$$

and

$$\vec{F}_R = m\left[\frac{d^2\vec{r}(t)}{dt^2}\right]_R = m\vec{a}_R. \tag{3.16b}$$

Obviously, $\vec{F}_I \neq \vec{F}_R$. Instead, we have

$$\vec{F}_I = m\left[\{\vec{\omega}\cdot\vec{r}(t)\}\,\vec{\omega} - \omega^2\vec{r}(t)\right] + \left[2m\vec{\omega}\times\frac{d\vec{r}(t)}{dt}\right]_R + m\left\{\left[\frac{d\vec{\omega}}{dt}\right]_R \times\vec{r}(t)\right\} + \vec{F}_R. \tag{3.17}$$

Therefore, the mass times acceleration term in the *rotating* frame of reference becomes

$$\vec{F}_R = \vec{F}_I + \vec{F}_{cfg} + \vec{F}_{\text{Coriolis}} + \vec{F}_{\text{LSC}}, \tag{3.18}$$

where,

$$\vec{F}_{cfg} = -m\vec{\omega}\times\{\vec{\omega}\times\vec{r}(t)\} = -m\left[\{\vec{\omega}\cdot\vec{r}(t)\}\vec{\omega} - \omega^2\vec{r}(t)\right], \text{ (called “centrifugal force”)}, \tag{3.19a}$$

$$\vec{F}_{\text{Coriolis}} = -\left[2m\vec{\omega}\times\frac{d\vec{r}(t)}{dt}\right]_R, \text{ (called “Coriolis force”), and} \tag{3.19b}$$

$$\vec{F}_{\text{LSC}} = -m\left\{\left[\frac{d\vec{\omega}}{dt}\right]_R \times\vec{r}(t)\right\}, \text{ (called “Leap Second Correction force”)}. \tag{3.19c}$$

The Coriolis force, $\vec{F}_{\text{Coriolis}}$, is named after Gaspard Gustave de Coriolis (Fig. 3.7).

Gaspard Gustave de Coriolis
1792–1843

Jean Bernard Léon Foucault
1819–1868

Fig. 3.7 Coriolis and Foucault (pronounced ‘Fookoh’) made important contributions to our understanding of changes in momentum of an object which is in motion, as observed on Earth.

- The equations, Eqs. 3.16a and 3.16b, are both correct. They both also have exactly the same form, but in different frames of reference. In both the frames they express, the Newtonian principle, that the rate of change of momentum of an object is equal to the force acting on it. Both of these equations express the causality principle that the acceleration resulting in the motion of an object is linearly proportional to the force that 'causes' it. The nature of the force is, however, completely different in the two frames of reference. In the inertial frame of reference U_I, the force $\vec{F}_I$ would be associated with a physical interaction between an agency and the target on which it acts. It would be identified with one of the fundamental interactions in the physical universe: gravity, electromagnetic-nuclear-weak, nuclear-strong (or a sum of some of these physical interactions). In the rotating frame U_R the force $\vec{F}_R$ comprises of a superposition (sum) of the physical interaction force $\vec{F}_I$ and one (or more) pseudo-force, defined in Eq. 3.19(a, b, and c).

It is now a matter of straightforward extension that if another frame of reference, say U_{LR}, has a linear acceleration $\vec{f}$ with respect to an inertial frame of reference (such as the frame U'' discussed in Section 3.1), and is *also* rotating with respect to the inertial frame (similar to the frame U_R, discussed above), then the equation of motion in the frame U_{LR} akin to Newton's second law would be

$$\vec{F}_{LR} = \vec{F}_I + \vec{F}_{cfg} + \vec{F}_{Coriolis} + \vec{F}_{LSC} - m\vec{f} = m\vec{a}_{LR}, \tag{3.20a}$$

where $\vec{a}_{LR}$ is the acceleration of the object that an observer in the frame U_{LR} would see. We note that all the forces, except for the term $\vec{F}_I$, in Eq. 3.20a are pseudo-forces.

A few points pertaining to judicious use of the pseudo-forces are in order:

- None of the pseudo-forces has any role to play if the analysis of motion is made completely in the inertial frame of reference. Eq. 3.16a comprehensively expresses the Newton's second law in the spirit of the principle of causality: the acceleration (effect) is linearly proportional to the physical cause (force).
- The pseudo-forces are unphysical; they are introduced as mathematical artefacts in the theory of mechanics only when motion is analyzed in an accelerated frame of reference. They are introduced when motion is observed in a frame of reference that is either linearly accelerated with respect to an inertial frame of reference (as in Eq. 3.6), or in a frame of reference that is rotating with respect to the inertial frame (as in Eq. 3.16b), or both (as in Eq. 3.20).
- The principle of causality and determinism, which states that the rate of change of momentum of an object is equal to the force (stimulus) acting on it, refers to a physical fundamental force (gravity, electromagnetic-nuclear-weak, nuclear-strong). It is tacitly valid only in an inertial frame of reference. In fact, an inertial frame is often *defined* as one as one in which Newton's laws hold. The forces invoked in Newton's fundamental laws are real forces resulting from *physical* interactions. The force acting on an object in the inertial frame is exactly equal to the rate of change of momentum of the object: $\vec{F} = \dfrac{d\vec{p}}{dt} = m\vec{a}$. Pseudo-forces are invoked in a frame of reference that is *accelerated*

with respect to an inertial frame of reference, only in order to express its acceleration ($\vec{a}''$ or $\vec{a}_R$) as proportional to a force ($\vec{F}''$ or $\vec{F}_R$), which is, however, a combination of physical forces and pseudo-forces. Pseudo-forces thereby enable an interpretation of the perceived acceleration of an object as resulting from the application of a force, albeit an unreal one. They enable us write an equation of motion in an *accelerated* frame of reference that is *isomorphous* to Newton's second law in an *inertial* frame of reference.

- The conservation of linear momentum expressed in Newton's third law (Chapter 1) involves action–reaction forces which represent a pair of *physical* interactions in an inertial frame of reference. Pseudo-forces are naturally irrelevant to the first and the third laws of Newton.

3.4 REAL EFFECTS THAT SEEK PSEUDO-CAUSES

We have already discussed the difference between the 'true' weight (gravitational force when you weigh yourself on ground) and the 'apparent' weight, as would be measured in an elevator accelerating upward or downward. The difference is entirely on account of the pseudo-force due to the linear acceleration of the elevator. The physical experience is as real as the difference in the reading actually seen on the weighing machine which can even be zero if the elevator is in a state of free fall. On a giant wheel rotating in a vertical plane, the effective weight of a person would be different at all orientations of the wheel. The effective weight at a position at some height on one side of the wheel that is going *up* would be different from that at the same height on the opposite side of the giant wheel that is moving *down.*

Rapid advances in technology make it necessary to develop comfort with pseudo-forces. It is no longer true that for most purposes we may regard Earth as an inertial frame of reference for the analysis of everyday phenomena. This supposition could possibly be made until a few decades ago. Now, air travel, rocket propulsion, television transmission, mobile telephones, other communication devices including the internet, connect people globally through satellite links. The very notion of what can be regarded as *everyday* phenomena has now changed drastically with advancing technology. Everyday tools and gadgets we use necessitate proper cognizance of every term in Eq. 3.18 (even Eq. 3.20a in some cases). Just like the effective weight actually *seen* on the scale in an accelerating elevator, all effects of all pseudo-forces are perceptible in the accelerated frame, they are real. However, their causes are *pseudo*, unphysical. They are only mathematical artefacts which compensate for the observer herself/himself being in an accelerated frame of reference.

We now consider the *real* effects due to the *pseudo*-forces $\vec{F}_{cfg}$, $\vec{F}_{\text{Coriolis}}$, and $\vec{F}_{\text{LSC}}$ in a rotating frame of reference. We consider two astronauts, Sunita (S) and Kalpana (K), in a satellite orbiting the Earth. A satellite is often imparted a rotational motion about an axis through it. Such a rotation provides stability, since it generates an angular momentum which cannot be changed without the application of a torque. The satellite orbiting the Earth is in a state of *free fall*. The satellite, along with everything within it, is accelerated continuously toward the Earth together. We consider a tool tossed by Kalpana toward Sunita in a plane orthogonal to Earth's gravity at the position of the satellite. (The names of these astronauts are chosen in honor of the first two Indian female astronauts, Kaplana Chawla and Sunita Williams). The only physical force acting on the tool after it is tossed away is Earth's gravity, $\vec{F} = m\vec{g}$, which would be exactly canceled by the term $m\vec{f}$ in Eq. 3.20a. We consider the rotational angular velocity $\vec{\omega}$ of the satellite (about its axis, considered to be along $\vec{g}$ in the present example) to be constant. The equation of motion in the frame U_A of the astronauts would then be, from Eq. 3.20a,

$$\vec{F}_A = \vec{F}_{cfg} + \vec{F}_{\text{Coriolis}}, + \vec{F}_{LSC} = m\vec{a}_A. \tag{3.20b}$$

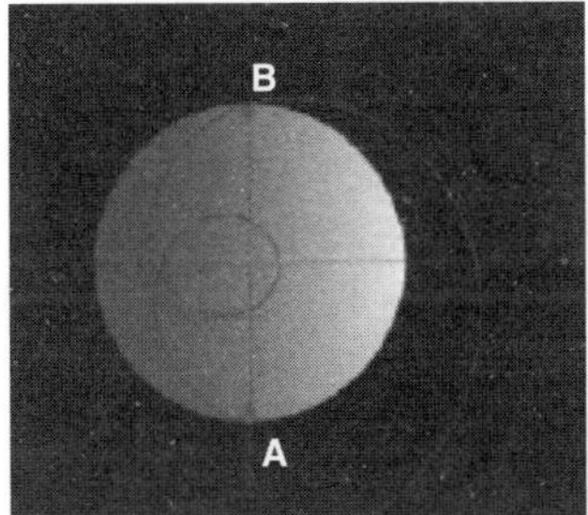

Fig. 3.8 Trajectory of a tool thrown by an astronaut toward another diametrically opposite to her in an orbiting satellite discussed in the text. The change in the momentum of the tool is only due to the pseudo-forces $\vec{F}_{cfg} + \vec{F}_{\text{Coriolis}}$ which are completely fictitious, and necessitated only because the motion is viewed by astronauts in an accelerated (rotating) frame of reference U_A [4]. The trajectories would be along various curves determined by the initial momentum imparted to the tool by the astronaut.

The trajectory of the tool, once thrown, would be essentially described as a projectile motion if it were viewed from an inertial frame of reference. In the astronauts' frame U_A, it would be along various curves in the plane orthogonal to $\vec{g}$, determined by the initial throw momentum. Details regarding the initial conditions that would generate the weird trajectory seen in Fig. 3.8 can be found in Reference [4].

A dramatic effect of the pseudo-force $\vec{F}_{\text{Coriolis}}$ is seen in the oscillations of the Foucault pendulum. Consider a simple pendulum oscillating in its plane and placed at the North Pole. Even as the plane of the pendulum is fixed, the Earth under it is rotating about the South–North axis. An observer on Earth at the North Pole will see that the plane of oscillation of the pendulum is turning. He will observe that it takes 24 hours for the plane of oscillation to complete a full turn. If the same experiment is carried out at the South Pole, an observer located there would see the same effect, except that the plane of rotation would turn in exactly the opposite direction, since she is an antipode of the observer on the North Pole. It would take 24 hours at the Poles for the plane of the motion of the simple harmonic oscillator to complete a full rotation through 2π, though the sense of rotation would be opposite. The period of rotation of the plane of oscillation, however, increases as one moves from the Poles to the Equator, where this period would be infinite, since at the Equator the plane of oscillation would not rotate at all. At the Equator, it would be some sort of an average behavior at the North and the South Poles, where the rotations are exactly opposite, so at the Equator, the plane of rotation therefore simply does not rotate. 'No rotation' corresponds to an 'infinite' time-period, so the period of rotation steadily increases from 24 hours at the Poles to 'no-rotation' at the Equator. It was Jean Bernard Léon Foucault (Fig. 3.7) who first demonstrated, in February 1851 at the Paris Observatory, that the plane of oscillation of the pendulum rotated clockwise, and took about 31.8 hours to complete one rotation.

We now proceed to determine a quantitative assessment of the real effects on motion of objects seen on Earth, considering the pseudo-forces in the earth-centered frame of reference, which rotates about its North–South axis. We begin by choosing an appropriate coordinate system, specific to the point on Earth from where the observation would be carried out. The particular coordinate system (Fig. 3.9) that is proposed below is simple, but important. It makes analysis of the pseudo-forces $\vec{F}_{cfg}$ and $\vec{F}_{\text{Coriolis}}$ easy. We shall employ a Cartesian (X, Y, Z) coordinate system, in which the position vector of any point is described in terms of a linear combination of the basis unit vectors $\{\hat{e}_x, \hat{e}_y, \hat{e}_z\}$ such that $\hat{e}_z$ is along the local vertical, determined by a plumb line at the particular point on earth.

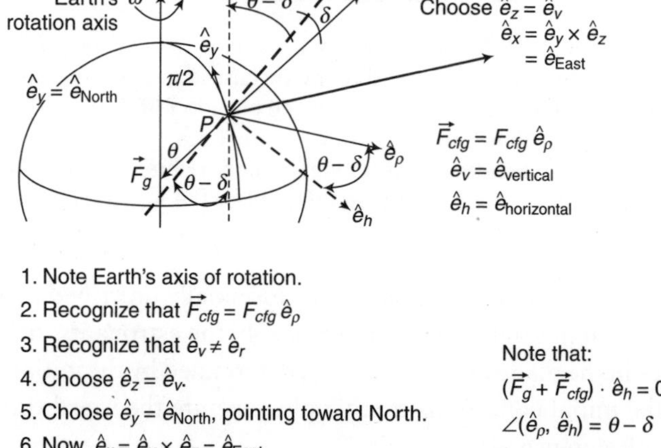

Fig. 3.9 Schematic diagram showing the choice of the Cartesian coordinate system employed at a point P on Earth to analyze the pseudo-forces $\vec{F}_{cfg}$ and $\vec{F}_{\text{Coriolis}}$ that must be considered due to Earth's rotation at an angular velocity $\vec{\omega}$ about the South–North axis. The angle $\measuredangle(\hat{e}_z, \hat{e}_r) = \delta$ is that between the local vertical along $\hat{e}_z$, and the radial outward line, along $\hat{e}_r$, from Earth's center. $\hat{e}_y$ is chosen to point toward the North Pole, and $\hat{e}_x = \hat{e}_y \times \hat{e}_z$ provides the direction of the local east. Note that $\hat{e}_Z = \hat{e}_V$ does not point toward Earth's center C. Also, note that the angle, $\lambda = \measuredangle(\vec{\omega}, \hat{e}_{\text{North}}) = \measuredangle(\vec{\omega}, \hat{e}_y)$. Note that we are assuming that Earth has a spherical shape.

A plumb line is just a piece of mass suspended at the bottom of a string held steady, as would be oriented if suspended from the roof of a room. It would be a simple harmonic oscillator if only it were set in small oscillations. The plumb line is, however, essentially steady in its equilibrium position. The line along which the plumb line is oriented is the 'local vertical'. If you looked up along the plumb line, you would therefore be looking in the direction of the local vertical. This is the direction along which the unit vector $\hat{e}_z$ is to be chosen first. Note that the unit vector $\hat{e}_z$ so chosen would not be along the radial line from Earth's center; since the mass suspended to the plumb line is being observed on the rotating Earth. Its equilibrium in the rotating frame is determined not just be gravity, but by the combined effect of gravity *and* the centrifugal force.

If the radial outward line from the center of Earth is along $\hat{e}_r$, then the angle $\measuredangle(\hat{e}_z, \hat{e}_r) = \delta$ would be zero only at the Poles, and at the Equator. At other points on Earth, it would be non-zero, even if rather small. We chose the direction $\hat{e}_y$ of the coordinate system in a plane orthogonal to $\hat{e}_z$ such that $\hat{e}_y$ points toward the North Pole on Earth. The third unit vector of the basis set of the Cartesian coordinate system, $\hat{e}_x$, is now chosen easily. It is simply given by the cross-product $\hat{e}_y \times \hat{e}_z$. It is defined with respect to the 'local vertical', and points toward the direction of the 'local east'. It cannot, of course, necessarily be exactly in the direction from which the Sun rises. The 'east' defined by the direction from which the Sun rises changes slightly over the year, since Earth's rotation axis makes an angle (~23.4°) with the ecliptic plane. The coordinate system $\{\hat{e}_x^{\,(\mathrm{East})}, \hat{e}_y^{\,(\mathrm{North})}, \hat{e}_z^{\,(\mathrm{Up})}\}$ we have introduced makes no reference to the direction from which the Sun rises. We can proceed to consolidate the formalism to quantify the real effects of pseudo-forces in the rotating Earth coordinate system.

An interesting upshot of Eq. 3.18 is that (i) a line in space defined by a radial line from Earth's center through its surface, (ii) a line along which a plumb line at rest is suspended, and (iii) a space curve along which a tiny mass is dropped above that point on the Earth's surface, would all be different. While line (i) is defined by geometry alone, the line (ii) would be seen by an observer on the Earth's surface to be pointing *away* from the Earth's center, on account of the centrifugal term, and the space-curve (iii) would require for its shape, seen by the observer on Earth the consideration of both the centrifugal and the Coriolis terms.

It is easy to see that in the Cartesian coordinate system so chosen, the angular velocity vector $\vec{\omega}$ would be in the YZ-plane, and hence given by

$$\vec{\omega} = (\vec{\omega} \cdot \hat{e}_y)\hat{e}_y + (\vec{\omega} \cdot \hat{e}_z)\hat{e}_z. \tag{3.21}$$

Let us now determine the real effect of the centrifugal pseudo-force.

$$\vec{F}_{cfg} = -m\vec{\omega} \times \{\vec{\omega} \times \vec{r}(t)\} = -m\omega^2 r\hat{e}_\omega \times (\hat{e}_\omega \times \hat{e}_r) = -m\omega^2 r\hat{e}_\omega \times (\sin\theta\, \hat{e}_\varphi) = m\omega^2 r \sin\theta\, \hat{e}_\rho. \tag{3.22}$$

The magnitude of the centrifugal force will consequently change with Earth's latitude, as shown in Fig. 3.10.

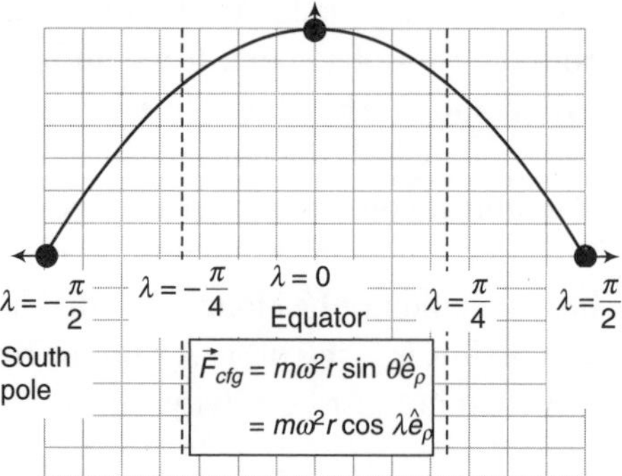

Fig. 3.10 The magnitude of the centrifugal force is largest at the Equator, and nil at the Poles. In between, it varies as the cosine of the latitude.

The effect of the centrifugal force in the earth-frame is that the equilibrium of every object that is at rest on Earth can be accounted for only by the combination of Earth's gravity, *and* the centrifugal force. This results in the object being radially pushed *away* in a direction that is orthogonal to Earth's axis of rotation. Of course, whatever is the normal reaction force on the object at rest that holds it at rest, must also be considered. A plumb line suspended on Earth's surface would therefore point in a direction *away* from the axis of rotation of Earth. The deflection angle δ (Fig. 3.9) is about 0.1° at a latitude of 45°.

The effect of the centrifugal force is perceptible; it results in the deflection of the plumb line away from Earth's axis of rotation. This is not because any real force is pushing it away, but only because we observe it in a rotating frame of reference. Clearly, whenever $\omega \neq 0$, $r \neq 0$ and $\theta \neq 0$ or π (see Eq. 3.22), an object with mass m would, in the earth-frame, appear to be pushed away. From the physical point of view, it is only the object's inertia that makes it appear as if the object is flung away.

An astronaut whose weight is mg on ground, will experience a centrifugal force of magnitude $m\omega^2 r \sin\theta$ (as per Eq. 3.22) radially outward if she/he is spun in a fast rotating machine, called a centrifuge. The acceleration experienced would correspond to $g_{\text{effective}} = \omega^2 r \sin\theta$. By spinning the astronauts rapidly, they are trained to experience high accelerations which they must sustain at the time of lift-off. In training, they often experience as much as $g_{\text{effective}} \sim 8g$. Usually they experience $\sim 3g$ when they are launched out on a space voyage in a rocket. *The physical experience experienced by an astronaut in a linear acceleration in a rocket launched upward, and in a high-speed rotating centrifuge, is essentially similar.* A laboratory centrifuge exploits exactly this to separate particles of different masses (or liquids of different densities). The more massive objects tend to retain their positions due to inertia, while lighter objects get easily centripetally accelerated, and thus move toward the center. This results in their separation.

We have seen above that while setting up a Newtonian equation of motion in the rotating frame of reference, one must include the three pseudo-forces, $\vec{F}_{cfg}$, $\vec{F}_{\text{Coriolis}}$ and $\vec{F}_{LSC}$. Now that we have learned how to estimate the effect of the centrifugal force, we focus on the Coriolis term alone. We shall neglect the centrifugal term, and also the leap-second correction. This is not a bad approximation, since Earth rotates about its axis at an angular speed of $\omega \approx 7.29 \times 10^{-5}$ radians/second. The centrifugal term is really small, being quadratic in Earth's rotational angular frequency.

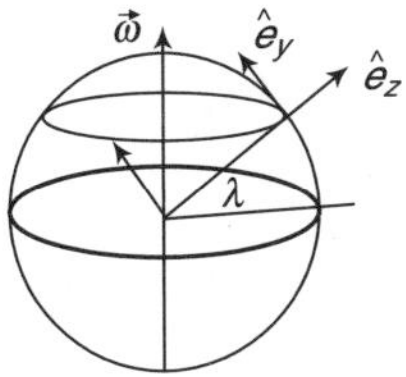

Fig. 3.11 This figure shows a simplification of the coordinate system that was employed in Fig. 3.9. First of all, for simplicity, the shape of Earth is regarded as spherical. The Earth rotates about its South–North axis. The local vertical at a point on Earth's surface at a latitude λ is taken to be along the Z-axis of the Cartesian coordinate system. The Y-axis is chosen to be point toward the North Pole, just as in Fig. 3.9. Since the centrifugal term is now ignored (to focus attention on the Coriolis term), the local vertical and the radial line are co-linear.

The Coriolis force is given by

$$\vec{F}_{\text{Coriolis}} = -2m[(\vec{\omega} \cdot \hat{e}_y)\hat{e}_y + [(\vec{\omega} \cdot \hat{e}_z)\hat{e}_z] \times [v_x\hat{e}_x + v_y\hat{e}_y + v_z\hat{e}_z]. \tag{3.23a}$$

The actual direction of this force is determined by the instantaneous velocity $\vec{v} = v_x\hat{e}_x + v_y\hat{e}_y + v_z\hat{e}_z$ in Earth's rotating frame.

It is easy to see that

$$\vec{F}_{\text{Coriolis}} = -2m\omega\,[\cos\lambda\hat{e}_y + \sin\lambda\hat{e}_z] \times [v_x\hat{e}_x + v_y\hat{e}_y + v_z\hat{e}_z], \tag{3.23b}$$

i.e., $$\vec{F}_{\text{Coriolis}} = -2m\omega\,[(\cos\lambda v_z - \sin\lambda v_y)\,\hat{e}_x + \sin\lambda v_x\hat{e}_y + (-\cos\lambda v_x)\hat{e}_z]. \tag{3.23c}$$

If you suspend a pendulum (Fig. 3.12) in a laboratory at the North Pole, and let it oscillate under the acceleration due to gravity, the floor of the laboratory under the pendulum would turn under it anticlockwise, along with the Earth (Fig. 3.13). For an observer in the laboratory, it is, however, rather the plane of the oscillation of the pendulum that would seem to turn clockwise. At the South Pole, the sense of rotation would be just the opposite. This rotation of the plane of oscillation would be seen at all latitudes, except at the Equator. The observer in the earth-fixed laboratory frame would of course wonder why the plane of oscillation of the pendulum is rotating. The change in the horizontal components of the momenta of the pendulum-bob cannot be accounted for by any physical interaction. Knowing in fact that there is no force in the horizontal (*XY*) plane, one can conclude by observing the rotation of the plane of oscillation of the simple pendulum that it is *actually* the *Earth* that must be rotating. In fact, the experiment carried out by Jean Bernard Léon Foucault at the Paris Observatory in 1851 demonstrated just that.

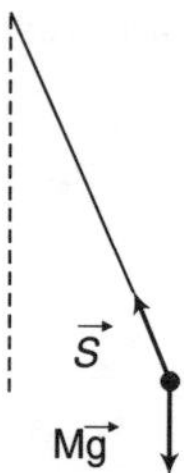

Fig. 3.12 Forces operating on a tiny mass m suspended by an inelastic string set into small oscillations, generating simple harmonic motion. The maximum angle is shown greatly exaggerated in this figure. Since the mass is in motion relative to the Earth, changes in its momentum as seen by an observer on Earth can be accounted for only after the inclusion of the Coriolis force, along with the gravitational attraction $m\vec{g}$ and the tension $\vec{S}$ in the string.

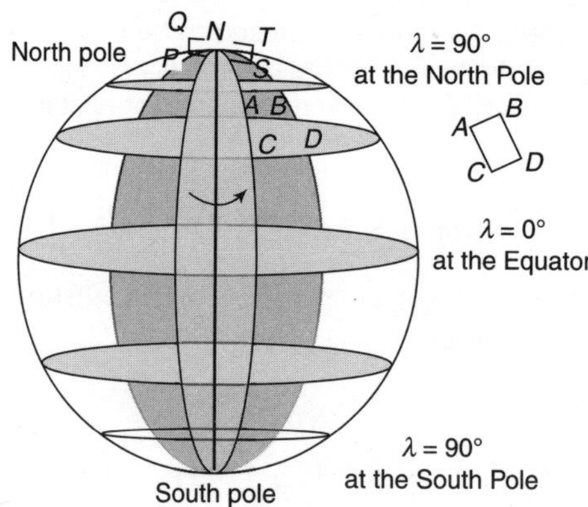

Fig. 3.13 Earth as a rotating frame of reference. If a simple pendulum experiment is performed in a laboratory at the North Pole on floor PSTQ (considered parallel to the equatorial plane, tangent to the earth at the North Pole), then the floor under the pendulum would turn anticlockwise. To the observer in the laboratory, it is the plane of oscillation of the pendulum that would appear to turn clockwise. Note also that if the laboratory floor, is at some other latitude on the surface of the earth, such as ABCD, then the edge AB would be turning slower than the edge CD, since the latter is farther from the axis of rotation.

The equation of motion of the bob shown in Fig. 3.12, in Earth's rotating frame of reference, is given by

$$m\ddot{\vec{r}}_R = m\vec{g} + \vec{S} - 2m\omega\,[(\cos\lambda v_z - \sin\lambda v_y)\hat{e}_x + \sin\lambda v_x\hat{e}_y + (-\cos\lambda v_x)\hat{e}_z)]. \tag{3.24}$$

The centrifugal term and the leap second correction term have been neglected in Eq. 3.23. In the Cartesian coordinate system $\{\hat{e}_x^{\,(\text{East})}, \hat{e}_y^{\,(\text{North})}, \hat{e}_z^{\,(\text{Up})}\}$ employed in Fig. 3.11, we write the tension in the string as

$$\vec{S} = S\hat{u} = S\,[\hat{e}_x\,(\hat{e}_x\cdot\hat{u}) + \hat{e}_y\,(\hat{e}_y\cdot\hat{u}) + \hat{e}_z\,(\hat{e}_z\cdot\hat{u})] = S[\hat{e}_x\cos\alpha + \hat{e}_y\cos\beta + \hat{e}_z\cos\gamma]. \tag{3.25}$$

Thus, the equation of motion of the bob becomes

$$m\ddot{\vec{r}}_R = -mg\hat{e}_z + S\begin{bmatrix}\hat{e}_x\cos\alpha + \\ \hat{e}_y\cos\beta + \hat{e}_z\cos\gamma\end{bmatrix} - 2m\omega\begin{bmatrix}(\cos\lambda v_z - \sin\lambda v_y)\hat{e}_x + \\ \sin\lambda v_x\hat{e}_y + (-\cos\lambda v_x)\hat{e}_z\end{bmatrix}. \quad (3.26)$$

The above equation is not easy to solve, even after neglecting the leap second correction and the centrifugal terms. We therefore contain ourselves by making some further approximations. The resulting solution, even if not fully accurate, provides an excellent account of the motion of the bob in Earth's frame of reference. Specifically, we make the following additional approximations:

- We assume that the magnitude of the tension $S \simeq mg$.
- We neglect the component v_z of the velocity of the bob.

Equation 3.26 is a vector equation, of which we now consider the equations for the X- and the Y- components:

$$m\ddot{x}_R = mg\cos\alpha + 2m\omega\sin\lambda v_y, \quad (3.27a)$$

$$m\ddot{y}_R = mg\cos\beta - 2m\omega\sin\lambda v_x. \quad (3.27b)$$

After cancelling out the mass of the bob on both sides of the equation above, we get the following coupled differential equations for the X- and Y- components of the bob's coordinates in Earth's frame of reference:

$$\ddot{x}_R = g\cos\alpha + 2\omega\sin\lambda\dot{y} = g\cos\alpha + 2\Omega\dot{y}, \quad (3.28a)$$

and

$$\ddot{y}_R = g\cos\beta - 2\omega\sin\lambda\dot{x} = g\cos\beta - 2\Omega\dot{x}. \quad (3.28b)$$

In the above equations, we have introduced $\Omega = \omega\sin\lambda$. (3.29)

The coupled differential Eqs. 3.28(a) and 3.28(b) can be solved easily by effecting the transformations

$$x = x'\cos(\Omega t) + y'\sin(\Omega t), \quad (3.30a)$$

$$y = -x'\sin(\Omega t) + y'\cos(\Omega t). \quad (3.30b)$$

Equations 3.30 give, on differentiation with respect to time, give

$$\dot{x} = -\Omega\,[x'\sin(\Omega t) - y'\cos(\Omega t)], \quad (3.31a)$$

$$\dot{y} = -\Omega\,[x'\cos(\Omega t) + y'\sin(\Omega t)]. \quad (3.31b)$$

Using the speeds $\dot{x}$ and $\dot{y}$ from Eq. 3.31 in 3.28, we get

$$\ddot{x} = g\cos\alpha - 2\Omega^2\,[x'\cos(\Omega t) + y'\sin(\Omega t)], \quad (3.32a)$$

$$\ddot{y} = g\cos\beta + 2\Omega^2\,[x'\sin(\Omega t) - y'\cos(\Omega t)]. \quad (3.32b)$$

Multiplying now Eq. 3.32a by $\cos(\Omega t)$ and Eq. 3.32(b) by $\sin(\Omega t)$, and then adding the two, we get

$$\ddot{x}\cos(\Omega t)+\ddot{y}\sin(\Omega t)=\begin{bmatrix} g\cos\alpha\cos(\Omega t)+ \\ g\cos\beta\sin(\Omega t) \end{bmatrix}+2\Omega^2\begin{Bmatrix} [x'\sin(\Omega t)-y'\cos(\Omega t)]\sin(\Omega t)- \\ [x'\cos(\Omega t)+y'\sin(\Omega t)]\cos(\Omega t) \end{Bmatrix}.$$

Now, we may neglect the quadratic terms $\Omega^2=\omega^2\sin^2\lambda$, which is absolutely fine, since the angular frequency of Earth's rotation is small, as noted above already.

Thus,

$$[\ddot{x}-g\cos\alpha]\cos(\Omega t)+[\ddot{y}-g\cos\beta]\sin(\Omega t)=0,$$

from which we immediately conclude that,

$$\ddot{x}-g\cos\alpha=0,\ \text{i.e.,}\ \ddot{x}-g\frac{x}{l}=0, \tag{3.33a}$$

and $$\ddot{y}-g\cos\beta=0,\ \text{i.e.,}\ \ddot{y}-g\frac{y}{l}=0. \tag{3.33b}$$

We conclude from Eqs. 3.33 that the bob's motion is described by a 2-dimensional oscillator. The superposition of these two simple harmonic orthogonal motions is described by an ellipse, which would, however, precess at an angular frequency $\Omega=\omega\sin\lambda$. We need to remind ourselves that our treatment has been approximate; we neglected $\dot{z}$, and also neglected terms quadratic in Earth's angular rotational speed. Also, we neglected the leap second correction. The result of the neglect of these terms is that the actual motion is more complicated. In particular, the 'plane' of oscillation of the bob is warped (curved) rather than flat, but the main conclusions we have drawn are very nearly accurate. Notwithstanding this minor difference, we shall therefore continue to refer to this as the rotation of the 'plane' of oscillation of the pendulum. The angular frequency of this rotation being Ω, the corresponding time period is

$$T=\frac{2\pi}{\Omega}=\frac{2\pi}{\omega\sin\lambda}=\frac{24\text{ hours}}{\sin\lambda}. \tag{3.34}$$

The least value of the time period for one full rotation of the plane of oscillation will be at the latitude $\lambda=\pm 90°$ (i.e., at the North and the South Poles) whence $\sin\lambda=\pm 1$. The negative sign at the South Pole denotes the fact that the sense of rotation of the plane would be opposite to that at the North Pole. At intermediate latitudes, the plane would take longer than 24 hours to complete one full rotation, and it would in fact take longer and longer as you go closer to the Equator, $\lambda=0°$, where the plane of oscillation of the bob would in fact remain essentially fixed.

There would be other tangible effects of the unreal pseudo-forces. For example, cyclonic storms spin counter-clockwise in the Northern Hemisphere, and clockwise in the Southern Hemisphere (Fig. 3.14).

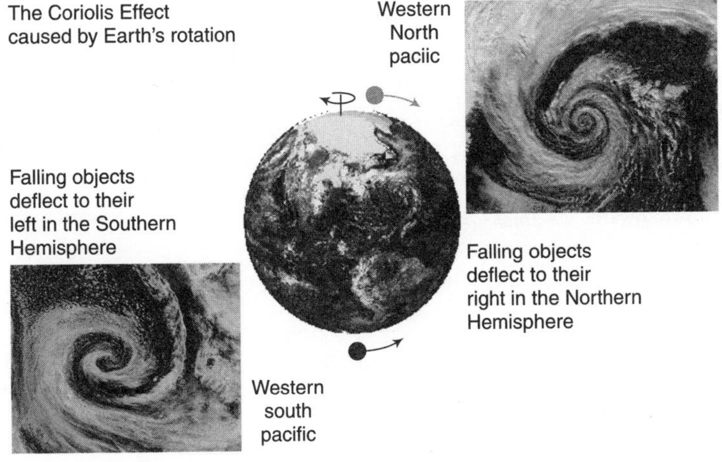

Fig. 3.14 Every object that is seen to be moving on Earth at a velocity $\vec{v}$ is seen to undergo a deflection, not because any physical interaction causes it, but because Earth itself is rotating. Using the coordinate system $\{\hat{e}_x^{(\text{East})}, \hat{e}_y^{(\text{North})}, \hat{e}_z^{(\text{Up})}\}$, described in the text, it is easy to see which way the deflection would result.

At the Equator, sin $\lambda = 0$ and cos $\lambda = 1$. Thus, if an object is dropped under gravity, its velocity would be $\vec{v} = -gt\hat{e}_z$ (i.e., $v_z = -gt$) and the Coriolis force acting on it would be

$$\vec{F}_{\text{Coriolis}} = -2m\omega\,[(\cos\lambda v_z)\hat{e}_x] = 2m\omega gt\cos\lambda\hat{e}_x, \tag{3.35}$$

which would provide the object a Coriolis acceleration equal to $2\omega gt\cos\lambda\hat{e}_x$, in the eastward direction. Over a time interval τ, the eastward acceleration of the object would impart to it an eastward speed, given by

$$v_x^{\text{East}} = \int_0^{\tau}(2\omega gt\cos\lambda)dt = 2\omega g\cos\lambda\int_0^{\tau} tdt = (\omega g\cos\lambda)\tau^2. \tag{3.36}$$

The object would therefore be deflected Eastward by an amount given by

$$\delta x^{\text{East}} = \int_0^{\tau} v_x dt = (\omega g\cos\lambda)\int_0^{\tau} t'^2 dt' = \omega g\cos\lambda\frac{\tau^3}{3}. \tag{3.37}$$

Table 3.1 The Coriolis effect, showing dramatic real effects of the pseudo-force.

Direction of an object's velocity on Earth in the coordinate system $\{\hat{e}_x^{(East)}, \hat{e}_y^{(North)}, \hat{e}_z^{(Up)}\}$	$\vec{F}_{Coriolis} = -2m\omega\begin{bmatrix}(\cos\lambda v_z - \sin\lambda v_y)\hat{e}_x + \\ \sin\lambda v_x\,\hat{e}_y + (-\cos\lambda v_x)\hat{e}_z\end{bmatrix}$ (from Eq. 3.23c)	Direction of Coriolis deflection
$-\hat{e}_z^{(Up)}$ (object falling down; v_z negative)	$-2m\omega[(\cos\lambda v_z)\hat{e}_x]$	$\hat{e}_x$ (toward east)
$\hat{e}_z^{(Up)}$ (object thrown up)	$-2m\omega[(\cos\lambda v_z)\hat{e}_x]$	$-\hat{e}_x$ (toward west)
$\hat{e}_x$ (object thrown toward east)	$-2m\omega v_x[\sin\lambda\hat{e}_y - \cos\lambda\hat{e}_z]$ $\sin\lambda > 0$ in the Northern Hemisphere $\cos\lambda > 0$	Toward south and also deflected upward
$-\hat{e}_x$ (object thrown toward west)	$-2m\omega v_x[\sin\lambda\hat{e}_y - \cos\lambda\hat{e}_z]\hat{e}$	Toward north and also deflected downward (since the object is thrown toward the west)
$\hat{e}_y$ (object thrown toward north)	$2m\omega\sin\lambda v_y\hat{e}_z$	Deflection toward east

Note: In the Southern Hemisphere, $\sin\lambda < 0$ and $\cos\lambda > 0$

We thus see from Eq. 3.37 that the longer an object falls, the more it gets deflected Eastward, as the cube of the time it falls through. Now, if the object falls through a height $h = \frac{1}{2}g\tau^2$, then $\tau = \sqrt{\frac{2h}{g}}$ and thus the Coriolis deflection of the object falling through that height would be $\delta x^{East} = \omega g\cos\lambda\left(\frac{2h}{g}\right)^{\frac{3}{2}}$. From the latitude, acceleration due to gravity, and Earth's rotational angular frequency, one can easily determine the Coriolis deflection. It is by no means small. For a fall through 100 meter at intermediate latitudes, it is of the order of a centimeter. Likewise, if you fire a bullet at a target that is 100 meter away, you would very well miss the bull's eye by a few centimeters. Not accounting for the fact that during the bullet's flight, the Earth under it would have rotated, can result in missing the target. At the Equator, the bullet would end up hitting *above* the bull's eye if you fired it toward the East. It would hit *below* the bull's eye if you fired it toward the west, as you can see from Table 3.1. The upward or downward deflection, effectively alters the acceleration due to gravity. This is called the Eötvös effect. Can you now imagine what would be the Coriolis deflection of an ICBM fired? The actual deflection depends on the latitude, speed of the ICBM and its velocity. It must now be obvious to you that the Coriolis effect is an extremely important one in navigation, communication, ballistic dynamics, etc. Unless one accounts for it thoroughly, modern lifestyle and needs of societies cannot even be sustained. A soccer ball kicked through 50 meters would deflect through a few centimeters during its flight. It could even miss the goal. Or, maybe, the shot barely may be successful if the Coriolis deflection favored the goal. During Falkland battles in the First World War, the British and the Germans shot canons at each other. The Falkland Islands are at a latitude of about 50° south of the Equator. Despite aiming well, the British shots landed about a hundred meters to the left of their intended targets due to the Coriolis effect.

The opposite sense of cyclonic winds in the Northern and the Southern hemispheres is already mentioned above. At the Equator, some people show tourists water currents in a funnel going in opposite directions just by crossing the Equator by a few steps on each side. However, these are unlikely to be accurate demonstrations of the Coriolis effect, since the funnel shape etc. is not calibrated. Notwithstanding such things, there are far-reaching, tangible atmospheric and oceanic effects whose dynamics is often studied using the 'Coriolis parameter', which is a measure of the horizontal component of the Coriolis force at a given latitude [5]. The force is unreal, of course, but the effect seen by us on Earth are dramatic which result in complex weather patterns and ocean currents, often producing havoc for human life.

We now turn our attention to the Leap Second Correction term in Eq. 3.19c. Astronomers have estimated that Earth's rotation is not constant [5]; $\dot{\vec{\omega}} = \left[\frac{d\vec{\omega}}{dt} \right] \neq \vec{0}$. This is on account of the Moon's gravitational interaction with the Earth, and especially how it affects the ocean tides. The Earth's rotational speed is affected also because of phenomena well *within* the Earth. While we live on the Earth's crust, deep inside is a very inhomogeneous mass distribution. The inner core is very hot and dense, surrounded by various shells, having materials having less density. Denser objects move toward the center. The rearrangement of mass is by no means complete. The tectonic plates often slide each other as heavier masses move inward. This causes earthquakes, and the redistribution of mass decreases the moment of inertia of the Earth. This results in increasing Earth's angular speed of rotation about its axis, since the angular momentum would be conserved. The length of the day has decreased by 1.8 microseconds as a result of the 2011 Japan earthquake (~9 Richter). The 2010 Chile earthquake (8.8 Richter) shortened the day by 1.26 microseconds, and the 2004 Sumatra quake (9.1 Richter) shortened the day by as many as 6.8 microseconds.

The change in the rotation speed of the Earth about its axis makes it necessary to apply the *Leap Second Correction* $\vec{F}_{LSC} = -m\left\{ \left[\frac{d\vec{\omega}}{dt} \right]_R \times \vec{r}(t) \right\}$ of Eq. 3.19c. The time taken by Earth to do one rotation differs from day to day, and from year to year. For example, the Earth was slower than atomic clocks by 0.16 second in 2005, by 0.30 second in 2006, by 0.31 second in 2007; and by 0.32 second in 2008. The unit of time, the 'second', is defined as the duration of 9192631770 periods of the radiation corresponding to the transition between two levels in the hyperfine structure of atomic Cesium-133. This does not agree with the commonly accepted definition of the second as $\frac{1}{24 \times 60 \times 60} = \frac{1}{86400}$ times the duration of a mean solar day. The difference demands re-calibration of the global time-keeping and affects functioning of network time protocol, GPS systems, etc. There are two time scales that are used for this purpose: (i) International Atomic Time (TAI), which is driven by about 200 high precision atomic clocks placed worldwide, and (ii) Universal Time (referred to as UT), also called the 'Astronomical Time', which is based on the time taken by the Earth to rotate once about its axis, i.e., the length of a day. Internet websites [6,7,8] provide continuous display of TAI and UT and shows the difference at any instant of time. Reference [9] provides an excellent read about 'The leap second: its history and possible future'. The Global Positioning System (GPS) time, is the atomic time scale employed using the atomic clocks placed at GPS ground control stations and at the GPS satellites. GPS time was set to zero at 0h on 6 January 1980 and no changes are made to it by the leap second. Thus GPS is currently ahead of the UTC (Coordinated Universal Time) by 18 second. *The UTC provides a 24-hour time standard and is adjusted using atomic clocks corrected by the Earth's rotation.* It is the International Earth Rotation and Reference System Service (IERS) which decides when to introduce a leap second in UTC. On an average day, the difference between the atomic clocks and Earth's rotation is around 0.002 second, or around 1 second every 1.5 years. The last leap second correction was made by adding one second at 23:59:60 to the UTC on 31 December 2016.

If the computer clocks are not adjusted for the leap second correction as and when needed, the electronic time-keeping system can go totally haywire. The consequences can be hazardous affecting the operations of cell phone networks, and even air traffic control system. The financial trading markets could fall out of rhythm and even breakdown if they did not agree on the *exact* time. Indeed, among other things, trading had to be suspended on 30 June 2012 on the National Association of Securities Dealers Automated Quotations (NASDAQ) and on the New York Stock Exchange (NYSE) to address this hazard.

Problems with Solutions

P3.1:

A young man of mass m is riding on 'Delhi's Eye Ferris Wheel'. The radius of the wheel is R and its angular speed is $\omega = \dot{\varphi}$. Determine the effective weight of the rider as a function of time as the wheel rotates.

***Solution*:**

Using Eq. 3.19a, the effective weight $\underset{\substack{\text{effective}\\\text{weight of}\\\text{the rider}}}{\vec{W}_{\text{effective}}}(t)$ is the net force on the rider in the downward direction.

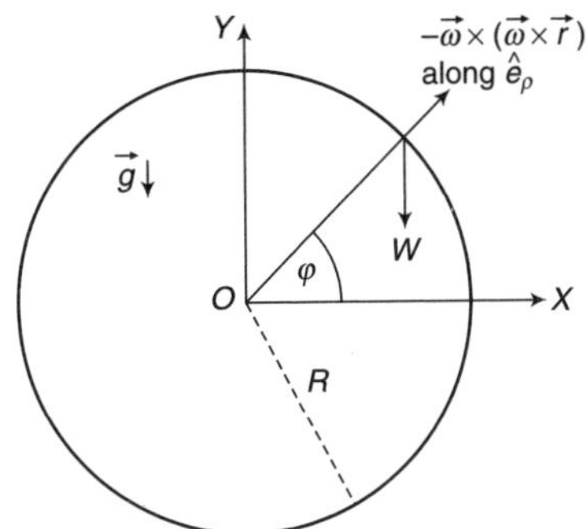

$$\vec{F}_{\substack{\text{effective}\\ \text{force on}\\ \text{the rider}}}(t) = -mg\hat{e}_y + m\vec{\omega}\times(\vec{\omega}\times\vec{r}(t)),$$

i.e., $$\vec{F}_{\substack{\text{effective}\\ \text{force on}\\ \text{the rider}}}(t) = -mg\hat{e}_y + mR\vec{\omega}\hat{e}_\rho(t).$$

Depending on where the passenger is located at the start of the rotation of the wheel, there is a phase factor which must be be accounted for. Considering this, we have:

$$\vec{F}_{\substack{\text{effective}\\ \text{force on}\\ \text{the rider}}}(t) = -mg\hat{e}_y + mR\vec{\omega}\{\cos(\omega t+\delta)\hat{e}_x + \sin(\omega t+\delta)\hat{e}_y\},$$

since $$\hat{e}_\rho(t) = \cos(\omega t+\delta)\hat{e}_x + \sin(\omega t+\delta)\hat{e}_y.$$

Therefore, $$\vec{W}_{\substack{\text{effective}\\ \text{weight of}\\ \text{the rider}}}(t) = -m[g - R\omega^2\sin(\omega t+\delta)\hat{e}_y.$$

Hence, $\vec{W}_{\substack{\text{effective}\\ \text{weight of}\\ \text{the rider}}}(t) = -m[g - \omega^2 R\sin(\omega t+\delta)\hat{e}_y$. Since the sinusoidal function varies between -1 and $+1$, the weight of the rider varies: $m(g-\omega^2R) \le mg \le m(g+\omega^2R)$. It is maximum at the bottom, and the least at the top.

P3.2:

A plumb line is suspended from the point P, from the roof of a room on the Earth, located at a latitude λ. In the rotating Earth's frame, as a result of the real effect of the centrifugal force on the mass suspended, the plumb line would not point at the Earth's center. It points slightly away from the center. Determine the angular departure δ between the local vertical along the plumb line and the radial line to the center of the Earth. As in the text, we assume that the earth's shape is perfectly spherical.

Solution:

Due to the centrifugal term, the plumb line is swung away from the Earth's axis of rotation, along $\breve{e}_\rho = \dfrac{\overrightarrow{OO'}}{\left|\overrightarrow{OO'}\right|}$, which is parallel to the equatorial (XY) plane. The point O' toward which the plumb line points is therefore somewhat displaced from the Earth's center O.

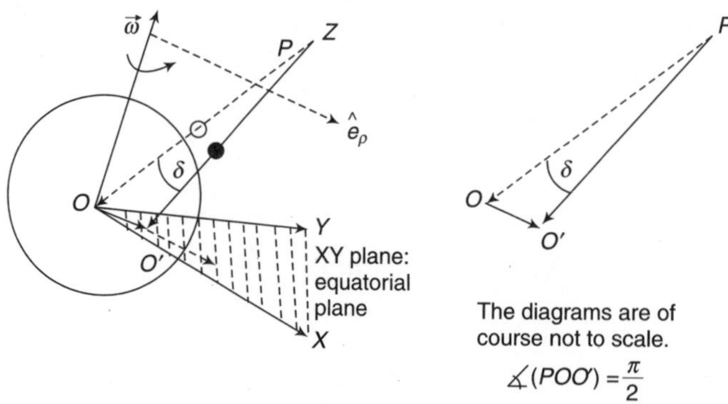

The plumb line points along *PO'*, and not toward the center *O* of the Earth. The force on the mass suspended at the bottom of the plumb line, by Eq. 3.18, is:

$$\vec{F}_R = \vec{F}_I + \vec{F}_{cfg} + \vec{F}_{Coriolis} + \vec{F}_{LSC} \simeq m\vec{g} + \vec{F}_{cfg}; \text{ i.e., } m\vec{g}' \simeq m\vec{g} + \vec{F}_{cfg}.$$

In the present case, the Coriolis term is zero, since the suspended mass has zero velocity, and we may ignore the leap second correction. The centrifugal term force is $-m\vec{\omega} \times (\vec{\omega} \times \vec{r})$.

Hence, the effective force on the mass suspended at the plumb line is $m\vec{g}' \simeq m\vec{g} - m\vec{\omega} \times (\vec{\omega} \times \vec{r})$.

We use the coordinate system employed in Fig. 3.11, and also use Eq. 3.23a, to analyze the above result.

Since $\vec{\omega} = [\omega \cos \lambda \hat{e}_y + \omega \sin \lambda \hat{e}_z]$, we have $\vec{\omega} \times \vec{r} = [\omega \cos \lambda \hat{e}_y + \omega \sin \lambda \hat{e}_z] \times R\hat{e}_z = \omega \cos \lambda \hat{e}_x$, and $\vec{\omega} \times (\vec{\omega} \times \vec{r}) = [\omega \cos \lambda \hat{e}_y + \omega \sin \lambda \hat{e}_z] \times \omega R \cos \lambda \hat{e}_x = -R\omega^2 \cos^2 \lambda \hat{e}_z + R\omega^2 \sin \lambda \cos \lambda \hat{e}_y$.

Hence, the centrifugal force is: $-m\vec{\omega} \times (\vec{\omega} \times \vec{r}) = mR\omega^2 \cos^2 \lambda \hat{e}_z - mR\omega^2 \sin \lambda \cos \lambda \hat{e}_y$. Note that this has a component along the vertical, but against the gravity. We therefore see that the effective weight of the mass suspended is now reduced; it is $mg - mR\omega^2 \cos^2 \lambda$.

From the triangle *POO'*, we get: $\tan\delta = \dfrac{-mR\omega^2 \sin \lambda \cos \lambda}{-mg + mR\omega^2 \cos^2 \lambda} = \dfrac{R\omega^2 \sin \lambda \cos \lambda}{g - R\omega^2 \cos^2 \lambda}$

$\tan\delta \simeq \delta = \dfrac{R\omega^2 \sin\lambda \cos\lambda}{g - R\omega^2 \cos^2 \lambda} \simeq \dfrac{R\omega^2 \sin\lambda \cos\lambda}{g}$. At latitude of 45°, this is about 0.1°.

P3.3:

A mass *m* is released with zero initial velocity from the point '*A*' shown in the figure, next to the Mantri Pinnacle Residential Tower, Bangalore, located at latitude of 12.972442°. The point '*A*' is at a height of 153 m. It is intended that the object dropped from '*A*' would hit the ground at the point '*B*', which is ***directly below*** the point '*A*'.

(a) Determine the amount of the sideways Coriolis deflection of the object dropped when it hits the ground.

(b) Sketch the Coriolis deflection as a function of latitude, if the latitude is varied from $-\dfrac{\pi}{2}$ to $+\dfrac{\pi}{2}$, for a fixed height.

(c) Plot the deflection against height from which the object is dropped, varying it from 100 m to 1.2 km.

Solution:

(a) Again we use the coordinate system employed in Fig. 3.11.

As before, we have $\vec{\omega} = \omega\cos(\lambda)\hat{e}_y + \omega\sin(\lambda)\hat{e}_z$. The Coriolis force is given by (Eqs. 3.19b and 3.23a):

$$F_{\text{Coriolis}} = -2m\left(\vec{\omega}\times\frac{d\vec{r}}{dt}\right)$$

Although the Coriolis force produces small velocity components in the $\hat{e}_x$, $\hat{e}_y$ directions, we may neglect $\dot{x}$ and $\dot{y}$, since the acceleration due to gravity is strong; the z-component of the velocity is the most important one.

Hence:

$$F_{\text{Coriolis}} \simeq -2m\begin{vmatrix}\hat{e}_x & \hat{e}_y & \hat{e}_z\\ 0 & \omega\cos\lambda & \omega\sin\lambda\\ 0 & 0 & -gt\end{vmatrix} = 2m\omega gt\cos\lambda\hat{e}_x$$

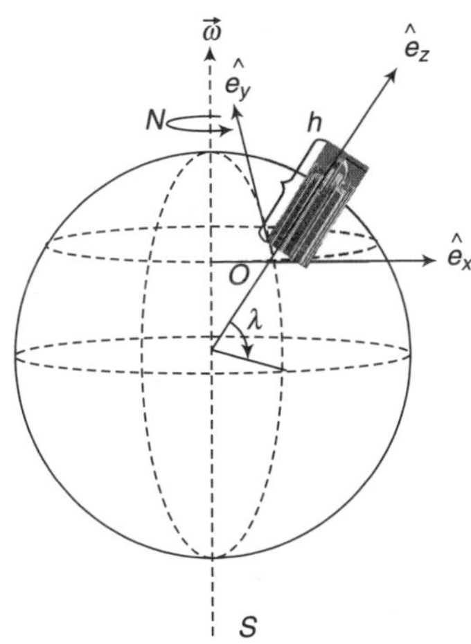

Integrating $\ddot{x} = 2\omega g t \cos(\lambda)$ and $\ddot{z} = -g$ twice with respect to time, and using the initial conditions at $t = 0$; we have $(x = 0, \dot{x} = 0)$ and $(z = h, \dot{z} = 0)$:

$$x(t) \cong \frac{1}{3}\omega g t^3 \cos\lambda \quad \text{and} \quad z(t) \simeq -\frac{1}{2}g t^2 .$$

The time interval the object takes to fall through the height h is therefore $t \simeq \sqrt{\frac{2h}{g}}$.

Hence, $x(t) \simeq \frac{1}{3}\omega g\left(\sqrt{\frac{2h}{g}}\right)^3 \cos\lambda = \frac{1}{3}\omega\left(\sqrt{\frac{8h^3}{g}}\right)\cos\lambda$

The deflection along the x direction is therefore, $d = \frac{1}{3}\omega\left(\sqrt{\frac{8h^3}{g}}\right)\cos\lambda = 0.0405 \text{ m} \approx 4 \text{ cm}$.

A point-sized object dropped from the point A would therefore miss the point B directly below it by ~4 cm. Note that this result is independent of the mass of the object dropped.

(b) Variation of the eastward deflection (in the Northern Hemisphere) as a function of latitude λ at constant height $h = 153$ m. From the above result, we know that the deflection will be maximum at the Equator, where $\cos\lambda = \cos 0 = 1$. Away from the Equator, it drops as cosine of the latitude, as shown in the adjacent figure.

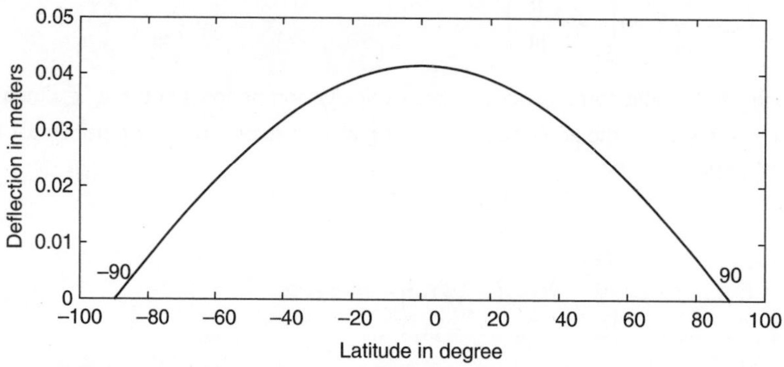

(c) Variation of the eastward deflection (in the Northern Hemisphere) as a function height, at constant latitude $\lambda = 12.973°$

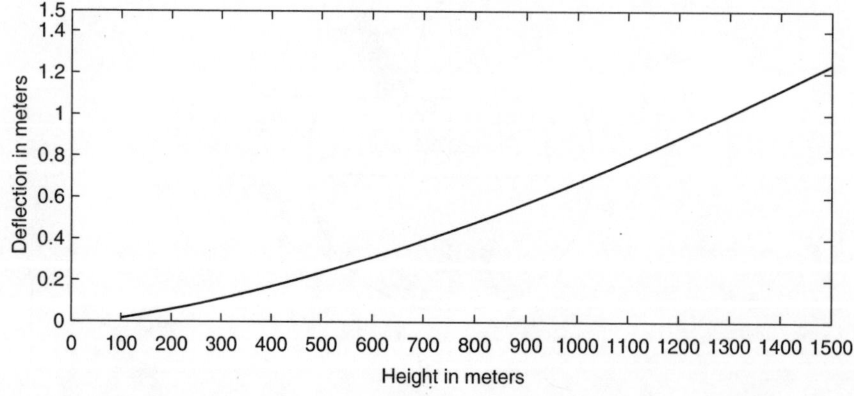

P3.4:

A metal ball of mass 50 kg falls freely on the surface of the Earth at the latitude of 12.973° from a height of 1 m. Determine the magnitudes of the gravitational, centrifugal, Coriolis and Leap-second-correction forces acting on it in the Earth's rotating frame of reference. Assume the Earth to be a perfect sphere of radius 6.4×10^6 m.

Solution:

Gravitational force: $\vec{F}_g = m\vec{g} \simeq 50 \times 10 = 500\ \text{N}(-\hat{e}_z);\quad |\vec{F}_g| = 500\ \text{N}$

Centrifugal force: $\vec{F}_{\text{Centrifugal}} = -m\vec{\omega} \times (\vec{\omega} \times \vec{r}') = m\omega^2 R \cos\lambda \hat{e}_p.$

with $\omega = 7.26 \times 10^{-5}$ rad/sec, and $R + h \simeq R$ and $\lambda = 12.973°$,

$\vec{F}_{\text{Centrifugal}} = 1.6164\ \hat{e}_p\ \text{N};\quad \left|\vec{F}_{\text{Centrifugal}}\right| = 1.6164\ \text{N}$

Coriolis force: $\vec{F}_{\text{Coriolis}} = -2\text{m}(\vec{\omega} \times \vec{v})\ -2m\omega gt \cos\lambda\hat{e}_x$

Now, time for free fall t is $t = \sqrt{\dfrac{2h}{g}} = \sqrt{\dfrac{2 \times 1}{10}} = 0.447\,\text{sec}$

We may use $g \simeq 10\ \text{ms}^{-2}$.

$$\vec{F}_C = -2 \times 50\ (7.26 \times 10^{-5}) \times 10 \times 0.447 \times \cos(12.973)\hat{e}_x \simeq 0.0316\ \text{N}\hat{e}_x;\ |\vec{F}_C| = 0.03161\ \text{N}$$

Leap second force: $\vec{F}_{\text{Leap}} = m(\vec{\dot{\omega}} \times \vec{R})$. We shall use $\vec{\dot{\omega}} = -0.05369 \times 10^{-10}\ \hat{e}_\omega$ rad/sec^2

$$\vec{F}_{\text{Leap}} = 50 \times (-0.05369) \times 10^{-10} \times 6.4 \times 10^6\ (\hat{e}_\omega \times \hat{e}_R) = -0.001718\ (\hat{e}_\omega \times \hat{e}_R)\text{N}$$

$$|\vec{F}_{\text{Leap}}| = 0.001674\ \text{N}$$

P3.5:

Two masses m_1 and m_2 are suspended from a point S by identical strings of length L. Both the masses rotate about the vertical line through the point of suspension S, at a constant angular velocity. Both the masses make the same angle ζ with the vertical. At the positions P_1 and P_2 respectively of the two masses, determine:

(a) the acceleration of either of the two masses as viewed from the point of suspension,

(b) the acceleration of the point of suspension as viewed from either of the two mass.

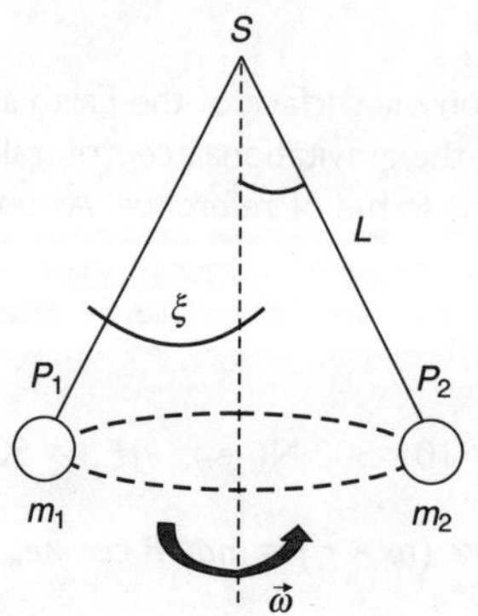

Solution:

Considering the symmetry of the problem, we shall use the cylindrical polar coordinates (ρ, φ, z).

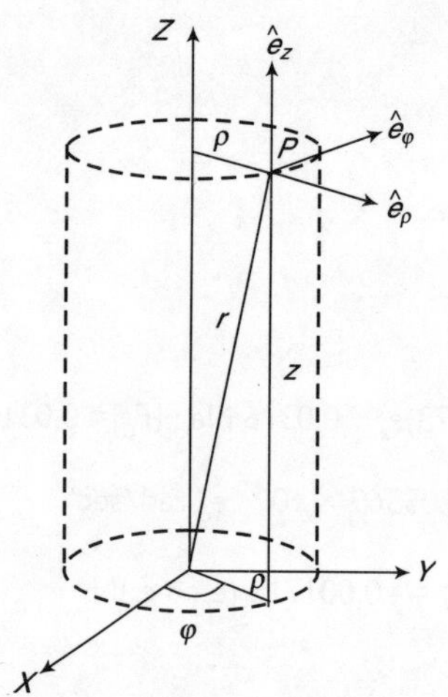

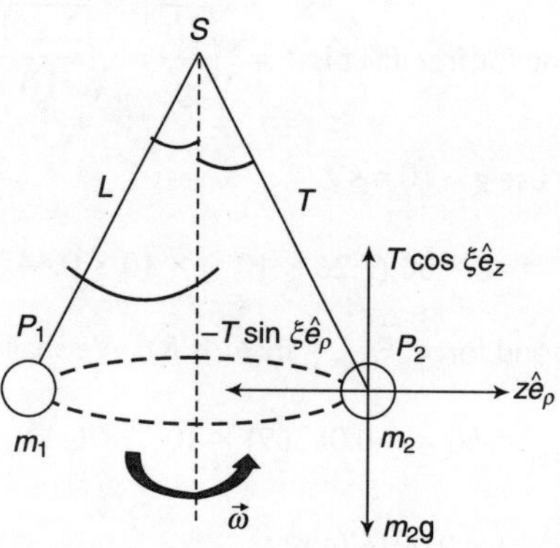

(a) Acceleration of the mass m_2, viewed from the point of suspension S.

From Newton's second law, $\vec{T} + m_2\vec{g} + m_2\vec{a}_R = \vec{0}$.

i.e., $-T \sin \theta \hat{e}_\rho + T \cos \theta \hat{e}_z - m_2 g \hat{e}_z + m_2 a_R \hat{e}_\rho = \vec{0}$.

The vertical acceleration is zero, hence: $T \cos \theta = m_2 g$.

Therefore: $T = \dfrac{m_2 g}{\cos \zeta}$ and $\vec{a}_R \hat{e}_\rho = g \tan \zeta \hat{e}_\rho$

(b) In the inertial frame of reference $\vec{a}_I = \vec{0}$ [fixed point of suspension].

Therefore, from Eq. 3.15e: $\vec{0} = \vec{a}_R + 2\vec{\omega} \times \vec{V}_R + \vec{\omega} \times (\vec{\omega} \times \vec{r})$ (i)

From Eq. 3.12 the velocity in inertial frame of reference is: $\vec{V}_I = \vec{V}_R + (\vec{\omega} \times \vec{r})$.

Since $\vec{V}_I = 0$, $\vec{V}_R = -(\omega \hat{e}_z \times \rho \hat{e}_\rho) = -(\omega L \sin \zeta)\hat{e}_\varphi$

Hence, from (i), we get: $\vec{0} = \vec{a}_R + 2\omega \hat{e}_z \times (-\omega L \sin \zeta)\hat{e}_\varphi - \omega^2 L \sin \zeta \hat{e}_\rho$

i.e., $\vec{0} = \vec{a}_R + 2\omega^2 L \sin \zeta \hat{e}_\rho - \omega^2 L \sin \zeta \hat{e}_\rho$. Thus: $\vec{a}_R = -\omega^2 L \sin \zeta \hat{e}_\rho$.

P3.6:

Determine the direction and magnitude of the Coriolis force acting on an object having a mass of 10 kg, at a latitude of 80° in the Southern Hemisphere, and moving at 3 m/s in the direction from the North Pole to the South Pole along a meridian on the Earth's surface.

Solution:

$$\vec{F}_{\text{Coriolis}} = -2m(\vec{\omega} \times \vec{v}) = -2m\omega\,(\cos\lambda\hat{e}_y + \omega\sin\lambda\hat{e}_z) \times \{3(-\hat{e}_y)$$

Hence: $\vec{F}_{\text{Coriolis}} = (-\hat{e}_x)1.44 \times 10^{-3}$ N (toward west)

Additional Problems

P3.7 If you were in a closed chamber moving at some acceleration along a straight line on the Earth's surface, how would you determine the acceleration of the chamber, given only a plumb line? Neglect the effects of the Earth's rotation, and its curvature. You can test this easily, especially when sitting in an aircraft, accelerating on the runway toward its take off.

P3.8 At (13° N, 79.5° E), a bullet is fired at a target 300 meters away due west. The bullet is shot at a horizontal velocity of 500 ms^{-1}. Determine the deflection of the bullet due to the Coriolis effect. Incidentally, when the Coriolis deflection is upward (see Table 3.1), it is commonly referred to as the Eötvös effect.

P3.9 Determine the time period of the Foucault pendulum at the following locations:

(a) Tirupati (13.63551° N, 79. 41989° E)

(b) Indore (22.71792° N, 75.8333° E)

(c) Moscow (55.7558° N, 37.6173° E)

P3.10 What should be the speed at which an object must be thrown vertically upwards from the ground at the Equator, so that when it returns to the ground it would fall 10 cm away from where it was thrown upward.

P3.11(a) What is an ocean gyre? How is it formed? (b) What is an Ekman spiral? How is it formed?

P3.12 Explain why the shape of the Earth is an oblate spheroid.

P3.13 A cylinder of mass M and radius R rolls at a uniform angular speed ω without slipping on a wooden plank. The wooden plank itself is accelerated horizontally at the rate λ ms^{-1}. Find the net linear acceleration of the cylinder and explain what causes it.

P3.14 Determine the angle between the vertical and the line along which a plumb line is oriented if the plumb line is suspended in an aircraft accelerating on the runway at α ms^{-2}. Also, determine the tension in the string.

P3.15 An iceberg of mass 3×10^5 tons at a latitude of 86° in the Southern Hemisphere is heading eastward at a speed of (¼) km per hour. Determine the Coriolis deflection it would suffer in one hour.

P3.16 A centrifuge spins at 36000 revolutions per minute. Determine the g-force a particle experiences at (a) 5 cm (b) 10 cm from the axis of the centrifuge.

P3.17 A bug is crawling radially outward on a turn-table. The bug's linear speed is uniform, at λ cm/sec. The turn-table's angular speed is also uniform at α rad/sec. (a) Determine the centrifugal and the Coriolis forces acting on the bug. (b) Determine the physical force acting on the bug. (c) How far can the bug crawl before it begins to slip? (d) What would happen if the angular speed of the turn-table is increased?

P3.18 Write the sum of components of all the force terms (including pseudo-forces) in Eq. 3.17 in (a) cylindrical polar coordinates and (b) spherical polar coordinated.

P3.19 A projectile is launched from latitude λ in the Southern Hemisphere toward the South at 45°. It is aimed at an object that is at a distance $\delta \ll R$, where R is the Earth's radius. Determine the deflection ξ it would suffer due to the Coriolis effect. Specifically, determine the dependence of ξ on the distance δ and on the latitude.

P3.20 The speed of the wind at latitude λ is v. A pressure-gradient force would be required to counter the Coriolis deflection of the wind to make it flow in a constant direction. Show that the required pressure-gradient force is $\vec{v}_R = \left[\frac{d}{dt}\right]_R \vec{r}$ where ρ is the density of air. Hint: consider the forces on a mass of air in a tiny volume element, τ.

References

[1] 'Science Demonstrations, NASA.' https://history.nasa.gov/SP-401/ch12.htm. Accessed on 13 October 2018.

[2] Ramasubramanian, K. M. D. S., M. D. Srinivas, and M. S. Sriram. 1994. 'Modification of the Earlier Indian Planetary Theory by the Kerala Astronomers (c. 1500 CE) and the Implied Heliocentric Picture of Planetary Motion.' *Current Science* 66(10): 784–790.

[3] Plofker, Kim. 2009. *Mathematics in India*. Princeton: Princeton University Press.

[4] Das, P. Chaitanya, G. Srinivasa Murthy, Gopal Pandurangan, and P. C. Deshmukh. 2004. 'The Real Effects of Pseudo-forces.' *Resonance* 9(6): 74–85.

[5] Omstedt, A., J. Elken, A. Lehmann, M. Lepparanta, H. E. M. Meier, K. Myrberg, and A.Rutgersson. 2014. 'Progress in Physical Oceanography of the Baltic Sea during the 2003–2014 Period.' *Progress in Oceanography* 128: 139–171.

[6] Perkins, Sid. 2016. 'Ancient Eclipses Show Earth's Rotation is Slowing.' *Science*. Science. https://www.sciencemag.org/news/2016/12/ancient-eclipses-show-earth-s-rotation-slowing. Accessed 26 April 2019.

[7] 'What is a Leap Second?' https://www.timeanddate.com/time/leapseconds.html. Accessed on 15 June 2017.

[8] http://www.leapsecond.com/java/gpsclock.htm. Accessed on 15 June 2017.

[9] Nelson, Robert Arnold, Dennis Dean McCarthy, Stephen Malys, Judah Levine, Bernard Guinot, Henry Frederick Fliegel, Ronald L. Beard, and T. R. Bartholomew. 2001. 'The Leap Second: Its History and Possible Future.' *Metrologia* 38(6): 509.

CHAPTER 4

Small Oscillations and Wave Motion

I will be the waves and you will be a strange shore.
I shall roll on and on and on, and break upon your lap with laughter.
And no one in the world will know where we both are.

—*Rabindranath Tagore*

4.1 BOUNDED MOTION IN ONE-DIMENSION, SMALL OSCILLATIONS

Rise and fall, ups and downs, profit and loss, success and failure, happiness and sorrow—almost every experience in life seems to be in a state of oscillation. It is the undulations of the sound waves in air which enable us to hear each other, and it is undulations of the electromagnetic waves in vacuum which bring energy (light) from the stars to us. Quantum theory tells us that there is a wave associated with every particle, and that leaves us with absolutely nothing in the physical world that is *not* associated with oscillations. Oscillations of what, where and how can only be a matter of details, but there is little doubt that the physical universe requires for its rigorous study an understanding of 'oscillations'. Oscillations are repetitive physical phenomena, which swing past some event, back and forth. They are ubiquitous, and they maintain things close to some mean behavior, at least most of the times. Whether it is the breeze jiggling the leaves, the rhythmic beating of the heart that is the acclamation of life itself, the gentle ripples on a river bed, or the mighty tides in the oceans, or for that matter fluctuations in the share market, we are always dealing with some expression of the simple harmonic oscillator, or a superposition of several of these, however, complex.

We have learned in earlier chapters that equilibrium, defined as motion at constant momentum, sustains itself when the object is left alone, completely determined by initial conditions. Departure from equilibrium is accounted for by Newton's equation of motion (second law), or equivalently (as we shall see in Chapter 6) by Lagrange's, or Hamilton's equations. In a 1-dimensional region of space, say along the X-axis, we consider a region in which the equilibrium of a particle may *change* because of spatial dependence of its interaction with the environment. This is represented by a space-dependent potential $V(x)$.

In such a region, different points of equilibrium are shown in Fig. 4.1. Left to itself with zero kinetic energy, a particle at a point, such as x_1 in this figure, would slide down toward the left. This is because the potential toward the left of that point is lower than that at its right. The potential *increases* at the point x_1 *from left to right*. This induces the object placed at that point to move in the *opposite* direction, from right to left. Greater the slope $\frac{dV}{dx}$, greater would be the stimulus that would make the object move toward the left. This stimulus is essentially the force in employed Newtonian dynamics:

$$F = -\frac{dV}{dx}. \tag{4.1}$$

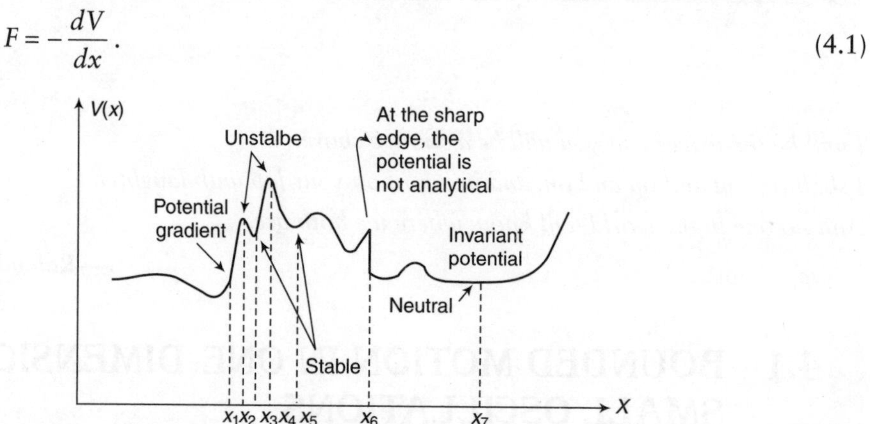

Fig. 4.1 Different kinds of common equilibria in one-dimension.

The minus sign in Eq. 4.1 is on account of the fact that the direction of the force is opposite to the direction in which the potential increases. The momentum of the particle would change in the direction opposite to that in which the potential increases. We have considered conservative fields here, which essentially means that the sum of whatever interaction(s) the particle has with its environment is represented by the function $V(x)$ sketched in Fig. 4.1. There is no other interaction an object has with anything else that is not included in $V(x)$. In other words, there are no unspecified degrees of freedom with which an object at that point may exchange energy. The slope $\frac{dV}{dx}$ is known as the gradient of the potential. This terminology will be further developed in Chapter 9 where we shall meet the 3-dimensional analogue of $\frac{dV}{dx}$.

At points such as x_3, or x_5, the particle would remain undisturbed, if left to itself with zero kinetic energy. An object, when placed at such points with no initial kinetic energy, would simply stay there forever, described by Galileo–Newton's law of inertia. In fact, a point mass can also be delicately balanced at points such as x_2 or x_4, where it can be left with zero kinetic energy, forever, in equilibrium. There is, however, an obvious difference in the nature of the equilibrium at the points x_2 and x_4 on one hand, and at x_3 and x_5 on the other. If displaced only slightly from the point x_3, the object would tend to *return* to the original point. The displacement of course will need to be effected by an external agent. Without an influence by an external agent, the object would only retain its equilibrium position of rest. On the other hand, a slight displacement from x_4, would only induce the object to move *further* away from the original point. Nonetheless, an external agent is necessary to trigger even this. Thus, the points x_3 and x_4 are respectively called 'stable' and 'unstable' equilibrium points. The point x_7, on the other hand, is also an equilibrium point, but a slight displacement from that point will leave the object yet again in a state of equilibrium, even at the new point. Such a point is called a point of 'neutral' equilibrium. At a cliff-like point such as x_6, the potential is not analytical. We know all too well what happens to an object depending on which side of the singularity it is placed.

In a region of space where the potential can change *differently* along *two independent degrees of freedom*, the nature of the equilibrium can be more complex. Fig. 4.2a shows a marble kept at the middle of a horse's saddle. Clearly, the marble would be in equilibrium. A slight displacement of the marble along the length of the saddle would tend to bring it *back* to the equilibrium point. However, a slight displacement along the width of the saddle would tend to move it further *away* from the equilibrium point. Such a point of equilibrium is called a *saddle point*. Another example of the saddle point equilibrium would be the midpoint of two equal electric charges, both positive (or negative). A positive (negative) test charge placed at their middle with zero kinetic energy would be in equilibrium. It would tend to get *back* to the original position if it is slightly displaced along the line joining the two charges, but it would tend to fly off if displaced slightly transverse to this line. We may ask if there can be a saddle point in a potential that is a function of just one variable. Fig. 4.2b illustrates one such possibility. The saddle point, just like a potential maximum or a minimum, is essentially an 'extremum'. The potential function is 'stationary' at that point. The derivatives of the potential along *all* degrees of freedom are all zero at that point.

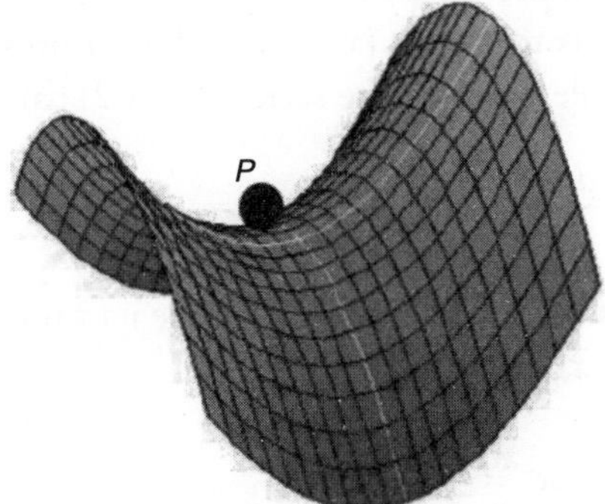

Fig. 4.2a Saddle point, when the potential varies differently along two independent degrees of freedom.

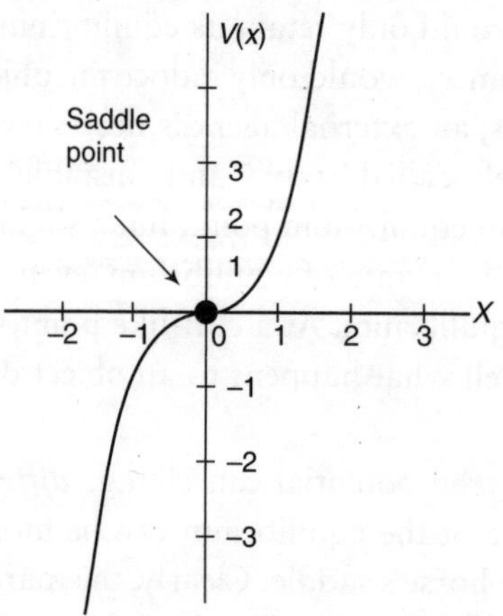

Fig. 4.2b A potential described by $V(x) = ax^3$. The origin is a 'saddle point', since the potential varies differently along the two sides of this point even if it depends only on a single degree of freedom.

Now, if the potential function $V(x)$ is continuous, and is differentiable to all orders, at a point of stable equilibrium, say x_0, then it can be written as a Taylor series at a point x that is close to x_0:

$$V(x) = V(x_0) + \left\{ \left.\frac{dV}{dx}\right|_{x_0} \right\}(x - x_0) + \left\{ \frac{1}{2!}\left.\frac{d^2V}{dx^2}\right|_{x_0} \right\}(x - x_0)^2 + \left\{ \frac{1}{3!}\left.\frac{d^3V}{dx^3}\right|_{x_0} \right\}(x - x_0)^3 + \ldots \quad (4.2)$$

We know that at a point of stable equilibrium, the potential is stationary, and hence its first derivative vanishes. If we therefore expand the potential near a stable equilibrium, we can always find a region of space on either side of the equilibrium for which the interval $(x - x_0)$ is small enough for us to *ignore* $(x - x_0)^3$. This ensures that all powers $(x - x_0)^n$ of $(x - x_0)$, for $n \geq 3$ are ignorable. The infinite series (Eq. 4.2) can then be truncated:

$$V(x) = V(x_0) + \left\{ \frac{1}{2!}\left.\frac{d^2V}{dx^2}\right|_{x0} \right\}(x - x_0)^2 \cdot \quad (4.3a)$$

Choosing the zero of the potential energy scale to be its value *at* x_0, and also choosing the zero of the X-scale such that $x_0 = 0$, we get

$$V(x) = \frac{1}{2!}\left\{ \left.\frac{d^2V}{dx^2}\right|_{x_0 = 0} \right\} x^2 = \frac{1}{2}\kappa x^2, \quad (4.3b)$$

In the above equation, we have introduced $\kappa = \left.\frac{d^2V}{dx^2}\right]_{x_0 = 0}$. (4.4)

It is known as the spring constant.

We have already seen that the stimulus (force) that would change the momentum of an object is given by $F = -\frac{dV}{dx}$ (Eq. 4.1), which in the present case would be

$$F = -\kappa x. \tag{4.5a}$$

The above relation describes a force that increases linearly as the displacement from the equilibrium point increases. The minus sign in Eq. 4.5a ensures that the force is always directed toward the equilibrium point. This was first stated as a Latin phrase, *ut tensio, sic vis*, which means *as the extension, so the force*, by Robert Hooke (1635–1703). He was a distinguished and brilliant physicist, contemporary of Isaac Newton and Robert Boyle. In fact, the well-known *Boyle's law* was discovered using an apparatus that was built by Hooke. Hooke was also one of the early physicists who also had the idea of the inverse square law of gravitational attraction. This law is of course credited to Newton, and for good reason. We shall discuss the gravitational interaction further in Chapter 8).

By carrying a series of experiments on elastic materials, Robert Hooke arrived at the law given in Eq. 4.5. The form expressed in this scalar relation is a simple one, but many materials have anisotropic properties which demand a *tensor* relationship between displacement and the restoring force. However, we shall refrain from these details. The Hooke's law is of great importance in studying material properties, and its applicability provides the limits of elastic behavior of these materials. It is applicable when the relationship between the restoring elastic force in a material is linearly proportional to the displacement from the equilibrium, and is always directed toward the equilibrium position. This property has earned this force its name; it is called the *restoring* force. Equation 4.5 is applicable when the *force–displacement* relationship is expressed using a simple scalar equation. This relationship is applicable only within a limiting value of the displacement. It is called the elastic limit. Equation 4.5 is a second order differential equation of motion. It is essentially the Newton's equation of motion for this system. Its solution turns out to be oscillatory, as described below. Since the displacements under this consideration are not large, this phenomenology is known as 'small' oscillations.

Common spring is one of the best examples of an elastic material. Springs are around us everywhere. It is hard to imagine appliances and machines where springs are not employed. Whether you compress it or elongate it, internal restoring forces get activated as soon as the spring is disturbed from its equilibrium configuration. The more the displacement from the equilibrium, the stronger is the restoring force. The mathematical relation between the displacement and the restoring force is linear, expressed as the aforementioned Hooke's law. An object is said to behave within its elastic limits when this linear relation holds. Our prototype of such a material is the mass–spring oscillator, shown in Fig. 4.3a.

Equation 4.5a gives

$$\frac{d^2x}{dt^2} + \frac{\kappa}{m}x = 0. \tag{4.5b}$$

A large number of systems in nature tend to preserve their equilibrium configuration. Even if many of us observe this in the phenomena around us, only the brilliant can extract a systematic law of nature from what we see. Galileo Galilei, in 1581, when he was just 17 years old, observed in a cathedral that a chandelier was swinging back and forth periodically.

Legend has it that he measured the periodicity of the oscillations of the chandelier by counting the swings against his biorhythm felt by checking his own pulse rate.

Fig. 4.3b shows a simple pendulum consisting of a tiny mass, m, suspended from a frictionless support by an inextensible string of length ℓ. When displaced from its mean position, the arc-length $s = \ell\theta$ provides a measure of the displacement of the bob. Its acceleration is therefore given by

$$a = \frac{d^2 s}{dt^2} = \ell \frac{d^2\theta}{dt^2}. \tag{4.6a}$$

The weight of the bob, mg, is resolved along two orthogonal components: one that is tangent to the arc along which the bob swings, pointed toward the equilibrium position, and the other that is perpendicular to it, along the tension in the string to which the bob is attached. The tension in the string cancels the component $mg \cos\theta$. The restoring force is $F = -mg \sin\theta$. This force imparts an acceleration

$$a = -g \sin\theta. \tag{4.6b}$$

$$\text{Thus, } \frac{d^2\theta}{dt^2} + \frac{g}{\ell}\sin\theta = 0. \tag{4.7}$$

Hence, we may write,

$$0 = \frac{d\theta}{dt}\left[\frac{d^2\theta}{dt^2} + \frac{g}{\ell}\sin\theta\right] = \frac{d\theta}{dt}\frac{d^2\theta}{dt^2} + \frac{g}{\ell}\frac{d\theta}{dt}\sin\theta; \tag{4.8a}$$

$$\text{i.e., } \frac{d}{dt}\left[\frac{1}{2}\left(\frac{d\theta}{dt}\right)^2 - \frac{g}{\ell}\cos\theta\right] = 0 \cos\theta\Bigg] = 0. \tag{4.8b}$$

From Eq. 4.8b, it follows that

$$\frac{1}{2}\left(\frac{d\theta}{dt}\right)^2 - \frac{g}{\ell}\cos\theta = \text{C, a constant.} \tag{4.8c}$$

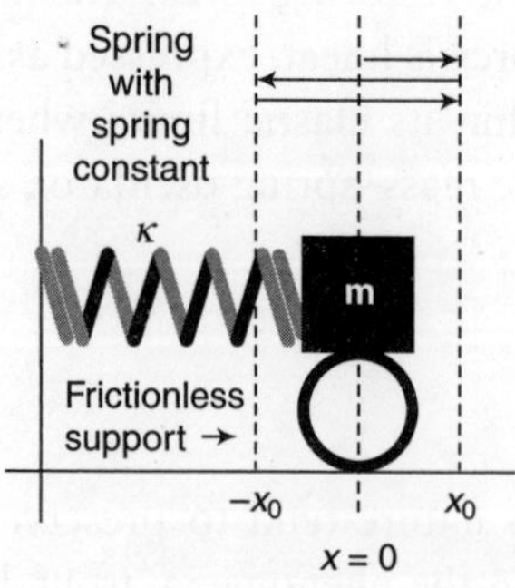

Fig. 4.3a A mass–spring oscillator, fixed at the left end and freely oscillating at the right, where the mass m is attached, without frictional losses.

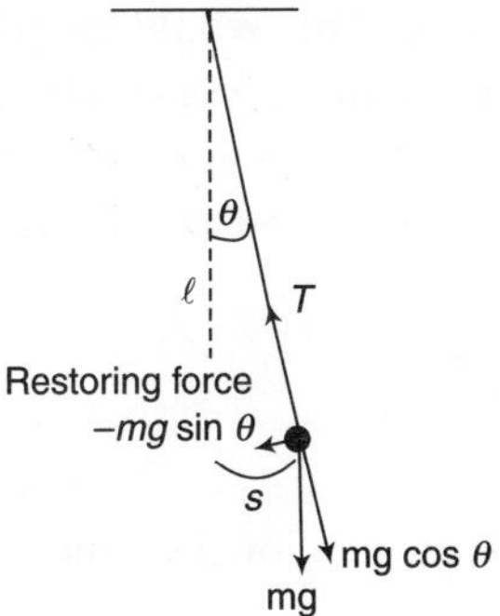

Fig. 4.3b A simple pendulum. The weight *mg* is resolved along the tangent to the arc toward the equilibrium, and orthogonal to it. Note that the restoring force is written with a minus sign since it is in a direction opposite to that in which the displacement angle and its sine increases.

We now use the initial conditions, at $t = 0$:

(i) position: $\theta\Big]_{t=0} = \theta_0$, the maximum angular displacement from which the bob is released to oscillate under gravity,

(iii) velocity $\left.\dfrac{d\theta}{dt}\right]_{t=0} = 0\,.$

The initial conditions give the value of $C = -\dfrac{g}{\ell}\cos\theta_0$, which gives

$$\left(\frac{d\theta}{dt}\right)^2 = \frac{2g}{\ell}(\cos\theta - \cos\theta_0), \text{ or, } \left(\frac{d\theta}{dt}\right) = \sqrt{\frac{2g}{\ell}}\sqrt{(\cos\theta - \cos\theta_0)}\,. \tag{4.9}$$

We consider only the positive square root, corresponding to 'real' velocity. The solutions are symmetric under time-reversal. Integrating Eq. 4.9, and recognizing that the angle swept by the pendulum bob from θ to θ_0 corresponds to a *quarter* of the time-period, T, we get

$$\int_0^{\theta_0}\left(\frac{d\theta}{\sqrt{(\cos\theta - \cos\theta_0)}}\right) = \sqrt{\frac{2g}{\ell}}\int_0^{T/4} dt = \frac{T}{4}\sqrt{\frac{2g}{\ell}}\,. \tag{4.10a}$$

Hence, the periodic time of the pendulum is given by

$$4 = \left(4\sqrt{\frac{\ell}{2g}}\right)\int_0^{\theta_0}\left(\frac{d\theta}{\sqrt{(\cos\theta - \cos\theta_0)}}\right). \tag{4.10b}$$

The pendulum described by Eq. 4.10 is not quite 'simple'. The periodic time clearly depends on the initial angle θ_0 at which the pendulum's bob is released. This angle could be anything from very small, to large. In other words, the periodic time will depend not merely on length of the pendulum ℓ and acceleration due to gravity g but also on the *amplitude* of oscillation, which is determined by θ_0. This would greatly restrict the pendulum's capacity to provide us with a device to measure time. Nonetheless, the happy word is that as long as the *maximum* angular displacement of the pendulum is *small* enough, so that for all angles we may approximate $\sin\theta \approx \theta$, we can write the following relation (for 'small' oscillations) instead of Eq. 4.7:

$$\frac{d^2\theta}{dt^2} + \frac{g}{\ell}\theta = 0. \tag{4.11a}$$

Since the arc-length S is merely proportional to the angle $\theta = \frac{s}{\ell}$ swept by the string of length ℓ, we also have

$$\frac{d^2s}{dt^2} + \frac{g}{\ell}s = 0. \tag{4.11b}$$

Equation 4.11b is identical to Eq. 4.5b if we identify $\frac{\kappa}{m} \leftrightarrow \frac{g}{\ell}$. We may therefore use essentially the same mathematical tools to address the pendulum corresponding to small oscillations, as we used for the mass–spring simple harmonic oscillator.

One of the strikingly beautiful aspects of the use of mathematics to describe physical phenomena is that we often come across essentially the *same* mathematical form, describing totally *different* physical situations. It is, as Wigner would say, a gift of nature that both the mass–spring oscillator and the simple-pendulum are described by exactly the *same* differential equation.

If the mass–spring oscillator whose dynamics is governed by the elastic properties of the material is not sufficiently distinct from a pendulum oscillating under gravity, consider an electrical circuit consisting of an inductor L and a capacitor C (Fig. 4.3c). The governing principle for the phenomenology that describes the dynamics of this electrical circuit is the Faraday–Lenz law. Its details would be discussed in a book on electrodynamics, but we shall visit it briefly in Chapter 12. The capacitor C is charged by the battery B, by turning the switch S_1 on. Switch S_1 is then turned off, and the switch S_2 is turned on to complete the LC circuit. The charge on the capacitor flows through the inductor, but as the current flows through the inductor, the changing magnetic flux through the inductor causes an induced EMF (electromotive force) which begins to oppose the current through it. This would cause a current to flow in the *opposite* direction, which would then recharge the capacitor. This sets in *oscillations* in the flow of the charge. Unlike what happens in a resistor, the current and the voltage in an inductance L, and in a capacitor C, do not *peak* together. According to the laws of electrodynamics, which we shall not detail here, the voltage across the capacitor is

$$V_C = \frac{Q}{C}, \tag{4.12a}$$

and that across the inductor is $V_L = -L\frac{dI}{dt} = -L\frac{d^2Q}{dt^2}$. (4.12b)

The current in the circuit is therefore given by

$$I = \frac{dQ}{dt} = \frac{d(CV)}{dt} = C\frac{dV}{dt}, \tag{4.12c}$$

and thus the current I is proportional to $\frac{dV}{dt}$, and not to V. It is only in the case of a resistor that the current I and the voltage V are proportional to each other (Ohm's law), and therefore would peak and fall together, in phase. In an electric circuit which has an inductor and a capacitor, the voltage *lags* the current in a *capacitor* by 90°, but *leads* the current in an *inductor* by the same amount. The voltages across the inductor and the capacitor are exactly out of phase, and represented therefore using a relative *minus* sign. Accordingly,

$$-V_L + V_C = 0. \tag{4.12d}$$

Therefore, we get $+L\frac{d^2Q}{dt^2} + \frac{Q}{C} = 0$, i.e., $\frac{d^2Q}{dt^2} + \frac{Q}{LC} = 0$. (4.13)

This equation again has exactly the same form as Eq. 4.5 for the mass–spring oscillator, or Eq. 4.11 for the simple pendulum. The identical forms of Eqs. 4.5, 4.11, and 4.13 immediately lead us to the following correspondence, $\frac{\kappa}{m} \leftrightarrow \frac{g}{\ell} \leftrightarrow \frac{1}{LC}$. One may develop the electro-mechanical analogues further, by associating $L \leftrightarrow m$ and $C \leftrightarrow \frac{1}{\kappa}$ since inductance plays the same role in electric circuits, as inertia does in mechanics. Likewise, capacitance plays the same role in electrical circuits as compliance (inverse of the spring constant) does for elastic materials. The electro-mechanical analogues are so powerful that Feynman referred to Eq. 4.13 as the Newton's law of electricity.

We now meet another, seemingly very different situation, which nevertheless will turn out to be equivalent to the situations described in Fig. 4.3a, b, and c. We see in Fig. 4.3d a particle released from rest at a point A in a parabolic frictionless bowl under gravity. The figure shows only a vertical planar section of the parabolic bowl, in which the motion of the particle takes place. At any point, the potential energy of the particle is mgh, h being the height from the base where the zero of the potential energy scale is set. The shape of the bowl being parabolic, the height at a point on the surface of the bowl has the functional form, $h = \sigma x^2$, which gives the potential energy $mgh = V(x) = mg\sigma x^2$. σ is an appropriate scaling constant which describes the parabolic shape of the bowl, and its dimension is L^{-1}. This allows us to write the conservative force on the object as

$$-\frac{dV}{dx} = F = -(2mg\sigma)x\,. \tag{4.14}$$

We immediately now recognize that $2mg\sigma$ plays a role as would allow us to consider it to be completely analogous to the spring constant κ of the mass–spring oscillator.

We may therefore actually write $2g\sigma = \frac{\kappa}{m}$. (4.15)

Motion of the tiny object in the parabolic bowl, considered to be frictionless, will then be given by the solution to the differential equation $m\frac{d^2x}{dt^2} + 2mg\sigma x = 0$, (4.16a)

i.e., $\frac{d^2x}{dt^2} + (2g\sigma)x = 0$ · (4.16b)

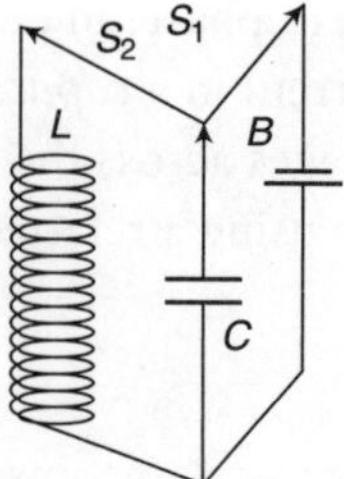

Fig. 4.3c An electric oscillator, also known as 'tank circuit', consisting of an inductor L and a capacitor C. Switch S_1 connects the capacitor to the battery, and S_2 to the inductor. Once the capacitor is charged by the battery, S_1 is switched off, and S_2 is switched on and electric current and charge are set up in the LC circuit through the switch S_2.

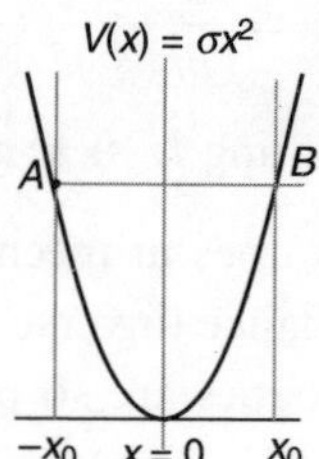

Fig. 4.3d An object released from point A in a parabolic container whose shape is described by the function ½kx^2 oscillates between points A and B, assuming there is no frictional loss of energy. All of its potential energy at point $-x_0$ gets transformed into kinetic energy at the equilibrium mid-point, which carries it up to the point B, when it turns around to oscillate.

We thus see that motion under gravity in a parabolic bowl is governed by exactly the same equation of motion as that in an elastic potential, and as that of the mass–spring oscillator. A and B are the turning points, just as the points at $x = \pm x_0$ in Fig. 4.3a.

Notwithstanding the fact that the differential equations 4.5, 4.11, 4.13, and 4.14 describe completely different physical situations, they all have the exact same form,

$$\ddot{q} + \alpha q = 0. \tag{4.17}$$

In the above equation, q represents the departure from equilibrium of the oscillating system, and α is an auxiliary constant determined by *intrinsic* properties of the system. For the mass–spring oscillator, $\alpha = \frac{\kappa}{m}$ and q is the displacement x from the equilibrium point, at the origin. The solutions are, however, equally applicable for *all* oscillators with an appropriate interpretation of κ and m.

Hence, we get: $-\alpha q = \ddot{q} = \dfrac{\delta v}{\delta t} = \dfrac{\delta\left(\dfrac{\delta q}{\delta t}\right)}{\delta t} = \left(\dfrac{\delta v}{\delta q}\right)\left(\dfrac{\delta q}{\delta t}\right) = v\dfrac{\delta v}{\delta q},$ (4.18a)

i.e., $-\alpha\int q dq = \int v dv$,

and hence, $-\alpha\dfrac{q^2}{2} + \varepsilon = \dfrac{v^2}{2}$, i.e., $\dfrac{v^2}{2} + \dfrac{\kappa}{m}\dfrac{q^2}{2} = \varepsilon$.

Hence, $m\dfrac{v^2}{2} + \kappa\dfrac{q^2}{2} = m\dfrac{\dot{q}^2}{2} + \kappa\dfrac{q^2}{2} = E = m\varepsilon.$ (4.18b)

We immediately recognize the two terms on the left hand side of the above equation respectively as the kinetic and the potential energy of the oscillator, and the right hand side as the total energy. ε has appeared as a constant of integration. The total energy of the system, $E = m\varepsilon$, must therefore be eventually determined from the initial conditions.

Since $v^2 = \dfrac{2}{m}\left(E - \kappa\dfrac{q^2}{2}\right)$, we get $v = \pm\sqrt{\dfrac{2}{m}\left(E - \kappa\dfrac{q^2}{2}\right)} = \pm\sqrt{\dfrac{2E}{m}}\sqrt{\left(1 - \dfrac{\kappa}{E}\dfrac{q^2}{2}\right)}$,

i.e., $\dfrac{dq}{dt} = \pm\sqrt{\dfrac{2E}{m}}\sqrt{\left(1 - \dfrac{\kappa}{E}\dfrac{q^2}{2}\right)}$. (4.19)

Hence, $\int\left(\dfrac{dq}{\sqrt{\left(1 - \dfrac{\kappa}{E}\dfrac{q^2}{2}\right)}}\right) = \pm\sqrt{\dfrac{2E}{m}}\int dt$. (4.20)

To solve this integral, we put

$$q = \sqrt{\frac{2E}{\kappa}}\cos\theta,\ \theta = \cos^{-1}\left(q\sqrt{\frac{\kappa}{2E}}\right),\ \text{hence } dq = -\sqrt{\frac{2E}{\kappa}}\sin\theta d\theta, \quad (4.21)$$

which gives $\int\left(\dfrac{-\sqrt{\dfrac{2E}{k}}\sin\theta d\theta}{\sqrt{(1-\cos^2\theta)}}\right) = \pm\sqrt{\dfrac{2E}{m}}\int dt$.

Therefore, $-\int\left(\dfrac{\sin\theta d\theta}{\sqrt{(1-\cos^2\theta)}}\right) = \pm\sqrt{\dfrac{k}{m}}\int dt = \pm\sqrt{\dfrac{\kappa}{m}}t + \eta$, where η is a constant of integration.

Thus, $-\int d\theta = \pm\sqrt{\frac{\kappa}{m}}t + \eta$, which gives (using Eq. 4.19), $\theta = \cos^{-1}\left(q\sqrt{\frac{\kappa}{2E}}\right) = \mp\sqrt{\frac{\kappa}{m}}t - \eta$

$$\text{Accordingly, } \left(q\sqrt{\frac{\kappa}{2E}}\right) = \cos\left(\mp\sqrt{\frac{\kappa}{m}}t - \eta\right) = \cos\left(\sqrt{\frac{\kappa}{m}}t + \phi\right). \tag{4.22a}$$

Above, instead of the constant of integration, η, we have used $\phi = \mp\eta$.

$$\text{Therefore, } q(t) = \left(\sqrt{\frac{2E}{\kappa}}\right)\cos\left(\sqrt{\frac{\kappa}{m}}t + \phi\right) = \lambda\cos\left(\sqrt{\frac{\kappa}{m}}t + \phi\right) = \lambda\cos(\omega_0 t + \phi), \tag{4.22b}$$

where $\omega_0 = \sqrt{\frac{\kappa}{m}}$ has dimension T^{-1}, and appropriate interpretations of $\frac{\kappa}{m}$ for the different oscillators described by Fig. 4.3a, b, c, and d. The maximum displacement $\lambda = \sqrt{\frac{2E}{\kappa}}$ from the equilibrium point is known as the *amplitude* of the oscillation. Since ω_0 for all the oscillators is determined completely by intrinsic properties of the different oscillators, it is known as the natural circular frequency of the simple harmonic oscillator. It is given by

$$\omega_0 = \left(\sqrt{\frac{\kappa}{m}} \leftrightarrow \sqrt{\frac{g}{\ell}} \leftrightarrow \frac{1}{\sqrt{LC}} \leftrightarrow \sqrt{2g\sigma}\right). \tag{4.23a}$$

The corresponding time period is given by

$$T_0 = \frac{2\pi}{\omega_0} = 2\pi\sqrt{\frac{m}{\kappa}} \leftrightarrow 2\pi\sqrt{\frac{\ell}{g}} \leftrightarrow 2\pi\sqrt{LC} \leftrightarrow \frac{2\pi}{\sqrt{2g\sigma}}. \tag{4.23b}$$

The velocity $\dot{q} = \frac{dq}{dt}$ of the oscillator would of course change at each instant of time, and is readily determinable from Eq. 4.22b. We observe that when $\phi \neq 0$, the displacement of the oscillator from the equilibrium would not be maximum at $t = 0$, and thus its potential energy also would not be a maximum. Likewise, its velocity would not be zero at $t = 0$, which means that the oscillator is set in motion at some intermediate point, but with some initial velocity. The sum of the initial potential energy at that stretch, together with the initial velocity, would determine the total energy of a system. In a conservative system, the nature of the energy of the system changes between its kinetic form and the potential form, the total remaining constant (Fig. 4.4b).

From Eq. 4.18b, we have,

$$\frac{\dot{q}^2}{2E/m} + \frac{q^2}{2E/\kappa} = \frac{\dot{q}^2}{2E/m} + \frac{q^2}{2E/m\omega_0^2} = 1, \tag{4.24a}$$

which is an equation to the ellipse in the position–velocity phase space of the oscillator, with semi-major axis $a = \sqrt{\frac{2E}{m}}$, (4.24b)

and semi-minor axis $b = \sqrt{\frac{2E}{\kappa}} = \sqrt{\frac{2E}{m\omega_0^2}}$. (4.24c)

The area of this ellipse is given by $A = \pi \times \sqrt{\frac{2E}{m}} \times \sqrt{\frac{2E}{m\omega_0^2}} = \frac{2\pi E}{\sqrt{m\kappa}}$. (4.24d)

It is a constant for a given oscillator of mass, m, stiffness constant, κ, and which is imparted a certain amount of energy, E , to set it in motion. Since the oscillator system dynamically 'evolves' under the restoring force to attain the state of motion described by the elliptic orbit in the phase space, the ellipse is often referred to as an 'attractor' in phase space. The region $q_1 < q < q_2$ shown in Fig. 4.5 depicts the potential to be quadratic in the neighborhood of the equilibrium point, q_0, and the phase space trajectory of the oscillator in this region will therefore be an ellipse.

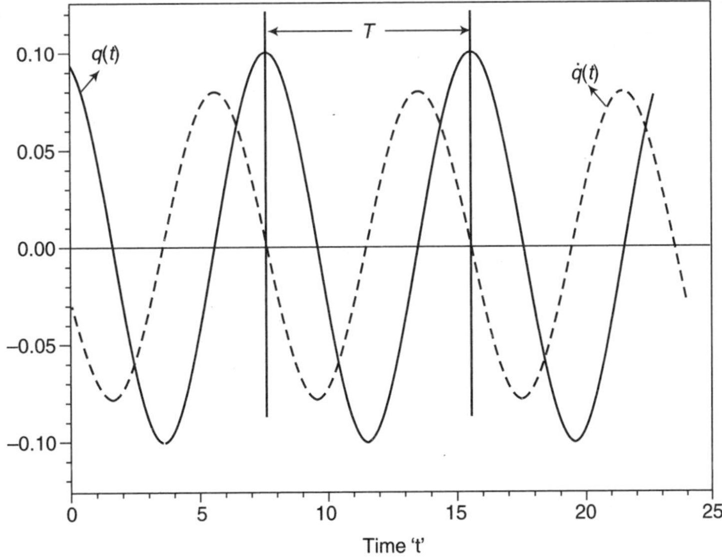

Fig. 4.4a Position $q(t)$ and velocity $\dot{q}(t)$ of a linear harmonic oscillator as a function of time, using Eq. 4.22. Plotted with $\kappa \cong 0.62 \times 10^{-3}$ N/Kg; $m = 10^{-3}$ kg; $E = 3.1 \times 10^{-6}$ J; $\lambda = 0.1$m; $\omega_0 \approx 0.79$ rad/sec; $T \cong 8$ sec and $\phi = 0.1\pi$ rad.

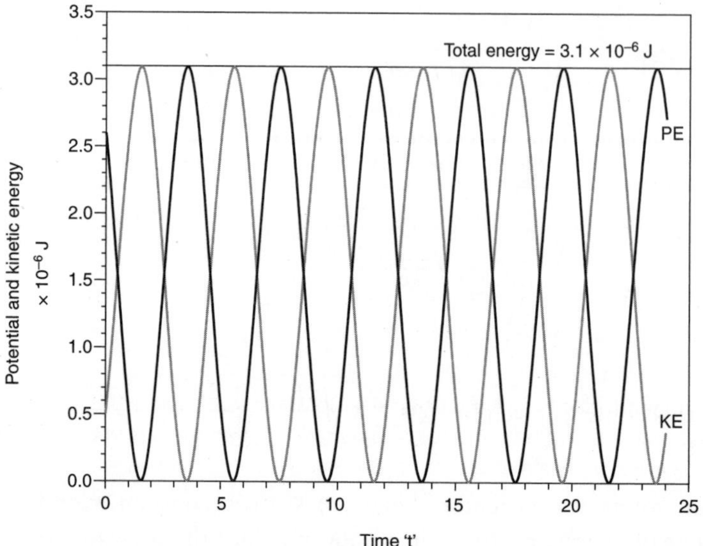

Fig. 4.4b Potential Energy (PE) and Kinetic Energy (KE) of the linear harmonic oscillator of Fig. 4.4a, as a function of time.

Motion of a particle at the point of stable equilibrium (such as q_0, shown in Fig. 4.5) carries the particle across the equilibrium point because of the particle's non-zero momentum at that instant. This motion carries the oscillator only as far as the *turning points*, where the entire energy becomes potential energy. At these points, the particle comes to a momentary halt, and reverses its direction of motion. At all points between the turning points, $q_1 < q < q_2$, there always are *two* possible values of the velocity $\dot{q}$, one having positive sign and the other having negative sign. This is so because at any of these intermediate points, the particle could be moving either toward left, or toward right, i.e., in the direction of decreasing value of q, or increasing.

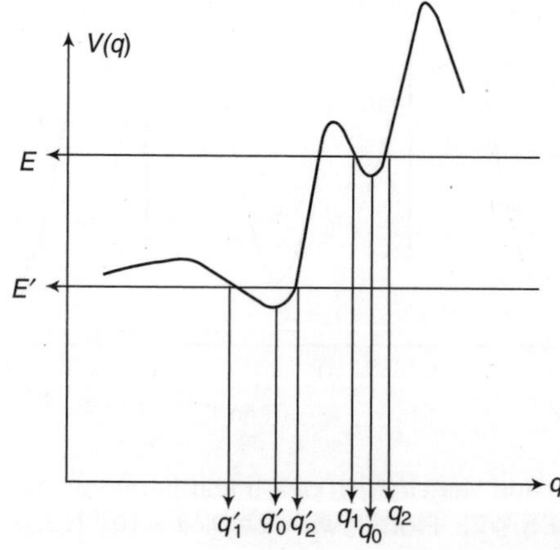

Fig. 4.5 Bound state motion about a point of stable equilibrium must remain in between the 'classical turning points'. A particle with total energy E' remains bound between q'_1 and q'_2 and one with total energy E remains bound between q_1 and q_2.

A particle released with zero $\dot{q}$ from the point q_2 remains *bound*, oscillating between the turning points q_2 and q_1. Since the motion is governed by a second order differential equation, the solution is completely symmetric with respect to the reversal of the sign of the velocity, $\dot{q}$. The time taken by the particle to go from q_2 to q_1 is essentially the same as that required for the reverse path q_1 to q_2. This is so not merely the case if the potential is strictly quadratic about the equilibrium point, but even if it is not so. Thus all motion within the confines of the turning points is strictly repetitive, periodic, generating closed orbits in the position–velocity phase space. The orbit in the position–velocity phase space is a perfect ellipse *only* if the potential is strictly quadratic between the turning points. When the turning points are farther away from the point of equilibrium into a region of the potential where it is no longer parabolic (such as the region $q_1' < q' < q_2'$ in Fig. 4.5), the trajectory in the phase space of the particle will no longer be a strict *ellipse*, even if it will still be a *closed* orbit. The reason for the departure from the perfect ellipse is due to the fact that while $a = \sqrt{2E/m}$ remains unaffected in the range $q_1' < q' < q_2'$, $b = \sqrt{2E/\kappa}$ cannot be defined in that range. This limitation comes from the fact that $\kappa = \left\{ \left. \dfrac{d^2V}{dx^2} \right|_{x_0 = 0} \right\}$ is valid *only* in the *quadratic* domain of Eq. 4.2. The region $q_1' < q' < q_2'$ clearly stretches beyond the quadratic region. Finally, we note the special case when the particle is placed at the point of stable equilibrium and is *not* imparted any initial velocity. Such a particle will just stay there; this is a special case of the ellipse, called a degenerate ellipse. It is effectively only a 'one point' orbit in the phase space.

The initial conditions for the oscillator can of course be different every time the system is set into *small* oscillations. By 'small oscillations', we essentially imply that the oscillations are confined to the region of space over which the potential can be considered essentially parabolic. The two parameters (λ, ϕ) in Eq. 4.22 have resulted as constants of integration, determined uniquely by the initial conditions on $(q_{t=0}, \dot{q}_{t=0})$. The solution given by Eq. 4.22 is the most general solution. In fact, without going through the steps (Eq. 4.18 and Eq. 4.21), we could have easily guessed that the solution to Eq. 4.15 would be of the form Eq. 4.20, since Eq. 4.15, by its very form, seeks a solution whose second derivative is proportional to itself. A function having this property is of course a rather familiar one. We know that the sine and the cosine functions are derivatives of each other (taking care of the sign). Both of these functions therefore have their *second* derivatives proportional to themselves. It really does not matter whether we consider the solution to be the sine function or the cosine, since they differ only in phase by $\pi/2$. It can be easily absorbed by shifting the phase in the argument of the function.

Alternative forms of the general solution result from a simple consideration that solution to a differential equation can always be written in any alternative, linearly independent, basis. Different choices of the basis set are all mathematically equivalent to each other. No matter which basis set is employed, the general solution is always a linear superposition of two independent functions. The coefficients in the superposition are always determinable from the two initial conditions provided by initial position and initial velocity. Some particular choice, however, could be more convenient to simplify the equations than another for a specific consideration. It is therefore useful to be acquainted with some of the more common alternative basis sets.

We have, from Eq. 4.22, the initial position and velocity given by

$$q_0 = q_{t=0} = \lambda \cos \phi = J, \quad \text{and} \quad v_0 = \dot{q}_{t=0} = \lambda \omega_0 \sin \phi = K\omega_0. \tag{4.25a}$$

From the initial conditions, (q_0, v_0), we get the two constants,

$$\lambda = \pm\sqrt{q_0^2 + \frac{v_0^2}{\omega_0^2}} \quad \text{and} \quad \phi = -\tan^{-1}\frac{v_0}{q_0\omega_0}. \tag{4.25b}$$

We may write the sinusoidal solution Eq. 4.22 in the following form:

$$q(t) = \lambda \cos(\phi)\cos(\omega_0 t) - \lambda \sin(\phi)\sin(\omega_0 t) = J\cos(\omega_0 t) + K\sin(\omega_0 t), \tag{4.25c}$$

$$\text{with } J = \lambda\cos(\phi) \quad \text{and} \quad K = -\lambda\sin(\phi). \tag{4.25d}$$

The constants $(J, K) = (\lambda \cos\phi, \lambda \sin\phi)$ are then immediately recognized as the coefficients in superposition of the linearly independent base pair, (cos $(\omega_0 t)$, sin $(\omega_0 t)$. They can be determined from the initial conditions on position and velocity.

The general solution can also be written as a superposition of another linearly independent base pair, $(e^{i\omega_0 t}, e^{-i\omega_0 t})$:

$$q(t) = Me^{i\omega_0 t} + M^* e^{-i\omega_0 t} \equiv (M, M^*), \tag{4.25e}$$

where the constants M and M^* are now complex numbers, conjugate of each other, and therefore determined by just two real numbers, which are the real and the imaginary parts of M, are given by

$$M_{rp} = \frac{M+M^*}{2} = \frac{J}{2} = \frac{\lambda\cos\phi}{2} \quad \text{and} \quad M_{ip} = \frac{M-M^*}{2i} = \frac{K}{2} = -\frac{\lambda\sin\varphi}{2}, \tag{4.25f}$$

which can also, of course, be determined from the initial condition on position and velocity, as one would expect. After all, there are no more, and no less, than two parameters to be fixed, no matter which basis set of linearly independent functions is employed.

We have already seen that the different physical situations described in Fig. 4.3a, b, c, and d are *all* described by the solution Eq. 4.22. Likewise, all of these situations can be pictorially depicted by a single reference circle shown in Fig. 4.6. It shows an object in circular motion in the ZX Cartesian plane at a constant angular velocity ω_0, with the center of the circle at the origin.

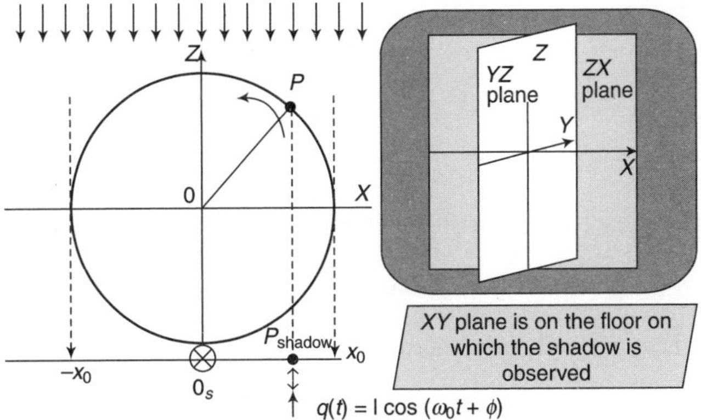

Fig. 4.6 The reference circle which can be related to each of the simple harmonic motion described in Fig. 4.3a, b, c, and d. The circular motion shown on the left side is in the *ZX* plane, but similar motion can take place in the *YZ* plane and the corresponding motion seen on the floor along the *Y*-axis.

We shall not worry about what would make a tiny object P go round at a uniform angular speed in a vertical plane. We restrict our attention to the *shadow*, P_{shadow}, of the point P on the ground as would be cast by parallel rays of light falling on it from the top. All light rays are considered to arrive along the direction of $-\hat{e}_z$. The motion of P_{shadow} as P goes the around the circle at uniform angular speed is described by Eq. 4.22. It is a simple harmonic motion (SHM) about the central equilibrium point, O_s. The circular motion of P and the SHM of P_{shadow} are thus *intimately* correlated. P_{shadow} performs SHM about the equilibrium point along the X-axis while the object P performs a circular motion in the ZX plane. *The reason this motion is called 'harmonic' is that composites of this fundamental oscillatory motion produce harmonious melodies.* In the next section we discuss how two (and more) concurrent simple harmonic motions are combined.

4.2 SUPERPOSITION OF SIMPLE HARMONIC MOTIONS

We now consider superposition of *two* simple harmonic motions. First, we restrict ourselves to the simple case when the two motions are collinear. One can learn the mathematical tools of constructing such superposition quite easily. The important thing to note in this context is that the principle of superposition is a fundamental principle and has a wide range of applicability. We have already seen that motion sufficiently close to a point of stable equilibrium is simple harmonic. Irrespective of the origin of the restoring force, whether material elasticity, or gravitational field, or anything else we require a detailed consideration of such phenomena. The principle of superposition holds in a wide range of situations. The techniques we discuss in this section are, however, limited to motion in what are called 'linear media'. In some real materials, the response of the medium to a disturbance is non-linear, and the behavior of the physical phenomena in such materials is more complex.

We now construct the superposition of two motions described respectively by $y_1(t)$ and $y_2(t)$:

$$y_1(t) = C_1 \cos(\omega\tau - \delta_1), \tag{4.26a}$$

$$y_2(t) = C_2 \cos(\omega\tau - \delta_2). \tag{4.26b}$$

The result of the (linear) superposition is then given by the sum of the two:

$$y(t) = (C_1 \cos\delta_1 + C_2 \cos\delta_2)\cos(\omega t) + (C_1 \sin\delta_1 + C_2 \sin\delta_2)\sin(\omega t). \tag{4.26c}$$

We see that this is completely equivalent to

$$y(t) = C\cos(\omega t - \delta) = C\cos(\omega t)\cos\delta + C\sin(\omega t)\sin\delta, \tag{4.26d}$$

with

$$C = \sqrt{C_1^2 + C_2^2 + 2C_1C_2\cos(\delta_2 - \delta_2)}, \tag{4.26e}$$

and

$$\delta = \tan^{-1}\frac{C_1 \sin\delta_1 + C_2 \sin\delta_2}{C_1 \cos\delta_1 + C_2 \cos\delta_2}. \tag{4.26f}$$

A few special cases are illustrated below, and the detailed mathematical constructions are left as exercises using the methods described above.

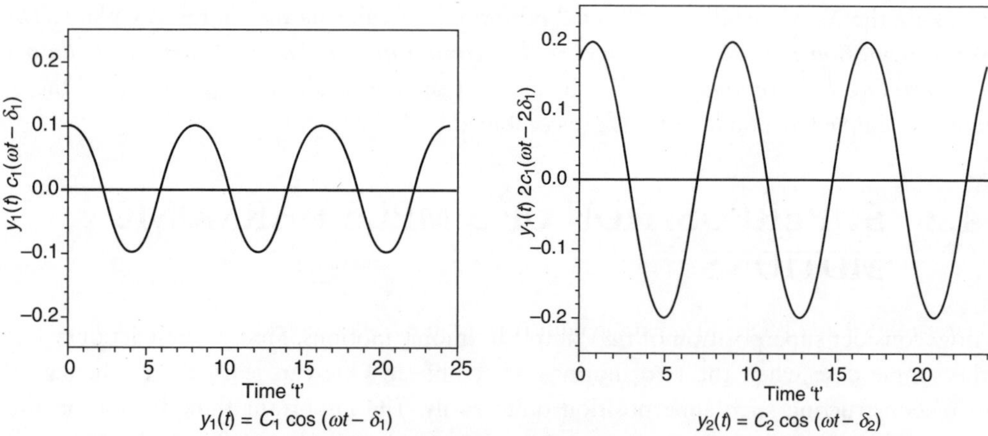

Fig. 4.7 Two simple harmonic motions at the same frequency, ω, but with different amplitudes, C_1 and C_2, and different phases, δ_1 and δ_2. In a linear medium, the two simply add up. Same angular frequency, ω, as in Fig. 4.4a has been used. Other parameters used for this figure are $C_1 = \lambda$, as was used in Fig. 4.4a; and $C_2 = 2C_1$, Also, $\delta_1 = \phi$ as was used in Fig. 4.4a, and $\delta_2 = 2\delta_1$. The phases, of course, determine the values of the displacements at the time 'ero', and thus at all values of the time.

If $\delta_2 = \delta_1$, then $C = C_1 + C_2$ and the amplitudes reinforce each other, the motion due to both the independent components being in phase. That is, the two oscillations are in step with each other. If $\delta_1 - \delta_2 = \pi$, then $C = |C_1 - C_2|$, which means that if the two amplitudes are exactly equal, the resultant would actually vanish. If the two amplitudes are not exactly equal, the net amplitude of the superposition would be diminished, being out of phase. This results in attenuation.

We now consider superposition of two collinear simple harmonic motions whose amplitudes and frequencies may *both* be different. Generally, this does not result in a periodic disturbance. In the special case when the amplitudes are equal, and the frequencies are rather close to each other, the resultant motion can nevertheless be considered to be periodic, even if only approximately in some sense.

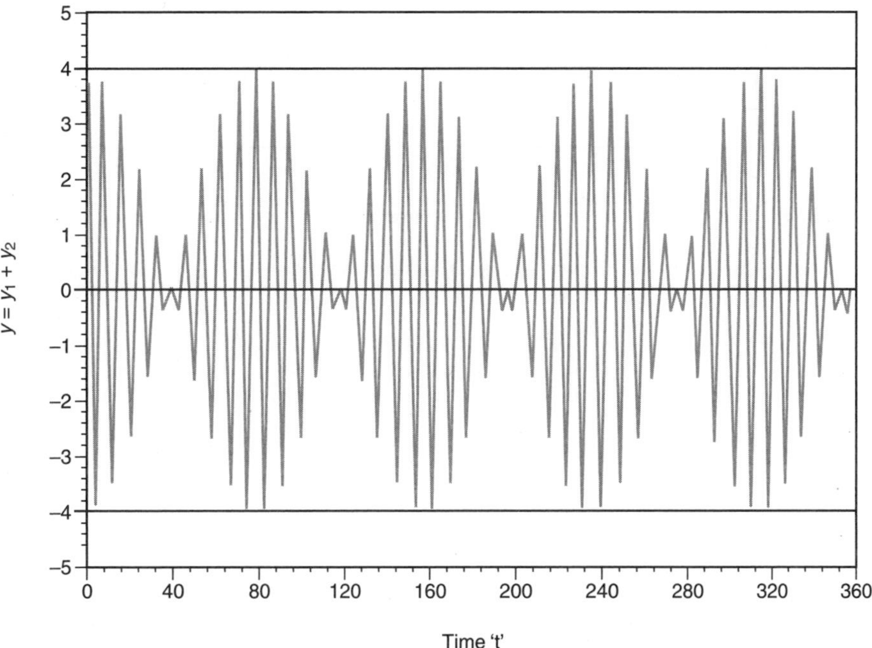

Fig. 4.8a If the frequencies are not vastly different, we get 'beats' and we can regard the resultant as 'amplitude modulated' harmonic motion. The reader is encouraged to discover the result using the superposition method described above.

$y_1 = A\cos(\omega_1 t)$; $y_2 = A\cos(\omega_2 t)$ $\omega_1 = 0.8$; $\omega_2 = 0.72$

$$y = 2A\cos\left(\frac{\omega_1 + \omega_2}{2}\right)t\ \cos\left(\frac{\omega_1 - \omega_2}{2}\right)t$$

$A = 2$

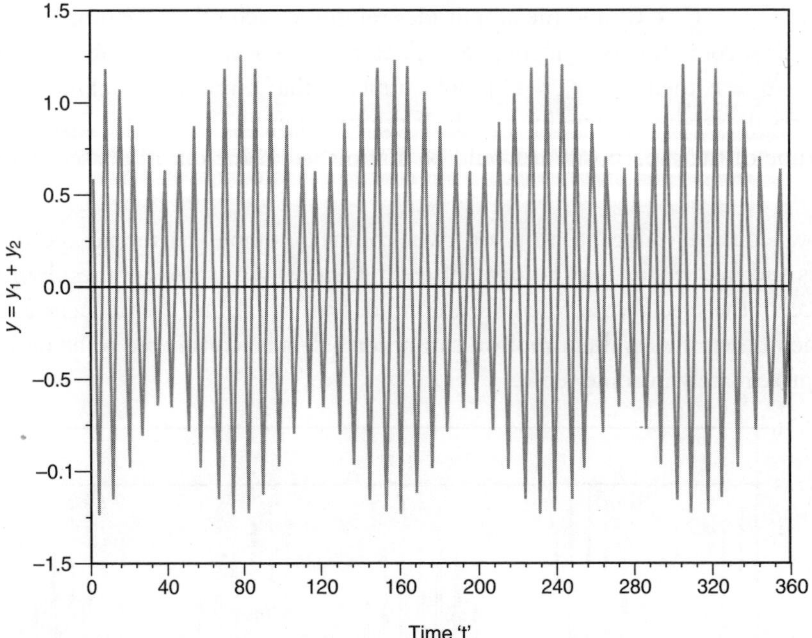

Fig. 4.8b Superposition of two collinear simple harmonic motions for which the frequencies and the amplitudes are both different.

$$y_1 = A_1 \cos(\omega_1 t);\ y_2 = A_2 \cos(\omega_2 t)$$

Let $A_1 = \cos\delta;\ A_2 = \sin\delta$

$$\delta = 0.314;\ \omega_1 = 0.8;\ \omega_2 = 0.72$$

$$y = \cos\delta \cos(\omega_1 t) + \sin\delta \cos(\omega_2 t)$$

$$y = \frac{\cos(\delta + \omega_1 t) + \cos(\delta - \omega_1 t)}{2} + \frac{\sin(\delta + \omega_2 t) + \sin(\delta - \omega_2 t)}{2}$$

When the simple harmonic motions are not collinear, obtains intriguing resultant patterns, generally known as Lissajous patterns, after Jules Antoine Lissajous (1822–1880), who constructed an apparatus to generate such superposition of waves. A large number of inventions (the television screen, for example) employ this construction, in one form or another.

Let us presume now that the vertical *ZX* plane in which the object *P* (of Fig. 4.6) is executing circular motion is no longer static. We now let the whole *ZX* plane oscillate along the *Y*-axis (which is orthogonal to the plane of the Fig.4.6) about the origin. Since *x* and *y* are independent degrees of freedom, P_{shadow} will continue to oscillate along the *X*-dimension, but will concurrently undergo oscillations along the *Y*-dimension. While the oscillations along the *X*-axis are between $-x_0$ and $+x_0$, those along the *Y*-axis would be between $-y_0$ and $+y_0$. The maximum displacement, i.e., the amplitude λ_y of the oscillations along the *Y*-axis, and that along the X-axis, λ_x, need not be the same, though the two certainly may be equal in some special cases. Likewise, the displacement from the equilibrium may not reach its maximum along the Y-axis at the same instant of time when the displacement along the X-axis reaches its respective maximum. That is, the oscillations need not be in phase. Thus ϕ_y and ϕ_x may or may not be equal. The pattern traced on the XY surface by P_{shadow} can thus be quite complicated, and the complexity can get only richer if the frequencies $\omega_y = 2\pi\nu_y$ and $\omega_x = 2\pi\nu_x$ are different; though these two also of course may be equal in some special case. The most interesting situations are those when these two independent frequencies are integer multiples of each other. These patterns on the XY plane are not just complex, but also beautiful, known as the Bowditch–Lissajous figures, named after the American mathematician Nathaniel Bowditch (1773–1838) and the French physicist Jules Antoine Lissajous (1822–1880), who studied these patterns extensively.

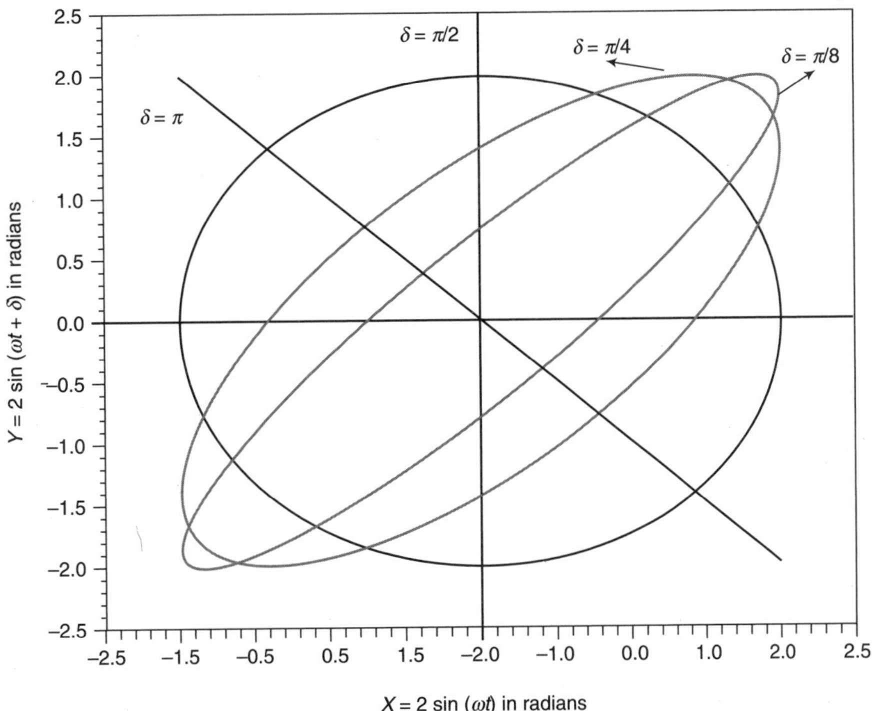

Fig. 4.9a Superposition of two simple harmonic motions, both having the same circular frequency, ω, same amplitude A, but different phases.

The value of $\omega = 0.8$ and $A = 2$ has been used here, as in Fig. 4.4a.

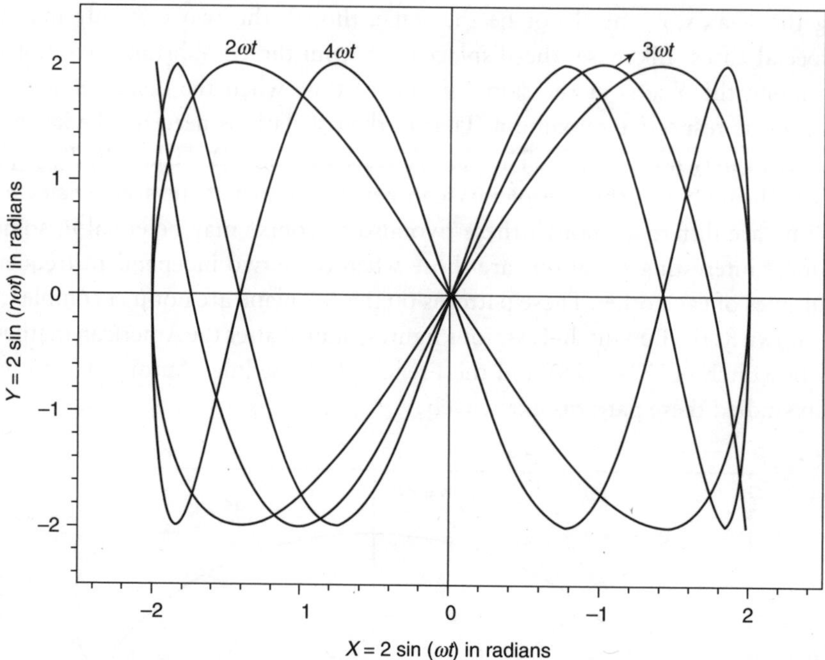

Fig. 4.9b Superposition of two simple harmonic motions which are orthogonal to each other, both with same amplitude A, no phase difference, but different circular frequencies.

The value of $\omega = 0.8$ and A = 2 has been used here, as in Fig. 4.4a.

The Bowditch–Lissajous patterns already are an example of superposition of SHM, but these were along two orthogonal (hence independent) degrees of freedom. We observe that the differential equation is a linear equation, since only the first powers of q and of $\ddot{q}$ appear in it. In fact, a differential equation for a variable q of any order is said to be linear if only the *first* powers of q and also of $\frac{d^n q}{dt^n}$ for *all* values of n appear in it. In fact, Eq. 4.17 is not merely linear, it is also homogenous. A linear homogenous equation has an interesting and important property: sum of any two of its solutions is also a solution. In the present case, therefore, if there are two solutions, $q_1(t)$ corresponding to one set of initial displacement and initial velocity, and $q_2(t)$ corresponding to another set of initial displacement and initial velocity, then their algebraic sum

$$q_3(t) = q_1(t) + q_2(t), \tag{4.27}$$

is also a solution to the same differential equation.

We have so far considered oscillations of an object about a point of stable equilibrium. There is also an energy associated with the 'disturbance' that is oscillating. In a frictionless environment, this energy is conserved, changing at every instant between kinetic and potential forms. The mean (i.e., average over one period) kinetic energy and mean potential energy are respectively given by:

$$\langle KE \rangle = \left\langle \frac{1}{2} m\dot{q}^2 \right\rangle = \frac{m}{2}\left\langle [-\omega_0 \lambda \sin(\omega_0 t + \phi)]^2 \right\rangle = \frac{m}{4}\lambda^2 \omega_0^{\ 2}, \tag{4.28a}$$

$$\langle PE \rangle = \left\langle \frac{1}{2} \kappa q^2 \right\rangle = \frac{\kappa}{2}\left\langle [\lambda \cos(\omega_0 t + \phi)]^2 \right\rangle = \frac{\kappa}{4}\lambda^2, \tag{4.28b}$$

where we have used $<\cos^2(\omega t)dt> = \frac{1}{2} = <\sin^2(\omega t)dt>$.

We note, however, that $<PE> = <KE>$, since $\omega_0^{\ 2} = \frac{\kappa}{m}$. (4.29)

The total energy of the oscillating disturbance, however, remains localized in space in a tiny region of space about its mean position.

4.3 WAVE MOTION

We shall now extend the idea of *displacement* of the oscillator about its mean position to a *disturbance* about the mean position. In order to develop a rather general analysis of the wave motion, we deliberately avoid the question about disturbance of just *what*, i.e., we talk about the disturbance without worrying about just what is being disturbed. Essentially, this will allows us a generalization of the vocabulary we have employed so far to accommodate electromagnetic fields. The oscillators we have talked about so far are essentially mechanical oscillators, like the mass attached to a spring, or a simple pendulum, executing small oscillations. The oscillations that we shall now talk about may be more generally referred to as a time-dependent *disturbance*, such as the amplitude of an electric and the magnetic vector of an electromagnetic wave. In what follows, the term *disturbance* would continue to include the displacement of a mechanical oscillator about its mean position, but will also include the amplitude of an electric and magnetic vector of an electromagnetic field. Such a formulation of the wave motion will make it easy to adapt it to for the understanding of the electromagnetic waves, which we shall consider later, in Chapter 13. There is energy associated also with the properties of these electric and magnetic vectors. Their magnitudes oscillate about an equilibrium point in a strictly sinusoidal manner, analogous to the oscillating displacement of the mechanical oscillators. In the case of electromagnetic waves, no material particle oscillates; it is just the magnitudes of the time-dependent electric and magnetic components of the wave which oscillate. I have already used the term wave, without even defining it, appealing to your intuitive idea that a wave associated with the electromagnetic wave transports energy. Essentially, it is the electromagnetic waves from Sun which bring energy to the Earth, and sustains life. Without this energy, there would be no life on Earth. *Transport of energy* through a region of space is an essential property that waves have, and the electromagnetic waves achieve this even through vacuum. Electromagnetic waves transport energy in the direction of $\vec{E}(t) \times \vec{B}(t)$ wherein $\vec{E}(t)$ is the oscillating electric disturbance, and $\vec{B}(t)$ is the oscillating magnetic disturbance. We shall discuss the electromagnetic waves in Chapter 13, but here we refer only to some of their properties of direct relevance to consider the wave phenomena.

Mechanical oscillators transport energy through a medium. Particles of the medium enable energy transport by nudging their respective neighboring particles. The energy gets transported, without any physical, material, transport of the oscillating particles. The material particles themselves remain confined to the tiny range of small oscillations about the mean position of each. The energy transport takes place in a direction that is either orthogonal to the motion of the disturbance (in which case we have transverse waves) or along the line of the oscillating disturbance (in which case we have longitudinal waves). Electromagnetic waves are transverse waves, since they transport energy in the direction of $\vec{E}(t) \times \vec{B}(t)$, as you will learn in Chapter 13.

The oscillations were described by a function of time alone, as in Eq. 4.22b. A wave transports *energy* through space, through the nudge of the neighboring oscillators. A wave must be therefore described by a function of *both* space and time. In a 1-dimensional problem, say along the X-axis of a coordinate system, the oscillating disturbance can be described therefore by a function, $y(x, t)$. Its magnitude is a measure of the disturbance at the point x, at the time t. The space-dependence of the amplitude $y(x, t)$ of the wave is related to the time-dependence, as one would expect. The following second ordered linear partial differential equation, called the 'wave equation', provides this relation:

$$\frac{\partial^2 \psi(x,t)}{\partial x^2} = \gamma \frac{\partial^2 \psi(x,t)}{\partial t^2}, \tag{4.30a}$$

wherein γ must have the dimensions $L^{-2}T^{-2}$, i.e., dimensions of the square of the inverse of speed. In fact, the wave equation is given by

$$\frac{\partial^2 \psi(x,t)}{\partial x^2} = \frac{1}{c^2} \frac{\partial^2 \psi(x,t)}{\partial t^2}, \tag{4.30b}$$

where c is the speed of the propagation of the wave along the X-axis. The letter c is most often used for the speed of light waves in vacuum, but we have used it to represent the speed of whatever wave that is under our immediate consideration, including the electromagnetic waves. The space and time dependence of the wave-amplitude can be factored as follows:

$$\psi(x, t) = \xi(x)\tau(t). \tag{4.31a}$$

Substituting Eq. 4.31a in the wave equation, we get

$$\tau(t)\frac{d^2\xi(x)}{dx^2} = \frac{\xi(x)}{c^2}\frac{d^2\tau(t)}{dt^2}. \tag{4.31b}$$

Therefore, $$\frac{c^2}{\xi(x)}\frac{d^2\xi(x)}{dx^2} = \frac{1}{\tau(t)}\frac{d^2\tau(t)}{dt^2}. \tag{4.31c}$$

The left hand side of Eq. 4.31c depends only on ζ, and the right hand side only on t. We therefore conclude that both the sides are *independent* of *both* ζ and t. Hence, both the sides are equal to a constant. We denote this constant by $-\omega^2$. This constant separates the ζ-dependence from t-dependence:

$$\frac{c^2}{\xi(x)}\frac{d^2\xi(x)}{dx^2} = -\omega^2 = \frac{1}{\tau(t)}\frac{d^2\tau(t)}{dt^2}. \tag{4.31d}$$

Thus, with the help of the constant of separation, $-\omega^2$, the wave equation separates into two different second order differential equations. One of these, the space part, is

$$\frac{d^2\xi(z)}{dx^2} + \frac{\omega^2}{c^2}\xi(x) = \frac{d^2\xi(x)}{dx^2} + k^2\xi(x) = 0. \tag{4.32a}$$

where we have introduced $k^2 = \frac{\omega^2}{c^2}$. Here, $k = \frac{\omega}{c} = \frac{2\pi\nu}{c} = \frac{2\pi}{\lambda}$ is called the 'wavenumber'. This k is of course different from the spring constant κ we have used earlier. The k used in Eq. 4.32 has dimension L^{-1}, whereas the spring constant κ has dimension MT^{-2}. When the relation between ω and k is linear (the proportionality being given by the speed c), the medium through which the wave propagates is said to be non-dispersive.

All wave solutions corresponding to different frequencies propagate at a fixed speed c, which is known as the phase velocity. Even if it is a minor matter related to mere nomenclature, it is important to note that it is not at all uncommon to have the wavenumber defined as $\frac{1}{\lambda}$ instead of $\frac{2\pi}{\lambda}$. Equation 4.32a is known as the 1-dimensional Helmholtz equation.

The time-dependence in the wave equation (the other part of Eq. 4.31d) is given by the solutions of the following equation, now separated from the wave equation (Eq. 4.30):

$$\frac{d^2\tau(t)}{dt^2} + \omega^2\tau(t) = 0. \tag{4.32b}$$

The solutions of Eq. 4.30, 4.31 are given by

$$\xi(x) = Ae^{ikx} + Be^{-ikx}, \tag{4.33a}$$

$$\text{and } \tau(t) = Ce^{i\omega t} + De^{-i\omega t}. \tag{4.33b}$$

The solution to the wave equation Eq. 4.30 can now be written, using 4.33:

$$\psi(x,t) = \xi(x)\,\tau(t) = \{Ae^{ikx} + Be^{-ikx}\}\{Ce^{i\omega t} + De^{-i\omega t}\},$$

$$\text{or, } \psi(x,t) = Me^{i(kx+\omega t)} + Ne^{i(kx-\omega t)} + Je^{-i(kx-\omega t)} + Ke^{-i(kx+\omega t)}. \tag{4.34a}$$

We see that the wave equation admits solutions expressible as linear superposition of disturbances in the form $f_1(kx - \omega t)$, $f_2(kx - \omega t)$ which travel at the speed c. The argument $(kx - \omega t)$ of f_1 ensures that the space-dependence varies linearly with x, just as the time-dependence varies linearly with t. This property retains the profile of the functions intact, as the disturbance propagates along $\hat{e}_x$, and the value of x increases. Similar argument holds for $f_2(kx + \omega t)$ which represents a wave that travels in the opposite direction, i.e., along $-\hat{e}_x$ at the same speed, c. The general solution (Eq. 4.34a) has four constants, M, N, J and K, two of which get fixed by the *initial* conditions $\psi(x, t)\big]_{t=0}$ and $\left.\frac{\partial \psi(x, t)}{\partial t}\right]_{t=0}$ and the other by the two *spatial* boundary conditions, such as (i) the values of $\psi(x, t)\big]_{x=0}$ and $\psi(x, t)\big]_{x=x_{\max}}$ if the function is restricted to the range $0 \leq x \leq x_{\max}$, (ii) or $\psi(x, t)\big]_{x=0}$ and $\left.\frac{\partial \psi(x, t)}{\partial x}\right]_{x=0}$.

Now, $k = \frac{\omega}{c}$ may belong to a discrete set, or to a continuous set, depending on what frequencies ω can be sustained under the imposition of the boundary conditions.

The fact that the equation is a linear partial differential equation is responsible for the fact that its different solutions can be linearly superposed to get *new* solutions. These solutions to the wave equation are *naturally* very well adapted to the simple harmonic sinusoidal oscillations about the mean equilibrium point. Such admixtures are complicated already. These solutions can also of course be written equivalently in terms of the trigonometric sine and cosine functions. Furthermore, all frequencies ω_j, = 1, 2, 3, ... admissible naturally by the oscillating system must be included in the most general solution, which adds only further complexity, and also richness, to the most general solution. The general solution must therefore be written as sum of sets of four terms, one set for each admissible frequency:

$$\psi(x, t) = \sum_j \{M_j e^{i(k_j x + \omega_j t)} + N_j e^{i(k_j x - \omega_j t)} + J_j e^{-i(k_j x - \omega_j t)} + K_j e^{-i(k_j x + \omega_j t)}\}. \tag{4.34b}$$

When no boundary conditions restrict the values of the wavenumber k (or equivalently, before the imposition of any boundary condition), we may therefore write the most general solution of the wave equation to be given by

$$\psi(x, t) = \int_{-\infty}^{\infty} \{M(k)e^{i(kx+\omega(k)t)} + N(k)e^{i(kx-\omega(k)t)} + J(k)e^{-i(kx-\omega(k)t)} + K(k)e^{-i(kx+\omega(k)t)}\}dk. \tag{4.34c}$$

By explicitly allowing $\omega = \omega(k)$, not necessarily given by the linear relation $\omega = kc$, we have allowed for the most general solution which accommodates a dispersive medium.

The terms $\theta_j = k_j x \pm \omega_j t$ in Eq. 4.34 represent the phase of the jth component in the traveling waves. When waves are traveling in a medium, the surface in the region of space on which the phase would be 'equal' (to each other) would represent the 'wave-front'. The movement of the wavefront contains full information about how the wave propagates in the medium. For an arbitrary set of points on this surface, the phase being equal, $\delta\theta_j = k_j \delta x \pm \omega_j \delta t = 0$. Hence, $\frac{\delta x}{\delta t} = \mp \frac{\omega_j}{k_j}$. The $\mp$ sign behind $\frac{\omega_j}{k_j}$ (which is a ratio of intrinsically positive quantities) therefore determines whether the position x of the wave-front decreases, or increases, as time advances, i.e., for $\delta t > 0$. This determines if the wave is propagating toward the left, or the right, respectively, along the x-axis.

For now, let us consider a simple linear combination of two of the multitude of linearly independent functions, corresponding to disturbances with natural frequency, ω_r, moving in opposite directions:

$$\psi_r(x, t) = Me^{+i(kx+\omega_r t)} + Ne^{+i(kx-\omega_r t)}. \tag{4.35a}$$

We further consider these two components to be of equal magnitude, i.e., $M = N$.

$$\psi_r(x, t) = N\left[e^{+i(kx+\omega_r t)} + e^{+i(kx-\omega_r t)}\right] = 2Ne^{ikx}\cos(\omega_r t). \tag{4.35b}$$

The real part of this solution is given by

$$\mathrm{Re}[\psi_r(x, t)] = 2N\cos(kx)\cos(\omega_r t). \tag{4.35c}$$

For all values of x for which (kx) is an odd multiple of $\frac{\pi}{2}$, the disturbance goes to zero and we have a permanent node at that point, regardless of time. Such solutions constitute '*standing waves*'. Between the nodes, the disturbance exhibits sinusoidal oscillations.

4.4 FOURIER SERIES, FOURIER TRANSFORMS, AND SUPERPOSITION OF WAVES

The sinusoidal trigonometric functions which appear in the solutions of the simple harmonic oscillator are of exceptional value in mathematical analysis. One important application they have in solving problems in a wide range of domains from mathematics to biology comes from methods developed by Jean Baptiste Joseph Fourier (1768–1830). He was the 9th child (out of 12) of a French tailor's second wife. Fourier was taught by Lagrange and Laplace, and he left his own mark amongst the greatest of all mathematicians, even as he served as science adviser to Napoleon's army. Fourier presented his work in 1807 on the propagation of heat in solid materials using rigorous mathematical methods he developed, but his work had first been rejected by Lagrange and Laplace both, among others including Poisson, but then it was accepted later. Fourier analysis provides a powerful tool, called Fourier transforms, to solve differential equations. The methodology draws its strength from the series expansion of a mathematical function, named after Fourier. Series expansions to represent functions are well known, such as the power series expansions which employ the derivatives of the function in the coefficients of various powers of the independent variable.

Fourier showed that when -

(i) a function is periodic, i.e., $\phi(x + L) = \phi(x)$, with $L = b - a$,

(ii) the function is single-valued (except possibly at a finite number of isolated discontinuities),

(iii) and the function has a finite number of maxima/minima within one period, and

(iv) if the integral $\int_a^b |\phi(x)|\,dx$ converges,

then one can find appropriate coefficients a_0, a_n and b_n for all values of n to expand the function in the following series:

$$\phi(x) = \frac{a_0}{2} + \sum_{n=1}^{\infty} a_n \cos\frac{2n\pi x}{L} + \sum_{n=1}^{\infty} b_n \sin\frac{2n\pi x}{L}. \tag{4.36}$$

This *assertion* holds good on account of the fact that sine and cosine functions are periodic. It would become *useful* if we can determine the coefficients a_n, b_n for all values of n. This can indeed be done, as shown below, which would establish that the assertion (Eq. 4.36) is both *valid*, and *useful*.

Integrating the function from a to b, we get

$$\int_a^b \phi(x)dx = \int_a^b \frac{a_0}{2}dx + \sum_{n=1}^{\infty} a_n \int_a^b \cos\frac{2n\pi x}{L}dx + \sum_{n=1}^{\infty} b_n \int_a^b \sin\frac{2n\pi x}{L}dx = \frac{a_0}{2}L.$$

Hence, $a_0 = \frac{2}{L}\int_a^b \phi(x)dx$.

Next, multiplying the function $\phi(x)$ by $\cos\frac{2m\pi x}{L}$ and integrating over the same range, we get

$$\int_a^b \phi(x)\cos\frac{2m\pi x}{L}dx = \int_a^b \frac{a_0}{2}\cos\frac{2m\pi x}{L}dx + \sum_{n=1}^{\infty} a_n \int_a^b \cos\frac{2n\pi x}{L}\cos\frac{2m\pi x}{L}dx + \\ + \sum_{n=1}^{\infty} b_n \int_a^b \sin\frac{2n\pi x}{L}\cos\frac{2m\pi x}{L}dx.$$

Now, using $\int_a^b \cos\frac{2n\pi x}{L}\cos\frac{2m\pi x}{L}dx = \frac{L}{2}\delta_{mn}$ and $\int_a^b \sin\frac{2n\pi x}{L}\cos\frac{2m\pi x}{L}dx = 0$, we get the coefficients a_m for all values of m.

Next, multiplying the function $\phi(x)$ by $\sin\frac{2m\pi x}{L}$ and integrating over the same range, we get the coefficients b_m for all values of m. Toward this, we use the fact that $\int_a^b \sin\frac{2n\pi x}{L}\sin\frac{2m\pi x}{L}dx = \frac{L}{2}\delta_{mn}$. Thus, we find all the coefficients necessary in Eq. 4.36, validating the claim that every periodic function is expressible as a superposition of harmonics of sine and cosine functions.

It is often convenient to use exponential functions, instead of the trigonometric functions. Thus, we may express the Fourier series as

$$\phi(x) = \frac{a_0}{2} + \sum_{n=1}^{\infty} (a_n - ib_n)e^{i\frac{2n\pi x}{L}} + \sum_{n=1}^{\infty} (a_n + ib_n)e^{-i\frac{2n\pi x}{L}}. \tag{4.36a}$$

By introducing new coefficients, c_n, which can be determined from (a_n, b_n) using the relations,

$$c_n = \begin{bmatrix} \frac{1}{2}(a_n - ib_n) & ; \; n > 0 \\ \frac{1}{2}(a_n) & ; \; n = 0 \, , \\ \frac{1}{2}(a_{-n} + ib_{-n}) & ; \; n < 0 \end{bmatrix} \tag{4.36b}$$

we may write, equivalently, $\phi(x) = \sum_{n=-\infty}^{\infty} c_n e^{i\frac{2n\pi x}{L}}$. (4.36c)

The coefficients, $c_n = c_{-n}^*$, can be determined from (a_n, b_n), or more directly, using the integral

$$c_n = \frac{1}{L}\int_a^b \phi(x) e^{-i\frac{2\pi nx}{L}} dx \, . \tag{4.37a}$$

The Fourier series thus provides a breakup of any periodic, well-behaved function, $\phi(x)$, such as in Fig. 4.10a, in terms of trigonometric sine and cosine functions. Alternatively, it provides a breakup of any periodic function $\phi(x)$ in terms of the exponential functions.

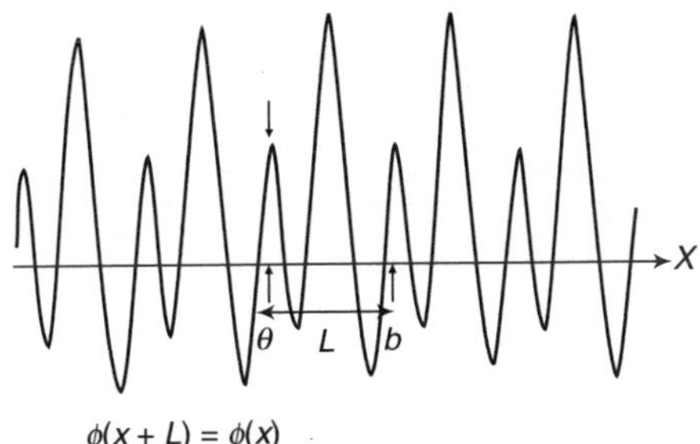

Fig. 4.10a Example of a periodic function which replicates its pattern infinitely both toward its right and toward its left.

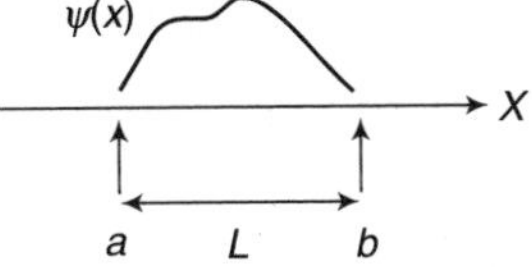

Fig. 4.10b The function $\psi(x)$ defined over the interval (a,b) is not periodic. However, it can be continued on both sides (as shown in Fig.4.10d and 4.10e) to make it periodic.

By a 'well-behaved' function, we mean that it is single-valued (except possibly at a finite number of isolated discontinuities), and that the integral $\int_a^b |\phi(x)| dx$ converges.

The capacity to use the Fourier series for the function $\psi(x)$ of Fig. 4.10b comes from a small trick. We replicate the function along the X-axis so that $\psi(x)$ becomes a part of a function, $\phi(x)$, which is periodic. The Fourier expansion can be applied to the function, $\phi(x)$, to get the coefficients in the expansion. The replication needs to take advantage of the periodicity of sine and cosine functions which are respectively odd and even functions of the argument. Hence, it does not help to compose a replication such as in Fig. 4.10c.

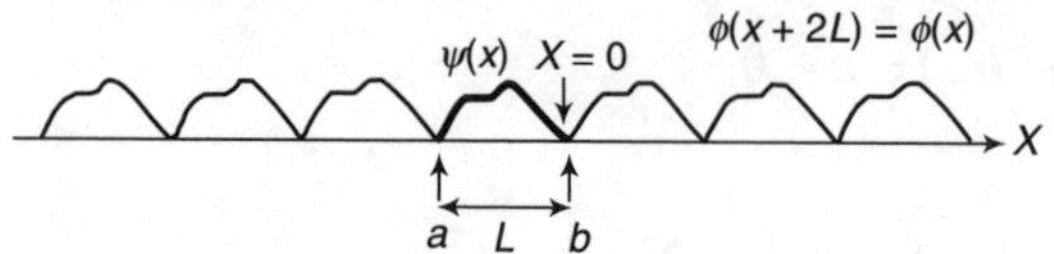

Fig. 4.10c The extended function $\phi(x)$ coincides with $\psi(x)$ in the region (a, b). It is periodic, but it is neither odd nor an even function of x.

On the other hand, replication such as in Fig. 4.10d and Fig. 4.10e is fruitful. Note not just the similarities in the repeated patterns in Fig. 4.10c, 4.10d and 4.10e, but also the differences.

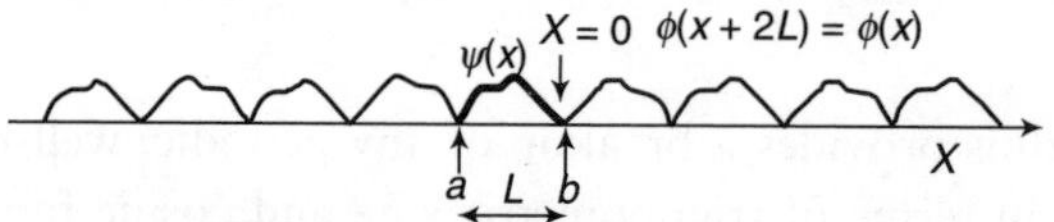

Fig. 4.10d The extended function $f(x)$ coincides with $\psi(x)$ in the region (a, b). It is periodic, and it is an even function of x.

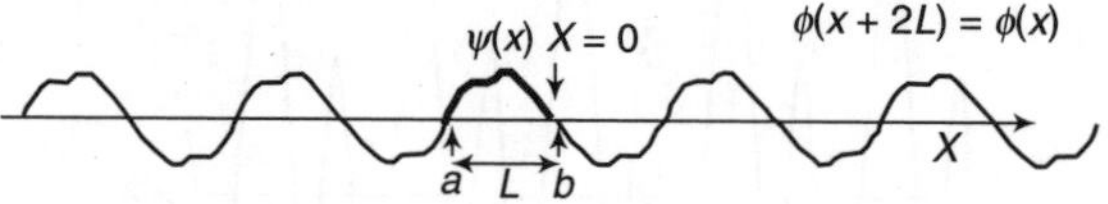

Fig. 4.10e The extended function $\phi(x)$ coincides with $\psi(x)$ in the region (a, b). It is periodic, and it is an odd function of x.

It is remarkable that this idea can be extended to any well-behaved function, $\psi(x)$, even if it is not periodic, such as the one shown in Fig. 4.10b.

We note, of course, that if the periodicity of the wave has a pattern such as in Fig. 4.10d or Fig. 4.10e, rather than how it is in Fig. 4.10a, then the function c_n will be given by an integral over the range 2L instead of L. Hence, c_n will be given by

$$c_n = \frac{1}{2L}\int_{-L}^{+L} \phi(x) e^{-i\frac{2\pi n x}{2L}} dx = \frac{1}{2L}\int_{-L}^{+L} \phi(x) e^{-i\frac{\pi n x}{L}} dx . \tag{4.37b}$$

Fig. 4.11 illustrates an odd-function, made of square periodic patterns with sharp edges, as in square waves, in terms of Fourier series. Only a few terms in the series are employed in this illustration. The final results we get are approximate, which improve as more terms are added.

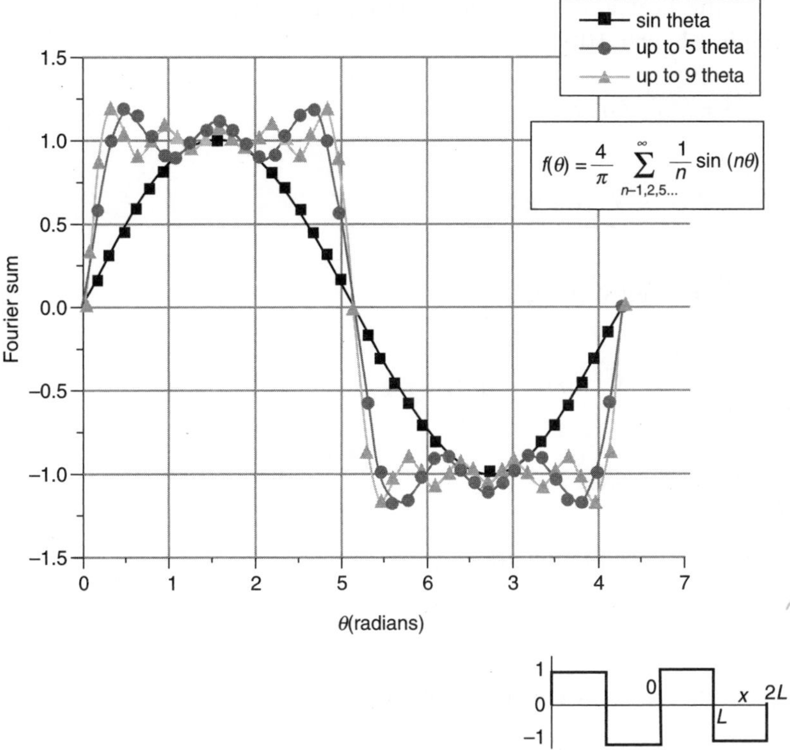

Three curves are plotted here, showing summations of terms up to $n\theta$ with $n = 1$, 5, and 9. Note how the approximation improves as more terms get added. Note also that only sine terms are included, the function being odd.

Fig. 4.11 The Fourier sum of a few terms to express the square wave $f(x)$ which is repetitive but has sharp edges. Mathematically the function is expressed in terms of the Heaviside step function H, defined below.

$$f(x) = 2\left[\mathrm{H}\left(\frac{x}{\mathrm{L}}\right) - \mathrm{H}\left(\frac{x}{\mathrm{L}} - 1\right)\right] - 1$$

when $x \in [0, 2\mathrm{L}]$.

$$\begin{aligned} \mathrm{H}(x) &= 0 \text{ for } x < 0, \\ &= 1 \text{ for } x > 0 \end{aligned}$$

The utility of the Fourier series is most exploited in solving different kinds of differential equations. Some examples where Fourier methods are useful include the differential equations which involve partial derivatives with respect to space and time, as they appear in the heat equation, and/or the diffusion equation, and also the wave equation.

Examples of the heat and/or the diffusion equation are (using *notation* that we shall introduce *later* in Chapter 10):

$$\frac{\partial T(\vec{r},t)}{\partial t} = k^2 \vec{\nabla}^2 T(\vec{r},t), \tag{4.38a}$$

$$\frac{\partial}{\partial t}[R(\vec{r})g(t)] = k^2 \vec{\nabla}^2 [R(\vec{r})g(t)], \tag{4.38b}$$

$$R(\vec{r})\frac{dg(t)}{dt} = k^2 g(t) \vec{\nabla}^2 R(\vec{r}), \tag{4.38c}$$

$$\frac{1}{g(t)}\frac{dg(t)}{dt} = \frac{k^2}{R(\vec{r})}\vec{\nabla}^2 R(\vec{r}) = -\lambda k^2, \tag{4.38d}$$

$$\frac{1}{g(t)}\frac{dg(t)}{dt} + \lambda k^2 = 0. \tag{4.38e}$$

Examples of the wave equation are

$$\vec{\nabla}^2 \Psi(\vec{r},t) = \frac{1}{c^2}\frac{\partial^2 \Psi(\vec{r},t)}{\partial t^2}, \tag{4.39a}$$

and the quantum mechanical Schrödinger equation

$$i\hbar \frac{\partial \Psi(\vec{r},t)}{\partial t} = \left[\frac{(-i\hbar\vec{\nabla})^2}{2m} + V(\vec{r},t)\right]\Psi(\vec{r},t). \tag{4.39b}$$

In the above equations, $\vec{\nabla}$ represents the gradient operator which we shall introduce in Chapter 10. Suffice it is at this point to state that it is made up of partial derivatives with respect to spatial coordinates. The wave equation contains the second order derivative with respect to space and also with respect to time, whereas the Schrödinger equation has only the first order derivative with respect to time. This property makes it somewhat analogous to the heat equation (Eq. 4.38), in some sense. Nevertheless, the Schrödinger equation falls in the category of the wave equation (Eq. 4.39) as it admits solutions similar to those of Eq. 4.39a. Both quantum mechanics, and a detailed study of differential equations, fall beyond the scope of this book, but we proceed to discuss the solution to the 1-dimensional wave equation Eq. 4.30.

Now, the coefficients n in the Fourier series Eq. 4.36 change in steps of unity, i.e., $\Delta n = 1$. Hence, we may write

$$\phi(x) = \sum_{n=-\infty}^{+\infty} c_n e^{+i\left(\frac{n\pi x}{L}\right)} \Delta n = \frac{L}{\pi}\sum_{n=-\infty}^{+\infty} c_n e^{+i\left(\frac{n\pi x}{L}\right)} \frac{\pi \Delta n}{L}. \tag{4.40}$$

We introduce a variable k having dimension of 'inverse length' by defining it as

$$k = \frac{n\pi}{L}, \tag{4.41a}$$

so that

$$\Delta n = \frac{L\Delta k}{\pi}. \tag{4.41b}$$

Equation 4.40 thus gives

$$\phi(x) = \frac{L}{\pi}\sum_{n=-\infty}^{+\infty} c_n e^{+i(kx)} \frac{\pi}{L}\frac{L\Delta k}{\pi} = \sum_{n=-\infty}^{+\infty}\left(\frac{L}{\pi}c_n\right)e^{+i(kx)}\Delta k \,. \tag{4.42}$$

Now, c_n depends on n, and hence on k. We therefore introduce the coefficient $A(k)$ which we shall use instead of c_n, defined by the following relation:

$$A(k) = \sqrt{2\pi}\left(\frac{Lc_n(k)}{\pi}\right) = \sqrt{2\pi}\left(\frac{L}{\pi}\right)\frac{1}{2L}\int_{-L}^{+L}\phi(x)e^{-i\frac{\pi nx}{L}}dx = \frac{1}{\sqrt{2\pi}}\int_{-L}^{+L}\phi(x)e^{-ikx}dx. \tag{4.43a}$$

By using Eq. 4.43, Eq. 4.42 becomes

$$\phi(x) = \sum_{n=-\infty}^{+\infty}\left(\frac{L}{\pi}\frac{\pi A(k)}{L\sqrt{2\pi}}\right)e^{+i(kx)}\Delta k = \sum_{n=-\infty}^{+\infty}\frac{A(k)}{\sqrt{2\pi}}e^{+i(kx)}\Delta k \to \frac{1}{\sqrt{2\pi}}\int_{-\infty}^{\infty}A(k)e^{+ikx}dk \,. \tag{4.43b}$$

Equations 4.43a and 4.43b are called *Fourier transforms* of each other. Extending these relations trivially to 3-dimensions, we get the corresponding 3-dimensional relations:

$$A(\vec{k}) = \frac{1}{(\sqrt{2\pi})^3}\iiint_{\substack{\text{whole}\\ \text{space}}}\phi(\vec{r},t)e^{-i\vec{k}\bullet\vec{r}}d^3\vec{r}\,, \tag{4.44a}$$

$$\phi(\vec{r}) = \frac{1}{(\sqrt{2\pi})^3}\iiint_{\substack{\text{whole}\\ \text{of }\vec{k}-\text{space}}}A(\vec{k})e^{+i\vec{k}\bullet\vec{r}}d^3\vec{k}\,. \tag{4.44b}$$

We have seen that the wave motion's time dependence is given by Eq. 4.33b. From the product of the space-function (Eq. 4.35b) and $e^{-i\omega t}$, we get the following solution to the wave equation:

$$\phi(\vec{r},t) = \frac{1}{(\sqrt{2\pi})^3}\iiint_{\substack{\text{whole}\\ \text{of }\vec{k}-\text{space}}}A(\vec{k})e^{+i(\vec{k}\bullet\vec{r}-\omega t)}d^3\vec{k}\,. \tag{4.45}$$

We now include in our discussion the consequence of the possibility of the medium being dispersive; i.e., $\omega = \omega(k)$. The relation between ω and (k) is not just linear in a dispersive medium. It is enough to discuss the 1-dimensional wave,

$$\phi(x,t) \approx \int_{-\infty}^{+\infty}a(k)e^{i(kx-\omega t)}dk\,. \tag{4.46a}$$

Essentially, we have in the above superposition, a number of waves having an infinite range of k. Often, we are interested in the superposition of waves having their wave numbers in a small range Δk centered about some particular value, such as k_0. Such a superposition is known as 'wave packet'. It is an approximation to Eq. 4.46a, and the integral in this relation is then restricted to the following:

$$\phi(x,t) \approx \int_{\Delta k} a(k) e^{i(kx-\omega t)} dk. \tag{4.46b}$$

The integration is only over a limited range of k. The medium in which these waves propagate may be dispersive. In the small range Δk we may write

$$\omega = \omega(k) = \omega(k_0) + \left[\frac{d\omega}{dk}\right]_{k_0} (k-k_0) + \mathrm{O}\{(k-k_0)^2\}. \tag{4.47}$$

The last term in Eq. 4.47 represents collectively all terms of order $(\Delta k)^2$, and smaller. These can be ignored in an approximation.

Thus,

$$\phi(x,t) = \int_{\Delta k} a(k) e^{i\left[kx-\omega(k_0)t-k\left[\frac{d\omega}{dk}\right]_{k_0} t+k_0\left[\frac{d\omega}{dk}\right]_{k_0} t\right]} \{e^{+ik_0x} \times e^{-ik_0x}\}\, dk, \tag{4.48a}$$

i.e., $$\phi(x,t) = \left[\int_{\Delta k} a(k) e^{i(k-k_0)[x-v_g t]} dk\right] e^{+i(k_0x-\omega_0 t)} = \tilde{A} e^{+i(k_0x-\omega_0 t)}, \tag{4.48b}$$

where $\omega_0 = \omega(k)]_{k=k_0}$, (4.49a)

$$v_g = \left.\frac{d\omega(k)}{dk}\right]_{k=k_0}, \tag{4.49b}$$

and $$\tilde{A} = \int_{\Delta k} a(k) e^{i(k-k_0)[x-v_g t]} dk. \tag{4.49c}$$

Equation 4.48 represents a wave of length $\lambda_0 = \frac{2\pi}{k_0}$ and frequency $\nu_0 = \frac{\omega_0}{2\pi}$, but whose amplitude $\tilde{A}$ is a function of both space and time. The dependence on space and time is not completely independent of each other, since x and t appear in the combination term $(x - v_g t)$. Therefore, $\tilde{A}$ is not the same at every point along the x-axis; but it is the same where $(x - v_g t) = \eta$, a constant. Equation 4.48 therefore represents a 'wave packet' which propagates in space such that $\delta\eta = 0 = (\delta x - v_r \delta t)$. The wave packet consists of a group of waves having wave numbers k in the range Δk. This group moves at the velocity v_g given by Eq. 4.49b. The components in this wave packet corresponding to different values of $k = \frac{2\pi}{\lambda_k} = \frac{2\pi\nu_k}{c_k} = \frac{\omega_k}{c_k}$ move at different individual phase velocities, c_k. The wave packet is thus not localized at a particular point on the x-axis. It is spread over a range Δx. To localize the wave packet, the spread Δk needs to be large. Δk and Δx therefore have a reciprocal relationship:

$$\Delta k \Delta x \approx 1. \tag{4.50}$$

The uncertainty relation Eq. 4.50 has stunning consequences in quantum mechanics which students would be introduced to in a later course. The essential reason that the wave packet is not localized at a particular point in space, but is spread over a region Δx, is the approximation we made in writing Eq. 4.46a as 4.46b. The wider the spread Δk we admit in Eq. 4.46b, the closer we approach the exact expression of Eq. 4.46a, and the more localized the wave packet gets. The reciprocal relationship between Δk and Δx is intrinsic to the formalism of Fourier transforms as we have found above. In quantum mechanics, Δk appears as the uncertainty in momentum in units of $\hbar = \frac{h}{2\pi}$, h being the Planck's quantum of 'action'. Action is defined later, in Chapter 6. We alert the young reader to refrain from directly interpreting Eq. 4.50 as Heisenberg's principle of uncertainty, since the quantum uncertainty requires a counter-intuitive approach. It must be defined in terms of operator formalism of dynamical variables, which is a totally non-classical concept and well beyond the scope of this book. The above formalism, however, is a useful step in the interpretation of the quantum uncertainty principle.

An important property of wave motion that we have used is the principle of superposition. This comes from the fact that the wave equation is a *linear* equation. Only the first powers of the wave-amplitude and that of any of its derivatives appears in the wave equation. This property allows us to construct sums of the solutions which are also solutions of the differential equation. Thus, if two different waves pass through a point at the same instant of time, the sum of their amplitudes is also a solution. The amplitude functions, of course, depend on space and time both, and also on the initial conditions which set an arbitrary additional phase in the arguments of the sinusoidal functions. Thus, at a particular point in space and at a particular instant of time, the net displacement of the oscillation in the wave is governed by the net sum of all individual wave passing criss-cross through that point at that time. Waves from a television set and an electronic radio would go through each other in a room carrying their respective characteristics without affecting each other, even if at each point, and at each instant, the net oscillation would be the sum of amplitudes from all the waves passing criss-cross. What we hear/see of the sound/light wave is because of the *energy* the wave delivers to the sensors. Energy is required to drive the sensor, such as required to cause vibrations in the diaphragms of our ear-drums. What provides a measure of this energy is the modulus-square of the amplitude of the wave, such as in Eq. 4.28, and at the point of detection any effect of the superposition of the waves criss-crossing each other would get washed out if the sensor is not fast and sensitive enough to register this effect before conditions at that spot change. Wave amplitudes at that spot change with each next moment, when the superposition could lead to either an addition or a subtraction of amplitudes, depending on the net phase $(\vec{k} \cdot \vec{r} \pm \omega\tau + \phi)$ of each component wave at each instant. The sources which generated each component wave may fall *out of step* at the next instant, causing the superposition effect to get washed out even before it is detected. However, if the sources are *coherent*, i.e., they produce waves which maintain constant phase relationship with each other, then the superposition effect at each spot where the resulting coherent waves superpose would generate a detectable interference pattern independent of time. In this case, the phase relationships between the superposed waves is *sustained* over a long-enough period. Thus, while superposition of waves always takes place whenever waves cross each other, they require coherence over a period *long-enough* and over a region that is large-enough to produce *detectable interference patterns*, called *fringes*, which are fluctuations in the total intensity of the superposed waves. There can thus be billions and trillions of waves that crisscross each other and get superposed at each instant of time where they cross, but pass through each other without producing fringes. An experiment conducted by Thomas Young (1773–1829) in 1801, now rated as one of the top ten most beautiful experiments ever, laid down the principles of interferometry which is an incredibly powerful technique that has been used to gain insights in the phenomenology of wave motion, whether of sound or water or light waves, or de Broglie quantum particle waves, or gravitational waves.

Young's experiment provides a neat way of providing two waves that are coherent such that they maintain a mutual constant phase relationship and thereby generate a detectable fringe pattern from their superposition. This is achieved by drawing the two wave trains that would superpose from a single wave generated at the source S by letting it pass through two slits S_1 and S_2 as shown in Fig. 4.12.

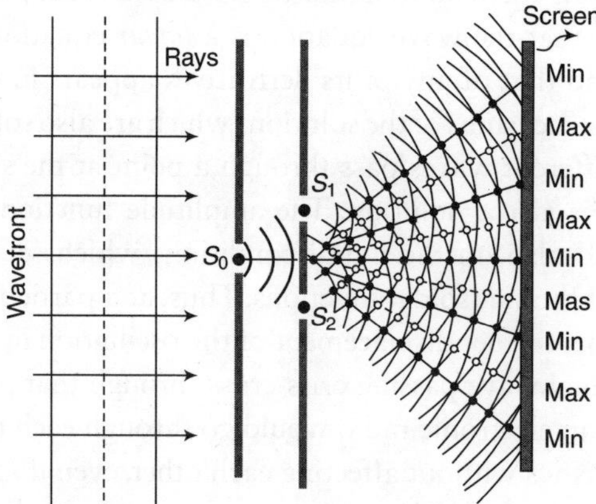

Fig. 4.12a Experimental setup to get interference fringes from superposition of two waves.

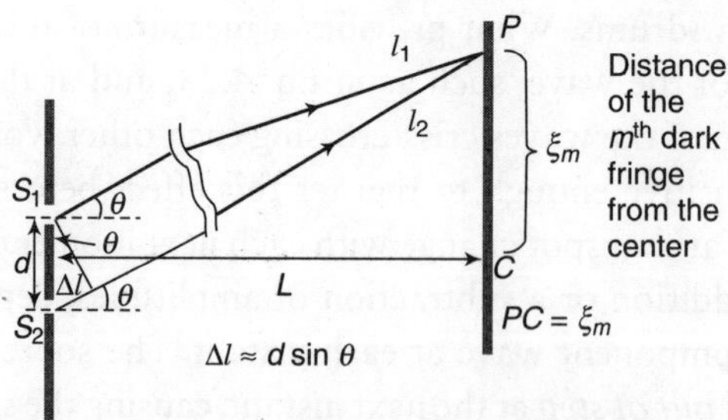

Fig. 4.12b Geometry to determine the fringe pattern in Young's double-slit experiment.

As seen from Fig. 4.12c, if the path length difference $\Delta\ell = d\sin\theta \approx d\frac{\xi}{L}$ accumulated at the detector point P by the waves traveling from the two slits is an integral multiple of λ, the wavelength of the waves, then the amplitudes of the two waves add up *in phase*, generating constructive interference (bright fringes). If they add up *out of phase*, the superposition results in destructive interference (dark fringes). Both the waves arrive at the point of dark fringes; *it is not that no wave gets there*. However, the waves arrive at that point out of step with each other and the resultant intensity diminishes creating an impression that no wave has reached there. This again is due to the fact that the energy is in the *square of the amplitude-modulus*, not in the amplitude itself. For the mth and the (m+1)th dark fringes, we have

$$\Delta\ell_m = d\sin\theta_m \approx d\frac{\xi_m}{L} = \left(m\lambda + \frac{\lambda}{2}\right) \quad \text{and} \quad \Delta\ell_{m+'} = d\sin\theta_{m+1} \approx d\frac{\xi_{m+1}}{L} = \left(m\lambda + \frac{3\lambda}{2}\right).$$

Thus,

$$\xi_m = \left(m\lambda + \frac{\lambda}{2}\right)\frac{L}{d} = \left(m + \frac{1}{2}\right)\frac{\lambda L}{d} \quad \text{and} \quad \xi_{m+1} = \left((m+1)\lambda + \frac{\lambda}{2}\right)\frac{L}{d} = \left(m + \frac{3}{2}\right)\frac{\lambda L}{d}$$

$$\text{Hence, } \xi_{m+1} - \xi_m = \frac{\lambda L}{d}. \tag{4.51}$$

Equation 4.51 gives the separation of adjacent dark fringes, which is $\frac{L}{d}$ times the wavelength of the wave. Since $\frac{L}{d} >>> 1$, measurement of separation of fringes enables determination of even very short wavelengths.

Note: *Large* oscillations are dealt with in the Chapter 6.

Problems with Solutions

P4.1:

Find the stationary points (maxima, minima, saddle point) of the function $f(x, y)$ given by:

$$f(x, y) = 2x^3 + 6xy^2 - 3y^3 - 150x.$$

***Solution*:**

Taking partial derivatives:

$$f_x = \frac{\partial f}{\partial x} = 6x^2 + 6y^2 - 150; \qquad f_y = \frac{\partial f}{\partial y} = 12xy - 9y^2$$

$$f_{xx} = \frac{\partial^2 f}{\partial x^2} = 12x; \qquad f_{yy} = \frac{\partial^2 f}{\partial y^2} = 12x - 18y$$

$$f_{xy} = \frac{\partial}{\partial x}\frac{\partial f}{\partial y} = 12y = \frac{\partial}{\partial y}\frac{\partial f}{\partial x} = f_{yx}$$

At the stationary points, we have $\frac{\partial f}{\partial x} = 6x^2 + 6y^2 - 150 = 0$, and also $\frac{\partial f}{\partial y} = 12xy - 9y^2 = 0$.

i.e., $x^2 + y^2 = 25$ and $y(4x - 3y) = 0$. The second of these two equations implies either that either $y = 0$ or $4x = 3y$.

If $y = 0$, then, from the first equation we get $x = \pm 5$. Hence the stationary points are: $(\pm 5, 0)$.

If $4x = 3y$, then, we may put $x = \frac{3}{4}y$ in the first equation and get $y = \pm 4$, for which $x = \pm 3$.

Hence the stationary points are: $(\pm 3, \pm 4)$.

Thus, there are four stationary points: (5, 0), (–5, 0), (3, 4),(–3, –4).

Hint for remaining part of the problem:

To determine if each of these stationary points is a maximum, minimum or a saddle point, we determine the second derivatives of the function. It is fruitful to construct the matrix:

$\mathbb{M} = \begin{bmatrix} f_{xx} & f_{xy} \\ f_{yx} & f_{yy} \end{bmatrix}$ and find the determinant $D = |\mathbb{M}|$ at each of the stationary points. The character of the stationary point is identified by the following conditions:

$D > 0$ and $f_{xx} > 0 \rightarrow$ local minimum; $D > 0$ and $f_{xx} < 0 \rightarrow$ local maximum.

$D > 0 \rightarrow$ saddle point; $D = 0 \rightarrow$ indeterminate.

P4.2:

A mass slides on a frictionless surface, held between two walls spaced at a distance '2a' by springs as shown in the figure. The surface on which it slides is not shown. It is just the plane of the figure, so the figure shows the 'top' view. The un-stretched length of each spring is 'a_0' and we need to stretch each spring to length 'a' in order to hook it to the mass, when in equilibrium. The mass can be set into oscillations between the wall by (a) moving it to a side, and let go, or (b) pulling it along a transverse direction, and let go. Determine the effective spring constant of the net simple harmonic motion in both the cases (a) and (b).

(c) Discuss the 'slinky' approximation when $\frac{a_0}{\ell} \ll 1$, where ℓ is the stretched length of the spring when the transverse displacement is y.

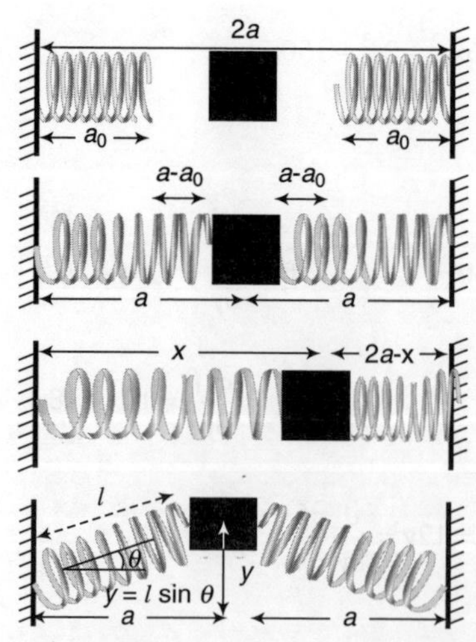

Solution:

The restoring force on the mass is the sum of the restoring forces of the two springs.

(a) $M\ddot{x} = F_{\substack{\text{restoring}\\ \text{force}}}(x) = -k(x - a_0) + k(2a - x - a_0) = -2k(x - a_0)$.

Effective spring constant = 2k.

(b) $M\ddot{y} = F_{\substack{\text{restoring}\\ \text{force}}}(y) = -2T\sin\theta = -2k\dfrac{(\ell - a_0)}{\ell}y = -2k\left(1 - \dfrac{a_0}{\ell}\right)y.$

Note: $\ell = \dfrac{y}{\sin\theta}$ is not independent of y. Since, $M\ddot{y} = -2k\left(1 - \dfrac{a_0}{\ell}\right)y$, the restoring force, is not linearly proportional to the displacement from the equilibrium. Hence, this equation does ***not*** describe a simple harmonic motion.

'Slinky' approximation: $a_0 \ll a < \ell : M\ddot{y} \simeq -2ky$. In this approximation, the motion is simple harmonic, with an effective spring constant = 2k, even if the length of the slinky is quite large.

P4.3:

Two masses are held together by two springs of spring constants k_1 and k_2 on a frictionless surface as shown, with a third spring in between, having a spring constant k_3, connecting the two masses.

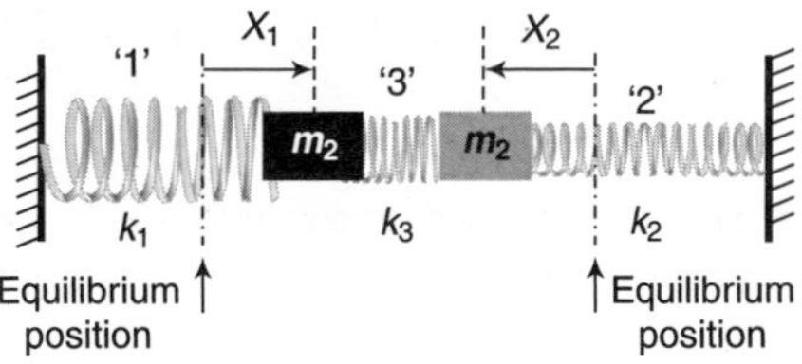

The three springs are in their relaxed, un-stretched ***equilibrium*** positions, shown by the dashed vertical lines. Positive direction of displacement of the spring '1', x_1, is to the right. The positive displacement of the spring '2', x_2, is considered toward the left. The spring '3' is therefore compressed by $(x_1 + x_2)$. Neglect friction on the surface. Determine the frequencies at which the mass m_1 can oscillate, and the frequencies at which the mass m_2 can oscillate.

Solution:

The equations of motion of the two masses are:

$$m_1\ddot{x}_1 = -k_1x_1 - k_3(x_1 + x_2) \text{ and } m_2\ddot{x}_2 = -k_2x_2 - k_3(x_1 + x_2)$$

Hence: $m_1\ddot{x}_1 + (k_1 + k_3)x_1 + k_3x_2 = m_1\ddot{x}_1 + k_{13}x_1 + k_3x_2 = 0,$ (i)

where $k_{13} = k_1 + k_3$.

Similarly, $m_2\ddot{x}_2 + (k_2 + k_3)x_2 + k_3x_2 = m_2\ddot{x}_2 + k_{23}x_2 + k_3x_1 = 0,$ (ii)

where $k_{23} = k_2 + k_3$.

If the third term were absent on the left hand sides of the equations (i) and (ii), they would have represented simple harmonic motions ***respectively*** at circular frequencies $\omega_{10} = \sqrt{\dfrac{k_{13}}{m_1}}$ and $\omega_{20} = \sqrt{\dfrac{k_{23}}{m_2}}$.

We look for solution of the type: $x_1 = A_1 e^{\lambda t}$ and $x_2 = A_2 e^{\lambda t}$, where A_1 and A_2 are constants. Substituting these in the equations (i) and (ii), we get two algebraic equations:

$$(\lambda^2 m_1 + k_{13})\, A_1 + k_3 A_2 = 0 \text{ and } (\lambda^2 m_2 + k_{23})\, A_2 + k_3 A_1 = 0.$$

Between the above two algebraic equations, we have three unknowns, A_1, A_2 and λ. We can therefore get the ratio $\frac{A_2}{A_1}$ in terms of other known quantities, and the unknown, λ:

Thus: $\frac{A_2}{A_1} = \frac{-m_1\lambda^2 + k_{13}}{k_3}$ and also $\frac{A_2}{A_1} = \frac{k_3}{m_2\lambda^2 + k_{23}}$.

Hence: $\frac{-m_1\lambda^2 + k_{13}}{k_3} = \frac{k_3}{m_2\lambda^2 + k_{23}}$, i.e., $m_1 m_2 \lambda^4 + (m_2 k_{13} + m_1 k_{23})\lambda^2 + (k_{13}k_{23} - k_3^2) = 0$,

which is a quadratic equation in λ^2 with two solutions λ_1^2 and λ_2^2 given by:

$$\lambda^2 = -\lambda_1^2 = -\left(\omega_{10}^2 + \frac{1}{2}\delta\omega^2\right) \text{ and } \lambda^2 = -\lambda_2^2 = -\left(\omega_{20}^2 + \frac{1}{2}\delta\omega^2\right), \text{ where}$$

$$\delta\omega = (\omega_{10}^2 - \omega_{20}^2)\left[\left(1 + \frac{4\kappa^4}{(\omega_{10}^2 - \omega_{20}^2)^2}\right)^{\frac{1}{2}} - 1\right], \text{ where } \kappa^2 = \frac{k_{13}}{\sqrt{m_1 m_2}}.$$

Accordingly, $\lambda = \pm i\lambda_1$ and $\lambda = \pm i\lambda_2$ which provide the solutions $x_1 = A_1 e^{\pm i\lambda_1 t}$ and $x_2 = A_2 e^{\pm i\lambda_2 t}$, and hence the frequencies at which the two masses can oscillate.

P4.4:

A string is wrapped several times around a cylinder of radius R, and then its free end is tied to a mass m. In equilibrium, the mass hangs at a distance l_0 vertically below cylinder, as shown in the figure. Find the potential energy if the pendulum has swung to an angle θ from the vertical. Show that for small angles, it can be written in the Hooke's law form as $U = \frac{1}{2}k\theta^2$.

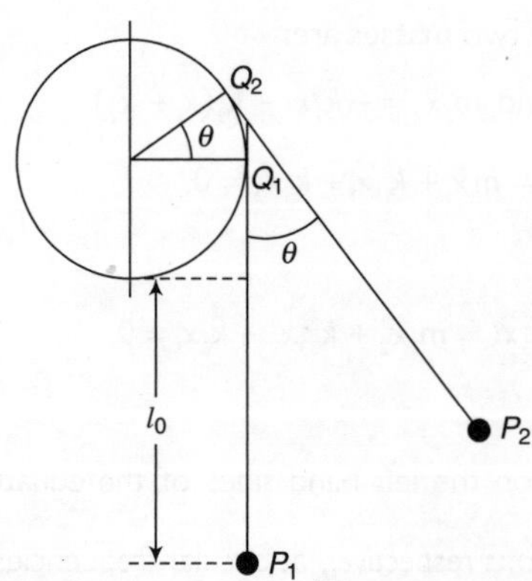

***Solution*:**

The arc length $Q_1Q_1 = R\theta$.

Length of the pendulum in the vertical position is $P_1Q_1 = R + l_0$,

Length of the pendulum when the mass is swung to the position P2:

$$P_2Q_2 = R + l_0 + R\theta.$$

Note that the point of suspension has now shifted from the point Q_1 to Q_2, and hence it has shifted upward, and also to the left.

The increase in the potential energy of the mass is due to the *lift* $P_1'R_1$.

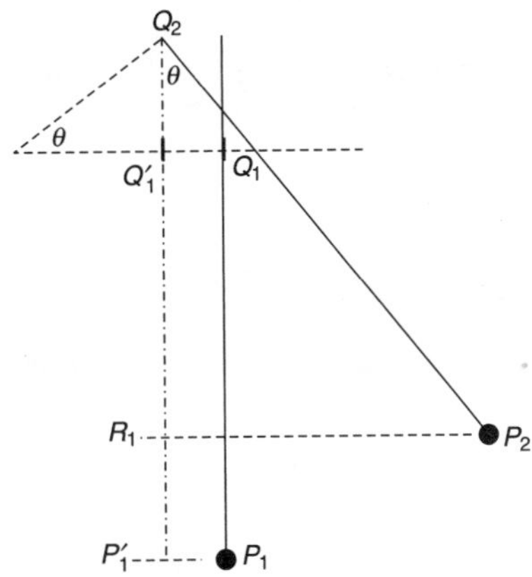

$$P_1'R_1 = P_1'Q_2 - R_1Q_2 = P_1Q_1 + Q_1'Q_2 - P_2Q_2\cos\theta \simeq P_1Q_1 + \text{arc} - \text{length}\ (Q_1Q_2) - P_2Q_2\cos\theta$$

$$P_1'R_1 = (R + l_0) + R\theta - (R + l_0 + R\theta)\cos = (R + l_0) + R\theta - (R + l_0 + R\theta)\left(1 - \frac{\theta^2}{2!} + \frac{\theta^4}{4!} - \dots\right).$$

$$P_1'R_1 \simeq (R + l_0) + R\theta - (R + l_0 + R\theta)\left(1 - \frac{\theta^2}{2!}\right) = (R + l_0)\frac{\theta^2}{2},\ \text{neglecting higher order terms.}$$

Hence: $U(\theta) = \frac{1}{2}mg(R + l_0)\theta^2 = \frac{1}{2}k\theta^2$, with $k = mg(R + l_0)$.

P4.5:

Determine the Fourier transform of the Gaussian function $\phi(x) = \frac{1}{\sqrt{2\pi\sigma^2}}e^{-\left(\frac{x^2}{2\sigma^2}\right)}$.

***Solution*:**

$$\tilde{\phi}(k) = \frac{1}{\sqrt{2\pi}}\int_{-\infty}^{+\infty}\frac{1}{\sqrt{2\pi\sigma^2}}e^{-\frac{x^2}{2\sigma^2}}e^{-ikx}dx = \frac{1}{\sqrt{2\pi}}\int_{-\infty}^{+\infty}\frac{1}{\sqrt{2\pi\sigma^2}}e^{-\left(\frac{x^2}{2\sigma^2} + ikx\right)}dx$$

$$\frac{x^2}{2\sigma^2} + ikx = \frac{1}{2\sigma^2}(x + ik\sigma^2)^2 + \frac{k^2\sigma^2}{2};\ \text{i.e.,}\ \frac{1}{2\sigma^2}(x + ik\sigma^2)^2 = \left(\frac{x^2}{2\sigma^2} + ikx\right) - \frac{k^2\sigma^2}{2}$$

$$\tilde{\phi}(k)=\frac{1}{\sqrt{2\pi}}\int_{-\infty}^{+\infty}\frac{1}{\sqrt{2\pi\sigma^2}}e^{-\left(\frac{1}{2\sigma^2}(x+ik\sigma^2)^2+\frac{k^2\sigma^2}{2}\right)}dx=\frac{1}{\sqrt{2\pi}}\left[\int_{-\infty}^{+\infty}\frac{1}{\sqrt{2\pi\sigma^2}}e^{-\left(\frac{1}{2\sigma^2}(x+ik\sigma^2)^2\right)}dx\right]e^{-\left(\frac{k^2\sigma^2}{2}\right)}$$

$\frac{1}{2\sigma^2}(x+ik\sigma^2)$; Put $u=(x+ik\sigma^2)$; $du=dx$;

$$\tilde{\phi}(k)=\frac{1}{\sqrt{2\pi}}\frac{1}{\sqrt{2\pi\sigma^2}}e^{-\left(\frac{k^2\sigma^2}{2}\right)}\left[\int_{-\infty}^{+\infty}e^{-\left(\frac{u^2}{2\sigma^2}\right)}du\right]$$

Essentially, we need to determine the integral $I(a)=\int_{-\infty}^{\infty}dx\,e^{-ax^2}$, with $a=\frac{1}{2\sigma^2}$

$$I^2(a)=\int_{-\infty}^{\infty}dx\,e^{-ay^2}\int_{-\infty}^{\infty}dy\,e^{-ay^2}=\int_{-\infty}^{\infty}dx\int_{-\infty}^{\infty}dy\,e^{-a(x^2y^2)}$$

Using plane polar coordinates is helpful here. Thus: $x^2+y^2=\rho^2$

The area of an elemental patch in the two coordinate systems is: $dxdy=(\rho d\varphi)(d\rho)=\rho d\rho d\varphi$

$$I^2(a)=\int_0^{\infty}\rho e^{-a\rho^2}d\rho\int_0^{2\pi}d\varphi=2\pi\int_0^{\infty}\rho e^{-a\rho^2}d\rho$$

We may use a change of integration variable, s instead of ρ, such that:

$s=-a\rho^2$, hence $ds=-2\alpha\rho d\rho$, and $\rho d\rho=-\frac{ds}{2a}$.

$$I^2(a)=2\pi\int_{-\infty}^{0}e^s\frac{ds}{2a}=\frac{\pi}{a}(e^0-e^{-\infty})=\frac{\pi}{a}\qquad I(a)=\sqrt{\frac{\pi}{a}}=\sqrt{\pi 2\sigma^2}$$

$$\tilde{\phi}(k)=\frac{1}{\sqrt{2\pi}}\frac{1}{\sqrt{2\pi\sigma^2}}e^{-\left(\frac{k^2\sigma^2}{2}\right)}\left[\left[\sqrt{\pi 2\sigma^2}\right]\right]=\frac{1}{\sqrt{2\pi}}e^{-\left(\frac{k^2\sigma^2}{2}\right)}$$

P4.6:

A particle undergoes two orthogonal simultaneous simple harmonic oscillations, described by $x=C\cos(\omega t)$ and $y=D\cos(2\omega t+\delta)$. Determine the superposed trajectory of the particle.

Solution:

We have, from the equation for x, $\frac{x}{C}=\cos(\omega t)$.

From the equation for y, we have:

$$\frac{y}{D}=\cos(2\omega t+\delta)=2(\cos^2\omega t-1)\cos\delta-2\sin(\omega t)\cos(\omega t)\sin\delta$$

Hence, $\frac{y}{D}=2\left(\frac{x^2}{C^2}-1\right)\cos\delta-2\left(\sqrt{1-\frac{x^2}{C^2}}\right)\left(\frac{x}{C}\right)\sin\delta.$

Squaring: $\left(\frac{y}{D}+\cos\delta\right)^2 - 4\left(\frac{y}{D}+\cos\delta\right)\left(\frac{x}{C}\right)^2\cos\delta + 4\left(\frac{x}{C}\right)^4\cos^2\delta = 4\left(1-\frac{x^2}{C^2}\right)\left(\frac{x}{C}\right)^2\sin^2\delta.$

i.e., $$\left(\frac{y}{D}+\cos\delta\right)^2 - 4\left(\frac{y}{D}+\cos\delta\right)\left(\frac{x}{C}\right)^2\cos\delta + 4\left(\frac{x}{C}\right)^4\cos^2\delta - 4\left(1-\frac{x^2}{C^2}\right)\left(\frac{x}{C}\right)^2\sin^2\delta = 0$$

i.e., $$\left(\frac{y}{D}+\cos\right)^2 + 4\frac{x^2}{C^2}\left(\frac{x^2}{C^2}-1-\frac{y}{D}\cos\delta\right) = 0$$

When $\delta = 0$: $\left(\frac{y}{D}+1\right)^2 + 4\frac{x^2}{C^2}\left(\frac{x^2}{C^2}-1-\frac{y}{D}\right) = 0$

i.e., $$\left(\frac{y}{D}+1-2\frac{x^2}{C^2}\right)^2 = 0\text{, i.e., } \left(\frac{y}{D}+1-2\frac{x^2}{C^2}\right)\left(\frac{y}{D}+1-2\frac{x^2}{C^2}\right) = 0$$

i.e., $\left(\frac{y}{D}+1-2\frac{x^2}{C^2}=0\right)$ $\rightarrow$ describes a parabola; i.e., or, $\frac{1}{D}\left(y+D-2D\frac{x^2}{C^2}\right) = 0$,

which is equivalent to the condition: $y = -D + \frac{2D}{C^2}x^2$ $\rightarrow$ also describes the same parabola.

Thus the trajectories are two coincident parabolas.

Additional Problems

P4.7 A wooden block of mass M 0.25 kg slides on a frictionless flat surface inclined at an angle $\theta = 30°$ with respect to the horizontal. It is hooked to a spring, having a spring constant $k = 100\ \text{Nm}^{-1}$, supported at the top. The relaxed, un-stretched, length of the spring is 25 cm. (a) Determine the length of the stretched spring at which the block stops sliding down. (b) From its equilibrium position, once it stops sliding, if the block is slightly pulled downward and let go, find the frequency of oscillations the block would undergo. Ignore friction on the surface of the inclined plane.

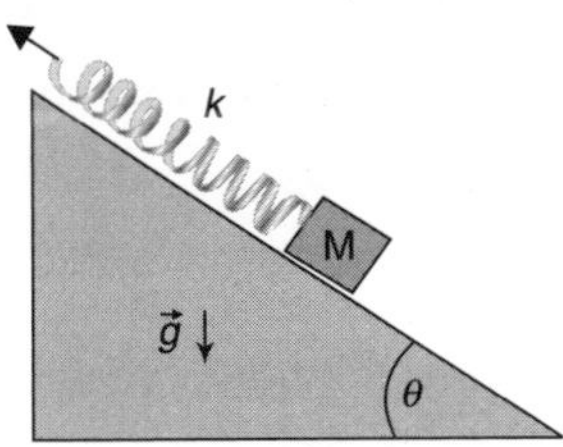

P4.8 An organ pipe which is 1 meter long is open at both ends. Determine the fundamental frequency this pipe will support and also determine the first two overtones. The speed of sound is ~345 ms^{-1}.

P4.9 A spring with spring constant $k = 100$ N/m has a mass 1kg tied to one of its end, while the other end is held fixed.

(a) Determine the angular speed, frequency, and the time period of the oscillation.

(b) If the initial position is $x_{t=0} = 0$ and the initial speed is $v_{t=0} = 20$ m/s, obtain the general solution for $x(t)$.

P4.10 A particle moving in a 2-dimensional space experiences restoring forces along both the x- and y- displacements: $F_x = -k_x x$ and $F_y = -k_y y$.

Determine the potential in which this particle is moving.

P4.11 The potential energy of a particle of mass m and at a distance r from the origin is given by $U(r) = U_0\left(\frac{r}{R} + \lambda^2 \frac{R}{r}\right)$, where U_0, R and λ are positive constants. Determine the position r_0 where the particle will be in equilibrium. Show that the particle's potential energy has the form as $U(x) = \varepsilon + \frac{1}{2}kx^2$, where ε is a constant. Find the angular frequency of oscillation if the particle is slightly displaced from the equilibrium.

P4.12 Determine the Lissajous pattern traced by a particle subjected to two simultaneous orthogonal simple harmonic motions, given by $x(t) = 2 \sin(2\pi t)$ and $x(t) = 3 \sin(\pi t)$

P4.13 A wave having a frequency ν_1 and wavelength λ_1 travels from a medium '1' to medium '2'. The speed of the wave in the medium '1' is half of the speed it has in the medium '2'. Determine the frequency and the wavelength of the wave in the medium '2'.

P4.14 Determine the beat frequency when two tuning forks at frequencies 485 Hz and 490 Hz are struck at the same time.

P4.15 A car approaching at 40 m/s blows its horn at a frequency of 450 Hz. The speed of sound may be taken to be 335 m/s. At what frequency will the sound of the horn be heard by an observer on the road? (Doppler effect)

P4.16 A sound burst has an intensity of 10^{-2} Wm^{-2} at a distance of 1m from the source. How far from the source can this burst be heard? Take into account the fact that the burst would spread evenly in all directions, and that when intensity of sound is less than 10^{-4} Wm^{-2} then it is not audible by the human ear. Also, determine the ratio of the amplitude at this distance to that at 1m from the source.

P4.17 A uniform rope of length 10m and mass 5 kg hangs from the roof. A block of 1 kg is attached to the rope at the bottom. A transverse pulse of wavelength 0.05 m is generated at the bottom of the rope. What is the wavelength of the pulse when it reaches the top of the rope?

P4.18 Write a small program to determine the following superposition $\phi(x)$ and plot $\phi(x)$ as a function of (x):

$$\phi_2(x) = \sqrt{\frac{1}{2}}[\sin(7x) + \sin(8x)]$$

$$\phi_3(x) = \sqrt{\frac{1}{3}}[\sin(7x) + \sin(7.5x) + \sin(8x)]$$

$$\phi_6(x) = \sqrt{\frac{1}{6}}[\sin(7x) + \sin(7.2x) + \sin(7.4x) + \sin(7.6x) + \sin(7.8x) + \sin(8x)]$$

$$\phi_6(x) = \sqrt{\frac{1}{11}}\begin{bmatrix} \sin(7x) + \sin(7.1x) + \sin(7.2x) + \sin(7.3x) + \sin(7.4x) + \sin(7.5x) + \\ \sin(7.6x) + \sin(7.7x) + \sin(7.8x) + \sin(7.9x) + \sin(8x) \end{bmatrix}$$

Observe the above four sketches. What do you observe? Analyze your observations.

P4.19(a) Determine the coefficients in the Fourier series expansions of the following periodic functions:

$$\phi(x) = |x| \text{ for } -2 < x < 2; \text{ and } \phi(x+4) = \phi(x)$$

$$\phi(x) = x \text{ for } 0 < x < 4; \text{ and } \phi(x+4) = \phi(x)$$

(b) Compare the above two series.

P4.20 Solve the following heat conduction problem using Fourier method:

$$2\frac{\partial^2\phi(x,t)}{\partial x^2} = \frac{\partial\phi(x,t)}{\partial t};\ 0 < x < 9;\ t > 0$$

Given: Boundary conditions: $\phi(0, t) = 0$ and $\phi(9, t) = 0$.

and Initial condition: $\phi(x,0) = 25\sin\left(\frac{\pi x}{3}\right) + 45\sin\left(\frac{4\pi x}{3}\right) - 12\sin\left(\frac{3\pi x}{3}\right)$.

CHAPTER 5

Damped and Driven Oscillations; Resonances

Lost time is never found again.

—*Benjamin Franklin*

5.1 DISSIPATIVE SYSTEMS

Energy dissipative systems prompt us to think of physical phenomena in which energy is not conserved, it is lost. We attribute these losses to 'friction', which is the common term used to describe energy dissipation. Now, in our everyday experience, we are primarily involved with the gravitational and electromagnetic interactions, and both of these are essentially conservative. What is it, then, that makes friction *non-conservative*? What does non-conservation of energy, or 'energy-loss', really mean? Fundamental interactions in nature allow energy to be changed from one form to another, but *not* created or destroyed. It therefore seems that the term 'energy loss' is used somewhat loosely. We must be really careful when we talk about dissipative phenomena.

Losing money is often a matter of concern, as also other things we sometimes 'lose' from time to time, including *time* itself. Time wasted does not ever come back, of course; but nor does the time well-spent. The difference between the two is that the latter is *accounted for* by the gains made, and for the former there is simply no account. Isn't it merely a matter of book-keeping? It is not at all uncommon that we plan to do something during the day, and end up not doing it. We then sit and wonder where lost track of time. Did the day just *skip* over the afternoon and ring in the evening? If that did not happen, where was the time 'lost'? You possibly remember everything that you did since morning, and you may be able to account for every hour you spent, except perhaps for what happened between 2:30 pm and 3 pm. That was the time you ransacked your house to find your mathematics text book, and did not realize that it took you half an hour to find it. The book had not really vanished, even if you suspected that it was lost. You had to look for it all over, from your Dad's room to your Sister's. As such, neither the book was lost, nor was half an hour scooped out of the day. Only that it was not *well accounted* for. Same thing can happen to money: you begin with Rs 100 in your purse, buy vegetables for Rs 40, fruits for Rs 45 and just cannot find the remaining Rs 15. It may well be that you did not keep track of the ball-pen you bought for that amount on the way; not that the money was actually lost. Even if the money accidentally dropped out of your purse, it would *not vanish*. Someone lucky may actually find it.

Energy loss due to friction is pretty much like that. Energy gets *dissipated*, or *lost* only in the sense that it gets transferred to *unspecified degrees of freedom* which you had *not* included in the equations of motion you may be solving. You may, for example, set up the equation of motion for a block sliding over a surface, but not include the energy transferred to the particles on the surface of the table by the block. These particles are associated with the unspecified degrees of freedom, like the lucky person who finds the Rs 15 fallen out of your purse, or the pen you bought, but just forgot. Dissipation, non-conservation of energy, energy lost in damping and friction, are all only a result of unaccounted book-keeping. The unspecified degrees of freedom, with which the object under study interacts, take away some of its energy.

In the case of a simple pendulum, for example, there may be friction at the point of suspension of the pendulum, or air resistance, that we had not considered. In the mass-spring oscillator, there is that heating of the spring due to its kinetic motion that we did not account for. This heat would get radiated away, and thus 'lost' in the above sense, even if the energy has only moved into the radiation field. In an electrical oscillator made up of an inductance and a capacitor, there is always a resistive component in the circuit elements, and this resistive element would get heated when a current flows in it, taking the energy away from the LC circuit.

In this chapter, we reconcile with the fact that a real oscillator, whether the mass-spring, or the pendulum, the electrical LC circuit, or any other, would have to interact with some unspecified degrees of freedom which would take away some of the energy from the oscillator system. The energy loss may be significant, or only minor, but very unlikely to be zero. The detailed structure of the unspecified degrees of freedom may be extremely complex. Just think of the number of air molecules a pendulum would interact with during its swing. The complexity of this many-body problem is so severe that it becomes prohibitively difficult to account for the pendulum's interaction, and the resulting energy exchanges, with these elements. Nonetheless, there is a lot of empirical data available using which one can lump together the net *effect* of all of these unspecified degrees of freedom in a *single* term. We can then insert this term in the equation of motion of the oscillator to account for the dissipation in an approximate manner. We introduce this term in the next section. Notwithstanding the fact that it is preposterous to hope that a single term would effectively mimic the collective effects of thousands, even millions, of interactions with unspecified objects in the surroundings, we shall savor the fact that this strategy actually works in most situations.

5.2 DAMPED OSCILLATORS

An oscillator is damped due to the dissipative interactions not included in the equation of motion, Eq. 4.15. This equation does not consider the unspecified degrees of freedom. Due to these untold, unspecified, degrees of freedom, the oscillator is no longer *simple* harmonic. Equations 4.27 and 4.28 breakdown due to the neglect of interactions of system with minute, and many, elements in its surroundings. The energy of the oscillator itself is not conserved; energy leaks *out of the oscillator* system, and *into unspecified degrees of freedom* in the oscillator's surroundings. The nature of the unspecified degrees of freedom is very complex. For example, dynamically changing air resistance to a simple pendulum, and friction at the point of suspension of the pendulum, both contribute to the energy dissipation. There is no easy way of including these effects at a fundamental level in the equation of motion. One may, however, include the effect of dissipation in the equation of motion, even if only approximately, using *empirical* knowledge.

Friction arises primarily due to the *movement* ($\dot{\vec{r}} \neq \vec{0}$) of the oscillator through the medium, (including at the point of support of a pendulum). This *movement* is a common aspect of dissipation in oscillators, including the electrical LC oscillator where the resistance to the electrical charges in motion causes obstruction to the flow of the electric current and causes energy dissipation. It is thus natural to expect that the damping effect can be represented, at least approximately, by an additional force in the equation of motion that would (a) oppose motion and (b) be proportional to the oscillator's *instantaneous speed*. We are of course free to suspect that the effective damping term may be proportional to the *square* of the instantaneous speed, or perhaps to some other polynomial function of the speed. It is, however, most common to consider the effective damping force to be linearly proportional to the instantaneous speed of the oscillator. It is also natural to expect that this term would be independent of the direction of motion of the oscillator, whether toward the equilibrium or away from it. Hence only the absolute magnitude of the instantaneous speed of the oscillator would matter.

For the 1-dimensional oscillator along the X-axis, we may therefore write the effective force due to the multitude of the unspecified degrees of freedom as:

$$F_{\text{friction}} = -bv = -b\dot{x}, \tag{5.1}$$

where $b > 0$, and is known as the damping coefficient. The negative sign in Eq. 5.1 implies that the frictional force *opposes* movement of the oscillator. The unspecified degrees of freedom would take away energy from the oscillator, and hence *slow* it down. The equation of motion for the damped oscillator is thus obtained by adding the speed-dependent damping force to the restoring 'spring' force in the equation of motion for the oscillator:

$$m\ddot{x} = \kappa x - b\dot{x}, \tag{5.2a}$$

$$\text{i.e., } \ddot{x} + 2\gamma\dot{x} + \omega_0^2 x = 0 \text{ with } \gamma = \frac{b}{2m} \text{ and } \omega_0^2 = \frac{\kappa}{m}. \tag{5.2b}$$

In real situations, damping cannot be avoided, and hence energy-loss is also unavoidable. Even if energy-loss seems undesirable, damping is not always ruinous. On the contrary, damping is often necessary. Consider, for example, an automobile which does not have shock absorbers which provide damping. The effect can be best seen in a bumpy ride in a video available on the YouTube at https://www.youtube.com/watch?v=5Mr-UgWr8-s (accessed on

8 June 2018). Various different kinds of mechanisms are therefore engineered to prevent damages due to unwanted oscillations. Damping techniques include employing mechanical damping plates, Coulomb damping arising out of electrostatic interactions between sliding surfaces, viscous damping used by immersing the oscillator in a viscous lubricant medium, magnetic damping which exploits eddy currents in some electrical devices, and even radiation damping due to energy dissipation by accelerated charged particles. These interactions seem undesirable if energy loss is to be avoided, but they also play a constructive role in situations such as driving on a bumpy road where damping can actually save a few of your bones.

The case of zero-damping represents the mathematical model of an ideal simple harmonic oscillator. A real physical oscillator must include interactions with the surroundings; if not in all details, at least approximately by an effective equivalent representation. The simplicity of the damping term hides the complexities in the multitude of the interactions the oscillator has with tens of thousands of particles in the surroundings. The interaction between the oscillator and each of these particles may be weak and negligible, but the cumulative effect of all such interactions is usually not ignorable. Note that $\gamma > 0$ since $b > 0$. I will like to alert you that since b and γ account for damping, *both* of them are generally known as the *damping coefficient*.

We know from Eq. 4.25e that solution to the free, *not*-damped, simple harmonic oscillator has the form $e^{\pm i\omega_0 t}$ where $\omega_0 = \sqrt{\frac{\kappa}{m}}$ for the mass-spring oscillator. For the other mechanical or electrical oscillators considered in Chapter 4, the natural frequency of the oscillator is given by other *intrinsic* physical properties of the system, such as $\omega_0 = \sqrt{\frac{g}{l}}$ for the simple pendulum, or $\omega_0 = \sqrt{2g\sigma}$ for the marble-in-a-parabolic-bowl oscillator, or $\omega_0 = \frac{1}{\sqrt{LC}}$ for the electrical oscillator made of an inductance and a capacitance.

The dynamics of a the damped oscillator is expected to be determined primarily by the restoring force in Eq. 5.2, when the additional influence from damping is relatively small. In such a case, damping would only lightly, or at least not too heavily, perturb the dominant term. It will therefore be natural to expect that the solution for the damped oscillator is *considerably similar* to that for the free, *not*-damped, oscillator; but it would also be *somewhat different*. The difference would of course be on account of the friction that is now included. We therefore seek a solution in the same *form* given in Eq. 4.25e, which is $e^{\pm i\omega_0 t}$, but also account for the *difference* from the free oscillator by choosing an exponent u that is somewhat different from ω_0. The problem of studying damping then reduces to determining an appropriate exponent u as would enable it to include the combined effect of the restoring force, and the damping force. The exponential form (Eq. 4.25e) offers much convenience with regard to taking the time-derivative of the exponential function e^{iut}; differentiation with respect to time can be effectively replaced by straight forward multiplication by iu. We must now determine the exponent u that would serve our purpose. Accordingly, we examine the consequences of employing the solution in the even *more* general, but very similar, form:

$$x(t) = Ae^{qt}. \tag{5.3}$$

By substituting Eq. 5.3 in Eq. 5.2, we get

$$q^2 + \frac{c}{m}q + \omega_0^2 = 0, \quad \text{i.e., } mq^2 + cq + \kappa = 0; \tag{5.4a}$$

$$\text{with } m\omega_0^2 = \kappa, \tag{5.4b}$$

the spring constant for the 'free' oscillator. Equation 5.4a is readily recognized as the usual quadratic equation whose solutions are given by the roots

$$q = \frac{-c \pm \sqrt{c^2 - 4m\kappa}}{2m}, \tag{5.5a}$$

$$\text{i.e., } q_1 = -\gamma + \sqrt{\gamma^2 - \omega_0^2}, \tag{5.5b}$$

$$\text{and } q_2 = -\gamma - \sqrt{\gamma^2 - \omega_0^2}, \tag{5.5c}$$

$$\text{where } \gamma = \frac{c}{2m}. \tag{5.5d}$$

The *general* solution to the equation of motion for the damped oscillator therefore is:

$$x(t) = A_1 e^{q_1 t} + A_2 e^{q_2 t}. \tag{5.6}$$

A_1 and A_2 are constants, determined by the initial conditions, at $t = 0$. The nature of the solutions lends itself to the following three possibilities:

Case 1: $\gamma > \omega_0$. In this case, $\sqrt{\gamma^2 - \omega_o^2} = \xi$ is a positive real number with $\xi < \gamma$, so the two roots (Eq. 5.5), $q_1 = -\gamma + \xi$ and $q_2 = \gamma - \xi$, are both real, and both negative. In fact, q_2 is even more negative than q_1. For reasons as would become clear shortly, an oscillator for which $\gamma > \omega_0$ is known as the *overdamped* oscillator.

Case 2: $\gamma = \omega_0$. The two roots become degenerate, i.e., they are equal. An oscillator for which $\gamma = \omega_0$ is called the *critically damped* oscillator.

Case 3: $\gamma < \omega_0$. The two roots become complex. The oscillator for which $\gamma < \omega_0$ is called the *underdamped* oscillator.

We now discuss the dynamics of the oscillator for the above mentioned three cases.

Case 1: $\gamma > \omega_0$ Overdamped oscillator, or, strong damping:

In this case, $x(t) = A_1 e^{q_1 t} + A_2 e^{q_2 t} = e^{-\gamma t}[A_1 e^{\xi t} + A_2 e^{-\xi t}]$. Both the terms approach zero as $t \to \infty$. The second of the two terms approaches zero faster than the first. If the system is *displaced* from the equilibrium position $x = 0$, then it would not return to the equilibrium position unless the first term also approaches it. The decay rate is therefore determined by the *first* term for which $q_1 = -\gamma + \xi$. Since both the terms would go to zero only asymptotically as $t \to \infty$, we inquire if the oscillator can, however, get to the equilibrium point in finite time under any circumstances. This could happen if the superposition of the two terms is zero at some finite time $\tau > 0$: $A_1 e^{\xi\tau} + A_2 e^{-\xi\tau} = 0$, i.e. if $\tau = \frac{1}{2\xi}\ln\left(-\frac{A_2}{A_1}\right)$. Since $0 < \xi < \gamma$, we can thus get a solution with $\tau > 0$ only with $\left(-\frac{A_2}{A_1}\right) > 1$.

The velocity of the oscillator is $\dot{x}(t) = -\gamma e^{-\gamma t}(A_1 e^{\xi t} + A_2 e^{-\xi t}) + e^{-\gamma t}(\xi A_1 e^{\xi t} - \xi A_2 e^{-\xi t})$.

The two constants, A_1, A_2 can be easily obtained in terms of the initial conditions on position and velocity:

$$x_0 = x(t=0) = A_1 + A_2 \text{ and } v_0 = \dot{x}(t=0) = A_1(\xi - \gamma) - A_2(\xi - \gamma), \tag{5.7a}$$

$$\text{which gives } A_1 = \frac{x_0(\xi + \gamma) + v_0}{2\xi} \text{ and } A_2 = \frac{x_0(\xi - \gamma) - v_0}{2\xi}. \tag{5.7b}$$

We cannot arbitrarily choose A_1 and A_2. They are totally determined by the initial values (x_0, v_0). The latter can nevertheless be chosen in a variety of ways. The condition $-\frac{A_2}{A_1} > 1$ requires, for example, $x_0 < 0$ and $v_0 > 0$. However, if from the very beginning we had interchanged the coefficients A_1 and A_2, we could have got a solution for $\tau > 0$ with $-\frac{A_1}{A_2} > 1$ which would require $x_0 > 0$ with $v_0 < 0$. In either case, we would need a displacement away from the equilibrium point, on either side, but the initial velocity to be such that it is essentially directed *toward* the equilibrium position. In such a case, the overdamped oscillator has the possibility of crossing the equilibrium mean point once, but *only* once.

The solution, in terms of the initial position and velocity, therefore is:

$$x(t) = e^{-\gamma t}\left[\left\{\frac{x_0(\xi + \gamma) + v_0}{2\xi}\right\}e^{\xi t} + \left\{\frac{x_0(\xi - \gamma) - v_0}{2\xi}\right\}e^{-\xi t}\right]. \tag{5.8}$$

We have seen in the previous chapter that the phase space trajectory of the not-damped simple harmonic oscillator is an ellipse. The phase space trajectory of the damped oscillator is expected to be different. For the overdamped oscillator, it is often rather complicated, though asymptotically it attains a simple form, as we illustrate now. As mentioned earlier, the asymptotic solution is dominated by the first of the aforementioned two terms, since the second term would go to zero faster than the first term as $t \to \infty$. Thus, $x(t \to \infty) = e^{-(\gamma-\delta)t}\left\{\frac{x_0(\xi + \gamma) + v_0}{2\xi}\right\}$, and the corresponding velocity would be $\dot{x}(t \to \infty) = -(\gamma - \xi)x(t \to \infty)$. Thus, as time progresses, the phase space trajectory of the overdamped oscillator would become *linear*, with a negative slope, but the approach to this asymptotic behavior from the initial point in the phase space would have a rather intricate time-dependence. On the other hand, if the initial conditions are such that $-x_0(\xi + \gamma) = v_0$, i.e., if $A_1 = 0$, then the second term is the only one that remains to be considered, and the displacement of the oscillator is then given by

$x(t) = e^{-(\gamma-\delta)t}\left\{\dfrac{x_0(\xi-\gamma)-v_0}{2\xi}\right\}$. The corresponding velocity is $\dot{x}(t) = -(\gamma+\xi)x(t)$, which provides the trajectory in the phase space to be *linear* yet again, but with a *different* negative slope.

Fig. 5.1 shows the time-dependence of the displacement from equilibrium of an overdamped oscillator for three different initial conditions. The mass and the spring constant of the oscillator is the same that we had considered in Fig. 4.4 of the previous chapter, except that we have now admitted damping. The oscillator therefore has the same intrinsic, natural, frequency, $\omega_0 \approx 0.79$, but damping is admitted now in the equation of motion for the oscillator, with $\gamma = 0.85$. The damping coefficient we have chosen is such that $\gamma^2 > \omega_0$, and the oscillator is therefore clearly *overdamped*. The three curves in Fig. 5.1 correspond to the following three initial conditions, chosen specially to illustrate different features of the dynamical behavior of the overdamped oscillator:

(i) $x(t = 0) = x_0 = 2.0$ m; $v(t = 0) = v_0 = 0.0$ m/s (continuous curve),

(ii) $x(t = 0) = x_0 = 2.0$ m; $v(t = 0) = v_0 = -11.71$ m/s (dotted curve),

(iii) $x(t = 0) = x_0 = -2.0$ m; $v(t = 0) = v_0 = 6.73$ m/s (dashed curve).

In the case (i), the oscillator is displaced at $t = 0$ to $x(t = 0) = x_0 = 2.0$ m and is released with zero initial velocity. As one would expect, the oscillator would take infinite time to return to its equilibrium position of zero-displacement, and without any overshoot. The case (ii) represents release of the oscillator from the same initial displacement, but with a negative velocity. The initial displacement being positive, and the initial velocity negative, we note that the oscillator is given an initial movement *toward* the equilibrium. This enables an overshoot once, but only once, since subsequent behavior involves an exponential decay of the amplitude. In the case (iii), the displacement at $t = 0$ is negative, but the initial velocity is positive, i.e., again *toward* the equilibrium, which again enables an overshoot once, and only once, for the same reason. Of course, this overshoot is from the opposite side. We find that the magnitude of the initial velocity in the case (ii) is greater than that in the case (iii). Hence the case (ii) oscillator overshoots past the equilibrium position in a *shorter time duration* than the case (iii) oscillator. Furthermore, we see that if at $t = 0$ the oscillator is given an initial negative displacement and is released with an initial velocity which is *also* negative, (i.e., *away* from the equilibrium), then initially the oscillator would move *farther away* from the equilibrium. The displacement would become even more negative. The oscillator would move in that direction due to the initial momentum imparted to it. Nonetheless, the restoring force and damping would both immediately get into play and reverse the direction of the motion. An overshoot would, however, not be possible; the oscillator would take infinite time to return to equilibrium.

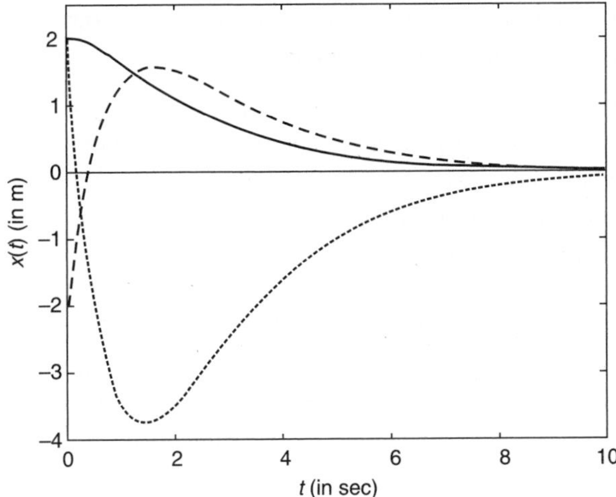

Fig. 5.1 Amplitude versus time plot for an 'overdamped' oscillator. Overshoot is generally not possible (continuous curve) in finite time, but in special cases (dashed, and continuous, curve) it is possible once, but no more than once.
The three curves correspond to different initial conditions (see text for details) of the position and velocity with which the oscillator is set into motion.
Oscillator parameters used for this plot are m = 1g,
$\kappa = 0.62 \times 10^{-3}$ N/kg,
$\omega_0 = 0.79$ rad/sec.
The damping coefficient chosen for this plot is $\gamma = 0.85$ rad/sec.

To get further insight into the dynamics of an overdamped oscillator, we rewrite Eq. 5.5 in a slightly different form:

$$q = \frac{-c \pm \sqrt{c^2 - 4m\kappa}}{2m} = \frac{-c \pm c\left(1 - \frac{4m\kappa}{c^2}\right)^{\frac{1}{2}}}{2m} \approx \frac{-c \pm c\left(1 - \frac{1}{2}\frac{4m\kappa}{c^2}\right)}{2m} = \frac{-c \pm c \mp \frac{1}{2}\frac{4m\kappa}{c}}{2m},$$

so that the two roots are

$$q_1 = \frac{-\frac{1}{2}\frac{4m\kappa}{c}}{2m} = -\frac{\kappa}{c}, \tag{5.9}$$

$$q_2 = \frac{-2c + \frac{1}{2}\frac{4m\kappa}{c}}{2m} = \frac{m\kappa - c^2}{mc} \approx -\frac{c}{m}. \tag{5.10}$$

Thus, when the solution is governed primarily by q_1, the mass of the oscillator does not matter. However, if the motion is mostly governed by the other root, then the restoring force does not matter much. It is the competition between the damping force and inertia that would influence the oscillator dynamics. The damping coefficient appears in the denominator in the first root, and in the numerator in the second. Hence the two terms affect the oscillator's approach to equilibrium very differently. Of course, the relative importance of the two terms is determined by the initial conditions. This is part of the reason why despite the fact that the solutions have relatively simple forms in the asymptotic ($t \to \infty$) domain, the phase trajectories have a complicated evolution in the initial stages.

Overdamping is often desirable. A classic example is that of an automatic door shutter which would shut after a button opens it for the passage of a person on a wheel chair, or in hospitals where the ward boys have to open the door to push a patient on a stretcher out of a gate. The door would shut automatically after giving enough time for the person(s) to pass through it. *Heavy* damping would provide for enough time, as the approach to equilibrium would be slow. However, this will not be useful for the shock absorbers of a car which must return to the equilibrium quickly. A push-button water tap is another such example. It is used to save water, considering the possibility that some people may forget to close the tap. Overdamped push-button taps would allow water to flow for enough time for use. The overdamped tap-shutter returns slowly to the shut-position, but would surely close it avoiding the possibility of uncontrolled wastage. A plane landing on autopilot would be under the risk of being fractured if an overshoot on landing is not prevented, and the principle of overdamping is useful to design the mechanical response. In fact, any object that must be stopped in a confined space will require overdamping which disables overshoot. The detailed mechanism of breaking in planes, or trains/buses, or elevators, or any such object is, however, too complicated to be discussed here. In many situations, the damping mechanism is tuned to be close to what is known as 'critical damping', discussed below as Case 2, although one could be on either side of criticality for optimal desired response.

Case 2: $\gamma = \omega_0$ Critical damping

In this case, $\gamma = \omega_0$ and the two roots are equal: $q_1 = q_2 = -\gamma$. This is therefore known as the case of repeated roots. We write the solution as $x(t) = Ae^{-\gamma t}$. This cannot, however, be the most general solution, which must have *two* arbitrary constants. We must therefore look for a second solution which is linearly independent of $Ae^{-\gamma t}$. One can look for such a solution by multiplying $e^{-\gamma t}$ by a polynomial function of time, $f(t) \neq$ *constant*. We explore the simplest such product, $t \times e^{-\gamma t}$, and construct the following superposition of these two linearly independent functions:

$$x(t) = Ae^{-\gamma t} + Bte^{-\gamma t} = (A + Bt)e^{-\gamma t}. \tag{5.11}$$

We immediately see that for some $t = \tau > 0$, if $(A + B\tau) = 0$, implying that the system can reach equilibrium position once. Subsequently, after the overshoot, the system can reverse its motion but will approach equilibrium only exponentially. It is obvious that A and B will have to have opposite signs. A has the dimension of length, and B has the dimension of velocity. From Eq. 5.11, we see that the velocity of the oscillator would be given by:

$$\dot{x}(t) = -\gamma(A + Bt)e^{-\gamma t} + Be^{-\gamma t}. \tag{5.12}$$

From the initial conditions, at $t = 0$, we get

$$x_0 = x(t = 0) = A,$$

and $v_0 = \dot{x}(t = 0) = -\gamma A + B = -\gamma x_0 + B.$

Hence, we get

$$x(t) = (x_0 + v_0 t + \gamma x_0 t)e^{-\gamma t}. \quad (5.13)$$

Thus, the system can return to equilibrium when

$$\tau = -\frac{x_0}{(v_0 + \gamma x_0)}.$$

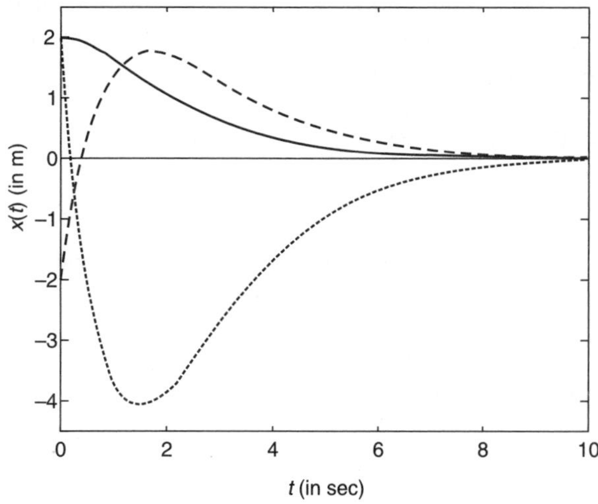

Fig. 5.2 The critically damped oscillator can overshoot the equilibrium once, and only once. Subsequently, it can never return to the equilibrium poisition in any finite time.
Oscillator parameters used for this plot are m = 1g,
$\kappa = 0.62 \times 10^{-3}$ N/kg,
$\omega_0 = 0.79$ rad/sec.
The damping coefficient chosen for this plot is $\gamma = \omega_0$.

Fig. 5.2 shows the displacement vs. time graph for a critically damped oscillator. The same set of three different initial conditions are used here as were used for the overdamped oscillator of Fig. 5.1. The damping coefficient chosen for this illustration was the same as the natural frequency of oscillation ($\gamma = 0.79 = \omega_0$), which is just what makes this oscillator *critically* damped. The three curves in Fig. 5.2 correspond to the same three initial conditions as were used in Fig. 5.1 for the overdamped oscillator:

(i) $x(t = 0) = x_0 = 2.0$ m; $v(t = 0) = v_0 = 0.0$ m/s (continuous curve)

(ii) $x(t = 0) = x_0 = 2.0$ m; $v(t = 0) = v_0 = -11.71$ m/s (dotted curve)

(iii) $x(t = 0) = x_0 = -2.0$ m; $v(t = 0) = v_0 = 6.73$ m/s (dashed curve)

It may not be apparent from the rather similar looking Fig. 5.1 and Fig. 5.2 that, for the same initial conditions, the critically damped oscillator can reach the equilibrium *quicker* than the overdamped oscillator. There is nothing counter-intuitive about this, since excess damping would only *further* impede the movement of the oscillator. The critically damped oscillator therefore returns fastest to the equilibrium without oscillations about it. Accordingly, when *damping-without-oscillations* is needed, the damping coefficient is often chosen closest to the critical condition.

Case 3: $\gamma < \omega_0$ Underdamped oscillator, or, subcritical damping

For this case, all the other parameters being the same as in the case (i) and case (ii), we have chosen $\gamma = 0.05$ rad/sec. This value makes the damping coefficient much smaller than the intrinsic natural frequency (ω_0 = 0.79 rad/sec) of oscillation of the system. Since $\gamma < \omega_0$, $\xi = \sqrt{\gamma^2 - \omega_0^2} = i\omega$ is now an imaginary number (ω : real and positive). The two roots (Eq. 5.5b,c) are therefore complex numbers:

$$q_{1,2} = -\gamma \pm \xi = -\gamma \pm i\sqrt{\omega_0^2 - \gamma^2} = -\gamma \pm i\omega, \tag{5.14a}$$

$$\text{where } \omega = \sqrt{\omega_0^2 - \gamma^2}\ . \tag{5.14b}$$

Observe that $\omega < \omega_0$, by an amount determined by γ. The general solution to the equation of motion for the damped oscillator is therefore given by

$$x(t) = A_1 e^{(-\gamma + i\omega)t} + A_2 e^{(-\gamma - i\omega)t}, \tag{5.15a}$$

$$\text{i.e., } x(t) = e^{-\gamma t}\{A_1 e^{+i\omega t} + A_2 e^{-i\omega t}\}, \tag{5.15b}$$

$$\text{or, } x(t) = e^{-\gamma t}\{(A_1 + A_2)\cos(\omega t) + i(A_1 - A_2)\sin(\omega t)\}. \tag{5.15c}$$

We introduce two new parameters, B and θ, *instead of* A_1 and A_2, by writing $A_1 + A_2 = B\sin\theta$ and $i(A_1 + A_2) = B\cos\theta$. Equivalently, $A_1 = -\dfrac{iBe^{+i\theta}}{2}$ and $A_2 = +\dfrac{iBe^{-i\theta}}{2}$. Use of the new parameters enables us to write the solution in the following discerning form:

$$x(t) = Be^{-\gamma t}\{\sin\theta\cos(\omega t) + \cos\theta\sin(\omega t)\}, \tag{5.15d}$$

$$\text{i.e., } x(t) = (Be^{-\gamma t})\sin(\omega t + \theta). \tag{5.15e}$$

Each form in Eqs. 5.15a–e represents the *most* general solution, which is a superposition of two linearly independent solutions with two arbitrary constants, which are obtainable from the initial conditions $x(t = 0)$ and $\dot{x}(t = 0)$. The equivalence of the forms employing trigonometric functions and the exponential functions is a great advantage in solving the equation of motion for the simple harmonic and/or the damped oscillators. All the forms are completely equivalent, but the form (Eq. 5.15e) lends itself to straightforward physical interpretation. It represents an oscillatory sinusoidal behavior. It is similar to the solution to the simple harmonic (not-damped) oscillator, but it is also somewhat different. It entails three additional considerations to reckon with:

(i) the amplitude ($Be^{-\gamma t}$) diminishes exponentially as time advances,

(ii) the oscillator will go through the mean position $x = 0$ at a regular period given by $T = \dfrac{2\pi}{\omega} > \dfrac{2\pi}{\omega_0}$, but with an amplitude which diminishes exponentially with time. The complete pattern (displacement vs. time) is therefore not exactly repetitive, but the zeros of the displacement are strictly repetitive with an exact time period $T = \frac{2\pi}{\omega}$, which is therefore known as the *period of the damped oscillator.*

(iii) at $t = 0$, $x(t = 0) = B \sin \theta$, i.e., the oscillation is *phase-shifted* by $\theta = \sin^{-1}\left(\frac{x(t=0)}{B}\right)$.

The solution in Eq. 5.15e has diminishing amplitude. It is called the *transient* solution since the oscillations cannot last long.

Fig. 5.3 shows the displacement vs. time graph for a critically damped oscillator. The damping coefficient chosen in this case corresponds to weak damping: $\gamma = 0.05 < \omega_0$ (underdamped oscillator). The three curves in Fig. 5.3 correspond to the very same set of the three initial conditions as were used in Fig. 5.1 and Fig. 5.2:

(i) $x(t = 0) = x_0 = 2.0$ m; $v(t = 0) = v_0 = 0.0$ m/s (continuous curve)

(ii) $x(t = 0) = x_0 = 2.0$ m; $v(t = 0) = v_0 = -11.71$ m/s (dotted curve)

(iii) $x(t = 0) = x_0 = -2.0$ m; $v(t = 0) = v_0 = 6.73$ m/s (dashed curve)

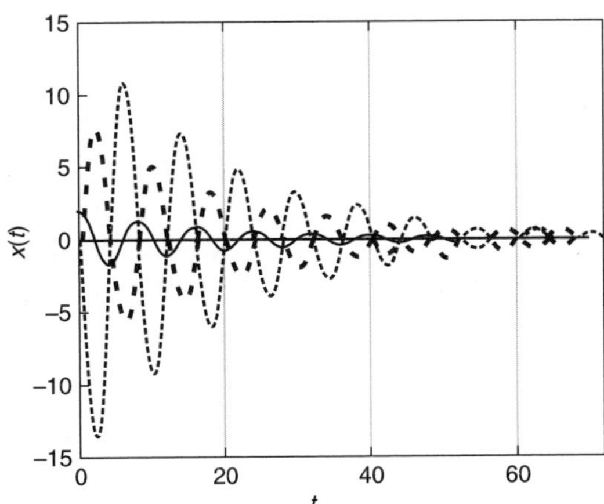

Fig. 5.3 Schematics sketch showing displacement of an under-damped oscillator as a function of time.
Oscillator parameters used for this plot are $m = 1$ kg,
$\kappa = 0.62 \times 10^{-3}$ N/kg,
$\omega_0 = 0.79$ rad/sec.
The damping coefficient chosen for this plot is $\gamma = 0.05$ rad/sec.

The number of oscillations of an under-damped oscillator in a small interval δt would be $N(\text{in } \delta t) = \frac{\delta t}{T} = \nu\delta t = \frac{\omega\delta t}{2\pi}$, with $\nu = \frac{1}{T}$, which is the frequency of the *zeros* of the displacement. In two successive periods, the amplitude falls in the ratio

$$\frac{B_{n+1}}{B_n} = \frac{Be^{-\gamma(t+T)}}{Be^{-\gamma t}} = e^{-\gamma T} = e^{-\varphi}. \tag{5.16}$$

Over a time-interval corresponding to N oscillations, $\delta t = NT$, the amplitude diminishes by the factor $\frac{B_{N+1}}{B_1} = e^{-\gamma NT} = e^{-N\varphi}$. Thus, when $\gamma = \frac{1}{NT}$, the amplitude decreases by a factor $\frac{1}{e}$. For this reason, γ is sometimes also known as the logarithmic decrement factor. Unlike the previous case of the 'overdamped oscillator', we have oscillations about the mean position. Damping is not so large that the oscillatory behavior is completely *killed* even as the amplitude of the oscillations diminishes. An oscillator with such properties is therefore known as the *underdamped* oscillator.

5.3 FORCED OSCILLATIONS

The equation of motion for an oscillator, including one with damping, which we have considered so far is essentially a homogeneous differential equation. The only thing one did to the oscillator was to give it an initial displacement, impart to it an initial velocity, and then let it go. The subsequent temporal evolution of the system was completely governed by properties of the oscillating system, including its natural surroundings, to which it would lose energy through damping. Damping can be caused by an *internal* element of the oscillator also, such as unavoidable circuit resistance to current in an electrical LC oscillator, or the internal friction between various parts of a spring as it oscillates and its internal layers slide over each other, causing friction. Damping can also be caused due to the medium in which the oscillator is placed, such as a viscous medium in which a mass spring may be oscillating, or the air in which a pendulum may be swinging. Regardless of the cause of damping, once the system is let go from an initial displacement and with an initial velocity imparted to it, the oscillator considered hitherto was left alone to itself, along with its natural internal and external surroundings that caused damping. This was accounted for in Eq. 5.2 which describes the damped oscillator.

We now add an inhomogeneous term to Eq. 5.2. My favorite example of this case is that of a mother rocking the cradle of her baby for the hand that rocks the cradle rules the world, as the famous proverb, after the poem by W. R. Wallace. Left to itself, the cradle, once set into oscillations would continue to oscillate forever, were it not for damping and friction at the supports (and that due to air). The effect of damping would cause the amplitude of oscillation to decay, as discussed in the previous section. However, the attentive mother would rock the cradle every once so often. This force on the oscillating cradle is completely external to the oscillating system. It is the mother who would decide the timing of this external force, its amplitude, and its frequency. The time at which the mother would rock the cradle could be when the cradle is moving toward her, or away from her, or anytime in between the cradle's oscillation. It would therefore have a phase difference with respect to the oscillator's motion. The (circular) frequency Ω at which the mother rocks the cradle may be different from ω_0, the *natural frequency* of oscillation of the cradle. It can even be different from the frequency $\omega = \sqrt{\omega_0^2 - \gamma^2}$, which is the frequency of the zeros of the *damped* oscillator. Such an oscillator's dynamics is then determined by the intrinsic restoring force, $-\kappa x$, the damping force $-c\dot{x}$, and the external driving force F_{dr}.

The equation of motion of the damped mass-spring oscillator which is subjected to an external driving force, then, is

$$m\ddot{x} = F = -\kappa x - c\dot{x} + F_{dr}, \tag{5.17}$$

i.e., $$\ddot{x} + \frac{c}{m}\dot{x} + \frac{\kappa}{m}x = \frac{F_{dr}}{m}. \tag{5.18a}$$

Likewise, the equation of motion for a simple pendulum with damping, which is given an external push every now and then, is given by

$$\ddot{\theta} + \frac{c}{ml}\dot{\theta} + \frac{g}{l}\theta = \frac{F_{dr}}{ml}. \tag{5.18b}$$

The electrical LRC oscillator which contains a source of alternating current is also a similar damped-driven oscillator. The equation of motion for the dynamics of charges that flow in it is given by a completely identical equation:

$$\ddot{Q} + \frac{R}{L}\dot{Q} + \frac{1}{LC}Q = \frac{V^{(\text{ext})}}{L}. \tag{5.18c}$$

We shall obtain the solutions of Eq. 5.18a for the mass-spring 'damped and driven' oscillator which we consider to be the prototype of all such systems. The actual form of the solution depends on the functional form of F_{dr}. We consider a periodic force whose magnitude is oscillatory, varying at a (circular) frequency, Ω. The instant at which the external driving force is set on, need not be the one when the driving force is at peak, or zero. In general, it can be anything in between. A phase angle θ in the expression for the driving force is the appropriate parameter which carries information about this. Accordingly, we consider a periodic driving force to be represented by

$$F_{dr} = F_0 e^{i(\Omega t + \theta)} = \widehat{F} e^{i(\Omega t)}, \text{ with } \widehat{F} = F_0 e^{i\theta}. \tag{5.19}$$

We have extracted the phase θ of the driving force and included it in the 'complex' amplitude $\widehat{F}$. The equation of motion for the damped and driven oscillator therefore is:

$$\ddot{x} + 2\gamma\dot{x} + \omega_0^2 x = \frac{\widehat{F}}{m}e^{i\Omega t}. \tag{5.20}$$

It is instructive to build the solutions in two stages. Just as we first considered the damped oscillator *not* subjected to any external force, we now consider a *driven* oscillator *without* damping. This would help us gain some insight into the dynamics of the damped and driven oscillator, which is the full physical system that we really want to study. We shall then insert the damping term later. The equation of motion for the driven oscillator with no damping is then given by

$$\ddot{x} + \omega_0^2 x = \frac{\widehat{F}}{m}e^{i\Omega t}, \tag{5.21a}$$

We look for the solution of Eq. 5.21 in the form

$$x(t) = \hat{x}e^{i\Omega t}, \tag{5.21b}$$

where the 'complex' amplitude includes the phase factor.

The use of complex amplitudes—whether for displacement, or for the applied force—need not trouble us. Complex numbers only enable us deal with two numbers together, both of which are essentially real. We shall secure information about physical parameters by examining the real parts of the complex functions. After all, both the real part and the imaginary part of a complex number are equally real. Since $\dot{x} = i\Omega x$, and $\ddot{x} = (i\Omega)^2 x$, the exponential form allows us to interpret the effect of differentiation with respect to time by the operator (d/dt), to be equivalent to multiplication by $(i\Omega)$.

We therefore get

$$(\omega_0^2 - \Omega^2)\hat{x} = \hat{F}/m, \tag{5.22a}$$

$$\text{i.e.,}\quad \hat{x} = \frac{\hat{F}}{\{m(\omega_0^2 - \Omega^2)\}}. \tag{5.22b}$$

It might concern some of you that when Ω, the driving frequency becomes equal to natural frequency, ω_0, the amplitude blows up to infinity. That is no worry, since we are considering dynamics of the system only within the oscillator's elastic limit, and hence so this formalism becomes inapplicable. Besides, real oscillators must include some unavoidable damping, which is ignored in Eq. 5.21.

We now address the complete equation of motion (Eq. 5.20) for the damped and driven oscillator. The solution will be somewhat similar to, but also somewhat different, from Eq. 5.21b. The difference would be necessitated essentially by damping. We therefore look for a solution having the following form:

$$x(t) = \hat{A}e^{i\Omega t}, \text{ with } \hat{A} = A_0 e^{i(\theta - \phi)}, \tag{5.23a}$$

$$\text{which gives } \dot{x} = \hat{A}(i\Omega)e^{i\Omega t} = i\Omega x, \tag{5.23b}$$

$$\text{and } \ddot{x} = \hat{A}(i\Omega)^2 e^{i\Omega t} = -\Omega^2 x. \tag{5.23c}$$

The angle ϕ above takes care of the phase lag of oscillation with respect to the driving force $\hat{F}e^{i\Omega t}$. Using Eq. 5.23 in Eq. 5.20, we get:

$$A_0 e^{i(\theta - \phi)} = \frac{\{F_0 e^{i\theta}/m\}}{\{(\omega_0^2 - \Omega^2) + i2\gamma\Omega\}}, \tag{5.24a}$$

$$\text{i.e.,}\quad e^{-i\phi} = \frac{\{F_0/(mA_0)\}}{(\omega_0^2 - \Omega^2) + i2\gamma\Omega}. \tag{5.24b}$$

Separating now the real and imaginary parts of the complex number by multiplying both the numerator and denominator on the right hand side of Eq. 5.24b by the complex conjugate of the denominator, we get

$$\cos\phi = \frac{\{F_0/(mA_0)\}(\omega_0^2 - \Omega^2)}{(\omega_0^2 - \Omega^2)^2 + 4\gamma^2\Omega^2}, \tag{5.25a}$$

and

$$\sin\phi = \frac{\{F_0/(mA_0)\}2\gamma\Omega}{(\omega_0^2 - \Omega^2)^2 + 4\gamma^2\Omega^2}, \tag{5.25b}$$

$$\text{with } \phi = \tan^{-1}\left\{\frac{2\gamma\Omega}{\omega_0^2 - \Omega^2}\right\}. \tag{5.25c}$$

Squaring and adding Eqs. 5.25a and 5.25b, we get $\sin^2\phi + \cos^2\phi = 1$, from which we get:

$$A_o(\Omega) = \frac{F_0}{m\sqrt{(\omega_0^2 - \Omega^2)^2 + 4\gamma^2\Omega^2}}. \tag{5.26}$$

Now, since our solution to the differential equation of motion for the damped and driven oscillator is given by $x(t) = \widehat{A}e^{i\Omega t}$, with $\widehat{A} = A_0 e^{i(\theta - \phi)}$, we see that the solution is:

$$x(t) = \widehat{A}e^{i\Omega t} = A_0 e^{i(\Omega t + \theta - \phi)t} = \frac{(F_0/m)}{\sqrt{(\omega_0^2 - \Omega^2)^2 + 4\gamma^2\Omega^2}} e^{i(\Omega t + \theta - \phi)}, \tag{5.27a}$$

with the external-frequency dependent amplitude of the oscillation given by

$$A_0 = A_0(\Omega) = \frac{(F_0/m)}{\sqrt{(\omega_0^2 - \Omega^2)^2 + 4\gamma^2\Omega^2}}. \tag{5.27b}$$

The solution to the damped and driven oscillator given in Eq. 5.27a is called the *steady state* solution. It is repetitive, at the frequency of the driving force, and remains *steady*, hence its name. The steady state solution is in contrast with the *transient* solution given in Eq. 5.15e. The transient solution is solution to the homogeneous differential equation (Eq. 5.2b), which does not contain the driving term. We observe that the damping coefficient γ plays a much bigger role in the transient solution than in the steady state solution. The complete solution to the damped and driven oscillator may now be written as a *sum* of the steady state solution and the transient solution. However, we may ignore the damping coefficient γ in the 'steady state part', but of course not in the 'transient part'. Accordingly, the general solution to the damped and driven oscillator is given by

$$x(t) = Be^{-\gamma t}\sin(\omega t + \delta) + \frac{(F_0/m)}{(\omega_0^2 - \Omega^2)} e^{i(\Omega t + \theta - \phi)}. \tag{5.28}$$

There are *three* frequencies of importance in the general solution. These are: (i) ω_0 the natural frequency of the oscillator, which is determined completely by the oscillators intrinsic properties (ii) ω, the frequency at which the zeros of the damped oscillator occur, which is determined by the effective damping coefficient, no matter what causes the damping, and (iii) Ω, the frequency of the driving oscillatory external force. This is tunable, which means that the external frequency Ω can be varied till the amplitude of oscillation is maximized at resonance.

The instantaneous displacement of the oscillator $x(t)$ is out of step with the driving force F_{driving} by the angle, ϕ. The amplitude of the oscillation is determined by the amplitude F_0 of the driving force, modulated further by the external-frequency dependent factor, $\frac{1}{\sqrt{(\omega_0^2-\Omega^2)^2+4\gamma^2\Omega^2}}$, and of course also by the inertia m. In particular, it depends on the proximity of the frequency Ω of the external driving force to the intrinsic natural frequency ω_0 of the oscillator. The nature of the solution also depends on the damping coefficient, γ. The results we have obtained are extremely important. Using them, fascinating applications in mechanical, electrical and various other physical systems have been developed by physicists and engineers.

5.4 RESONANCES, AND QUALITY FACTOR

The amplitude A_0 of the oscillation (Eq. 5.27) depends on (i) the intrinsic natural frequency of the oscillating system, and also (ii) Ω, the frequency of the driving force. We can therefore control the frequency Ω of the external driving force to actually augment the amplitude, $x(t)$. The amplitude would be an extremum at the frequency $\Omega = \Omega_r$ of the driving force for which $\frac{dA_0(\Omega)}{d\Omega} = 0$.

$$\text{From } \frac{dA_0}{d\Omega} = \frac{-\frac{1}{2}\frac{F_0}{m}\{2(\omega_0^2-\Omega^2)(-2\Omega)+8\gamma^2\Omega\}}{\{(\omega_0^2-\Omega^2)^2+4\gamma^2\Omega^2\}^{3/2}}, \tag{5.29}$$

we see that the numerator of $\frac{dA_0}{d\Omega}$ is zero when $\Omega^2 = \omega_0^2 - 2\gamma^2$. Hence the frequency of the external driving force when amplitude of the oscillation would be a maximum is:

$$\Omega_r = \sqrt{\omega_0^2 - 2\gamma^2}\,. \tag{5.30a}$$

This frequency is known as the *resonance* frequency, and the subscript in the notation Ω_r denotes just that. We recognize that

$$\Omega_r = \sqrt{\omega_0^2\left(1-\frac{2\gamma^2}{\omega_0^2}\right)} = \omega_0\left(1-\frac{2\gamma^2}{\omega_0^2}\right)^{\frac{1}{2}} \approx \omega_0\left(1-\frac{\gamma^2}{\omega_0^2}\right) = \omega_0 - \frac{\gamma^2}{\omega_0}\,. \tag{5.30b}$$

We know that when damping is present, the zeros of the *underdamped* oscillator do not occur at the intrinsic natural frequency of the oscillator. Therefore, Eqs. 5.30a, and 5.30b for the resonance frequency involve the intrinsic natural frequency ω_0. It is also useful to express it in a form involving the frequency of zeros of the damped oscillator, which is $\omega = \sqrt{\omega_0^2 - \gamma^2}$. Using Eq. 5.14b and Eq. 5.30a, the resonance frequency is

$$\Omega_r = \sqrt{\left(\omega^2+\gamma^2\right)-2\gamma^2} = \sqrt{\omega^2-\gamma^2} = \left\{\omega^2\left(1-\frac{\gamma^2}{\omega^2}\right)\right\}^{1/2} \simeq \omega\left(1-\frac{\gamma^2}{2\omega^2}\right). \tag{5.30c}$$

Using Eq. 5.30a in Eq. 5.26, we get the amplitude of the oscillation *at* the resonance frequency. This is the maximum amplitude the oscillation can possibly have:

$$A_0(\Omega)_{\text{MAX}} = \frac{F_0/m}{\sqrt{(\omega_0^2-(\omega_0^2-2\gamma^2))^2+4\gamma^2(\omega_0^2-2\gamma^2)}}, \tag{5.31a}$$

$$\text{i.e., } A_0(\Omega)_{\text{MAX}} = \frac{F_0/m}{2\gamma\sqrt{\omega_0^{\,2}-\gamma^2}}. \tag{5.31b}$$

Using Eq. 5.31b, the amplitude (given in Eq. 5.27b) at an arbitrary frequency of the external driving force can be written in terms of the maximum amplitude that occurs at resonance:

$$A_0(\Omega) = \frac{2\gamma A_0(\Omega)_{\text{MAX}}\sqrt{\omega_0^{\,2}-\gamma^2}}{\sqrt{(\omega_0^2-\Omega^2)^2+4\gamma^2\Omega^2}}. \tag{5.32a}$$

We note that $\omega_0^2-\Omega^2 = (\omega_0-\Omega)(\omega_0+\Omega) = (\omega_0-\Omega)(2\omega_0+(\Omega-\omega_0))$. The second factor in this product, $(2\omega_0 + (\Omega - \omega_0))$, is nearly equal to $2\omega_0$, since $(\Omega - \omega_0) << 2\omega_0$. We therefore ignore the term $(\Omega - \omega_0)$ in this factor. In other words, $\omega_0^2-\Omega^2 \simeq (\omega_0-\Omega)(2\omega_0)$. Thus close to the resonance, and when $\gamma << \omega_0$, we can write

$$A_0(\Omega) \simeq \frac{A_0(\Omega)_{\text{MAX}}\,2\gamma\omega_0}{\sqrt{\{(\omega_0-\Omega)(2\omega_0)\}^2+4\gamma^2\omega_0^{\,2}}} = \frac{A_0(\Omega)_{\text{MAX}}}{\sqrt{\dfrac{(\omega_0-\Omega)^2}{\gamma^2}+1}}. \tag{5.32b}$$

At the *resonance frequency* $\Omega_r \simeq \omega_0$, the amplitude (Eq. 5.32) is the *largest*. The use of the approximate equality is on account of the approximation $\omega_0^2-\Omega^2 \simeq 2\omega_0(\omega_0-\Omega)$ that was used above.

When $\Omega = \Omega_r \pm \gamma$, we shall have

$$A_0(\Omega) \simeq \frac{A_0(\Omega)_{\text{MAX}}}{\sqrt{\left\{\dfrac{(\omega_0-\Omega_r\pm\gamma)^2}{\gamma^2}+1\right\}}} = \frac{A_0(\Omega)_{\text{MAX}}}{\sqrt{\left\{\dfrac{(\pm\gamma)^2}{\gamma^2}+1\right\}}} = \frac{A_0(\Omega)_{\text{MAX}}}{\sqrt{2}}. \tag{5.33}$$

The range of frequency values between $\Omega = \Omega_r - \gamma$ and $\Omega = \Omega_r + \gamma$ is of special interest. Within this range of the frequencies, the amplitude of the oscillation is reduced by a factor $\frac{1}{\sqrt{2}}$ of times the maximum amplitude (Eq. 5.30), *or less*. Since the energy is proportional to square of the amplitude, it would be reduced in this range by $\frac{1}{2}$, *or less*. Thus, for frequencies separated at the most by 2γ about the resonance frequency, the energy imparted to the oscillator by the driving agency reduces at the most to half the maximum at resonance.

The prototype of a damped and driven oscillator we shall consider is the electrical circuit shown in Fig. 5.4. The inductance L and the capacitor C constitutes the oscillator having an intrinsic natural frequency $\omega_0 = \sqrt{\frac{1}{LC}}$, the resistor R provides the damping, and an alternating electric voltage source $V^{\text{ext}}(t)$ placed in the circuit provides a periodic driving agency which operates at a tunable frequency Ω. In Chapter 4, we have already seen that the correspondence $\frac{\kappa}{m} \leftrightarrow \frac{g}{\ell} \leftrightarrow \frac{1}{LC} \leftrightarrow 2g\sigma$ will enable us to employ the analysis to *any* of the oscillators described in Fig.4.3a,b,c,d. The intrinsic natural frequency of these oscillators is:

$$\omega_0 = \sqrt{\frac{\kappa}{m}} \leftrightarrow \sqrt{\frac{g}{\ell}} \leftrightarrow \sqrt{\frac{1}{LC}} \leftrightarrow \sqrt{2g\sigma} = 2\pi\nu_0 = \frac{2\pi}{T_{\text{period}}}, \tag{5.34a}$$

and the corresponding periodic time is

$$T_{\text{period}} = 2\pi\sqrt{\frac{m}{\kappa}} \leftrightarrow 2\pi\sqrt{\frac{\ell}{g}} \leftrightarrow 2\pi\sqrt{LC} \leftrightarrow 2\pi\sqrt{\frac{1}{2g\sigma}} = \frac{1}{\nu_0}. \tag{5.34b}$$

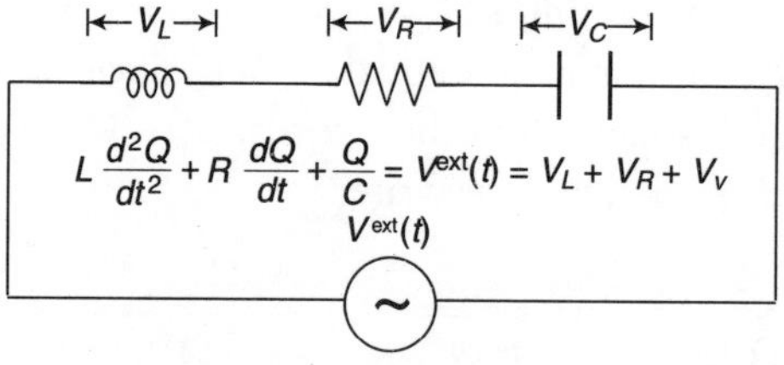

$$L\frac{d^2Q}{dt^2} + R\frac{dQ}{dt} + \frac{Q}{C} = V^{\text{ext}}(t) = V_L + V_R + V_v$$

Fig. 5.4 This figure shows an electric LRC circuit which is our prototype of a damped and driven oscillator. The voltage $V^{\text{ext}}(t)$ generated by an external source is distributed across the inductance, resistor, and the capacitance.

The equation of motion for the time-dependent charge Q in the LRC circuit is already given in Eq. 5.18c. The electro-mechanical analogues $L \leftrightarrow m, C \leftrightarrow \frac{1}{\kappa}$, and $\frac{R}{2L} \leftrightarrow \gamma$ establish the equivalence between the electrical LRC oscillator and the damped mass-spring oscillator. The electrical *inductance* L plays the same role in an electrical circuit as the *inertia* m plays in a mechanical device, and the *capacitance* C plays the same role as that of *compliance*, which is the inverse of the spring-constant in a mass-spring oscillator. The role of the *damping coefficient* $\gamma = \frac{b}{2m}$ is now played by $\frac{R}{2L}$, i.e., the resistor plays the same role as the coefficient b: $R \leftrightarrow b$. It should cause no confusion that both b and γ are commonly referred to as damping coefficient. The equation of motion for the charge q in the damped and driven LRC circuit is:

$$L\frac{d^2q}{dt^2} + R\frac{dq}{dt} + \frac{1}{C}q = V^{\text{ext}}(t), \tag{5.35a}$$

$$\text{i.e.,} \quad \frac{d^2Q}{dt^2} + \frac{R}{L}\frac{dQ}{dt} + \frac{1}{LC}Q = \frac{1}{L}V^{\text{ext}}(t) \tag{5.35b}$$

Toward the end of the Section 5.3, we have seen that for frequencies Ω of the external driving agency separated at the most by 2γ about the resonance frequency $\Omega_r \simeq \omega_0$, i.e., in the range, $\Omega_r - \gamma \leq \Omega \leq \Omega_r + \gamma$, the energy pumped into the LRC circuit by the driving agency reduces at the most to *half of the maximum*. For this reason, 2γ is known as the *resonance width*. It is used to define a ratio, known as the quality factor, defined as

$$Q = \frac{\Omega_r}{2\gamma} \approx \frac{\omega_0}{2\gamma}, \tag{5.36a}$$

$$\text{i.e.,} \quad Q = \omega_0 \times \frac{L}{R} = \sqrt{\frac{1}{LC}} \times \frac{L}{R} = \frac{1}{R}\sqrt{\frac{L}{C}}. \tag{5.36b}$$

In an electric circuit containing a voltage source alternating at a frequency ω, and an inductance L, the product ωL impedes the flow of current. The impedance offered to an alternating circuit is similar to the ohmic resistance offered by the conducting elements in the circuit, collectively symbolized by R in Fig. 5.4. The quality factor is then the ratio of the inductive impedance to the Ohmic resistance. A lower resistance results in a *narrower resonance width* $2\gamma \leftrightarrow \frac{R}{L}$, resulting the maximum energy transfer over a narrower frequency range of the external driving agency. It is a desirable quality in electrical communication devices so that one can tune the circuit close to a particular frequency. It is for this reason that Q is called the *quality factor*. The amplitude of the oscillation as a function of frequency of the driving agency is shown in Fig. 5.5. You can see that lower the damping, sharper is the resonance, and better the quality of the oscillator to respond resonantly to a particular frequency of the driving agency. Essentially, maximum energy transfer over a narrower, sharper, range of frequency is achieved by *matching* the frequency of the driving agency Ω with the intrinsic natural frequency $\sqrt{\frac{1}{LC}}$ of the oscillator. This can be achieved by varying the frequency of the alternating external voltage, or, for that matter, by changing the inductance L, or the capacitance C in the circuit. In radio-wave communications, it is easy to change the capacitance by adjusting, for example, the distance between the plates of a parallel-plates capacitor. This can be done easily by controlling a knob. This is just what tuning a radio is about.

Every object, not just a mass-spring system or a simple pendulum, has an intrinsic natural frequency. For complex shapes and materials, it is not easy to determine it using first principles, but it is possible to determine it using the principle of superposition of oscillators and applying sophisticated theories which deal with coupled oscillators. These systems would accept maximum energy from an external disturbance if the driving forces stimulate the natural frequencies of the system. It is said that a distinguished Indian classical 'dhrupad' music vocal singer, Pandit Baijnath Mishra (or Baijanath Nayak?), better known

as 'Baiju', who lived in Chanderi (Madhya Pradesh, India) in the late sixteenth century, could sing generating harmonics which would cause resonances in glass creating oscillations large enough to shatter it. In recent times, such a feat has been performed in front of television cameras, and you can read about it in the book 'Raise Your Voice' by Jamie Vendera (ISBN: 978-0-974911-5-8), and see it on the YouTube:
(https://www.youtube.com/watch?v=IZD8ffPwXRo,
https://www.youtube.com/watch?v=Oc27GxSD_bI, accessed on 21 October 2018).

Soldiers break their step while walking on a bridge to avoid causing its resonant collapse. The collapse of the Tacoma Narrows Bridge on 7 November 1940 has often been attributed to such a resonant amplification of its oscillations, but it is argued that a 'negative' damping factor has been a more important factor (*The Failure of the Tacoma Bridge: A Physical Model* by Daniel Green and William G. Unruh, Am. J. Phys. 74 –8, page 206, August 2006) .

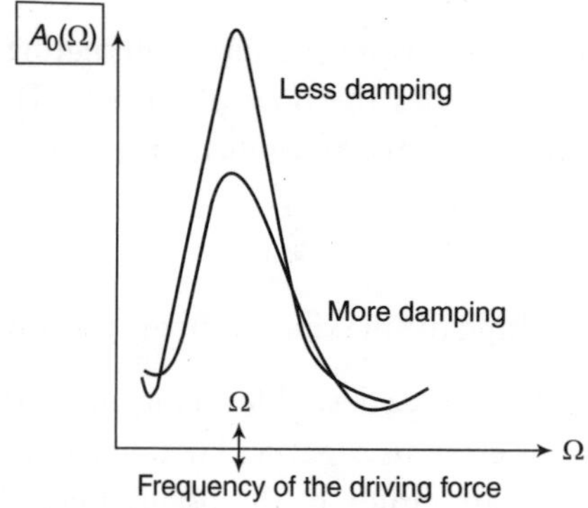

Fig. 5.5 This figure shows the amplitude of oscillation as a function of the frequency of the driving agency. It peaks at the resonance energy. The resonance is sharper when damping is less, which enhances the quality factor.

The phenomenology concerning oscillations, positive and negative damping, forced oscillations, resonances, and superposition of oscillations, is of paramount importance. Its applications encompass a gigantic range, from classical to quantum mechanics, from an analysis of bio-rhythms to those of fluctuations in the share market. Technology employs it, from sophisticated applications in communication devices to medical applications in diagnostics, including fMRI (functional Magnetic Resonance Imaging).

Problems with Solutions

P5.1:

Consider a periodic function of time, $\phi(t) = \cos(\omega t)$, with time period τ. (a) Determine if the average value of ϕ over a single period is equal to the average over an arbitrary time interval. (b) Show that the average over a large time interval is almost equal to the average over unit time period.

Solution:

The average of $\phi(t)$ over one time period is $\langle\phi\rangle_\tau = \frac{1}{\tau}\int_0^\tau \phi(t)dt = \frac{1}{\tau}\int_0^\tau \cos(\omega t)dt = 0$. Essentially, the positive and negative contributions of the cosine function cancel over a full time period. However, if you take the average value over the first quarter of the time period, the cosine function is always positive and there is no cancellation: $\langle\phi\rangle_{\frac{\tau}{4}} = \frac{4}{\tau}\int_0^{\frac{\tau}{4}} \phi(t)dt = \frac{4}{2\pi}$, which is of course different from the average over one time period.

The average over a long time interval T is $\langle\phi\rangle_{T\to\infty} = \frac{1}{T}\int_0^T \phi(t)dt = \frac{1}{T}\left[\int_0^{n\tau}\phi(t)dt + \int_{n\tau}^T \phi(t)dt\right]$, and hence $\langle\phi\rangle_{T\to\infty} = \frac{1}{T}[0+R] = \frac{R}{T} \to 0$. [R: remainder, from the integral from $n\pi$ to T]

P5.2:

In a somewhat crude, but very fruitful, model of the atom, each electron in it is considered to be bound to an equilibrium position by a restoring force $-k\vec{r}(t)$, where $\vec{r}(t)$ denotes its displacement from the equilibrium position at the instant t. The constant k is therefore akin to the spring constant of a restoring force. The electron is also subject to a damping force, $\Gamma_d m\dot{\vec{r}}(t)$, where m is the mass of the electron, Γ_d is the damping coefficient. The damping force is proportional to the instantaneous velocity of the electron. The damping is due to the unspecified degrees of freedom; for example, it could be due to the electron radiating away some electromagnetic energy as it oscillates, and loose energy thereby. We consider the oscillations of the electron to be driven by an electric field $\hat{\varepsilon}eE_0e^{-i(\Omega t+\theta)}$, where Ω is the frequency of the driving electric field having polarization direction $\hat{\varepsilon}$, and e is the electron charge. θ is a phase factor.

(a) Show that $\vec{r}(t) = \breve{\varepsilon}\frac{e}{m}\frac{E_0}{\omega_0^2 - \Omega^2 - i\Gamma_d\Omega}e^{-i(\Omega t+\theta)}$ is a solution to the differential equation of motion of the electron.

(b) Use your result to obtain the atomic polarizability, α, which is just the ratio of the electron dipole moment to the applied electric field, given by $\alpha = \frac{|e\vec{r}|}{|\vec{E}|}$.

(c) How will you determine the average power pumped into the atomic system by the applied electric field?

Solution:

(a) The equation of motion for the atomic electron in this (semi-classical) model is:

$$m\ddot{\vec{r}} = -k\vec{r}(t) - m\Gamma_d\dot{\vec{r}}(t) + \hat{\varepsilon}eE_0e^{-i(\Omega t+\theta)}. \text{ Hence } \ddot{\vec{r}} + \omega_0^2\vec{r}(t) + \Gamma_d\dot{\vec{r}}(t) = \breve{\varepsilon}\frac{e}{m}E_0e^{-i(\Omega t+\theta)}.$$

We have to verify that $\vec{r}(t) = \breve{\varepsilon}\frac{e}{m}\frac{E_0}{\omega_0^2 - \Omega^2 - i\Gamma_d\Omega}e^{-i(\Omega t+\theta)}$ is a solution to the equation of motion.

Therefore, we must first determine the velocity and the acceleration.

We immediately get: $\dot{\vec{r}}(t) = -i\Omega\vec{r}(t)$, and $\ddot{\vec{r}}(t) = -\Omega^2\vec{r}(t)$. Direct substitution of the position vector, velocity and acceleration immediately confirms that the equation of motion is satisfied by the proposed solution.

(b) We note that the displacement of the electron would be zero if the applied electric field vanishes. The displacement of the electron from the equilibrium point is therefore on account of the driving force. It results in an induced oscillating dipole moment, $\vec{p}(t) = e\vec{r}(t)$.

The atomic polarizability therefore is:

$$\alpha = \frac{|e\vec{r}|}{|\vec{E}|} = \frac{\left| e\breve{\varepsilon}\frac{e}{m}\frac{E_0}{\omega_0^2 - \Omega^2 - i\Gamma_d\omega} e^{-i(\Omega t+\theta)} \right|}{\left|\breve{\varepsilon}E_0 e^{-i(\Omega t+\theta)}\right|} = \frac{e^2}{m}\frac{1}{\omega_0^2 - \Omega^2 - i\Gamma_d\Omega}$$

(c) Average power, say $\langle Q \rangle$, pumped into the atomic system by the EM field is the real part of the average time-rate of work-done by the applied field in displacing the electron from the equilibrium point:

$$\langle Q \rangle = \frac{1}{T}\int_0^T \frac{dW}{dt}dt = \frac{1}{T}\int_0^T \frac{\vec{F}\cdot d\vec{r}}{dt}dt = \frac{1}{T}\int_0^T \left[\left(\frac{\vec{F}}{e}\right)\cdot (e\dot{\vec{r}}) \right] dt$$

Note that using the complex variables is a fruitful mathematical tool even if we are interested only in its real part.

Hence:

$$\langle Q \rangle_{\substack{\text{Real}\\ \text{part}}} = \frac{1}{T}\int_0^T \left[\text{Re}\left(\frac{\vec{F}}{e}\right)\cdot \text{Re}(e\dot{\vec{r}}) \right] dt$$

The above integral can be evaluated to determine the power pumped into the atomic system by the applied electric field.

[Caution. *The model employed above is the classical model for absorption of energy by an atomic electron from an oscillating electric field, which may be the electric component of an electromagnetic field. The actual quantum mechanical model does not permit the use of classical ideas like position and velocity at the same time. The classical model is nonetheless useful to develop fruitful notion about what is called the spectral oscillator strength in atomic physics.*]

P5.3:

Consider a 1-dimensional damped harmonic oscillator driven by a periodic square-wave external force. Find the relative amplitude of the first three terms of the solution for the displacement $x(t)$. It is given that the natural frequency of the oscillator $\omega_0 = 3\omega$ where ω is the angular frequency of the sinusoidal terms in the Fourier expansion of the periodic square wave, given by $\phi(t) = \frac{4}{\pi}\left[\sin(\omega t) + \frac{1}{3}\sin(3\omega t) + \frac{1}{5}\sin(5\omega t) + \ldots\right]$. Also given is the Quality Factor, $Q = 100$.

Solution:

When the driving force is $F_{dr} = F_0 e^{i(\Omega t+\theta)}$, consisting of a single driving frequency Ω, the solution to the damped driven oscillator is: $x(t) = \frac{F_o}{m\sqrt{(\omega_o^2 - \Omega^2)^2 + 4\gamma^2\Omega^2}} e^{i(\Omega t+\theta-\phi)}$.

Under the influence of the periodic square wave driving force, the solution therefore is:

$$x(t) = A_1 \sin(\omega t - \phi_1 + \theta) + A_3 \sin(3\omega t - \phi_3 + \theta) + A_5 \sin(5\omega t - \phi_5 + \theta) + \ldots$$

with $A_1 = \dfrac{F_o^1}{m\sqrt{(\omega_o^2 - \omega^2)^2 + 4\gamma^2\omega^2}}$,

$$A_3 = \frac{F_o^3}{m\sqrt{(\omega_o^2 - 9\omega^2)^2 + 4\gamma^2 9\omega^2}}, \text{ and } A_5 = \frac{F_o^5}{m\sqrt{(\omega_o^2 - 25\omega^2)^2 + 4\gamma^2 25\omega^2}}$$

From the coefficients in the Fourier series, we get: $F_o^1 = \dfrac{4}{\pi}$, $F_o^3 = \dfrac{4}{3\pi}$ and $F_o^5 = \dfrac{4}{5\pi}$.

It is given that $\omega_0 = 3\omega$ and also that the quality factor is $Q = 100$. We therefore conclude that $\gamma = \dfrac{\omega_o}{2Q} = \dfrac{3\omega}{2Q}$. Hence, $\gamma = \dfrac{\omega_o}{200} << \omega_o$. From this, we determine that the oscillator is *weakly* damped.

Substituting $\omega_0 = 3\omega$ and $\gamma = \dfrac{\omega_o}{200}$ we can readily determine the amplitudes: A_1, A_3, and A_5 using the above relations. Note that the strongest contribution will come from A_3 for which the denominator will be minimum.

P5.4:

Consider the steady state solution of a driven damped oscillator. Explain how the power delivered to the oscillator by the driving source is expended.

Solution:

The total energy of the oscillator is $E = \dfrac{1}{2}m\dot{x}^2 + \dfrac{1}{2}\kappa x^2$. Therefore, $\dfrac{dE}{dt} = \dot{x}(m\ddot{x} + \kappa x)$.

From Eq. 5.17, we have: $m\ddot{x} + \kappa x = -c\dot{x} + F_{dr}$. Now, in the steady state, $\dfrac{dE}{dt} = 0$.

Hence, $0 = \dot{x}(m\ddot{x} + \kappa x) = \dot{x}(-c\dot{x} + F_{dr}) = F_{dr}\dot{x} - c\dot{x}^2$. Since $F_{dr}\dot{x} = F_{dr}\dfrac{\delta x}{\delta t}$ is the rate at which power is delivered to the oscillator, we see that it is expended in overcoming damping.

P5.5:

The damping coefficient of an oscillator, whose intrinsic natural frequency is ω_0, is γ. It is driven by a sinusoidal force at a variable frequency ω. Show that the amplitude of the oscillator's displacement is maximum when $\omega = \sqrt{\omega_0^2 - 2\gamma^2}$.

Solution:

From Eq. 5.27, we have: $A_0 = A_0(\Omega) = \dfrac{(F_0/m)}{\sqrt{(\omega_0^2 - \Omega^2)^2 + 4\gamma^2\Omega^2}}$. This would be a maximum, when

$A_0^2 = A_0^2(\Omega) = \dfrac{(F_0/m)^2}{(\omega_0^2 - \Omega^2)^2 + 4\gamma^2\Omega^2}$ is a maximum, i.e., when $\frac{d[A_0^2(\Omega)]}{d\Omega} = 0$. Taking the derivative, it immediately follows that the condition is satisfied when the driving frequency is $\sqrt{\omega_0^2 - 2\gamma^2}$.

P5.6:

Analyze the damped and periodically driven oscillator using the Fourier series. Express the periodic force $\phi(t) = \sum_{n=0}^{\infty} f_n \cos(n\omega t)$ as $\phi = \mathrm{Re}(g)$, where the complex function g is $g(t) = \sum_{n=0}^{\infty} f_n e^{in\omega t}$. Show that the real solution to the oscillator's equation of motion is $x = \mathrm{Re}(z)$, where $z(t) = \sum_{n=0}^{\infty} c_n e^{in\omega t}$ with the coefficients given by $c_n = \frac{f_n}{\omega_0^2 - n^2\omega^2 + 2i\gamma n\omega}$.

Solution:

The equation of motion is $\ddot{x} + 2\gamma\dot{x} + \omega_0^2 x = \phi$. In terms of complex variables and complex function, this equation can be written as: $\ddot{z} + 2\gamma\dot{z} + \omega_0^2 z = g$.

Hence: $\sum_{n=0}^{\infty} c_n(-n^2\omega^2)e^{in\omega t} + \sum_{n=0}^{\infty} 2\gamma c_n(in\omega)e^{in\omega t} + \sum_{n=0}^{\infty} \omega_0^2 c_n e^{in\omega t} = \sum_{n=0}^{\infty} f_n e^{in\omega t}$.

Therefore: $\sum_{n=0}^{\infty} c_n(\omega_0^2 - n^2\omega^2 + 2i\gamma n\omega)e^{in\omega t} = \sum_{n=0}^{\infty} f_n e^{in\omega t}$. Equating now the corresponding Fourier coefficients: $c_n(\omega_0^2 - n^2\omega^2 + 2i\gamma n\omega) = f_n$. Hence: $c_n = \frac{f_n}{\omega_0^2 - n^2\omega^2 + 2i\gamma n\omega}$.

Now, introducing the phase $\delta_n = \tan^{-1}\left(\frac{2\gamma n\omega}{\omega_0^2 - n^2\omega^2}\right)$, we get:

$$\mathrm{Re}[z(t)] = \mathrm{Re}\left[\sum_{n=0}^{\infty} c_n e^{in\omega t}\right] = \mathrm{Re}\left[\sum_{n=0}^{\infty} \frac{f_n}{\omega_0^2 - n^2\omega^2 + 2i\gamma n\omega} e^{in\omega t}\right]$$

$$= \mathrm{Re}\left[\sum_{n=0}^{\infty} \frac{f_n}{\sqrt{(\omega_0^2 - n^2\omega^2)^2 + 4\beta^2 n^2\omega^2}} e^{i(n\omega t - \delta_n)}\right]$$

i.e., $$\mathrm{Re}[z(t)] = \sum_{n=0}^{\infty} \frac{f_n}{\sqrt{(\omega_0^2 - n^2\omega^2)^2 + 4\beta^2 n^2\omega^2}} \cos(n\omega t - \delta_n) = \sum_{n=0}^{\infty} x_n(t) = x(t)$$

Additional Problems

P5.7 A 1-dimensional mass-spring oscillator, oscillating about the origin of the X-axis, and having a spring constant k and damping constant γ has a natural frequency of 0.5 Hz. Its amplitude diminishes to half the initial value in 2 seconds. Determine the damping coefficient and the spring constant in terms of the oscillator's mass.

P5.8 Consider a damped oscillator, with natural frequency ω and damping constant γ, both fixed. It is driven by a force $F(t) = F_0 \cos(\omega_d t)$.

(a) Find the rate $P(t)$ at which $F(t)$ does work and show that the average rate $<P>$ over any number of cycles is $m\gamma\omega_d^2 A^2$ where A is the amplitude of the oscillations. Verify that this is the same as the average rate at which energy is lost to the resistive force.

(b) Show that as ω_d is varied $<P>$ is maximum when $\omega_d = \omega$; that is, the resonance of the power occurs exactly at $\omega_d = \omega$.

P5.9 A damped oscillator has a damping constant $\gamma = \frac{\omega_0}{2}$. By what factor does the oscillator's amplitude diminish in one period?

P5.10 While driving a car on a road, the bumps cause the wheels and the connecting axle to oscillate on the springs.

(a) When four 80 kg persons climb into the car, the body sinks by a couple of centimeters. Estimate the spring constant k of each of the four springs.

(b) If an axle assembly (axle plus two wheels) has total mass 50 kg, what is the natural frequency of the assembly oscillating on its two springs?

(c) If the bumps on a road are 80 cm apart, at about what speed would these oscillations go into resonance?

P5.11 A damped oscillator is being driven close to resonance i.e., $\omega = \omega_0$.

(a) Show that the oscillator's total energy is $E = \frac{1}{2} m\omega^2 A^2$.

(b) Find out the energy dissipated ΔE_{dis} during a single cycle by the damping force.

(c) Show that $Q = 2\pi \frac{E}{\Delta E_{dis}}$.

P5.12 When at resonance, show that (a) a series L, C, R circuit behaves as a purely resistive circuit, (b) below the resonance it is capacitive and (c) above the resonance it is inductive. (d) How would it be for a parallel L, C, R circuit? (Also, examine the electro-mechanical analogues for this case).

P5.13 A large Foucault pendulum that hangs in many science museums can swing for many hours before it damps out. Taking the decay time to be about 8 hours and the length to be 30 meters, find the quality factor Q.

P5.14 An undamped oscillator has period 1 second. When weak damping is added, it is found that the amplitude of oscillation drops by 50% in one period, where this new period can be defined as the time between two successive maxima. Find the ratio β/ω_0. What is the period of the damped oscillation?

P5.15 A single degree of freedom viscosity damped system has a spring stiffness of 6000 N/m, critical damping constant 0.3 Ns/mm, and a damping ratio of 0.3. If the system is given an initial velocity of 1 m/sec, determine the maximum displacement of the system.

P5.16 A damped harmonic oscillator is described by the equation $\frac{d^2x}{dt^2} + 4\frac{dx}{dt} + 5x = 0$.

(a) Determine if the motion is underdamped, overdamped or critically damped.

(b) Find the angular frequency of the damped oscillations.

P5.17 A damped and driven oscillator satisfies the equation $\ddot{x} + 2C\dot{x} + \omega^2 x = F(t)$, where C and ω are positive constants. Find the equation of motion of the oscillator when the force $F(t)$ is given by the saw tooth form $F(t) = F_0 t$ for $-\pi < t < \pi$, and $F(t)$ is periodic with period 2π.

P5.18 A driven oscillator satisfies the equation $\ddot{x} + \omega^2 x = F_0 \cos[\omega(1 + \varepsilon)t]$, where ε is a positive constant. Show that the solution that satisfies the initial conditions $x = 0$ and $\dot{x} = 0$ when $t = 0$ is:

$$x(t) = \frac{F_0}{\varepsilon\left(1+\frac{1}{2}\varepsilon\right)\omega^2} \sin\frac{1}{2}\varepsilon\omega x(t) \sin\omega\left(1+\frac{1}{2}\varepsilon\right)t.$$

Sketch the graph of this solution for the case in which ε is small.

P5.19 Solve the differential equation for damped driven harmonic oscillator by a damped harmonic force: $F_d(t) = F_0 e^{-\alpha t} \cos(\omega t)$. Comment on its behavior as a function of time.

P5.20 The amplitude of a damped harmonic oscillator drops to e^{-1} of its initial value after n oscillations. Show that the ratio of the period of oscillation to the period of oscillation with no damping is approximately $1+\frac{1}{8\pi^2 n^2}$.

CHAPTER 6

The Variational Principle

When I was in high school, my Physics teacher—whose name was Mr Bader—called me down one day after physics class and said, 'You look bored; I want to tell you something interesting'. Then he told me something which I found absolutely fascinating, and have since then, always found fascinating. Every time the subject comes up, I work on it... The subject is this—the principle of least action.

—*Richard P. Feynman*

6.1 THE VARIATIONAL PRINCIPLE AND EULER–LAGRANGE'S EQUATION OF MOTION

In the previous chapters, we have worked with the Newtonian formulation of classical mechanics. Its central theme relies on the use of 'force' as the very cause of change in momentum. The cornerstone of Newtonian mechanics is this principle of causality. It is expressed in Newton's second law as a linear relation between the acceleration and the force. It is the result of the *equality* between the force and the rate of change of momentum. The relation between force and momentum is at the very heart of Newtonian formulation of classical mechanics. It turns out that classical mechanics has an *alternative* but *equivalent* formulation, based on what is known as the '*variational principle*', or '*Hamilton's principle of variation*'. In many universities, the principle of variation [1, 2] is introduced after a few years of college education in physics, and after a few courses on mechanics, including electrodynamics. However, there have been a few proposals [3, 4, 5] which recommend an *early* exposure in college curriculum to this fascinating approach. In fact, Richard Feynman was introduced to the principle of variation by his high school teacher, Mr Bader. Feynman went on to develop the path integral approach to the quantum theory based on the principle of variation. The path integral approach to quantum mechanics provides an *alternative* formulation of the quantum theory; it is equivalent to Heisenberg's uncertainty principle, and the Schrödinger equation. It has the capacity to describe a mechanical system and to account for how it evolves with time. The variational principle can be adapted to provide a *backward integration* of classical mechanics as an approximation toward the development of quantum theory. Newtonian formulation is not suitable for this purpose.

More need not be said to provide motivation introducing this topic even in a first college course on mechanics. We shall therefore jump into the subject right away. For parts this first section, we shall closely follow the Reference [6].

It nearly feels like a deep conspiracy of nature that the *principle of causality and determinism,* and the *variational principle*, produce results that are completely equivalent. The former makes no use of the 'variation principle', and the latter makes no use of 'force'. The '*variational principle*' is also known as the '*principle of extremum action*'. The general mathematical framework for the development and application of this technique is the 'calculus of variation'. Its beginnings can be traced to the solution provided by Isaac Newton to the famous brachistochrone problem, which was posed by Johann (also known as Jean or John) Bernoulli in 1696. I hasten to add, nevertheless, that the formulation of the principle of variation has a rich and intense history which dates back to periods well before the brachistochrone problem was posed by Bernoulli, as following: *Given two points S and F in a vertical plane* (Fig. 6.1), *what is the curve traced out by a particle acted on by gravity, which starts at S and reaches F, in the shortest time?* In Greek, 'brachistos' means 'the shortest', and 'chronos' means 'time'. The curve along which the object traverses in the least time is therefore known as the 'brachistochrone'.

The history of the principle of variation is as romantic as it is fascinating. In one of its earliest forms, it can be traced to the works of Pierre Fermat (1601–1665) who explained how a ray of light takes the path it does in reflection and refraction, when it meets the boundary of an interface between two media. The 'principle of extremum action', likewise explains how objects move the way they do. It explains how a stream of water running down a hill would take the path of steepest descent. It also accounts for the trajectories of any mechanical object with reference to initial conditions, and governed by an equation of motion that is *derivable* from the principle of extremum action. Rich contributions following Fermat's work by Maupertuis (1698–1759) and Euler (1707–1783) culminated in the works of Lagrange (1736–1813) and Hamilton (1805–1865), providing a robust framework for the entire discipline of classical mechanics.

Declared Bernoulli: "*I, Johann Bernoulli, address the most brilliant mathematicians in the world. Nothing is more attractive to intelligent people than an honest, challenging problem, whose possible solution will bestow fame and remain as a lasting monument. Following the example set by Pascal, Fermat, etc., I hope to gain the gratitude of the whole scientific community by placing before the finest mathematicians of our time a problem which will test their methods and the strength of their intellect. If someone communicates to me the solution of the proposed problem, I shall publicly declare him worthy of praise.*" *Bernoulli knew the solution, but he challenged other mathematicians to tackle this problem. Five solutions to this problem are famous, those by (1) Isaac Newton (1642–1727), (2) Johann's younger brother Jacob Bernoulli, also known as James or Jacques Bernoulli* (1655–1705), (3) G. W. Leibniz (1646–1716), (4) Guillaume François Antoine, Marquis de l'Hôpital (1661–1704), and (5) Johann Bernoulli himself. Newton's solution was published anonymously by the Royal Society (with the help of Charles Montague), but Johann Bernoulli immediately recognized that it was Newton's. He said: "we know the lion by his claw."

The variational principle rests on the premise similar to Galileo–Newtonian mechanics that an isolated mechanical state of the system with one degree of freedom is represented by the pair $(q, \dot{q})$, with q representing its position, and $\dot{q}$ its velocity. However, the position and velocity in this scheme have a broader sense which accommodates the following two distinctive features:

(i) q is the instantaneous coordinate of the object, called its 'generalized coordinate'. It provides the *essential* information about the position of the object under study, after accounting for any constraints which provide partial information about the position.

(ii) $\dot{q}$ provides the instantaneous velocity, called the 'generalized velocity'. It is the time-derivative of the generalized coordinate.

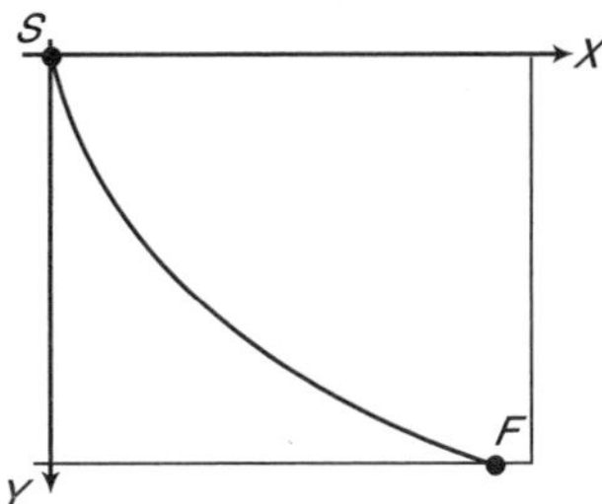

Fig. 6.1 The solution to the brachistochrone problem is *not* along the shortest distance, along the straight line, from the start-point S to the finish-point F. It is, rather, along a curved path, SF. The trajectory is best understood in the framework of the calculus of variation.

To understand the difference between the generalized coordinates and the commonly employed coordinates, such as Cartesian, or spherical polar or parabolic cylindrical coordinates (Chapter 2) in a Euclidean 3-dimensional space, we first consider a two-particle system, P_1 and P_2, located respectively at position vectors $\vec{r}_1$ and $\vec{r}_2$ · (Fig. 6.2). If these two particles were to move independently of each other, then we have to work with their six degrees of freedom, three for each of the two particles. These could be, for example, their Cartesian coordinates, (x_1, y_1, z_1) for P_1, and (x_2, y_2, z_2) for P_2. However, if there is a constraint that requires them to move in such a way that the distance d between these two particles is held *fixed*, by the condition $(\vec{r}_1 - \vec{r}_2)\cdot(\vec{r}_1 - \vec{r}_2) = d^2$, then the *degrees of freedom* reduce from six to five. The constraint that the two particles are at a fixed distance from each other, contains in it some information about the coordinates of the two particles, making it unnecessary to specify all the six coordinates. The positions of the pair of particles is in fact completely specified by just the five degrees of freedom $\{x_1, y_1, z_1, \lambda, \text{and } \delta\}$, as shown in Fig.6.2. Of these five degrees of freedom which determine the exact positions of both the particles P_1 and P_2, the first three have the dimensions of length, and the next two are angles. Furthermore, if a third particle P_3 is now to be placed such that its distances ℓ from P_1, and t from P_2, both are also fixed, then P_3 can only be on the intersection of a sphere centered at P_1 with radius ℓ, and another sphere which is centered at P_2 with radius t. The third particle can be placed anywhere on the circle of intersection of the two spheres, for example, at points P_3, P_3', etc., shown in Fig. 6.2. The 3-particle system, constrained as described above, therefore has only one degree of freedom corresponding to the point on the circle of intersection of the two spheres just described.

The difference between the generalized coordinates and the 'usual' physical configuration coordinates in the Euclidean space becomes important when we consider an N-particle system subjected to *constraints*. If there are m constraints, the number of degrees of freedom reduces from 3N to (3N-m). These independent degrees of freedom are then represented by (3N-m) 'generalized coordinates', as in the case of the 2-particles, and the 3-particles, systems depicted in Fig. 6.2. The degrees of freedom may not have the dimension of length, as pointed out earlier. There are two types of constraints, called holonomic, and non-holonomic. We shall not get into these details here. The objective of this chapter is not a detailed study of the dynamics of constrained many-particle systems using the generalized coordinates. Instead, we address the dynamics of a single point particle to only elucidate the *principle of variation* itself, which is a fundamentally different alternative to the Newtonian formulation of mechanics. This alternative approach produces completely equivalent results. As mentioned above, this formulation is well adapted to develop quantum theory, as was done by Richard Feynman. This is of course not to suggest that the variational method provides a derivation of quantum mechanics from the classical. That is simply not possible, but that is a much larger issue we shall not discuss here.

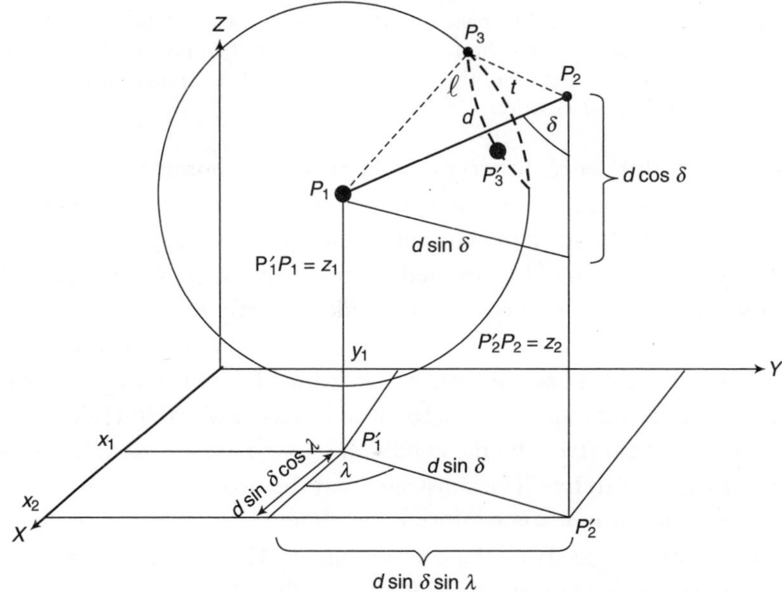

Fig. 6.2 The positions of both the particles, P_1 and P_2, are completely specified by just the five parameters, $\{x_1, y_1, z_1, \lambda, \text{and } \delta\}$, one less than the set of six Cartesian coordinates one would write for the two particles. The five variables are sufficient to completely specify the exact location of both the particles, since the distance between them is fixed. The degrees of freedom of the two-particle system are therefore five. Now, if a third particle is to be placed such that it is at a fixed distance from the first and another fixed distance from the second, we have only one degree of freedom left for this third particle.

We have already observed that the generalized coordinate q need not have the dimension of L. Its time-derivative, the generalized velocity $\dot{q}$ therefore does not necessarily have the dimensions LT^{-1}. The 'generalized momentum' in this formalism is defined as $p = \frac{\partial L}{\partial \dot{q}}$ where L is the *Lagrangian*, defined below. The generalized momentum is also therefore not necessarily the 'mass times velocity' as in Newtonian mechanics, although it may well be just that as a possible special case.

A definite integral over time, from time t_1 to t_2, of the *difference* between the kinetic energy T, and the potential energy of a system, i.e., $\int_{t_1}^{t_2} (T - V)\,dt$, can be constructed. The integrand $(T - V)$ is called the *Lagrangian*, after Joseph-Louis Lagrange (1736–1813). It is denoted by $L(q, \dot{q})$. The integral we have introduced is usually denoted by the letter S, given by

$$S = \int_{t_1}^{t_2} (T - V)\,dt = \int_{t_1}^{t_2} L(q, \dot{q})\,dt, \qquad (6.1a)$$

where $L(q, \dot{q}) = T - V$. (6.1b)

In the early days of variational calculus, the integral $S = \int_{t_1}^{t_2} L(q, \dot{q})\,dt$ used to be called 'Hamilton's principal function'. Following Feynman, it is now called 'action' [8]. Being the product 'energy × time', it is obvious that it has dimensions same as those of the angular momentum, ML^2T^{-1}. The well-known quantum of action, known as the *Planck's constant* (or, as the Planck–Einstein–Bose constant), is commonly represented by the letter h. It is one of the fundamental building blocks of the quantum theory.

The *principle of variation* is best understood as an *ansatz*, that a mechanical system evolves in such a way that '*action*' is an extremum. Often, this is a minimum, and hence the principle is also referred to as the principle of least action. Nevertheless, it is important to note that it can also be a maximum. In essence, 'action' is an extremum. The choice of the integrand of action, namely the Lagrangian L, as $T - V$ is a smart one, since it explicitly depends on the position q (as $V = V(q)$), and on the velocity $\dot{q}$, (as $T = T(\dot{q}^2)$. The quadratic dependence on velocity makes this choice particularly suitable for isotropic space. The temporal evolution of the mechanical system $(q, \dot{q})$, is of central concern in mechanics. The Lagrangian $L(q, \dot{q})$ has the complete information about the mechanical state of the system, since its argument $(q, \dot{q})$ characterizes the state completely, as we are already familiar with from the previous five chapters. You would expect, naturally, that the equation of motion which would describe the temporal evolution of the system is obtainable from the Lagrangian, as per the law of nature. Newtonian mechanics, and the principle of variation, both address essentially the same question: How does a mechanical system, described by $(q, \dot{q})$, evolve with time? Both aim at accounting for the trajectory taken by the system in the phase space from its start time at t_1, to the finish time t_2, of a time interval. The principle of variation is, however, in sharp contrast with the Newtonian explanation of this process. In Newtonian mechanics, the condition of sustenance of equilibrium is accounted for in terms of the inertia of the system (Galileo–Newton law of inertia), and it explains the departure from equilibrium as the linear response of the system to a stimulus. The qualitative and the quantitative statement of this approach is stated in Newton's second law, $\vec{F} = \frac{d\vec{p}}{dt} = m\vec{a}$. This law interprets the *rate* of change of momentum to be the very force (stimulus) which causes the change in momentum, resulting in an acceleration $\vec{a}$. The linear proportionality between the acceleration and the force is the mass (inertia) of the system. For a given mass, for greater the force, greater will be its acceleration. Thus the *linear stimulus–response* is central to Newtonian formulation of mechanics. On the other hand, the idea of stimulus (force) is *irrelevant* to the alternative formulation of classical mechanics which is based on the variational principle.

The following points highlight the similarities, and also the differences, between Newtonian formulation and the alternative exposition of classical mechanics based on the principle of variation:

- Both *Newtonian formulation of mechanics, and that based on the principle of variation* describe the mechanical state of the system by its position and momentum (or velocity) in an inertial frame of reference. Such a state is depicted by a point $\{(q, p)$ or $(q, \dot{q})\}$ in the phase space of that object in *both* the formulations. In the formulation based on the variational principle, the state of the mechanical system is represented equivalently by the Lagrangian, $L(q, \dot{q})$ which is completely determined by $(q, \dot{q})$.
- Newton's laws attribute the trajectory of a mechanical system in the phase space to (i) its inertia alone *when the mechanical state remains in equilibrium*, and (ii) when the object's equilibrium *changes*, to the force that acts on it. Thus, 'force' plays a pivotal role in accounting for the temporal evolution of the mechanical system.
- The principle of variation attributes the evolution of the mechanical system over a period of time to 'action' being an 'extremum'. 'Force' has absolutely no role in this formulation.

For 'action' to be an extremum, any *variation* in it when determined along *alternative* trajectories must be zero; i.e., $\delta S = 0$. This is akin to the derivative of a function of one-variable being zero at an extremum (maximum or minimum). We therefore ask the following question: what is the variation, or change, in the 'action' S, if we imagine the system to evolve along *alternative* paths? These paths are in the position–velocity phase space, from the initial time t_i, to the end time, (i.e., the final time) t_f, of the time-interval under consideration. Fig. 6.3 will make this discussion clear. The continuous line in this figure represents the actual path the mechanical system takes in the configuration space from the start position q_i at time t_i to reach the finish point q_f at time t_f. This is the path governed by the laws of nature which the object takes. It is of course possible to think of alternative paths, such as the one shown by the dashed line in this figure. We ask just 'how' can we describe unambiguously the particular (continuous) path the system takes, and not one of the alternative (dashed or dotted) paths? The principle of variation actually does *not* tell us *why* the system takes the path it does. Rather, it *describes precisely the path nature selects* according to its laws. Thus the principle describes 'what' is the law of nature and 'how' it operates, and not 'why' the law of nature is what it is. Newtonian mechanics described the law of nature using the 'stimulus–response' arguments; the principle of variation does so by stating that the evolution of the mechanical system is along the path that makes 'action' (Eq. 6.1) an extremum. Physics teaches us what the laws of nature are and how they operate, not why the laws are what they are.

In a more advanced course, you will learn how the description of a state of a system by a point in phase space has to be refined further, since the laws of nature are even more complex, and even largely counter-intuitive. Nature demands a more accurate description. The improved model to describe the law of nature is of course the quantum theory. Richard Feynman's, 'path integral approach to quantum theory' is based on a suggestion due to Paul Adrien Maurice Dirac. Feynman's approach also accounts for the fact that the path projected by the Lagrange–Hamilton 'classical' variational principle is such an *excellent approximation* for most systems we deal with in our day-to-day life, even if classical mechanics is strictly untenable. It is interesting to observe that the formulations of classical mechanics by Newton, Lagrange, and Hamilton are equivalent to each other,and that of quantum mechanics by Heisenberg, Schrodinger and Feynman also are equivalent to each other.

There are infinite different possibilities for alternative paths. Only two of these variants are shown by the dashed and the dotted curves in Fig. 6.3. These are only hypothetical paths that can be mathematically conjectured. The mechanical system would *not* evolve along these alternative paths, but nothing stops us from imagining alternative pictorial trajectories and then ask which of these alternative paths is selected by nature for the system to evolve. We can think of these alternative paths generated by a variation in some control parameter, γ. Each alternative path can then be construed as determined by one of the infinite different values that γ can take. The classical principle of variation tells us that nature selects the path, for some particular value of γ, for which the 'action, S' is an *extremum*.

For the path selected by nature, action being 'stationary', a *change* in it is zero,

$$\text{i.e.,} \quad \delta S = \int_{t_1}^{t_2} L(q+\delta q, \dot{q}+\delta\dot{q})\,dt - \int_{t_1}^{t_2} L(q,\dot{q})\,dt = 0. \tag{6.2}$$

From Eq. 6.2, it follows that

$$0 = \delta S = \int_{t_1}^{t_2} \left\{ \frac{\partial L}{\partial q}\delta q + \frac{\partial L}{\partial \dot{q}}\underline{\delta\dot{q}} \right\} dt = \int_{t_1}^{t_2} \left\{ \frac{\partial L}{\partial q}\delta q + \underline{\frac{\partial L}{\partial \dot{q}}\delta\left(\frac{d}{dt}q\right)} \right\} dt. \tag{6.3a}$$

In the last term (which is underlined only to identify it for our discussion), the two operations, $\frac{d}{dt}$ and δ commute, being independent of each other. The first operator provides the derivative with respect to time, whereas the second obtains variation of the path in the phase space. Hence,

$$0 = \int_{t_1}^{t_2} \left\{ \frac{\partial L}{\partial q}\delta q + \frac{\partial L}{\partial \dot{q}}\frac{d}{dt}(\delta q) \right\} dt = \int_{t_1}^{t_2} \left\{ \frac{\partial L}{\partial q}\delta q \right\} dt + \int_{t_1}^{t_2} \left\{ \left(\frac{\partial L}{\partial \dot{q}}\right)\left(\frac{d(\delta q)}{dt}\right) \right\} dt. \tag{6.3b}$$

In the last term, we integrate the product of two factors in the integrand using integration by parts. The result is

$$0=\int_{t_1}^{t_2}\left\{\frac{\partial L}{\partial q}\delta q\right\}dt+\left[\frac{\partial L}{\partial \dot{q}}\delta q\right]_{t_1}^{t_2}-\int_{t_1}^{t_2}\left\{\frac{d}{dt}\left(\frac{\partial L}{\partial \dot{q}}\right)\delta q\right\}dt, \tag{6.3d}$$

$$\text{i.e.,}\quad 0=\int_{t_1}^{t_2}\left\{\frac{\partial L}{\partial q}\delta q\right\}dt+\left\{\left[\frac{\partial L}{\partial \dot{q}}\delta q\right]_{t2}-\left[\frac{\partial L}{\partial \dot{q}}\delta q\right]_{t1}\right\}-\int_{t_1}^{t_2}\left\{\frac{d}{dt}\left(\frac{\partial L}{\partial \dot{q}}\right)\delta q\right\}dt. \tag{6.3e}$$

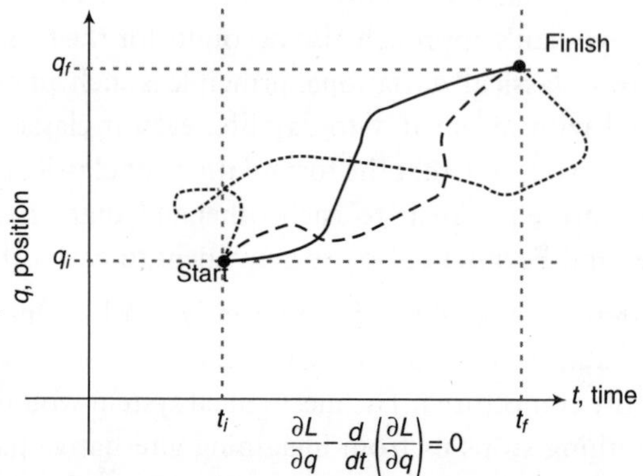

Fig. 6.3 We consider temporal evolution of the mechanical system from the initial time t_i to the final time t_f over a time interval $(t_f - t_i)$. The dark continuous line shows the trajectory which the system takes under laws of nature. The other two curves show hypothetical alternative paths of reaching the same destination from the same initial state. The end-points of these trajectories are therefore fixed, but the paths between the end-points can be varied.

The middle term vanishes, since $[\delta q]_{\underset{t_2}{\text{at}}} = 0 = [\delta q]_{\underset{t_1}{\text{at}}}$. After all, we are considering alternative paths between essentially the same initial point at the start, and the same destination point at the end of the interval between t_1 and t_2 (Fig. 6.3).

$$\text{Hence, we get } 0=\int_{t_1}^{t_2}\left\{\frac{\partial L}{\partial q}-\frac{d}{dt}\left(\frac{\partial L}{\partial \dot{q}}\right)\right\}\delta q dt. \tag{6.4}$$

Now, δq is an *arbitrary* change in the path. Before we discuss (below) more formally the mathematical methodology to discuss how this change is effected using a parameter γ, we note that the *condition* under which Eq. 6.4 would hold for *arbitrary* δq is

$$\frac{\partial L}{\partial q}-\frac{d}{dt}\left(\frac{\partial L}{\partial \dot{q}}\right)=0. \tag{6.5}$$

Equation 6.5 is known as the Euler–Lagrange's equation, or simply as the Lagrange equation. It is an 'equation of motion', just like the one in Newton's second law, but for the *degree of freedom* q rather than for each physical coordinate. If there are N generalized coordinates, *one for each degrees of freedom*, $\{q_i, i = 1, 2, 3, ..., N\}$, then each q_i satisfies the Lagrange's equation corresponding to it. In such a case, q in Eq. 6.5 must be replaced by q_i.

Leonhard Euler (1707–1783) is one of the most celebrated mathematician–physicist of all times respected for his ground-breaking contributions to a large number of disciplines, including mechanics, optics, fluid dynamics, musicology, etc., apart from fundamental mathematics. We now provide a more *formal* basis for the result in Eq. 6.5, *by explicitly considering a parameter γ which can be varied to generate alternative paths for the system's evolution.*

We consider a function $\phi_\gamma(\xi)$ of an independent variable ξ. Just how the function $\phi_\gamma(\xi)$ depends on ξ is construed to be controlled by a continuous parameter γ, which therefore appears as subscript. For example, we may write

$$\phi_\gamma(\xi) = \phi_{\gamma=0}(\xi) + \gamma\chi(\xi). \tag{6.6a}$$

We now consider another function ψ_γ of ξ which may have a rather complicated dependence on ξ. Nonetheless, we consider this dependence to be expressible in a simple manner, as a combination of three terms:

(i) direct dependence of ψ_γ on ξ, as some simple function of ξ,

(ii) dependence of ψ_γ on a *function* of ξ, namely $\phi_\gamma(\xi)$. The function $\phi_\gamma(\xi)$, in turn, depends on ξ in a manner that is determined by the control parameter γ,

(iii) a dependence of ψ_γ on the derivative, $\frac{\partial}{\partial\xi}[\phi_\gamma(\xi)]$, which is the rate of change of $\phi_\gamma(\xi)$ with respect to ξ.

We can therefore write $\psi_\gamma(\xi)$ in terms of the above 3 *arguments*:

$$\psi_\gamma = \psi_\gamma\left(\phi_\gamma(\xi), \frac{\partial}{\partial\xi}[\phi_\gamma(\xi)], \xi\right). \tag{6.6b}$$

The third argument of ψ_γ listed in Eq. 6.6b indicates the *explicit* dependence on ξ of ψ_γ, which as mentioned above is written as a simple function of ξ. In addition to such an explicit dependence of ψ_γ on ξ, the first two arguments in Eq. 6.6b stand for an *implicit* dependence of ψ_γ on ξ, through its dependence on $\phi_\gamma(\xi)$ and on $\phi'_\gamma = \frac{\partial\phi_\gamma}{\partial\xi}$. All the three arguments are necessary to express the *complete* dependence of ψ_γ on ξ. Leaving out one or the other would limit our capacity to know ψ_γ in terms of ξ. However, not all of the three arguments may *always* be required; in some particular cases, one or the other arguments may not be required.

We now consider the following definite integral of $\psi_\gamma(\xi)$ over the independent variable, ξ:

$$I_\gamma = \int_{\xi_1}^{\xi_2} \psi_\gamma\left(\phi_\gamma(\xi), \frac{\partial}{\partial\xi}[\phi_\gamma(\xi)], \xi\right) d\xi. \tag{6.6c}$$

The Euler equation is a mathematical condition that must be satisfied for the integral I_γ to be an extremum with regard to the variations in its value for different γ. The parameter γ influences how $\phi_\gamma(\xi)$ depends on ξ. The extremum can be either a maximum or a minimum, or even a saddle point. Specifically, it is the stationary property of the extremum that is underscored in this analysis. We set the zero of the γ scale where I_γ has an extremum. Any *change* in the value of γ (from $\gamma = 0$) would therefore result in increasing (or decreasing) the value of I_γ, if the extremum is a minimum (or maximum). As we change γ, the rate at which the integral I_γ changes with respect to γ is given by

$$\frac{\partial I_\gamma}{\partial \gamma} = \frac{\partial}{\partial \gamma}\int_{\xi_1}^{\xi_2} \psi_\gamma\left(\phi_\gamma(\xi), \frac{d\phi_\gamma(\xi)}{d\xi}, \xi\right) d\xi = \int_{\xi_1}^{\xi_2}\left[\frac{\partial}{\partial \gamma}\psi_\gamma\left(\phi_\gamma(\xi), \frac{\partial \phi_\gamma(\xi)}{\partial \xi}, \xi\right)\right] d\xi. \tag{6.6d1}$$

Hence,

$$\frac{\partial I_\gamma}{\partial \gamma} = \int_{\xi_1}^{\xi_2}\left[\frac{\partial \psi_\gamma}{\partial \phi_\gamma}\frac{\partial \phi_\gamma}{\partial \gamma} + \frac{\partial \psi_\gamma}{\partial \phi'_\gamma}\frac{\partial \phi'_\gamma}{\partial \gamma}\right] d\xi, \tag{6.6d2}$$

wherein, we have denoted the derivative with respect to ξ by prime.

From Eq. 6.6a we have

$$\frac{\partial \phi_\gamma}{\partial \xi} = \frac{\partial}{\partial \xi}\phi_{\gamma=0}(\xi) + \gamma\frac{\partial \chi}{\partial \xi},$$

$$\text{i.e.,} \quad \phi'_\gamma = \frac{\partial \phi_{\gamma=0}(\xi)}{\partial \xi} + \gamma\chi'. \tag{6.6e1}$$

Hence, differentiating ϕ_γ' with respect to the continuous parameter γ, which is the control parameter in Eq. 6.6a, we get

$$\frac{\partial}{\partial \gamma}\phi'_\gamma(\xi) = \chi'(\xi). \tag{6.6e2}$$

We also have, from Eq. 6.6(a), $\frac{\partial \phi_\gamma}{\partial \gamma} = \chi(\xi)$. (6.6e3)

Accordingly, using Eqs. 6.6e2 and 6.6e3 in Eq. 6.6d2, we get

$$\frac{\partial I_\gamma}{\partial \gamma} = \int_{\xi_1}^{\xi_2}\left[\frac{\partial \psi_\gamma}{\partial \phi_\gamma}\chi(\xi) + \frac{\partial \psi_\gamma}{\partial \phi'_\gamma}\chi'(\xi)\right] d\xi. \tag{6 .6f1}$$

We therefore get using Eq. 6.6e3 and then interchanging the derivation operator with respect to ξ and γ in Eq. 6.6f1,

$$\frac{\partial I_\gamma}{\partial \gamma} = \int_{\xi_1}^{\xi_2}\left[\frac{\partial \psi_\gamma}{\partial \phi_\gamma}\chi(\xi)\right] d\xi + \left[\frac{\partial \psi_\gamma}{\partial \phi'_\gamma}\chi(\xi)\right]_{\xi_1}^{\xi_2} - \int_{\xi_1}^{\xi_2}\left[\frac{d}{d\xi}\left(\frac{\partial \psi_\gamma}{\partial \phi'_\gamma}\right)\chi(\xi)\right] d\xi, \tag{6.6f2}$$

wherein we have integrated the second terms on the right hand side of Eq. 6.6f1 by parts.

Now, $\chi(\xi_2) = 0 = \chi(\xi_1)$, ξ_1 and ξ_2 being the fixed end-points of the range of the definite integral. Hence the term in the middle of the right hand side of Eq. 6.6f2 is zero.

$$\text{Hence, } \frac{\partial I_\gamma}{\partial \gamma} = \int_{\xi_1}^{\xi_2} \left[\frac{\partial \psi_\gamma}{\partial \phi_\gamma} - \frac{d}{d\xi}\left(\frac{\partial \psi_\gamma}{\partial \phi'_\gamma} \right) \right] \chi(\xi)\, d\xi . \tag{6.6f3}$$

Now, for I_γ to be an extremum for arbitrary $\chi(\xi)$, the necessary condition is $\frac{\partial I_\gamma}{\partial \gamma} = 0$. Hence,

$$\left[\frac{\partial \psi_\gamma}{\partial \phi_\gamma} - \frac{d}{d\xi}\left(\frac{\partial \psi_\gamma}{\partial \phi'_\gamma} \right) \right] = 0. \tag{6.6g}$$

Eq. 6.6g is a necessary condition for the integral I_γ to have a stationary value, and is known as the Euler equation.

Identifying the independent variable ϕ (Eq. 6.6a) as the q (generalized coordinate), ξ as time t, ψ (Eq. 6.6b) as the Lagrangian L, and I (Eq. 6.6c) as the action S, we get the Euler–Lagrange equation, which is Eq. 6.5.

6.2 THE BRACHISTOCHRONE

We now address the brachistochrone problem mentioned in Section 6.1 using the Lagrange–Euler formalism. Let us take a quick look at Fig. 6.1. We consider a mass m to go down under gravity, but *constrained* to move along frictionless curves. For example, the object may be a point mass sliding down along a curved surface, or it may be a bead sliding down on a curved wire. One can of course think about infinite different curves between the start point S and the finish point F. The challenge posed by Bernoulli was to determine the specific curve that would enable the object of mass m to reach the point F, in the *least time*, if it is released under gravity from the start point S at zero initial speed. We use the method of variational calculus described above to determine this path. We take the zero of the energy scale to be given by the energy that the particle has at rest, at the start point S (Fig. 6.1). As the particle falls under gravity, it would pick up kinetic energy, while its gravitational potential energy would become *negative;* the total energy remaining constant. The potential energy at a point where the particle's coordinate is y is then $V(y) = -mgy$, and the kinetic energy would be $T(y) = \frac{1}{2}mv^2 = -V(y) = mgy$. Note that the Y-axis is pointing downward in Fig. 6.1. The particle's speed at any instant is $v = \sqrt{2gy}$. The time interval δt taken by the particle to traverse a tiny (infinitesimal) distance δs is then

$$\delta t = \frac{\delta s}{v} = \frac{\sqrt{\delta x^2 + \delta y^2}}{v} = \frac{\delta x\sqrt{1 + \left(\frac{\delta y}{\delta x} \right)^2}}{\sqrt{2gy}}. \tag{6.7}$$

From the above equation, we see that the shape of the curve along which the particle would traverse under gravity in the least time can be parametrized in terms of the height function y, instead of the time parameter t. Since the Y-axis is pointing downward, you might wish to call y as the *depth* rather than the height. The *total time* t_F taken by the particle to reach the point F at which the x-coordinate is x_F, when it is released at t = 0 at the start-point S (Fig. 6.1) with zero velocity, therefore is

$$\tau = \int_{t_S=0}^{t_F} dt = \int_{x_S=0}^{x_F} \left(\frac{\sqrt{1+\left(\frac{dy}{dx}\right)^2}}{\sqrt{2gy}} \right) dx = \int_0^{x_F} \left(\frac{\sqrt{1+y'^2}}{\sqrt{2gy}} \right) dx, \quad \left\{ \text{with } y' = \frac{dy}{dx} \right\}. \tag{6.8a}$$

Hence,
$$\tau = \int_0^{x_F} \Psi(y, y')\, dx, \tag{6.8b}$$

where
$$\Psi(y, y') = \frac{\sqrt{1+y'^2}}{\sqrt{2gy}}. \tag{6.8c}$$

The advantage of writing the time-interval in terms of the integral over the X-coordinate is that it enables us parametrize the time-interval in terms of a relationship between x and y involving the derivative $\frac{dy}{dx}$ which has all the information about the shape function, which is exactly what we wish to determine. As the particle descends, its x-coordinate *increases* (toward the right in Fig. 6.1) at each instant of time (since the X-axis is pointing toward the right in Fig. 6.1). We therefore examine the x-dependence of the integrand $\Psi(y, y')$ of Eq. 6.8b:

$$\frac{d}{dx}[\Psi(y,y')] = \frac{\partial \Psi}{\partial y}\frac{dy}{dx} + \frac{\partial \Psi}{\partial y'}\frac{dy'}{dx} = \frac{\partial \Psi}{\partial y} y' + \frac{\partial \Psi}{\partial y'} y''. \tag{6.9}$$

Thus, bringing all the terms in Eq. 6.9 to one side, we get

$$0 = \frac{d\Psi}{dx} - \frac{\partial \Psi}{\partial y'} y'' - y' \frac{\partial \Psi}{\partial y}, \tag{6.10a}$$

i.e.,
$$0 = \frac{d}{dx}\left[\Psi - y' \frac{\partial \Psi}{\partial y'} \right]. \tag{6.10b}$$

We therefore conclude that $\left[\Psi - y' \frac{\partial \Psi}{\partial y'} \right] = \kappa$, a constant. $\quad$ (6.11)

Equation 6.11 will provide the equation to the trajectory for the fastest path for the particle to get to the finish point F from the start point S under gravity, as detailed below.

Now,
$$\frac{\partial \Psi}{\partial y'} = \frac{\partial}{\partial y'} \frac{\sqrt{1+y'^2}}{\sqrt{2gy}} = \frac{y'}{\sqrt{2gy\,(1+y'^2)}}. \tag{6.12}$$

Hence,
$$\frac{\sqrt{1+y'^2}}{\sqrt{2gy}} - y' \left(\frac{y'}{\sqrt{2gy\,(1+y'^2)}} \right) = \kappa, \tag{6.13a}$$

i.e., $$\frac{(1+y'^2)-y'^2}{\sqrt{2gy\,(1+y'^2)}}=\frac{1}{\sqrt{2gy\,(1+y'^2)}}=\kappa. \tag{6.13b}$$

Hence, $$y(1+y'^2)=y+yy'^2=\frac{1}{2g\kappa^2}=2\beta,\ \text{where } 2\beta \text{ is another constant.} \tag{6.13c}$$

Hence, $$y'=\sqrt{\frac{2\beta-y}{y}}. \tag{6.14}$$

The derivative $y'=\frac{dy}{dx}$ must remain positive, as there is no interaction that can reverse it. Hence, we have the inequality,

$$2\beta \geq y \leq 0. \tag{6.15a}$$

We parametrize the relation given in Eq. 6.14 by changing the variable $y \to \theta$ by defining

$$y=\beta-\beta\cos\theta=\beta(1-\cos\theta). \tag{6.15b}$$

At $y = 0$, when the particle is released at the start point S, θ must be zero.

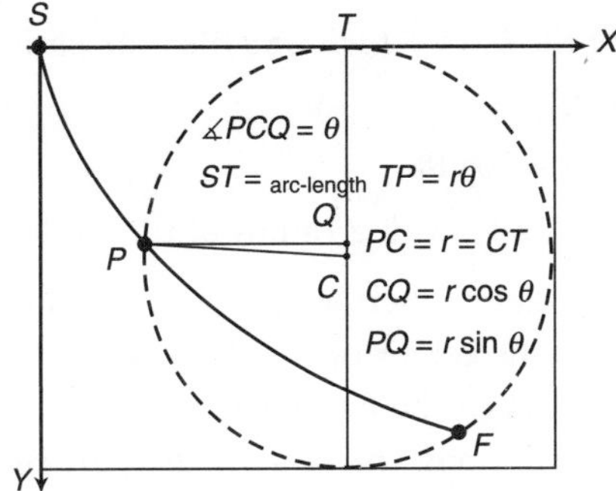

Fig. 6.4 The solution to the brachistochrone problem turns out to be a cycloid, shown in this figure. The (x, y) coordinates are given by Eq. 6.17 in terms of the angle θ.

Using Eq. 6.14 and Eq. 6.15b,

$$\frac{dy}{dx}=y'=\sqrt{\frac{2\beta-(\beta-\beta\cos\theta)}{(\beta-\beta\cos\theta)}}=\sqrt{\frac{\beta+\beta\cos\theta}{\beta-\beta\cos\theta}}=\sqrt{\frac{(1+\cos\theta)}{(1-\cos\theta)}}=\sqrt{\frac{\cos^2\frac{\theta}{2}}{\sin^2\frac{\theta}{2}}}=\frac{\cos\frac{\theta}{2}}{\sin\frac{\theta}{2}}. \tag{6.16a}$$

i.e., $$dy=\frac{\cos\frac{\theta}{2}}{\sin\frac{\theta}{2}}\,dx. \tag{6.16b}$$

Likewise from Eq. 6.15b, we get

$$dy = \beta \sin \theta \, d\theta = 2\beta \sin \frac{\theta}{2} \cos \frac{\theta}{2} \, d\theta. \tag{6.16c}$$

We therefore get

$$2\beta \sin^2 \frac{\theta}{2} \, d\theta = dx, \tag{6.17a}$$

i.e.,

$$dx = \beta(1 - \cos \theta) d\theta. \tag{6.17b}$$

Integrating the above expression, we get,

$$x = \beta (\theta - \sin \theta) + C, \tag{6.17c}$$

where C is a constant of integration. From the initial condition $\theta = 0$, when $x = 0$, we get $C = 0$. The other constant to be determined is β, and this is also easily determined since the path must pass through the point F for which the coordinates are known, say (x_F, y_F).

Hence, $x_F = \beta (\theta_F - \sin \theta_F)$,

and $y_F = \beta (1 - \cos \theta_F)$.

The equation to the path of the *brachistochrone* (i.e., the path along which the time taken is the least) is therefore given by

$$x = r (\theta - \sin \theta); \; y = r (1 - \cos \theta), \tag{6.18}$$

which defines a 'cycloid' (Fig. 6.4), with a radius of the circle given by $r = \beta$.

A cycloid, as is well known, can be described as the locus of a point on the rim of a circle, such as the bicycle wheel, while the center of the circle itself traverses along a straight line. From Fig. 6.4, we see that $\theta = 0$ when the point P is at the origin S. As the circle rolls down through a horizontal distance ST along the X-axis, the particle would slide under gravity along the arc SP. The arc-length PT of the circle is of length $r\theta$, since the radius of the circle is $r = \beta$. The Cartesian coordinates of the point P are given by Eq. 6.18. The coordinates (x_C, y_C) of the center of the circle C, are given by $x_C = ST = (arc)TP = r\theta$, equal to the arc-length on the circle, and $y_C = CT = r$ (the radius of the circle). The trajectory of the particle along which it reaches the finish-point F fastest under gravity in the least time, is the cycloid curve SPF (Fig. 6.4).

To illustrate an experimental verification of the solution to the brachistochrone problem, we refer to the demonstration described in Reference [6]. In Fig. 6.5 is shown a small brachistochrone (cycloid), along with similarly fabricated three other slopes from an acrylic $(C_5O_2H_8)_n$ sheet. The apparatus so fabricated are shown in Fig. 6.5. It is easy to build these models. Some students may wish to fabricate such apparatus in their college laboratories and workshops. To machine the brachistochrone, the curvature of an acrylic sheet must be contoured to produce the exact cycloid. The apparatus can then be fabricated using, for example, a CNC–VMC (Computerized Numerical Controller–Vertical Milling Machine).

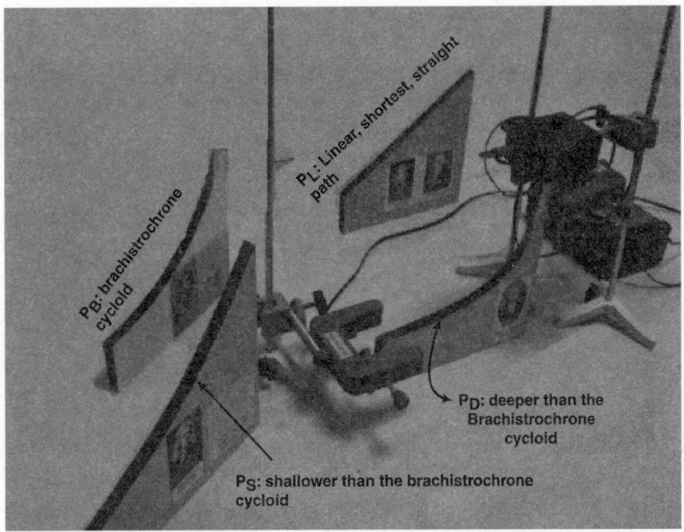

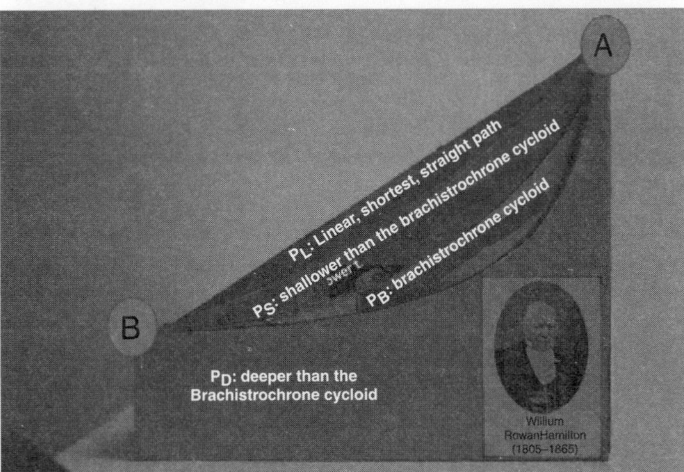

Fig. 6.5 The experimental setup (top panel) to perform the brachistochrone experiment. The time intervals taken by an object to come down under gravity alone from identical points '*A*' to '*B*' (right panel) are measured using two photogates, seen in the top panel (also seen in Fig. 6.6). The experiment is done on four different paths, P_L, P_S, P_B and P_D as described in the text. The experiment is described in the text. It is based on Ref.[6].

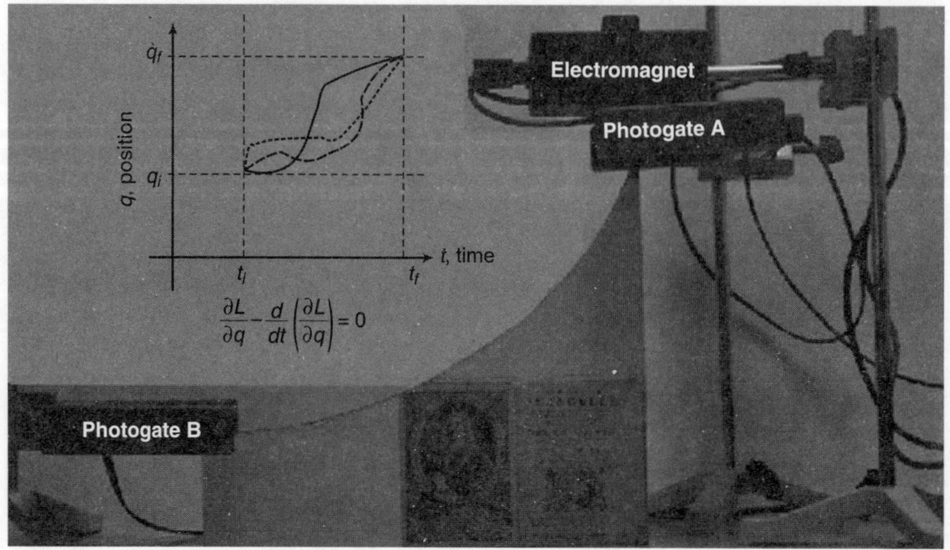

Fig. 6.6 The *brachistochrone model* placed for the conduct of the experiment to determine the time-interval δt to measure the time taken by a mass m to traverse under gravity from the point 'A' to 'B' shown in Fig. 6.5. A metallic object is released from rest (zero initial speed) by an electromagnet at the point A. The photogates A and B respectively record the start and end instants of time from whose difference the interval δt is determined.

The three additional slides, fabricated with different curvatures for comparison are labelled as (i) P_L, (ii) P_S and (iii) P_D, in Fig. 6.5. They represent different curvatures, respectively (i) the linear (straight) path between the points 'A' and 'B', (ii) a path shallower than the cycloid, and (iii) a path deeper than the cycloid. One can then drop a small mass at the top, from point A shown in Fig. 6.5, at zero velocity and record the time it takes to reach the point B under gravity. The time interval can be recorded using the photogate, as shown in Fig. 6.6. The experiment would reveal that the time taken for the mass to roll down under gravity would be the shortest for the cycloid path.

In the physical world around us, every object takes its paths along which its 'action' is minimum (rather, an 'extremum'). Two such trajectories (also from Reference [6]) are shown in Fig. 6.7. These may be called the '*brachistochrone carom boards*'. If a striker strikes the parabolic reflector along a line parallel to the axis of the parabola, then after reflection it passes through the focus of the parabola. Likewise, if a striker is hit from one of the two foci of an elliptic carom board, then after bouncing off the boundary, it passes through the other focus of an elliptic carom board, as shown in Fig. 6.7. The brachistochrone carom boards can also be easily fabricated in a college workshop.

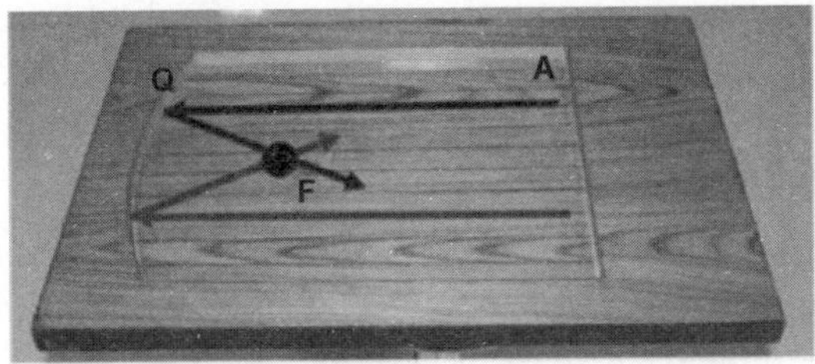

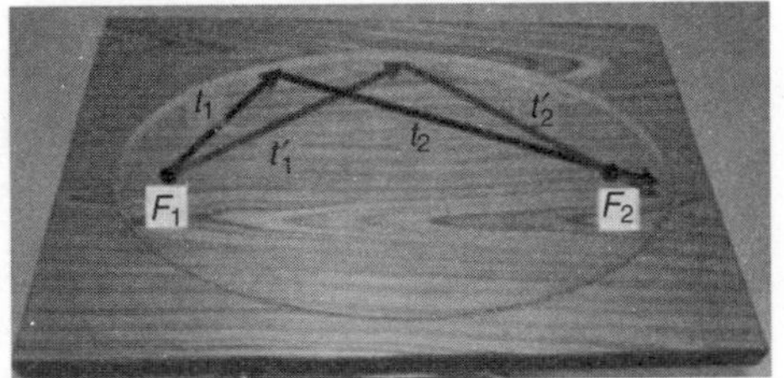

Fig. 6.7 The brachistochrone carom boards. Carom board striker incident along *any* line parallel to AQ in the parabolic reflector (on the left) is reflected back by the parabolic surface through the focus F. In the elliptic reflector (on the right), a striker struck *from* the focus F_1 always gets reflected through the second focus F_2 regardless of which point on the curved boundary it was aimed at.

The results of the carom board experiments, and of the brachistochrone experiment, are akin to the path taken by a ray of light, as is well known in the laws of optical reflection and refraction. These are completely determined by the variational principle of extremum action, as discussed above.

6.3 SUPREMACY OF THE LAGRANGIAN FORMULATION OF MECHANICS

Even as the variational principle makes no use of the idea of force, it can be easily related to the calculus of variation, since it is derivable from the Lagrangian:

$$\frac{\partial L}{\partial q} = -\frac{\partial V}{\partial q} = F, \tag{6.19a}$$

Likewise, the second term in the Euler–Lagrange's equation (Eq. 6.5) is

$$\frac{d}{dt}\left(\frac{\partial L}{\partial \dot{q}}\right) = \frac{dp}{dt} = \dot{p}, \tag{6.19b}$$

$$p = \frac{\partial L}{\partial \dot{q}}, \text{ being the 'generalized momentum'.}$$

Essentially, Eqs. 6.5 and 6.19 together affirm the equivalence of the Lagrange's equation with that of Newton's second law, namely, that the force is equal to the rate of change of momentum. This is more than a happy coincidence. Would it not be terrible if the equation of motion based on Newton's principle of causality and determinism were to be incompatible with the Lagrange–Hamilton formulation of classical mechanics, based on the principle of variation? It not amazing that these two approaches, based on completely different ideas, produce completely equivalent equations of motion for a classical mechanical system? Though equivalent, the variational method offers three impressive advantages over Newtonian method:

(a) The equation of motion resulting from the principle of variation is applicable to the 'generalized coordinates', after eliminating constraints, if any. The generalized coordinates have been introduced above, but a detailed study of 'holonomic' and 'non-holonomic' constraints is left for a slightly more advanced study of this subject.

(b) The formulation based on the principle of variation is well adapted to the development of quantum mechanics. It provides the foundation for the path integral approach to the quantum theory, due to Richard Feynman [7,8].

(c) The equation of motion, namely the Lagrange's equation which we obtained from the variational principle readily illustrates the powerful connection between symmetry and a conservation principle. This connection, capsuled in Noether's theorem, provided the central theme for Chapter 1. In particular, we see that when the potential is independent of the coordinate, it follows from Eq. 6.5 (and using Eq. 6.19), that the momentum is conserved. We see this from the fact that time-derivative of the momentum is zero: $\dot{p} = \frac{d}{dt}\left(\frac{\partial L}{\partial \dot{q}}\right) = 0$ *because* $\frac{\partial L}{\partial q} = 0$. The invariance in the value of potential (and hence in the value of the Lagrangian) as the coordinate is changed, is a categoric expression of symmetry.

When the Lagrangian is independent of a coordinate q that coordinate is called a *cyclic coordinate*. Since the momentum p is defined with reference to each particular coordinate by the relation $p = \frac{\partial L}{\partial \dot{q}}$, the pair $(q, p) = \left(q, \frac{\partial L}{\partial \dot{q}}\right)$ is called a pair of canonically conjugate variables. The independence of the Lagrangian with respect to a coordinate expresses a symmetry (invariance) with respect to any change in that coordinate. The conclusion embedded in point (c) above is thus an expression of the essence in Noether's theorem: associated with every symmetry, there is a conservation principle, and vice versa.

We now illustrate the power of the Lagrangian formulation by solving the equation of motion for a few dynamical systems. In particular, we shall use the Lagrange's equation of motion to solve the equation of state for (a) a cycloidal pendulum, also known as Huygens' pendulum, and (b) a bead riding on a circular hoop. The examples chosen have a large occurrence in problems of interest in physics and engineering, and also in natural systems that we encounter.

(a) Cycloidal pendulum

We have seen in Chapter 4 that the time period for a simple pendulum is $T_0 = 2\pi\sqrt{\frac{l}{g}}$. This expression for the periodic time of the pendulum is valid only for *small* oscillations. i.e., if the amplitude of oscillation is not large. It is valid when the largest angle the pendulum's suspension string makes with the vertical is rather small, for which the approximation $\sin\theta \approx \theta$ can be applied. For a mass-spring oscillator, the corresponding expression is valid when the cubic (and higher) terms in Eq. 4.2 are ignored. If θ is not so small that powers of the θ of the order 3 or more in the power series expansion of $\sin\theta$ are cannot be ignored, then the periodic time of oscillation will not be given by the simple expression for T_0. Instead, it will be given by

$$T = T_0\left(1 + \frac{\theta_0^2}{16} + \frac{11\theta_0^4}{3072} + \frac{173\theta_0^6}{737280} + \cdots\right). \tag{6.20}$$

The result presented in Eq. 6.20 is manifestly complicated. It is derived using 'elliptic integral' in the appendix to this chapter (see Eq. 6A.13b). Now, T_0 is useful to device a clock. It is independent of the initial angular displacement, θ_0. For *not-so-small* oscillations, the time-period would be given by Eq. 6.20, and hence it would depend on the actual magnitude of θ_0. The pendulum will need to be calibrated separately for the magnitude of the angular displacement *every time* it is set into oscillations. The re-calibration will be required *even at the same location on Earth*. Christian Huygens (1629–1695), a Dutch physicist–mathematician who was a contemporary of Isaac Newton, recognized the fact that the trajectory of the pendulum bob in *small* oscillations is a *tiny* arc of an exact circle. He realized that it is exactly this property that makes it possible to employ the approximation $\sin\theta \approx \theta$ which is crucial in getting the time period of the oscillator to be determined *only* by its length, and the acceleration due to gravity, independent of the amplitude. Huygens paid attention to the fact that the smallness of the amplitude was necessary to validate the approximation, $\sin\theta \approx \theta$. Now, since the bob of the pendulum traverses on a circle, the angle between the tangent to its trajectory and the horizontal is essentially the same between the pendulum's string and the vertical. This would *not* be the case if the bob were to move on a curve which is *not* circular. Huygens therefore concluded that if the time period is *not* to change *even* if the amplitude of the oscillation is *large* and the approximation $\sin\theta \approx \theta$ would *not* be applicable, then the arc on which the bob moves must *not* be circular.

Huygens himself made many clocks. He used to develop clock designs. Clockmakers at that time had already learned to *constrain* the pendulum string's motion by installing short metal (or wooden) plates on either side of the string, near the point of suspension. The plates, called chops, constrained and altered the string's movement, and thereby controlled the periodic time. Huygens invented a design in which the chops used were two semi-cycloids, shown in Fig. 6.8. When a pendulum is suspended from the cusp between the two chops, the effective point of suspension slides on the surface of the chops. The pendulum's bob then traverses on a curve, which is a congruent cycloid, since the *involute of a cycloid is a cycloid that is congruent to the original cycloid, but shifted from the original.*

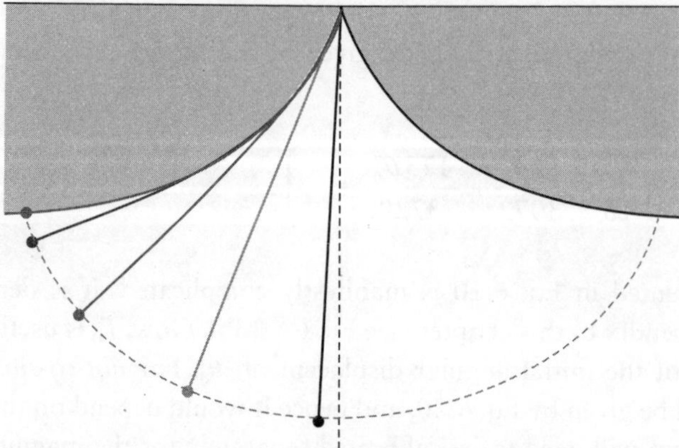

Fig. 6.8a Huygens' cycloidal pendulum [9]. The five pendula shown in this figure, each, have the same string-length but different amplitudes. The strings wrap around, gently over, the cycloid chops. The five pendula shown in this figure are shown at an instant when each is farthest in its respective oscillation from the equilibrium. Note how the effective length of the five different pendula drops the farther the bob moves away from the equilibrium.

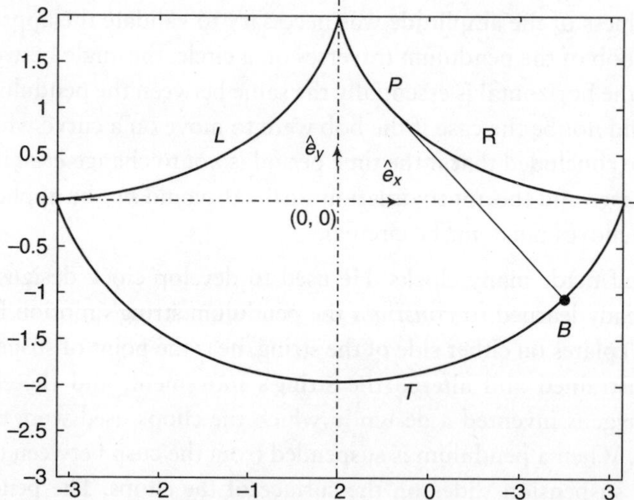

Fig. 6.8b The point of suspension P slides on the right arc R when the bob swings toward the right of the equilibrium. It slides on the left arc L when it oscillates and moves toward the left of the equilibrium. The bob B itself oscillates on the cycloid T. The net effect of this ingenious design is that the periodic time of the pendulum, remains exactly the same even if the amplitude goes well beyond the regime of the approximation $\sin\theta \approx \theta$.

Huygens had thus recognized that if the objective is to design a pendulum whose periodic time would be *independent* of the amplitude, then one must gradually *shift* and *slide* the very point of suspension using chops of appropriate shape. The shape of the curved surface cannot be arbitrary. It turns out that the arcs L, R and even T (Fig. 6.8b) are *all* cycloids. The pendulum's bob in Fig. 6.8 traverses on the cycloid, whose equation is given by

$$x = r(\theta - \sin\theta), \text{ and } y = r(\cos\theta - 1). \tag{6.21}$$

This is the same relationship as discussed in the context of Fig. 6.4, except that the sign of the y-coordinate is now reversed. The reversal in sign is on account of the fact that the reference circle which would roll to the right (so that a point on its rim would generate the cycloid-arc Fig. 6.8) corresponds to $y \le 0$. The reference circle in Fig. 6.6, on the other hand, would have $y \ge 0$. The angle θ fixes both the Cartesian coordinates which are related to each other by the equation to the cycloid. Hence, we may now proceed to set up the Lagrange's equation of motion using the generalized coordinate, θ, and the generalized velocity, $\dot{\theta}$. We may write the Lagrangian for the bob using the Cartesian coordinates, or equivalently in the polar coordinates (Chapter 2):

$$L = T - V = \frac{1}{2}m(\dot{x}^2 + \dot{y}^2) - mgy, \tag{6.22a}$$

i.e., $$L = mr^2\dot{\theta}^2(1 - \cos\theta) - mgr(\cos\theta - 1). \tag{6.22b}$$

To use Lagrange's equation (Eq. 6.5), we need the following derivatives of the Lagrangian:

$$\frac{\partial L}{\partial \dot{\theta}} = 2mr^2\dot{\theta} - 2mr^2\dot{\theta}\cos\theta, \tag{6.22c}$$

$$\frac{d}{dt}\frac{\partial L}{\partial \dot{\theta}} = 2mr^2\ddot{\theta} - 2mr^2\ddot{\theta}\cos\theta + 2mr^2\dot{\theta}^2\sin\theta, \tag{6.22d}$$

and, $$\frac{\partial L}{\partial \theta} = mr^2\dot{\theta}^2\sin\theta + mgr\sin\theta. \tag{6.22e}$$

Now, using Eq. 6.22 in Eq. 6.5, we get

$$\ddot{\theta} = \frac{(gr - r^2\dot{\theta}^2)\sin\theta}{2r^2(1 - \cos\theta)} = -\frac{(r\dot{\theta}^2 - g)\sin\theta}{2r(1 - \cos\theta)}$$

$$= -\frac{(r\dot{\theta}^2 - g)\sqrt{1 + \cos\theta}}{2r\sqrt{1 - \cos\theta}} = -\frac{(r\dot{\theta}^2 - g)\cos\dfrac{\theta}{2}}{2r\sin\dfrac{\theta}{2}}, \tag{6.23a}$$

i.e., $$-2r\ddot{\theta}\sin\frac{\theta}{2} - r\dot{\theta}^2\cos\frac{\theta}{2} = -g\cos\frac{\theta}{2}. \tag{6.23b}$$

We now use the fact that $\frac{d}{dt}\left(\cos\frac{\theta}{2}\right) = \left(-\sin\frac{\theta}{2}\right)\frac{1}{2}\dot{\theta},$ (6.24a)

and $$\left(\frac{d}{dt}\right)^2\left(\cos\frac{\theta}{2}\right) = \left(-\sin\frac{\theta}{2}\right)\frac{1}{2}\ddot{\theta} - \left(\frac{1}{2}\right)^2\dot{\theta}^2\cos\frac{\theta}{2}, \tag{6.24b}$$

i.e., $$4r\left(\frac{d}{dt}\right)^2\left(\cos\frac{\theta}{2}\right) = 2r\left(-\sin\frac{\theta}{2}\right)\ddot{\theta} - r\dot{\theta}^2 \text{ co} \tag{6.24c}$$

Equation 6.23b and Eq. 6.24c immediately give: $\left(\frac{d}{dt}\right)^2\left(\cos\frac{\theta}{2}\right) = -\left(\frac{g}{4r}\right)\left(\cos\frac{\theta}{2}\right).$

(6.25)

Equation 6.25 has exactly the same form as Eq. 4.17. Its solution is essentially a simple harmonic motion, with a periodic time that is completely independent of the amplitude. It is given by

$$T = 2\pi\sqrt{\frac{4r}{g}} = 4\pi\sqrt{\frac{r}{g}}. \tag{6.26}$$

This is an amazing result, since it has been obtained *even* if the amplitude of oscillation is well beyond the applicability of the $\sin\theta \cong \theta$ approximation. It is thus valid even for *large* oscillations. This has been achieved by constraining the motion of the pendulum by using cycloid-shaped chops (Fig. 6.8). Huygens' cycloidal pendulum [9] is also known as an *isochronous pendulum*, since it always has the same periodic time, no matter what the amplitude of oscillation is. We have been able to get this result easily using Lagrangian method. The same result can of course be obtained using Newtonian method, but the procedure would be rather complicated.

(b) Bead on a circular hoop

We consider a circular hoop in a vertical YZ plane on which a bead can ride without any friction. We consider the zero of the bead's potential energy at the bottom of the circle. We shall consider a right-handed Cartesian reference coordinate system in which the X-axis is into the plane of Fig. 6.9, and the Z-axis points downward. If this hoop is spun about its Z-axis, the bead would tend to retain its original position due to its inertia, but the forces of constraints (normal reaction) would pick up a component against gravity. Newtonian equation of motion for the bead is determined by the combination of the force due to gravity, and the force of constraint due to the normal reaction which is along the radial line. The bead's motion can only be tangential to the hoop. Hence the normal reaction on the bead due to the hoop cannot do any work on the bead. The bead's distance from the center remains fixed, at $r = R$, which is the radius of the hoop. The bead therefore has two degrees of freedom, viz., its spherical polar coordinate θ, and its azimuthal coordinate φ.

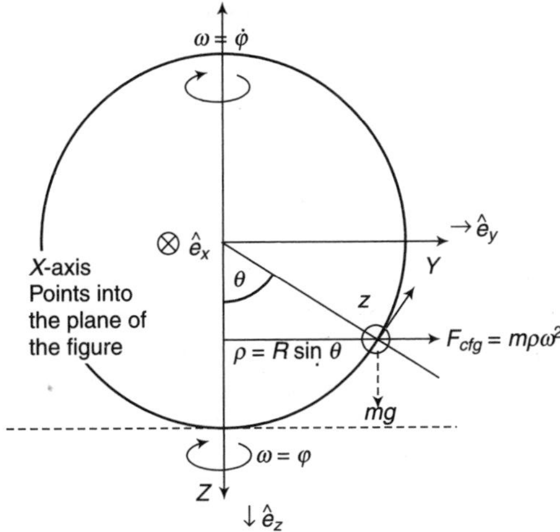

Fig. 6.9 A circular hoop in the vertical plane. Note that the Cartesian reference coordinate system has its *X*-axis into the plane of the paper and the *Z*-axis points downward, in the direction of gravity.

Newtonian formulation of mechanics would require the forces of constraints to be considered to study the motion of the bead, whereas the Lagrangian formulation enables us solve the problem by eliminating forces of constraints altogether from the beginning. This is achieved by setting up Lagrange's equation of motion in terms of the generalized coordinates, which in the present case are (θ, φ). The position and the velocity of the bead are respectively given by

$$\vec{r} = r\hat{e}_r \rightarrow \underset{\{\dot{r}=0\}}{\overset{\{r=R,\ \text{constant}\}}{=}} \rightarrow R\hat{e}_r \tag{6.27a}$$

and $\vec{v} = R\dot{\theta}\hat{e}_\theta + R\sin\theta\,\dot{\varphi}\hat{e}_\varphi.$ (6.27b)

The Lagrangian for the bead can therefore immediately be written as

$$L = T - V = \frac{1}{2}m\,(R^2\dot{\theta}^2 + R^2\sin^2\theta\,\dot{\varphi}^2) - mg\,(R - R\cos\theta). \tag{6.28}$$

The derivatives of the Lagrangian for use in Eq. 6.5 are

$$p_\theta = \frac{\partial L}{\partial\dot{\theta}} = mR^2\dot{\theta}, \text{ (which is the generalized momentum)}, \tag{6.29a}$$

which gives $\dfrac{d}{dt}\dfrac{\partial L}{\partial\dot{\theta}} = \dfrac{d}{dt}p_\theta = mR^2\ddot{\theta}.$ (6.29b)

We also need the partial derivative of the Lagrangian with respect to the angular displacement,

$$\frac{\partial L}{\partial\theta} = mR^2\dot{\varphi}^2\sin\theta\cos\theta - mgR\sin\theta = mR^2\sin\theta\left(\dot{\varphi}^2\cos\theta - \frac{g}{R}\right). \tag{6.29c}$$

Using Eqs. 6.29b and 6.29c, the Lagrange's equation becomes

$$mR^2\ddot{\theta} = mR^2\dot{\varphi}^2 \sin\theta\cos\theta - mgR\sin\theta, \tag{6.30a}$$

i.e.,
$$\ddot{\theta} = \left(\cos\theta - \frac{g}{R\omega^2}\right)\omega^2 \sin\theta, \tag{6.30b}$$

where we have used $\omega = \dot{\varphi}$.

Equation 6.30b cannot be solved easily, but it has a beautiful form that we could arrive at elegantly using the Lagrangian formulation. It is enormously useful, since we can infer from it several details about the conditions under which the bead can be in equilibrium ($\ddot{\theta} = 0$; $\dot{\theta} \rightarrow$ constant) while the hoop is rotating. The bead can of course be at equilibrium when $\omega = 0$ and the hoop is simply not spinning. Interestingly, the bead can be at equilibrium *even* when the hoop is spinning about the Z-axis. This can happen under the following conditions:

(i) $\theta = 0$, (bottom of the hoop), since it makes $\sin\theta = 0$,

(ii) $\theta = \pi$ (top of the hoop), since this also makes $\sin\theta = 0$,

and

(iii) $\cos\theta = \frac{g/R}{\omega^2}$, with $\frac{g}{R} \leq \omega^2$ since we must always have $\cos\theta \leq 1$. Essentially, this is possible when the rotation speed of the hoop $\omega \geq \omega_c$, where, the critical frequency ω_c is

$$\omega_c = \sqrt{g/R}. \tag{6.31}$$

When $\omega < \omega_c$, the bead can be at equilibrium at the bottom of the hoop, or at its top.

When $\omega > \omega_c$, there are *three* points at which the bead can be at equilibrium. These include the two points $\theta = 0$ and $\theta = \pi$ (as before), and also $\theta_0 = \cos^{-1}\left(\frac{g/R}{\omega^2}\right) = \cos^{-1}\left(\frac{\omega_c^2}{\omega^2}\right)$.

Consider a small departure from the equilibrium point, $\delta(t) = \theta(t) - \theta_0$.

$$\sin\theta = \sin(\theta_0 + \delta(t)) = \cos\delta(t)\sin\theta_0 + \sin\delta(t)\cos\theta_0 \cong \sin\theta_0 + \delta(t)\cos\theta_0 + O(\delta^2), \tag{6.32a}$$

$$\cos\theta = \cos(\theta_0 + \delta(t)) = \cos\delta(t)\cos\theta_0 - \sin\delta(t)\sin\theta_0 \cong \cos\theta_0 - \delta(t)\sin\theta_0 + O(\delta^2). \tag{6.32b}$$

Thus, the product of the two trigonometric functions is given by

$$\sin\theta\cos\theta = [\sin\theta_0 + \delta(t)\cos\theta_0 + O(\delta^2)]\,[\cos\theta_0 - \delta(t)\sin\theta_0 + O(\delta^2)],$$

i.e., $\sin\theta\cos\theta = [\sin\theta_0\cos\theta_0 + \delta(t)\cos^2\theta_0 + \cos\theta_0 O(\delta^2) - \delta(t)\sin^2\theta_0 + O(\delta^2)$,

i.e., $\sin\theta\cos\theta \cong \sin\theta_0\cos\theta_0 + \delta(t)\{\cos^2\theta_0 - \sin^2\theta_0\}$, i.e., $\sin\theta\cos\theta \cong \sin\theta_0\cos\theta_0 + \delta(t)\{2\cos^2\theta_0 - 1\}$. (6.33)

The equation of motion, Eq. 6.30, using Eq. 6.31, Eq. 6.32, and Eq. 6.33 becomes

$$\ddot{\theta} - \omega^2[\sin\theta_0 + \cos\theta_0 + \delta(t)\{2\cos^2\theta_0 - 1\}] + \omega_c^2[\sin\theta_0 + \delta(t)\cos\theta_0] = 0. \quad (6.34)$$

i.e., $\ddot{\theta} + [\omega^2 - 2\omega^2\cos^2\theta_0 + \omega_c^2\cos\theta_0]\,\delta(t) = 0.$ (6.35)

The three equilibrium points mentioned above, $\theta_0^{(i)} = 0$, $0_0^{(ii)} = \pi$, and $\theta_0^{(iii)} = \cos^{-1}\left(\frac{\omega_c^2}{\omega^2}\right)$, can now be analyzed in further details using Eq. 6.35 which expresses conditions in the neighborhood of an equilibrium point.

(i) $\theta_0^{(i)} = 0$. For this case, Eq. 6.35 gives: $\ddot{\theta} = [\omega^2 - \omega_c^2]\,\delta(t)$.

This implies that $\ddot{\theta} < 0$ for $\omega < \omega_c$, and $\theta_0^{(i)} = 0$ is therefore a point of *stable* equilibrium if the hoop is spun at a speed *below* the critical angular velocity. On the other hand, $\ddot{\theta} > 0$ for $\omega > \omega_c$, and this makes $\theta_0^{(i)} = 0$ a point of *unstable* equilibrium when the hoop is spun at a speed *above* the critical angular velocity. As you increase the speed past the critical speed, the nature of the equilibrium changes from 'stable' to 'unstable'.

(ii) $\theta_0^{(ii)} = \pi$. For this case, Eq. 6.35 gives: $\ddot{\theta} = [\omega_c^2 + \omega^2]\,\delta(t)$.

This implies that $\ddot{\theta} > 0$ hence the equilibrium is 'unstable' for both $\omega < \omega_c$ (low speed), for $\omega > \omega_c$ (high speed).

(iii) $\theta_0^{(iii)} = \cos^{-1}\left(\frac{\omega_c^2}{\omega^2}\right)$. This equilibrium point is accessible to the bead only at speeds above the critical angular speed, $\omega > \omega_c$, since $\cos\theta_0^{(iii)} = \left(\frac{\omega_c^2}{\omega^2}\right)$ must be less than or equal to 1.

For this case, Eq. 6.35 gives: $\ddot{\theta} + \left[\omega^2 - 2\omega^2\left(\frac{\omega_c^4}{\omega^4}\right) + \omega_c^2\left(\frac{\omega_c^2}{\omega^2}\right)\right]\delta(t) = 0,$

i.e., $\ddot{\theta} = \frac{(-\omega^4 + 2\omega_c^4 - \omega_c^4)}{\omega^2}\delta(t) = \frac{(\omega_c^4 - \omega^4)}{\omega^2}\delta(t.)$ This third point has to be a point of *stable* equilibrium, since we know that for this case the hoop *must* be spun at a speed *above* the critical speed, i.e., $\omega > \omega_c$. As the angular speed of the hoop is increased past the critical speed, the point at the bottom of the hoop changes its character from stable to unstable equilibrium. Note that $\theta_0 = \cos^{-1}\left(\frac{\omega_c^2}{\omega^2}\right)$ represents two, and not just one, points of stable equilibrium since the critical angle represents a cone with vertex at the center of the hoop, which intersects the hoop at two symmetric points on the opposite side of the vertical axis. This is an example of what is known as 'bifurcation', as one point of stability disappears, and two other stable points appear. This is akin to the bifurcation seen in the orbit of the logistic map that will be discussed in Chapter 9.

It is of course possible to study this problem using Newtonian method. However, it would be far more tedious, and it would be rather difficult to solve the problem in the laboratory (or inertial) frame of reference. Some convenience results from employing the rotating frame of reference in which the spinning hoop would be static, but then one would have to employ a pseudo-force. The centrifugal force would be pointed *away* from the axis of the spinning hoop. One can then resolve the centrifugal and the gravitational forces along the radial and the tangential components, to find the sum of the components and then determine the conditions for equilibrium, Using the Lagrangian method above is much neater, and provides insights into the nature of the equilibrium points without employing pseudo-forces.

6.4 HAMILTON'S EQUATIONS AND ANALYSIS OF THE S.H.O. USING VARIATIONAL PRINCIPLE

In general, one may write the Lagrangian to include an explicit time dependence, $L = L(q(t), \dot{q}(t), t)$, and not merely the implicit time dependence, $L = L(q(t), \dot{q}(t))$. The latter would be the case for isolated systems not interacting with surroundings. The total energy of the system would be expressible as the sum of its kinetic and potential energy described by the collection of all generalized coordinates and generalized velocities, of which the Lagrangian is a function. On the other hand, an explicit time-dependence has to be included when there are *unspecified* degrees of freedom with which the system may interact, and exchange energy with. In such a case, we shall have the total time-derivative of the Lagrangian $L(q, \dot{q}, t)$ to be given by

$$\frac{dL}{dt} = \left(\frac{\partial L}{\partial q}\right)\dot{q} + \left(\frac{\partial L}{\partial \dot{q}}\right)\ddot{q} + \frac{\partial L}{\partial t}. \tag{6.36a}$$

Now, when Eq. 6.5 holds, $\dfrac{\partial L}{\partial q} = \dfrac{d}{dt}\left(\dfrac{\partial L}{\partial \dot{q}}\right)$, and we have

$$\frac{dL}{dt} = \left(\frac{\mathrm{d}}{\mathrm{dt}}\left\{\frac{\partial L}{\partial \dot{q}}\right\}\right)\dot{q} + \left(\frac{\partial L}{\partial \dot{q}}\right)\ddot{q} + \frac{\partial L}{\partial t} = \frac{d}{dt}\left(\left\{\frac{\partial L}{\partial \dot{q}}\right\}\dot{q}\right) + \frac{\partial L}{\partial t}. \tag{6.36b}$$

Hence, $$\frac{d}{dt}\left(\left\{\frac{\partial L}{\partial \dot{q}}\right\}\dot{q} - L\right) = -\frac{\partial L}{\partial t}. \tag{6.37}$$

In other words, when $\dfrac{\partial L}{\partial t} = 0$, i.e., when the Lagrangian has *no* explicit time-dependence, we conclude from Eq. 6.37 that $\left\{\dfrac{\partial L}{\partial \dot{q}}\right\}\dot{q} - L$ is a constant with respect to time. This quantity is called the *Hamiltonian's Principal Function, H*:

$$H = \left[\left\{\frac{\partial L}{\partial \dot{q}}\right\}\dot{q} - L\right] = p\dot{q} - L(q, \dot{q}), \tag{6.38a}$$

where $p = \frac{\partial L}{\partial \dot{q}}$ is the generalized momentum. It is easy to see, especially for a particle in one-dimension motion with one degree of freedom, that

$$H = p\dot{q} - L(q, \dot{q}) = m\dot{q}^2 - L(q, \dot{q}) = 2T - (T - V) = T + V. \tag{6.38b}$$

Thus, when the Lagrangian is *not* an explicit function of time, the Hamilton's principal function is nothing but the total conserved energy of the isolated system. Having introduced the generalized momentum as $p = \frac{\partial L}{\partial \dot{q}}$, we must not forget that the state of the system with one degree of freedom is characterized by only two dynamical variables: $(q, \dot{q})$ or (q, p). The Hamiltonian in Eq. 6.38 must therefore be written only in terms of two variables, *no more*. Thus, dependence of all quantities on $(q, \dot{q})$ must be written in terms of dependence on (q, p). This is achieved using what is known as Legendre transformations. The Hamiltonian is then written in terms of (q, p). Thus, we have the Lagrangian $L = L(q, \dot{q})$, a function of the generalized coordinate and the generalized velocity, but the Hamiltonian $H = H(q, p)$, a function of the generalized coordinate and the generalized momentum. Even though we have considered a simple system with only one degree of freedom represented by $(q, \dot{q})$ or equivalently by (q, p), the above formalism can be easily generalized to a system having N degrees of freedom. In particular, Eq. 6.38 would then be generalized to the following:

$$H\,(q_1, q_2,.., q_{N-1}, q_N; p_1, p_2,.., p_{N-1}, p_N) = \sum_{i=1}^{N} p_i\,\dot{q}_i$$
$$- L(q_1, q_2,.., q_{N-1}, q_N; \dot{q}_1, \dot{q}_2,.., \dot{q}_{N-1}, \dot{q}_N). \tag{6.39}$$

Equating the differential increment dH in H obtained from the left hand side of Eq. 6.39 to the same obtained from the right hand side of the same equation, we get

$$\sum_k \frac{\partial H}{\partial q_k} dq_k + \sum_k \frac{\partial H}{\partial p_k} dp_k = \sum_k p_k d\dot{q}_k + \sum_k \dot{q}_k dp_k - \sum_k \frac{\partial L}{\partial q_k} dq_k - \sum_k \frac{\partial L}{\partial \dot{q}_k} d\dot{q}_k, \tag{6.40a}$$

$$\text{i.e.,}\quad \sum_k \frac{\partial H}{\partial q_k} dq_k + \sum_k \frac{\partial H}{\partial p_k} dp_k = -\sum_k \frac{\partial L}{\partial q_k} dq_k + \sum_k \dot{q}_k\, dp_k. \tag{6.40b}$$

Since the differential increments dq_i and dp_i are independent for every i, their respective coefficients on the two sides of Eq. 6.40b must be necessarily equal. Hence,

$$\forall\, k, -\frac{\partial L}{\partial q_k} = \frac{\partial H}{\partial q_k};\ \text{i.e.,}\ \frac{d}{dt}\frac{\partial L}{\partial \dot{q}_k} = \dot{p}_k = -\frac{\partial H}{\partial q_k}, \tag{6.41a}$$

$$\text{and}\quad \forall k,\ \dot{q}_k = \frac{\partial H}{\partial p_k}. \tag{6.41b}$$

Equations 6.41 tell us the rates at which the generalized coordinates q_κ and the generalized momenta p_κ evolve with time. They are, therefore, the *equations of motion.* These equations are called *Hamilton's equations*. They have exactly the same information about the temporal evolution of the classical system as is contained in *Lagrange's equation*, Eq. 6.5. The Lagrange's equation is, however, a second order differential equation and has to be integrated twice to get the solution. It would need two constants of integration, given by the initial conditions for the position $q(t = 0)$ and the initial velocity, $\dot{q}(t = 0)$. The Hamilton's equations, on the other hand, are *two* first order differential equations. We have to integrate both of them, but only once. This again requires two constants of integration, which are given by the two initial conditions for the position $q(t = 0)$ and the momentum $p(t = 0)$. We emphasize the fact that irrespective of whether one plans on employing the Lagrange's equation or the Hamilton's equations, one must *always begin with the Lagrangian*, expressed in terms of the generalized coordinates and the generalized velocities. To get to Hamilton's equations, one must determine the generalized momenta from the Lagrangian, and then express the Hamilton's principal function in terms of the generalized coordinates and the generalized momenta. Hamilton's equations can then be set up and solved, if the initial position and momentum are known.

The classical equations of motion, whether Newton's, Lagrange's, or Hamilton's, remain invariant under time-reversal, i.e., when time t goes to $-t$. This is because of the fact that Newton's and Lagrange's equation involve the *second* order derivative with respect to time. The sign-reversal under time-reversal therefore happens *twice*, leaving the overall sign invariant. In the Hamilton's equations, only the *first* order time derivative operator is employed, but there is a sign asymmetry in the *two* first order equations of motion (Eq. 6.41). This ensures that Hamilton's equations are also symmetric under time-reversal. Thus, if we know the position and velocity (or position and momentum) of an object, then we can determine its position and velocity (or position and momentum) at any other instant of time, both *prior* to that instant, or *later*. We can thus trace the history of the trajectory and also predict it in the future, in the position–velocity (or position–momentum) phase space.

We illustrate the method of solving problems in mechanics using Lagrange's and Hamilton's equations using a very simple example, that of the mass-spring simple harmonic oscillator, whose mass is m and the elastic spring constant is κ. The simple harmonic oscillator is characterized by the quadratic potential $\frac{\kappa}{2}q^2$ which describes it, where q represents the displacement of the oscillator from its equilibrium position. We have already studied the physics of the oscillator in some details in Chapters 4 and 5. As mentioned above, we must begin by first setting up the Lagrangian for the simple harmonic oscillator:

$$L = L(q, \dot{q}) = T - V = \frac{m}{2}\dot{q}^2 - \frac{\kappa}{2}q^2. \tag{6.42}$$

From the Lagrangian, the generalized momentum is readily obtained: $p = \frac{\partial L}{\partial \dot{q}} = m\dot{q}$. Using now the Lagrangian, the Lagrange's equation (Eq. 6.5) gives

$$-\kappa q - m\dot{q} = 0, \tag{6.43}$$

which essentially is the same as Newton's equation of motion for the simple harmonic oscillator. The Hamiltonian for this system is readily obtained by using the Lagrangian (Eq. 6.42) and writing the Hamiltonian in terms of (q, p) instead of $(q, \dot{q})$. This gives

$$H = \left[\left\{\frac{\partial L}{\partial \dot{q}}\right\}\dot{q} - L\right] = m\dot{q}^2 - \left(\frac{m\dot{q}^2}{2} - \frac{\kappa q^2}{2}\right) = \frac{m\dot{q}^2}{2} + \frac{\kappa q^2}{2} = \frac{p^2}{2m} + \frac{\kappa q^2}{2}. \tag{6.44}$$

Hamilton's equations are obtained by taking the partial derivatives of the Hamiltonian (6.44) with respect to the generalized coordinate, and with respect to the generalized momentum. Thus,

$$\dot{q} = \frac{\partial H}{\partial p} = \frac{p}{m}, \tag{6.45a}$$

and $$\dot{p} = -\frac{\partial H}{\partial q} = -\kappa q. \tag{6.45b}$$

While Eq. 6.45a is just the Newtonian velocity, Eq. 6.45b is essentially the same as Newton's equation of motion.

As mentioned earlier, the central problem in mechanics is this: how do you describe a mechanical system and how does it evolve with time? Newtonian mechanics invokes the linear stimulus-response formalism, subsumed in Newton's second law, $\vec{F} = \dot{\vec{p}}$, to account for this when equilibrium is disturbed. The variational principal provides an answer to the same question that is completely equivalent to that offered by Newtonian mechanics. However, the variational principle makes no use of the idea of 'force'. Our limited goal in this chapter is to provide an early exposure to the variational method, which accounts for how *alternative* trajectories in the phase space are optimized.

If a life-guard wishes to run and save a drowning person in a river, it is best that he approaches the bank along a path along which he would reach the accident-victim fastest. This must be optimized considering the fact that the speed at which he would run on land to approach the bank would be different from the speed at which he would swim after jumping in the waters. Clearly, the solution would be given by the variational principle, and the most rapid path would be given as per the Fermat's principle.

We end this chapter by reminding that the variational principle is best adapted to interpret the connections between symmetry and conservation principles which are at the heart of Noether's theorem, mentioned in Chapter 1. Besides, this formulation is best suited for development of the quantum theory.

APPENDIX: Time Period for Large Oscillations

From Eq. 4.10b, the time period of a pendulum is given as

$$T = \left(4\sqrt{\frac{\ell}{2g}}\right)\int_0^{\theta_0}\left(\frac{d\theta}{\sqrt{(\cos\theta - \cos\theta_0)}}\right). \tag{6A.1a}$$

In the above equation, the angle $\theta]_{t=0} = \theta_0$ is the maximum angular displacement from which the bob of a simple pendulum is released to oscillate under gravity. The pendulum therefore oscillates in the angular range $0 \le \theta \le \theta_0$, on each side of the equilibrium. The periodic time, in the above equation, is written in terms of the integration over the angular motion in this range in the above equation. As explained in Chapter 4, the factor '4' in Eq. 6A.1 comes from the fact that the angle swept by the pendulum bob from 0 to θ_0 corresponds to a *quarter* of the time-period T. For 'small' oscillations, the largest angular displacement θ_0 is small enough to approximate $\cos\theta_0 \approx 1 - \frac{\theta_0^2}{2}$. This approximation will of course hold good for all smaller angles. Using this approximation in Eq. 6A.1, we get

$$T_0 = \left(4\sqrt{\frac{\ell}{2g}}\right)\int_0^{\theta_0}\left(\frac{d\theta}{\sqrt{\left(\left(1-\frac{\theta^2}{2}\right)-\left(1-\frac{\theta_0^2}{2}\right)\right)}}\right),$$

i.e.,
$$T_0 = \left(4\sqrt{\frac{\ell}{2g}}\right)\sqrt{2}\int_0^{\theta_0}\frac{d\theta}{\sqrt{\theta_0^2-\theta^2}}. \tag{6A.1b}$$

Now, we know that $\sin^{-1}\left(\frac{x}{a}\right) = \int \frac{1}{\sqrt{a^2-x^2}}dx$, and hence

$$T_0 = \left(4\sqrt{\frac{\ell}{g}}\right)\left[\sin^{-1}\frac{\theta}{\theta_0}\right]_0^{\theta_0} = 2\pi\sqrt{\frac{\ell}{g}}. \tag{6A.2}$$

Equation 6A.2 is independent of the initial angular displacement, θ_0. As long as the initial displacement is not large, it does not matter what its actual value is. This property makes it very useful to use the simple pendulum as a clock. For large oscillations, however, Eq. 6A.1a cannot be solved using the approximation $\cos\theta \approx 1 - \frac{\theta^2}{2}$. We therefore have to search for a more elaborate solution.

Using the identity $\cos\theta = 1 - 2\sin^2\frac{\theta}{2}$, in Eq. 6A.1a, we get

$$\cos\theta - \cos\theta_0 = 2\left(\sin^2\frac{\theta_0}{2} - \sin^2\frac{\theta}{2}\right), \text{ thereby giving}$$

$$T = \left(4\sqrt{\frac{\ell}{2g}}\right)\int_0^{\theta_0}\left(\frac{d\theta}{\sqrt{2\left(\sin^2\frac{\theta_0}{2} - \sin^2\frac{\theta}{2}\right)}}\right). \tag{6A.3}$$

It is fruitful to change the variables from (θ_0, θ) to (ξ, u) by defining

$$\xi = \sin\left(\frac{\theta_0}{2}\right) \text{ and } \sin u = \frac{\sin\left(\frac{\theta}{2}\right)}{\sin\left(\frac{\theta_0}{2}\right)}. \tag{6A.4a}$$

We therefore have $\sin\left(\frac{\theta}{2}\right) = \xi \sin u.$ (6A.4b)

Differentiating both sides of Eq. 6A.4b,

$$\frac{1}{2}\cos\left(\frac{\theta}{2}\right) d\theta = \xi \cos u \, du, \tag{6A.5a}$$

and $d\theta = \frac{2\xi(1 - \sin^2 u)^{1/2}}{(1 - \xi^2 \sin^2 u)^{1/2}} du.$ (6A.5b)

As time varies from 0 to $\frac{T}{4}$, u varies from 0 to $\frac{\pi}{2}$.

The time period given in Eq. 6A.3 therefore can be written as

$$T = \left(4\sqrt{\frac{\ell}{2g}}\right)\int_0^{\frac{\pi}{2}}\left(\frac{2\xi(1 - \sin^2 u)^{1/2}}{(1 - \xi^2 \sin^2 u)^{1/2}\,(2\,(\xi^2 - \xi^2 \sin^2 u))^{1/2}}\right) du, \tag{6A.6a}$$

i.e., $T = \left(4\sqrt{\frac{\ell}{g}}\right)\int_0^{\frac{\pi}{2}} \frac{du}{(1 - \xi^2 \sin^2 u)^{1/2}}.$ (6A.6b)

The above integral is an *elliptic integral of the first kind*.

Expanding the term $(1 - \xi^2 \sin^2 u)^{-1/2}$ using the binomial theorem we get

$$(1 - \xi^2 \sin^2 u)^{-1/2} = 1 + \frac{1}{2}\xi^2 \sin^2 u + \frac{3}{8}\xi^4 \sin^4 u + \frac{5}{16}\xi^6 \sin^6 u + \ldots \tag{6A.7}$$

Accordingly,

$$T = \left(4\sqrt{\frac{\ell}{g}}\right)\int_0^{\pi/2}\left(1 + \frac{1}{2}\xi^2 \sin^2 u + \frac{3}{8}\xi^4 \sin^4 u + \frac{5}{16}\xi^6 \sin^6 u + \ldots\right) du. \tag{6A.8}$$

Using now the identity $\sin^2 u = \left(\frac{1-\cos 2u}{2}\right)$, and integrating the terms, we get

$$T=\left(4\sqrt{\frac{\ell}{g}}\right)\left\{\begin{array}{l}[u]_0^{\pi/2}+\frac{1}{2}\xi^2\int_0^{\pi/2}\left(\frac{1-(\cos 2u)}{2}\right)du+\frac{3}{8}\xi^4\int_0^{\pi/2}\left(\frac{1-\cos 2u}{2}\right)^2 du \\ +\frac{5}{16}\xi^6\int_0^{\pi/2}\left(\frac{1-\cos 2u}{2}\right)^3 du+\ldots\end{array}\right\}. \quad (6A.9a)$$

We are now able to write the periodic time in terms of an infinite series in powers of $\xi^2 = \sin^2\left(\frac{\theta_0}{2}\right)$:

$$\text{i.e., } T=\left(4\sqrt{\frac{\ell}{g}}\right)\left[\frac{\pi}{2}+\frac{1}{2}\xi^2\left(\frac{1}{2}\right)\left(\frac{\pi}{2}\right)+\frac{3}{8}\xi^4\left(\frac{1}{4}\right)\left(\frac{3\pi}{4}\right)+\frac{5}{16}\xi^6\left(\frac{1}{8}\right)\left(\frac{5\pi}{4}\right)+\ldots\right], \quad (6A.9b)$$

$$\text{i.e., } T = 2\pi\sqrt{\frac{\ell}{g}}\left[1+\left(\frac{1}{2}\right)^2\xi^2+\left(\frac{(1)\,(3)}{(2)\,(4)}\right)^2\xi^4+\left(\frac{(1)\,(3)\,(5)}{(2)\,(4)\,(6)}\right)^2\xi^6+\ldots\right], \quad (6A.9c)$$

$$\text{i.e., } T = 2\pi\sqrt{\frac{\ell}{g}}\left[1+\frac{1}{4}\xi^2+\frac{9}{64}\xi^4+\frac{225}{2304}\xi^6+\ldots\right]. \quad (6A.9d)$$

In terms of the initial angular displacement, θ_0, the periodic time therefore is

$$\text{i.e., } T = 2\pi\sqrt{\frac{\ell}{g}}\left[1+\frac{1}{4}\sin^2\left(\frac{\theta_0}{2}\right)+\frac{9}{64}\sin^4\left(\frac{\theta_0}{0}\right)+\frac{225}{2304}\sin^6\left(\frac{\theta_0}{2}\right)+\cdots\right]. \quad (6A.9e)$$

The periodic time of a pendulum when it is set into motion with somewhat larger initial displacement is not so simple. It is larger than the periodic time for the pendulum having small oscillations by the following factor:

$$\frac{T}{T_0} = 1+\frac{1}{4}\sin^2\left(\frac{\theta_0}{2}\right)+\frac{9}{64}\sin^4\left(\frac{\theta_0}{2}\right)+\frac{225}{2304}\sin^6\left(\frac{\theta_0}{2}\right)+\ldots. \quad (6A.10a)$$

For small oscillations, only the first term on the right hand side of Eq. 6A.10a is good enough. When the magnitude of the oscillation is beyond the domain of the approximation $\cos\theta \approx 1-\frac{\theta^2}{2}$, the time-period depends on the actual value of the initial angular displacement of the pendulum.

Introducing an auxiliary variable $\left(\frac{\theta_0}{2}\right) = \alpha$, we get

$$\frac{T}{T_0} = 1 + \frac{1}{4}\sin^2\alpha + \frac{9}{64}\sin^4\alpha + \frac{225}{2304}\sin^6\alpha + \dots . \tag{6A.10b}$$

We can write the above expression in a more compact form, by writing $\xi^2 = \sin^2\alpha$:

$$\frac{T}{T_0} = 1 + \frac{1}{4}\xi^2 + \frac{9}{64}\xi^4 + \frac{225}{2304}\xi^6 + \dots \tag{6A.10c}$$

Now, we note that

$$\xi^2 = \sin^2\alpha = \frac{1 - \cos 2\alpha}{2}, \tag{6A.11a}$$

and hence, using $\cos\alpha = 1 - \frac{\alpha^2}{2!} + \frac{\alpha^4}{4!} - \frac{\alpha^6}{6!} + \dots,$

and therefore $1 - \cos 2\alpha = \frac{4\alpha^2}{2!} - \frac{16\alpha^4}{4!} + \frac{64\alpha^6}{6!} - \dots,$
we get

$$\xi^2 = \alpha^2 - \frac{\alpha^4}{3} + \frac{2\alpha^6}{45} - \dots \tag{6A.12a}$$

Furthermore, recognizing that $\xi^4 = \sin^4\alpha = \frac{(1 - \cos 2\alpha)^2}{4}$, (6A.12b)

and using the same trick, we can see that $\xi^4 = \alpha^4 - \frac{2}{3}\alpha^6$, (6A.12c)

and $\xi^6 = \frac{(1 - \cos 2\alpha)^3}{8}$, likewise, can be simplified as:

$$\xi^6 = \alpha^6. \tag{6A.12d}$$

Substituting the value of ξ^2, ξ^4, ξ^6 in Eq. 6A.10c, we get

$$\frac{T}{T_0} = 1 + \frac{1}{4}\left(\alpha^2 - \frac{\alpha^4}{3} + \frac{2\alpha^6}{45} - \cdots\right) + \frac{9}{64}\left(\alpha^4 - \frac{2}{3}\alpha^6\right) + \frac{225}{2304}(\alpha^6) + \cdots \tag{6A.13a}$$

We can now write the above result in terms of the angular displacement θ_0, which is just 2α:

$$\frac{T}{T_0} = 1 + \frac{\theta_0^2}{16} + \frac{11\theta_0^4}{3072} + \frac{173\theta_0^6}{737280} + \cdots . \tag{6A.13b}$$

Hence, $T = T_0\left(1 + \frac{\theta_0^2}{16} + \frac{11\theta_0^4}{3072} + \frac{173\theta_0^6}{737280} + \cdots\right)$, which is the result we have used above in Eq. 6.20. (6A.13b)

Thus, when the oscillations are not small, the expression for the periodic time is rather complicated, and it is not independent of the initial displacement. This would make such a pendulum unsuitable to be used as a clock, as it will need to be calibrated even for the same length and at the same location on Earth to account for the actual value of the initial value of the angular displacement θ_0 on which the periodic time depends. The cycloidal pendulum designed by Huygens discussed in Section 6.3 removes the need for such re-calibration, by making the bob of the pendulum not on the arc of a circle, but on the arc of a cycloid.

Problems with Solutions

P6.1:

Two mass points m_1 and m_2 ($m_1 \neq m_2$) are connected by a string of length l passing through a hole in a horizontal table. The string and the mass m_1 move without friction on the table and the mass m_2 moves freely on a vertical line.

(a) What initial velocity must m_1 be given so that m_2 will remain motionless at a distance d below the surface of the table?

(b) If m_2 is slightly displaced in a vertical direction, small oscillations are set up in the system. Determine the time-period of these oscillations by setting up and solving Lagrange's equations.

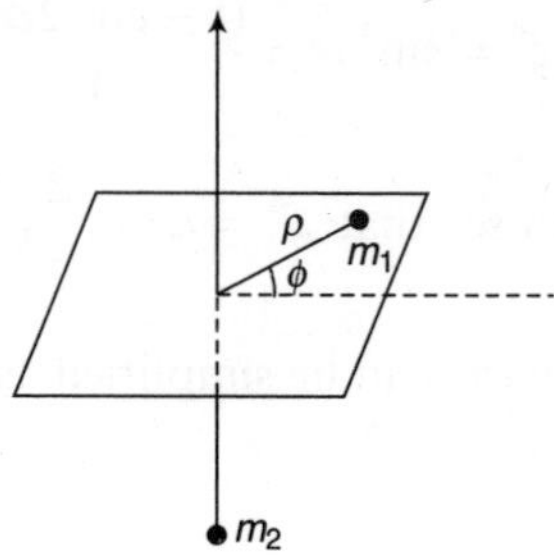

***Solution*:**

(a) The mass m_1 must have a velocity perpendicular to the string such that the centripetal force on it is equal to the gravitational force on m_2. Therefore, $\frac{m_1 v^2}{l-d} = m_2 g$, and $v = \sqrt{\frac{m_2(l-d)g}{m_1}}$.

(b) The mass m_2 is located at the z-coordinate $-l + r$ and thus has velocity $\dot{\rho}$. The Lagrangian for the system is: $L = T - V = \frac{1}{2}m_1(\dot{\rho}^2 + \rho^2\dot{\phi}^2) + \frac{1}{2}m_2\dot{\rho}^2 + m_2 g(l-\rho)$.

Lagrange's equations: $m_1\rho^2\dot{\phi}^2 = \lambda$, a constant, and $(m_1 + m_2)\ddot{\rho} - m_1\rho\dot{\phi}^2 = m_2 g = 0$.

At $t = 0$, $\rho = l - d$, $v = \sqrt{m_2(l-d)g/m_1} = v_0$, say, so $\dot{\phi}_0 = \frac{v_0}{l-d} = \sqrt{\frac{m_2}{m_1}\frac{g}{l-d}}$.

Hence: $m_1\rho^2\dot{\phi}^2 = m_1\ (l-d)^2\ \dot{\phi}_0 = m_1\sqrt{\frac{m_2}{m_1}(l-d)^3 g}$.

Therefore, $\rho\dot{\phi}^2 = \frac{\rho^4\dot{\phi}^2}{\rho^3} = \frac{m_2}{m_1}\left(\frac{l-d}{r}\right)^3 g$ and $(m_1+m_2)\ddot{\rho} - m_2\left(\frac{l-d}{r}\right)^3 g + m_2 g = 0$.

Let $\rho = (l-d) + h$, where $h \ll (l-d)$

Then: $\ddot{\rho} = \ddot{h}$, and $\rho^{-3} = (l-d)^{-3}\left(1+\frac{h}{l-d}\right)^{-3} \approx (l-d)^{-3}\left(1-\frac{3h}{l-d}\right)$.

The above equation gives: $\ddot{h} + \frac{3m_2 g}{(m_1+m_2)(l-d)}h = 0$, which is an equation to a linear harmonic oscillator. Hence h oscillates about O, i.e., ρ oscillates about the value $(l-d)$, at an angular frequency $\omega = \sqrt{\frac{3m_2 g}{(m_1+m_2)(l-d)}}$. The corresponding time-period is $T = 2\pi\sqrt{\frac{(m_1+m_2)(l-d)}{3m_2 g}}$.

P6.2:

A uniform solid cylinder of radius **R** and mass **M** rests on a horizontal plane. An identical cylinder rests on it, as shown in figure. The upper cylinder is given an infinitesimal displacement so that both cylinders roll without slipping.

(a) What is the Lagrangian of the system?

(b) What are the constants of the motion?

(c) Show that as long as the cylinder remain in contact,

$$\dot{\theta}^2 = \frac{12g(1-\cos\theta)}{R(17+4\cos\theta - 4\cos^2\theta)},$$

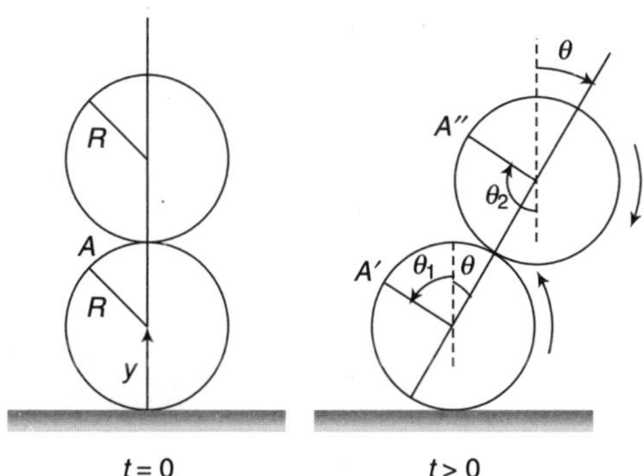

where θ is the angle which the plane containing the axes makes with the vertical.

Solution:

(a) The system possesses two degrees of freedom, described by two generalized coordinates. Toward this, we shall employ θ_1, the angle of rotation of the lower cylinder, and θ, the angle made by the plane containing the two axes of the cylinders and the vertical.

Initially, the plane containing the two axes of the cylinders is vertical. At a later time, this plane makes an angle θ with the vertical. The original point of contact, A, now moves to A′ on the lower cylinder and to A″ on the upper cylinder. From figure, we have $\theta_1 + \theta = \theta_2 - \theta$, or $\theta_2 = \theta_1 + 2\theta$.

Using Cartesian coordinates (x, y) in the plane perpendicular to the plane that contains the axes of the cylinders, and ***through their centers of mass***, we have,

at $t > 0$, for the lower cylinder: $x_1 = -R\theta_1$, $y_1 = R$,

and for the upper cylinder: $x_2 = x_1 + 2R\sin\theta$ and $y_2 = 3R - 2(R - \cos\theta) = R + 2R\cos\theta$.

Corresponding velocity: $\dot{x}_1 = -R\dot{\theta}_1$, $\dot{y}_1 = 0$ and $\dot{x}_2 = -R\dot{\theta}_1 + 2R\dot{\theta}\cos\theta$, $\dot{y}_2 = -2R\dot{\theta}\sin\theta$

The kinetic energy of the lower cylinder: $T_1 = \frac{1}{2}m\dot{x}_1^2 + \frac{1}{2}MR^2\dot{\theta}_1^2 = \frac{3}{4}MR^2\dot{\theta}_1^2$,

and that of the upper cylinder: $T_2 = \frac{1}{2}m(\dot{x}_1^2 + \dot{y}_1^2) + \frac{1}{4}MR^2\dot{\theta}_2^2$, i.e., $T_2 = \frac{1}{2}MR^2(\dot{\theta}_1^2 - 4\dot{\theta}_1\dot{\theta}\cos\theta$

$+ 4\dot{\theta}^2) = \frac{1}{4}MR^2(\dot{\theta}_1^2 + 4\dot{\theta}_1\dot{\theta} + 4\dot{\theta}^2) = \frac{1}{4}MR^2[3\dot{\theta}_1^2 + 4\dot{\theta}_1\dot{\theta}(1 - 2\cos\theta) + 12\dot{\theta}^2)$

The potential energy (with its zero at the horizontal plane): $V = Mg(y_1 + y_2) = 2MR(1 + \cos\theta)g$.

Hence the Lagrangian: $L = T - V = \frac{1}{2}MR^2[3\dot{\theta}_1^2 + 2\dot{\theta}_1\dot{\theta}(1 - 2\cos\theta) + 6\dot{\theta}^2] - 2MR(1 + \cos\theta)g$.

(b) Total mechanical energy is constant:

$$E = T + V = \frac{1}{2}MR^2[3\dot{\theta}_1^2 + 2\dot{\theta}_1\dot{\theta}(1 - 2\cos\theta) + 6\dot{\theta}^2] + 2MR(1 + \cos\theta)g.$$

When $\frac{\partial L}{\partial q_i} = 0$, it follows from the Lagrange's equation that $\frac{\partial L}{\partial \dot{q}_i}$ is conserved.

Hence: $\frac{\partial L}{\partial \dot{\theta}_1} = MR^2[3\dot{\theta}_1 + \dot{\theta}(1 - 2\cos\theta)] = \text{constant}.$

(c) Results of (b) hold when the cylinders remain in contact. Initially, $\theta = 0$, $\dot{\theta}_1 = \dot{\theta} = 0$, hence,

$$\frac{1}{2}MR^2[3\dot{\theta}_1^2 + 2\dot{\theta}_1\dot{\theta}(1 - 2\cos\theta) + 6\dot{\theta}^2] + 2MR(1 + \cos\theta)g = 4MRg,$$

$MR^2[3\dot{\theta}_1^2 + \dot{\theta}(1 - 2\cos\theta)] = 0$. Using these relations:

$$\dot{\theta}^2[18 - (1 - 2\cos\theta)^2] = \frac{12}{R}(1 - \cos\theta)g, \quad \text{i.e.,} \quad \dot{\theta}^2 = \frac{12(1 - \cos\theta)g}{R(17 + 4\cos\theta - 4\cos^2\theta)}.$$

P6.3:

Find the differential equation of the geodesic on the surface of an inverted cone with semi-vertical angle ξ.

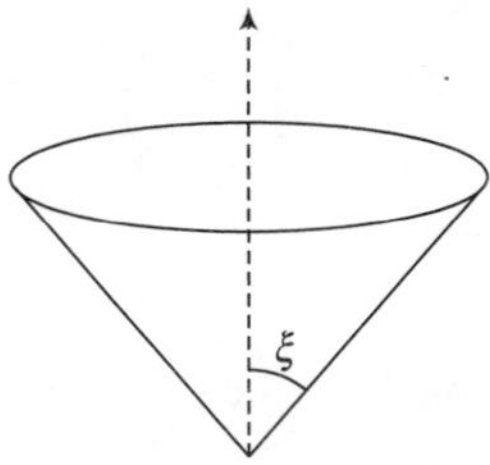

Solution:

Surface of the cone is characterized by: $x^2 + y^2 = z^2 \tan^2 \xi$, with ξ: constant

Thus: $x = ar\cos\phi$, $y = ar\sin\phi$ and $z = br$, where $a = \sin\xi$ and $\beta = \cos\xi$ are constants.

The metric $ds^2 = dx^2 + dy^2 + dz^2$ on the surface is: $ds^2 = dr^2 + a^2r^2\, d\phi^2$. Hence, the total length of the curve $\phi = \dot{\phi}(r)$ on the surface of the cone is given by: $s = \int dr\sqrt{1 + a^2r^2\dot{\phi}^2}$; $\dot{\phi} = \dfrac{d\phi}{dr}$.

The length s is stationary if the integrand $f = \sqrt{1 + a^2r^2\dot{\phi}^2}$ satisfies the Euler–Lagrange's equation:

$\dfrac{\partial f}{\partial \phi} - \dfrac{d}{dr}\left(\dfrac{\partial f}{\partial \dot{\phi}}\right) = 0$; i.e., $\dfrac{d}{dr}\left(\dfrac{a^2r^2\dot{\phi}}{f}\right) = 0$. Solving for ϕ' we get: $\dfrac{d\phi}{dr} = \dfrac{c_1}{ar\sqrt{a^2r^2 - c_1^2}}$, where c_1 is a constant. This is the required differential equation of geodesic. The geodesic on the surface of the cone is obtained by integrating the last equation:

$$\phi = \int \frac{c_1}{ar\sqrt{a^2r^2 - c_1^2}}dr + \alpha = \frac{1}{a}\sec^{-1}\left(\frac{ar}{c_1}\right) + \alpha. \text{ Hence, } r = \frac{c_1}{a}\sec[a(\phi - \alpha)].$$

P6.4:

Find the curve which extremizes the functional $I(y(x)) = \int_0^{\frac{\pi}{4}} (y''^2 - y^2 + x^2)dx$,

under the conditions that $y(0) = 0$, $y'(0) = 1$ and $y\left(\dfrac{\pi}{4}\right) = y'\left(\dfrac{\pi}{4}\right) = \dfrac{1}{\sqrt{2}}$

Solution:

The condition for the functional $I(y(x)) = \int_0^{\frac{\pi}{4}} (y''^2 - y^2 + x^2)dx$ to be an extremum is that the integrand $f = (y''^2 - y^2 + x^2)$ satisfies the Euler–Lagrange's equation:

$$\frac{\partial f}{\partial y} - \frac{d}{dx}\left(\frac{\partial f}{\partial y'}\right) + \frac{d^2}{dx^2}\left(\frac{\partial f}{\partial y''}\right) = 0. \text{ Hence, } -2y + \frac{d^2}{dx^2}(2y'') = 0 \text{ . i.e., } \frac{d^4y}{d^4x} - y = 0.$$

The solution of this last equation is: $y = ae^x + be^{-x} + c\cos x + d\sin x$, where a, b, c, d are constants of integration, to be determined.

$$y(0) = 0 \Rightarrow a + b + c = 0,$$

$$y\left(\frac{\pi}{4}\right) = \frac{1}{\sqrt{2}} \Rightarrow ae^{\frac{\pi}{4}} + be^{-\frac{\pi}{4}} + \frac{1}{\sqrt{2}}c + \frac{1}{\sqrt{2}}d = \frac{1}{\sqrt{2}},$$

$$y'(0) = 1 \Rightarrow a - b + d = 1$$

$$y'\left(\frac{\pi}{4}\right) = \frac{1}{\sqrt{2}} \Rightarrow ae^{\frac{\pi}{4}} - be^{-\frac{\pi}{4}} - \frac{1}{\sqrt{2}}c + \frac{1}{\sqrt{2}}d = \frac{1}{\sqrt{2}}$$

Solving these equations, the required curve is $y = \sin x$.

P6.5:

In 1662, Fermat proposed that the propagation of light obeyed the generalized principle of least time. In optics, Fermat's principle, or the principle of least time, is the principle that the path taken between two points by a ray of light is the path that can be traversed in the least time. Using Fermat's principle and the calculus of variation, prove the Snell's law of refraction of light.

Solution:

We use Cartesian coordinates, as in the adjacent figure. Light travels from the point $P_1(0, y_1, 0)$ to the point $P_2(x_2, -y_2, 0)$. The light beam intersects the plane glass interface at the point $Q(x, 0, z)$. We know that the velocity of light in any medium is given by $v = \frac{c}{n}$, where n is the refractive index of the medium and c is the velocity of light in vacuum. Hence, the transit time from the point P_1 to the point P_2 is: $t = \int_1^2 dt = \int_1^2 \frac{ds}{v} = \frac{1}{c}\int_1^2 n\,ds = \frac{1}{c}\int_1^2 n(x, y, z)\sqrt{1 + (x')^2 + (z')^2}$.

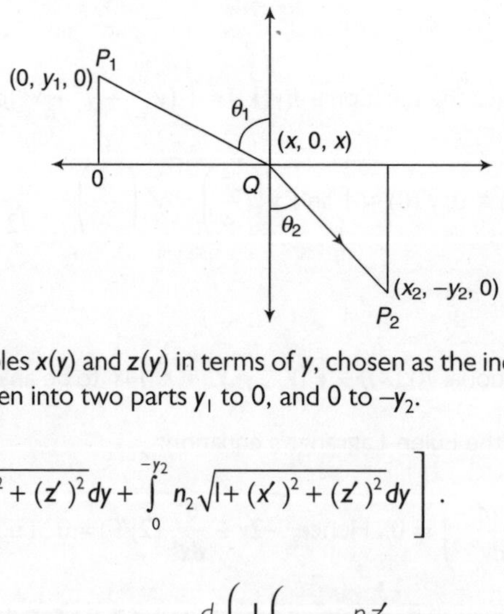

(0, y_1, 0) P_1
θ_1
(x, 0, x)
0
Q
θ_2
(x_2, $-y_2$, 0)
P_2

We parametrize the variables $x(y)$ and $z(y)$ in terms of y, chosen as the independent variable. The range of integration can be broken into two parts y_1 to 0, and 0 to $-y_2$.

$$t = \frac{1}{c}\left[\int_{y_1}^{0} n_1\sqrt{1 + (x')^2 + (z')^2}dy + \int_{0}^{-y_2} n_2\sqrt{1 + (x')^2 + (z')^2}dy\right].$$

The Euler's equation for z simplifies to: $0 + \frac{d}{dy}\left(\frac{1}{c}\left(\frac{n_1 z'}{\sqrt{1 + x'^2 + z'^2}} + \frac{n_2 z'}{\sqrt{1 + x'^2 + z'^2}}\right)\right) = 0$

Hence, $z' = 0$, and therefore z is a constant. The initial and the final values were chosen to be $z_1 = z_2 = 0$, therefore we have at the interface $z = 0$.

Similarly, the Euler's equation for **x** gives: $0+\frac{d}{dy}\left(\frac{1}{c}\left(\frac{n_1 x'}{\sqrt{1+x'^2+z'^2}}+\frac{n_2 x'}{\sqrt{1+x'^2+z'^2}}\right)\right)=0$.

Now, $x' = \tan\theta_1$, and $x' = -\tan\theta_2$, and $z' = 0$.

Hence: $0+\frac{d}{dy}\left(\frac{1}{c}\left(\frac{n_1\tan\theta_1}{\sqrt{1+(\tan\theta_1)^2}}-\frac{n_2\tan\theta_2}{\sqrt{1+(\tan\theta_2)^2}}\right)\right)=\frac{d}{dy}\left(\frac{1}{c}\left(n_1\sin\theta_1-n_2\sin\theta_2\right)\right)=0$.

Therefore, $\frac{1}{c}(n_1\sin\theta_1-n_2\sin\theta_2)=Z$, a constant, whose value must be zero, since when $n_1 = n_2$, we must have $\theta_1 = \theta_2$. Thus, Fermat's principle leads to Snell's law: $\frac{\sin\theta_2}{\sin\theta_1}=\frac{n_1}{n_2}$.

P6.6:

A sphere of radius *a* and mass *m* rests on the top of a fixed rough sphere of radius *b*. The first sphere is slightly displaced so that it rolls without slipping. Obtain the equation of motion for the rolling sphere.

Solution:

The constraint: $b\,d\xi = a\,d\psi$, i.e., $b\,d\xi - a\,d\psi = 0$.

The Lagrangian for the system is: $L = T_1 + T_2 - V$.

T_1, T_2: kinetic energies of two spheres and V is the potential energy of the small sphere.

Here, $T_1=\frac{1}{2}m(a+b)^2\dot{\xi}^2$ and $T_2=\frac{1}{2}I\omega^2$,

where $\omega = \dot{\xi} + \dot{\psi}$, and I: moment of inertia. Hence,

$$L=\frac{1}{2}m(a+b)^2\dot{\xi}^2+\frac{1}{2}I\omega^2-mg(a+b)\cos\xi$$

Using Euler–Lagrange equation for ξ: $\frac{d}{dt}\left(\frac{\partial L}{\partial\dot{\xi}}\right)-\frac{\partial L}{\partial\xi}=\lambda b$.

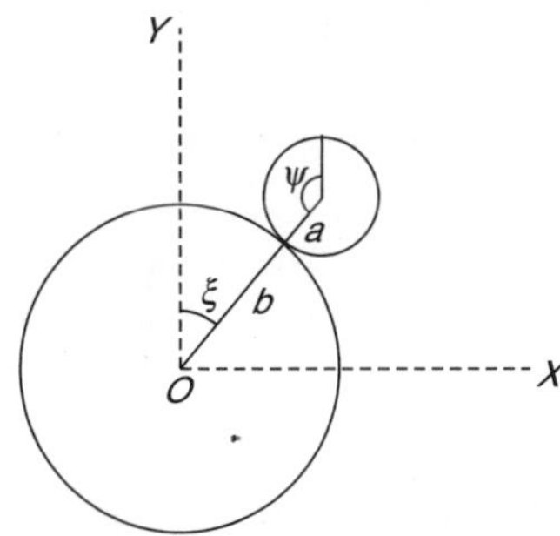

Hence: $\frac{d}{dt}[m(a+b)^2\dot{\xi}+I(\dot{\xi}+\dot{\psi})]+mg(a+b)\sin\dot{\xi}=\lambda b$. (i)

Using the Euler–Lagrange equation for ψ: $\frac{d}{dt}\left(\frac{\partial L}{\partial \dot{\psi}}\right)-\frac{\partial L}{\partial \psi}=-\lambda a$.

$$\frac{d}{dt}[I(\dot{\xi}+\dot{\psi})]=-\lambda a. \quad \text{(ii)}$$

From the equations (i) and (ii):

$$m(a+b)^2\ddot{\xi}+I(\ddot{\xi}+\ddot{\psi})+mg(a+b)\sin\xi=-\frac{I(\ddot{\xi}+\ddot{\psi})b}{a},$$

i.e., $m(a+b)^2\ddot{\xi}+I(\ddot{\xi}+\ddot{\psi})\left[\frac{a+b}{a}\right]+mg(a+b)\sin\xi=0$.

Hence: $(a+b)^2\left[m+\frac{I}{a^2}\right]\ddot{\xi}+mg(a+b)\sin\xi=0$, because $ad\psi-bd\xi=0$ and hence $\ddot{\psi}=\frac{b}{a}\ddot{\xi}$.

Thus: $\ddot{\xi}=-\frac{5g\sin\xi}{7(a+b)}$ (since $I=\frac{2}{5}ma^2$).

Additional Problems

P6.7 Consider a medium for which the refractive index is $n=\frac{a}{\rho^2}$, where α is a constant and ρ is the distance from the origin. Use Fermat's Principle to find the path of a ray of light traveling in a plane containing the origin. (Hint: use plane polar coordinates. Parametrize with $\varphi=\varphi(\rho)$ and show that the resulting path is a circle passing through the origin).

P6.8 Find the dimensions of a parallelepiped having maximum volume circumscribed by a sphere of radius R.

P6.9 Find the shortest path between two points whose Cartesian coordinates are (0, –1, 0) and (0, 1, 0) on the conical surface $z=1-\sqrt{x^2+y^2}$. What is the length of this path?

P6.10 Find the extremum of the functional $J(x)=\int_0^{\pi}(2x\sin t-\dot{x}^2)dt$ that satisfies $x(0)=x(\pi)=0$.

Show that this extremum provides the global maximum of J.

P6.11 Consider a mass m on the end of a spring of natural length l and spring constant k. It is suspended from the roof of a room. The spring has a mass suspended at its bottom. Let ξ be the instantaneous length of the spring from the point of suspension. The spring can also oscillate like a pendulum, and as it oscillates, it makes an instantaneous angle ψ at the point of suspension with the vertical line.

(a) Show that the Lagrangian for this system is: $L=\frac{1}{2}m(\dot{\xi}^2+\xi^2\dot{\psi}^2)-\frac{1}{2}k(\xi-l)^2+mg\xi\cos\psi$

(b) Write down the Lagrange equation of motion.

(c) Determine the equilibrium configurations, and identify the points of stable and unstable equilibrium.

(d) At each equilibrium point, approximate the Lagrange equations near the equilibrium to the first order and obtain the solution of the resulting first order differential equation.

(e) Determine the oscillation frequencies near the points of stable equilibrium.

P6.12 Suppose a string is tied between two fixed points ($x = 0, y = 0$) and ($x = l, y = 0$). Let $y(x, t)$ be a small transverse displacement of the string from its equilibrium at a position $x \in (0, l)$ and $t > 0$. The string has a linear mass density μ per unit length, and constant tension F. Show that the kinetic and potential energies are respectively given by:

$T = \frac{\mu}{2}\int_0^l y_t^2 dx$ and $V = F\int_0^l (\sqrt{1 + y_x^2} - 1)dx$. The subscripts in the integrands denote partial derivatives. You may neglect the effect of gravity. If the displacement y is small; $|y_x| \ll 1$, show that the Lagrangian can be approximated by an expression quadratic in y, and find the Euler–Lagrange equation for the approximate Lagrangian.

P6.13 Assume that the cost of flying an aircraft at height z is e^{-kz} per unit distance of flight-path, where k is a positive constant. Consider that the aircraft flies in the (x, z)-plane from the point $(-a, 0)$ to the point $(a, 0)$ where $z = 0$ corresponds to ground level. Assuming $ka < \frac{\pi}{2}$, determine the extremal for the problem that would minimize the total cost of the flight.

P6.14 (Dido's problem) Given a rope of a fixed length L, what shape should the rope take so that the isoperimetric area enclosed by the rope and a straight line segment is maximized?

P6.15 Consider a metal rope that is hanging between two posts. The length of the rope is L, and it has uniform linear mass density ρ. Determine the shape the rope must take that would minimize its total gravitational potential energy.

P6.16 Show that the shape of the curve must be a catenary if it minimizes the surface area that it would sweep when revolved about the z-axis.

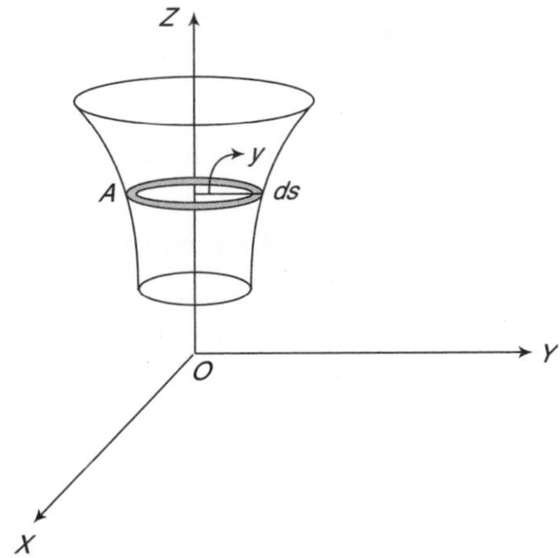

P6.17 Determine, using variational calculus, the shortest path between (0, 1) and (1, 1) that goes first to the horizontal axis $y = 0$ (i.e., the X-axis) and bounces back. Show that the best path treats this axis like a mirror: angle of incidence = angle of reflection.

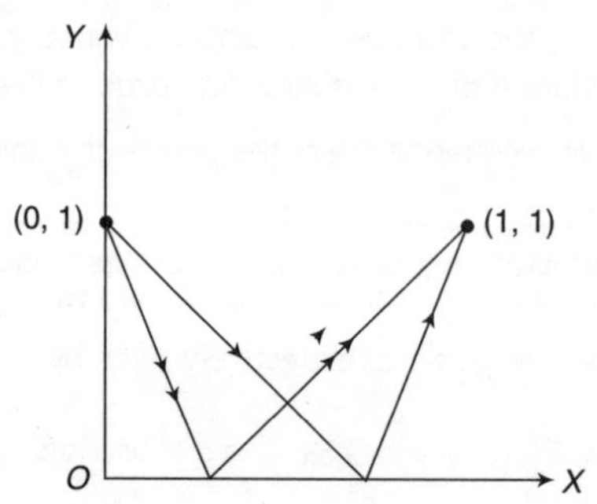

P6.18 (a) A regular helix trajectory around a cylinder has $x = \cos\theta$, $y = \sin\theta$, $z = u(\theta)$. Its length is $L = \int\sqrt{dx^2 + dy^2 + dz^2} = \int\sqrt{1 + (u')^2}\,d\theta$. Show that u' = constant satisfies Euler's equation.

(b) Show that the geodesic on a right circular cylinder is a helix.

P6.19 If f satisfies the Euler–Lagrange's equation, i.e., if $\frac{\partial f}{\partial y} - \frac{d}{dx}\left(\frac{\partial f}{\partial y'}\right) = 0$, then show that f is the total derivative $\frac{dg}{dx}$ of some function of x and y. Prove also the converse of this result.

P6.20 Find the extremum of the functional $J(x) = \int_1^2 \frac{\dot{x}^2}{t^3}dt$ that satisfies $x(1) = 3$ and $x(2) = 18$. Show that this extremum provides the global minimum of J.

References

[1] Goldstein, Herbert, Charles Poole, and John Safko. 2002. *Classical Mechanics*. 3rd Edition. New York: Pearson Education, Inc.

[2] Mukunda, N., and E. C. G. Sudarshan. 1974. *Classical Dynamics*. New York: John Wiley and Sons.

[3] Moore, Thomas A. 2004. 'Getting the Most Action Out of Least Action: A Proposal.' *Am. J. Phys.* 72(4): 522–527.

[4] Hanca, J., E. F. Taylor, and S. Tulejac. 2004. 'Deriving Lagrange's Equations Using Elementary Calculus.' *Am. J. Phys.* 72(4): 510–513.

[5] Hanca, J., and E. F. Taylor. 2004. 'From Conservation of Energy to the Principle of Least Action: A Story Line'. *Am. J. Phys.* 72(4): 514–521.

[6] Deshmukh, P. C., Parth Rajauria, Abiya Rajans, B. R. Vyshakh, and Sudipta Dutta. 2017. 'The Brachistochrone.' *Resonance* 22(9): 847–866.

[7] Derbes, David. 1996. 'Feynman's Derivation of the Schrödinger Equation.' *Am. J. Phys.* 64(7): 881–884.

[8] Feynman, R. P., R. B. Leighton, and M. Sands. 1964. *The Feynman Lectures on Physics*. 2: 19–2 Massachusetts: Addison-Wesley Publishing Company Inc. Reading.

[9] https://phys.libretexts.org/TextBooks_and_TextMaps/Classical_Mechanics/Book%3A_Classical_Mechanics_(Tatum)/19%3A_The_Cycloid/19.09%3A_The_Cycloidal_Pendulum. Accessed 6 June 2018.

CHAPTER 7

Angular Momentum and Rigid Body Dynamics

The fragrance of flowers spreads only in the direction of the wind, but the goodness of man spreads in all directions.

—*Chanakya*

7.1 THE ANGULAR MOMENTUM

We have already discussed in Chapters 1 and 6 the important connections between symmetry and conservation principles. We know that linear momentum is conserved in translationally invariant space. Translation invariance means that the properties of space remain exactly the same if you take a step away from where you are standing. This must be regardless of the size of the step, as long as you stay throughout the step taken within the medium under consideration. When space is *translationally* invariant, it is called *homogeneous*. We now consider a different kind of geometric invariance, namely rotational invariance, in our usual 3-dimensional Euclidean space. Properties of space may or may not be the same in all directions. Wind, for example, can flow in any one of the infinite possible directions. It carries the fragrance of flowers in the direction in which it flows. However, some properties may not change, no matter in which direction of space you may look at. An example of this is the strength $\frac{q}{r}$ of the Coulomb potential of a point charge q. It decreases as the inverse of the distance r from the charge. Such properties are *rotationally invariant*, and are called *isotropic*.

As one would expect, in line with the Noether's theorem (Chapters 1 and 6), there ought to be a conservation principle associated with the isotropic, rotational invariance. The physical quantity that is conserved in isotropic space is the *angular momentum*. A point particle whose instantaneous position vector is $\vec{r}$ with respect to a point O, and is moving with a linear momentum $\vec{p}$ has an angular momentum $\vec{\ell} = \vec{r} \times \vec{p}$ about the point O. The angular momentum, being the cross-product of two polar vectors, is an axial vector. Just the way the linear momentum is conserved in *homogenous* space, the angular momentum is conserved in *isotropic* space. The linear momentum of a system changes if we apply a force on it, and at a rate that is equal to the applied force, $(\dot{\vec{p}} = \vec{F})$. Likewise, the angular momentum $\vec{\ell}$ of that object would *change* if we applied a *torque*,

$$\vec{\tau} = \vec{r} \times \vec{F}, \tag{7.1a}$$

on it.

The rate of change of angular momentum is equal to the torque acting on it:

$$\dot{\vec{\ell}} = \vec{\tau}. \tag{7.1b}$$

Equation 7.1 explains why the angular momentum is conserved in *isotropic* space. In such a space, the force would have the same value in all directions; i.e., it would be radial, and have a component only along $\hat{e}_r$. It will have no component along $\hat{e}_\theta$ or $\hat{e}_\varphi$. The cross product of the two vectors in Eq. 7.1a is therefore zero, ensuring that $\dot{\vec{\ell}}$ is zero and thus the constancy of the angular momentum, $\vec{\ell}$. The analogy between the equation of motion that gives the rate of change of the *linear* momentum with that which gives the equation of motion for the rate of change of the *angular momentum* is often useful. However, it should be used with care. The need for such a care becomes extremely important as one progresses to more advanced applications in physics, especially using quantum theory. The linear momentum can be considered for all algebraic purposes to be a 'free' vector, in the sense that two linear momenta of equal magnitudes and same direction are *equal* (in Euclidean 3d space), no matter where they are placed; i.e., no matter where the tip or the tail of the arrows which represent them. The same is not true, in general, about two angular momenta, since the *very definition* of the angular momentum $\vec{\ell} = \vec{r} \times \vec{p}$ involves the position vector $\vec{r}$ which is referenced to a particular point in space. This reference point must therefore *always* be specified while analysing the angular momentum. For this reason, the angular momentum is often called as the *moment of momentum*. Likewise, any change in it can be effected only by the *moment of the force*, $\vec{\tau} = \vec{r} \times \vec{F}$, also about the *same* point.

The above definition is readily extended to a set of discrete particles, or even to a continuous medium. The angular momentum of a system of discrete particles about a point, like that of the asteroids in the belt between Mars and Jupiter about the Sun, is given by their sum, $\sum_i \vec{\ell}_i = \sum_i \vec{r}_i \times \vec{p}_i$, where $\vec{r}_i$ is the position vector of the center of mass of the ith asteroid with respect to the Sun, and $\vec{p}_i$ is the product of the mass of the asteroid with instantaneous linear velocity of the center of mass. For continuous mass distributions, these summations are easily extended into integrals over a line, a surface or a volume, as appropriate.

In the case of a linear mass distribution that has a mass density $\lambda(\vec{r})$ at a point whose position vector is $\vec{r}$ relative to a point O, the mass of an infinitesimal line segment ds located at $(\vec{r})$ is $\lambda(\vec{r})ds$. The angular momentum of the extended 1-dimensional slim linear object, like a wire, is therefore $\vec{\ell}_{1d-\text{wire}} = \int \vec{r} \times \lambda(\vec{r})\, \vec{v}(\vec{r})\, ds$, where $\vec{v}(\vec{r})$ is the instantaneous velocity of the tiny mass element $\lambda(\vec{r})ds$. For a surface mass distributions of a 2-dimensional flat-plate, having a surface mass density $\sigma(\vec{r})$, and for a 3-dimensional mass distribution at volume mass density $\rho(\vec{r})$, the corresponding expressions for the angular momentum about the point O are, respectively, $\vec{\ell}_{2d\text{-Flat-Plate}} = \iint \vec{r} \times \sigma(\vec{r})\vec{v}(\vec{r})\, da$ and $\vec{\ell}_{3d\text{-Object}} = \iiint \vec{r} \times \rho(\vec{r})\vec{v}(\vec{r})\, dV$. Note that the *position* in the mass distribution of a point in these expressions is given by the vector $\vec{r}$, which is the position vector with respect to some fixed point O. These definitions are general, since they can be used even when the mass density is *not* uniform, and also when the shape over which the mass is distributed—whether over a line, surface or volume—is irregular. In this chapter we shall consider rotational motion of rigid bodies, (i) about a point or an axis *within* itself, commonly called the *spin* motion, and/or (ii) about a point or an axis *outside* that object, called the *orbital* motion.

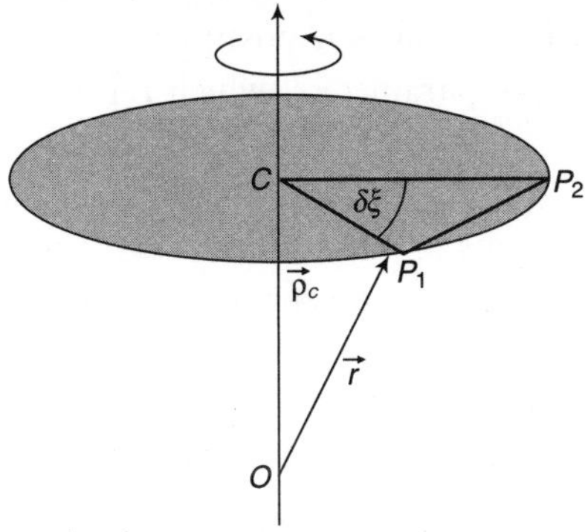

Fig. 7.1 In rotational motion, a point on a body moves from point P_1 to P_2 in an infinitesimal time interval δt over an arc length δs = arc (P_1 P_2). This is an arc of a circle of radius $CP_1 = \rho_C$. It sweeps an angle $\delta\xi$ subtended at the point C by the arc (P_1 P_2).

In the rotational motion, an object sweeps an angle about a point (or an axis) in an arc, as shown in Fig. 7.1. The magnitude of instantaneous linear velocity with which a point moves during this rotation is therefore

$$v = \lim_{\delta t \to 0} \frac{\delta s}{\delta t} = \lim_{\delta t \to 0} \frac{\rho_C \delta\xi}{\delta t} = \rho_C \left(\lim_{\delta t \to 0} \frac{\delta\xi}{\delta t} \right) = \rho_C \omega, \tag{7.2a}$$

wherein we define $\omega = \lim_{\delta t \to 0} \frac{\delta\xi}{\delta t} = \frac{d\xi}{dt}$, (7.2b)

as the angular speed of rotation. It gives us the angle swept per unit time. While $\omega = \frac{d\xi}{dt}$ is just a scalar, we promote the idea of a vector $\vec{\omega}$, which shall be called the angular velocity. This would have a *direction* given by the right-hand-screw's forward motion, if it is turned the same way as in this rotation. Since the instantaneous linear velocity of the point is in the direction P_1 to P_2 in the limit $\delta t \to 0$, we see from Fig. 7.1 that

$$\vec{v} = \vec{\omega} \times \vec{r} = \vec{\omega} \times (\overrightarrow{OC} + \vec{\rho}_C) = (\vec{\omega} \times \overrightarrow{OC}) + (\vec{\omega} \times \vec{\rho}_C) = \vec{\omega} \times \vec{\rho}_C. \tag{7.2c}$$

We readily see from Eq. 7.2c, since $\vec{v}$ and $\vec{\rho}_C$ are polar vectors, that the angular velocity is in fact an axial vector, also called a pseudo-vector (see Chapter 2), just as the angular momentum vector itself is.

Now, the characteristic, defining, property of a rigid body is that the distance d_{ij} between any of its pair of particles (i, j) does not change, no matter what net force acts on it. As opposed to this, particles in a heap of sand would move relative to each other, as would molecules of a fluid. A rigid body having N particles in 3-dimensions would need 3N spatial Cartesian coordinates (and likewise in any other coordinate system, such as the spherical polar or the cylindrical polar coordinates). However, the constraint that d_{ij} = constant between every pair of particles in the rigid body makes it possible to specify a point P anywhere in the rigid body by specifying just six parameters, which represent the generalized coordinates that describe the *position* and *orientation* of the rigid body. From 3N, down to just six, is an extra ordinary reduction in the amount of information needed. This reduction is enabled by taking advantage of the constraints d_{ij} = constant as explained in the discussion on Fig. 6.2 of Chapter 6. This figure illustrates the six generalized coordinates of an arbitrary point P in the rigid body. Three of these coordinates are with respect to a frame $\{x, y, z\}$ which is rigidly fixed to a point O *within* the rigid body which itself. The remaining three coordinates are those of the point O in the laboratory frame of reference $\{X, Y, Z\}$.

The frame $\{x, y, z\}$ is the *body-fixed* frame, and the frame $\{X, Y, Z\}$ as the *space-fixed* frame, or the laboratory frame, which we shall usually consider to be an *inertial* frame of reference. The body-fixed frame could have a linear motion, or a rotational motion, or a combination of both, with respect to the space-fixed frame. The origin of the body-fixed frame is best chosen to be its center of mass which will not change no matter how the body is turned or tossed, since the body is rigid. For a rigid body, unlike a bowl of sand or a fluid, redistribution of mass does not occur during any of its motion. It is therefore natural to analyze the motion of a rigid body in terms of (i) its *spin* angular momentum, i.e., the angular momentum of the whole rigid body about an axis *through* its center of mass, and (ii) the *orbital* motion of the center of mass itself about a point that is exterior to the object. We shall take the origin of the laboratory frame of reference $\{X, Y, Z\}$ to be this external point.

For simplicity, we first consider a flat 2-dimensional rigid body, shown in Fig. 7.2, which is spinning at angular velocity $\vec{\omega} = \omega\hat{e}_z$ about an axis through its center of mass C, *and* perpendicular to the plane of the body. The mass density of this body may or may not be uniform. It is fruitful to employ a cylindrical polar coordinates system with its Z-axis along the axis of rotation of the flat body, with its origin at the center of mass of the body. Thus, the mass δm_P in an elemental area δa located at the point P (Fig. 7.2) is given by $\sigma(\vec{r})\delta a$, where $\sigma(\vec{r})$ is the surface mass density *at* the point P. In Fig. 7.2, we employ a subscript O to reference a variable to the origin O of the laboratory coordinate system. Likewise, we use the subscript C to reference the coordinates to the center of mass C of the object.

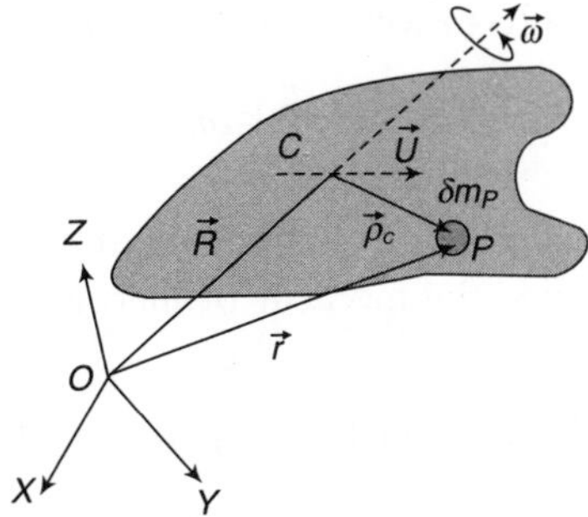

Fig. 7.2 A 2-dimensional flat mass distribution revolves at angular velocity $\vec{\omega}$ about the center of mass of the disk, at C. The center of mass itself is moving at the velocity $\vec{U}$ in the laboratory frame. $\delta m_P = \sigma(P)\delta a$ is a tiny mass in elemental area δa at the point P located at the position vector $\vec{\rho}_C$ with respect to the center of mass, and $\vec{r}$ with respect to the laboratory frame. The position vector of the center of mass of the disk with respect to O is $\vec{R}$.

The angular momentum of the flat-disk with respect to O is

$$\vec{\ell}_O = \iint \vec{r} \times \sigma(\vec{r})\vec{v}_O(P)da, \tag{7.3}$$

where $\vec{v}_O(P)$ is the instantaneous velocity of a tiny element δm_P at the point P. This velocity is the result of (i) movement at velocity $\vec{U}$ of the *entire flat-plate* with respect to the laboratory frame, and (ii) rotation of the flat-plate about its center of mass.

Thus, $$\vec{v}_O(P) = \vec{U} + \vec{v}_C(P). \tag{7.4}$$

where $\vec{v}_C(P)$ is the instantaneous velocity of the point P with respect to the center of mass, C, of the flat body.

Hence, $$\vec{\ell}_O = \iint (\vec{R} + \vec{\rho}_C) \times \delta m_P(\vec{v}_C(P) + \vec{U}), \tag{7.5a}$$

i.e., $$\vec{\ell}_O = \iint \vec{R}\times \delta m_P\vec{v}_C(P) + \iint \vec{R}\times \delta m_P\vec{U} + \iint \vec{\rho}_C \times \delta m_P\vec{v}_C(P) + \iint \vec{\rho}_C \times \delta m_P\vec{U}. \tag{7.5b}$$

We now examine the four surface integrals in Eq. 7.5b.

The first term is: $$\iint \vec{R}\times \delta m_P\vec{v}_C(P) = \iint \vec{R} \times \overrightarrow{\delta p_C}(P) = \vec{R}_O \times \vec{p}_C^{\,\text{Net}} = \vec{0}, \tag{7.6a}$$

since net linear momentum $\vec{p}_C^{\,\text{Net}}$ of the traveling body in its *own* frame referred to its center of mass is, of course, zero.

The second term is: $$\iint \vec{R} \times \delta m_P\vec{U} = \vec{R} \times \vec{U} \iint \delta m_P = \vec{R} \times M\vec{U} = \vec{L}_0, \tag{7.6b}$$

which is the *orbital angular momentum* of the object, denoted by subscript 'o', about the center of the laboratory frame of reference.

The third term is $\iint \vec{\rho}_C \times \delta m_P\vec{v}_C(P) = \iint \rho_C(P)\hat{e}_{\rho C} \times \delta m_P\, \rho_C(P)\dot{\varphi}_C\, \hat{e}_{\varphi C},$

wherein we have used Eq. 2.12b to write $\vec{v}_C(P) = \rho_C(P)\dot{\varphi}_C\hat{e}_{\varphi C}$.

Hence, $\iint \vec{\rho}_C \times \delta m_P \vec{v}_C(P) = \hat{e}_z\dot{\varphi}_C \iint [\rho_C(P)]^2 \delta m_P = \hat{e}_z\dot{\varphi}_C I_{zC} = \hat{e}_z\omega_C I_{zC}$, (7.6c)

where I_{zC} is the 'moment of inertia' of the disc, about an axis normal to the disc through its center of mass. We shall define the moment of inertia more precisely a bit later, in this chapter.

Finally, the fourth term is $\iint \vec{\rho}_C \times \delta m_P \vec{U} = \left[\iint \vec{\rho}_C\ \delta m_P\right] \times \vec{U} = \vec{0}$, (7.6d)

since, $\left[\iint \vec{\rho}_C\ \delta m_P\right]$ gives the position vector of the center of mass in its *own* frame, and is obviously $\vec{0}$.

Combining the four terms in Eq. 7.6a, b, c, and d and putting them back in Eq. 7.5b, we get

$$\vec{\ell}_O = \vec{L}_0 + \hat{e}_z\omega_C I_{zC}. \tag{7.7}$$

Equation 7.7 neatly enables us to separate the net angular momentum of the rigid body when it is (i) spinning about an axis through it, and (ii) also traveling along an orbit, about an external point. The 'spinning' motion about an axis generates the *spin angular momentum* $\hat{e}\omega_C I_{zC}$ of the object. Similarly, the trajectory that the center of mass has about an external point generates its *orbital angular momentum* $\vec{L}_0$. The separation of the angular momentum of a rigid body into the spin part and the angular part is extremely useful. In fact, the total *kinetic energy* of the rigid body also neatly separates into the spin part and the orbital part, as we now see. We write the kinetic energy of the rigid body in the frame of reference centered at the external point O (Fig. 7.2) as the integration over the kinetic energy of each mass element, $\delta m_P = \sigma(\vec{r})da$, in the body. The total kinetic energy then is

$$T = \iint \frac{1}{2}\sigma(\vec{r})\,[\vec{v}_O(P)\cdot\vec{v}_O(P)]\,da, \tag{7.8a}$$

i.e., $$T = \iint \frac{1}{2}\sigma(\vec{r})\,[\{\vec{U} + \vec{v}_C(P)\}\cdot\{\vec{U} + \vec{v}_C(P)\}]\,da. \tag{7.8b}$$

Hence,

$$T = \frac{1}{2}\vec{U}\cdot\vec{U}\iint \sigma(\vec{r})\,da + \frac{1}{2}\iint \sigma(\vec{r})\,\{\vec{v}_C(P)\cdot\vec{v}_C(P)\}\,da + \vec{U}\cdot\iint \sigma(\vec{r})\vec{v}_C(P)\,da. \tag{7.8c}$$

The last term vanishes, since it involves the scalar product of the velocity of center of mass with the *net liner momentum* of the body with respect to its *own* center of mass. $\iint \sigma(\vec{r})\vec{v}_C(P)\,da$ is nothing but the term $\vec{p}_C^{\,\text{Net}}$ in Eq. 7.6c. We already know that it is zero.

Thus, $T = \left\{ \frac{1}{2} \vec{U} \cdot \vec{U} \iint \sigma(\vec{r}) da \right\} + \left\{ \frac{1}{2} \iint \sigma(\vec{r}) \{ \vec{v}_C(P) \cdot \vec{v}_C(P) \} \, da \right\}$. (7.8d)

Writing $\vec{v}_C(P)$ in plane polar coordinates centered at the center of mass C, we see that

$$T = \frac{1}{2}(\vec{U} \cdot \vec{U})M + \iint \frac{1}{2} \sigma(\vec{r}) [(\dot{\rho}\hat{e}_\rho + \rho\dot{\varphi}\hat{e}_\varphi) \cdot (\dot{\rho}\hat{e}_\rho + \rho\dot{\varphi}\hat{e}_\varphi)] (\rho d\rho d\varphi). \tag{7.8e}$$

Hence, $T = \frac{MU^2}{2} + \frac{1}{2}\dot{\varphi}^2 \iint \rho^2 (\sigma(\vec{r}) \, \rho d\rho d\varphi) = \frac{MU^2}{2} + \frac{I\dot{\varphi}^2}{2} = \frac{MU^2}{2} + \frac{I\omega^2}{2}$. (7.8f)

Hence, the net kinetic energy of the orbiting *and* spinning rigid body is

$$T_{\text{net KE}} = T_{\text{translational}} + T_{\text{rotational}}. \tag{7.9}$$

We see that just as the separation of the total angular momentum of the body in its orbital part (i.e., instantaneous *translational* part) and its *rotational* (spin) part, the kinetic energy of an object in a state of linear and rotational motion can also be written as an arithmetic sum of corresponding components. Equations 7.7 and 7.9 provide a very elegant classification of the total motion of the rigid body in terms of the translational motion of its center of mass, and its rotational motion about that center. This terminology, stemming from the classification of the rigid body's physical properties in terms of its translational (orbital) motion and rotational (spin) motion, is carried into the quantum theory as well. The analogy is extremely useful, but also very deceptive, since quantum theory involves rather counter-intuitive and challenging ideas which are fundamentally different, and well-beyond classical analogies.

We are often confronted with the analysis of motion that is usually complicated. For example, consider the Earth's orbit around the Sun. We can express it as an elliptic orbit which may be considered to be in the XY plane. The Earth's spin about its North–South axis through its center of mass is, however, such that the spin axis is *tilted* with respect to the Z-axis, at an angle ~23.5°. There are other complications as well. For example, the mass–density inside the Earth is anisotropic. Besides, the mass density changes with time, not merely because of earthquakes but also because rivers flow, and occasionally even large land-masses slide. Therefore, to deal with more complex situations, it is necessary to develop a more powerful formalism of the rotational motion of a rigid body. Toward that end, we introduce in our formalism physical entities called the 'moment of inertia tensor', 'Euler angles', etc.

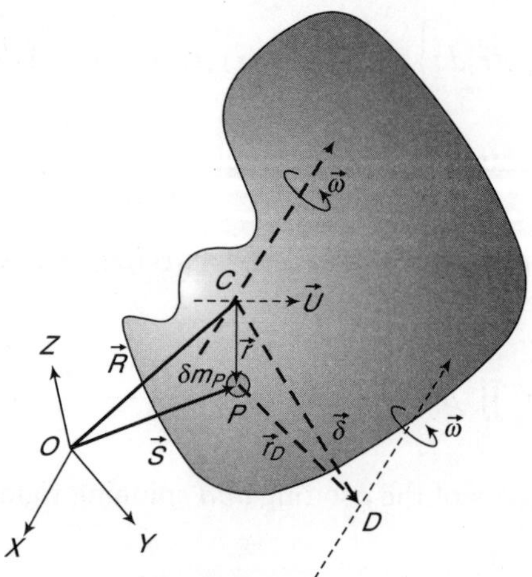

Fig. 7.3 A 3-dimensional mass distribution that revolves at angular velocity $\vec{\omega} = \omega_x \hat{e}_X + \omega_y \hat{e}_Y + \omega_z \hat{e}_z$ about an axis shown by a dark dashed line through the center of mass at the point C of the object. The center of mass C itself is located at the tip of the position vector $\vec{R}$ with respect to the laboratory frame of reference and has a linear velocity $\vec{U}$ in this frame.

$\delta m_P(\vec{r}) = \rho(P)\delta V$ is a tiny mass in an elemental volume δV at the point P, located at the position vector $\vec{r}$, with respect to the center of mass. Its position vector with respect to the origin of the laboratory frame is $\vec{S}$. To get the result for the parallel axis theorem, we consider the object to be spinning about an axis through a point D, displaced from P, by $\overrightarrow{CD} = \vec{\delta}$. The axis of rotation (light dashed line) through the point D is parallel to the one considered earlier through the center of mass C.

We can now write an expression for the total angular momentum of an object, whose center of mass is moving at a velocity $\vec{U}$ with respect to the laboratory frame centered at O and which is also spinning about an axis through its center of mass at an angular velocity $\vec{\omega}$. If the instantaneous velocity of a tiny mass $\delta m_P = \rho(P)\delta V$ at P, in an infinitesimal volume element δV, is $\vec{v}_C(P)$ with respect to the center of mass of the object, then the angular momentum of the object about O is given by

$$\vec{\ell}_O = \iiint (\vec{R} + \vec{r}) \times \delta m_P\, (\vec{v}_C(P) + \vec{U}) = \iiint (\vec{R} + \vec{r}) \times \rho(\vec{r})\, (\vec{v}_C(P) + \vec{U}) dV. \tag{7.10a}$$

In the above equation, $\vec{R}$ is the position vector of the center of mass with respect to O, and $\vec{r}$ is the position vector of the point P with respect to the center of mass. $\rho(\vec{r})$ is density of the material at the point P. Thus,

$$\vec{\ell}_O = \vec{R} \times \iiint \vec{v}_C(P)\delta m_P + \iiint \vec{r} \times \vec{v}_C(P)\delta m_P + \vec{R} \times \vec{U} \iiint \delta m_P + \left(\iiint \vec{r}\delta m_P\right) \times \vec{U}. \tag{7.10b}$$

Let us now analyze the above four terms. In the first term, the cross product of $\vec{R}$ is with an integral which gives the total linear momentum of the moving body in its *own* center of mass frame. The latter, we already know, is $\vec{p}_C^{\,\text{Net}}$; it is essentially zero. Hence the first term vanishes. Likewise, the last term involves the cross product with the velocity $\vec{U}$ with the position of center of mass in its own frame, which is also of course zero. Thus, only the second and the third terms survive.

$$\vec{\ell}_O = \iiint \vec{r} \times \vec{v}_C(P)\delta m_P + (\vec{R} \times M\vec{U}), \tag{7.11a}$$

i.e., $$\vec{\ell}_O = \iiint \vec{r} \times (\vec{\omega} \times \vec{r})\delta m_P + \vec{\ell}_O^{\,\text{orbital}}. \tag{7.11b}$$

The first term above comes from the internal rotational motion of the body. It contributes to the total angular momentum. This part of the angular momentum is due to the object's *spin* about an axis through itself, and is therefore called the *spin angular momentum*, $\vec{\ell}_C^{\,\text{spin}}$. The second term in Eq. 7.11 is just the cross product of position vector of the center of mass of the object with the net linear momentum of the object moving at the velocity of $\vec{U}$ in the laboratory frame. This net motion describes the trajectory, i.e., the *orbit*, of the object regardless of any additional internal motion the object has about its own axis. It is therefore the *orbital angular momentum* $\vec{\ell}_O^{\,\text{orbital}}$ of the body about the origin O. Equation 7.11 thus expresses the *total* angular momentum of the object as a sum of the spin part $\vec{\ell}_C^{\,\text{spin}}$ and the orbital part $\vec{\ell}_O^{\,\text{orbital}}$.

Expanding the vector triple product $\vec{r} \times (\vec{\omega} \times \vec{r})$ in Eq. 7.11b, we get

$$\vec{\ell}_C^{\,\text{spin}} = \iiint r^2 \vec{\omega}\delta m_P(r) - \iiint (\vec{r} \cdot \vec{\omega})\vec{r}\,\delta m_P(r). \tag{7.12}$$

Equation 7.12 is simple, but it contains an interesting result: It shows that the spin angular momentum $\vec{\ell}_C^{\,\text{spin}}$ and the spin angular velocity $\vec{\omega}$ are not necessarily parallel to each other, even if $\vec{\ell}_C^{\,\text{spin}}$ is on account of $\vec{\omega}$. This is almost counter-intuitive, since quite commonly, the angular velocity and the angular momentum are parallel to each other. We discuss the consequences of this peculiar result in the next section.

7.2 THE MOMENT OF INERTIA TENSOR

The complex distribution of mass inside rigid body is an important factor to reckon with to understand why $\vec{\ell}_C^{\,\text{spin}}$ and $\vec{\omega}$ are not necessarily parallel to each other. Of the two integrals on the right hand side of Eq. 7.12, the first one is always parallel to $\vec{\omega}$, but not the second. We cannot hope that the vectors in the second integrand in different directions will necessarily cancel each other. We therefore have a close look at Eq. 7.12, which we rewrite as

$$\vec{\ell}_C^{\,\text{spin}} = (\omega_x \hat{e}_x + \omega_y \hat{e}_y + \omega_z \hat{e}_z) \iiint r^2 \delta m_P - \iiint (x\omega_x + y\omega_y + z\omega_z)(x\hat{e}_x + y\hat{e}_y + z\hat{e}_z)\delta m_P. \tag{7.13}$$

As you will soon discover, uncomplaining *term-by-term* expansion of Eq. 7.13 will soon return rich dividends.

Thus,

$$\left.\begin{aligned}\vec{\ell}_C^{\text{spin}} &= (\omega_x\hat{e}_x \iiint r^2\delta m_P + \omega_y\hat{e}_y \iiint r^2\delta m_P + \omega_z\hat{e}_z \iiint r^2\delta m_P)\\ &- \iiint (x\omega_x + y\omega_y + z\omega_z)(x\hat{e}_x)\delta m_P - \iiint (x\omega_x + y\omega_y + z\omega_z)(y\hat{e}_y)\delta m_P\\ &- \iiint x\omega_x + y\omega_y + z\omega_z\,(z\hat{e}_z)\delta m_P.\end{aligned}\right\} \tag{7.14a}$$

Assembling now the components along the unit vectors $\{\hat{e}_x, \hat{e}_y, \hat{e}_z,\}$, we get

$$\left.\begin{aligned}\vec{\ell}_C^{\text{spin}} = \hat{e}_x&\left[\omega_x\iiint (x^2+y^2+z^2)\delta m_P - \iiint (x^2\omega_x + xy\omega_y + zx\omega_z)\delta m_P\right] +\\ \hat{e}_y&\left[\omega_y\iiint (x^2+y^2+z^2)\delta m_P - \iiint (xy\omega_x + y^2\omega_y + yz\omega_z)\delta m_P\right] +\\ \hat{e}_z&\left[\omega_z\iiint (x^2+y^2+z^2)\delta m_P - \iiint (zx\omega_x + yz\omega_y + z^2\omega_z)\delta m_P\right].\end{aligned}\right\} \tag{7.14b}$$

In each Cartesian component, we observe that the symmetric quadratic terms (i.e., those involving $x^2\omega_x$, $y^2\omega_y$ and $z^2\omega_z$) pair up with opposite signs, and hence cancel each other. We therefore get

$$\left.\begin{aligned}\vec{\ell}_C^{\text{spin}} = \hat{e}_x&\left[\omega_x\iiint (y^2+z^2)\delta m_P - \omega_y\iiint xy\delta m_P - \omega_z\iiint zx\delta m_P\right] +\\ \hat{e}_y&\left[\omega_y\iiint (x^2+z^2)\delta m_P - \omega_x\iiint xy\delta m_P - \omega_z\iiint yz\delta m_P\right] +\\ \hat{e}_z&\left[\omega_z\iiint (x^2+y^2)\delta m_P - \omega_x\iiint zx\delta m_P - \omega_y\iiint yz\delta m_P\right].\end{aligned}\right\} \tag{7.14c}$$

We see how the spinning mass distribution generates an angular momentum that need *not* be parallel to the angular velocity of the spin. The following simplified *notation* for each of the nine terms in Eq. 7.14c will enable us write the above result in an impressive and useful form:

$$I_{xx} = \iiint (y^2+z^2)\delta m_P, \tag{7.15a}$$

$$I_{xy} = \iiint xy\,\delta m_P = I_{yx} \quad (\leftarrow \text{note the symmetry}), \tag{7.15b}$$

$$I_{xz} = \iiint xz\,\delta m_P = I_{zx} \quad (\leftarrow \text{note the symmetry}), \tag{7.15c}$$

$$I_{yy} = \iiint (z^2+x^2)\delta m_P, \tag{7.15d}$$

$$I_{yz} = \iiint yz\,\delta m_P = I_{zy} \quad (\leftarrow \text{note the symmetry}), \tag{7.15e}$$

$$\text{and}\quad I_{zz} = \iiint (x^2+y^2)\delta m_P. \tag{7.15f}$$

The 6+3 equations, Eqs. 7.15a–f (inclusive of the three equations corresponding to the symmetry $I_{yx} = I_{xy}$, $I_{zx} = I_{xz}$ and $I_{zy} = I_{yz}$), give the nine elements of the *moment of inertia tensor*'. It is denoted by 'I', or better by '$\mathbb{I}_{C,(x,y,z)}$'. The fat letter $\mathbb{I}$ emphasizes the fact that it is a tensor. It is a tensor of rank 2, with its 9 elements denoted by elements of a

3×3 matrix, using the notation introduced in the appendix to Chapter 2. The subscripts in $\mathbb{I}_{C, (x, y, z)}$ are often omitted for brevity. They are included here only to emphasize that the form of the inertia tensor is *specific* to the point about which it is defined, and *also* to the coordinate frame to which it is referenced. The inertia tensor changes from one coordinate system to another as the values of the integrals in Eq. 7.15a–f will change. These integrals involve the mass distribution relative to a *particular* coordinate system.

The inertia tensor in two coordinate systems may in some cases bear a simple relationship with each other. For example, we consider the spin about an axis through a different point, such as the point D in Fig. 7.3. We also consider this axis to be parallel to the previous one. In this case, the expressions in Eqs. 7.15a–f get only slightly modified. If we denote the position vector of the point P relative to the point D as $\vec{r}\,'$, or $\vec{r}_D$ and the vector $\overrightarrow{DC} = \vec{\delta}$ then we have $\vec{r}\,' = \vec{r} + \vec{\delta}$ and hence $x' = x + \delta_x$, $y' = y + \delta_y$, and $z' = x + \delta_x$. The elements of the moment of inertia tensor about a parallel axis through the point D will then be given by similar expressions:

$$(I_{x'x'})_D = \iiint (y'^2 + z'^2)\delta m_P = \iiint ((y + \delta_y)^2 + (z + \delta_z)^2)\delta m_P$$

i.e., $$(I_{x'x'})_D = \iiint (y^2 + z^2)\delta m_P + (\delta_y^2 + \delta_z^2) \iiint \delta m_P + 2\delta_y \iiint y\delta m_P + 2\delta_z \iiint z\delta m_P$$

The last two terms drop off, they being zero from the very definition of the center of mass.

Thus, $$(I_{x'x'})_D = (I_{xx})_C + M(\delta_y^2 + \delta_z^2). \tag{7.15g}$$

Likewise, $$(I_{x'y'})_D = (I_{xy})_C + M(\delta_x \delta_y), \text{ etc.} \tag{7.15h}$$

As we can see from Eqs. 7.15g, 7.15h (etc.), the moment of inertia tensor about an axis through the point D, when the axes through C and D are parallel, can be determined simply by adding a few simple terms, such as $M(\delta_y^2 + \delta_z^2)$, $M(\delta_x \delta_y)$ (etc.). This result is called the *generalized parallel axis theorem.*

Due to symmetry, each of the equation numbers 7.15b, 7.15c and 7.15e actually represent a *pair* of equations; called the *products of inertia*. Using Eq. 7.15, the Eq. 7.14c can now be written in a pronouncedly revealing form:

$$\vec{\ell}_C^{\text{spin}} = \begin{Bmatrix} \hat{e}_x\left[\omega_x I_{xx} - \omega_y I_{xy} - \omega_z I_{xz}\right] + \\ \hat{e}_y\left[-\omega_x I_{yx} + \omega_y I_{yy} - \omega_z I_{yz}\right] + \\ \hat{e}_z\left[-\omega_x I_{zx} - \omega_y I_{zy} + \omega_z I_{zz}\right] \end{Bmatrix} = \begin{Bmatrix} \hat{e}_x \ell_{C,x}^{\text{spin}} + \\ \hat{e}_y \ell_{C,y}^{\text{spin}} + \\ \hat{e}_z \ell_{C,z}^{\text{spin}} \end{Bmatrix}, \tag{7.16}$$

i.e., $$\begin{bmatrix} \ell_{C,x}^{\text{spin}} \\ \ell_{C,y}^{\text{spin}} \\ \ell_{C,z}^{\text{spin}} \end{bmatrix} = \begin{bmatrix} I_{xx} & -I_{xy} & -I_{xz} \\ -I_{yx} & I_{yy} & -I_{yz} \\ -I_{zx} & -I_{zy} & I_{zz} \end{bmatrix} \begin{bmatrix} \omega_x \\ \omega_y \\ \omega_z \end{bmatrix}. \tag{7.17a}$$

As in the notation of the appendix to Chapter 2, we identify the three components of the column matrix as a vector, and rewrite Eq. 7.17a as

$$\vec{\ell}_C^{\,\text{spin}} = \mathbb{I}_{C,(x,y,z)}\,\vec{\omega}, \tag{7.17b}$$

$$\text{with } \mathbb{I}_{C,(x,y,z)} = \begin{bmatrix} I_{xx} & -I_{xy} & -I_{xz} \\ -I_{yx} & I_{yy} & -I_{yz} \\ -I_{zx} & -I_{zy} & I_{zz} \end{bmatrix}. \tag{7.18}$$

Equation 7.18 represents a rank-2 tensor (see Section 2.4), called the *inertia tensor, about the center of mass* C and *with respect to the basis* $\{\hat{e}_x, \hat{e}_y, \hat{e}_z\}$. The symmetry emphasized in Eqs. 7.15a, b, and c demonstrates that the inertia tensor is a *symmetric tensor*; the elements of the matrix (Eq. 7.18) which are equidistant from the diagonal of the matrix are equal. The diagonal elements given by Eqs. 7.15a, d, and are summations of squares, and are thus always positive. The off-diagonal elements (product of inertia), can be either positive or negative, or even zero. Their values, as we can see from the nature of the integrands in Eq. 7.15, depend on the mass distributions about a *specific* point, and referenced to a *particular orientation* of the coordinate axes. The inertia tensor therefore contains information about *how the mass of the object is distributed about its center of mass and with respect to a particular orientation of the coordinate system.*

The matrix multiplication in Eq. 7.17 elegantly reveals in a compact form that the angular momentum of a rigid body is *not necessarily* parallel to the angular velocity. The elements of the inertia tensor provide a measure of the components of the spin angular momentum in terms of the mass distribution of the spinning body, and the components of the angular velocity.

The inertia tensor thus enables us write the relation between the spin angular momentum and the spin angular velocity in a neat form. It also helps write the relationship between the spin angular speed and the rotational kinetic energy in a simple form. In Eq. 7.8f, and Eq. 7.9, the rotational kinetic energy was identified as $T_{\substack{\text{rotational}\\\text{kinetic}\\\text{energy}}} = \frac{I\omega^2}{2}$. It has a quadratic dependence on the angular speed. The moment of inertia is, however, a tensor, given by Eq. 7.18. Retaining the same form, the rotational kinetic energy is therefore written in terms of the inertia tensor, using matrix multiplication rules, as

$$T_{\substack{\text{rotational}\\\text{kinetic}\\\text{energy}}} = \frac{1}{2}\{[\omega_x \quad \omega_y \quad \omega_z]_{1\times 3}\}\left\{\begin{bmatrix} I_{xx} & -I_{xy} & -I_{xz} \\ -I_{yx} & I_{yy} & -I_{yz} \\ -I_{zx} & -I_{zy} & I_{zz} \end{bmatrix}_{3\times 3}\begin{bmatrix} \omega_x \\ \omega_y \\ \omega_z \end{bmatrix}_{3\times 1}\right\} = \frac{1}{2}\vec{\omega}^{\dagger}\mathbb{I}\vec{\omega}]. \tag{7.19}$$

We can also get the same result by following the slightly lengthy procedure such as in Eq. 7.8. This would require writing the kinetic energy as an integral over the elemental kinetic energy of the tiny mass elements in the complete mass-distribution of the body, and then identify the elements of the moment of inertia tensor, as we did in Eq. 7.15. However, we got the result expressed in Eq. 7.19 by exploiting how tensor products are neatly contracted to give a tensor of a lower rank. Above, we got a scalar through this product, namely the rotational kinetic energy, $T_{\substack{\text{rotational}\\\text{kinetic}\\\text{energy}}} = \frac{1}{2}\vec{\omega}\cdot\vec{\ell}$, by taking (half of) the tensor product $\vec{\omega}^{\dagger}\mathbb{I}\vec{\omega}$ using matrix-multiplication.

For the remaining portion of this chapter, it is best to use the framework of the algebra of matrices. The brief review of matrix algebra that is summarized in the appendix to Chapter 2 is adequate for the purpose of this chapter.

Using the matrix diagonalization method, the inertia tensor can be referenced to a particularly convenient frame of reference, oriented along what are called the *principal axes* of the spinning body. The inertia tensor has a diagonal form when referenced to the principal axes. This is easily understood by considering a 3 × 3 square matrix $\mathbb{T}$ which would transform the inertia tensor matrix $\mathbb{I}_{C,(x,y,z)}$, and bring it into its diagonal form given by

$$\mathbb{I}_{C,(x',y',z'),d} = \mathbb{T}\,\mathbb{I}_{C,(x,y,z)}\,\tilde{\mathbb{T}}. \tag{7.20a}$$

In the above equation, $\tilde{\mathbb{T}}$ (also written alternatively as $\mathbb{T}^T$) is the transpose of $\mathbb{T}$.

Writing out the matrix elements in the above equation, we have

$$\mathbb{I}_{C,(x',y',z'),d} = \begin{bmatrix} T_{11} & T_{12} & T_{13} \\ T_{21} & T_{22} & T_{23} \\ T_{31} & T_{32} & T_{33} \end{bmatrix} \begin{bmatrix} I_{xx} & -I_{xy} & -I_{xz} \\ -I_{yx} & I_{yy} & -I_{yz} \\ -I_{zx} & -I_{zy} & I_{zz} \end{bmatrix} \begin{bmatrix} T_{11} & T_{21} & T_{31} \\ T_{12} & T_{22} & T_{32} \\ T_{13} & T_{23} & T_{33} \end{bmatrix} = \begin{bmatrix} i_1 & 0 & 0 \\ 0 & i_2 & 0 \\ 0 & 0 & i_3 \end{bmatrix}. \tag{7.20b}$$

The eigenvalues of the inertia tensor given by the diagonal elements, (i_1, i_2, i_3) are called the *principal moments*, as they give the moments of inertia of the object about its three principal axes. The principal axes are not along the vectors in the basis $\{\hat{e}_1, \hat{e}_2, \hat{e}_3\}$, but along the rotated basis $\{\hat{e}'_1, \hat{e}'_2, \hat{e}'_3\}$, respectively, which we refer to as the principal basis. The same matrix $\mathbb{T}$ would also transform the components of the angular momentum vector $\vec{\ell}_C^{\text{spin}}$. The result of this transformation would yield the components of the spin angular momentum in the rotated basis:

$$\left[\vec{\ell}_C^{\text{spin}} \text{ in } \{\hat{e}_{x'}, \hat{e}_{y'}, \hat{e}_{z'}\}\right] \equiv \begin{bmatrix} \ell_{C,x'}^{\text{spin}} \\ \ell_{C,y'}^{\text{spin}} \\ \ell_{C,z'}^{\text{spin}} \end{bmatrix} = \begin{bmatrix} T_{11} & T_{12} & T_{13} \\ T_{21} & T_{22} & T_{23} \\ T_{31} & T_{32} & T_{33} \end{bmatrix} \begin{bmatrix} \ell_{C,x}^{\text{spin}} \\ \ell_{C,y}^{\text{spin}} \\ \ell_{C,z}^{\text{spin}} \end{bmatrix}. \tag{7.21a}$$

Likewise, the components of the angular velocity in the rotated bases are

$$[\vec{\omega} \text{ in } \{\hat{e}_{x'}, \hat{e}_{y'}, \hat{e}_{z'}\}] \equiv \begin{bmatrix} \omega_{x'} \\ \omega_{y'} \\ \omega_{z'} \end{bmatrix} = \begin{bmatrix} T_{11} & T_{12} & T_{13} \\ T_{21} & T_{22} & T_{23} \\ T_{31} & T_{32} & T_{33} \end{bmatrix} \begin{bmatrix} \omega_x \\ \omega_y \\ \omega_z \end{bmatrix}. \tag{7.21b}$$

In the principal basis set of unit vectors $\{\hat{e}'_1, \hat{e}'_2, \hat{e}'_3\}$, the Eq. 7.17a takes the following form:

$$[\vec{\ell}_C^{\text{spin}} \text{ in } \{\vec{e}_x', \vec{e}_y', \vec{e}_z'\}] \equiv \begin{bmatrix} \ell_{C,x'}^{\text{spin}} \\ \ell_{C,y'}^{\text{spin}} \\ \ell_{C,z'}^{\text{spin}} \end{bmatrix} = \begin{bmatrix} i_1 & 0 & 0 \\ 0 & i_2 & 0 \\ 0 & 0 & i_3 \end{bmatrix} \begin{bmatrix} \omega_x' \\ \omega_y' \\ \omega_z' \end{bmatrix} = \begin{bmatrix} i_1\omega_x' \\ i_2\omega_y' \\ i_3\omega_z' \end{bmatrix}. \tag{7.22}$$

Even if Eq. 7.22 employs the diagonal form of the inertia tensor, its diagonal elements, in general, need not be equal to each other. We may therefore omit the explicit tensor (matrix) notation and simply write

$$[\vec{\ell}_C^{\text{spin}} \text{ in } \{\vec{e}_x', \vec{e}_y', \vec{e}_z'\}] = i\vec{\omega}, \tag{7.23}$$

if and only if $i_1 = i_2 = i_3 = i$; *or*, if any two of the three components of the angular velocity in the basis $\{\hat{e}_1', \hat{e}_2', \hat{e}_3'\}$ are zero. This would happen if the body is rotating essentially *about* one of the principal axes. In this case, only one of the components of the angular velocity would be non-zero.

In general, $\vec{\ell}_C^{\text{spin}}$ and $\vec{\omega}$ are therefore *not* parallel, with an innate consequence that their cross product would not be the null vector. This situation will soon be found to have important and interesting ramifications.

Using Eq. 7.21a and Eq. 7.17a, we have

$$\begin{bmatrix} \ell_{C,x'}^{\text{spin}} \\ \ell_{C,y'}^{\text{spin}} \\ \ell_{C,z'}^{\text{spin}} \end{bmatrix} = \begin{bmatrix} T_{11} & T_{12} & T_{13} \\ T_{21} & T_{22} & T_{23} \\ T_{31} & T_{32} & T_{33} \end{bmatrix} \times \left\{ \begin{bmatrix} I_{xx} & -I_{xy} & -I_{xz} \\ -I_{yx} & I_{yy} & -I_{yz} \\ -I_{zx} & -I_{zy} & I_{zz} \end{bmatrix} \begin{bmatrix} \omega_x \\ \omega_y \\ \omega_z \end{bmatrix} \right\}. \tag{7.24}$$

We now resolve the unit matrix as

$$\begin{bmatrix} 1 & 0 & 0 \\ 0 & 1 & 0 \\ 0 & 0 & 1 \end{bmatrix} = \mathbb{T}^T\mathbb{T} = \begin{bmatrix} T_{11} & T_{21} & T_{31} \\ T_{12} & T_{22} & T_{32} \\ T_{13} & T_{23} & T_{33} \end{bmatrix} \begin{bmatrix} T_{11} & T_{12} & T_{13} \\ T_{21} & T_{22} & T_{23} \\ T_{31} & T_{32} & T_{33} \end{bmatrix}, \tag{7.25}$$

and insert it just before the column matrix representing the angular velocity in Eq. 7.24. We then get

$$\begin{bmatrix} \ell_{C,x'}^{\text{spin}} \\ \ell_{C,y'}^{\text{spin}} \\ \ell_{C,z'}^{\text{spin}} \end{bmatrix} = \begin{bmatrix} T_{11} & T_{12} & T_{13} \\ T_{21} & T_{22} & T_{23} \\ T_{31} & T_{32} & T_{33} \end{bmatrix} \begin{bmatrix} I_{xx} & -I_{xy} & -I_{xz} \\ -I_{yx} & I_{yy} & -I_{yz} \\ -I_{zx} & -I_{zy} & I_{zz} \end{bmatrix} \begin{bmatrix} T_{11} & T_{21} & T_{31} \\ T_{12} & T_{22} & T_{32} \\ T_{13} & T_{23} & T_{33} \end{bmatrix}$$

$$\times \begin{bmatrix} T_{11} & T_{12} & T_{13} \\ T_{21} & T_{22} & T_{23} \\ T_{31} & T_{32} & T_{33} \end{bmatrix} \begin{bmatrix} \omega_x \\ \omega_y \\ \omega_z \end{bmatrix}. \tag{7.26a}$$

We can interpret Eq. 7.26a by bracketing the product matrices from a slightly different point of view, without changing their order, by using the associative property of matrix multiplication. Accordingly, we express Eq. 7.26a as

$$\begin{bmatrix} \ell_{C,x'}^{\text{spin}} \\ \ell_{C,y'}^{\text{spin}} \\ \ell_{C,z'}^{\text{spin}} \end{bmatrix} = \left\{ \left\{ \begin{bmatrix} T_{11} & T_{12} & T_{13} \\ T_{21} & T_{22} & T_{23} \\ T_{31} & T_{32} & T_{33} \end{bmatrix} \begin{bmatrix} I_{xx} & -I_{xy} & -I_{xz} \\ -I_{yx} & I_{yy} & -I_{yz} \\ -I_{zx} & -I_{zy} & I_{zz} \end{bmatrix} \begin{bmatrix} T_{11} & T_{21} & T_{31} \\ T_{12} & T_{22} & T_{32} \\ T_{13} & T_{23} & T_{33} \end{bmatrix} \right\} \times \left\{ \begin{bmatrix} T_{11} & T_{12} & T_{13} \\ T_{21} & T_{22} & T_{23} \\ T_{31} & T_{32} & T_{33} \end{bmatrix} \begin{bmatrix} \omega_x \\ \omega_y \\ \omega_z \end{bmatrix} \right\} \right\}. \tag{7.26b}$$

Equation 7.26b has the following straight forward interpretation in the (rotated) principal basis:

$$[\vec{\ell}_C^{\text{spin}} \text{ in } \{\vec{e}_x, \vec{e}_y, \vec{e}_z\}] = \{\mathbb{I}_{C,(x',y',z'),d}\} \{[\vec{\omega}]_{\{\hat{e}_x', \hat{e}_y', \hat{e}_z'\}}\}. \tag{7.27}$$

Equation 7.27 is essentially the same as Eq. 7.22, which expresses the relationship between angular momentum and angular velocity, in the principal basis $\{\hat{e}_x', \hat{e}_y', \hat{e}_z'\}$.

We now discuss the consequence of the fact that angular momentum and angular velocity vectors need not be parallel, by considering the example of a cube of uniform mass density ρ. Its moment of inertia needs to be referenced (i) with respect to a point about which the cube rotates, *and* (ii) with respect to the *specific orientation* of the coordinate frame that is referenced. The orientation of the coordinate system determines the *products of inertia*, and also all other integrals given in Eq. 7.15. This is only to be expected. Any tensor whose rank is greater than 0 has components which *do* change when referenced to a different orientation of the frame of reference; only a scalar is excluded from this.

We first determine the inertia tensor (Eq. 7.15) of the cube considered to be rotating at constant angular velocity $\vec{\omega} = \omega\hat{e}_x$ about the axis pointing in the direction $\hat{e}_x$. This unit vector belongs to the basis set $\{\hat{e}_x, \hat{e}_y, \hat{e}_z\}$ of the laboratory frame of reference, oriented along the edges of the cube, and centerd at the corner (point O, Fig. 7.4).

Thus, using Eq. 7.15a, we have

$$I_{xx} = \iiint (y^2 + z^2)\delta m_P = \rho \int_{x=0}^{a} \int_{y=0}^{a} \int_{z=0}^{a} y^2 dxdydz + \rho \int_{x=0}^{a} \int_{y=0}^{a} \int_{z=0}^{a} z^2 dxdydz$$

i.e., $$I_{xx} = \frac{M}{a^3}(a)\left(\frac{a^3}{3}\right)(a) + \frac{M}{a^3}(a)(a)\left(\frac{a^3}{3}\right) = \frac{2Ma^2}{3}.$$

Similarly, evaluating all the other integrals in Eq. 7.15, we get

$$\mathbb{I}_{C,(x,y,z)} = Ma^2 \begin{bmatrix} \frac{2}{3} & -\frac{1}{4} & -\frac{1}{4} \\ -\frac{1}{4} & \frac{2}{3} & -\frac{1}{4} \\ -\frac{1}{4} & -\frac{1}{4} & \frac{2}{3} \end{bmatrix}. \tag{7.28}$$

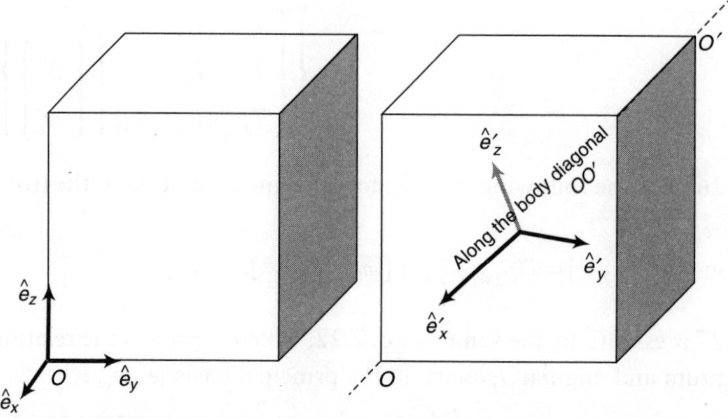

Fig. 7.4 The figure on the left shows the Cartesian coordinate frame with base vectors $\{\hat{e}_x, \hat{e}_y, \hat{e}_z\}$, centered at the corner O. The unit vectors are respectively parallel to the cube's edges, each side being of length 'a'. The figure on the right shows the axis along the body-diagonal OO' and a rotated coordinate frame of reference $\{\hat{e}'_x, \hat{e}'_y, \hat{e}'_z\}$. The unit vector $\hat{e}'_x$ is along $\overrightarrow{O'O}$, and the other two unit vectors $\{\hat{e}'_y, \hat{e}'_z\}$ are arbitrarily oriented in the plane orthogonal to $\hat{e}'_x$.

Since the cube is considered to spin about the axis through the origin and parallel to $\hat{e}_x$, we get

$$\vec{\ell}_C^{spin} = \mathbb{I}_{C,(x,y,z)}\,\vec{\omega} = Ma^2 \begin{bmatrix} \frac{2}{3} & -\frac{1}{4} & -\frac{1}{4} \\ -\frac{1}{4} & \frac{2}{3} & -\frac{1}{4} \\ -\frac{1}{4} & -\frac{1}{4} & \frac{2}{3} \end{bmatrix} \begin{bmatrix} \omega \\ 0 \\ 0 \end{bmatrix} = M\omega a^2 \begin{bmatrix} \frac{2}{3} \\ \frac{-1}{4} \\ \frac{-1}{4} \end{bmatrix}, \tag{7.29a}$$

$$\text{i.e.,} \quad \vec{\ell}_C^{\text{spin}} = M\omega a^2 \left(\frac{2}{3}\hat{e}_x - \frac{1}{4}\hat{e}_y - \frac{1}{4}\hat{e}_z \right). \tag{7.29b}$$

Clearly, $\vec{\ell}_C^{\text{spin}}$ is not parallel to $\vec{\omega} = \omega\hat{e}_x$. We now diagonalize the inertia tensor to obtain the transformation matrix which will give us the unit base vectors along the principal axes of the cube.

Now, the inertia tensor $\mathbb{I}_{C\ni(x,y,z)}$ in Eq. 7.28 is only a multiple of the matrix $\mathbb{A}$ that we had used in Eq. 2A.17 (Appendix to Chapter 2):

$$\mathbb{I}_{C\ni(x,y,z)} = Ma^2\mathbb{A}. \tag{7.30}$$

We can therefore adapt the analysis of Chapter 2 (between Eq. 2A.17 and Eq. 2A.27) and write the inertia tensor in its diagonal form, $\mathbb{I}_{C\ni(x',y',z')}$, referred to its principal axes:

$$\mathbb{I}_{C\ni(x',y',z')} = \begin{bmatrix} \frac{1}{\sqrt{3}} & \frac{1}{\sqrt{3}} & \frac{1}{\sqrt{3}} \\ -\frac{1}{\sqrt{2}} & \frac{1}{\sqrt{2}} & 0 \\ -\frac{1}{\sqrt{6}} & -\frac{1}{\sqrt{6}} & \frac{2}{\sqrt{6}} \end{bmatrix} [Ma^2\mathbb{A}] \begin{bmatrix} \frac{1}{\sqrt{3}} & -\frac{1}{\sqrt{2}} & -\frac{1}{\sqrt{6}} \\ \frac{1}{\sqrt{3}} & \frac{1}{\sqrt{2}} & -\frac{1}{\sqrt{6}} \\ \frac{1}{\sqrt{3}} & 0 & \frac{2}{\sqrt{6}} \end{bmatrix}, \tag{7.31a}$$

and hence,

$$\mathbb{I}_{C\ni(x',y',z')} = (Ma^2) \begin{bmatrix} \frac{1}{6} & 0 & 0 \\ 0 & \frac{11}{12} & 0 \\ 0 & 0 & \frac{11}{12} \end{bmatrix}. \tag{7.31b}$$

The transformation $\tilde{\mathbb{V}}_{3\times 3}$ (of Eq. 2A.27) diagonalizes the inertia tensor. It also rotates the laboratory basis $\{\hat{e}_x, \hat{e}_y, \hat{e}_z\}$ to a new basis $\{\hat{e}'_x, \hat{e}'_y, \hat{e}'_z\}$ that has unit vectors respectively parallel to the principal axes of the cube. Hence,

$$\begin{bmatrix} \hat{e}'_x \\ \hat{e}'_y \\ \hat{e}'_z \end{bmatrix} = \tilde{\mathbb{V}}_{3\times 3} \begin{bmatrix} \hat{e}_x \\ \hat{e}_y \\ \hat{e}_z \end{bmatrix} = \begin{bmatrix} \frac{1}{\sqrt{3}} & \frac{1}{\sqrt{3}} & \frac{1}{\sqrt{3}} \\ -\frac{1}{\sqrt{2}} & \frac{1}{\sqrt{2}} & 0 \\ -\frac{1}{\sqrt{6}} & -\frac{1}{\sqrt{6}} & \frac{2}{\sqrt{6}} \end{bmatrix} \begin{bmatrix} \hat{e}_x \\ \hat{e}_y \\ \hat{e}_z \end{bmatrix} = \begin{bmatrix} \frac{1}{\sqrt{3}}\hat{e}_x + \frac{1}{\sqrt{3}}\hat{e}_y + \frac{1}{\sqrt{3}}\hat{e}_z \\ -\frac{1}{\sqrt{2}}\hat{e}_x + \frac{1}{2}\hat{e}_y \\ -\frac{1}{\sqrt{6}}\hat{e}_x - \frac{1}{\sqrt{6}}\hat{e}_y + \frac{2}{\sqrt{6}}\hat{e}_z \end{bmatrix}. \tag{7.32}$$

The principal axes of the cube are therefore along the following directions, referenced to the laboratory frame:

$$\tilde{e}'_x = \frac{1}{\sqrt{3}}\begin{bmatrix} 1 \\ 1 \\ 1 \end{bmatrix}; \quad \hat{e}'_y = \frac{1}{\sqrt{2}}\begin{bmatrix} -1 \\ 1 \\ 0 \end{bmatrix}; \quad \hat{e}'_z = \frac{1}{\sqrt{6}}\begin{bmatrix} -1 \\ -1 \\ 2 \end{bmatrix}. \tag{7.33}$$

We see that $\hat{e}'_x$ is along the body diagonal of the cube. The other two unit vectors of the principal basis, $\hat{e}'_y$ and $\hat{e}'_z$, are mutually orthogonal to each other, but arbitrarily oriented in a plane orthogonal to $\hat{e}'_x$. We already know that only if the cube is spun about one of the principal axes will it have an angular momentum that is parallel to its angular velocity.

Note that with reference to the 'principal' basis $\{\hat{e}'_x, \hat{e}'_y, \hat{e}'_z\}$ of the unit vectors, the following holds:

(i) the angular velocity vector expressed in this basis has only one non-zero component, since the cube is spun about one of the three principal axes,

(ii) the angular momentum vector expressed in this basis also has only one non-zero component. This renders the angular momentum vector parallel to the angular velocity vector,

(iii) the inertia tensor is diagonal, given by Eq. 7.31.

The nomenclature of the particular basis, $\{\hat{e}'_x, \hat{e}'_y, \hat{e}'_z\}$ as the 'principal' basis is therefore very appropriate.

7.3 TORQUE ON A RIGID BODY: EULER EQUATIONS

The diagonalization of the inertia tensor provides us an orientation of the coordinate system and gives us the 'principal' basis. In general, it would not be congruent with the laboratory frame of reference. Fig. 7.5 shows, for a cylinder, the relative orientations of the space-fixed frame of reference (X_1, X_2, X_3), and the body-fixed frame (X'_1, X'_2, X'_3) wherein the axes are along the principal axes of the body. Now, the advantage of the space-fixed (inertial) frame of reference lies in the fact that Newton's laws are valid in this frame; the cause-effect relationship holds. The cause that changes the equilibrium is a genuine physical interaction. In the rotating, body-fixed frame, on the other hand, we would need to introduce various pseudo-forces (Chapter 3). Yet the advantage of the body-fixed rotating frame of reference is that it is in *this* frame that the inertia tensor is diagonal. The inertia tensor referenced to the space-fixed, on the other hand, is messy; i.e., it has off-diagonal non-zero elements. One should therefore expect to pay some price for the simple form of the inertia tensor in the rotating frame, just as pseudo-forces had to be introduced in Chapter 3.

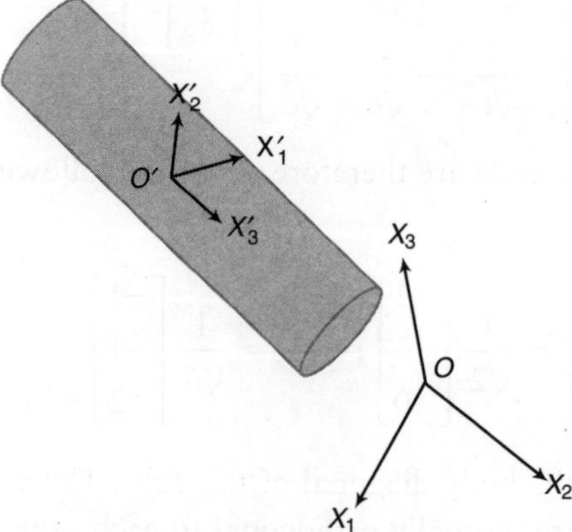

Fig. 7.5 This figure shows the relative orientation at a given instant between the space-fixed inertial laboratory frame (X_1, X_2, X_3) and the body-fixed rotating frame of reference (X'_1, X'_2, X'_3).

We have learned that angular momentum of a system would change if and only if a torque is applied on it. This is true only in an inertial frame of reference. In such a frame, the rate of change of angular momentum about a certain point (or an axis) is exactly equal to the *physical* torque about that point (or the axis). This cannot hold in a rotating frame of reference. This result is connected intimately with the relationship we have studied in Chapter 3. We have seen in Eq. 3.13 that the time-derivative operator in the inertial (space-fixed) frame is *not* equal to that in the rotating (body-fixed). These two rates are related by

$$\left[\frac{d}{dt}\right]_I \equiv [\vec{\omega}\times]+\left[\frac{d}{dt'}\right]_R. \tag{7.34a}$$

Applying the above time-derivative operator on the angular momentum vector, we get

$$[\vec{\tau}]_{\substack{\text{physical}\\\text{torque}}} = \left[\frac{d\vec{L}_I}{dt}\right] \equiv [\vec{\omega}\times\vec{L}]_{\substack{R,\\\text{body}\\\text{fixed}}} + \left[\frac{d\vec{L}_R}{dt'}\right]. \tag{7.34b}$$

The left hand side is the torque due to a physical interaction experienced by the body; it is responsible for the change in its angular momentum. The two terms on the right-hand side can be called *pseudo-torques*, in analogy with the pseudo-force used in Chapter 3. They are consequences of employing the rotating frame of reference.

Thus,

$$\begin{bmatrix}\tau_1\\\tau_2\\\tau_3\end{bmatrix}_{\substack{\text{physical}\\\text{torque}}} = \begin{bmatrix}\dfrac{dL_1}{dt}\\\dfrac{dL_2}{dt}\\\dfrac{dL_3}{dt}\end{bmatrix}_{\{\hat{e}'_1,\hat{e}'_2,\hat{e}'_3\}} \equiv [\vec{\omega}\times\vec{L}]_{\{\hat{e}'_1,\hat{e}'_2,\hat{e}'_3\}} + \frac{d}{dt'}\left(\begin{bmatrix}L'_1\\L'_2\\L'_3\end{bmatrix}_{\{\hat{e}'_1,\hat{e}'_2,\hat{e}'_3\}}\right). \tag{7.34c}$$

Equation 7.34b and 7.34c are known as the Euler's equation.

Now, the angular velocity in the body-fixed (rotating) frame of reference $\{\hat{e}'_1, \hat{e}'_2, \hat{e}'_3\}$ is

$$\vec{\omega}' = (\hat{e}'_1 \cdot \vec{\omega}')\hat{e}'_1 + (\hat{e}'_2 \cdot \vec{\omega}')\hat{e}'_2 + (\hat{e}'_3 \cdot \vec{\omega}')\hat{e}'_3, \tag{7.35a}$$

i.e., $$\vec{\omega}' = \vec{\omega}'_1\hat{e}'_1 + \vec{\omega}'_2\hat{e}'_2 + \vec{\omega}'_3\hat{e}'_3. \tag{7.35b}$$

Hence,

$$[\vec{\omega}\times\vec{L}]_{\{\hat{e}'_1,\hat{e}'_2,\hat{e}'_3\}} = \begin{vmatrix}\hat{e}'_1 & \hat{e}'_3 & \hat{e}'_3\\ \omega'_1 & \omega'_2 & \omega'_3\\ \ell'_1 & \ell'_2 & \ell'_3\end{vmatrix} = \begin{bmatrix}\hat{e}'_1(\omega'_2\ell'_3-\omega'_3\ell'_2)\\ +\hat{e}'_2(\omega'_3\ell'_1-\omega'_1\ell'_3)\\ +\hat{e}'_3(\omega'_1\ell'_2-\omega'_2\ell'_1)\end{bmatrix}. \tag{7.36a}$$

Using Eq. 7.27, we get

$$\frac{d}{dt'}\left(\begin{bmatrix} L'_1 \\ L'_2 \\ L'_3 \end{bmatrix}\right) = \frac{d}{dt'}\begin{bmatrix} i_1\omega'_1 \\ i_2\omega'_2 \\ i_3\omega'_3 \end{bmatrix} = \begin{bmatrix} i_1\dot{\omega}'_1 \\ i_2\dot{\omega}'_2 \\ i_3\dot{\omega}'_3 \end{bmatrix}. \tag{7.36b}$$

Therefore, using Eqs. 7.36a and 7.36b in Eq. 7.34, we get

$$\begin{bmatrix} \tau_1 \\ \tau_2 \\ \tau_3 \end{bmatrix} = \begin{bmatrix} \omega'_2\ell'_3 - \omega'_3\ell'_2 \\ \omega'_3\ell'_1 - \omega'_1\ell'_3 \\ \omega'_1\ell'_2 - \omega'_2\ell'_1 \end{bmatrix} + \begin{bmatrix} i_1\dot{\omega}'_1 \\ i_2\dot{\omega}'_2 \\ i_3\dot{\omega}'_3 \end{bmatrix}, \tag{7.37a}$$

i.e.,
$$\begin{bmatrix} \tau_1 \\ \tau_2 \\ \tau_3 \end{bmatrix} = \begin{bmatrix} \omega'_2 i_3\omega'_3 - \omega'_3 i_2\omega'_2 \\ \omega'_3 i_1\omega'_1 - \omega'_1 i_3\omega'_3 \\ \omega'_1 i_2\omega'_2 - \omega'_2 i_1\omega'_1 \end{bmatrix} + \begin{bmatrix} i_1\dot{\omega}'_1 \\ i_2\dot{\omega}'_2 \\ i_3\dot{\omega}'_3 \end{bmatrix} = \begin{bmatrix} \omega'_2\omega'_3(i_3 - i_2) \\ \omega'_3\omega'_1(i_1 - i_3) \\ \omega'_1\omega'_2(i_2 - i_1) \end{bmatrix} + \begin{bmatrix} i_1\dot{\omega}'_1 \\ i_2\dot{\omega}'_2 \\ i_3\dot{\omega}'_3 \end{bmatrix}, \tag{7.37b}$$

which gives us the *three* Euler equations, in the scalar form. These equations are non-linear, because of the appearance of the *products* of angular frequencies. We will therefore not be able to use the principle of superposition to combine the elementary solutions of Eq. 7.37 to get composite solutions of the same. The inertia elements that appear in the Euler's equations are actually the *eigenvalues* of the inertia tensor. They are not the complicated integrals of Eq. 7.15. The *detailed* mass distribution that is involved in the integrands (in Eq. 7.15) is therefore not important in using the Euler's equations. Only the eigenvalues of the inertia tensor in its eigenbasis are of importance. Thus, *bodies even with different shapes that somehow have the same principal axes and the same eigenvalues of the inertia tensor are described by the same set of Euler's equations.* The simplest geometrical shape that has three principal moments of inertia is a homogeneous ellipsoid, so problems on rigid body rotations often reduce to finding an *equivalent* ellipsoid. This technique was developed by Cauchy and Poinsot in the period roughly between 1825 and 1840.

In particular, for a rigid body spinning about its own axis and subjected to no *physical* torque, we have $[\vec{\tau}]_{I,\text{ physical torque}} = \vec{0}$. Therefore, from Eq. 7.37b, we get

and
$$\begin{bmatrix} i_1\dot{\omega}'_1 \\ i_2\dot{\omega}'_2 \\ i_3\dot{\omega}'_3 \end{bmatrix} = \begin{bmatrix} \omega'_2\omega'_3(i_2 - i_3) \\ \omega'_3\omega'_1(i_3 - i_1) \\ \omega'_1\omega'_2(i_1 - i_2) \end{bmatrix} = \begin{bmatrix} \frac{dL'_1}{dt} \\ \frac{dL'_2}{dt} \\ \frac{dL'_3}{dt} \end{bmatrix}_{\{\vec{e}'_1,\, \vec{e}'_2,\, \vec{e}'_3\}} . \tag{7.38}$$

An observer in the rotating frame must employ, therefore, a pseudo-torque to be acting on the body, even if there isn't such a physical torque present. Such a pseudo-torque alone would account for the change in angular momentum with time in her/his frame. Causality and determinism, which are central principles in Newtonian mechanics, can then be used in the rotating frame, albeit with a pseudo-torque. This is completely similar to the 'centrifugal push' you feel in a turning car; it is an effect of the analysis in a rotating frame of reference. In Chapter 3 we discussed the real effects of pseudo-forces. The real effects we considered there were real changes in linear momenta which an experimentalist in an accelerated frame observes. In the present case, the experimentalist would observe real changes in the angular momenta in her/his frame of reference, but not because a physical torque acted on the system, but because the observer is in a rotating frame of reference.

A symmetric top has $i_1 = i_2 \neq i_3$, and this reduces its Euler equations (7.38) in a *force-free* region (i.e., a region in which even gravity is switched off) to

$$\begin{bmatrix} \dfrac{dL'_1}{dt'} \\ \dfrac{dL'_2}{dt'} \\ \dfrac{dL'_3}{dt'} \end{bmatrix} = \begin{bmatrix} i_1\dot{\omega}'_1 \\ i_2\dot{\omega}'_2 \\ i_3\dot{\omega}'_3 \end{bmatrix} = \begin{bmatrix} \omega'_2\omega'_3(i_1 - i_3) \\ \omega'_3\omega'_1(i_3 - i_1 \\ 0 \end{bmatrix}. \tag{7.39}$$

Equation 7.39 is of great interest since it can describe the motion of a spinning planet in space, under appropriate conditions. Since $i_3\dot{\omega}'_3 = 0$, ω'_3, itself must be constant, since $i_3 \neq 0$. Thus,

$$\vec{\omega}'_3 = (\hat{e}'_3 \,.\, \vec{\omega}') = \chi, \text{ a constant.} \tag{7.40a}$$

Again, since $\omega'_2\omega'_3(i_1 - i_3) = i_1\omega'_1$, we have $\dot{\omega}'_1 = \omega'_2\omega'_3\dfrac{(i_1 - i_3)}{i_1} = \omega'_2(r\omega'_3)$,

i.e., $\dot{\omega}'_1 - \omega'_2\,(r\omega'_3) = 0,$ (7.40b)

and likewise, $\dot{\omega}'_2 - \omega'_1\,(r\omega'_3) = 0,$ (7.40c)

where we have used a dimensionless number, $\dfrac{(i_1 - i_3)}{i_3} = r = \dfrac{(i_1 - i_3)}{i_2}$. (7.41)

Equations 7.40b and 7.40c are coupled differential equations. These are best solved using the algebra of complex numbers. The complex numbers use the *imaginary* number $i = \sqrt{-1}$, but that really helps us to handle two *real* numbers together. This is very nice, since both the real part and the imaginary part of the complex numbers are real. We can thus combine Eqs. 7.40b and 7.40c in a single equation by using $i = \sqrt{-1}$ to multiply the second of these. Thus,

$$\{\dot{\omega}'_1 - \omega'_2\,(r\omega'_3)\} + i\{\dot{\omega}'_2 + (r\omega'_3)\,\omega'_1\} = 0,$$

i.e., $(\dot{\omega}'_1 - i\dot{\omega}'_2) + i(r\omega'_3)\,(\omega'_1 + i\omega'_2) = 0.$ (7.42)

We now simplify the notation by introducing

$$\Omega = (\omega'_1 + i\omega'_2); \quad \dot{\Omega} = (\dot{\omega}'_1 - i\dot{\omega}'_2). \tag{7.43}$$

The solution to the differential equation for Ω is

$$(\omega_1' + i\omega_2') = \Omega(t) = \omega_\alpha e^{-i(r\omega_3')t} = [\omega_\alpha \cos(r\omega_3' t) - i\omega_\alpha \sin(r\omega_3' t)], \tag{7.44}$$

where ω_α is a constant, complex in general, which we may write as $\omega_\alpha = |\omega_\alpha| e^{i\theta_\alpha}$. It includes a phase factor, θ_α, which can be taken as zero. Equating the real and the imaginary parts of Eq. 7.44, we get the components of the angular velocity vector in the body-fixed, rotating, frame of reference:

$$\omega_1' = \omega_\alpha \cos(r\omega_3' t) = (\hat{e}_1' \cdot \vec{\omega}'), \text{ and } \omega_2' = -\omega_\alpha \sin(r\omega_3' t) = (\hat{e}_2' \cdot \vec{\omega}'). \tag{7.45}$$

The magnitude of the angular velocity vector is

$$|\vec{\omega}| = \sqrt{(\omega_1')^2 + (\omega_2')^2 + (\omega_3')^2} = \sqrt{\omega_\alpha^2 + \chi^2} = \rho, \text{ a constant.} \tag{7.46}$$

$\omega_\alpha^2 + \chi^2 = \rho^2$ is an equation to a circle traced in a plane orthogonal to $\hat{e}_3'$ by the tip of the angular velocity vector $\vec{\omega}' = \overrightarrow{OP'}$. This circle subtends a cone at the origin, shown in Fig. 7.6a. The angular momentum vector $\vec{L} = \overrightarrow{OQ'}$ is *not* parallel to the angular velocity vector, but the vectors $(\hat{e}_3', \vec{\omega}, \vec{L})$ are all in a single plane through the axis X_3'. This plane rotates, as seen in the body-fixed rotating frame of reference, about the axis X_3' at the *circular* frequency, $(r\omega_3')$. The circular frequency $(r\omega_3')$ has a sign which depends on whether the ratio r is positive or negative. In turn, this depends, on whether $(i_2 = i_1) < i_3$ or $(i_2 = i_1) > i_3$, according to Eq. 7.41. An object for which $(i_2 = i_1) < i_3$ is called 'oblate' (flattened), while that with $(i_2 = i_1) > i_3$ is called 'prolate' (elongated).

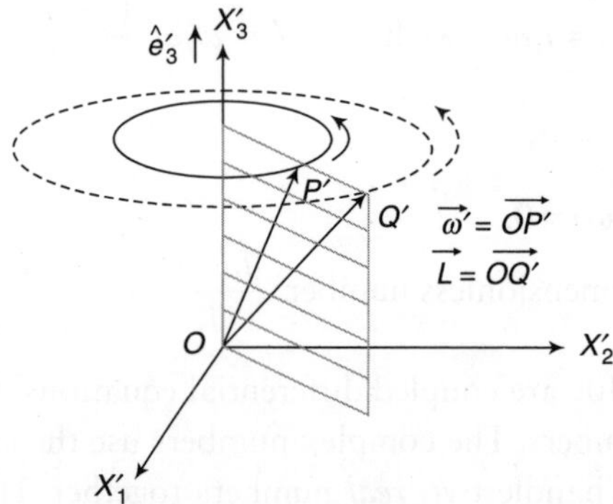

Fig. 7.6a In the body-fixed spinning frame, angular velocity and angular momentum vectors precess about the X_3' axis. The component of the angular velocity vector along $\hat{e}_3'$ is constant, and given by Eq. 7.40a. The other two components are time-dependent, given by Eq. 7.45, which causes the precession of the vector that represents the angular velocity.

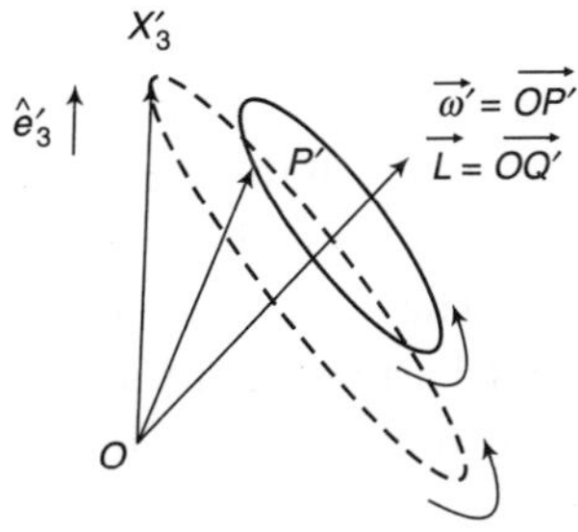

Fig. 7.6b In the space-fixed laboratory (inertial) frame of reference, the angular momentum is a constant, since the symmetric top is in a force free region. $(\hat{e}_3', \vec{\omega}, \vec{L})$ are, however, always in the same plane and hence both the angular velocity vector and the axis X_3' must precess about the angular velocity vector as shown above. The angular momentum vector in Fig.7.6b is the same as in Fig. 7.6a.

The components of both $\vec{\omega}'$ and $\vec{L}$ rotate on the surface of the cones traced respectively by the continuous and the dashed circles shown in Fig. 7.6a. In our consideration of the force-free symmetric top, the angular momentum in space-fixed laboratory frame of reference must be constant, shown in Fig. 7.6b. In this frame, both the angular velocity vector and the axis OX_3' would precess, since these three lie in the same plane, *no matter which frame of reference is used to view them.*

The above analysis has direct applicability in modern navigation devices, in addressing mysteries of the past, and making predictions in the distant future. For example, researchers rely on angular momentum analysis to estimate when some events in ancient periods may have taken place by studying the precession of the Earth's axis of rotation. To appreciate this, we examine how the axis of rotation of a rigid body precesses. To simplify the discussion on the precession of Earth's axis, we first consider a symmetric top, spinning about its symmetry axis through its pivot O and center of mass C (Fig. 7.7). The top we are describing now is no longer in a gravity-free region considered earlier. It is often known as a heavy-symmetrical top.

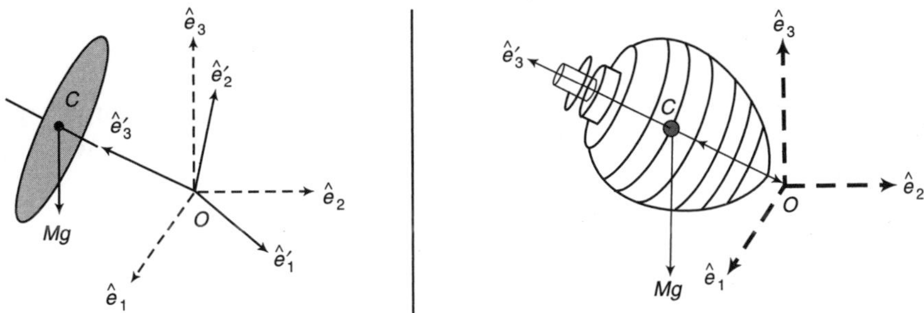

Fig. 7.7 This figure shows a top spinning about its symmetric axis. It experiences a torque due to Earth's gravity. Its weight $\vec{W} = m\vec{g}$ acts through its center of mass C as it spins about $\hat{e}_3'$, but precesses about $\hat{e}_3$. The space-fixed laboratory frame of reference is described by the basis set of orthonormal vectors $\{\hat{e}_1, \hat{e}_2, \hat{e}_3,\}$ and the top's rotating frame of reference by the basis $\{\hat{e}_1', \hat{e}_2', \hat{e}_3'\}$. The vector $\vec{R} = \overrightarrow{OC}$ is the position vector of the center of mass of the top with respect to the pivot at O.

Now, when you spin a top, you might want the axis about which it spins to be aligned with the local vertical. You expect that $\hat{e}_3'$ and $\hat{e}_3$ would be essentially parallel. Nevertheless, due to some imperfections at the surface on which the top is pivoted (or due to some inadvertent slip in how the top was set into motion), it could happen that at some instant, the direction $\hat{e}_3'$ would *not* be parallel to $\hat{e}_3$ (Fig. 7.7). The motion we discuss below is the one *subsequent* to this; because it is only thereafter that Earth's gravity would exert a torque on the top. This torque is otherwise zero, when the position vector of the center of mass is collinear with the vertical.

The top spins about its principal axis OC (Fig. 7.7), and therefore has its angular momentum parallel to the angular velocity:

$$\vec{L} = i_3\vec{\omega} = i_3\omega\hat{e}_3'. \tag{7.47}$$

The torque on the top exerted by gravity about the point O is

$$\vec{\tau} = \overrightarrow{OC} \times \vec{W} = \vec{R} \times M\vec{g} = \frac{d\vec{L}}{dt}, \tag{7.48a}$$

where $\vec{R}$ is the instantaneous position vector of the center of mass with respect to the pivot point, fixed at O. Both the magnitude and the direction of the angular momentum vector can change on application of the torque:

$$\vec{\tau} = \frac{d}{dt}(\hat{L}) = L\frac{d\hat{L}}{dt} + \left(\frac{dL}{dt}\right)\hat{L}. \tag{7.48b}$$

In Eq. 7.48b, we have factored the angular momentum vector into its magnitude and direction, given by a unit vector, and we consider the time-dependence of both the magnitude and the direction. However, a change in magnitude of the angular velocity of the top would also involve a change in its rotational kinetic energy. Hence, if the torque is weak, it is less likely to change the *magnitude* of the angular momentum. To an excellent approximation, therefore, we may write

$$\vec{\tau} \approx L\frac{d\hat{L}}{dt} = L\frac{d\hat{e}_3'}{dt} = L\dot{\hat{e}}_3', \tag{7.48c}$$

resulting in a change in the direction of $\hat{e}_3'$. We know from Chapter 2 that *change* in the direction of a unit-vector is *always* orthogonal to the direction of that vector. We see that the torque is, (since $\vec{R} = \overrightarrow{OC}$, in Fig. 7.7)

$$L\dot{\hat{e}}_3' = \vec{\tau} = \vec{R} \times M\vec{g} = Z_C' \, Mg[\hat{e}_3' \times (-\hat{e}_3)], \tag{7.49a}$$

where $Z_C' = \vec{R} \cdot \hat{e}_3'$, and hence,

$$\dot{\hat{e}}_3' = \frac{Z_C' Mg}{L}[\hat{e}_3' \times (-\hat{e}_3)] = \frac{Z_C' Mg}{i_3\omega}[\hat{e}_3 \times \hat{e}_3']. \tag{7.49b}$$

We conclude from Eq. 7.49 that when a weak torque acts on a spinning object, the direction of the spin angular velocity vector $\hat{e}_3'$ *precesses* about the direction $\hat{e}_3$. The *change* in its direction due to the torque is always orthogonal to both $\hat{e}_3$ and $\hat{e}_3'$.

We can now appreciate how this analysis can help us even determine when some ancient events may have occurred, for example, events described in the great epic Mahabharata. We note that Earth's equatorial plane is tilted at an angle ~23.5° with respect to the ecliptic (i.e., orbital) plane. We consider $\hat{e}_3$ to point in the direction of the ecliptic North, and $\hat{e}_3'$, the direction of Earth's angular velocity vector (about which Earth rotates), to point toward Polaris, the North Pole star. In fact, Earth's axis points *almost* toward Polaris, but not exactly. We shall, however, ignore the small difference. The angle between $\hat{e}_3$ and $\hat{e}_3'$ does not change appreciably over Earth's (one-year) orbit around the Sun. In fact, it does not change *much* even over *hundreds* of such orbits, but then, over longer periods the change becomes significant enough to reckon with. The result of the torque due to the tug exerted by the Sun and the Moon on the Earth accumulates over somewhat longer periods, making $\hat{e}_3'$ precess about $\hat{e}_3$, just as the heavy symmetrical top (Fig. 7.7) precesses.

Detailed calculations show that the direction of Earth's axis $\hat{e}_3'$ takes about ~26000 years to complete one full cycle of precession about the direction $\hat{e}_3$ of the ecliptic North. The direction in which the axis points in the celestial sphere therefore takes a little over 72 years to change by 1°. The change in the direction of $\hat{e}_3'$, over a period of ~5000 years is therefore *quite* significant, considering that the full rotation cycle takes only a little over five times that. Different distant stars are therefore the 'pole' stars in different periods during these 26000 years. About 5000 years ago, the axis pointed toward the star Thuban, which was the *then* pole star. About 13000 years from now, it would be Vega that would become the pole star. We can map the angular spacing between the different stars seen from Earth, and compare that with available records of which star *trailed* a neighboring star through the night. Comparison of a *simulation* of rotation of the celestial sphere about an axis pointing toward Thuban would test the assumption that events in the Mahabharata took place ~5000 years ago, since the great epic has striking descriptions, in beautiful poetry, of stellar patterns seen at that time. These patterns were in fact used in those periods as night-time navigation maps. We do not present the results of such studies, even if plausible findings seem to be available. Such studies are fascinating, isn't it? One could use similar analysis, in conjunction with available records of other ancient events, to *date* the construction of the pyramids or the sphinx, for example.

Earth's polar axis also wobbles. This effect is *different* from the 26,000 years of precession of the axis discussed above. The wobbling is due to an effect of the mass distribution *within* the Earth; it depends on its moment of inertia. The principal moments of inertia of the Earth along its axes in the equatorial plane are smaller, due to the fact that Earth has an oblate shape, than the moment of inertia along the polar axis: $(i_2 = i_1) < i_3$. We would therefore expect this to cause a precession of the axis of rotation, described by Eq. 7.45, at a circular frequency, $(r\omega_3')$, with r given by Eq. 7.41. For the Earth, it turns out to be, $|r| = \frac{i_1 - i_3}{i_1} \approx \frac{1}{300}$. This causes a small wobbling of the axis of rotation. One has to make some corrections, considering the interior dynamics in the Earth, inclusive of the fact that the tectonic plates slide over each other. The Earth is not quite a rigid body, and the wobbling effect is therefore better approximated by using $|r| \approx \frac{1}{400}$. The wobbling takes place at a circular frequency $\sim \frac{\omega}{400}$ and is very small, but it exists; it is real. It is called the *Chandler wobble*, after the astronomer Seth Chandler who discovered it in 1891. The Chandler wobble changed phase dramatically in the year 2005, but a discussion on that is beyond the scope of this book.

7.4 THE EULER ANGLES AND THE GYROSCOPE MOTION

The right handed Cartesian basis sets, $\{\hat{e}_1, \hat{e}_2, \hat{e}_3\}$ and $\{\hat{e}'_1, \hat{e}'_2, \hat{e}'_3\}$ (Fig. 7.7) are rotated with respect to each other. They both, however, have common origins, at the same point. Equation 2.12 (Chapter 2) tells us how the components $\{V_1, V_2, V_3\}$ of a vector $\vec{V}$ in the basis $\{\hat{e}_1, \hat{e}_2, \hat{e}_3\}$ are related to the components $\{V'_1, V'_2, V'_3\}$ of the same vector in the rotated basis $\{\hat{e}'_1, \hat{e}'_2, \hat{e}'_3\}$:

$$V'_i = \sum_{j=1}^{3} (\hat{e}'_1 \cdot \hat{e}_1)\, V_j;\ \text{with } i = 1, 2, 3, \tag{7.50a}$$

i.e.,

$$\begin{bmatrix} V'_1 \\ V'_2 \\ V'_3 \end{bmatrix} = \begin{bmatrix} \hat{e}'_1 \cdot \hat{e}_1 & \hat{e}'_1 \cdot \hat{e}_2 & \hat{e}'_1 \cdot \hat{e}_3 \\ \hat{e}'_2 \cdot \hat{e}_1 & \hat{e}'_2 \cdot \hat{e}_2 & \hat{e}'_2 \cdot \hat{e}_3 \\ \hat{e}'_3 \cdot \hat{e}_1 & \hat{e}'_3 \cdot \hat{e}_2 & \hat{e}'_3 \cdot \hat{e}_3 \end{bmatrix} \begin{bmatrix} V_1 \\ V_2 \\ V_3 \end{bmatrix}. \tag{7.50b}$$

Or, written in compact matrix notation using 'fat' symbols, we have:

$$\mathbb{V}'_{3\times 1} = \mathbb{R}_{3\times 3}\, \mathbb{V}_{3\times 1}. \tag{7.50c}$$

$\mathbb{V}'_{3\times 1}$ and $\mathbb{V}_{3\times 1}$ are 3 × 1 column matrices, and $\mathbb{R}_{3\times 3}$ is a 3 × 3 square matrix, which in the present context, is called the *rotation matrix*.

The length of the vector is $\sqrt{\vec{V} \cdot \vec{V}} = \sqrt{\mathbb{V}'^{\dagger}\mathbb{V}'} = \sqrt{\mathbb{V}^{\dagger}\mathbb{V}}$ since the length of the vector is independent of rotation of the coordinate axes. This guarantees that $\mathbb{R}^{\dagger}\mathbb{R} = \mathbb{I}_{3\times 3}$, which is a unit matrix. Essentially, this implies that $\sum_{k=1}^{3} \mathbb{R}^{\dagger}{}_{ik}\mathbb{R}_{kj} = \sum_{k=1}^{3} \mathbb{R}_{ki}\mathbb{R}_{kj} = \delta_{ij}$. We have used the fact that all elements of the transformation matrix, being direction cosines, are essentially real numbers. The transformation matrix is an orthogonal matrix; a name it derives from the above property in which the Kronecker delta appears. We see that its inverse is the same as its transpose, a property that is sometimes used as its characteristic signature. Now, the determinant of an orthogonal matrix, in general, is $|\mathbb{R}| = \pm 1$. When the determinant $|\mathbb{R}| = +1$, we deal with a *special* subset of the orthogonal transformations which is in fact constituted by pure *rotations* (SO(3), page 38). On the other hand, when the determinant of the orthogonal transformation matrix is $|\mathbb{R}| = -1$, then the transformation is not a *pure* rotation; it would involve a reflection or an inversion. *Our present interest is in pure rotations, for which,* $|\mathbb{R}| = +1$.

Now, we have nine elements in the rotation transformation matrix. All of these nine elements are, however, not independent, since they are inter-related through the six constraining relations $\sum_{k=1}^{3} \mathbb{R}_{ki}\mathbb{R}_{kj} = \delta_{ij}$ for $i, j = 1, 2, 3$. Three of the nine such relations repeat, and we are left with six constraints. Thus only 9 – 6 = 3 degrees of freedom provide the *complete* specification for an arbitrary *orientation* of the rigid body. The required set of three parameters can be chosen in a few alternate ways. Fig. 6.2 illustrated one way of achieving this. Here, we introduce an *alternative* method to make that choice. The alternative uses three angles, known as the *Euler angles* (φ, θ, ψ). These are defined below. The transformation from the coordinate frame $\{X, Y, Z\}$ to $\{X', Y', Z'\}$ consists of a *translation* of the origin of the coordinate frame $\{X, Y, Z\}$ to that of the frame $\{X', Y', Z'\}$, *followed by three rotations* as discussed below. The coordinates of center of mass of the rigid body provide for three degrees of freedom, and the remaining three degrees of freedom of the rigid body are provided by what are known as the Euler angles (φ, θ, ψ). Now, *the most general dislocation of a rigid body is a combination of a translational displacement and a rotation*, which is an important result, known as the *Chasles' theorem*. In fact, Chasles also proved that it is possible to choose the center of the body-fixed frame of reference, and the axis of rotation of the body, in such a way that the axis of rotation is in the same direction as the translational displacement. One could therefore reach any point on the rigid body via a 'screw motion'. The resulting symmetry is called *screw symmetry* and has many applications in the study of crystal structures. For our purpose, in order to focus attention on the *orientation* of the reference frame, we do away with the translational displacement of the origins of the two frames of reference. We shall therefore concentrate on the relative orientation of the frame of reference $\{X', Y', Z'\}$ with respect to the original frame $\{X, Y, Z\}$, shown in Fig. 7.8a and b. These two frames therefore have a common origin; they are co-centric.

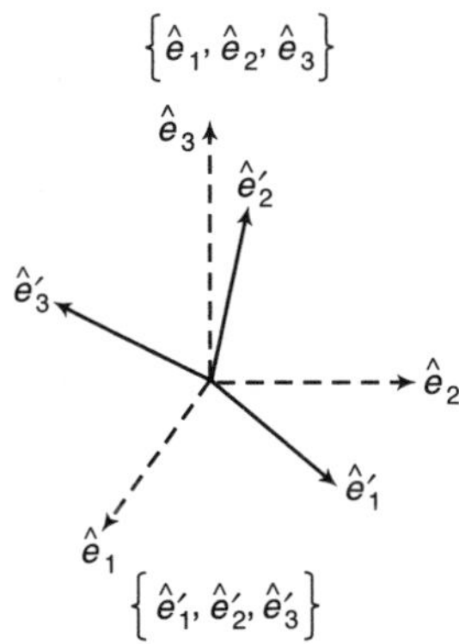

Fig. 7.8a This figure shows two orthonormal basis sets, $\{\hat{e}_1, \hat{e}_2, \hat{e}_3\}$ and $\{\hat{e}'_1, \hat{e}'_2, \hat{e}'_3\}$, which are respectively along the Cartesian frames $\{X, Y, Z\}$ and $\{X', Y', Z'\}$. Both have a common origin. An arbitrary orientation of $\{X', Y', Z'\}$ with respect to $\{X, Y, Z\}$ is expressible as a net result of three successive rotations through the Euler angles (φ, θ, ψ) described in Fig. 7.8b.

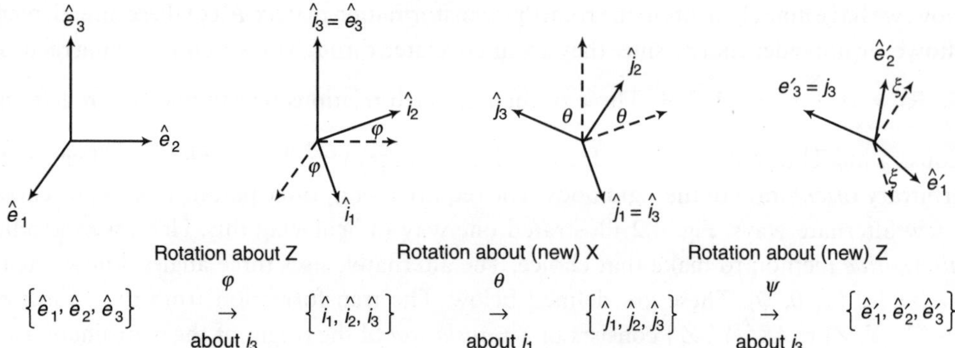

Fig. 7.8b This figure shows the three operations which describe the three Euler angles, (φ, θ, ψ). Two co-centered Cartesian orthonormal basis sets $\{\hat{e}_1, \hat{e}_2, \hat{e}_3\}$ and $\{\hat{e}'_1, \hat{e}'_2, \hat{e}'_3\}$, no matter what their relative orientation is, can be described by the three rotations respectively described by the three Euler angles, (φ, θ, ψ), described in the text.

To appreciate the Euler angles (φ, θ, ψ), we consider the following operations, with reference to Figs. 7.8a and 7.8b:

(i) Rotation of the $\{X, Y, Z\}$ frame about Z through an angle φ to get $\{X_i, Y_i, Z_i = Z\}$. The angle φ is called the angle of precession.

(ii) Rotation of the $\{X_i, Y_i, Z_i\}$ frame about X_i through an angle θ to get $\{X_j = X_i, Y_j, Z_j\}$. The angle θ is called the angle of nutation.

(iii) Rotation of the $\{X_j, Y_j, Z_j\}$ frame about Z_j through an angle ψ to get $\{X', Y', Z' = Z_j\}$. The angle ψ is called the spin angle.

We note from Fig. 7.8 that the direction $\hat{i}_1 = \hat{j}_1$ is orthogonal to $\hat{e}_3$, and also to $\hat{e}'_3$. A straight line through $\hat{i}_1$ is called the *line of node*.

It is easy to see that the three rotations described above are respectively represented by the following matrices:

$$\text{(i)}\quad \mathbb{R}_Z(\varphi) = \begin{pmatrix} \cos\varphi & \sin\varphi & 0 \\ -\sin\varphi & \cos\varphi & 0 \\ 0 & 0 & 1 \end{pmatrix}, \tag{7.51a}$$

$$\text{(ii)}\quad \mathbb{R}_X(\theta) = \begin{pmatrix} 1 & 0 & 0 \\ 0 & \cos\theta & \sin\theta \\ 0 & -\sin\theta & \cos\theta \end{pmatrix}, \tag{7.51b}$$

and

$$\text{(iii)}\quad \mathbb{R}_Z(\psi) = \begin{pmatrix} \cos\psi & \sin\psi & 0 \\ -\sin\psi & \cos\psi & 0 \\ 0 & 0 & 1 \end{pmatrix}. \tag{7.51c}$$

When the three rotations (φ, θ, ψ) are carried out *in the sequence* described above, the transformation convention is referred to as the Z-X-Z convention, and the net transformation is represented by the product of the above three matrices:

$$\mathbb{A}(\varphi, \theta, \psi) = \mathbb{R}_Z(\psi)\, \mathbb{R}_X(\theta)\, \mathbb{R}_Z(\varphi), \tag{7.52a}$$

i.e.,

$$\mathbb{A}(\varphi,\theta,\psi)=\begin{pmatrix} \boxed{\cos\psi\cos\varphi-\cos\theta\sin\varphi\sin\psi} & \boxed{\cos\psi\sin\varphi+\cos\theta\cos\varphi\sin\psi} & \boxed{\sin\theta\sin\psi} \\ \boxed{-\sin\psi\cos\varphi-\cos\theta\sin\varphi\cos\psi} & \boxed{-\sin\psi\sin\varphi+\cos\theta\cos\varphi\cos\psi} & \boxed{\sin\theta\cos\psi} \\ \boxed{\sin\varphi\sin\theta} & \boxed{-\cos\varphi\sin\theta} & \boxed{\cos\theta} \end{pmatrix}. \tag{7.52b}$$

Thus, the components of the position vector of a point $\{X, Y, Z\}$ in the coordinate frame whose orthonormal basis set of vectors is $\{\hat{e}_1, \hat{e}_2, \hat{e}_3\}$, are related to the components $\{x', y', z'\}$ of the same vector in the basis $\{\hat{e}_1', \hat{e}_2', \hat{e}_3'\}$, by the following matrix equation:

$$\begin{bmatrix} x' \\ y' \\ z' \end{bmatrix}_{3\times 1} = \mathbb{A}(\varphi,\theta,\psi)_{3\times 3}\begin{bmatrix} x \\ y \\ z \end{bmatrix}_{3\times 1}. \tag{7.53}$$

The transformation expressed by Eq. 7.53 is called the Euler Transformation.

The Euler transformations are not unique as there are alternative schemes of rotations which describe the net orientation of $\{X', Y', Z'\}$ relative to $\{X, Y, Z\}$ shown in Fig. 7.8a. Different conventions are commonly used in literature. The Euler angles (φ, θ, ψ) are commonly called respectively as the angles of ('precession', 'nutation', 'spin'). In the context of aerodynamics, they are referred to respectively as the ('yaw', 'pitch', 'roll') angles. The terminology is not unique, since it is developed by different mathematicians, physicists and engineers over hundreds of years. Galileo, Huygens, Newton, and Foucault all studied rotation dynamics of various objects. Foucault observed that a spinning wheel maintained its orientation, on account of the conservation of angular momentum, regardless of the Earth's rotation. The motion of a device, which you may have used as a toy in your childhood, called the 'gyroscope' (*with 'g' pronounced as in 'giraffe', not as in 'Google'*) is best described using the Euler angles. The name gyroscope was given to this device by Foucault, by combining the Greek words 'gyros', which means 'rotation', and 'skopien', which means to view, as it enables viewing rotations about different axes.

To illustrate the gyroscope dynamics, we revert to the motion of the top considered in Fig. 7.7. The torque exerted by gravity on the top about the point O using Eq. 7.49, is

$$\vec{\tau} = MgZ_C' \sin\theta\hat{\tau}. \tag{7.54}$$

In the above equation, the angle $\theta = \angle(\hat{e}_3, \hat{e}_3')$ and Z_C' is the coordinate (in the rotating 'top' frame) of the center of mass of the top with respect to the pivot at O (Fig. 7.7). The top can have *three* independent rotations – corresponding to $\dot{\varphi}$ (precession), another corresponding to $\dot{\theta}$ (nutation), and a third which is rotation (spin) about its own axis, coming from $\dot{\psi}$. The principal axes of the top are aligned about $\{\hat{e}'_1, \hat{e}'_2, \hat{e}'_3\}$ and the whole top can rotate about the corresponding axes OX'_1, OX'_2, OX'_3 with respective angular speeds $\{\omega'_1, \omega'_2, \omega'_3\}$.

Equation 7.22 would apply, and the components of angular momentum of the torque in the $\{\hat{e}'_1, \hat{e}'_2, \hat{e}'_3\}$ basis would be proportional to those of angular velocity in the same basis, the proportionality being given by the scalar eigenvalues of the inertia tensor. We can relate the components of the angular velocity with respect to the final frame to the time-derivatives of the Euler angles. The latter are the ratios, of the tiny angles (i) $\delta\varphi$ swept about $\hat{e}'_3$, (ii) $\delta\theta$ about $\hat{j}_2$ and (iii) dy about $\hat{e}'_3$, to a tiny time interval δt, in the limit $\delta t \to 0$. These are in different directions, *not* orthogonal to each other, but since infinitesimal rotations are vectors, they can in fact be added up.

Thus, the angular velocity will be given by the following sum of three vectors:

$$\vec{\omega} = \lim_{\delta t\to 0}\frac{\delta\varphi}{\delta t}\hat{\varphi} + \lim_{\delta t\to 0}\frac{\delta\theta}{\delta t}\hat{\varphi} + \lim_{\delta t\to 0}\frac{\delta\psi}{\delta t}\hat{\psi} = \dot{\varphi}\hat{\varphi} + \dot{\theta}\hat{\theta} + \dot{\psi}\hat{\psi}. \tag{7.55a}$$

It is clear that the unit vectors $(\hat{\varphi}, \hat{\theta}, \hat{\psi})$ are *not* mutually orthogonal, but each can be expressed in the common orthonormal body centered basis $\{\hat{e}'_1, \hat{e}'_2, \hat{e}'_3\}$. Accordingly, we get

$$\vec{\omega} = \dot{\varphi}\left\{\mathbb{A}(\varphi,\theta,\psi)\begin{bmatrix}0\\0\\1\end{bmatrix}\right\} + \dot{\theta}\left\{\mathbb{R}_Z(\psi)\begin{bmatrix}1\\0\\0\end{bmatrix}\right\} + \dot{\psi}\left\{\begin{bmatrix}0\\0\\1\end{bmatrix}\right\}. \tag{7.55b}$$

Using Eqs. 7.51 and 7.52, we get

$$\vec{\omega} = \dot{\varphi}\begin{bmatrix}\sin\theta\sin\psi\\ \sin\theta\cos\psi\\ \cos,\end{bmatrix} + \dot{\theta}\begin{bmatrix}\cos\psi\\ -\sin\psi\\ 0\end{bmatrix} + \dot{\psi}\begin{bmatrix}0\\0\\1\end{bmatrix}. \tag{7.55c}$$

The components of the angular velocity vector in the final body-fixed frame of reference in terms of the Euler angles are therefore given by

$$\vec{\omega} = \omega'_1\hat{e}'_1 + \omega'_2\hat{e}'_2 + \omega'_3\hat{e}'_3 = \begin{bmatrix}\omega'_1\\ \omega'_2\\ \omega'_3\end{bmatrix} = \begin{bmatrix}\dot{\varphi}\sin\theta\sin\psi + \dot{\theta}\cos\psi\\ \dot{\varphi}\sin\theta\cos\psi - \dot{\theta}\sin\psi\\ \dot{\varphi}\cos\theta + \dot{\psi}\end{bmatrix}. \tag{7.56}$$

Most often, for objects like the top, or more generally for a gyroscope, $|\dot{\varphi}| \ll |\dot{\theta}| \ll |\dot{\psi}|$.

The gyroscope has an axis of symmetry, which we shall refer to as the *longitudinal* axis. Symmetry ensures that the two transverse (T) moments of inertia are equal to each other, but most likely *different* from the longitudinal (L) moment. We shall therefore use the notation $i_1 = i_T = i_2$ for the transverse components, and $i_3 = i_L$, for the longitudinal component. Because of the azimuthal symmetry about the longitudinal axis, *any* pair of orthogonal directions in the plane at 90^0 to the longitudinal axis serves as the two transverse axes. From Fig. 7.8, we see that the Euler angle θ is a rotation about the axis $\hat{j}_1 = \hat{i}_1$, and the final Euler rotation through ψ is about the axis $\hat{j}_3 = \hat{e}'_3$. The direction $\hat{j}_1 = \hat{i}_1$ is therefore orthogonal to $\hat{e}'_3$, and we are free to choose this as the direction of *one* of the two transverse principal axes. This makes $\psi = 0$ at *each* instant. We therefore analyze the motion of the gyroscope in a rotating frame of reference, with its axes respectively aligned with the principal axes of the gyroscope.

With this choice of the frame of reference, the angular velocity of the rigid body becomes

$$\vec{\omega} = \omega_1' \hat{e}_1' + \omega_2' \hat{e}_2' + \omega_3' \hat{e}_3' = \begin{bmatrix} \dot{\theta} \\ \dot{\varphi}\sin\theta \\ \dot{\varphi}\cos\theta + \dot{\psi} \end{bmatrix}. \tag{7.57}$$

The angular momentum then becomes

$$\vec{L} = i_T\omega_1' \hat{e}_1' + i_T\omega_2' \hat{e}_2' + i_L\omega_3' \hat{e}_3' = \begin{bmatrix} i_T\dot{\theta} \\ i_T\dot{\varphi}\sin\theta \\ i_L(\dot{\varphi}\cos\theta + \dot{\psi}) \end{bmatrix}. \tag{7.58}$$

The torque is then given by

$$\left.\begin{aligned} \frac{d\vec{L}}{dt} = i_T\frac{d\dot{\theta}}{dt}\hat{e}_1' + i_T\frac{d(\dot{\varphi}\sin\theta)}{dt}\hat{e}_2' + i_L\frac{d(\dot{\varphi}\cos\theta + \dot{\psi})}{dt}\hat{e}_3' + \\ i_T\dot{\theta}\frac{d\hat{e}_1'}{dt} + i_T\dot{\varphi}\sin\theta\frac{d\hat{e}_2'}{dt} + i_L(\dot{\varphi}\cos\theta + \dot{\psi})\frac{d\hat{e}_3'}{dt} \end{aligned}\right\}. \tag{7.59}$$

The *changes* in the unit vectors are *always* orthogonal to them, as we have learned in Chapter 2. We can therefore resolve the *time-derivatives* of the unit vectors in the pair of unit vectors which are orthogonal to the unit vector whose time-derivative is under consideration. We can then collect all the terms together. The detailed derivation is simple; only a little bit laborious. It is left as an exercise. The result is

$$\left.\begin{aligned} \frac{d\vec{L}}{dt} = \{i_T\ddot{\theta} - i_T\dot{\varphi}^2\sin\theta\cos\theta + i_L(\dot{\varphi}\cos\theta + \dot{\psi})(\dot{\varphi}\sin\theta)\}\hat{e}_3 + \\ \{i_T\ddot{\varphi}\sin\theta + 2i_T\dot{\varphi}\dot{\theta}\cos\theta - i_L(\dot{\varphi}\cos\theta + \dot{\psi})\dot{\theta}\}\hat{e}_2 + i_L\frac{d(\dot{\varphi}\cos\theta + \dot{\psi})}{dt}\hat{e}_3 \end{aligned}\right\} \tag{7.60}$$

Equating the component along $\hat{e}_1'$ of the torque in Eq. 7.54, to that in Eq. 7.60, we get:

$$\{i_T\ddot{\theta} - i_T\dot{\varphi}^2\sin\theta\cos\theta + i_L(\dot{\varphi}\cos\theta + \dot{\psi})(\dot{\varphi}\sin\theta)\} = MgZ_C'\sin\theta, \tag{7.61}$$

and canceling $\sin\theta$ on both sides, we have

$$i_T\ddot{\theta} - i_T\dot{\varphi}^2\cos\theta + i_L(\dot{\varphi}\cos\theta + \dot{\psi})\dot{\varphi} = MgZ_C'. \tag{7.62a}$$

Furthermore, setting the components of the torque along $\hat{e}_2'$ and $\hat{e}_3'$ to be zero, we get

$$i_T\ddot{\varphi}\sin\theta + 2i_T\dot{\varphi}\dot{\theta}\cos\theta - i_L(\dot{\varphi}\cos\theta + \dot{\psi})\dot{\theta} = 0, \tag{7.62b}$$

$$\text{and} \quad i_L \frac{d(\dot{\varphi}\cos\theta + \dot{\psi})}{dt} = 0. \tag{7.62c}$$

The *gyroscope equation(s) of motion* (Eq. 7.62) can be further simplified by considering $\dot{\theta} = 0$, which would describe a precessing gyroscope *without nutation*. However, in addition to precession (due to $\dot{\varphi}$), the gyroscope can nutate, due to $\dot{\theta}$, and the general solution must include this effect. The resulting motion seems to defy gravity; yet it is in fact determined completely by the torque exerted by gravity itself. The motion is spectacular, which makes the gyroscope a popular toy that fascinates children. The gyroscope is, however, not a mere toy; it is an extremely sophisticated navigation and orientation-guidance device with applications in navigation, GPS, aerodynamics, satellite trajectory control, etc. If the gyroscope is mounted in a moving vehicle, any sudden acceleration of the vehicle (such as skidding of a car) would make it necessary to take into account the pseudo-forces which, even if unreal, have real effects, described above, in line with the discussion in Chapter 3. The gyroscope can therefore be used as a sensitive sensor in moving vehicles such as automobiles and aeroplanes to carefully track their rotations. Feedback electro-mechanical control mechanisms are installed in these vehicles as critical safety devices. Together with accelerometers (which are sensitive to *linear* accelerations), the gyroscopes are extremely useful in detecting the smallest of linear/rotational accelerations. This behavior of the gyroscope makes it a very useful and sophisticated device for guidance and control mechanisms. Complex feedback mechanisms are installed in addition to the gyroscope to adjust the position, orientation, and trajectories of skidding motor vehicles, and/or turning aeroplanes (Fig. 7.8). Aeroplane movements are then controlled by the 'ailerons' which generate the roll, the 'elevators', which regulate the pitch, and by the 'rudder' which governs the yaw.

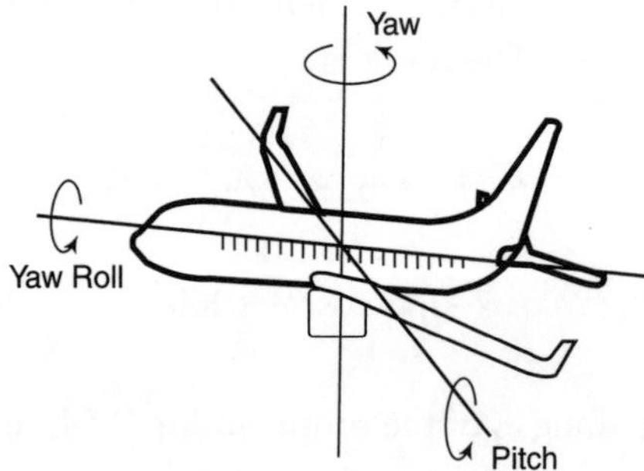

Fig. 7.8c The gyroscope is the perfect device that is used as a sensor to monitor and regulate yaw, roll, and the pitch of an aeroplane in flight. Modern MEMS (Micro-Electro-Mechanical Systems) technology uses sophisticated devices which sense and regulate rotational accelerations to control stability of vehicles. Other applications include governing camera dynamics to get good quality images by getting feedback on the orientations of cameras mounted on three orthogonal axes. The MEMS gyro-technology relies on advanced electro-mechanical coupling mechanisms.

Problems with Solutions

P7.1:

(a) Consider a long rod of uniform mass density having total mass M and length L. Show that its moment of inertia I_{CM} about the axis that is ***perpendicular*** to the rod ***through its center of mass is one fourth*** of its moment of inertia I_{End} about an axis that is perpendicular to the rod at one end.

(b) Determine I_{End} using the parallel axis theorem.

Solution:

The axis is perpendicular to the rod at one end of the rod, as shown in the figure. We lay the rod along the X-axis with one end at $x = 0$. The axis about which the moment of inertia is of interest is the Y-axis. Consider a tiny element of the rod, having a mass δm at a distance x (corresponding to the point P) on the rod from the origin.

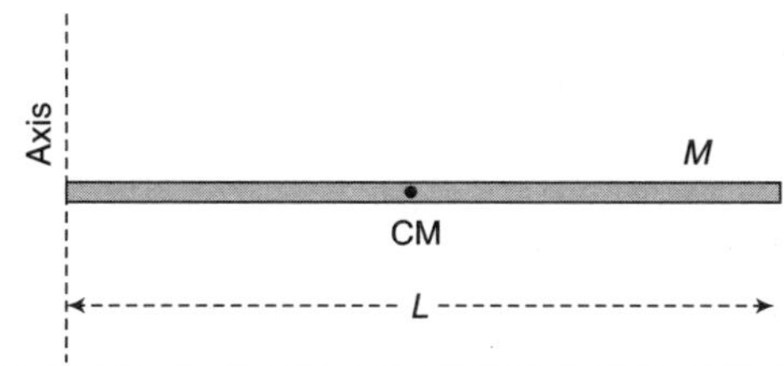

The rod is set up as a 2-dimensional flat mass distribution (Section 7.1): $I_{zC} = \int\int [\rho_C(P)]^2 \delta m_P$

In the present case, $z = y$, C corresponds to the origin, and we have a linear mass distribution. The linear mass density of the rod is $\lambda = M / L$ since the rod has a uniform mass distribution. Thus, the moment of inertia of the rod about an axis perpendicular to it at one end is

$$I_{End} = \int_0^L x^2 \delta m = \int_0^L x^2 \lambda dx = \frac{M}{L}\int_0^L x^2 dx = \frac{ML^2}{3}.$$

To find the moment of inertia of the rod about its center of mass, let us shift the origin of the coordinate system to its center. Then, we have

$$I_{CM} = \int_{-L/2}^{+L/2} x^2 \lambda dx = \frac{M}{L}\int_{-L/2}^{+L/2} x^2 dx = \frac{1}{12}ML^2. \text{ Thus: } I_{CM} = \frac{I_{End}}{4}.$$

(b) According to the parallel axis theorem given by Eq. 7.15g, the moment of inertia of an object of mass M about an axis passing through a point D can be written as the sum of moment of inertia of the object about a parallel axis passing through its center of mass, and the product of the mass M and the square of the distance of the center of mass from D. Therefore,

$$I_{End} = I_{CM} + M\left(\frac{L}{2}\right)^2 = \frac{ML^2}{12} + \frac{ML^2}{4} = \frac{ML^2}{3}.$$

P7.2:

Determine the moment of inertia of a solid cone having uniform mass density, base radius R, and height h, about (a) its azimuthal axis of symmetry through its base and (b) about the diameter of its circular base.

Solution:

We consider the coordinate system shown in the figure below.

(a) the moment of inertia with respect to the azimuthal axis of symmetry through its base (Z-axis) is given by (cf. Eq. 7.15f)

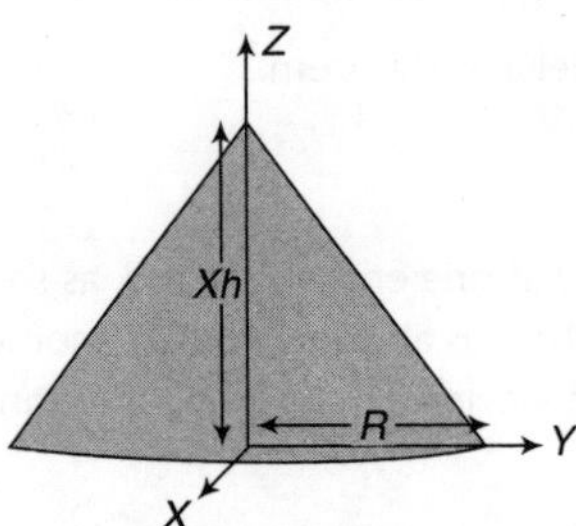

$$I_{zz} = \int (x^2 + y^2)\delta m = \frac{M}{\frac{1}{3}\pi R^2 h}\int_0^h dz\int_0^{2\pi} d\varphi\int_0^r \rho d\rho\rho^2 = \frac{M}{\frac{1}{3}\pi R^2 h}2\pi\int_0^h dz\frac{r^4}{4}.$$

Since $r = \frac{R}{h}z$, $I_{zz} = \frac{6M}{R^2 h}\cdot\frac{R^4}{4h^4}\cdot\frac{h^5}{5} = \frac{3}{10}MR^2$.

(b) The moment of inertia with respect to the diameter of the base (such as along the X-axis),

$$I_{xx} = \int (y^2 + z^2)\,\delta m = \frac{M}{\frac{1}{3}\pi R^2 h}\int_0^h dz\int_0^{2\pi} d\varphi\int_0^r \rho d\rho(\rho^2\sin^2\varphi + z^2),$$

$$= \frac{3M}{\pi R^2 h}\int_0^h dz\int_0^{2\pi} d\varphi\left(\frac{r^4}{4}\sin^2\varphi + \frac{r^2}{2}z^2\right) = \frac{3M}{\pi h^3}\int_0^h dz\int_0^{2\pi} d\varphi\left(\frac{R^2}{h^2}\frac{z^4}{4}\sin^2\varphi + \frac{z^4}{2}\right),$$

$$= \frac{3M}{\pi h^3}\left(\frac{1}{2}\frac{h^5}{5}\right)\left(\frac{\pi}{2}\cdot\frac{R^2}{h^2} + 2\pi\right) = \frac{3}{20}M(R^2 + 4h^2).$$

With respect to the diameter of the base, the moment of inertia therefore is:

$$I_{base} = \frac{3}{20}MR^2 + \frac{3}{5}Mh^2 + M\left(\frac{h}{4}\right)^2 = \frac{3}{20}MR^2 + \frac{53}{80}Mh^2.$$

P7.3:

Consider a symmetric spinning top as illustrated in the figure below. (a) Determine the Lagrangian of this system. (b) Write down the equations of motion for θ, φ and ψ, and identify the constants of motion. (c) Analyze the implication of the constancy of θ. (d) Use the expression for the total energy of the system for the analysis of the precession of the top, with nutation. (e) Sketch the effective potential experienced by the top and determine the boundary conditions of the angle θ. Discuss the effects of the angular momentum L_3 on the precession frequency Ω.

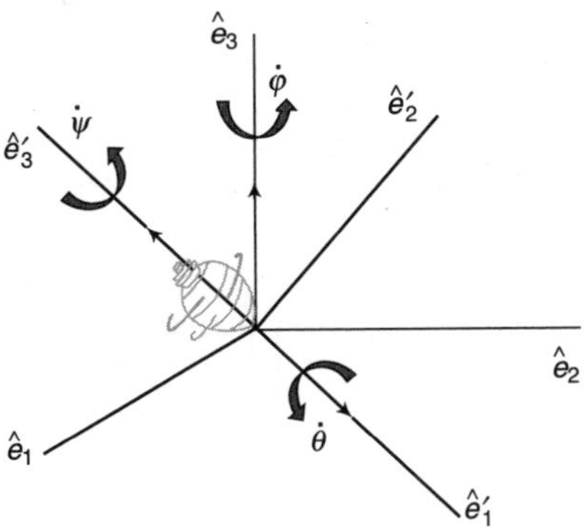

Solution:

(a) The kinetic energy of the spinning top is given as: $T = \frac{1}{2} i_1(\omega_x^2 + \omega_y^2) + \frac{1}{2} i_3\omega_z^2$,

where the components of angular velocity in terms of the Euler angles (Eq. 7.56) are:

$$\omega_x = \dot{\varphi} \sin\theta \sin\psi + \dot{\theta} \cos\psi;\ \omega_y = \dot{\varphi} \sin\theta \cos\psi - \dot{\theta} \sin,\ \psi, \text{ and } \omega_z = \dot{\varphi} \cos\theta + \psi.$$

Therefore, $T = \frac{1}{2} i_1(\dot{\theta}^2 + \dot{\varphi}^2 \sin^2\theta) + \frac{1}{2} i_3(\dot{\psi} + \dot{\varphi}\cos\theta)^2$.

For a symmetric top, $i_1 = i_2 \neq i_3$, where i_1, i_2, i_3 are the moments of inertia along the axes x, y and z respectively; φ, θ, and ψ are the Euler angles, namely the 'precession', 'nutation', and 'spin' respectively.

The potential energy is: $V = MgR \cos\theta$ where R is the distance of the center of mass of the top from the pivot. The Lagrangian of the top is:

$$\mathcal{L} = T - V = \frac{1}{2} i_1(\dot{\theta}^2 + \dot{\varphi}^2 \sin^2\theta) + \frac{1}{2} i_3(\dot{\psi} + \dot{\varphi}\cos\theta)^2 - MgR\cos\theta \ .$$

(b) The Lagrange equation for θ is: $\frac{d}{dt}\left(\frac{\partial \mathcal{L}}{\partial \dot{\theta}}\right) = \frac{\partial \mathcal{L}}{\partial \theta}$.

Hence: $i_1\ddot{\theta} = i_1\dot{\varphi}^2 \sin\theta \cos\theta - i_3 (\dot{\psi} + \dot{\varphi} \cos\theta)\dot{\varphi} \sin\theta + MgR \sin\theta$.

The equation for the cyclic coordinate φ gives: $p_\varphi = \frac{\partial \mathcal{L}}{\partial \dot{\varphi}} = i_1\dot{\varphi} \sin^2\theta + i_3(\dot{\psi} + \dot{\varphi} \cos\theta)\cos\theta$, which is a constant. There is no torque about the Z-axis; the z-component of the angular momentum is a constant of motion. Similarly, p_ψ is also a constant: $p_\psi = \frac{\partial \mathcal{L}}{\partial \dot{\psi}} = i_3(\dot{\psi} + \dot{\varphi} \cos\theta)$: constant.

Hence, the angular momentum with respect to the body frame is a constant of motion.

(c) We have, $L_3' = p_\psi = i_3\omega_3'$, from Eq. 7.47, where $\omega_3' = \dot{\psi} + \dot{\varphi}\cos\theta$. The rate at which the top precesses around the z-axis is given by the θ equation. If θ is constant, then, the left hand side $i_1\ddot{\theta} = 0$. Let us relabel $\dot{\varphi}$ as Ω. Then, we have: $i_1\Omega^2\cos\theta - i_3\omega_3'\Omega + MgR = 0$. The roots for this equation are given by $\Omega_\pm = \dfrac{i_3\omega_3'}{2i_1\cos\theta}\left(1 \pm \sqrt{1 - \dfrac{4Mi_1gR\cos\theta}{i_3^2\omega_3'^2}}\right)$. For large ω_3', the slow precession frequency is given by: $\Omega_- = \dfrac{i_3\omega_3'}{2i_1\cos\theta}\left[1 - \left(1 - \dfrac{1}{2}\cdot\dfrac{4Mi_1gR\cos\theta}{i_3^2\omega_3'^2}\right)\right] \approx \dfrac{MgR}{i_3\omega_3'}$. The larger root of Ω, Ω_+, is given by: $\Omega_+ = \dfrac{i_3\omega_3'}{2i_1\cos\theta}\left[1 + \left(1 - \dfrac{1}{2}\cdot\dfrac{4Mi_1gR\cos\theta}{i_3^2\omega_3'^2}\right)\right] \approx \dfrac{(i_3\omega_3')}{i_1\cos\theta}$.

Note that Ω_+ does not depend on g.

(d) The total energy of the top: $E = \dfrac{1}{2}i_1(\dot{\theta}^2 + \dot{\varphi}^2\sin^2\theta) + \dfrac{1}{2}i_3(\dot{\psi} + \dot{\varphi}\cos\theta)^2 + MgR\cos\theta$.

To determine $\dot{\varphi}$, consider the angular momentum for a symmetric spinning top:

$$\vec{L} = i_1\dot{\theta}\,\hat{e}_1' + i_2\dot{\varphi}\sin\theta\,\hat{e}_2' + i_3(\dot{\psi} + \dot{\varphi}\cos\theta)\,\hat{e}_3', \text{ and } \hat{e}_3 = \cos\theta\,\hat{e}_3' + \sin\theta\,\hat{e}_2'.$$

Therefore, $L_3 = \vec{L}\cdot\hat{e}_3 = i_1\dot{\varphi}\sin\theta\sin\theta + i_3(\dot{\psi} + \dot{\varphi}\cos\theta)\cos\theta = i_1\dot{\varphi}\sin^2\theta + L_3'\cos\theta$.

Hence, $\dot{\varphi} = \dfrac{L_3 - L_3'\cos\theta}{i_1\sin^2\theta}$.

Now the total energy is: $E = \dfrac{1}{2}i_1\dot{\theta}^2 + \dfrac{1}{2}\dfrac{(L_3 - L_3'\cos\theta)^2}{i_1\sin^2\theta} + \dfrac{L_3'^2}{2i_3} + MgR\cos\theta = \dfrac{1}{2}i_1\dot{\theta}^2 + U_{eff}(\theta)$, where the effective potential energy is: $U_{eff}(\theta) = \dfrac{1}{2}\dfrac{(L_3 - L_3'\cos\theta)^2}{i_1\sin^2\theta} + \dfrac{L_3'^2}{2i_3} + MgR\cos\theta$.

(e) The first term corresponds to a U-shaped potential as shown in the figure below. It goes to infinity as $\theta \to 0$ or π, which are the extreme values of θ in this case. We note from the above result that the total energy being $\dfrac{1}{2}i_1\dot{\theta}^2 + U_{eff}(\theta)$, we must have $E \geq U_{eff}(\theta)$, as $\dfrac{1}{2}i_1\dot{\theta}^2$ must be non-negative.

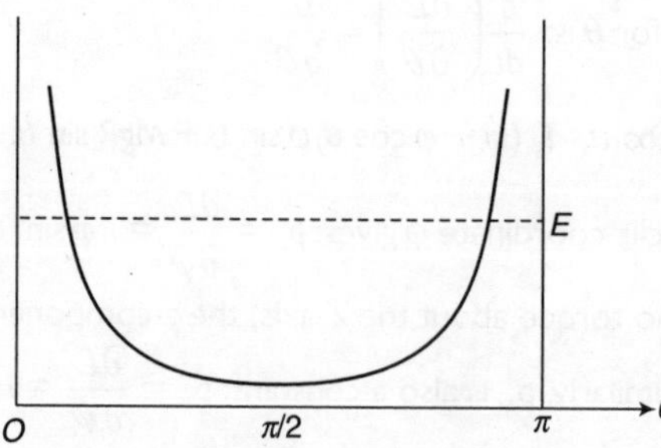

Thus the angle θ is confined between the two turning points. In fact, θ can oscillate and the top can undergo nutation when it precesses. From the expression for $\dot{\varphi}$, we see that when L_3 is greater than L_3', $\dot{\varphi}$ cannot change sign; it will either decrease or increase. Thus, the top precesses in one direction whereas the angle θ varies periodically between two values determined by the boundary condition. If, on the other hand, $L_3 < L_3'$, then $\dot{\varphi}$ can vanish at an angle given by $L_3 - L_3' \cos\theta_0 = 0$. If this angle lies outside the range specified by the boundary condition, $\dot{\varphi}$ will not vanish, but only nutate. If the angle lies within the limits, then the sign of $\dot{\varphi}$ will change twice during an oscillation of θ.

P7.4:

Consider a free rotating rigid body with equal moments of inertia i about two of its axes of rotation (say, $\hat{x}_1$ and $\hat{x}_2$), whereas the third moment of inertia about the $\hat{x}_3$ axis is given as I_3. Show that with respect to the body frame of reference, the angular frequency $\vec{\omega}$ and the angular momentum $\vec{L}$ precess with frequency $\Omega \equiv \omega_3 (i_3 - i)/i$. Also, show that the axis $\hat{x}_3$ and the angular velocity vector $\vec{\omega}$ precess around the angular momentum vector $\vec{L}$ at a frequency $\frac{\vec{L}}{i}$ with respect to laboratory frame of reference. Justify that the two results are consistent.

Solution:

From the Euler's equations of motion (Eq. 7.37b), dropping the primes on ω, (since the principal moments of inertia are aligned with the body frame of reference), we have

$$\tau_1 = i_1\dot{\omega}_1 + (i_3 - i_2)\omega_3\omega_2;\ \tau_2 = i_2\dot{\omega}_2 + (i_1 - i_3)\omega_1\omega_3 \text{ and } \tau_3 = i_3\dot{\omega}_3 + (i_2 - i_1)\omega_2\omega_1.$$

Since $i_1 = i_2$, say both equal to i, we have

$$0 = i\dot{\omega}_1 + (i_3 - i)\omega_3\omega_2,\quad 0 = i\dot{\omega}_2 + (i - i_3)\omega_1\omega_3;\quad \text{and}\quad 0 = i_3\dot{\omega}_3.$$

Thus, ω_3 is constant. From the first two of the above equations, $0 = \dot{\omega}_1 + \Omega\omega_2$ & $0 = \dot{\omega}_2 - \Omega\omega_1$, where $\Omega \equiv \omega_3 (i_3 - i)/i$. Hence, $\ddot{\omega}_1 + \Omega^2\omega_1 = 0$ (and similar relation for ω_2).

The solutions to the second order differential equations are: $\omega_1 = A\cos(\Omega t + \theta)$, & $\omega_2 = A\sin(\Omega t + \theta)$.

Thus we see that $\vec{\omega}$ traces a cone around the axis $\hat{x}_3$ with an angular frequency Ω.

Since $\vec{L} = (L_1, L_2, L_3) = (i\omega_1, i\omega_2, i_3\omega_3)$, we see that $\vec{L}$ also precesses around $\hat{x}_3$ with the frequency Ω.

With reference to the body frame, $\vec{\omega} = \omega_1\hat{x}_1 + \omega_2\hat{x}_2 + \omega_3\hat{x}_3$,

$$\vec{L} = i(\omega_1\hat{x}_1 + \omega_2\hat{x}_2) + i_3\omega_3\hat{x}_3 = i(\vec{\omega} - \omega_3\hat{x}_3) + i_3\omega_3\hat{x}_3 = i(\vec{\omega} + \Omega\hat{x}_3).$$

Thus, $\vec{\omega} = \left(\frac{\vec{L}}{i} - \Omega\breve{x}_3\right)$. Since $\vec{L}$, $\vec{\omega}$, and $\hat{x}_3$ are related through a linear relationship, they share a common plane. Since there are no torques acting on the system, the angular momentum $\vec{L}$ remains constant. Taking this to be along the z-direction of the lab-fixed frame, we see that the $\hat{x}_3$ and $\vec{\omega}$ precess about $\vec{L}$. Let $\vec{\omega}'$ be the angular frequency of precession. Then the rate of change of $\hat{x}_3$ is:

$\frac{dx_3}{dt} = \vec{\omega}' \times \breve{x}_3 = \left(\frac{\vec{L}}{i} - \Omega\breve{x}_3\right) \times \breve{x}_3 = \left(\frac{L}{i}\right)\breve{L} \times \breve{x}_3$. Thus the frequency of precession of $\hat{x}_3$ is given by L/i. From the body-fixed frame of reference, $\vec{L}$ and $\vec{\omega}$ precess about $\hat{x}_3$, at a frequency Ω.

P7.5:

Determine the location of the center of mass of (a) a solid hemisphere and (b) a hemispherical shell of radius R.

Solution:

The position of the center of mass of a rigid body of mass M defined with respect to a coordinate system is given by $\vec{R}_{CM} = \frac{1}{M}\int \vec{r}dm$, where $\vec{r}$ gives the position vector of the tiny mass elements dm in the rigid body.

(a) For solid hemisphere: Let us choose a coordinate system as shown in the figure. The x and y components of the center of mass are zero because of the symmetry of the mass distribution about these axes. If ρ is the density of the material of the sphere, the z component is:

$$z_{CM} = \frac{\rho}{M}\int_0^R dr\, r^2 \int_0^{\pi/2} d\theta \sin\theta \int_0^{2\pi} d\phi\, z = \frac{3R}{8},$$

where $z = r\cos\theta$, and $M = \frac{2}{3}\pi R^3 \rho$.

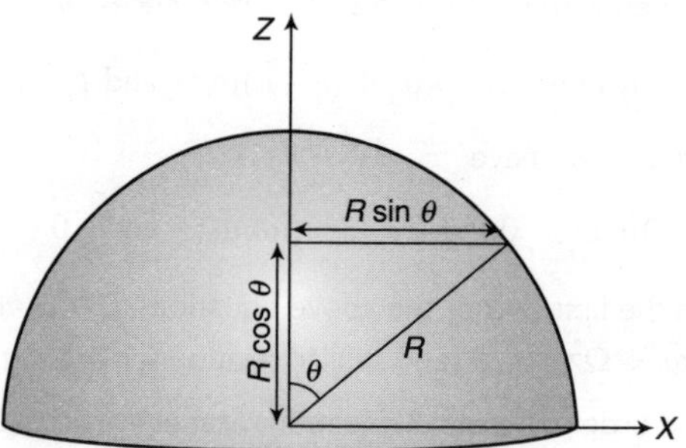

(b) For hemispherical shell: Again, from symmetry, the x and y components of the center of mass are zero. The z component is: $z_{CM} = \frac{\sigma}{M}R^2 \int_0^{\pi/2} d\theta \sin\theta \int_0^{2\pi} d\phi\, z = \frac{R}{2}$, since $z = r\cos\theta$ and $M = 2\pi R^2$.

P7.6:

A gyroscope wheel is attached at one end of an axle of length d. The other end of the axle is suspended from a string of length l. The wheel is set into motion so that it executes uniform precession in the horizontal plane. The wheel has mass m and moment of inertia about its center of mass I_0. Its spin angular velocity is ω_s. Neglect the masses of the shaft and the string. Find the angle β that the string makes with the vertical. Assume that β is small; and $\sin\beta \approx \beta$.

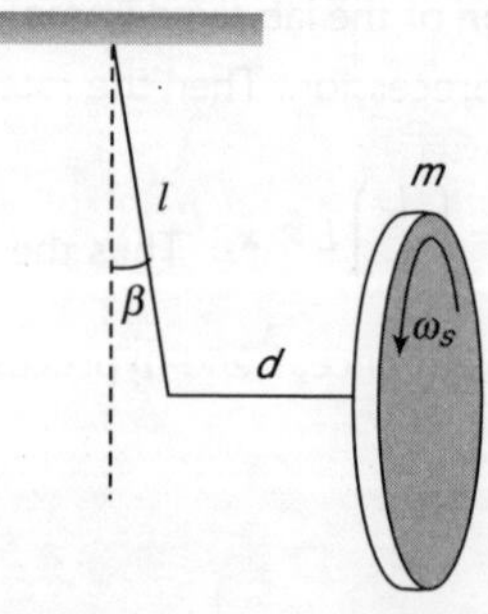

Solution:

Let the *normal* distance from the point of contact of the string at the support, to the center of mass of the disc of the gyroscope, is r; i.e., $r = l \sin \beta + d \simeq l\beta + d$. The weight mg of the disc is balanced by the vertical component of tension in the string T, hence: $T \cos \beta = mg$. Hence, $T = mg$ ($\cos \beta \approx 1$). Centripetal force due to precession at an angular frequency (Ω) is provided by the horizontal component of the tension. Hence: $T \sin \beta = m\Omega^2 r$; i.e., $T\beta = m\Omega^2 r$.

The torque on the disc: $\tau = mgd = \frac{d}{dt}$ (*angular momentum*) $= \Omega \times$ *angular momentum* $= \Omega I_0 \omega_s$.

Therefore, $\Omega = \frac{mgd}{I_0\omega_s}$. Since $r = l\beta + d$, and $\beta = \Omega^2 r / g$, we get $\beta = \frac{\Omega^2}{g}(l\beta + d)$, and hence $\beta = \frac{d}{g/\Omega^2 - l}$

where Ω is given by $\Omega = \frac{mgd}{I_0\omega_s}$.

P7.7:

Consider a cube of side a and mass m placed on the top of a table by balancing it on its edge. If the cube is left unsupported, it would topple, and then rotate till it lands on the table on its face. If the cube does not slip, determine the angular velocity just when its face hits the table surface.

Solution:

When the cube is balanced on its edge, the center of mass of the cube is at a height of $a/\sqrt{2}$, and the potential energy of the cube is $mga/\sqrt{2}$. When the cube lands on the table's surface on its face, its potential energy is $mga/2$. The difference in the potential energy gives the kinetic energy of rotational. Thus, we have $\frac{1}{2}I\omega^2 = mga\left(\frac{1}{\sqrt{2}} - \frac{1}{2}\right)$. The moment of inertia of the cube about its axis is known to be $\frac{2}{3}ma^2$. Substituting the value, we get $\omega = \sqrt{\frac{3g}{a}\left(\frac{1}{\sqrt{2}} - \frac{1}{2}\right)}$.

P7.8:

Consider two-point masses of mass m traveling in a circular trajectory of radius r at a circular frequency ω. The two particles are diametrically opposite to each other. The circle of motion is centered at the point $(0, 0, z_0)$, and the plane of the circle is parallel to the xy-plane. What is the angular momentum of the system? If one of the particles is removed from the system, what would be the new angular momentum of the system?

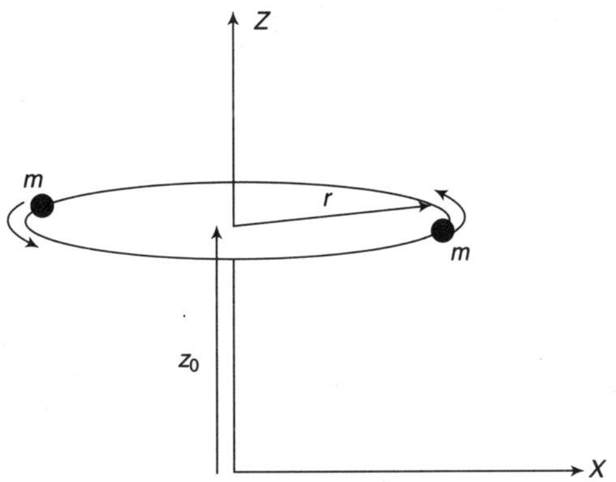

Solution:

The angular momentum is

$$\vec{L} = \mathbb{I}\vec{\omega} = \begin{bmatrix} \int dm(y^2+z^2) & -\int dm(xy) & -\int dm(xz) \\ -\int dm(yx) & \int dm(z^2+x^2) & -\int dm(yz) \\ -\int dm(zx) & -\int dm(zy) & \int dm(x^2+y^2) \end{bmatrix} \begin{bmatrix} 0 \\ 0 \\ \omega_z \end{bmatrix}.$$

Note that, in this case, only the last column of the moment of inertia tensor contributes to the angular momentum of the system. The integrals $\int dn(xz)$ and $-\int dm = \breve{e}_z 2m\omega r^2$ are zero, as the x and the y coordinates of one particle has signs opposite to those of the other particle, the particles being diametrically opposite to each other on the circular orbit.

Hence, $\vec{L} = \mathbb{I}\vec{\omega} = \breve{e}_z\omega\int dm(x^2+y^2) = \breve{e}_z\omega r^2\int dm = \breve{e}_z 2m\omega r^2 a.$

If one of the particles is removed from the system, then the integrals $-\int dm(xz)$, and $-\int dm(yz)$, do not vanish, and the angular momentum has non-zero components along the x and y directions. It is

$$\vec{L} = \mathbb{I}\vec{\omega} = m\omega \begin{bmatrix} -xz_0\breve{e}_x \\ -yz_0\breve{e}_y \\ r^2\breve{e}_z \end{bmatrix}.$$

The angular momentum and the angular velocity are not parallel to each other in this case.

Additional Problems

P7.9 Show that the kinetic energy of rotation of a rigid body is given by $T = \frac{1}{2}\vec{\omega}\cdot\vec{L}$.

P7.10 Equal masses are placed at the vertices of an isosceles triangle which forms the base of a pyramid with height of 4 cm. Where is the center of mass of this system?

P7.11 Determine the moment of inertia of a solid sphere of radius **R** about an axis through its center.

P7.12(a) Consider a solid cylinder of mass **m** and radius **R** rolling on a floor without slipping with speed **v**. Show that the total kinetic energy of the body is equal to the rotational KE of the body with respect to any point in the body which is instantaneously stationary. [moment of inertia of a cylinder $I_{zz} = \frac{1}{2}mR^2$].

(b) The above-mentioned solid cylinder rolls without slipping down a plane inclined at an angle 30°. Show that its acceleration is independent of the dimensions of the cylinder.

P7.13 A small homogeneous sphere of mass **m** and radius **r** rolls without sliding on the outer surface of a larger stationary sphere of radius **R** as shown in the adjacent figure. Let θ be the instantaneous angle of the small sphere with respect to the z-axis, as shown. The smaller sphere starts from rest at the top of the larger sphere ($\theta = 0$).

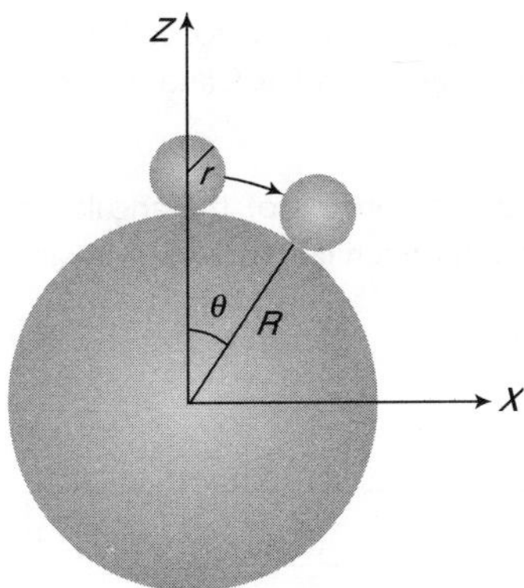

(a) Determine the velocity of the center of the small sphere as a function of q.

(b) Determine the angle at which the small sphere would fly off the large one.

P7.14 A mass m travels perpendicular to a stick of mass m and length l, which is initially at rest. At what location should the mass collide elastically with the stick, so that the mass and the center of the stick move with equal speeds after the collision?

P7.15 A mass m moves at a speed v_0 perpendicular to a stick of mass m and length ℓ, which is initially at rest. The mass collides completely inelastically with the stick at one of its ends; it sticks to it. Determine the resulting angular velocity of the system by choosing

(a) the center mass of the system after collision,

(b) the center mass of the stick before collision.

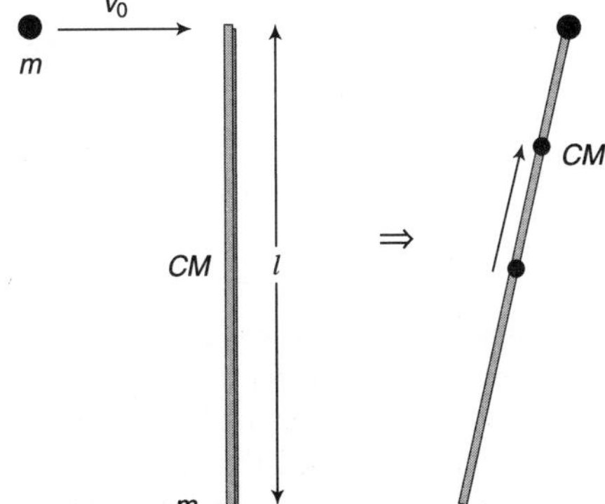

P7.16(a) Determine the inertia tensor of a hollow cube of mass M, and side L, with the coordinate axes parallel to the edges of the cube. The origin is at the center of the cube. Also, determine the angular momentum of the cube about Z-axis, when it rotates about it with an angular velocity ω. Assume now that the cube is flattened to form a perfect square in the XY-plane. What would be its angular momentum with respect to the origin if it rotates in the XY-plane?

(b) In the above question, assume that during flattening, the box is distorted into some arbitrary planar shape in the XY-plane. Which of the elements in the inertia tensor will be non-vanishing? Show that $I_{zz} = I_{xx} + I_{yy}$.

P7.17 Use the moment of inertia tensor matrix to determine the moment of inertia of (a) solid sphere, and, also a hollow sphere of radius R and (b) a solid, and also a hollow, cylinder of mass M, radius R and height h.

P7.18 Determine the x, y, and z components of the angular momentum of a rectangular block of edges a, b, and c, respectively, when it is rotating with an angular velocity ω about either of its edges.

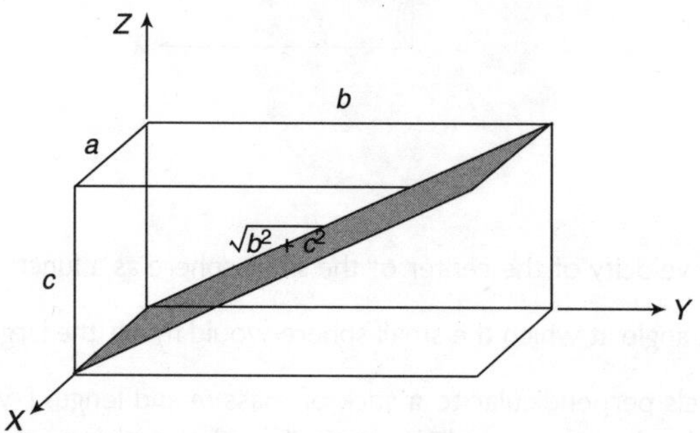

P7.19 Consider a sheet in the form of an isosceles triangle of mass M, vertex angle 2θ and common side length L. Determine the moment of inertia about an axis passing through its vertex and normal to its plane.

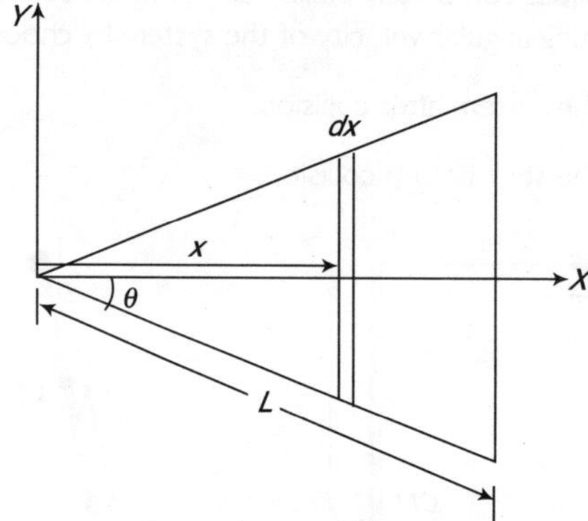

P7.20 The Earth may be approximated as a spinning top. Because of its equatorial bulge, its moment of inertia about the polar axis is slightly greater than the other two moments. Use $i_1 = i_2 = i$ and $i_3 = 1.00327i$. Also, the polar axis of the Earth is 0.2° tilted with respect to its axis of rotation. Show that the precession of the axis of the Earth about its polar axis (Chandler wobble) is 305 days. Also, show that the period of this wobble should be about a day when seen from the space frame.

CHAPTER 8

The Gravitational Interaction in Newtonian Mechanics

Because there is a law such as gravity, the universe can and will create itself from nothing.

—Stephen Hawking

8.1 CONSERVED QUANTITIES IN THE KEPLER–NEWTON PLANETARY MODEL

To an ancient man who watched the sky without nuances of smoke, dust and light contaminations in the atmosphere, the sky would have looked many times more beautiful and brighter than it does now. With amazing regularity, night after night, the sky would turn around his village. With no television to dissipate his time, he would admire and wonder, what is it that makes the sky look so very nearly the same each night, and yet that tad different. What a wonder it would seem that the world turned around him, every single day. A keen observer would notice, however, that amid the twinkling stars, there were some bright objects that seemed to be wandering a little bit in space, here today, and there tomorrow. Over days, weeks, and months, they would drift even far apart from the group of stars they were first sighted with. Astronomy is in some sense the mother of both physics and mathematics, ever since the curious man explored reasons to account for his observations. In studying astronomy, man hit on the very method of science, which would require geometry, trigonometry and eventually differential and integral calculus. Early models included imaginary forces, driven often by mythological gods and daemons, stories of whom fascinate children the world over even today. The myth, however, obliterates reason. Sections of the society sadly even now continue to be driven by superstition, rather than the knowledge earned by man over centuries. Today, much is known, and even as mindboggling questions continue to challenge physicists, superstition is thankfully becoming increasingly dispensable.

Star-gazing and analyzing motion of the planets, first considered as wandering stars, thus reveals the very method of scientific exploration. Human curiosity demands a model to be developed in order to account for the observed phenomena. One may trust the model as long as it does not lead to any discrepancy or contradiction, or internal inconsistency. A few early deductions known to some Indian and Greek astronomers have turned out to be quite accurate. For example, Aryabhatta, in the fifth century CE, had inferred and proclaimed that Earth has a spherical shape, not flat (as some people seem to want to believe even today). Brahmagupta, in the seventh century, had estimated the circumference of Earth to be ~36,210 kms, remarkably close to 40,075 kms as we now know. The Kerala astronomers, in the fifteenth century CE, even before Copernicus, were very much aware of the Sun's central position in our solar system, even though this knowledge is often termed as the 'Copernicus revolution'. A scientific model is tested against observations and experiments. Its ability to *predict* phenomena is always challenged. If the predictions bear out right, the model gains further support; not otherwise.

Inadequacies, and inconsistencies, in the models require modifications of the models, improvisations, or even total replacements. Our understanding of the laws of nature today has undergone a myriad of brilliant contributions from thinkers and philosophers. Amongst these were some outstanding observers, mathematicians, astronomers, and physicists. Tycho Brahe (14 Dec 1546–24 Oct 1601) was one of them. His life was extra-ordinary. Stories about interesting episodes about his life ranging from how he lost his nose in a fight with a friend to decide whose mathematics was right to his affair with the queen of Denmark can be found on the internet. Brahe made systematic observations on the wandering stars—the planets in our solar system—which he catalogued carefully. His laboratory was generously funded by the King of Denmark.

Johannes Kepler (15 Dec 1571–15 Nov 1630), a German mathematician and astronomer, sought Tycho Brahe's data for analysis, but Brahe refused to share it with Kepler. Kepler admitted later that he usurped Brahe's data taking advantage of his fateful illness. The planetary model [1] had already undergone refinements and modifications, since the earliest model due to Ptolemy (100–170 CE), which considered the Earth at the center of the universe. Copernicus' (1473–1543) model had the Sun at the center, and the planets going around it in perfect circles. Tycho Brahe's (1546–1601) model had the Sun and the Moon go around the Earth, and the rest of the planets go around the Sun. Amongst Brahe's most famous discoveries was the supernova in 1572, in the constellation *Cassiopeia* (known as *Sharmishtha* in the school of astronomy that developed in India).

Johannes Kepler (1571–1630) disagreed with Brahe's planetary model and adopted the Copernicus model, but modified it to have the planets go around the Sun in *elliptic* orbits, rather than in Copernican circles. Kepler thought that planetary motion was driven by some kind of radiating *invisible spokes* from the Sun (Fig. 8.1) which held the planets and guided their motion. The idea of the gravitational interaction was established systematically by Isaac Newton, in the seventeenth century, well after Kepler.

Kepler's model was able to account for planetary positions with much greater accuracy than any of the previous models. The elliptic shapes Kepler considered were mostly based on clever conjectures, by smart-guessing the trajectory from Brahe's careful data of the positions of the wandering stars, i.e., the planets. Kepler orbits (almost) correctly predicted the positions of the planets. Accurate predictions of time-period of the planets to complete their revolution about the Sun could be made on the basis of Kepler orbits. The developments of the planetary models through the works of Ptolemy, Copernicus, Tycho Brahe, and Kepler, and parallel developments at the Kerala school of Astronomy in India [2], which were in fact frequently *ahead* of the developments in Europe, make fascinating reading for a student interested in the history of science.

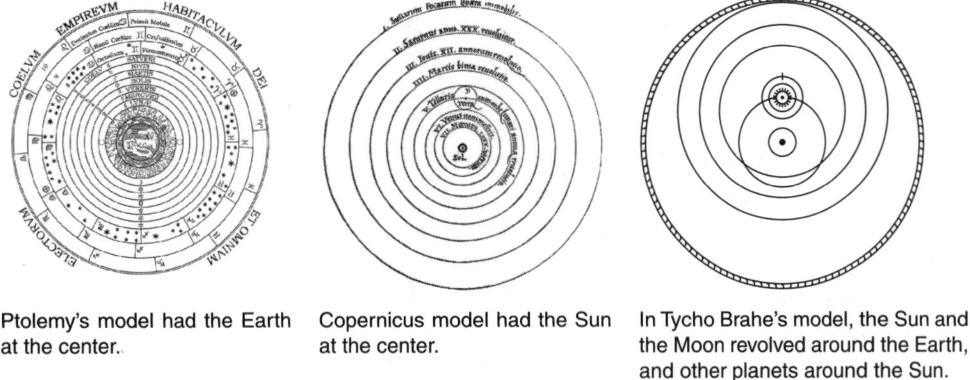

Ptolemy's model had the Earth at the center.

Copernicus model had the Sun at the center.

In Tycho Brahe's model, the Sun and the Moon revolved around the Earth, and other planets around the Sun.

Fig. 8.1 The planetary models of Ptolemy, Copernicus, and Tycho Brahe [1].

The Kepler problem can be cast as a two-body problem. It is exactly solved using analytical methods within the framework of the Newtonian (Chapters 1–3), or Lagrangian mechanics (Chapter 6). However, to get accurate solutions, influences of other celestial bodies also need to be accounted for. Slight inaccuracy in knowing the initial conditions can, however, render the planetary motion chaotic, albeit only on a very long time scale. We shall discuss the dynamics of chaos in the next chapter. Besides, for accurate description of celestial motion, one has to go *beyond* Newtonian mechanics. One must include relativistic corrections, which we shall discuss in Chapters 13 and 14.

We now proceed to formulate the Kepler problem within the framework of the Galileo–Newtonian mechanics. We shall refer to the two-body interaction problem between the Sun and one of its planets as the Kepler–Newton problem. Our prototype of this interaction would be the gravitational force of attraction between the Sun and the Earth. Fig. 8.2 schematically shows the center of mass coordinate system we shall employ for our analysis of the two-body problem. Depending on the relative masses of the two interacting bodies, their center of mass would be somewhere between them. If one of the masses is also much bigger and larger than the other, the center of mass could be physically located even *inside* the bigger body. Since the Earth's mass is about 3,33,000 times less than that of the Sun, the center of mass of the Earth–Sun system is well *inside* the Sun. On the other hand, the center of mass of the Jupiter–Sun system is a few tens of thousands of kilometers *above* the Sun's surface, since the Sun is only about 1080 times more massive than Jupiter.

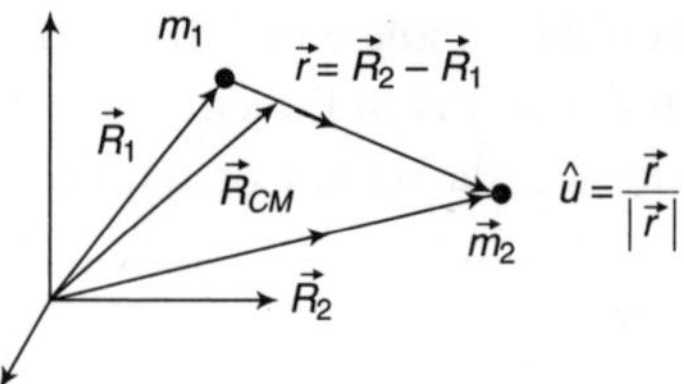

Fig. 8.2 The center of mass of the two masses m_1 and $m_1$is between them, always closer to the larger mass, so much that if $m_1 \gg m_2$ and is also big sized, the center of mass can be physically located inside the body with mass, m_1. In this figure, the center of mass is located at the tip of the vector $\vec{R}_{CM}$.

Discovery of the nature of interaction between the two masses makes an exceptionally fascinating story [3]. Edmund Halley, in 1684, had proposed that the force of attraction between a planet and the Sun decreases as the inverse square of the distance between them. Christopher Wren had also arrived at the same conclusion around the same time. This conjecture, however, needed a proof, which Robert Hooke claimed he had, but did not deliver. Halley asked Newton what would be the shape of a planet's orbits if the force between them diminished as the inverse square of the distance between the planet and the Sun. Newton responded that the orbit would be an ellipse (See Problem 8.4, based on *Feynman's Lost Lecture*). This was believed to be so, until then, only from the Brahe–Kepler empirical result, or from unproven models claimed by Hooke. Newton's answer was based on a calculation using *new* mathematics which he had developed. Newton had to reconstruct the proof over the next few months. It is this great piece of work that was published as Newton's *Philosophiae Naturalis Principia Mathematica*. Isaac Newton was, by some, described as the man who wrote a book that neither he nor anybody else understood. It was the first explanation of motion of an object in terms of the forces that act on it. The explanations employed in this model would accurately stand true for all objects, small and big, colored or not, and separated by distances whether small or astronomical. It thus got to be known as one of the first universal law of nature.

The governing relations that determine the dynamics of planetary motion, then, are the prodigious Newton's laws for the forces by the masses m_1 and m_2 on each other.
The force by object '1' on '2' is given by

$$\vec{F}_{\text{by\_1\_on\_2}} = m_2\ddot{\vec{R}}_2 = -G\frac{m_1 m_2}{\left|\vec{R}_2 - \vec{R}_1\right|^2}\hat{u} \tag{8.1a}$$

and the force by '2' on '1' is

$$\vec{F}_{\text{by\_2\_on\_1}} = m_1\ddot{\vec{R}}_1 = G\frac{m_1 m_2}{\left|\vec{R}_2 - \vec{R}_1\right|^2}\hat{u} \tag{8.1b}$$

The unit vector $\hat{u}$ is shown in Fig. 8.2.

The center of mass of the system is given by

$$\vec{R}_{CM} = \frac{m_1\vec{R}_1 + m_2\vec{R}_2}{m_1 + m_2}, \; m_1 \gg m_2. \tag{8.2a}$$

The position vector of '2' relative to '1' is

$$\vec{r} = \vec{R}_2 - \vec{R}_1 = \vec{r}_2 - \vec{r}_1. \tag{8.2b}$$

The acceleration of the reduced mass, $\mu = \frac{m_1 m_2}{m_1 + m_2} \approx m_2 = m_{\text{planet}}$, with reference to the center of mass of the 2-body system is given by

$$\ddot{\vec{r}} = \ddot{\vec{R}}_2 - \ddot{\vec{R}}_1 = -G(m_1 + m_2)\frac{\vec{r}}{|\vec{r}\,|^3} = -\kappa\frac{\vec{r}}{|\vec{r}\,|^3}, \tag{8.3a}$$

where the constant $\kappa = G(m_1 + m_2) \approx Gm_1 = Gm_{\text{sun}}$, (8.3b)

since $m_1 \gg m_2$, as is the case for the Sun–Earth system. The dimensions of κ are L^3T^{-2}.

The equation of motion for the reduced mass $\mu \simeq m_{\text{planet}}$, is therefore for all practical purposes the equation of motion for the Earth itself, and given by

$$\ddot{\vec{r}} + \kappa\frac{\vec{r}}{|\vec{r}\,|^3} = \vec{0}. \tag{8.4}$$

This equation of motion describes the 'relative motion' of the smaller mass relative to the larger mass, assuming that the difference in the masses is huge. It contains sum and substance of the law of nature that governs the dynamics between two objects. Effectively, we have reduced the two-body problem to a one-body problem.

We shall now invoke the connections between *laws of nature* and *conservation principles* which we discussed in Chapters 1 and 6. We had discovered intimate connections between the two, as much as their derivability from each other. To determine what physical quantities are conserved in the Sun–Earth gravitational Kepler–Newton problem (Eq. 8.4), all we have to do is carry out some simple algebra [4].

Taking the dot product of the velocity with the equation of motion, we get

$$\dot{\vec{r}} \cdot \ddot{\vec{r}} + \frac{\kappa}{|\vec{r}\,|^3}\dot{\vec{r}} \cdot \vec{r} = 0. \tag{8.5a}$$

Using $\dot{\vec{r}} = \frac{d}{dt}\vec{r} = \frac{d}{dt}(r\hat{u}) = \dot{r}\hat{u} + r\dot{\hat{u}}$, (8.5b)

we find that $\dot{\vec{r}} \cdot \ddot{\vec{r}} = \vec{v} \cdot \dot{\vec{v}} = v\dot{v}$, (8.5c)

and that $\dot{\vec{r}} \cdot \vec{r} = (\dot{r}\hat{u} + r\dot{\hat{u}}) \cdot (r\hat{u}) = r\dot{r}$. (8.5d)

Thus, $v\dot{v} + \frac{\kappa}{r^3}r\dot{r} = 0$, (8.6a)

i.e., $$v\lim_{\delta t \to 0}\frac{\delta v}{\delta t} = -\frac{\kappa \lim_{\delta t \to 0}\frac{\delta r}{\delta t}}{r^2}. \tag{8.6b}$$

Integrating Eq. 8.6b with respect to time, we get

$$\frac{v^2}{2} = \frac{\kappa}{r} + E, \text{ i.e., } \frac{v^2}{2} - \frac{\kappa}{r} = E. \tag{8.7}$$

The dimensions of the constant of integration E are $[E] = L^2T^2$. We can see from Eq. 8.7 that E is nothing but the specific (*i.e., per unit mass*) mechanical energy. You must appreciate how a constant of motion, a conserved quantity, the energy, has emerged from the integration with respect to *time*. This is very much in accord with the connection between symmetry and the invariance principle that we discussed in the previous chapters. It should not surprise you that it is the integration with respect to *time* that has resulted in the conservation of *energy*, since these two physical quantities are canonically conjugate, as discussed in Chapter 6.

Above, we took the dot product of the equation of motion with velocity, and discovered a conserved quantity, the energy E. Now, taking the cross product of the position vector with the equation of motion, we get

$$\vec{r} \times \ddot{\vec{r}} + \vec{r} \times \kappa \frac{\vec{r}}{|\vec{r}|^3} = \vec{0}, \text{ i.e., } \vec{r} \times \ddot{\vec{r}} = \frac{d}{dt}(\vec{r} \times \dot{\vec{r}}) = \vec{0}. \tag{8.8}$$

Yet again, a conserved quantity has emerged, since the time-derivative of $(\vec{r} \times \dot{\vec{r}})$ is zero. The conserved quantity is essentially the specific (i.e., *per unit mass*) angular momentum. The very reason that angular momentum is conserved is the fact that the force between the Sun and the Earth is collinear with the radial position $\vec{r}$ of the Earth with respect to the Sun, i.e., the force has a radial symmetry. It is this symmetry which makes the term $(\vec{r} \times \ddot{\vec{r}} = \vec{0})$, and has been is used to discover that $(\vec{r} \times \dot{\vec{r}})$ is invariant with respect to time. We shall denote the 'specific angular momentum' by $\vec{H}$:

$$\vec{H} = \vec{r} \times \vec{v} = \vec{r} \times \frac{\vec{p}}{m} = \frac{\vec{\ell}}{m}, \tag{8.9a}$$

$\vec{\ell}$ being the angular momentum, $\vec{r} \times \vec{p}$, itself.

We now use the results from Chapter 2, in which we discussed how unit vectors in plane polar coordinates change with time. We see readily that

$$\vec{H} = \vec{\rho} \times \frac{d}{dt}(\rho\hat{e}_\rho) = \vec{\rho} \times \left[\rho\frac{d\hat{e}_\rho}{dt} + \hat{e}_\rho\dot{\rho}\right] = \rho\hat{e}_\rho \times \rho\dot{\hat{e}}_\rho = \rho^2\hat{e}_\rho \times \left(\frac{d\varphi}{dt}\right)\hat{e}_\varphi \tag{8.9b}$$

$$\text{i.e., } \vec{H} = \omega r^2\hat{e}_z = \omega\rho^2\hat{e}_z, \text{ where } \omega = \dot{\varphi}. \tag{8.9c}$$

Fig. 8.3 schematically shows two positions of the Earth, E and E', on its orbit around the Sun. The two points could be extremely close to each other, though in the figure their separation is exaggerated. We see that the magnitude of specific angular momentum vector is *twice* $\frac{\delta A}{\delta t}$, where the area $\delta A = \frac{1}{2}\rho^2\delta\varphi$ is swept by the Earth on its orbit while moving from the position E to a neighboring position E', through the angle $\delta\varphi$. The constancy of the angular momentum is thus nothing but the mathematical precise expression of Kepler's empirical law that the planets sweep equal areas in equal time. Kepler had concluded this from Tycho Brahe's compilation of observed positions of the planets. Kepler's law thus finds a rationale in the constancy of angular momentum in a spherically symmetric force-field. The rate of change of the area swept by the planet on its elliptic orbit simply is

$$\dot{A} = \lim_{\delta t \to 0} \frac{\delta A}{\delta t} = \frac{1}{2}\rho^2 \lim_{\delta t \to 0} \frac{\delta \varphi}{\delta t} = \frac{1}{2}\rho^2 \dot{\varphi} = \frac{1}{2}\rho^2 \omega = \frac{H}{2} = \frac{\ell}{2m_p}, \text{ with} \tag{8.9d}$$

$\ell = m_p H$ being the angular momentum of the planet about the focus. Yet again we see the connection between symmetry and a conservation law. These are beautiful illustrations of the Noether's theorem which we have mentioned in the earlier chapters.

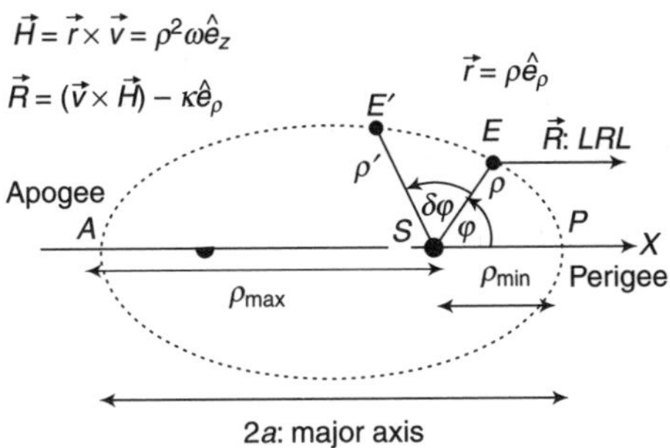

Fig. 8.3 E and E' are two neighboring positions of the Earth on its elliptic orbit about the Sun, shown at one of the two foci of the ellipse. The plane polar coordinate system, described in Chapter 2, is best suited to describe the position of the Earth on its orbit around the Sun. The angular momentum is orthogonal to the plane containing $(\hat{e}_\rho, \hat{e}_\varphi)$. It is along $\hat{e}_z$ unit vector of the cylindrical polar coordinate system. In this coordinate system, distance r is denoted by ρ.

The eccentricity of the elliptic (closed) orbits is between 0 and 1; $\varepsilon = 0$ being that of a circle. The eccentricity of a parabola is unity, and that of hyperbola is greater than unity. The two-body Kepler–Newton problem provides for solutions with *each* of these possibilities. The eccentricity of the Halley's comet, whose period is 78.1 years, is 0.967, and that of comet Hyakutake, whose period is ~40000 years, is 0.9998. An object in hyperbolic orbit, with eccentricity about 1.12, called 'Oumuamua' was discovered by a Canadian astronomer Robert Weryk on 19 October 2017. It is traveling in inter-stellar space, and fleeting through the periphery of our solar system.

The constancy of angular momentum implies that motion is confined to a plane which is orthogonal to $\vec{\ell}$. We may therefore use plane polar coordinates, (ρ, φ). Also, using the constancy of angular momentum, we can now get a different, new, conserved quantity (i.e., a constant of motion). This will lead us to further insights in the Kepler–Newton problem. We proceed by taking the cross product of the specific angular momentum with the equation of motion,

$$\vec{H} \times \ddot{\vec{r}} + \vec{H} \times \left(\kappa \frac{\vec{r}}{|\vec{r}|^3} \right) = \vec{0}, \tag{8.10a}$$

$$\text{i.e.,}\quad \left[\vec{H}\times\ddot{\vec{r}}+\frac{\kappa}{|\vec{r}\,|^3}(\vec{r}\times\vec{v})\times\vec{r}\right]=\left[\vec{H}\times\ddot{\vec{r}}+\frac{\kappa}{|\vec{r}\,|^3}\{(\vec{r}\cdot\vec{r})\vec{v}\boxtimes(\vec{r}\cdot\vec{v})\vec{r}\}\right]=\vec{0}. \tag{8.10b}$$

$$\text{Thus, } \vec{H}\times\ddot{\vec{r}}+\frac{\kappa}{|\vec{r}\,|^3}(r^2\vec{v}-r\dot{r}\vec{r})=\vec{0}. \tag{8.11}$$

Now, $\frac{d}{dt}(\vec{H}\times\vec{v})=\frac{d}{dt}(\vec{H}\times\dot{\vec{r}})=\vec{H}\times\ddot{\vec{r}}$, since $\frac{d\vec{H}}{dt}=\vec{0}$. This enables us to see that:

$$\frac{d}{dt}(\vec{H}\times\vec{v})+\frac{\kappa}{|\vec{r}\,|^3}(r^2\vec{v}-r^2\dot{r}\hat{e}_{\rho})=\frac{d}{dt}(\vec{H}\times\vec{v})+\kappa\left(\frac{\vec{v}}{r}-\frac{\dot{r}}{r^2}\vec{r}\right)=\vec{0}. \tag{8.12}$$

It then follows that $\frac{d}{dt}(\vec{H}\times\vec{v})+\kappa\frac{d}{dt}\left(\frac{\vec{r}}{r}\right)=\vec{0}$, i.e.,

$$\frac{d}{dt}\left[(\vec{H}\times\vec{v})+\kappa\left(\frac{\vec{r}}{r}\right)\right]=\vec{0}, \tag{8.13a}$$

$$\text{i.e.,}\quad \frac{d}{dt}\left[(\vec{H}\times\vec{v})+\kappa\left(\frac{\vec{\rho}}{\rho}\right)\right]=\vec{0}, \tag{8.13b}$$

where we have recognized the unit vector $\hat{e}_{\rho}=\frac{\vec{\rho}}{\rho}$ of the plane polar coordinates explicitly to emphasize that the motion is essentially planar. Eq. 8.13 has a null *time-derivative* of a vector, which must therefore be a constant of motion. The conserved (being a constant) physical vector, which we have just obtained is

$$(\vec{H}\times\vec{v})+\kappa\hat{e}_{\rho}=-\vec{R}. \tag{8.14a}$$

It represents the third physical quantity, after 'energy' and 'angular momentum', which is conserved in the Kepler–Newton analysis of the Earth's motion around the Sun. Both the magnitude and direction of the vector remain unchanged, no matter where the Earth is on its orbit around the Sun. It is important to underscore that this constant of motion has emerged, yet again, by carrying out simple vector algebra with, and hence from, *the equation of motion*. A conservation principle, yet again, results from the laws of nature; and vice versa.

The constant vector,

$$\vec{R}=(\vec{v}\times\vec{H})-\kappa\hat{e}_{\rho}, \tag{8.14b}$$

is called as the Laplace–Runge–Lenz Vector (LRL) (Fig. 8.4). The name of Wilhelm Lenz is associated with this vector for his work on the Coulomb potential between the electron and the proton in an atom of Hydrogen, which also produces the inverse square law for the force [5,6], just as in the gravitational Kepler–Newton case. Sometimes, the Laplace–Runge–Lenz vector is defined after scaling it with reduced mass for which the equation of motion is set up, and/or, with a sign that is *opposite* to what we have used in Eq. 8.14a and b. These differences, of course, do not alter the interesting physics that its analysis leads us to. Scaled by the reduced mass $\mu = \dfrac{m_{\text{sun}}\, m_{\text{planet}}}{m_{\text{sun}} + m_{\text{planet}}} \approx m_{\text{planet}}$, the LRL vector is

$$\vec{R}' = \frac{\vec{p} \times \vec{l}}{\mu} - \mu\kappa\hat{e}_\rho. \tag{8.14c}$$

Pierre-Simon Laplace
1749–1827

Carl David Tolmé Runge
1856–1927

Wilhelm Lenz
1888–1957

Fig. 8.4 The vector $\vec{R} = (\vec{v} \times \vec{H}) - \kappa\hat{e}_\rho$ is named as the Laplace–Runge–Lenz vector.

8.2 DYNAMICAL SYMMETRY IN THE KEPLER–NEWTON GRAVITATIONAL INTERACTION

Let us quickly recapitulate what we have learned above. Essentially, beginning with the physical (inverse square) *law*, namely, $\ddot{\vec{r}} + \kappa\dfrac{\vec{r}}{|\vec{r}\,|^3} = \vec{0}$, we deduced, by carrying out simple vector-algebraic operations on the equation of motion, that (a) the energy is conserved, (b) the angular momentum is conserved, and (c) the Laplace–Runge–Lenz (LRL) vector is conserved. In line with our discussion in earlier chapters (especially Chapter 1 and 6) on the relationship between symmetry and conservation laws, we see that, in the spirit of the Noether's theorem, (i) the conservation of energy is associated with symmetry with respect to temporal evolution, and (ii) the conservation of angular momentum is associated with the fact that the gravitational force field has a central (i.e., spherical) symmetry. Shouldn't then we, therefore, expect that there is some symmetry corresponding to the constancy of the LRL vector? In order to determine this symmetry, let us ask ourselves under what conditions is the LRL vector conserved; i.e., what physical properties determine its constancy.

Now, the constancy of the LRL vector is expressed by the condition that its time-derivative is zero:

$$\vec{0} = \frac{d\vec{R}}{dt} = \left(\frac{d\vec{v}}{dt} \times \vec{H} + \vec{v} \times \frac{d\vec{H}}{dt} \right) - \kappa \frac{d\hat{e}_\rho}{dt}, \tag{8.15a}$$

$$\text{i.e.,} \quad \vec{0} = \frac{d\vec{v}}{dt} \times (\vec{r} \times \vec{v}) - \kappa \hat{e}_\varphi \dot{\varphi}, \tag{8.15b}$$

since we already know that the (specific) angular momentum $\vec{H}$ is a constant for the spherically symmetry potential. Therefore, for Eq. 8.15 to hold, the law $\vec{F} = \frac{d\vec{p}}{dt} = m\frac{d\vec{v}}{dt}$ needs to get *pinned down* to a *specific form* as would make $\frac{d\vec{v}}{dt}$ to be exactly what it takes to satisfy Eq. 8.15. It is easy to see that if the force (per unit mass) is given by $\frac{d\vec{v}}{dt} = -\frac{\kappa}{\rho^2}\hat{e}_\rho$, then, and only then, the relation

$$\vec{0} = \frac{d\vec{R}}{dt} = \left(-\frac{\kappa}{\rho^2} \hat{e}_\rho \times (\rho^2 \dot{\varphi} \hat{e}_z) \right) - \kappa \dot{\varphi} \hat{e}_\varphi = \kappa \dot{\varphi} \hat{e}_\varphi - \kappa \dot{\varphi} \hat{e}_\varphi, \tag{8.15c}$$

is satisfied. In concluding this, we have used the fact that the unit vectors $\{\hat{e}_\rho, \hat{e}_\varphi, \hat{e}_z\}$ constitute a right handed cylindrical polar coordinate system (see Chapter 2 for details). That the *form* of the force has to be given by the inverse square of the distance is a critical factor in making the LRL vector a constant of motion. Any other central field force, other than the inverse square force, would not satisfy Eq. 8.15. In contrast to this, the angular momentum would be a constant of motion for any form of the force having spherical symmetry; including forces that depend on any polynomial function of the distance from the center, as long as there is spherical symmetry. In the context of the Noether's theorem about the intimate relations between symmetry and conservation laws, the symmetry associated with the conservation (constancy) of the LRL vector is called the *dynamical symmetry*. This terminology comes from the fact that it is the *dynamics* (rather than geometry), namely the specific *form* of the (inverse-square) force, that determines the conservation of the LRL vector.

From the Laplace–Runge–Lenz vector, we can even actually determine the exact geometrical shape of trajectory described by the planet in the force-field of the Sun. More commonly, this shape, which would be the equation to the orbit of the planet, would be obtained by solving the differential equation of motion for the planet. It was for this purpose that Newton invented differential calculus and figured out that the solution to this equation is given by an ellipse. In doing so, Newton established the law of gravitational attraction between any two masses, and also the principle of causality contained in his second law that must be used in the equation of motion. This method is discussed in Section 8.3. For now, we demonstrate a *completely different* way of getting the equation of to the orbit, using the LRL vector.

We first take the scalar product between the position vector, $\vec{r} = r\hat{e}_\rho$ with the Laplace–Runge–Lenz vector $\vec{R}$:

$$(\vec{v} \times \vec{H}) \cdot \vec{r} - \kappa r = \vec{R} \cdot \vec{r}. \tag{8.16a}$$

On interchanging the dot and the cross in the scalar triple product, and on reversing the sign throughout the equation, we get $\vec{H} \cdot (\vec{v} \times \vec{r}) + \kappa r = -\vec{R} \cdot \vec{r}$, i.e., $-H^2 + \kappa\rho = -\vec{R}\cdot\vec{\rho}, = -R\rho\cos\varphi,$ (8.16b)

where $\varphi = \angle(\vec{R}, \vec{\rho})$. Equation 8.16b provides the relation (ρ, φ) which therefore gives the equation to the orbit:

$$\rho = \frac{H^2}{\kappa + R\cos\varphi} = \frac{(H^2/\kappa)}{1 + (R/\kappa)\cos\varphi} = \frac{\lambda}{1 + \varepsilon\cos\varphi}. \tag{8.16c}$$

We recognize Eq. 8.16c to be the equation to an ellipse, only expressed in plane polar coordinates. The numerator in Eq. 8.16c, $\lambda = \frac{H^2}{\kappa}$, is identified as the latus rectum, and the eccentricity of the conic section is $\varepsilon = \frac{R}{\kappa}$. Note that the LRL vector $\vec{R}$ has a magnitude which is nothing but the eccentricity of the orbit (in units of κ). It is therefore also called the *eccentricity vector*. The rigorous geometric figure that emerges from this solution confirms Kepler's guess; it provides a mathematical proof that the planetary orbits are ellipses. The proof given above using the LRL vector is a robust alternative to the historic proof first given by Newton using the differential equation of motion, which we deal with in the next Section. In order to write the equation to the ellipse (Eq. 8.16c) in the familiar Cartesian coordinate system, we note that

$$\rho + \rho\varepsilon\cos\varphi = \lambda,$$

i.e., $\rho + \varepsilon x = \lambda,$ (8.16d)

since the Cartesian x-coordinate is $x = \rho\cos\varphi$.

This gives

$$x^2 + y^2 = \rho^2 = (\lambda - \varepsilon x)^2 = \lambda^2 - 2\lambda\varepsilon x + \varepsilon^2 x^2,$$

i.e., $y^2 + (1-\varepsilon^2)\left(x + \frac{\lambda\varepsilon}{1-\varepsilon^2}\right)^2 = \frac{\lambda^2}{1-\varepsilon^2}.$

Dividing throughout by $\frac{\lambda^2}{1-\varepsilon^2}$, we get the equation to the ellipse, in the celebrated Cartesian form:

$$\frac{\left[x + \frac{\lambda\varepsilon}{1-\varepsilon^2}\right]^2}{a^2} + \frac{y^2}{b^2} = 1, \tag{8.17a}$$

with $a^2 = \frac{\lambda^2}{(1-\varepsilon^2)^2}$ and $b^2 = \frac{\lambda^2}{1-\varepsilon^2}$ i.e., $a = \frac{\lambda}{1-\varepsilon^2}$ and $b = \frac{\lambda}{\sqrt{1-\varepsilon^2}}$. (8.17b)

Since $\frac{2b^2}{a} = 2\frac{\lambda^2}{1-\varepsilon^2} \times \frac{1-\varepsilon^2}{\lambda} = 2\lambda = 2\frac{H^2}{\kappa}$, we see that the latus rectum of the ellipse (Fig. 8.3) is, *indeed* (as claimed above), given by

$$\lambda = \frac{H^2}{\kappa} = \frac{b^2}{a} = a(1-\varepsilon^2). \tag{8.17c}$$

The path of the orbit, recognized as ellipse has been obtained above from the properties of the LRL vector. Equation 8.17a describes the ellipse in the rather commonly used Cartesian coordinates, with the focus of the ellipse located at the Cartesian coordinates $\left(-\frac{\lambda\varepsilon}{1-\varepsilon^2}, 0\right)$.

The LRL vector not only helps determine rigorously that the trajectory of the planetary orbits would be ellipses, but also that the orbits would be *fixed*; i.e., the ellipse does not precess (Fig. 8.5). We shall validate this conclusion by showing that direction of the constant LRL vector is *along* the major axis of the ellipse; and hence *its constancy prevents precession* of the ellipse. We can determine the direction of the LRL vector, when the planet is at the perigee of the ellipse (for example). We immediately see that when the planet is at the perigee, both the terms on the right hand side of Eq. 8.14, and hence the LRL vector itself, must be essentially *along* the major axis of the ellipse. This direction must remain constant, and its conservation ensures that the ellipse itself would get fixed, i.e., it would not undergo any rosette motion (Fig. 8.5).

Fig. 8.5 This figure shows the precession of an ellipse while remaining in its plane. This can happen without violating the constancy of the angular momentum, which only requires the motion to stay confined within the plane. The motion would look, from a distance, somewhat like the petals of a rose and is therefore called rosette motion.

Constancy of angular momentum only confines the motion of the planet to stay within a plane. It is the constancy of the LRL vector, mandated by the dynamical symmetry stemming from the strict *inverse square force law*, which ensures that the ellipse does not precess.

It is, however, a matter of further details that when we take into account the gravitational influences of the other planets in the solar system, the *net* field is not expressible strictly in terms of the inverse square of a planet's distance from its center of mass with the sun. Other planets cause a perturbation of the inverse square law force, and hence the planetary orbit would be expected to precess. Indeed, over long time periods, the planetary trajectories can become chaotic. We shall discuss 'chaos' in the next chapter. In our solar system, the motion of planets is indeed seen to somewhat precess, and most of this is accounted for by taking into account influence of other planets *within* the framework of Newtonian mechanics. However, there would remain a small unaccounted precession which requires going beyond Newtonian mechanics. It demands a far more sophisticated theory of gravity, namely the General Theory of Relativity (GTR), developed by Albert Einstein [7]. We shall discuss the GTR component of planetary precession in Chapter 14.

The dynamical symmetry associated with conservation of the Laplace–Runge–Lenz vector is a beautiful example of the intimate connection between symmetry and conservation principles. This is one of the cornerstones of modern physics. One can appreciate how crucial this connection is by reflecting on Eugene Wigner's comment: "*It is now natural for us to try to derive the laws of nature, and to test their validity by means of the laws of invariance, rather than to derive the laws of invariance from what we believe to be the laws of nature.*" The importance of dynamical symmetry embedded in the constancy of the Laplace–Runge–Lenz vector for a two-body interaction goes *beyond* the domain of planetary motion in classical mechanics, for the two-body interaction between the proton and the electron in the Hydrogen atom is also governed essentially by the same dynamical symmetry [5,6].

8.3 KEPLER–NEWTON ANALYSIS OF THE PLANETARY TRAJECTORIES ABOUT THE SUN

It is an unparalleled turning point in the development of mathematics, astronomy, and physics, that that one man, Isaac Newton, recognized that the force between an object falling on Earth, whether an apple or a coconut, and one that keeps the Earth in its orbit around the Sun has essentially the identical form, namely the inverse square law of gravity. Furthermore, Newton established that the solution that describes the trajectory of the object obtained on solving the equation of motion, $\vec{F} = \frac{d\vec{p}}{dt} = m\frac{d\vec{v}}{dt}$, using the inverse square law is an ellipse. It is humbling to note that calculus did not even exist before Newton; in fact, he invented it. In doing so, and applying it to the force of attraction between the Sun and the Mars, or the Sun and any other planet, and to the force between any two masses, Newton essentially introduced what we now regard as the *universal* law of nature. Essentially, the very same law of gravity holds between the Sun and one planet, or another, or for that matter between *any* mass, and another. Could there have been a more beautiful synthesis of geometry, differential calculus, laws of physics, and astronomy?

The Earth's rotation about the Sun is a fascinating phenomenon. The orbital trajectory, the average distance the Earth has from the Sun, etc., are *just* appropriate to make life on Earth so special. Ours is the only planet in our solar system that supports life, even if the possibility of life on many exoplanets is a well-established, distinct, possibility. Earth's orbit is very special. The orbit takes a full year to complete. Major events on this orbit, such as the Earth's closest approach to the Sun, or the farthest, are periodic. The closest approach is *not* in summer on the Northern Hemisphere, contrary to what one might imagine. It is in fact in winter. The seasons on the Earth are not quite determined by its *distance* from the Sun; they are instead determined mostly by the *tilt* the Earth's axis has with respect to the plane of revolution around the Sun. The summer solstice occurs around 21 June every year, and the winter solstice around 21 December. On 21 June in the Northern Hemisphere, the duration of the day (i.e., the duration from sunrise to sunset) is the longest. On 21 December it is the shortest. In the Southern Hemisphere, these are reversed. These dates are quite different from when the Earth is at the perihelion on its orbit, or when it is at the aphelion. The words perihelion and aphelion come from Greek, in which 'peri' means 'close', and 'apo' means 'far', while 'Helius' is the Sun-God. The Earth is at its perihelion and aphelion about two weeks *after* the winter solstice, and the summer solstice, respectively. However, this changes little bit from year to year, due to various astronomical events. The seasons are therefore not determined by the Earth's timing at the perihelion or at the aphelion, but by the tilt (which is ~23.5^0) of the Earth's axis of rotation with respect to its orbital plane.

We shall, however, not pursue the interesting topic of how the Earth's seasons are determined. Instead, we shall re-affirm, without using the LRL vector this time, that the trajectory of the Earth's orbit around the Sun is described by the equation to an ellipse. We shall do it by solving Newton's equation of motion. Essentially, the vector equation of motion (Eq. 8.4), written in terms of the plane polar components along the base vectors ($\hat{e}_\rho$, $\hat{e}_\varphi$) is

$$(\ddot{\rho} - \rho\dot{\varphi}^2)\hat{e}_\rho + (2\dot{\rho}\dot{\varphi} + \rho\ddot{\varphi})\hat{e}_\varphi + \kappa\frac{\hat{e}_\rho}{\rho^2} = \vec{0}. \tag{8.18a}$$

Hence, $$\ddot{\rho} - \rho\dot{\varphi}^2 + \frac{\kappa}{\rho^2} = 0, \tag{8.18b}$$

and $$2\dot{\rho}\dot{\varphi} + \rho\ddot{\varphi} = 0. \tag{8.18c}$$

It is now convenient to introduce inverse distance as $\xi = \frac{1}{\rho}$, which gives, from Eq. 8.9c,

$$\dot{\varphi} = H\xi^2 \text{ and } \ddot{\varphi} = 2H\xi\dot{\xi}. \tag{8.19a}$$

We also get

$$\dot{\rho} = -\frac{1}{\xi^2}\dot{\xi} = -\frac{1}{\xi^2}\frac{d\xi}{d\varphi}\frac{d\varphi}{dt} = -\frac{1}{\xi^2}\frac{d\xi}{d\varphi}\dot{\varphi} = -H\frac{d\xi}{d\varphi}, \tag{8.19b}$$

and $$\ddot{\rho} = -H\frac{d^2\xi}{d\varphi^2}\dot{\varphi} = -H^2\xi^2\frac{d^2\xi}{d\varphi^2}. \tag{8.19c}$$

Hence, from the radial component of the equation of motion (Eq. 8.18b) we get

$$-H^2\xi^2\frac{d^2\xi}{d\varphi^2} - \frac{1}{\xi}(H\xi^2)^2 + \kappa\xi^2 = 0. \tag{8.20a}$$

Dividing throughout by $-H^2\xi^2$, we get

$$\frac{d^2\xi}{d\varphi^2} + \xi - \frac{\kappa}{H^2} = 0. \tag{8.20b}$$

The solution to this equation is

$$\xi = A\cos(\varphi - \varphi_0) + \frac{\kappa}{H^2}, \tag{8.21a}$$

i.e., $$\rho = \frac{\dfrac{H^2}{\kappa}}{1 + \left(\dfrac{AH^2}{\kappa}\right)\cos(\varphi - \varphi_0)}, \tag{8.21b}$$

with $A\left(\dfrac{H^2}{\kappa}\right) = A\lambda = \varepsilon$, which is the eccentricity of the orbit, and λ the latus rectum. This is completely equivalent to Eq. 8.16, which we had obtained using the LRL vector.

We now turn our attention to the determination of the minimum and the maximum distance of the Earth from the center of the force. These points are also called *apses*. At an apse, $\dot{\rho} = 0$, which implies, from Eq. 8.19b that $\left(\dfrac{d\xi}{d\varphi}\right) = 0$.

Now, from Eq. 8.16c (and from Fig. 8.3) we see that the perigee of the ellipse is at a distance

$$\rho_p = \frac{\lambda}{1+\varepsilon} = \frac{H^2/\kappa}{1+\varepsilon}, \tag{8.22a}$$

and that the apogee is at a distance

$$\rho_a = \frac{\lambda}{1-\varepsilon} = \frac{H^2/\kappa}{1-\varepsilon}. \tag{8.22b}$$

We now consider the potential field of the Sun in which the Earth revolves around it in bound orbits. The conservation of energy equation is given by

$$\frac{1}{2}v^2 + V(\rho) = E, \tag{8.23a}$$

i.e., $$\frac{1}{2}(\dot{\rho}^2 + \rho^2\dot{\varphi}^2) + V(\rho) = E. \tag{8.24}$$

Therefore, $$\frac{1}{2}\left[H^2\left(\frac{d\xi}{d\varphi}\right)^2 + H^2\xi^2\right] + V(\rho) = E. \tag{8.25}$$

The potential energy function for an attractive inverse square field is $V(\rho) = \dfrac{-\kappa}{\rho} = -\kappa\xi$, and hence, we can write

$$\frac{\dot{\rho}^2}{2} + \frac{H^2}{2\rho^2} - \frac{\kappa}{\rho} = E. \tag{8.26}$$

We now interpret the above 1-dimensional equation as the statement of conservation of energy, with the first term correspond to the kinetic energy (per unit mass). The *effective* 1-dimensional (radial) potential is then given by

$$U_{\text{eff}} = \frac{H^2}{2\rho^2} - \frac{\kappa}{\rho}. \tag{8.27a}$$

We can get this result also using an alternative method. Toward this, we first differentiate the equation to the ellipse (Eq. 8.16c) with respect to time, and get

$$\dot{\rho} = -\frac{\lambda}{(1+\varepsilon\cos\theta)^2}(-\varepsilon\sin\varphi\dot{\varphi}) = \frac{\lambda\varepsilon\sin\varphi\dot{\varphi}}{(1+\varepsilon\cos\varphi)^2} = \frac{\rho^2\lambda\varepsilon\sin\varphi\dot{\varphi}}{\lambda^2} = \frac{\varepsilon\rho^2\sin\varphi\dot{\varphi}}{\lambda},$$

which gives, $\dot{\rho}^2 = \dfrac{\varepsilon^2\rho^4\dot{\varphi}^2\sin^2\varphi^2}{\lambda^2} = \dfrac{\varepsilon^2\lambda\kappa\sin^2\varphi}{\lambda^2} = \dfrac{\varepsilon^2\kappa\sin^2\varphi}{\lambda}.$

We can now define the effective radial potential by subtracting the radial kinetic energy from the total energy:

$$V_{\text{eff}} = \frac{\kappa(\varepsilon^2-1)}{2\lambda} - \frac{1}{2}\dot{\rho}^2 = \frac{\kappa(\varepsilon^2-1)}{2\lambda} - \frac{\varepsilon^2\kappa\sin^2\varphi}{2\lambda}$$

$$= \frac{\kappa(\varepsilon^2\cos^2\varphi - 1)}{2\lambda} = \frac{\kappa(\varepsilon\cos\varphi + 1)(\varepsilon\cos\varphi - 1)}{2\lambda}.$$

Using now the fact that $\varepsilon\cos\varphi - 1 = \dfrac{\lambda}{\rho} - 2 = \dfrac{\lambda - 2\rho}{\rho}$, we get

$$V_{\text{eff}} = \frac{\kappa}{2\lambda}\frac{\lambda}{\rho}\frac{\lambda-2\rho}{\rho} = \frac{\kappa\lambda}{2\rho^2} - \frac{\kappa}{\rho} = \frac{H^2}{2\rho^2} - \frac{\kappa}{\rho}, \tag{8.27b}$$

which is the same as Eq. 8.27a.

It is instructive to sketch the effective radial potential as a function of the distance from the center of the force, which is the center of mass of the Sun–Earth system. We plot, in Fig. 8.6, both the components of the effective radial potential, $-\dfrac{\kappa}{\rho}$ and $\dfrac{H^2}{2\rho^2}$, and also their sum, which gives the effective radial potential $V_{\text{eff}}(\rho)$ as a function of the distance ρ. Fig. 8.6 shows this plot in the range $0 \le \rho < (\rho_a + \rho_p)$. The component $\dfrac{H^2}{2\rho^2}$ is of course not a *physical* potential; it is a pseudo-potential, and hence called the centrifugal potential. It has appeared only because we have projected the 2-dimensional motion into a 1-dimensional radial equation.

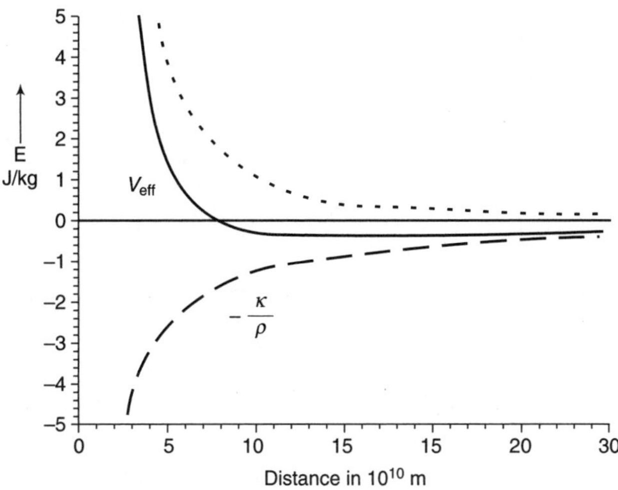

Fig. 8.6a The physical *one-over-distance* (longer dashes, always negative), the centrifugal (shorter dashes, always positive), and the effective radial potential (continuous line) which appear in Eq. 8.27b.

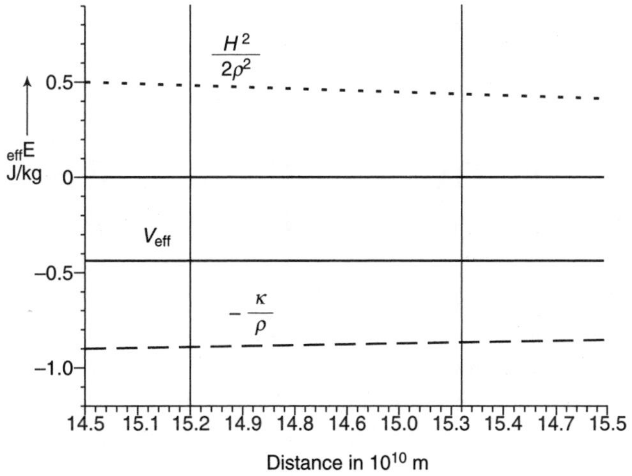

Fig. 8.6b A magnification of Fig. 8.6a in a rather small range of the distance. On this magnification scale, the effective potential curve only appears to be parallel to the E = 0 horizontal line, but it is of course not so. A further magnification, shown in Fig.8.6c, shows the shape of the effective potential clearly.

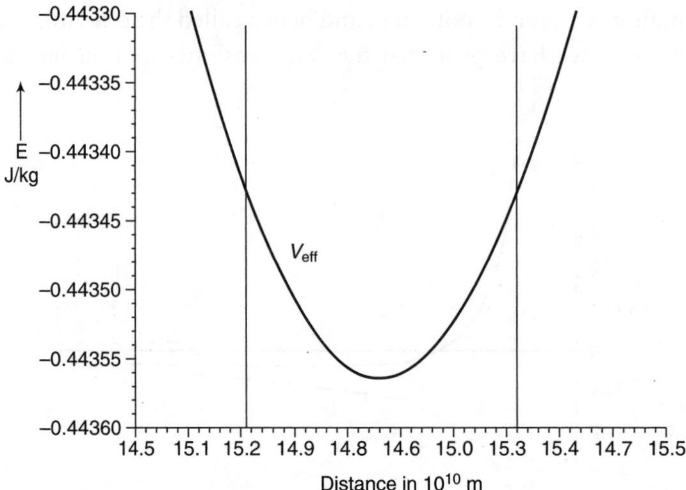

Fig. 8.6c The effective potential has a minimum, which is not manifest on the scales employed in Fig. 8.6a and 8.6b.

The radial effective potential has a minimum (Fig. 8.6c) at a distance that would correspond to the distance at which the planet would be in a circular orbit. Equation 8.26, however, has two values of the distance at which $\dot{\rho} = 0$. From this condition, we get the two apses, from the solutions of the quadratic equation, $2E\rho^2 + 2\kappa\rho - H^2 = 0$. The solutions give us the distances of the perigee and the apogee:

$$\rho_{p,a} = -\frac{\kappa \pm \sqrt{\kappa^2 + 2EH^2}}{2E}. \tag{8.28}$$

8.4 MORE ON THE MOTION IN THE SOLAR SYSTEM, AND IN THE GALAXY

The eccentricities of the orbits of various planets in our solar system are given in Table 8.1, from Reference [9], along with other planetary data for all the planets in our solar system. Even though this table includes information for Pluto, discovered in 1930 (by Clyde Tombaugh using Kepler–Newton laws,) Pluto lost its status as a planet in 2006 when the General Assembly of the International Astronomical Union (IAU) adopted new criteria to define a planet. For an object to be called a planet, the criteria introduced then required that the object -

(a) is in orbit around the Sun,

(b) has sufficient mass for its self-gravity to overcome rigid body forces so that it assumes a hydrostatic equilibrium (nearly round) shape,

(c) has cleared the neighborhood around its orbit.

Pluto has many other objects in the Kuiper belt in its vicinity, and does not satisfy the criterion (c).

Table 8.1 Planetary data for our solar system.

Planetary Data [9]	MERCURY	VENUS	EARTH	MOON	MARS	JUPITER	SATURN	URANUS	NEPTUNE	PLUTO
Mass (10^{24}kg)	0.330	8.87	8.972	0.073	0.642	1898	568	88.8	102	0.0146
Diameter (km)	4879	12,104	12,756	3475	6792	142,984	120,536	51,118	49,528	2370
Density (kg/m^3)	5427	5243	5514	3340	3933	1326	687	1271	1638	2095
Gravity (m/s^2)	3.7	8.9	9.8	1.6	3.7	23.1	9.0	8.7	11.0	0.7
Escape Velocity (km/s)	8.3	10.4	11.2	2.4	8.0	59.5	38.5	21.3	23.5	1.3
Rotation Period (hours)	1407.6	−5832.5	23.9	658.7	28.6	9.9	10.7	−17.2	18.1	−153.3
Length of Day (hours)	4222.6	2802.0	28.0	705.7	28.7	9.9	10.7	17.2	18.1	153.3
Distance from Sun (10^6 km)	57.9	105.2	149.6	0.384*	227.9	778.6	1433.5	2872.5	4498.1	5905.4
Perihelion (10^6 km)	48.0	107.5	147.1	0.363*	205.6	740.5	1352.6	2741.3	4448.5	4438.8
Aphelion (10^6 km)	69.8	105.9	152.1	0.406*	249.2	818.6	1518.5	3003.6	4548.7	7378.9
Orbital Period (days)	88.0	228.7	368.2	27.3	687.0	4331	10,747	30,589	59,800	90,560
Orbital Velocity (km/s)	47.4	38.0	29.8	1.0	28.1	13.1	9.7	8.8	8.4	8.7
Orbital Inclination (degrees)	7.0	3.4	0.0	8.1	1.9	1.3	2.5	0.8	1.8	17.2
Orbital Eccentricity	0.208	0.007	0.017	0.085	0.094	0.049	0.087	0.046	0.011	0.244
Obliquity to Orbit (degrees)	0.034	177.4	23.4	8.7	28.2	3.1	28.7	97.8	28.3	122.5
Mean Temperature (C)	167	464	15	−20	−65	−110	−140	−195	−200	−225
Surface Pressure (bars)	0	92	1	0	0.01	Unknown*	Unknown*	Unknown*	Unknown*	0.00001
Number of Moons	0	0	1	0	2	67	62	27	14	5
Ring System?	No	No	No	No	No	Yes	Yes	Yes	Yes	No
Global Magnetic Field?	Yes	No	Yes	No	No	Yes	Yes	Yes	Yes	Unknown

The total time period taken by a planet to go around the Sun once can be easily obtained from the total area $\Theta = \int_0^\tau \frac{dA}{dt}\,dt$ swept by it, which gives, using Eq. 8.9, $\pi ab = \frac{H}{2}\tau$, and

hence $$\tau = \frac{2\pi ab}{H} = \frac{2\pi a^2\sqrt{1-\varepsilon^2}}{H}. \tag{8.29a}$$

We immediately recognize this result to correspond to Kepler's third law, since,

$$\tau^2 = \frac{4\pi^2 a^4(1-\varepsilon^2)}{H^2} = \frac{4\pi^2 a^4}{H^2}\,\frac{\left(\frac{H^2}{\kappa}\right)}{a} = \frac{4\pi^2}{\kappa}a^3 = \frac{4\pi^2}{G(m_1+m_2)}a^3. \tag{8.29b}$$

The third law of Kepler is also called the harmonic law, as Kepler found in it a certain musical pattern, and thus harmony, in the celestial dynamics of planetary motion. We have used, above, the value of κ from Eq. 8.3b. If m_1 is the mass of the Sun, and m_2 that of the planet, we can approximate, when $m_1 \gg m_2$, that $m_1 + m_2 \simeq m_{\text{sun}}$. Now, on its elliptical trajectory, the speed v of the planet changes from point to point, but we can get an approximate idea about it by ignoring the eccentricity of the ellipse. Assuming that the orbit is circular, we may write approximately that, $\tau = \frac{2\pi\rho}{v}$, giving us the following relation (from Eq. 8.29b) between the *square* of the speed a planet, and its *mean* distance from the center of mass:

$$\left(\frac{2\pi\rho}{v}\right)^2 = \frac{4\pi^2}{G(m_{\text{sun}}+m_{\text{planet}})}\rho^3, \tag{8.29c}$$

i.e., $$v^2 = \frac{G(m_{\text{sun}}+m_{\text{planet}})}{\rho}. \tag{8.30}$$

We obtained the result in Eq. 8.30 from Kepler's third law, but we have assumed *circular*, rather than *elliptical* orbits. This assumption is, however, not required, since the relation Eq. 8.30 is readily obtainable also from a general theorem due to R. E. Clausius, known as the virial theorem [10]. This theorem is applicable whenever we are dealing with a mechanical system of N interacting particles whose positions, nor speeds, blow up; i.e., they have an upper bound. The theorem enables you to obtain the *long-time-average* of the potential energy of the system, $\langle V\rangle$, from the long-time-average of its kinetic energy $\langle T\rangle$, or vice-versa.

The virial theorem states that when the potential is of the form $V\,\alpha\,\frac{1}{r^s}$, r being the distance between the interacting particles, then:

$$2\langle T\rangle = -s\langle V\rangle. \tag{8.31}$$

The virial theorem, Eq. 8.31, for the gravitational interaction can be easily established considering a quantity C_{virial}, which Clausius called the 'virial'. It is constructed from the sum of scalar products of momentum $\vec{p}$ with position vector $\vec{r}$ for each of the N particles. Clasius defined the virial of an interacting system as:

$$c_{\text{virial}} = \sum_{i=1}^{N} \vec{p}_i \cdot \vec{r}_i. \tag{8.32}$$

Thus, when the potential is given by $V(r) = -\frac{\kappa}{r^s}$, we get the time-derivative of the virial as:

$$\dot{c}_{\text{virial}} = \mu\dot{\vec{v}} \cdot \vec{r} + \mu\vec{v} \cdot \vec{v} = \vec{F} \cdot \vec{r} + 2T = \left(-\frac{\partial V}{\partial r} \hat{e}_r \right) \cdot \vec{r} + 2T = sV + 2T. \tag{8.33a}$$

The time average, of the time-derivative of the virial, over a *large* time τ is:

$$\frac{1}{\tau}\int_0^{\tau} \frac{dc_{\text{virial}}}{dt}\, dt = \langle \dot{c}_{\text{virial}} \rangle = s\langle V \rangle + 2\langle T \rangle, \tag{8.33b}$$

i.e.,
$$\frac{1}{\tau}\left[c_{\text{virial}}^{(t=\tau)} - c_{\text{virial}}^{(t=0)} \right] = \langle \dot{c}_{\text{virial}} \rangle = s\langle V \rangle + 2\langle T \rangle. \tag{8.33c}$$

Therefore, for *large* time durations, since $\tau \gg \left[c_{\text{virial}}^{(t=\tau)} - c_{\text{virial}}^{(t=0)} \right]$, we get $s\langle V \rangle + 2\langle T \rangle = 0$ which is the virial theorem.

The virial for a Sun–planet two-body system in the center of mass frame of reference is simply

$c_{\text{virial}} = \vec{p} \cdot \vec{r} = \mu\vec{v} \cdot \vec{r}$ where the reduced mass is $\mu = \frac{m_{\text{sun}} m_{\text{planet}}}{m_{\text{sun}} + m_{\text{planet}}}$. For the gravitational

(*and also for the Coulomb*) interaction we have $s = 1$, hence $2\langle T \rangle = -\langle V \rangle$, which is the same as Eq. 8.30. Thus the restriction of employing circular orbits is *not* required for the conclusion expressed as Eq. 8.30.

For various planets in our solar system, we now plot the relationship between the *mean* orbital speed of the planet, and its *mean* distance from the center of mass, about which it is rotating. This plot is called the *velocity curve* or as the *rotation curve*. It can be plotted for the planets in our solar system using the data presented in Table 8.1. It is shown in Fig. 8.7. The rotation curve for our solar system is seen to be completely in accordance with the virial theorem (Eq. 8.31, with $s = 1$).

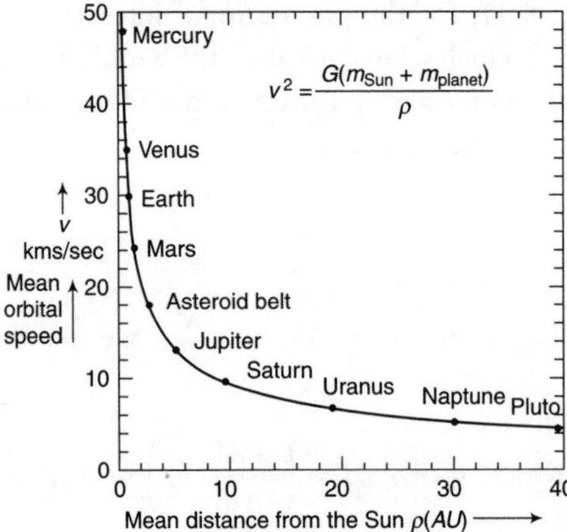

Fig. 8.7 Graph showing the relation between a planet's mean speed and its mean distance from the center of mass in our solar system. Included in this graph is Pluto (which lost its status as a planet), but also the asteroid belt, which consists of a very large number of objects, some of which, such as Ceres, are cognizable, which orbit the Sun at a *mean* distance between 2.2 and 3.2 AU.

The Sun being enormously massive compared to all the planets, the center of mass would practically be at the center of the Sun, but strictly speaking it would actually *dynamically* move as planets revolve around the Sun. We see from Fig. 8.7 and from Eq. 8.26 that *farther* the orbiting object is from the center of mass, the *slower* it would move. From the rotation curve, and Eq. 8.30, one can estimate the mass of the planet.

We now ask if the Kepler–Newton model's conclusion depicted in Fig. 8.7 accounts for other objects in the cosmos, to mass distributions in glaxies, looking into the Milky Way, and even beyond, into the millions of galaxies in the universe. We expect all matter to be governed by the gravitational interaction. Our impression of a galaxy bears some analogy to the planetary system of our Sun; millions of stars in a galaxy go around a supermassive black hole at its center, much like planets go around the Sun. Our Milky Way is one of the greatest spectacle on a clear moonless night. If you have not seen it, I recommend that you go well away from the city light and pollution to enjoy its breath-taking view. Our Milky Way, also called "Aakaash Ganga" (the Ganges in the sky), has a spiral structure that spans about 120,000 light-years across. In astronomy, distances are measured in units of kiloparsec (kpc), which is 3,262 light-years. Our solar system is located at about 8 kpc (i.e., about 27,000 light years) from the Milky Way's center. There are perhaps 400 billion stars of various sizes and brightness in it, perhaps much more. It is estimated that there are perhaps ~200 billion galaxies in the observable universe, stretching out into a region of space that is possibly 13.8 billion light-years far from us, in all the directions. Furthermore, there are possibilities that there are multiverses, rather than just one universe. It is mindboggling to think about how many exoplanets have intelligent life, or for that matter how much mass exists in each galaxy, and how to determine it.

At the center of our Milky Way galaxy is a supermassive black-hole, Sagittarius A* (Sgr A*). We visualize the billions of stars in the galaxy to orbit this black hole, much the same way planets orbit the Sun in our solar system. It is now natural to ask if the rotation curve (velocity curve) for stellar motion in the galaxy is similar to that for the planets, shown in Fig. 8.7. Now, the rotation curves for various galaxies in the visible universe have been studied by a number of astronomers, beginning with the Dutch astronomer Jan Oort in the 1930's using stellar Doppler shifts. From such measurements, it is now concluded using the mass/luminosity ratios that the rotation curves for the galaxies [11] are *not* the Newton–Keplerian curves of Fig. 8.7. Instead, they are as shown in Fig. 8.8. The difference is striking: instead of the velocities diminishing with distance from the center of the galaxies, the velocities seem to remain unchanged even as you go a large distance away from the center. The velocity curves in Fig. 8.8 are therefore not in accordance with the form, $v^2 \simeq \frac{GM}{\rho}$, as in Eq. 8.30.

To *empirically* account for the fact that the velocities *flatten out with distance*, rather than decline, Eq. 8.30 can, however, be modified to allow the mass M in the numerator on the right hand side to actually *increase* with the distance ρ, giving

$$\{v(\rho)\}^2 = \frac{G\{M(\rho)\}}{\rho}. \tag{8.34a}$$

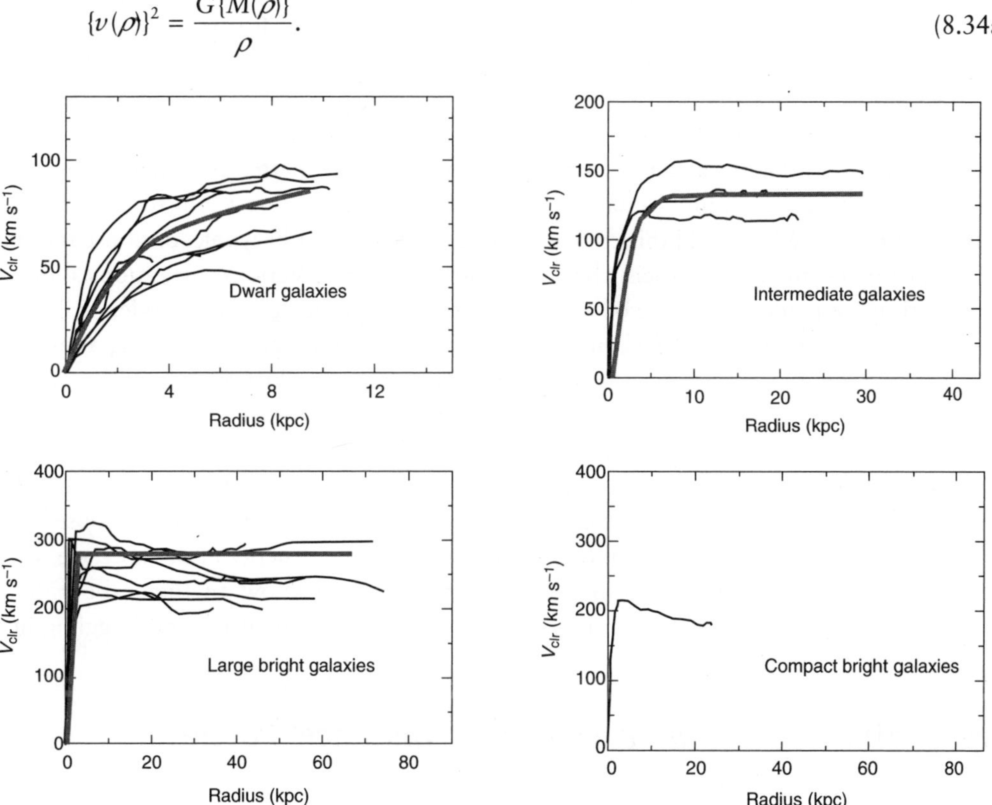

Fig. 8.8 The rotation curves for galaxies [10] are not Keplerian (as in Fig. 8.7). Unlike the Keplerian drop seen in rotation curve for the planets in our solar system these rotation curves flatten out, suggesting that the mass that keeps the stars in motion is *not* centerd as it is in our solar system. It is possibly smeared out, even if it is not seen in the observations carried out. This suggests that there is a different kind of matter out there which is neither luminous, nor does it interact with light.

Equation 8.34a provides for a mass in the galaxy that spreads out far into the space, becoming larger as one goes away from the center. It is in clear departure from the Newton–Keplerian planetary model of the mass distribution in our planetary system. From Eq. 8.34a, we find that

$$M(\rho) = \frac{\rho v(\rho)^2}{G}. \tag{8.34b}$$

This relation suggests that the mass, normally interpreted as the quantity of matter, actually *increases linearly*, since the velocities (as per Fig. 8.8) remain more or less constant, as you go large distances from the center of the galaxy. The mass of the galaxy, in our naïve thinking, comes from that of the stars, which are of course luminous, or from gases and planetary objects in the galaxy, which would scatter the galactic light. We expect the mass then to stretch out from the center of the galaxy. However, as much light as would account for the spread-out mass $M(\rho)$ of Eq. 8.34b is not observed. The rotation curves point toward a stretched-out distribution of mass in the galaxies, but it is not seen. The gravitational *effect* of this unseen matter is manifest, but the matter itself is not seen. It is therefore called *dark matter*. It does not glow, i.e., it is not self-luminous like the stars. More surprisingly, it does not even interact with stellar light, at least not in the manner we know ordinary matter does. To reconcile the galactic rotation curves with the amount of light seen scattered in our galaxy, it would need dark matter as much as a 100 billion times the matter as is there in our Sun. It is estimated that the amount of dark matter in the galaxies is almost four to five times the amount of matter. What could this dark matter be made of? Is it baryonic, or non-baryonic? Is it made up of something physicists have not known hitherto? Various suggestions about its composition are made [11], some of which go even beyond the Standard Model of Physics. We have only given the name *dark matter* to what we know practically nothing about. We have hit on a question whose answer would provide a huge breakthrough in understanding our universe. In Chapter 14, we shall also meet *dark energy*, another unknown. Even as we encounter these fascinating and challenging questions in this introductory course on classical mechanics, their answers must wait, not just for more advanced courses, but perhaps also for discoveries and advances yet to take place. It is estimated that matter that is observable and accounted for constitutes just about 5% of the total mass-energy in the universe. Dark matter (~27%), and dark energy (~68%), is what we do not know anything about. Since Copernicus, Galileo and Newton, in just these few hundred years, it is mind-boggling how much we have learned, yet so little.

Appendix: The *Repulsive* 'One-Over-Distance' Potential

The Coulomb interaction is governed by a potential that has exactly the same form as the gravitational potential discussed in this chapter. While the gravitational interaction between two masses is attractive, the Coulomb interaction between two charges is attractive if the two charges are unlike, and repulsive when the charges have the same sign. The repulsive Rutherford scattering experiment was one of the pioneering ones which established the basic atomic structure. Fascinating accounts of this experiment, and commentary on the role it played in understanding the atomic structure, can be easily found elsewhere in the literature. Our limited interest in this appendix is to discuss the Coulomb–Rutherford scattering, since the form of the potential which determines the results is exactly the same as the gravitational interaction we have discussed in this chapter. Within this limited scope, we shall, in this appendix, consider the repulsive 'one over distance' Coulomb potential,

$$V_+(\rho) = -\frac{\kappa}{\rho}, \tag{8A.1}$$

which has exactly the same *form* as the attractive gravitational potential. For the Coulomb case, we *define*

$$\kappa = -\frac{1}{m_\alpha}\frac{Qq}{4\pi\varepsilon_0}. \tag{8A.2}$$

In the gravitational case, we had $\kappa_g \approx Gm_1$ (we have added here the subscript 'g' for gravity) with the dimensions of $[\kappa] = L^3T^{-2}$. In Eq. 8A.2, we have defined the constant κ in terms of m_α, which represents the mass of the α-particles, the charge Q, which is the charge of the of each nucleus in a metal foil which would scatter off an incident beam of α- particles, and q is the charge of the α- particle. The constant, κ, so chosen, has the same dimensions as the constant used to describe the gravitational interaction. The minus sign on the right hand side of Eq. 8A.2 takes care of the fact that the potential given by Eq. 8A.1 is *repulsive*.

Our prototype of the Coulomb repulsive interaction is the Rutherford–Coulomb scattering of α particles of mass m_α, and charge $+q$, by positively charged atomic nuclei having charge $+Q$ in a thin foil of gold. The incident particle is repelled by the Coulomb force,

$$f(\rho) = \frac{1}{4\pi\varepsilon_0}\frac{Qq}{\rho^2}. \tag{8A.3}$$

The differential equation of the orbit is then given by

$$\frac{d^2\xi}{d\varphi^2} + \xi = \frac{\kappa}{H^2}, \text{ with } \rho = \frac{1}{\xi}, \text{ and } \kappa = -\frac{1}{m_\alpha}\frac{Qq}{4\pi\varepsilon_0}. \tag{8A.4a}$$

We have already seen that the solution of the above equation is $\xi = \delta\cos(\varphi - \varphi_0) - \frac{\kappa}{H^2}$

i.e., $$\rho = \frac{1}{\delta\cos(\varphi - \varphi_0) - \frac{\kappa}{H^2}}$$

or, $$\rho = \frac{\lambda}{\varepsilon\cos(\varphi - \varphi_0) - 1}, \tag{8A.4b}$$

where $\lambda = \frac{H^2}{\kappa}$ and $\varepsilon = \sqrt{1 + \frac{2EH^2}{\kappa^2}}$. (8A.4c)

At an initial velocity, v_i, the total *specific* energy (i.e., energy per unit mass) of an incident α-particle incident at an impact parameter s is $E = \frac{1}{2}v_i^2$ and hence $v_i = \sqrt{2E}$. The magnitude of the *specific* angular momentum about the target nucleus is, $H = |\vec{H}| = |\vec{r} \times \vec{v}_i| = rv_i \sin\Theta = sv_i$. Fig. 8A.1 shows a schematic representation of the Rutherford–Coulomb scattering. It shows the trajectories of mono-energetic α-particles scattered by the positive nuclei in a thin metal foil. We employ azimuthal symmetry around the axis of incidence of the α-particles, so all the particles incident at the same impact parameter in an annular ring would be scattered off at the same scattering angle as shown. The scattered particles would exit the reaction zone through a tiny strip on the surface of a sphere, centerd at the target nucleus. Since we have $H^2 = 2Es^2$, the eccentricity becomes, $\varepsilon = \sqrt{1 + \frac{4E^2s^2}{\kappa^2}}$. We see that the eccentricity is the square root of an essentially positive quantity added to 1. Hence, the $\varepsilon > 1$, and the orbit is a hyperbola.

We consider, in Fig. 8A.1, scattering of incident α particles in the ring 'A' of radius s and thickness ds. Particles in this ring get scattered at an angle φ_s and escape through the ring 'B'. Essentially, since no particles are lost, or created; they are only scattered. Hence, *all* the particles in the ring 'A' (at impact parameters between s and $s + ds$) escape through the ring 'B'. If N is the number of particles per unit area in the incident beam, the number of particles in the ring 'A' is $N2\pi\, sds$, and *each one of these particles* is scattered essentially through the ring 'B' at the scattering angles between φ_s and $\varphi_s + d\varphi_s$ (see panel on the left side of Fig. 8A.1).

Now, from the equation to the hyperbola, $\rho = \frac{\lambda}{\varepsilon\cos(\varphi - \varphi_0) - 1}$, we find that ρ is minimum when $\cos(\varphi - \varphi_0) = 1$, that is when $\varphi = \varphi_0$. The trajectory (ρ, φ) of the α-particles is shown in Fig. 8A.2. We choose the axis with respect to which the angle φ is measured such that $\rho \to \infty$ at $\varphi = 0$. Now, $\rho \to \infty$ when $\varepsilon\cos(\varphi - \varphi_0) = 1$, i.e. when $\cos(\varphi - \varphi_0) = \frac{1}{\varepsilon}$ which is of course less than unity (since $\varepsilon > 1$). Having chosen $\rho \to \infty$ at $\varphi = 0$, we have $\cos(-\varphi_0) = \frac{1}{\varepsilon}$ when $\rho \to \infty$, which guarantees that we also have $\rho \to \infty$ when $\cos(+\varphi_0) = \frac{1}{\varepsilon}$. This would happen at $\varphi = 2\varphi_0$ when $\cos(\varphi - \varphi_0) = \cos(+\varphi_0) = \frac{1}{\varepsilon}$. Hence, the scattering angle, which is the angle between the two asymptotes respectively at $\varphi = 0$ and $\varphi = 2\varphi_0$, is given by $\varphi_s = \pi - 2\varphi_0$. It then follows that $\varphi_0 = \frac{\pi}{2} - \frac{\varphi_s}{2}$, which is shown in Fig. 8A.2.

Now, the relation $\dfrac{1}{\varepsilon}=\cos(\varphi_0)$ translates to $\dfrac{1}{\sqrt{1+\dfrac{2EH^2}{\kappa^2}}}=\cos\left(\dfrac{\pi}{2}-\dfrac{\varphi_s}{2}\right)=\sin\left(\dfrac{\varphi_s}{2}\right)$,

which gives $1+\dfrac{2EH^2}{\kappa^2}=1+\cot^2\dfrac{\varphi_s}{2}$,

and hence: $$\cot\frac{\varphi_s}{2}=\frac{(\sqrt{2E})H}{\kappa}=\frac{(\sqrt{2E})sv_i}{\kappa}=\frac{(\sqrt{2E})s(\sqrt{2E})}{\kappa}=\frac{2Es}{\kappa}. \tag{8A.5}$$

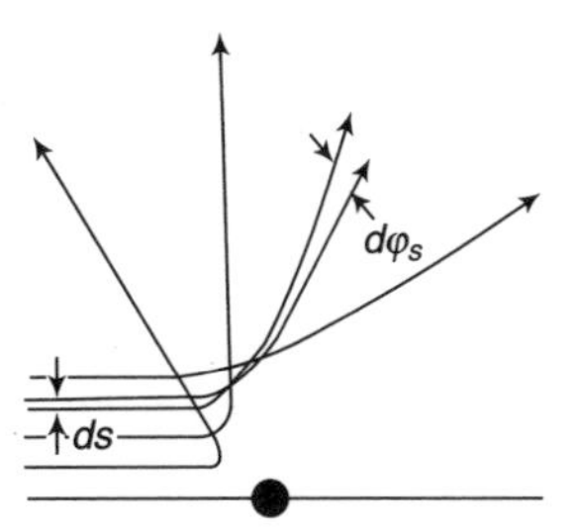

Particles with the least impact parameter are scattered back. Those at the largest impact parameters are scattered least.

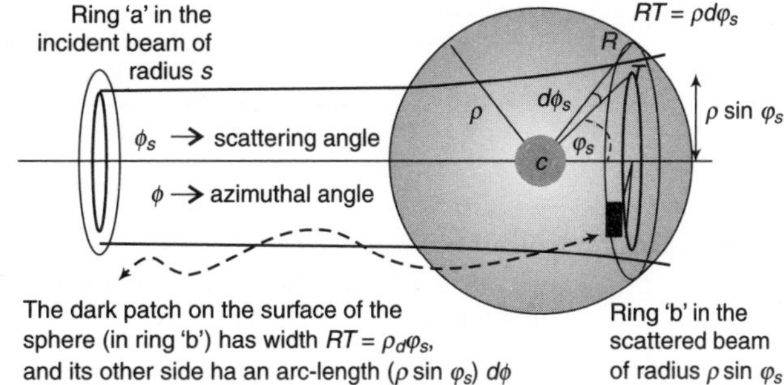

The scattered particles exit the scattering zone through a strip on the surface of the sphere of radius ρ centered at the target nucleus. The arc-length of the dark tiny patch, shown on the Ring 'b', is $\rho\sin\varphi_s d\phi$, and its width is $RT=\rho d\varphi_s$.

The area of the dark patch on the strip (ring 'b'), is

$\rho d\varphi_s\times\rho\sin\varphi_s d\phi=\rho^2\sin\varphi_s d\varphi_s\,d\phi$.

Fig. 8A.1 Coulomb–Rutherford scattering of α particles. All particles having the same impact parameter undergo identical deviation.

Accordingly, $$s=\frac{\kappa}{2E}\cot\frac{\varphi_s}{2}=\frac{\kappa}{2E}\sqrt{\frac{1+\cos\varphi_s}{1-\cos\varphi_s}}=\frac{Q}{4\pi\varepsilon_0(2E)}\sqrt{\frac{1+\cos\varphi_s}{1-\cos\varphi_s}}. \tag{8A.6}$$

The solid angle subtended by the small elemental area shown shaded on the ring 'b' in Fig. 8A.2 at the center 'C' of the sphere is $d\omega=\sin\varphi_s d\varphi_s d\phi$, where ϕ is the azimuthal angle about the axis along which the projectiles are incident on the target. Note the difference in the font φ and ϕ which have both been used. The complete ring 'b' therefore subtends an angle of $d\Omega=2\pi\sin\varphi_s d\varphi_s$ at the center 'C'. The angle $d\Omega$ results simply from sweeping that shaded area on the ring 'b' through one full circle about the axis.

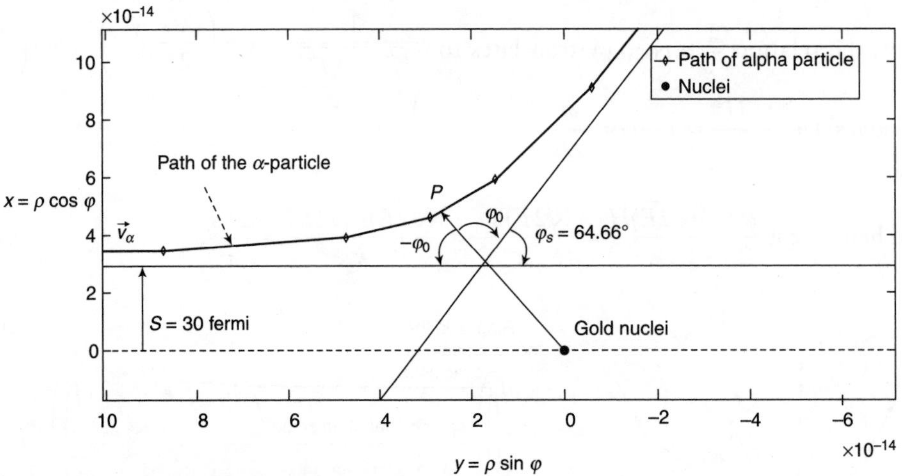

Fig. 8A.2 Typical trajectory of α particles arriving at an impact parameter ϕs and scattered at an angle ϕs by the charge Q. Note that the Cartesian axes are rotated in this figure for the sake of clarity.

The number dN of the incident particles that are deflected through a given angle φ_s is proportional to the number N of incident particles, to the number n of scattering centers per unit area of the target foil, and to the solid angle $d\Omega$ about the angle φ_s corresponding to scattering at *all* angles between φ_s and $\varphi_s + d\varphi_s$. The proportionality is called the differential scattering cross-section, and it is a function of the angle φ_s:

$$dN = \sigma(\varphi_s)\, Nnd\Omega = \sigma(\varphi_s) Nn2\pi \sin \varphi_s d\varphi_s. \tag{8A.7a}$$

This number must be equal to the number of particles in the ring 'a' of Fig. 8A.1, since all the particles in ring 'a' escape the scattering zone through the ring 'b'.

Thus, $$dN = Nn2\pi s\, ds, \tag{8A.7b}$$

or, $$\frac{dN}{N} = n\sigma(\varphi_s)\, d\Omega. \tag{8A.8}$$

Substituting the value of $d\Omega$ we get,

Thus, $$\sigma(\varphi_s) = \frac{1}{nd\Omega}\frac{dN}{N} = \frac{1}{nd\Omega}\frac{Nn2\pi s\, ds}{N} = \frac{2\pi s\, ds}{d\Omega} = \frac{2\pi s\, ds}{2\pi \sin\varphi_s d\varphi_s} = \frac{s}{\sin\varphi_s}\frac{ds}{d\varphi_s}. \tag{8A.9a}$$

Accordingly, we write $$\sigma(\varphi_s) = \frac{s}{\sin \varphi_s}\left|\frac{ds}{d\varphi_s}\right|. \tag{8A.9b}$$

To find the scattering cross section for charged particles, we differentiate $\cot\frac{\varphi_s}{2} = \frac{2Es}{\kappa}$ with respect to φ_s. Thus, $\dfrac{1}{2\sin^2\left(\dfrac{\varphi_s}{2}\right)} = \dfrac{2E}{\kappa}\left|\dfrac{ds}{d\varphi_s}\right|$, which gives

$$\left|\frac{ds}{d\varphi_s}\right| = \frac{\kappa}{2E}\frac{1}{2\sin^2\left(\frac{\varphi_s}{2}\right)}. \tag{8A.10}$$

Hence, $\sigma(\varphi_s) = \frac{s}{\sin\varphi_s}\frac{\kappa}{2E}\frac{1}{2\sin^2\left(\frac{\varphi_s}{2}\right)},$ (8A.11)

$$= \frac{s}{2\sin\left(\frac{\varphi_s}{2}\right)\cos\left(\frac{\varphi_s}{2}\right)}\frac{\kappa}{2E}\frac{1}{2\sin^2\left(\frac{\varphi_s}{2}\right)} = \frac{\kappa^2}{16E^2}\frac{1}{\sin^4\left(\frac{\varphi_s}{2}\right)}.$$

Finally, using $s = \frac{\kappa}{2E}\cot\frac{\varphi_s}{2}$, we get

$$\sigma(\varphi_s) = \frac{\kappa^2}{16E^2}\frac{1}{\sin^4\left(\frac{\varphi_s}{2}\right)}. \tag{8A.12}$$

The above relation is famously known as the Rutherford scattering formula. It is obviously valid for both attractive and repulsive 'one-over-distance' potentials. Interestingly, using quantum mechanical scattering theory also we get an identical expression for the scattering cross-section. An illustrative hyperbolic path is plotted for $x = \rho\cos\varphi$ versus $y = \rho\cos\varphi$ in Fig. 8A.2. For the purpose of this sketch, semi-latus-rectum $\lambda = \frac{H^2}{\kappa}$ has been calculated considering an alpha particle incident at an impact parameter of $s = 30$ fm. Using the kinetic energy $\frac{1}{2}mv_i^2 = 6MeV$, we get the initial velocity v_i is found to be 0.658×10^7 m/s. We have used the mass of alpha particle to be $m_\alpha = 6.645 \times 10^{-27}$ kg. Thus, its angular momentum is $l = mH = msv_i = 131.17 \times 10^{-35}$ kgm^2/s.

Also, we have, $\kappa = \frac{1}{m_\alpha}\frac{Qq}{4\pi\varepsilon_0} = \frac{1}{6.645\times10^{-27}}\frac{79\times2\times(1.602\times10^{-19})^2\mathrm{C}^2}{4\pi\times8.85419\times10^{-12}\mathrm{N}^{-1}\mathrm{C}^2m^{-2}}$

$$= 547.82\times10^{-2}\,\mathrm{Nm^2Kg^{-1}}$$

and hence the scattering angle for the chosen impact parameter turns out to be given by:

$$\cot\frac{\varphi_s}{2} = \frac{2sE}{\kappa} = \frac{2\times30\times10^{-15}\times6\times10^{6}\times1.6\times10^{-19}}{3640.32\times10^{-29}} = 1.582,$$

and $\tan\frac{\varphi_s}{2} = 0.633.$

Using these values, we get the scattering angle: $\varphi_s = 64.66°$. (8A.13a)

It follows that

$$\varphi_0 = \frac{\pi}{2} - \frac{\varphi_s}{2} = 90° - \frac{64.66°}{2} = 57.67°, \tag{8A.13b}$$

and $$\varepsilon = \sqrt{1+\left(\cot\frac{\varphi_s}{2}\right)^2} = \sqrt{1+(\cot 32.33°)^2} = 1.871. \tag{8A.13c}$$

Problems with Solutions

P8.1:

A satellite encircling the Earth on a circular orbit of radius r_1 is to be lifted into a concentric circular orbit of radius $r_2 > r_1$ by means of two rocket impulses. The required amount of fuel is optimized if the transition is made via an elliptic orbit with the Earth at one of the foci. Find out the required velocity increment for the transition, in terms of r_1 and r_2.

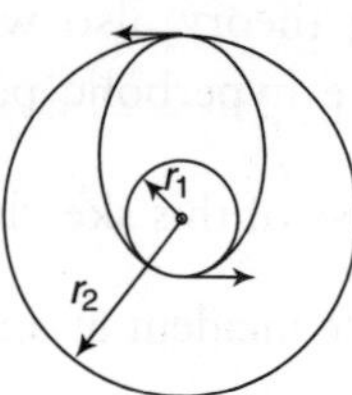

Solution:

The velocities of the circular orbits are $v_{r_1} = \sqrt{\frac{\mu}{r_1}}$ and $v_{r_2} = \sqrt{\frac{\mu}{r_2}}$, where μ is the proportionality in the inverse square law for the gravitational force $f = -\mu\frac{m}{r^2}$. In the gravitational field of the Earth, $\mu = GM_{Earth} = 3986 \times 10^2\ \text{km}^2\text{s}^{-2}$. Now, the transition from the lower orbit to the higher is made via an elliptical orbit. The velocity v is a maximum or minimum respectively at an apsis when $r \to r_1 = a(1 - e)$ and $r \to r_2 = a(1 + e)$, where e is the eccentricity of the ellipse. From the conservation of energy and angular momentum, the velocity of a particle at position r, moving in an elliptic orbit having major axis as $2a$ is given by, $v^2 = \frac{2\mu}{r} - \frac{\mu}{a}$. Hence, $v_{max}^2 = \frac{2\mu}{a(1-e)} - \frac{\mu}{a} = \frac{\mu}{a}\frac{1+e}{1-e}$ and $v_{min}^2 = \frac{2\mu}{a(1+e)} - \frac{\mu}{a} = \frac{\mu}{a}\frac{1-e}{1+e}$.

For a circular orbit (e = 0) of radius r, we have $v = \sqrt{\frac{\mu}{r}}$.

Since $\frac{r_{min}}{r_{max}} = \frac{1-e}{1+e} = \frac{r_1}{r_2}$ and $r_1 + r_2 = 2a$, we get $a = \frac{1}{2}(r_1 + r_2)$ so that,

$$v_{max} = \sqrt{\frac{\mu(1+e)}{a(1-e)}} = \sqrt{\frac{r_2}{r_1}\frac{\mu}{a}} = \sqrt{\frac{2\mu r_2}{(r_1+r_2)r_1}} \text{ and } v_{min} = \sqrt{\frac{2\mu r_1}{(r_1+r_2)r_2}}$$

Hence, the required velocity increment is: $\Delta v = \sqrt{\frac{2\mu r_2}{(r_1+r_2)r_1}} - \sqrt{\frac{\mu}{r_1}} + \sqrt{\frac{\mu}{r_2}} - \sqrt{\frac{2\mu r_1}{(r_1+r_2)r_2}}$.

P8.2:

A particle of mass m moves under the action of a central force whose potential is $V(r) = Kmr^3$ ($K > 0$). Then,

(i) for what kinetic energy and angular momentum of the particle will the orbit be a circle of radius R about the origin?

(ii) determine the period of the ensuing circular motion.

Solution:

Given, $V(r) = Kmr^3$ ($K > 0$). Therefore, $F = -\frac{\partial V}{\partial r} = -3Kmr^2$.

(i) For the motion to be circular, we must have $F = -\frac{mv^2}{r} = -3Kmr^2$. Hence, the kinetic energy is: $\frac{1}{2}mv^2 = \frac{3}{2}Kmr^3$ and the angular momentum is: $mvr = mr(3Kr)^{1/2}$, since $v^2 = 3Kr^3$.

(ii) Since, $v = r\omega = \frac{2\pi r}{T}$, we get $v^2 = \left(\frac{2\pi r}{T}\right)^2 = 3Kr^3$, and $T = \frac{2\pi}{\sqrt{3Kr}}$.

P8.3:

A huge cloud of gas in the Sun starts collapsing under its own force of gravity releasing its energy in the form of electromagnetic radiation. If this were to be a plausible explanation of energy production in the Sun, determine the age of the Sun from the total amount of energy radiated till now, assuming that the Sun has been radiating with the same luminosity $L_\odot$ during its entire life time. Given: mass of the Sun $M_\odot = 1.989 \times 10^{30}$ kg, radius of the Sun $R_\odot = 6.95 \times 10^5$ km, and luminosity $L_\odot = 3.828 \times 10^{26}$ W. Assume that the cloud is spherically symmetric, with radius R and mass M.

Solution:

Total energy of such a cloud: E = kinetic energy + potential energy = $T + V$

We have, from Eq. 8.31 (virial theorem): $2\langle T\rangle = -\langle V\rangle$. Hence, $E = -\frac{1}{2}V + V = \frac{V}{2}$

We must therefore determine only the potential energy. The contraction of the cloud initiates at the ***outermost*** layer, and subsequently the radius of the cloud of gas would decrease. This process continuous as the radius of the spherical cloud diminishes.

Let us consider the potential dV of a tiny amount of mass dm of the cloud in the ***outermost*** layer (outermost spherical shell), at a distance r from the center. We need to consider the gravitational potential energy of the mass dm in the outermost shell in the field of the total mass $M(r)$ distributed at a mass–density $\rho(r)$ inside, up to the radius r.

Thus $V = -4\pi G\int_0^R M(r)\rho(r)rdr$. The density generally has a radial dependence, but we will assume it to be uniform $\rho = \frac{M}{\frac{4}{3}\pi R^3}$; hence: $M(r) = \frac{4}{3}\pi r^3\rho$.

Accordingly, $V = -4\pi G\left(\frac{M}{\frac{4}{3}\pi R^3}\right)^2 \frac{4}{3}\pi\int_0^R r^4 dr = -\frac{3GM^2}{5R}$; Hence: $E = \frac{V}{2} = -\frac{3GM^2}{10R}$.

Assuming that the cloud has been collapsing from an infinite distance at the beginning, i.e., $E_{r=\infty} = 0$, we get:

The amount of energy radiated by the Sun till now is $E_{\text{radiated}} = E_{r=\infty} - E_{r=R_\odot} = \frac{3GM_\odot^2}{10R_\odot}$

Therefore $E_{\text{radiated}} = \frac{3GM_\odot^2}{10R_\odot} \simeq 1.1\times 10^{41}\,\text{J}$. This gives the age of the Sun to be

$\Delta t = \frac{E_{\text{radiated}}}{L_\odot} = \frac{1.1\times 10^{41}}{3.828\times 10^{26}}\text{s} \simeq 0.335\times 10^{15}\text{s} \simeq 10^7$ years. [Note: Notwithstanding the assumptions made in this problem, the age of the Sun is estimated to be ~4.6×10^9 years. Also, the energy yield from the Sun is mostly due to nuclear fusion of hydrogen nuclei into helium.]

P8.4:

Preamble: This problem must be appreciated in the context of its historical context. Edmond Halley, Robert Hooke and Christopher Wren once raised the following question to Isaac Newton over a cup of coffee: *If the force between a planet and the Sun is assumed to vary as the inverse square of the distance between the Sun and the planet, would it result in the orbit of the planet to be an ellipse?* As is mentioned in this chapter already, Johannes Kepler had already conjectured, semi-empirically from the data compiled by Tycho Brahe, that the shape would be an ellipse. The question raised looked for a geometric determination of the orbit's shape to find if it would indeed be an ellipse. Newton answered, authoritatively, that yes, it would in fact be an ellipse, and he offered a proof. Newton's proof was reconstructed by Feynman in one of his lectures, now known as *Feynman's Lost Lecture*. It is an amazing example of Feynman's insightful lectures, which were lost but the contents were thankfully preserved by David L. Goodstein and Judith R. Goodstein (See '*Engineering and Science*', Number 3, page 15, 1996). A very impressive video based on David L. Goodstein and Judith R. Goodstein's notes has been produced by Grant Sanderson for his 3*Blue*1*Brown* channel, available online on the YouTube (https://www.youtube.com/watch?v=xdljYBtnvZU). The following problem is essentially based on Feynman's Lost Lecture, for which the best source is the narrative by Goodstein and Goodstein, and the YouTube video by Grant Sanderson.

P8.4(a):

Assume that a planet goes round the Sun, sweeping equal area in equal time. If it is also assumed now that the planet is attracted to the Sun by a force which goes as the inverse square of the distance between the two, show that the planet's velocity vector sweeps a circle in the velocity space as the planet goes round the Sun in the direct space.

P8.4(b):

Use the fact that the velocity vector sweeps a circle in the velocity space to show that the planet's orbit about the Sun must be an ellipse.

Solution:

(a) The planet sweeps equal area in equal time (Kepler's second law; conservation of the angular momentum). The areal velocity, $\frac{\Delta A}{\Delta t} = \text{constant}$. Therefore, as the planet traverses its orbit, it sweeps a larger area, the more time it spends on its trajectory. Hence, Δt and ΔA are linearly proportional: $(\Delta t)\alpha\ (\Delta A)$. We now use a geometric construction to solve this problem. Draw a circle with radius *R* centered at *O*. From an *eccentric* point *A*, draw a straight line connecting this eccentric point *A* to an arbitrary point *P* on the circumference of the circle (Fig. P8.4.1). Draw the perpendicular bisector of *AP*. *MQ* is the perpendicular bisector (Fig. P8.4.2). Then, *QP* = *QA*, since the triangle (*QMP*) is congruent with the triangle (*QMA*). Note that *OP* = *OQ* + *QP* = *OQ* + *QA*, which is a constant. We get different points *Q*, corresponding to different points P_0, P_1, P_2, P_3, ... on the circumference of the circle. The different points *Q* would then essentially trace an ellipse (Fig. P8.4.3), since in each case *OQ* + *QA* would be constant, essentially equal to the radius *OP*. This constancy is in fact just the signature property you would have used to construct an ellipse, by tracing the locus of a point whose sum of distance from two points is a constant.

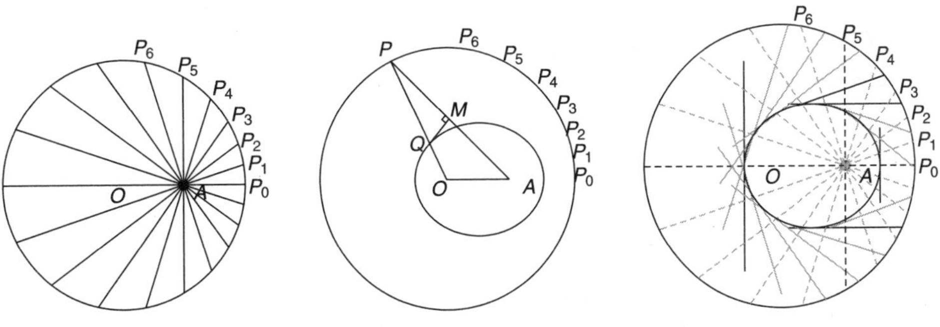

Fig. P8.4.1 **Fig. P8.4.2** **Fig. P8.4.3**

Notice that except for the point *Q* which is on the ellipse, every *other* point *Q′* on the perpendicular bisector will make (*OQ′* + *Q′A*) > (*OQ* + *QA*). We conclude therefore that the perpendicular bisector is tangent to the ellipse. It thus follows that these bisectors are along the velocity vector of the planet. ***Note that the bisector MQ is tangent to the enclosed ellipse at the point Q through which the radial line OP passes.***

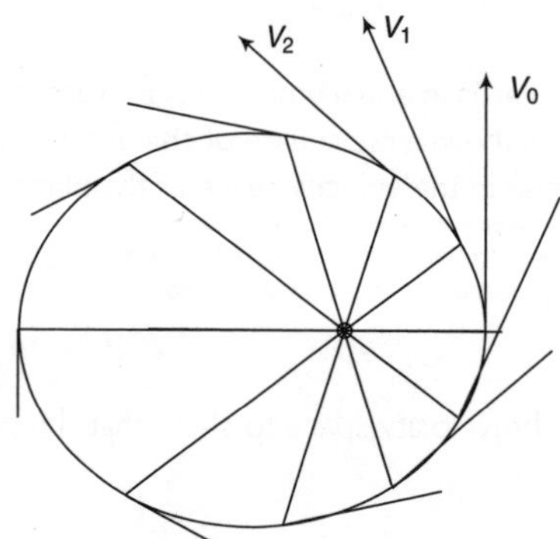

Fig. P8.4.4

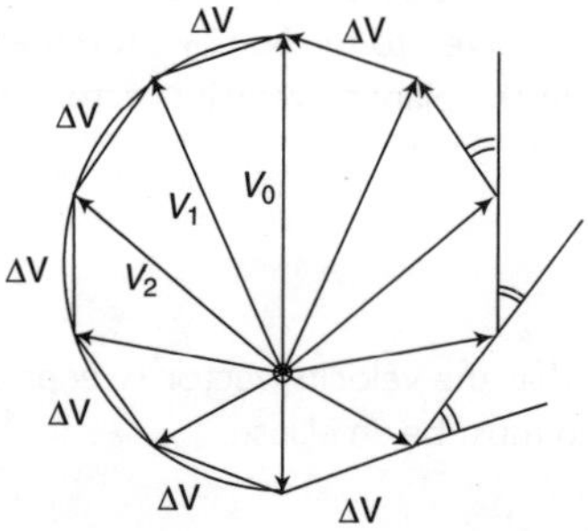

Fig. P8.4.5a Velocity space figure.

The velocity vectors corresponding to the different points $P_0, P_1, P_2, P_3, \ldots$ (Fig. P8.4.2) are of course different. These would be respectively $\vec{V}_0, \vec{V}_1, \vec{V}_2, \vec{V}_3, \ldots$ (Fig. P8.4.4) etc. Note that the speed of the planet would be highest at the point closest to the Sun (i.e., at the perigee of the orbit). Hence the corresponding velocity vector would have the largest length, representing its magnitude.

Let us now drag all the velocity vectors, in the velocity space, so that their tails are at a common point (Fig. P8.4.5a). We look at the velocity vectors at the start, and at the end, of a slice on the planet's orbit (Fig. P8.4.1). We look at such slices for which adjacent slices subtend equal angles at the Sun. The shape of the slices, for infinitesimal movements, would be a triangle. The area of the triangles is $\frac{1}{2}(\text{arc-length}) \times (\text{distance})$, which is equal to $\frac{1}{2}\rho d\varphi \times \rho = \frac{1}{2}\rho^2 d\varphi$. Essentially, we find that the area of the infinitesimal triangular slice is proportional to the square of the distance from the Sun: $(\Delta A)\ \alpha\ (\rho^2)$. We already had $(\Delta t)\ \alpha\ (\Delta A)$ (Kepler's second law).

Hence, $(\Delta t)\alpha\,(\rho^2)$.

If you now look at the ***difference*** in the speeds at neighboring points, $(\Delta V) = a(\Delta t)$, since the acceleration is $a = \lim_{\Delta t \to 0} \frac{\Delta V}{\Delta t}$. Furthermore, as per the assumption suggested by Halley, Hooke and Wren, if the force (hence the acceleration) is assumed to be given by the inverse square law, we shall have $(\Delta V) = a(\Delta t)\ \alpha \left(\frac{1}{\rho^2}\right)(\rho^2)$. Hence, distance-square in the numerator, and in the denominator, cancels. The ***change*** in velocity is therefore constant, independent of the distance, for adjacent slices.

In Fig. P8.4.5a, you see that all the sides of the polygons have the same length (ΔV) as per this conclusion. It is the assumption that the acceleration (i.e., the force) is proportional to the inverse square of the distance that has led us to the circle traced out by the tip of the velocity vectors in the velocity space. Now, a regular polygon (cyclic polygon) having all sides of equal length has an inscribed circle ('***incircle***'). This circle is tangent to every side at the midpoint; i.e., a regular polygon is a ***tangential polygon***. The tip of the velocity vectors therefore traces a circle (Fig. P8.4.5a) in the velocity space. The external angles of the velocity-polygon which inscribes the circle of course must all be equal to each other. Note that in this part of the solution, we have presumed that the planet sweeps equal area in equal time, but we have <u>not</u> presumed that the shape of the orbit is an ellipse, which is in fact what we must now to show. We know that Kepler's second law is essentially a statement of the constancy of the orbital angular momentum, which is a property of ***any*** radial force. This property follows from the radial symmetry, and is not based on the shape of the orbit to be an ellipse.

(b) We shall see now that the above fact helps us establish that the planet's orbit in the real space must be an ellipse, thereby answering the question raised by Edmond Halley, Robert Hooke, and Christopher Wren. Now, as the planet's position advances through an angle θ with respect to an axis through the position of the Sun, the tip of its velocity vectors (Fig. P8.4.5b) would sweep the same angle θ on the arc of the circle in the velocity space.

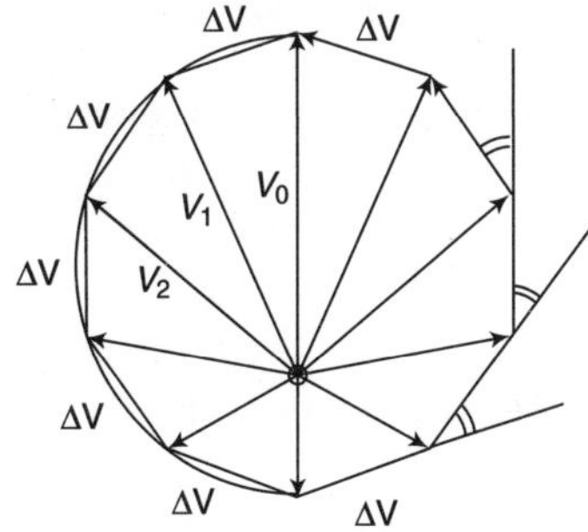

Fig. P8.4.5b Velocity space figure.

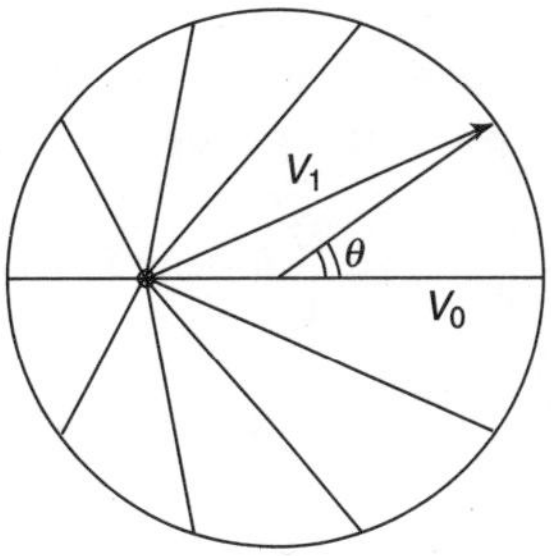

Fig. P8.4.6 Velocity space figure.

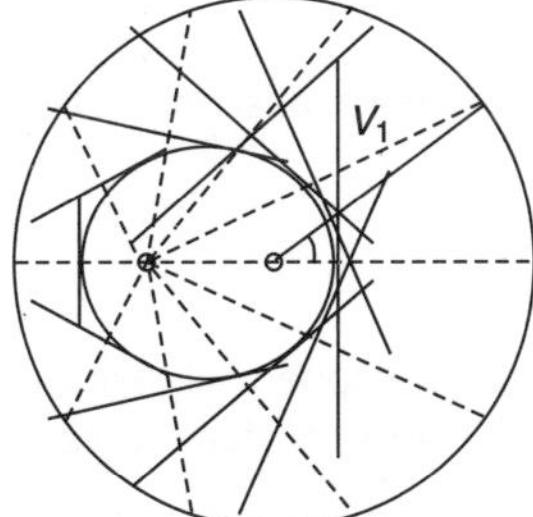

Fig. P8.4.7 Velocity space figure.

In the real space, the velocity vectors are along the tangent to whatever is the shape of the curve traced by the planet around the Sun. Therefore, we now ask what the shape of the planet's orbit must be like, if the tip of its tangent velocity vectors sweep out a circle in the velocity space. We are looking for such a curve that the velocity tangent to each point on it is in such a direction that the corresponding velocity vectors would trace a circle in the velocity space. In order to determine the shape of the planet's orbit, we now first rotate the whole circle circumscribing the polygon in Fig. P8.4.5b by 90°, resulting in Fig. P8.4.6. Fig. P8.4.5b is in fact the same as Fig. P8.4.5a, but it is deliberately repeated to place it just before Fig. P8.4.6. This makes it easy to see how the velocity VI (in Fig. P8.4.5a,b), rotated clockwise, appears in Fig. P8.4.6. Subsequently, we take each of the individual velocity vectors in Fig. P8.4.6 and rotate them ***anticlockwise*** about their ***respective*** mid-points through 900, as shown in Fig. P8.4.7. We see now, after this 90° rotation, that the elliptic orbit has emerged.

P8.5:

[Note: ***In the previous problem, we used the method explained in Feynman's reconstruction of Newton's geometrical method. In the present problem, you need to integrate the equation of motion to get the shape of the orbit.***] Using the Kepler–Newton equation of motion for a planet's orbit in the Sun's gravitational inverse-square force field, show that the velocity vector of the planet sweeps out a circle, and determine the radius of the circle.

Solution:

The gravitational force in the Kepler–Newton problem is $\vec{F} = m\frac{d\vec{v}}{dt} = -\frac{\kappa}{\rho^2}\breve{e}_\rho$.

From Eq. 2.11, we have $\breve{e}_\rho = -\frac{1}{\dot{\varphi}}\frac{d\breve{e}_\varphi}{dt}$, and hence: $\frac{d\vec{v}}{dt} = \frac{\kappa}{m\rho^2}\frac{1}{\dot{\varphi}}\frac{d\breve{e}_\varphi}{dt} = \frac{\kappa}{m\rho^2\dot{\varphi}}\frac{d\breve{e}_\varphi}{dt} = \frac{\kappa}{\ell}\frac{d\breve{e}_\varphi}{dt}$,

where $\vec{\ell}$ is the orbital angular momentum (Eq. 8.9). In an infinitesimal time interval, dt, therefore, the velocity of the planet changes by:

$\vec{dv} = \frac{\kappa}{\ell}d\breve{e}_\varphi = \frac{\kappa}{\ell}d(\breve{e}_\varphi) = \frac{\kappa}{\ell}d(-\sin\varphi\breve{e}_x + \cos\varphi\breve{e}_y)$; (Eq. 2.8a used).

and hence: $\vec{dv} = -\frac{\kappa}{\ell}\cos\varphi d\varphi\breve{e}_x - \frac{\kappa}{\ell}\sin\varphi d\varphi\breve{e}_y$; i.e., $dv_x = -\frac{\kappa}{\ell}\cos\varphi d\varphi$ and $dv_y = -\frac{\kappa}{\ell}\sin\varphi d\varphi$.

We now integrate the above differential increments in the components of the velocity vector, using the conditions $v_x = 0$ and $v_y = 0$ at $\varphi = 0$: $v_x = -\frac{\kappa}{\ell}\sin\varphi$ and $v_y = \frac{\kappa}{\ell}(\cos\varphi - 1) + v$.

Hence: $v_x^2 + \left(v_y - v + \frac{\kappa}{\ell}\right)^2 = \left(\frac{\kappa}{\ell}\right)^2$, which describes a circle of radius $\frac{\kappa}{\ell}$.

P8.6:

(a) Set up the Lagrangian for a particle in a spherically symmetrical potential and from the Lagrange's equations, obtain the differential equation of that would describe the particle's trajectory.

(b) A particle, moving in a central force field located at $\rho = 0$ describes a spiral $r = e^{-\varphi}$. Prove that the magnitude of force is inversely proportional to ρ^3.

Solution:

(a) Since the orbital angular momentum of a particle in spherically symmetric potential is conserved, we know that the motion of that particle must be confined to a plane that is orthogonal to the angular momentum. We thus employ plane polar coordinates in the plane of the orbit. The Lagrangian for the system is: $\mathcal{L} = T - V = \frac{1}{2}m(\dot{\rho}^2 + \rho^2\dot{\varphi}^2) - V(\rho)$.

The Lagrange's equation for the azimuthal angle: $\frac{d}{dt}\left(\frac{d\mathcal{L}}{d\dot{\varphi}}\right) - \frac{d\mathcal{L}}{d\varphi} = 0$. $\therefore \frac{d}{dt}(m\rho^2\dot{\varphi}) = 0$.

i.e., $\ell = m\rho^2\dot{\varphi}$, the orbital angular momentum, is a constant – just as we know.

Again Lagrange's equation for ρ coordinate is: $\frac{d}{dt}\left(\frac{d\mathcal{L}}{d\dot{\rho}}\right) - \frac{d\mathcal{L}}{d\rho} = 0$.

Hence: $\frac{d}{dt}(m\dot{\rho}) - m\rho\dot{\varphi}^2 + \frac{\partial V}{\partial \rho} = 0$. i.e., $m\ddot{\rho} - m\rho^2\dot{\varphi} = f(\rho)$. Hence: $m\ddot{\rho} - \frac{\ell^2}{m\rho^3} = f(\rho)$.

Now, $\dot{\rho} = \frac{d\rho}{dt} = \frac{d\rho}{d\varphi}\frac{d\varphi}{dt} = \frac{d\rho}{d\varphi}\dot{\varphi} = \frac{\ell}{m\rho^2}\frac{d\rho}{d\varphi}$.

Hence: $\ddot{\rho}=\frac{d}{dt}\left(\frac{\ell}{m\rho^2}\frac{d\rho}{d\varphi}\right)=\frac{\ell}{m\rho^2}\frac{d}{d\varphi}\left(\frac{d\rho}{d\varphi}\right)\frac{d\varphi}{dt}=\left(\frac{\ell}{m\rho^2}\right)^2\frac{d}{d\varphi}\left(\frac{d\rho}{d\varphi}\right)$.

A change of variable is now useful. Use $u=\frac{1}{\rho}$ instead of ρ. This gives $\frac{du}{d\varphi}=-\frac{1}{\rho^2}\frac{d\rho}{d\varphi}$.

Accordingly, $\ddot{\rho}=\left(\frac{\ell}{m\rho^2}\right)^2\frac{d}{d\varphi}\left(\frac{d\rho}{d\varphi}\right)=\frac{\ell^2u^4}{m^2}\left(\frac{1}{u^2}\right)\frac{d}{d\varphi}\left(\frac{du}{d\varphi}\right)=-\frac{\ell^2u^2}{m^2}\frac{d}{d\varphi}\left(\frac{du}{d\varphi}\right)$.

Hence $-\frac{\ell^2u^2}{m}\frac{d^2u}{d\varphi^2}-\frac{\ell^2u^3}{m}=f\left(\frac{1}{u}\right)$, i.e., $\frac{d^2u}{d\varphi^2}+u=-\frac{m}{\ell^2u^2}f\left(\frac{1}{u}\right)$.

(b) Since $\rho=e^{-\varphi}$, $u=\frac{1}{\rho}=e^{\varphi}$. The differential equation of the orbit, using $u=e^{\varphi}$ in the result of part (a): $e^{\varphi}+e^{\varphi}=-\frac{m}{\ell^2}e^{-2\varphi}f\left(\frac{1}{u}\right)$; i.e., $f\left(\frac{1}{u}\right)=-2\frac{\ell^2}{m}e^{3\varphi}$. Or: $f(\rho)=-\frac{2\ell^2}{m}\frac{1}{\rho^3}$.

Additional Problems

P8.7 A particle of mass m moves in a central repulsive force field that varies with the radial distance ρ from the force center as $f(\rho)=\frac{\kappa}{\rho^3}$, with $\kappa>0$. In the adjacent figure, the mass m approaches the center of the force at O from a great distance at a velocity v_0; and the impact parameter is p. Determine the closest distance of approach of the particle to the point O as the incident particle gets repelled away by the central field.

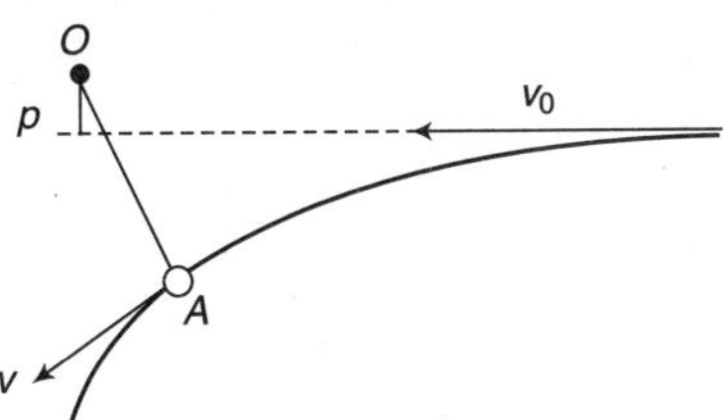

P8.8 A spacecraft, encircling the Earth in a circular orbit of radius r_c, was inserted into an elliptic orbit by firing a rocket. If the speed of the spacecraft was increased by 10% by a sudden blast of rocket motor, what is the equation of the new orbit? Determine the distance to the apogee.

P8.9 The interaction between two particles is described by a potential $V(r)=\frac{\kappa e^{-\alpha r}}{r}$ (called 'Yukawa potential'), where $\kappa>0$ and $\alpha>0$. Determine the force derivable from this potential. Obtain an expression for the energy and angular momentum of a particle of mass m that moves in circular orbits in the Yukawa potential.

P8.10 It is given that the potential in which a satellite is moving is $\frac{\kappa}{r}$. Show that when the satellite moves in an elliptic orbit, its speed v is given by $v^2 = \kappa\left(\frac{2}{r} - \frac{1}{a}\right)$; and when it is in a parabolic orbit, the velocity is expressed by $v^2 = \frac{2\kappa}{r}$.

P8.11 A comet is seen at a distance r_0 from the Sun. It is moving with a speed v_0 and its direction of motion makes an angle φ with the radius vector from the Sun. Determine the eccentricity of the comet's orbit.

P8.12 Show that the energy of a planet in an elliptic orbit can be written as $E = -\frac{GMm}{r_{max} + r_{min}}$, and the time period of the orbit is $T = \frac{2\pi GM}{\left(-\frac{2E}{m}\right)^{3/2}}$.

P8.13 At an apsidal distance a, a particle of mass m is injected, at a velocity $v = \frac{k}{a}$, into an orbit of an attractive central force, per unit mass, given by:

$$\frac{f}{m} = -\kappa^2[2(a^2 + b^2)u^5 - 3a^2b^2u^7].$$

Here, $u = \frac{1}{r}$, and a, b, κ are constants. Show that the orbit is described by:

$$\rho^2 = a^2 \cos^2 \varphi + b^2 \sin^2 \varphi.$$

P8.14 A steamer is cruising around a light house. Its velocity v relative to the water is always perpendicular to the line connecting it to the light house. The velocity of the water current in the sea water is $u < v$, in an arbitrary direction. Determine the orbit of the steamer.

P8.15 A particle is describing a parabola about a center of force which attracts according to the inverse square of the distance. If the speed of the particle is suddenly made one half of its previous value, without changing the direction of motion of the particle when the particle is at one end of the latus rectum, prove that the new path is an ellipse with eccentricity $e = \sqrt{\frac{5}{8}}$.

P8.16 The eccentricity of the Earth's orbit is $e = 0.0167$. Show that the time intervals spent by the Earth on the two sides of the minor axis are $\left(\frac{1}{2} \pm \frac{e}{\pi}\right)$ of year. Determine the difference in these time intervals.

P8.17 A planet moves on an orbit of a central force (per unit mass) $\frac{f}{m} = -\left(\frac{\mu}{r^2} + \frac{\lambda}{r^3}\right)$; it is given that λ is small. Show that the perturbation term $\frac{\lambda}{r^3}$ leads to a precession of the apsides. What is the condition for the stability of the planet on a circular orbit of radius a? Assuming that this condition is satisfied, determine the resulting orbit, and the angles of the apsides. [Note: Adding such a perturbation term is fruitful to understand the precession of planetary orbits which in fact occurs due to the relativistic effects ignored in Kepler–Newton mechanics. The relativistic correction is discussed in Chapter 14.]

P8.18 Determine the differential scattering cross-section and the total scattering cross-section for the scattering of a particle by a rigid elastic sphere.

P8.19 Find the differential scattering cross-section for the scattering of particles by the potential $V(r)$, where $V(r) = \alpha\left(\frac{1}{r} - \frac{1}{R}\right)$ for $r < R$ and $V(r) = 0$ for $r > R$.

P8.20 Determine the differential scattering cross section for α-particles by lead (Pb, Z = 82) nuclei, provided that the initial energy of a-particles is 11×10^{-13} Joule and the scattering angle is 30°. Also, determine the value of the impact parameter.

References

[1] https://faculty.history.wisc.edu/sommerville/351/351-182.htm. Accessed on 25 June 2017.

[2a] Ramasubramanian, K. M. D. S., M. D. Srinivas, and M. S. Sriram. 1994. 'Modification of the Earlier Indian Planetary Theory by the Kerala Astronomers (c. 1500 CE) and the Implied Heliocentric Picture of Planetary Motion.' *Current Science* 66(10): 784–790.

[2b] Srinivas, M. D. 2012. *Kerala School of Astronomy and Mathematics. Mathematics Newsletter.* 21(4) and 22(1): 118.

[3] 'Birth of a Masterpiece.' http://www.pbs.org/wgbh/nova/newton/principia.html. Accessed on 25 June 2017.

[4] Deshmukh, P. C., and Shyamala Venkataraman. 2011. 'Obtaining Conservation Principles from Laws of Nature – and the Other Way Around.' *Bulletin of the Indian Association of Physics Teachers* 3: 143–148.

[5] Deshmukh, P. C., Aarthi Ganesan, N. Shanthi, Blake Jones, James Nicholson, and Andrea Soddu. 2014. 'The "Accidental" Degeneracy of the Hydrogen Atom is No Accident. *Canadian Journal of Physics,* 93(3): 312–317. doi *10.1139/cjp-2014-0300.*

[6] Deshmukh, P. C., and J. Libby.

(a) 2010. 'Symmetry Principles and Conservation Laws in Atomic and Subatomic Physics-1.' *Resonance* 15: 832.

(b) 2010. 'Symmetry Principles and Conservation Laws in Atomic and Subatomic Physics-2.' *Resonance* 15: 926.

[7] Deshmukh, P. C., Kaushal Jaikumar Pillay, Thokala Soloman Raju, Sudipta Dutta, and Tanima Banerjee. 2017. 'GTR Component of Planetary Precession.' *Resonance.* 22(6). 577–596.

[8] Williams, David R. 2018. *Planetary Fact Sheet-Metric.* https://nssdc.gsfc.nasa.gov/planetary/factsheet/. Accessed on 4 July 2017.

[9] Fraternali, Filippo. Università di Bologna. https://www.unibo.it/sitoweb/filippo.fraternali/en. Accessed on 12 July 2017.

[10] Ladera, Celso L., and Eduardo Alomá y Pilar León. 2010. 'The Virial Theorem and its Applications in the Teaching of Modern Physics.' *Lat. Am. J. Phys. Educ.* 4(2): 260–266.

[11] Zioutas, Konstantin, Dieter H. H. Hoffmann, Konrad Dennerl, and Thomas Papaevangelou. 2004. 'What is Dark Matter Made Of' *Science* 306(5701): 1485–1488.

CHAPTER 9

Complex Behavior of Simple System

I am convinced that chaos research will bring about a revolution in natural sciences similar to that produced by quantum mechanics.

—*Gerd Binnig*[*]

9.1 LEARNING FROM NUMBERS

The equations of motion of classical mechanics, whether Newton's, Lagrange's, or Hamilton's, have us believe that given the state of the system at a particular time, one can always predict what its state would be any time later, or, for that matter, what it was any time earlier. This is because of the fact that the equations of motion are symmetric with respect to time-reversal: $(t)\rightarrow(-t)$. Classical mechanics relies on the assumption that position q and momentum p of a mechanical system are simultaneously knowable. Together, the pair (q, p) provides a signature of the *state of the system*. Their time-dependence, i.e., $(\dot{q}, \dot{p})$, provided by the equations of motion, accurately describes their temporal evolution. For macroscopic objects, this is an excellent approximation, and the classical laws of mechanics are stringently deterministic. This is stringently correct, but there is an important caveat, expressed succinctly by Stephen Hawking: "*Our ability to predict the future is severely limited by the complexity of the equations, and the fact that they often have a property called chaos... a tiny disturbance in one place, can cause a major change in another.*" The difficulty Hawking alludes to has nothing to do with the quantum principle of uncertainty, but to a challenge *within* the framework of the fully deterministic *classical* theory. The solution to the temporal evolution may become *chaotic*, *even* as they remain deterministic, due to *extreme sensitivity to the initial conditions* that are necessary to obtain the solution to the equation of motion. Careful admission of the previous remark would prepare you to embark your journey on the exciting field of chaos. Along the way, you will also meet objects having weird fractional dimensions. The field covered by chaos theory is vast, though relatively young. It is a very rich field and can be introduced from a variety of perspectives.

[*] Inventor of the Scanning Tunneling Microscope; winner of 1986 Nobel Prize.

The general field of chaos theory is often regarded as a study of a 'dynamical system' which is just about any quantity which changes, and one has reasons to track these changes and the sequence of values it may take, for example, over a time interval. The change in the state of the physical system may not merely be in response to a change in time, but on account of the system's sensitivity to any physical property on which the system depends. The range of applications of the study of dynamical systems include the study of weather patterns, discord in traffic, behavior of market economy, operations of electromagnetic oscillators and other electronic devices, including computers. Other applications include fields as diverse as manufacturing scheduling, communication systems, social behavior, medical diagnostics, and cell multiplication, to name just a few. The study of chaotic dynamics has opened up a mind-boggling range of applications in sciences, engineering, medical disciplines, and also in financial analysis, population dynamics, etc. Study of fractals in engineering, biology, medicine, financial markets, internet traffic, cell multiplication, is not just captivating, it is necessitated by the changing needs of the society. Chaos theory has emerged as a topic of central concern in all physical, chemical, mathematical and biological sciences with beneficial applications in all branches of human endeavor.

Despite the fact that the field of chaos theory is relatively young, its range of applications is so vast that it is necessary to pierce into its very *method*, and find its place in the larger domain of the practices of scientific inquiries into the physical laws. Traditionally, a physicist's inquiry into the laws of nature rides on one of the following two methods (and their combinations):

(a) <u>Observations and analysis:</u>

Some of the classic examples of this method are:

(i) Galileo's discovery of the periodic oscillations of the pendulum on watching the swing of the chandelier at a cathedral in Pisa,

(ii) his discovery of the law of inertia by performing various experiments, described in Chapter 1,

(iii) Rydberg's discovery of a formula that provides the regularities in the frequencies in the spectrum of the hydrogen atom,

(iv) Raman's formulation of molecular scattering theory based on detailed observations of light scattered by molecules.

(b) <u>Modeling the regularities in natural phenomena:</u>

This is best represented by:

(i) intuitive development by Isaac Newton of the principle of causality and determinism (Newton's second law), and his discovery of the conservation of momentum, implied by Newton's third law,

(ii) formulation of the principle of variation which completely accurately describes the evolution of the state of a classical system, *without* any reference to Newtonian cause-effect proportionality,

(iii) the quantum theory, to which stalwarts like Louis de Broglie, Erwin Schrodinger, Albert Einstein, S. N. Bose, Werner Heisenberg, Niels Bohr, Paul Dirac, Wolfganag Pauli, John von Neumann, Richard Feynman, etc., made pioneering contributions,

(iv) Maxwell's development of electrodynamics (Chapter 12), and Einstein's theory of relativity (Chapters 13 and 14), also fall in this category.

A third method, perhaps unexpected in the domain of investigations of physical laws of nature, stems from a detailed study of arithmetic, geometry and algebra, or more generally, study of *mathematics*, even as this method also remains firmly embedded in the study of natural sciences. The amazement about the role of mathematics in studying natural phenomena sparked by Wigner's remark, placed at the top of Chapter 2, is only further ignited by the study of deterministic chaos. This field is relatively young, gathering momentum since the works of Henry Poincare in response to a mathematics competition held to celebrate (in 1889) the 60th birthday of King Oscar II of Sweden (and Norway). The challenging question for which Poincare won the prize—a gold medal and Swedish Kroners 2,500—has come to be known as the *restricted 3-body* problem. It would address central scenarios such as the stability of the *Sun–Earth–Moon* system.

While the mathematical genesis of the theory of chaos is not so old, scientists had already developed skills much earlier to learn about the laws of nature from *numbers* (or from *sequence*, and from *arrangements* of numbers). Examples of this method are-

(i) π, originally defined as the ratio of the circumference to the diameter of the circle, now has equivalent and alternative definitions, including based on number theory, and also in the theory of randomness (statistics). It appears in a large number of formulae in physics, including the time period of a simple harmonic oscillator, Heisenberg's principle of uncertainty, and even Einstein's field equation of the general theory of relativity (Chapter 14).

(ii) e, the Euler's number is the base of the natural logarithms, invented by John Napier, and hence sometimes also called Napier's number. This appears in the expressions for radioactive decay, population dynamics, rate factors in charging and discharging of capacitors, flow-rates of water from an orifice, etc.

(iii) The Varahamihira's triangle (also known as Pascal's triangle), discussed previously in Chapter 2.

(iv) The Fibonacci (Leonardo of Pisa—*Liber Abaci*, 1202) sequence of numbers. This sequence of numbers was in fact known [1] to Indian mathematicians Gopala, around 1135, and to Hemachandra c. 1150, from their study of rhythmic beats, before Fibonacci. The Fibonacci sequence is best understood by counting the number of pairs of rabbits as the population of rabbits increases in discrete and uniform steps as time progresses. We shall consider a specific *rabbit-reproduction model* to estimate this population. In this model, we begin at the zeroth month with a new born pair of rabbits, a male and a female.

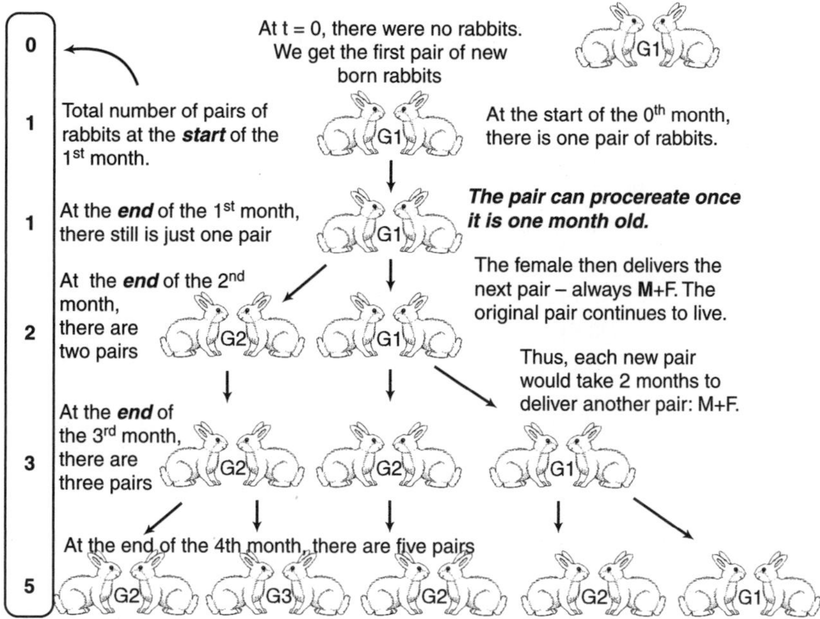

Fig. 9.1 This figure shows the number of rabbits at the start of the first month, and then at the end of the first, second, third, and the fourth month if their population grew as per the hypothetical model discussed in the text.

G1 is the first generation of the pair of rabbits. Any, and every, pair born to G1 is labelled G2. Any, and every, pair born to G2 is labelled G3, and so on. The number of the pairs of rabbits grows every month, and is given by the Fibonacci (infinite) sequence:

$F_S = \{0, 1, 1, 2, 3, 5, 8, 13, 21, 34,55,89,144,233, ...\}$

The pair would take a month to be able to procreate, and another month after that to deliver the next generation, which would also be essentially a male and a female. This pattern would continue with exactly the same evolution orderliness at each generation. Fig. 9.1 shows the number of pairs of rabbits for the first few generations. One can see from this figure that the number of pairs of rabbits grows in the sequence $0, 1, 1, 2, 3, 5, 8, 13, 21, 34, 55, 89, 144, 233, ...$ It should be clear from this sequence that it has an interesting property that subsequent numbers in it can be built as *the sum of the two previous numbers*. The sequence continues, and has infinite number of terms.

(v) The *golden ratio*, which is the limiting value of the ratio of consecutive numbers in the Fibonacci sequence. We see that the ratio of successive terms F in the Fibonacci sequence is given by: $\frac{1}{1} = 1$, $\frac{2}{1} = 2$, $\frac{3}{2} = 1.5$, $\frac{5}{3} = 1.666...$, $\frac{8}{5} = 1.6$, $\frac{13}{8} = 1.625$, $\frac{21}{13} = 1.61538...$, $\frac{34}{21} = 1.61904...$, $\frac{55}{34} = 1.617646...$, ... , and the limiting value is

$$\lim_{n \to large} \frac{F_{n+1}}{F_n} = 1.6180339887... = \phi \text{, the golden ratio.} \tag{9.1}$$

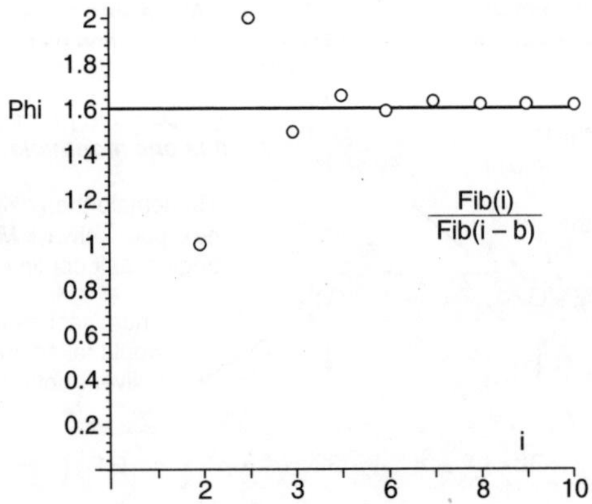

Fig. 9.2a The golden ratio is the limiting value of the ratio of successive Fibonacci numbers.

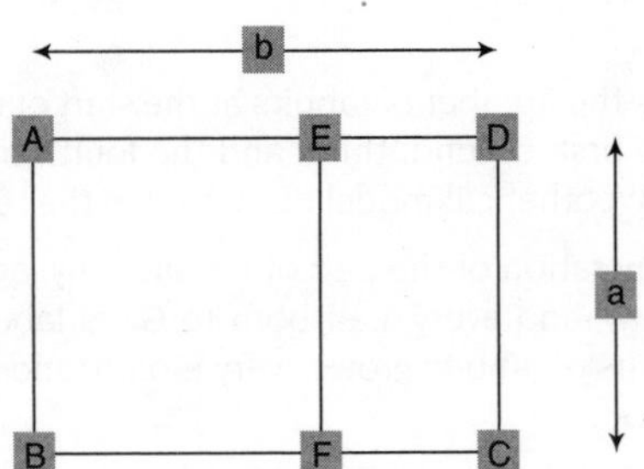

Fig. 9.2b A rectangle that has lengths of its sides in the golden ratio is called the golden rectangle.

Fig. 9.2a shows how the ratio of the successive terms in the Fibonacci sequence settles down to the golden ration. A rectangle ABCD (Fig. 9.2b) in which the sides of the rectangle are in this ratio is called the golden rectangle. If we drop a line EF parallel to DC such that AE = AB, then ABFE is a perfect square. The rectangles ABCD and FCDE are similar, which means that their sides have essentially the same ratio:

$$\frac{AD}{AB} = \frac{EF}{ED} = \phi. \tag{9.2a}$$

Then, we see that

$$\phi = \frac{b}{a} = \frac{a}{ED} = \frac{a}{b-a} = \frac{a}{a\left(\frac{b}{a} - 1\right)} = \frac{1}{(\phi - 1)}. \tag{9.2b}$$

Equation 9.2b is essentially a quadratic equation in ϕ:

$$\phi^2 - \phi - 1 = 0. \tag{9.2c}$$

Its positive solution is, $\phi = \dfrac{1+\sqrt{5}}{2} = 1.6180339877...$ (9.2d)

$$\phi^2 - \phi - 1 = 0.$$

One can also express the golden ratio as a continued fraction, since

$$\phi^2 = \phi + 1 = \phi\left(1 + \frac{1}{\phi}\right), \tag{9.3}$$

which immediately gives

$$\phi = 1 + \frac{1}{\phi} = 1 + \cfrac{1}{1 + \cfrac{1}{\phi}} = 1 + \cfrac{1}{1 + \cfrac{1}{1 + \cfrac{1}{\phi}}} = 1 + \cfrac{1}{1 + \cfrac{1}{1 + \cfrac{1}{1 + \cfrac{1}{\phi}}}} = \ldots \tag{9.4}$$

On observing the continuation of the endless fraction above, this continued fraction, one cannot escape the sense of seeing more of the same (*deja vu*), which is a common property of what are called fractals, discussed further in Section 9.4.

The Fibonacci sequence enables us to generate a spiral curve, called the *Fibonacci spiral* (Fig. 9.3a), which closely resembles a large number of curves found in nature (Fig. 9.3b). The construction of the Fibonacci spiral is quite simple. You begin with a square whose diagonal is AB as shown in Fig. 9.3a. On the immediate right side of this square, you place another square of the same size, whose diagonal would be BC, as shown. Now, join the points A, B, C by a smooth curve, which would go through the points D, E, F, and G (as shown), such that D, E, F, and G are the diagonally opposite corners of successive squares, whose sides increase in proportions (as shown in Fig. 9.3a) corresponding to the Fibonacci sequence. Thus, the squares with the diagonals AB, BC, CD, DE, EF, FG, ... have sides exactly in proportion to the Fibonacci sequence, and are thus respectively given by: 1, 1, 2, 3, 5, 8, 13, ... The resemblance of the Fibonacci spiral to curves occurring in nature is reason strong enough to study the Fibonacci sequence. One can only be awestruck to discover how these numbers give us insights into innate properties of nature.

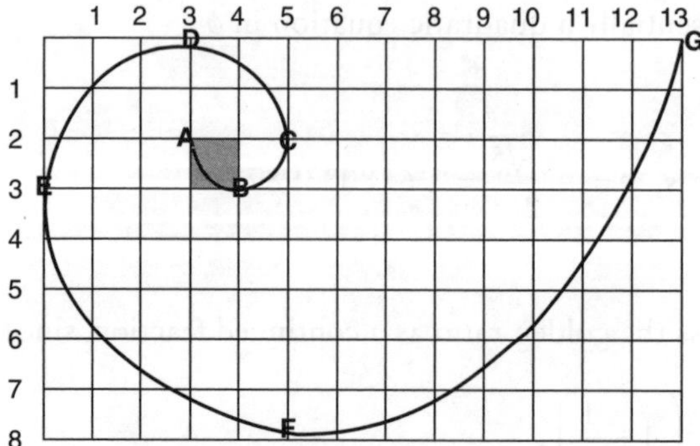

Fig. 9.3a Fibonacci spiral. It is drawn from the corner A of a square, joining it to the diagonally opposite point B and then to the next diagonally opposite point C, and then to D, etc.

Fig. 9.3b Fibonacci spirals in nature. Several patterns in nature show formations as in the Fibonacci spiral. The internet is rich with images and videos of such patterns.

9.2 CHAOS IN DYNAMICAL SYSTEMS

We have seen above amazing connections between the dance of numbers and physical properties in nature. A classic consequence of this situation is displayed in Fig. 9.4. This sketch illustrates two possible scenarios. They show possible evolution of a dynamical system from two alternative initial states which are *extremely* close to each other. In Fig. 9.4a, we see that the state of the system after long-enough time is not much different, no matter which of the two initial states it evolved from.

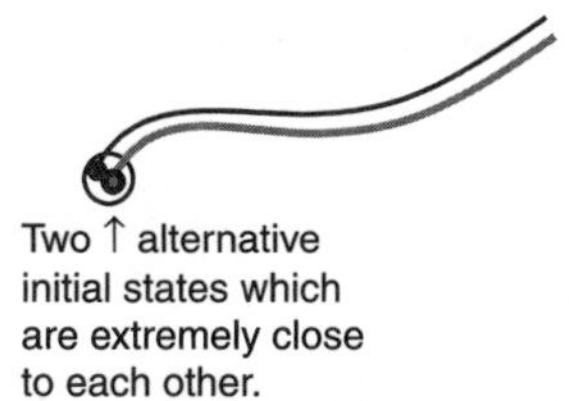

Fig. 9.4a Similar evolution trajectories with initial conditions slightly different from each other.

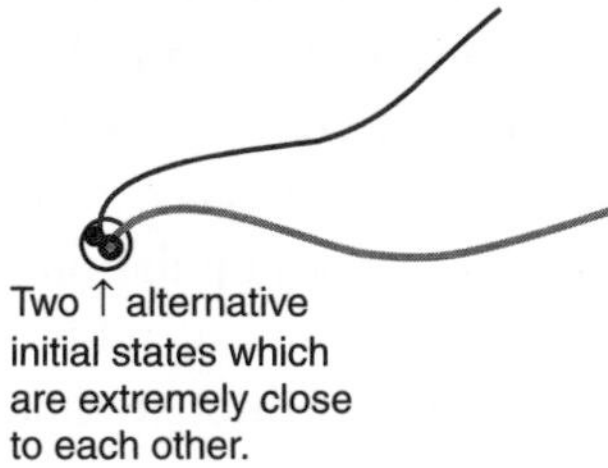

Fig. 9.4b Diverging evolution trajectories with initial conditions slightly different from each other.

However, Fig. 9.4b shows that even if the initial states are very close to each other, the states after long-enough time can be quite different from each other. The trajectories shown above can be in the *phase* space, or for that matter, in any other *parameter* space which may be of relevance in order to describe the system.

For example, we may be interested in (a) how many radioactive atoms would remain after a certain passage of time, since many of them would decay after a certain time, depending on the decay constant. We may also be interested in (b) the amount of charge that would remain on a capacitor after a certain time, if it is getting discharged at a certain rate. There are many such situations in nature that are of interest, including the (c) size of a population after a certain time, if this size is increasing (or decreasing) at a certain rate. In each of these situations, the *rate of change* $\left(\frac{dN}{dt}\right)$ of the physical quantity at any instant of time is directly proportional to the size $N(t)$ of the species *at that* instant. Such a relationship was used by Thomas R. Malthus (in 1798) to quantify the rate at which population of a biological species would grow, or decay. The rate of change of the population in the Malthus model is given by

$$\frac{dN}{dt} = rN, \tag{9.5}$$

where the coefficient r is *control parameter* which determines the growth (or decay) rate. Equation 9.5, however, does not account for any restraining elements, such as food-shortage, diseases, or the species being killed by predators. Pierre Verhulst (Belgian, 1838) introduced a modification of the Malthus model to account for any debilitating factor which would impede the growth rate by introducing a factor 'K' called the 'carrying capacity'.

The relationship that provides the rate equation for the rate of change of population that includes Verhulst's modification is

$$\frac{dN}{dt} = \left[rN\left(1 - \frac{N}{K}\right) \right]. \tag{9.6}$$

Equation 9.6 is non-linear. It is called the *logistic equation*. We observe that $\dot{N} = 0$ when $N = 0$, or when $N = K$. These two values of N are the equilibrium values of the population. The population would be stable, unchanged, once the population reaches either of these two values. The population will become zero in the case of decay, and it will stabilize at the value K in the case of growth. Now, Malthus and Verhulst models, regardless of their differences, assumed that the population is a continuous function of the time. Hence we could employ the derivative of the population with respect to time in both Eqs. 9.5 and 9.6. The requirement of continuous dependence of the population on time makes the mathematical models inapplicable to species such as the pairs of rabbits that we considered in Fig. 9.1, because the population of rabbits changed at discrete time intervals, and not continuously. We cannot define the derivative with respect to time in such cases. In fact, many physical properties, including the population of some biological species in the $(n + 1)^{\text{th}}$ generation, depend on the size of that physical property after n iterations, but the dependence may be *discrete*, rather than *continuous*. The Verhulst logistic equation must therefore be modified. Thus, we may replace Eq. 9.6 by

$$\frac{N((n+1)\delta t) - N(n\delta t)}{\delta t} = rN(n\delta t)\left[1 - \frac{N(n\delta t)}{K}\right]. \tag{9.7}$$

Equation 9.7 is similar to Eq. 9.6; it includes the debilitating factor K just as in Eq. 9.6. It is also non-linear, but the crucial difference is that it is a discrete equation in which the population changes discretely, whereas in Eq. 9.6 it changes continuously.

The rate at which populations of some species change across different generations provides an excellent (and a rather popular) example of the development of chaos. Let us consider the population of lions in a certain jungle, and preys on gazelles. Now, the population x_n of the lions in the nth generation depends on what it was in the previous generation, so we may expect a relation between them to be expressible as $x_{n+1} = rx_n$, where r is a fecundity (fertility) coefficient. You will note that now the population is changing discretely, unlike the Malthus model (Eq. 9.5). However, the logistic factor incorporated in the Verhulst model must be incorporated now. As lions prey on the gazelles, the population of gazelles would diminish, and the population of the lions would no longer grow at the same rate. The growth rate would *diminish* somewhat. The simplest modification that would incorporate the corresponding *reduction* of the fecundity coefficient can be written as

$$x_{n+1} = rn_x(1 - x_n). \tag{9.8}$$

We shall use Eq. 9.8 to represent the family of the discretized dependencies of the kind expressed in Eq. 9.7. This equation can be generalized to represent changes in the size of any species; it can be, for example, the concentration of a chemical in a reactive mixture.

It declares the fact that the value of a certain physical property x at the $(n + 1)^{th}$ iteration depends on the value of that property at the nth iteration. The value of x changes from one generation to the next *discretely*. The change is influenced by a fecundity coefficient, modulated by a debilitating factor. Equation 9.8 is called the *logistic map*. More generally, the relationship between the value of a physical parameter at some iteration and at the next one is expressed as

$$x_{n+1} = f(x_n, r). \tag{9.9}$$

> "I urge that people be introduced to the logistic equation early in their mathematics equation."
> — Robert M. May, 'Simple Mathematical Models with very Complicated Dynamics'
> *Nature* 261 (1976): 459-467

In order to predict the values of physical quantities asymptotically (i.e., after a long time, or after several iterations of the generation index n in Eq. 9.8 and 9.9), it is of great importance to understand the difference between the rate equation (Eq. 9.6) and the discrete difference equation (Eq. 9.8, 9.9). This difference would enable us account for the vast difference in a system's asymptotic evolution suggested in the sketch in Fig. 9.4b. It would account for the fact that even a small difference in the initial condition can produce a huge difference eventually in the state of the system, causing the situation to be chaotic, even if predictable. This is famously called the *butterfly effect* which alludes to a flutter of a butterfly in one part of the world, which only mildly changes the local atmospheric conditions when the butterfly flutters. The flutter may be a very minor event, but it may trigger a cascade of disturbances in the arrangements of molecules around the butterfly. The sequence of atmospheric effects triggered by the butterfly's flutter could amplify to such a magnitude that it would result in a storm in another part of the world. It is like a real mountain coming out of molehill.

Fig. 9.5 shows how a small difference in the value of the coefficient r in the deterministic logistic equation (Eq. 9.8) beginning with $x_1 = 0.02$, can make to the result x_{n+1} after just a few iterations on the number n. The three curves in Fig. 9.5 correspond to three *nearby* values of the coefficient r: $r_1 = 3.69$(dashed), $r_2 = 3.70$(continuous) and $r_2 = 3.71$(dotted). Even though the three curves start out together, they separate out after a few iterations on n. The result becomes *chaotic* even if the r values employed in the three curves are *not much different* from each other. It is this phenomenology, akin to making a huge mountain out of a small molehill, that chaos is about. The vast difference in the three curves (Fig. 9.5) after about 15 iterations, regardless of a rather small change in the value of r at the start, justifies the term *butterfly effect* mentioned above. Note that the result is not *random*, since it is totally deterministic. It is accurately *predictable* from the nature of the iterative equation. It is important to keep this difference in mind: the predictions of the logistic equation can be *chaotic*, but they are not random.

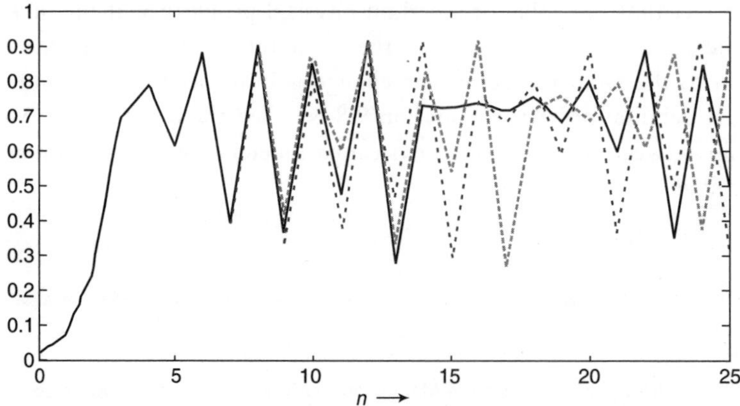

Fig. 9.5 In this figure, $x_{n+1} = rx_n(1-x_n)$ is plotted against n, beginning with $x_1 = 0.02$ for three different values of r:
$r_1 = 3.69$ (dashed),
$r_2 = 3.70$ (continuous) and
$r_2 = 3.71$ (dotted)

The difference (of 0.01) in the value of the coefficient r in the three curves shown in Fig. 9.5 is very small. Chaotic behavior was historically seen first for even a *hundred times smaller* a difference, in 1961, by Edward Norton Lorenz. He had studied mathematics at Harvard, and was working as a meteorologist with the U.S. Army. Lorenz was predicting the value of a physical property of the atmosphere using a deterministic equation. The equation contained a certain physical parameter, which we refer to as ρ. He had done this calculation earlier, and was only repeating it. However, this time around, he did so by initiating the calculation with a value $\rho = 0.506$ instead of $\rho = 0.506127$ which was used earlier. The difference 0.000127 is far smaller compared to the difference between the r values for the three curves in Fig. 9.5. Lorenz had thought that the tiny difference would not matter. He expected the difference to get ironed out in the calculation. Instead, the result of the calculation *diverged* chaotically (Fig. 9.6). It was vastly different. Lorenz went on to become one of the pioneering contributors to the theory of chaos.

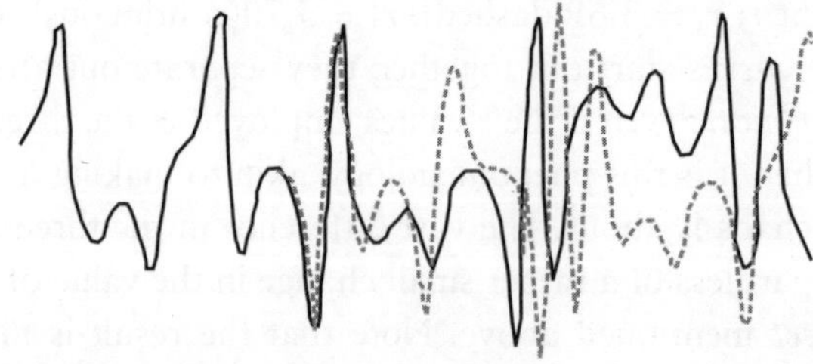

Fig. 9.6 Data showing Lorenz's original simulation results overlaid with the attempt to reproduce the *same* result. At first, the traces are similar, but after a few cycles they begin to differ and they are soon completely unrelated. Reference: https://fractalfoundation.org/OFC/OFC-6-5.html (accessed on 26 October 2018).

Clearly, we need to understand the butterfly effect in Fig. 9.5 and 9.6. To develop some insight into the chaotic behavior, we therefore consider a simple 1-dimensional *iterative, logistic map*, defined by Eq. 9.8. The sequence of values $\{x_{n+1}\}$, obtained by iterative applications of Eq. 9.8, by progressively increasing the iteration index, $\{n = 0, 1, 2, 3, ...\}$, provides the logistic map. The integer n thus keeps track of the number of iteration index, which is the 'generation number' index in the context of discrete population changes. We shall consider a particularly simple model of population dynamics in which the population of a certain species is determined by its size in the previous generation, through a linear relation,

$$y_{n+1} = ry_n. \tag{9.10a}$$

One must remember that the changes in the population take place *discretely*; as the generation index n increases through integer steps. Real numbers in-between the integers are of *no* relevance. y_n versus n represents discrete mapping; it is not an analytical function. We incorporate a restraining factor on the population, discussed above, by modifying Eq. 9.10a, and replace it by the relation

$$y_{n+1} = ry_n - ty_n^2. \tag{9.10b}$$

By scaling the parameter y_n by the factor $\frac{t}{r}$, we rewrite the population dynamics in terms of the parameter $x_n = \frac{t}{r}y_n$, and get the *difference equation* (Eq. 9.8), viz., the *logistic map* equation. One begins with a seed value x_o which belongs to the domain $x_o \epsilon [0, 1]$ and with a value of the control parameter r which belongs to the interval $r \epsilon [0, 4]$. The control factor in the logistic Eq. 9.8 plays an extremely crucial role in chaos theory. It impacts the result of the *map* in a dramatic manner. For small differences in the value of the control factor r, the result of iterations on Eq. 9.8 can be vastly different. We illustrate the evolution of the map given by Eq. 9.8 with few more examples of different values of the control parameter r. However, we shall always employ the same start-value $x_1 = 0.02$ (in arbitrary units. The Fig. 9.7a shows the results of the logistic map (Eq. 9.8) for $r = 1.5$ and 2.9, the Fig. 9.7b for $r = 3.1$ and 3.4, the Fig. 9.7c for $r = 3.45$ and 3.48 and the Fig. 9.7d shows the results of the logistic map for $r = 3.59$ and 3.70.

Fig. 9.7a shows that the logistic map evolves to a *steady state* solution for both the values of the control parameter, $r = 1.5$ and 2.9.

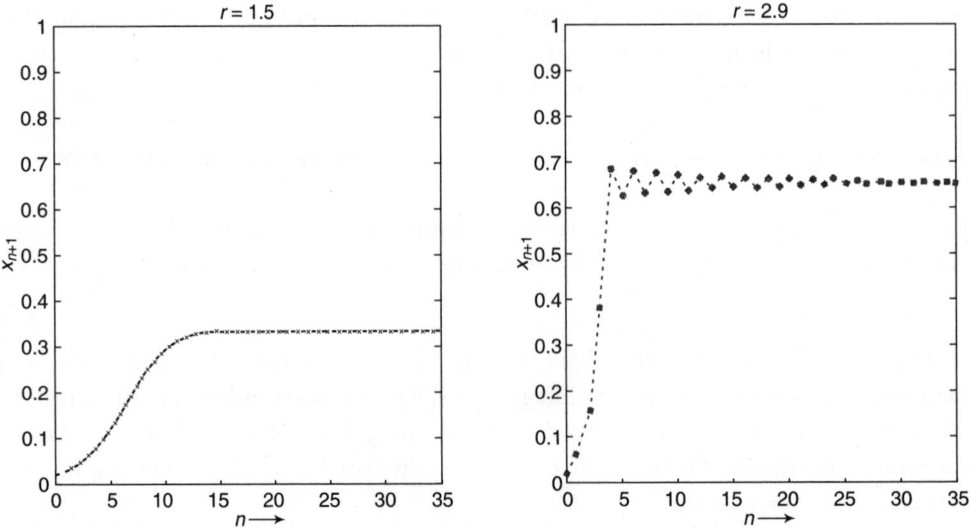

Fig. 9.7a The logistic map $x_{n+1} = rx_n (1 - x_n)$ for $r = 1.5$ and 2.9, starting with $x_1 = 0.02$.

However, as we go from $r = 2.9$ (second panel in Fig. 9.7a) to $r = 3.1$ (first panel in Fig. 9.7b), we no longer have a steady state. Instead, the results of successive iterations *alternate* between *two* values, labeled '1' and '2'. This pattern is called a *period two* oscillation. The values labeled as '1' and '2' repeat, all the way, but only the first three pairs are labeled for $r = 3.1$. For $r = 3.4$ also, the period 2 oscillations are seen. The first and the smallest, value of the control parameter for which bifurcation occurs is $r = 3$. For subsequent values of r, for example, for $r = 3.4$ shown in the second panel of Fig. 9.7b we again have period 2 oscillation.

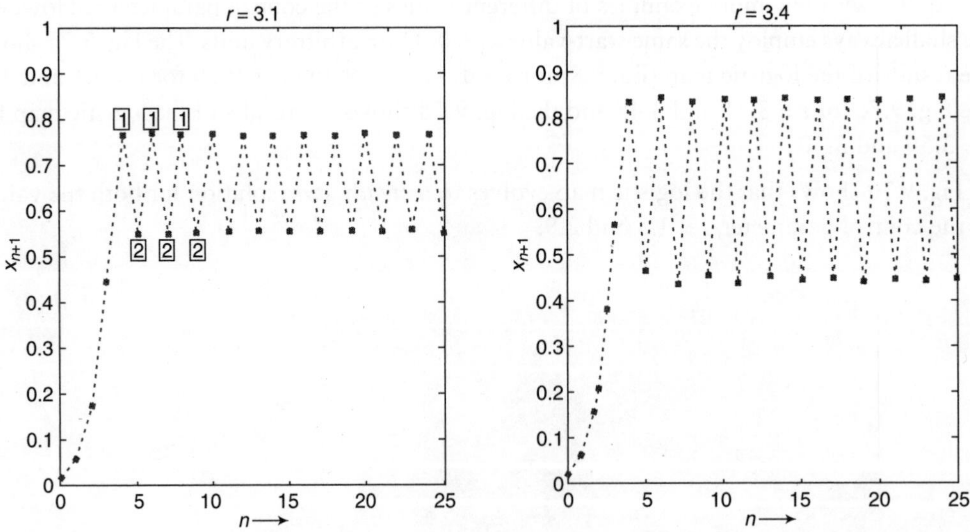

Fig. 9.7b The logistic map $x_{n+1} = rx_n (1 - x_n)$ for $r = 3.1$ and 3.4.

Bifurcation continues for higher values of r as we increment it through tiny numbers, till each branch bifurcates *yet again* at r = 3.449489... We show the bifurcation of each branch for two values of r in Fig. 9.7c.

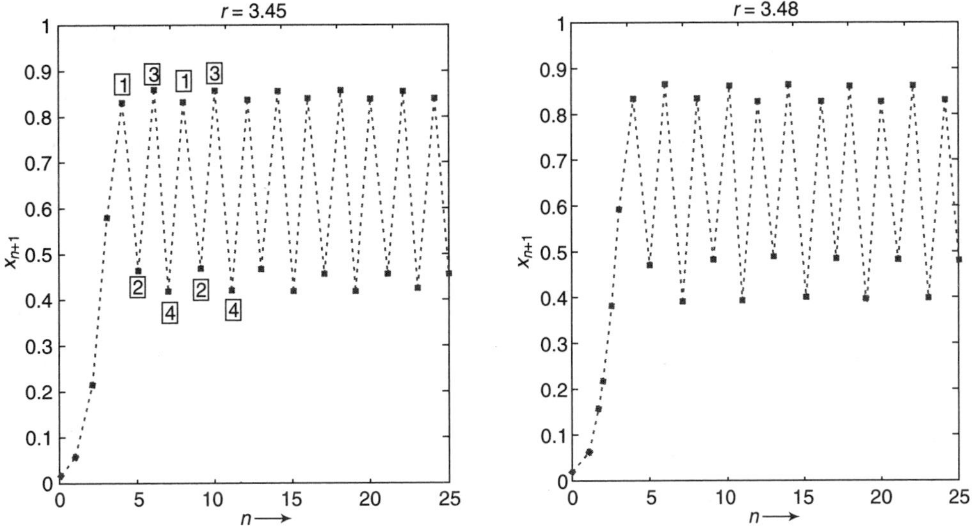

Fig. 9.7c The logistic map $x_{n+1} = rx_n(1 - x_n)$ for r = 3.45 and 3.48.

As we let the value of the control parameter go across $r = 1 + \sqrt{6} = 3.449489\ldots$, Fig. 9.7c shows *period four* oscillations. Results for r = 3.45 and r = 3.48 are shown in this figure. Successive results of the iterations on the map recur in *sets* of *four different values*, labeled as '1', '2', '3', '4'. Only the first two sets of four values in the case of r = 3.45 are labeled, but the period four oscillations continue for subsequent generations. For rather small changes in the value of the control parameter r, the nature of the solutions changes from the steady state (Fig. 9.7a), to period two oscillations (Fig. 9.7b), to period four oscillations (Fig. 9.7c). The bifurcation of each branch continues. Except for the value r = 3 for which bifurcation occurs first, subsequent values of r at which bifurcations take place are irrational, transcendental numbers. As the control parameter goes past r = 3.569946..., we no longer can describe the results in terms of 'period doubling' or 'bifurcation'. The result is chaotic, even if fully predictable, deterministic. Fig. 9.7d illustrates this for r = 3.59 and 3.70.

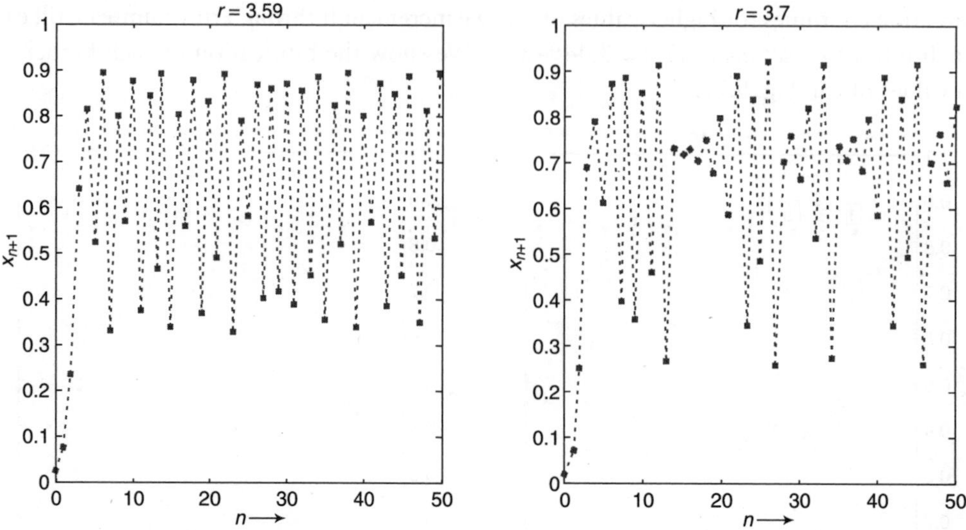

Fig. 9.7d The logistic map $x_{n+1} = rx_n(1 - x_n)$ for $r = 3.59$ and 3.70. We always get exactly the same value with a given start-value for x_1 after n number of iterations. The results after n iterations are therefore not random, but they wander wildly. *Chaos* is this *wild wandering* of the orbit generated by the asymptotic value of the physical parameter, x_n.

Fig. 9.8 shows the asymptotic values of the logistic map after four hundred iterations of the map (Eq. 9.5b) for a *range* of values of r, from $r = 2.8$ to $r = 4$, incremented in steps of 0.01. Note that Fig. 9.5 and Fig. 9.7 showed x_{n+1} versus n (i.e., population changes across the different successive generations). Each of the Fig. 9.5 through 9.7 corresponds to a selected value of the control parameter r. On the other hand, Fig. 9.8 shows x_{n+1} after a large number of iterations for a range of *different* values of the control parameter. Fig. 9.8 therefore shows the *asymptotic values* of the population x_n obtained from the logistic map. The asymptotic value(s) of the map is called an *attractor*. It is the value, or a set of values, to which the system evolves asymptotically, over a passage of time. The name 'attractor' is very appropriate, since it represents the ultimate value (or a set of values) which nature guides the physical property to attain asymptotically. For some values of the control parameter r, bifurcations are seen in the logistic map (Fig. 9.8). They look like tines of a fork along which the population x_n evolves asymptotically. Depending on the value of the control parameter r, the attractor can be a *steady state*, (as for values of r less than 3), or the attractor can be one of two possible values, as for $3 < r < 3.449489...$ When bifurcation takes place, the attractor is said to have a 'period 2' oscillation. Table 9.1 describes the nature of the attractor of the logistic map. The attractor generates an 'orbit' of the logistic map as the control parameter is changed.

Table 9.1 Nature of the attractor (asymptotic behavior) orbit of the logistic map showing the period doubling route to chaos.

$r < 3$: Steady state orbits. The asymptotic value of the attractor is unique. Once the steady state is reached for some value of the iteration n, it does not change after that. If the population were to start at this value, it would remain *fixed*, unchanged. However, the fixed point could be a *stable* fixed point, or an *unstable* one. For a stable fixed point, the population would be attracted toward the fixed point if it were only slightly different from it at the start. For an unstable fixed point, it will tend to run away from it if the starting value was only slightly different. This situation is analogous to stable and unstable equilibrium points in a potential.
$3 < r < 3.449489..$: In this range of the control parameter, first period doubling occurs. The attractor is *period 2* oscillation. The asymptotic value is predictably one of possible 2 values which can be determined uniquely for each value of r.
$3.449489.. < r < 3.544090..$: second period doubling occurs; the attractor is a *period* 2^2 oscillation. It is a *set* of 4 alternative, but predictable, values.
$3.544090.. < r < 3.564407..$: We have the third bifurcation in this range; *period* 2^3 oscillation. The solution *set* now consists of 8 alternatives, but predictable, values.
$3.564407.. < r < 3.568759..$: This is the region of the fourth period doubling. The asymptotic solution consists of a *set* of possible 16 values; *period* 2^4 oscillation
$3.568759.. < r < 3.569691..$: In this range of the control parameter, we have the 5th period doubling: *period* 2^5 oscillation. One can predict to which value among a larger set, consisting of a possible 32 values, the system will evolve to.
$3.569691.. < r < 3.569891..$: This is the regime of the 6th period doubling: The asymptotic solution belongs to a set of 64 possible values: *period* 2^6 oscillation.
$3.569891.. < r < r'$: Region of the 7th period doubling: *period* 2^7 oscillation. Depending on the value of r, the solution set contains 128 values. At r', the region of the eighth bifurcation would begin.
Bifurcations continue. In general, the attractor is a set of 2^N predictable values.
$3.569946.. < r < r_3$. For these values of the control parameter, the attractor shows chaos. In the chaotic regime, the solution set remains determinable and bounded, but we cannot identify any pattern in it. This is the regime of the strange attractor. The range of values of the control parameter r goes up to some value, which we shall refer to as r_3 of the control parameter at which period 3 oscillations appear, after the chaotic regime. For values of $r > r_3$, each of the three tines of the attractor tines remains steady up to some value of r beyond which each tine begins to bifurcate. The whole of the previous pattern is replicated (however, with a different aspect ratio) as each tine branches out.

If you zoom-in on any part of the chaotic regime of the logistic map diagram (Fig. 9.8), it will be similar to the original diagram, except that the x-dimension and the y-dimension of the plot would be scaled differently. It therefore has *approximate* self-similarity, and is called '*self-affine*'.

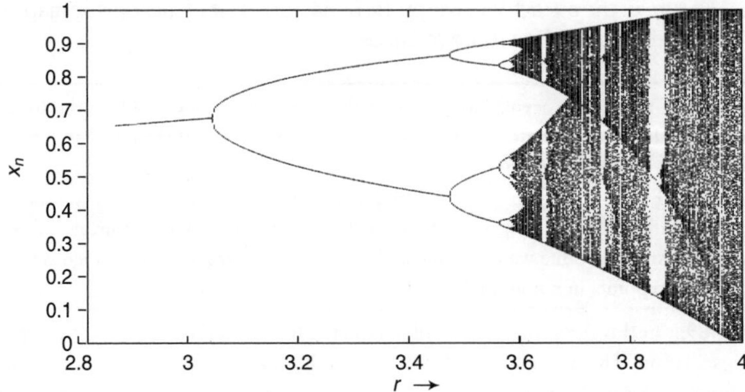

Fig. 9.8 The orbit of a physical quantity is a set of values it takes asymptotically, depending on the value of some control parameter. The orbit of the logistic map is made up of the bifurcation route to chaos.

Zooming-in involves stretching and scaling the x-dimension, as well as the y-dimension. Exact self-similarity would involve essentially the same scaling factor in both the directions, which is not the case with self-affine diagrams. In self-affine systems, scaling is *anisotropic*, i.e., the aspect ratio is not maintained. Many objects in nature, for example, a cauliflower and a fern, are self-affine. Several applets are easily accessible on the internet which show the self-affine structure of the logistic map. Furthermore, if you zoom-in further *on* the chaotic regime of the previously *zoomed-in* map, you will find a self-affine logistic map yet again. This would happen, in fact, again, and again and again, as you keep zooming. The self-affine property of the orbit of the logistic map is so *complete* that all the details, such as the period-three oscillations (seen above the value 3.82 in Fig. 9.8) re-appear on zooming. The period-three oscillations appear well after the onset of chaos, but then each of the three tines would undergo period-doubling, and then again, and again and again and again, and then get chaotic.

9.3 FRACTALS

The self-affine, or self-similarity, is the characteristic property of objects called *fractals*. This term was introduced by Benoit Mandelbrot (1924–2010), recognized as the father of fractal geometry. Fractals are objects having *fractional* dimensions. When we encounter this term for the first time, we may feel dismayed. In high-school geometry, we learn that a point has no dimension, a straight-line is 1-dimensional, a flat-surface has two dimensions, and a cube has three-dimensions. We are thus used only to objects having integer dimensions. Fractional dimension is best understood using the so-called Hausdorff–Besicovitch definition. We illustrate the course to the definition of *fractal*, which is an object having *fractional dimension*, using a straight line (Fig. 9.9a), a square (Fig. 9.9b), and a cube (Fig. 9.9c) as our elemental objects of discussion. Let us divide *each* dimension of these three objects into n equal parts. Thus, we divide the straight line into n parts, and also *each side* of the square

(Fig. 9.9b) and of the cube (Fig. 9.9c) into n parts. The number of *self-similar* pieces of the original object we get is

$$N = n^d, \tag{9.11a}$$

where the dimension of the original object is $d = 1$ for the straight line, $d = 2$ for the square, and $d = 3$ for the cube. From Eq. 9.11a, we can, therefore, *define* the dimension of the object as

$$d = \frac{\log N}{\log n}, \tag{9.11b}$$

since $\log N = d \log n$.

The dimension of an object defined by Eq. 9.11b is called its Hausdorff–Besicovitch dimension. Its defining criterion is not determined by the perception of space as guided by physical observations. Its characteristic defining attribute is inspired by the principle of *self-similarity* that is at the heart of Eq. 9.11a. It naturally accommodates the possibility of objects having fractional dimensions, i.e., objects being *fractals*.

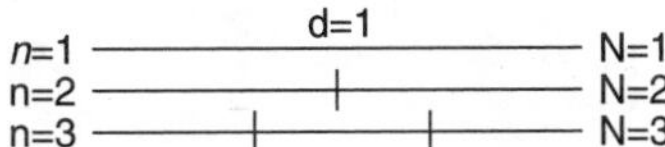

Fig. 9.9a

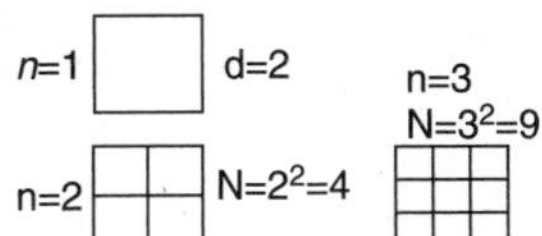

Fig. 9.9b

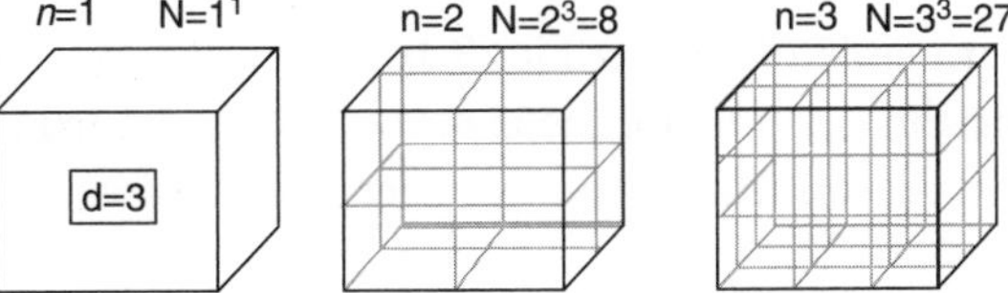

Fig. 9.9c

We exemplify the notion of fractal dimension defined by Eq. 9.11b using the 'Sierpinski carpet', and the 'Menger sponge'. If we divide *each side* of a square into 3 parts, we get exactly self-similar 9 small squares (Fig. 9.10). On removing the central square among these 9 (Fig. 9.10a), we are left with 8 squares. We begin with this object, having 8 squares, and iterate on it, i.e., we do to each of the 8 square as we did to the original square. On continuing such iterations, we shall get the Sierpinski carpet, shown in Fig. 9.10b.

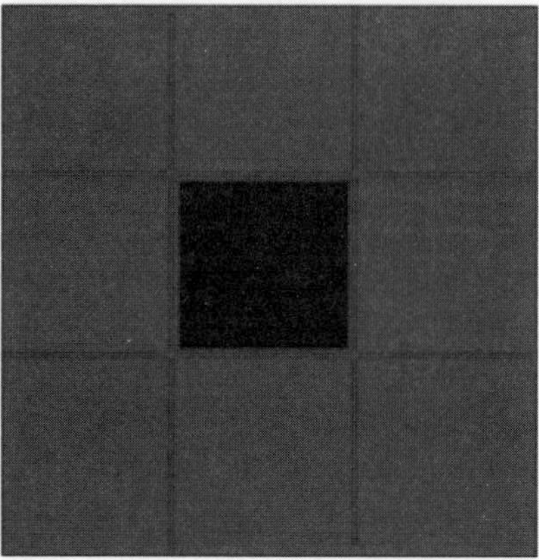

Fig. 9.10a Sierpinski carpet; removal of the central square.

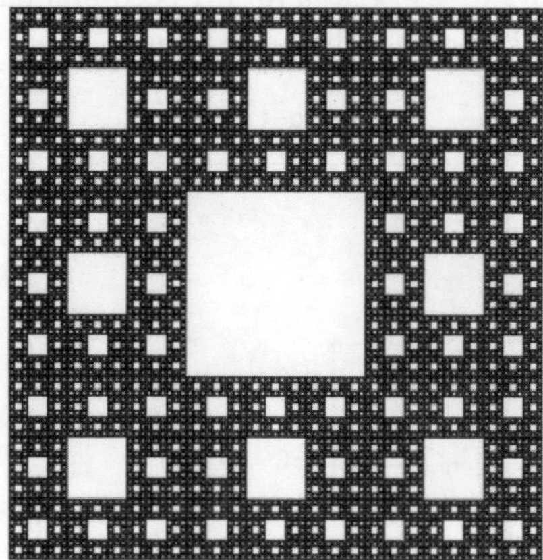

Fig. 9.10b Continuing the removal of the central square in the Sierpinski carpet.

Note that the guiding principle of getting Fig. 9.10b from 9.10a is *self-similarity*, and it is now amenable to define the fractional dimension of this object using Eq. 9.11b. From the nature of its construction (described above), the fractal dimension of the Sierpinski carpet is a fraction, given by

$$d = \frac{\log N}{\log n} = \frac{\log 8}{\log 3} \simeq 1.89... \tag{9.12a}$$

Likewise, on dividing *each side* of a cube into 3 parts, we can see that it can be thought of as made $3^3 = 27$ self-similar small cubes. If we now *remove* the central cube from *each* of the 6 sides of the cube, and also remove the seventh cube that is at the center of the full cube, we shall then be left with 20 small cubes. This cube is shown in Fig. 9.11.

Fig. 9.11 The Menger sponge.

We now iterate on each of the 20 small cubes in Fig. 9.11, i.e., we do the same thing to each of the 20 small cubes as we did to the first original cube, beginning with dividing each side of each cube into 3 equal parts. We continue the process, iterating the procedure on each *smaller-still* cube. The resultant object we get is called Menger sponge. Using Eq. 9.11b, determined by the criterion of self-similarity, the fractal dimension of the Menger sponge is

$$d = \frac{\log N}{\log n} = \frac{\log 20}{\log 3} \simeq 2.7268.. \tag{9.12b}$$

Another example of this kind is the object we get by constructing self-similar triangles, on each side of an equilateral triangle, as shown in Fig. 9.12a. When you place a self-similar equilateral triangle of side 1/3 of the original triangle, and place it on the middle of each side, the number of length elements on that side grows from 1 to 4 pieces, but each has a size one-third of the original side. The added triangle, has a sticks out of each side. While its base merges with the side of the original triangle, the two sides which stick out, along with the residual two sides in the wings of the central triangle, produce a total of 4 length elements.

Iterating by generating self-similar constructions of equilateral triangles on the central $(1/3)^{rd}$ part of each triangle would generate the shape shown in Fig. 9.12b. Its fractal dimension is

$$d = \frac{\log N}{\log n} = \frac{\log 4}{\log 3} \simeq 1.261.. \tag{9.12c}$$

Fig. 9.12a Construction of smaller equilateral triangles, each of side one-third the size of the previous triangle and erected on the middle of each side.

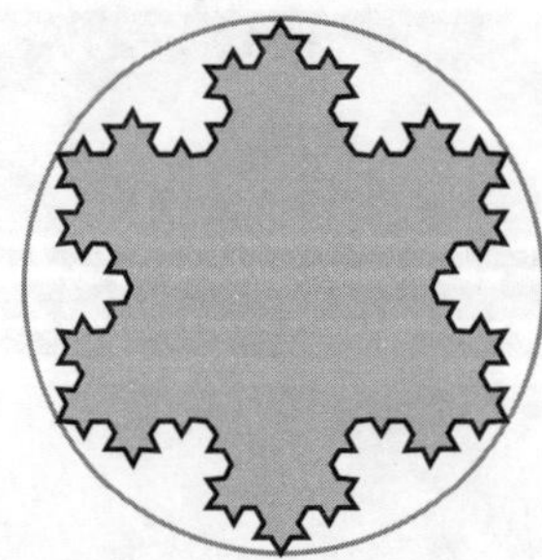

Fig. 9.12b The perimeter of the above figure obtained by constructing the Koch curve on each side of the triangle would have infinite length, but the whole pattern is enclosed in a finite area.

The perimeter of the 'triangle-on-triangle' construction described above will have infinite length, though it is bounded in a finite a area of the circle which circumscribes it. Fig. 9.13 shows how the length of this perimeter tends to infinity. The perimeter is called the Koch curve. It is a fractal, of dimensionality given by Eq. 9.12c.

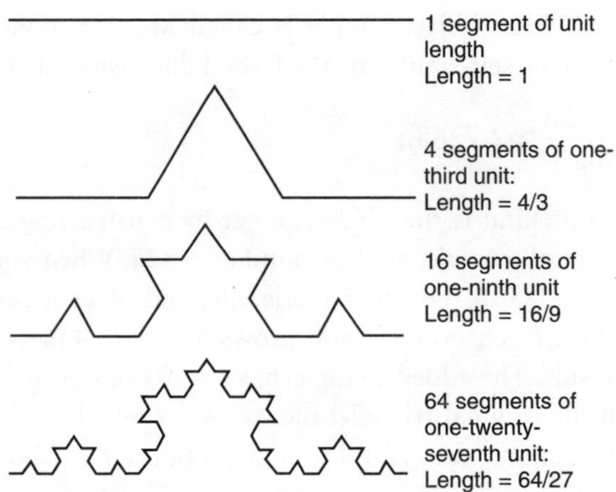

Fig. 9.13 The Koch curve has infinite length, as *each* length element increases by a factor of $\frac{4}{3} \simeq 1.33$ at every iteration.

The mathematical objects of Fig. 9.9, 9.10, 9.11, 9.12, and 9.13 each show self-similarity and are fractals. The basic elements in each of these figures are well-defined geometrical objects, like straight lines, squares, cubes and triangles. These objects were of utmost interest to Galileo Galilei who would say ".... the universe ... cannot be understood unless one first learns to comprehend the language in which it is written. It is written in the language of mathematics, and its characters are triangles, circles and other geometric figures." Self-similarity of the objects in Figs. 9.9 through 9.13 could be discussed in terms of such geometric objects. Fractals are therefore of great interest in objects occurring in nature. Other than cauliflowers and ferns, there are many striking examples of fractals in nature: clouds, lightning, broccoli, sea-shells, coastline, stalagmites and stalactites, etc. All of these objects have a spectacular self-affine property, though *not exact* self-similarity. Thus, arguing against Galileo, Mandelbrot's contrary view would be: "Clouds are not spheres, mountains are not cones, coastlines are not circles, and bark is not smooth ..." Each of these objects in nature shows self-similarity, but with anisotropic aspect ratio. These fractals are *self-affine*, rather than *self-similar*.

We return to the self-affine property of the orbit of the logistic map (Fig. 9.8). As discussed earlier, zooming on any part of the chaotic regime reveals its self-affine fractal structure. The recognition of the self-affine fractal structure of the logistic map enables us to generalize the notion of fractals and make the concept applicable to the orbit of the population expressed by the logistic map. Thus, the generalization extends the applications from objects like the Sierpinski carpet and the Menger sponge, to attractors of populations, or of solutions to any physical property whose successive values can be expressed using a map equation, such as Eq. 9.9, not just the logistic (quadratic) map of Eq. 9.8.

The notion of fractal is very powerful, and can be extended to relationships more complicated than Eq. 9.9. It is fruitful in situations involving more physical parameters, not merely the one represented by the variable x so far. The additional variables may be represented, for example, by y, z, etc., which may correspond to different physical properties, such as temperature, wind velocity, ... or whatever. In fact, in the work of Lorenz, which we referred to earlier in the discussion on Fig. 9.6, he was analyzing three atmospheric physical properties, which we shall call as x, y, and z (without getting into the details of what exactly they represent). These physical quantities satisfied the following set of coupled time-dependent differential equations:

$$\frac{dx}{dt} = -\sigma x + \sigma y, \tag{9.13a}$$

$$\frac{dy}{dt} = \rho x - y - xz, \tag{9.13b}$$

and $$\frac{dz}{dt} = xy - \beta z. \tag{9.13c}$$

For $\sigma = 10, \rho = 28, \beta = \frac{8}{3}$, if the asymptotic solutions (i.e., for large enough time t) for x, y, and z are plotted on three orthogonal axes in a 3-dimensional graphics plot, the result looked like the pattern in Fig. 9.14. The solution is chaotic, even if bounded. It is chaotic within the box, though predictable, and remains bound within the box. There is thus some order within the disorder, or disorder within the order, howsoever you describe it, in the nature of the asymptotic attractor. It is a strange attractor, called the Lorenz attractor. The chaotic dynamical system's asymptotic behavior is thus described over the *strange attractor*. The solution does not converge to a stead state, nor to any simple structure like the period-doubling, etc., of Fig. 9.8.

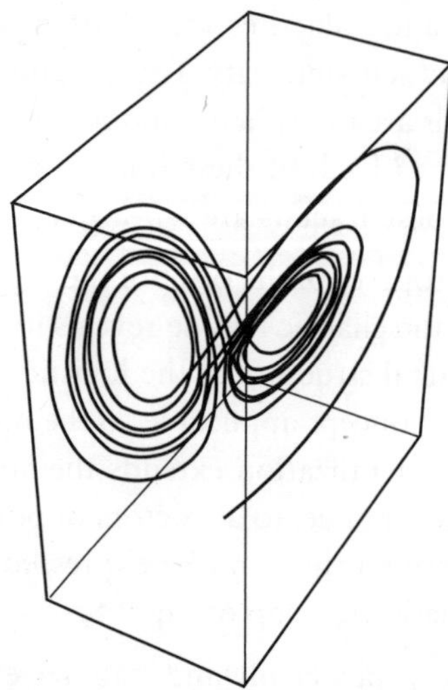

Fig. 9.14 The strange attractor is named after Lorenz, and is called the Lorenz attractor. The asymptotic orbit solution of Eq. 9.13 evolves to the strange attractor.

The shape of the strange attractor in Fig. 9.14, strangely enough, alludes to the wings of a butterfly whose mere flutter is capable of causing chaos.

We have seen above that on each zoom in the chaotic regime of the logistic map (Fig. 9.8), we find a bifurcation pattern at various values of the control parameter r. Several values of the control parameter at which bifurcation takes place are given in Table 9.1, above. Mitchell Jay Feigenbaum (born 1944) defined a ratio from the values of the control parameter at the nth bifurcation:

$$\delta_n = \frac{r_{n+1} - r_n}{r_{n+2} - r_{n+1}}. \tag{9.14a}$$

He found that its limit,

$$\delta = \lim_{n \to \infty} \delta_n = 4.66920160910299067185\ldots \tag{9.14b}$$

is a constant, irrespective of the initial value x_1 we employed to generate the logistic map. The limiting value given in Eq. 9.14b is called the Feigenbaum (first) constant. Its constancy is seen not merely with respect to the logistic map, but also with respect to other maps, such as the sine map defined by

$$x_{n+1} = r \sin(\pi x_n). \tag{9.15}$$

By changing the control parameter r, the orbit of the limiting values of the sine map x_n (Eq. 9.15) shows bifurcation and period doubling similar to the logistic map. Determining the limiting value of the ratio δ by applying Eq. 9.14 to the bifurcations of the sine map, it turns out that the numerical value turns out to be exactly the same as in Eq. 9.14b. This is really amazing. There are in fact a large number of maps, called the 'quadratic maps' (and not just the logistic map and the sine map) which show period-doubling. They *all* have the property that the values of their respective control parameter at which consecutive bifurcations take place conform exactly to the numerical value when ratios as per Eq. 9.14a are taken. The $\lim_{n\to\infty} \delta_n$ is therefore a universal constant, called the Feigenbaum (first) constant. This transcendental number ranks as one of the fundamental numbers in nature. On account of its universal applicability in all quadratic maps, it enjoys the same status as other fundamental number constants, such as π and e. There is also a second universal constant, denoted by α, which is also of importance in the context of the period-doubling route to chaos. It is called Feigenbaum's second constant, denoted by α. It is defined in terms of the width between the pair of tines following a bifurcation. It is given by the limiting value of another ratio,

$$\alpha = \lim_{n\to\infty} \frac{\xi_n}{\xi_{n+1}} = \pm 2.502907875095\ldots, \tag{9.16}$$

where ξ_n is the width between the tines after the nth bifurcation. The plus sign is used if the upper $(n+1)^{\text{th}}$ tine is referenced in the denominator, and the minus sign if it is the lower $(n+1)^{\text{th}}$ tine that is referenced.

Above, we have discussed the 'attractor' of the logistic map. It represents the eventual value which the parameter under investigation will attain as the system evolves. The system's evolution may be with reference to changes in time, or some other independent parameter, or with reference to the iterations of the equation to a map. We have already seen that the attractor may be a steady state, or a set of values. It can also be a set of values which repeat themselves, in which case the attractor is known as a limit cycle. For example, the attractor of a linear harmonic oscillator is a repetitive orbit (called the 'limit cycle') defined by an ellipse in the phase space:

$$\frac{p^2}{2m} + \frac{1}{2}\kappa q^2 = E, \tag{9.17}$$

whereas the attractor of a damped oscillator would spiral into a fixed point, as the system is dissipative and looses energy. The attractors of the linear harmonic oscillator are shown in Fig. 9.19a, and that for the damped oscillator in Fig. 9.19b.

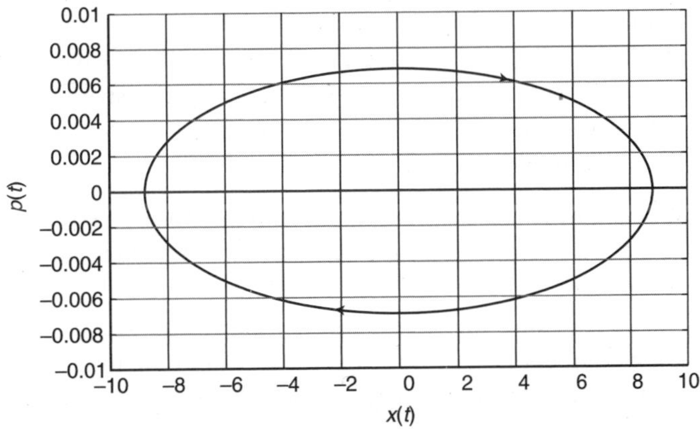

Fig. 9.15a Phase space trajectory of a simple harmonic oscillator.

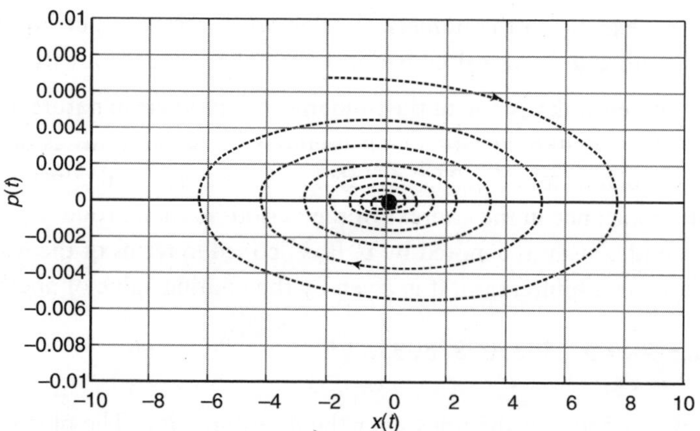

Fig. 9.15b Phase space trajectory of a damped oscillator.

We now consider the attractor of a map, defined using *complex numbers*, $z = x + iy$, where x and y are real numbers and $i = \sqrt{-1}$. We consider the map, defined by the sequence

$$z_{n+1} = z_n^2 + c, \tag{9.18}$$

where each z_n is a complex number that is iterated upon, and c is a constant complex number. It serves as a 'control parameter' for the orbit in the complex plane that the iterations with $n = 0, 1, 2, 3, \ldots$ would generate, starting with some value of z_0. One may get a variety of results when Eq. 9.18 is iterated upon, depending on what value of the control parameter c is chosen, and also what initial value z_0 is chosen. We may therefore change the real and/or imaginary part(s) of z_0, for a chosen complex number c, or change the real and/or imaginary part(s) of c, for a chosen initial value z_0. To keep track of the alternate possibilities, we shall write the complex numbers z_0 and c using different algebraic representatives of their real and imaginary parts: $z_0 = a + ib$, and $c = x + iy$. The set of pairs of complex numbers (z_0, c) therefore constitute a 4-dimensional parameter space, with the possibility that the 4 real numbers (a, b, x, y) take different values. To illustrate the orbit of values which z_{n+1} can take beginning with a choice of (z_0, c), let us consider a particular c with $(x = -1, y = 0)$, and z_0 determined by $(a_0 = 0, b_0 = 0)$. On applying Eq. 9.18, we see that we get z_1 with $(a_1 = -1, b_1 = 0)$, $z_2(a_2 = 0, b_2 = 0)$, and z_3 with $(a_3 = -1, b_3 = 0)$. Successive results of the iteration continue to alternate between $(a_{n,\text{ even}} = 0, b_{n,\text{ even}} = 0)$ and $(a_{n,\text{ odd}} = -1, b_{n,\text{ odd}} = 0)$. Essentially, we get a period 2 oscillation in this case.

However, if we begin with the same value of c, i.e., with $(x = -1, y = 0)$, but start the iterations with a different value of z_0, viz., $(a_0 = 0, b_0 = 0)$, then successive iterations of Eq. 9.18 yield real numbers which quickly grow to large values; they are not bounded. For various pairs (z_0, c) constituting the 4-dimensional parameter space, we get two types of sequences of complex numbers $\{z_n\}$, those with $|z_n| \leq 2$ and remain trapped in this bounded space, i.e., within a 'cage-set', in the complex plane, and those which go outside this trap. Sequences which slip outside this trap as $n \to \infty$ typically have absolute magnitudes which run-away limitlessly. Usually, a small number of iterations is sufficient to determine if the sequence would belong to the 'cage-set' or to the 'run-away' set.

The set of complex numbers c with $(z_0 = 0 + i0, c)$ for which the sequence $\{z_n\}$ remains within the bounds $|z_n| \leq 2$ is called the *Mandelbrot set*, after Benoit Mandelbrot. The Mandelbrot set has a fixed value of z_0, which is $z_0 = 0 + i0$. There are other sets of complex numbers of interest. A closely related set of complex number, for example, is the Julia set, named after the French mathematician, Gaston Julia. It is obtained on varying z_0, but keeping c fixed. Interesting relationships between the Julia set and the Mandelbrot set are established in some fascinating theorems which you will find in more specialized books on the theory of chaos and fractals, such as Ref. [2, 3, 4, 5]. The Mandelbrot set is shown in Fig. 9.16. This set has a very interesting relationship with the logistic map. The Mandelbrot set M intersects with the real line at (–2, 0) and (0.25, 0). If the control parameter r of the logistic map is associated with Re(z) of the Mandelbrot plot, we have the following one-one correspondence with the logistic map as follows:

$$z_{n+1} = z_n^2 + c \rightarrow r z_n (1 - z_n), \tag{9.19}$$

where r is in the space of the control parameter of the logistic map, i.e., r is in the interval [1, 4]. To see the relationship between the Mandelbrot set and the logistic map, we set

$$c = 0.5r\,(1 - 0.5r). \tag{9.20}$$

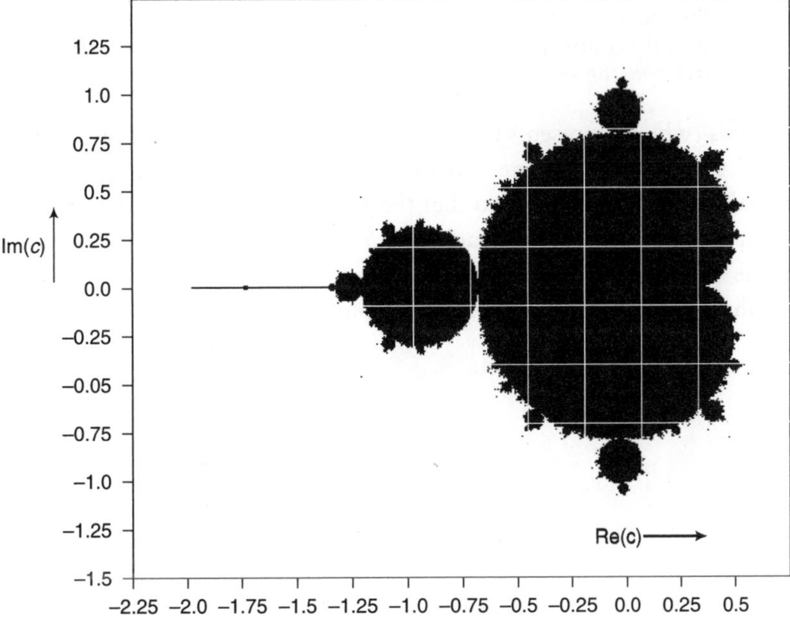

Fig. 9.16 The Mandelbrot set is the set of complex numbers, c such that starting from $z_0 = 0 + i0$, $z_{n+1} = z_n^2 + c$ remains bounded, with $|z_n| \leq 2$, as $n \rightarrow \infty$. The Cartesian coordinates of the points plotted are (Re(c), Im(c)), for all points c belonging to the Mandelbrot set.

As the control parameter r of the logistic map varies from 1 to 4, the parameter c (in Eq. 9.18) varies from 0.25 to −2. Fig. 9.17 shows the logistic map plotted using Eq. 9.19, 9.20, where c was calculated using in [1, 4] with $z_0 = 0 + i0$. Fig. 9.18 shows the correspondence between the *real axis* of the Mandelbrot set, and the *control parameter space* of the logistic map.

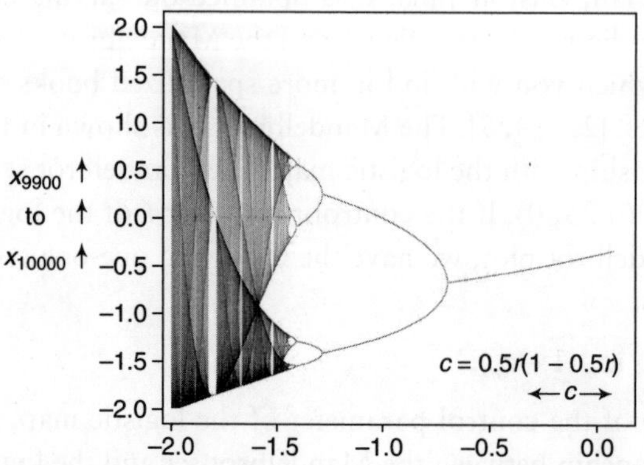

Fig. 9.17 The logistic map plotted using Eq. 9.19, with r in the range [1, 4]. Calculations are made with $z_0 = 0$.
$x_{n+1} = x_n^2 + c$ is plotted with $x_0 = 0$, calculated for 10,000 iterations. Out of the 10,000 iterations, the last 100 are plotted on the y-axis, from x_{9900} to x_{10000}, against the control parameter c between −2.0 and 0.25, which are the values of Re(z) where the set M intersects the real axis.

The original y-axis (which represents the values $x_{n+1} = x_n^2 + c$) of the logistic map in Fig. 9.17 lies between [−2.0, + 2.0]. The figure was first shifted to the y-range [0, 4], by adding +2. Then, it was scaled by a factor 0.225 so that the y-values would lie in the range [0,0.9]. The figure was then shifted again by subtracting 1.5 out of the y-values, so that the y-values would fall in the range [−1.5, −0.6]. This scaling is done to *squash* the logistic map, so that it can be seen with the Mandelbrot set in the same plot, in Fig. 9.18.

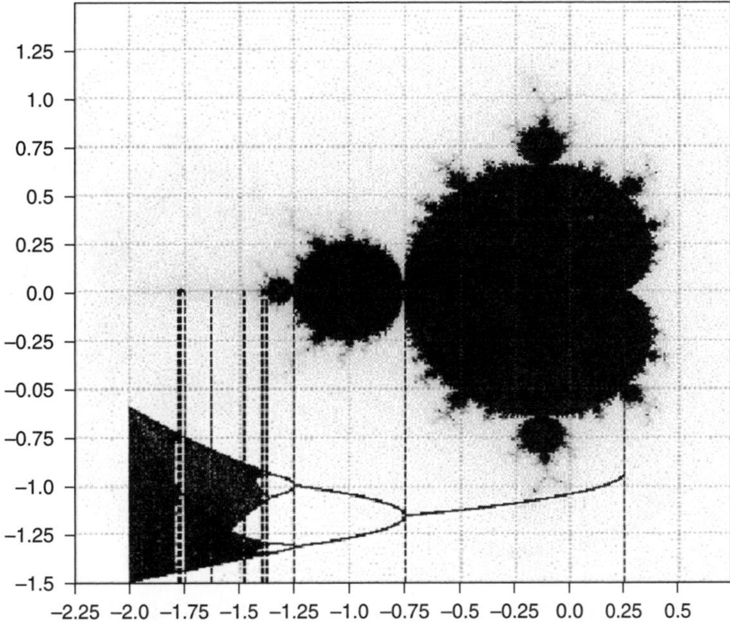

Fig. 9.18 The y-axis of logistic map is rescaled. It is displaced and squashed as explained in the text. This rescaling helps visualize clearly the relationship between the Mandelbrot set and the logistic map. There are therefore two *different* y-axis in this figure since the logistic map is squashed as described in the text. The calibration on the y-axis corresponds to the imaginary axis of the Mandelbrot set.

The Mandelbrot set is a fractal; it contains infinite self-affine copies of itself which are discovered by progressively zooming in the central region of the set, shown in Fig. 9.19(a), (b), (c), (d). This self-affine structure is the characteristic property of the fractals. The more you zoom in, the more copies of the original pattern are discovered, as one also would by zooming into a cauliflower. Very beautiful and mesmerizing images, even videos, in which different parts of the Mandelbrots appear in lovely colors and decorations, can be easily found on the internet. Most of these are made by smart undergraduate students; may be you can upload your results too?

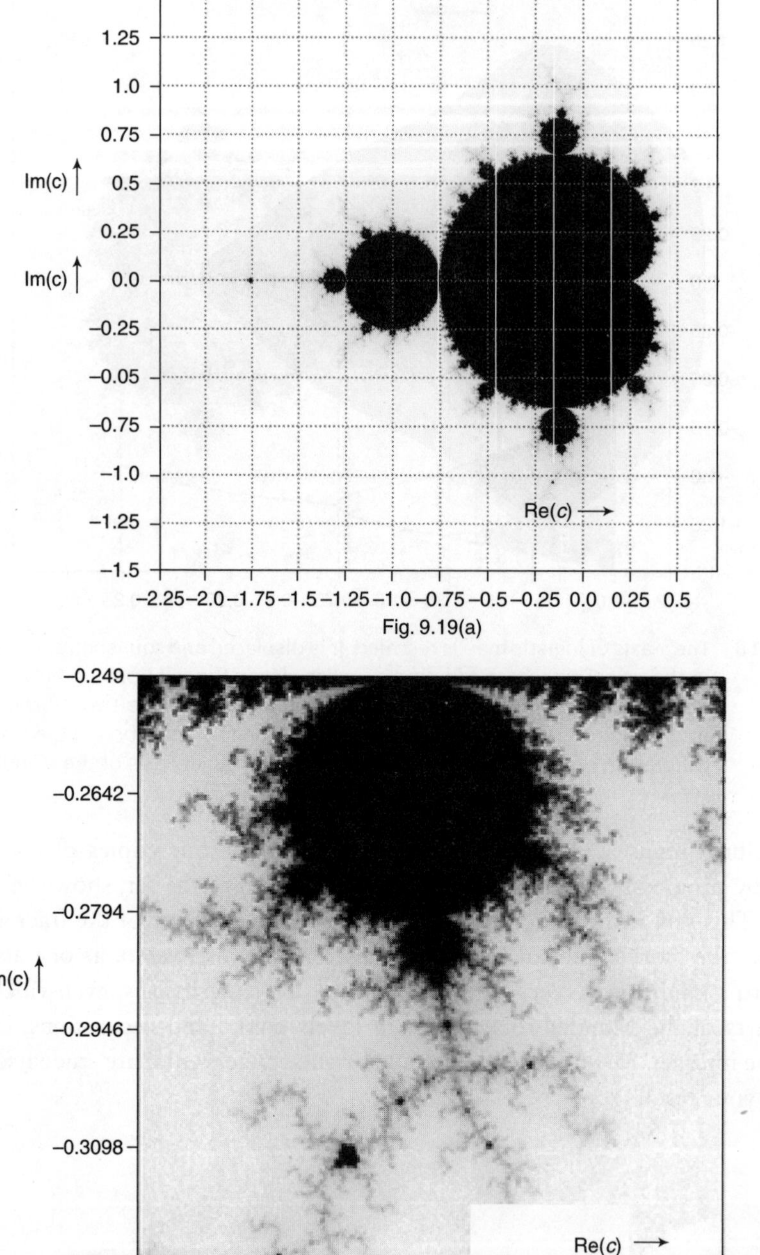

Fig. 9.19(a)

Fig. 9.19(b)

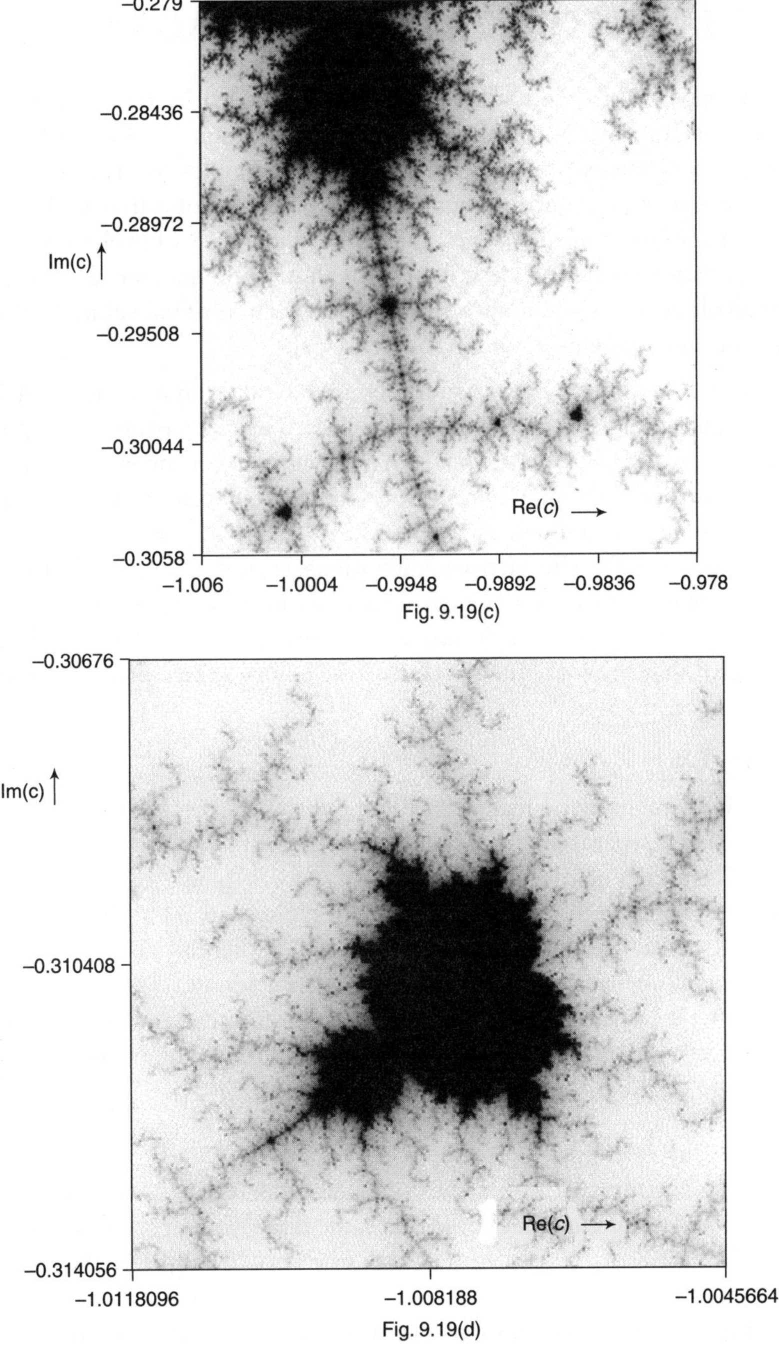

Fig. 9.19 The Mandelbrot set is a fractal. There are self-affine Mandelbrots *within* each Mandelbrot. As you keep zooming, these Mandelbrot *copies within the Mandelbrot* develop and become manifest. Figs 9.19(a), (b), (c), and (d) are from the *same* plot. Each subsequent of these figures is obtained on zooming the previous.

9.4 CHARACTERIZATION OF CHAOS

A measure of sensitivity to the initial conditions is provided by the Lyapunov coefficient, named after the Russian Aleksandr Mikhailovich Lyapunov (1857–1918), who was possessed with applications of some models in fluid dynamics to problems in astronomy. Lyapunov made significant contributions to the development of differential calculus and to the understanding of dynamical systems, and also to the theory of probability and potential theory. The Lyapunov coefficient describes, through a single number, the divergence in the evolution of an observable on account of a slight change in its initial value. We discuss it now, following its introductory elucidation in Reference [6].

In the chaotic regime, the values that an observable takes, in a sequence of iterations, is different for neighboring seed values, however, close. Their departure could grow with the number of iterations linearly, or faster through a polynomial dependence on some coefficient, or even faster, having in fact an exponential behavior. We consider two start-values, x_0 and $x_0' = x_0 + \varepsilon_0$. The sequence of numbers generated by the map $x_{n+1} = f(x_n, r)$ and $x_{n+1}' = f(x_n', r)$ could depart exponentially with increasing iterations, even if $\varepsilon_n = |x_n' - x_n|$ for $n = 0$ is very tiny: $\varepsilon_0 = \varepsilon$. To what extent x_n' different from x_n, i.e., how large the difference ε_n is, would depend on the control parameter, r. It may well be that for some value of r, the difference is minor (even zero), but for some other value of the control parameter, the difference could actually grow exponentially:

$$\varepsilon_n = |x_n' - x_n| = \varepsilon_0 e^{n\lambda_r}. \tag{9.21}$$

λ_r is called the Lyapunov exponent. Typically, the 'distance', (i.e., 'difference'), $\varepsilon_n = |x_n' - x_n|$, for each value of n, is a separation of the state of the system at the nth iteration *in the phase space* of the system. The notion of 'distance' in the phase space is to be used with care. We expect it to represent how separated the mechanical states of the system are at the nth iteration, if they were separated at a distance $\varepsilon_0 = |x_0' - x_0|$ at the start. Whether this distance would diverge, or converge, or remains unchanged, would determine if the dynamical system would turn chaotic or not. Pictorially, the possibility of an invariant separation, and that of diverging (chaotic, like the 'butterfly effect') separation, is shown in Fig. 9.4a and 9.4b. Now, the Euclidian distance d between two points whose position vectors $\vec{r}_1$ and $\vec{r}_2$ is $d = |\vec{r}_2 - \vec{r}_1| = \sqrt{(\vec{r}_2 - \vec{r}_1)\cdot(\vec{r}_2 - \vec{r}_1)}$. In a 2-dimensional Cartesian plane, this metric would be $d = \sqrt{(x_2 - x_1)^2 + (y_2 - y_1)^2}$. The reason this is straight forward is that the X-axis and the Y-axis both have the dimensions of length. In a position–momentum phase space, the dimension of the distance between two points is obviously different. Nonetheless, we ask what would happen to the distance between two neighboring start points in the phase space for a system as it dynamically evolves. If the Lyapunov exponent is negative, this distance would diminish, and become zero after a few iterations. If the Lyapunov coefficient is zero, then this distance in the phase space would remain constant, no matter after how many iterations. The system would, however, become chaotic if the Lyapunov coefficient is positive. The system may of course evolve differently, even for the same distance if the phase space points at the start x_0 and x_0' are *oriented differently* in the phase space. Accordingly, there may be several different Lyapunov coefficients for the chaotic system. The largest of these, is called the principal Lyapunov exponent. Unless stated otherwise, it is the *principal* Lyapunov exponent that is usually referenced.

The subscript r in λ_r (Eq. 9.21) would remind us that the rate of departure, even if exponential, would depend on some control parameter r. Beginning with its initial value, the observable would take a sequence of values as we iterate on it using the difference map. Depending on the start-value, x_0 or $x_0' = x_0 + \varepsilon_0$, successive iterations would generate the following two alternative sequences:

$$\{X\} = \{x_0, x_1, x_2, x_3, \ldots\ldots\ldots, x_n, x_{n+1}, \ldots\}, \tag{9.22a}$$

and $$\{X'\} = \{x_0', x_1', x_2', x_3', \ldots\ldots\ldots, x_n', x_{n+1}', \ldots\}. \tag{9.22b}$$

The difference between the successive values after the first iteration is:

$$\varepsilon_1 = x_1' - x_1 = f(x_0 + \varepsilon, r) - f(x_0, r) \simeq \left[\frac{df}{dx}\right]_{x_0} \varepsilon. \tag{9.23}$$

After n iterations, the difference is: $\varepsilon_n = |x_n' - x_n| = \varepsilon_0 e^{n\lambda_r}$. (9.24)

Thus, we may write:

$$x_n = f(x_{n+1}, r) = f(f(x_{n+2}, r), r) = f(f(f(x_{n+3}, r), r), r) = \ldots = f^n(x_0, r), \tag{9.25a}$$

$$x_n' = f(x_{n+1}', r) = f(f(x_{n+2}', r), r) = \ldots f^n(x_0', r) = f^n(x_0 + \varepsilon, r). \tag{9.25b}$$

The difference after the nth iteration will be:

$$\varepsilon_n = x_n' - x_n = f^n(x_0 + \varepsilon, r) - f^n(x_0, r) = \varepsilon e^{n\lambda_r}. \tag{9.26}$$

Hence, $$\frac{f^n(x_0 + \varepsilon, r) - f^n(x_0, r)}{\varepsilon} = e^{n\lambda_r},$$

i.e., $$\ln\left|\left[\frac{f^n(x_0 + \varepsilon, r) - f^n(x_0, r)}{\varepsilon}\right]\right| = n\lambda_r.$$

Hence, $$\lambda_r = \frac{1}{n}\ln\left|\left[\frac{f^n(x_0 + \varepsilon, r) - f^n(x_0, r)}{\varepsilon}\right]\right| = \frac{1}{n}\ln\left|\left[\frac{df^n}{dx}\right]_{x_0}\right|. \tag{9.27}$$

Now, the *first* derivative of the iterate '2' is easily obtained, using the 'chain rule':

$$\frac{d}{dx}\psi(\phi(x)) = \frac{d\psi}{d\phi}\frac{d\phi}{dx}. \tag{9.28}$$

Hence, we get:

$$\left[\frac{df^2(x)}{dx}\right]_{x=x_0} = \frac{d}{dx}f(f(x)) = \left[\frac{d}{df(x)}f(x)\right]_{f(x)=f(x_1)} \times \left(\frac{df}{dx}\right)_{x=x_0}$$

$$= \left(\frac{df}{dx}\right)_{x=x_1}\left(\frac{df}{dx}\right)_{x=x_0}$$

$$\text{i.e.,}\quad \left[\frac{df^n(x)}{dx}\right]_{x=x_0} = \left[\frac{df(x)}{dx}\right]_{x=x_{n-1}} \left[\frac{df(x)}{dx}\right]_{x=x_{n-2}} \left[\frac{df(x)}{dx}\right]_{x=x_{n-3}\ldots} \left[\frac{df(x)}{dx}\right]_{x=x_0},$$

$$\text{or,}\quad \left[\frac{df^n(x)}{dx}\right]_{x=x_0} = \left[\frac{d}{dx}f^n(x)\right]_{x=x_0} = \left[\frac{d}{dx}f(f(f\ldots f(x_0,r)))\right]_{x=x_0}. \tag{9.29}$$

Now, using the chain rule, we get

$$\begin{aligned}\left[\frac{df^n(x)}{dx}\right]_{x=x_0} &= \left[\frac{df(x)}{dx}\right]_{x=x_{n-1}} \left[\frac{df(x)}{dx}\right]_{x=x_{n-2}} \left[\frac{df(x)}{dx}\right]_{x=x_{n-3}\ldots} \left[\frac{df(x)}{dx}\right]_{x=x_0} \\ &= \prod_{i=0}^{n-1}\left[\frac{df(x)}{dx}\right]_{x=x_i}\end{aligned} \tag{9.30}$$

$$\text{Hence,}\quad \lambda_r = \frac{1}{n}\ln\left|\left[\frac{df^n}{dx}\right]_{x_0}\right| = \frac{1}{n}\ln\left|\prod_{i=0}^{n-1}\left[\frac{df(x)}{dx}\right]_{x=x_i}\right| = \frac{1}{n}\sum_{i=0}^{n-1}\ln\left|\left[\frac{df}{dx}\right]_{x_i}\right|. \tag{9.31a}$$

For a large number of iterations, in the limiting case, we have

$$\lambda_r = \lim_{n\to\infty}\frac{1}{n}\sum_{i=0}^{n-1}\ln\left|\left[\frac{df}{dx}\right]_{x_i}\right|. \tag{9.31b}$$

After the nth generation, the difference, $\varepsilon_n = x_0' - x_n = \varepsilon e^{n\lambda_r}$, could be very small, even zero if $\lambda_r < 0$. It can also remain the same as it was at the very beginning, if $\lambda_r = 0$. However, when $\lambda_r > 0$, the difference can be huge, leading to the 'butterfly effect'.

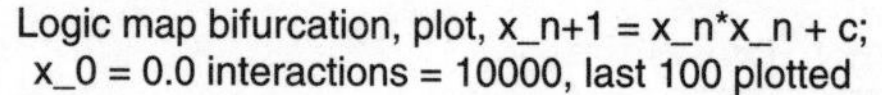

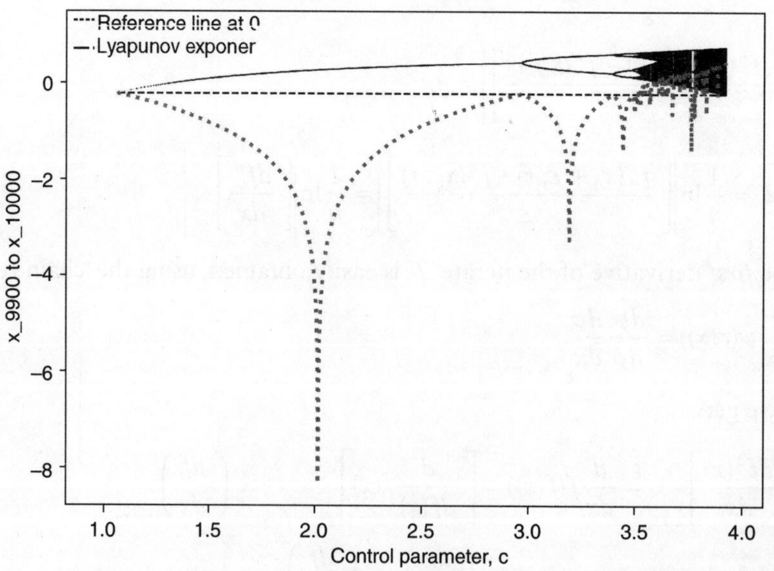

Fig. 9.20a The logistic map is shown here along with a plot of the Lyapunov exponent for the same.

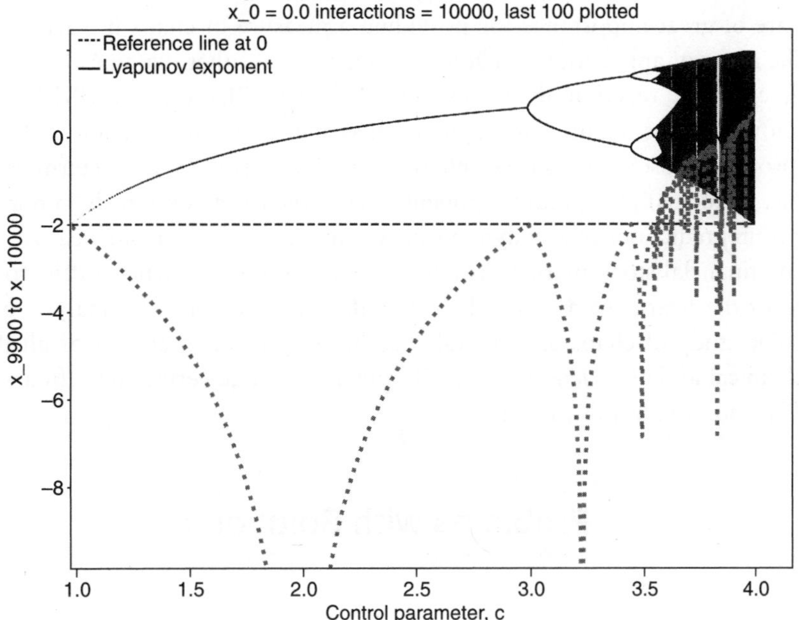

Fig. 9.20b Magnification of Fig. 9.20a.

We illustrate the method of calculating the Lyapunov exponent by taking the case of the logistic map. For 2000 values of r between 1 and 4, spaced at 0.0015, we calculate $x_{n+1} = rx_n(1 - x_n)$ for each value of r. For the plots shown in Fig. 9.20, an initial value of $x_0 = 0.6$ was used. This initial value is unimportant, as long as it is not 0. For each value of r, 10,000 values of x_n can be generated iteratively, using, for each n, obtaining x_{n+1} from x_n using the logistic map $x_{n+1} = f(x_n, r) = rx_{\mathrm{n}} - rx^2_{\mathrm{n}}$.

Correspondingly, $\dfrac{df}{dx} = r - 2rx = r(1-2x)$ (9.32)

can be calculated for each value of n, thereby generating a sequence of 10,000 values of $\dfrac{df}{dx}$.

The Lyapunov coefficient λ_r was then obtained, for each value of r, by using Eq. 9.31b. Fig. 9.20a shows the plot of λ_r versus r. Fig. 9.20b shows a magnification of this graph. When the Lyapunov exponent is negative, we get a steady state attractor. When the Lyapunov exponent is zero, or near-zero, the difference $|x_{n+1} - x_n|$ between successive values remains invariant, indicative of a period two oscillation. We see in Fig. 9.20 that the Lyapunov exponent goes to zero at those value of r where we see period doubling, and at subsequent bifurcation points of the logistic map. At negative values of the Lyapunov exponent, we find asymptotic stability of the logistic map. Finally, the eventual chaotic behavior of the logistic map is well accounted for by the positive value of the Lyapunov coefficient, which makes the difference $|x_{n+1} - x_n|$ chaotic. The plot for the Lyapunov exponent shows, at a glance, a huge amount of information about the dynamical system, its steady state and bifurcation patterns, and its chaotic behavior.

The techniques introduced in this chapter, including the calculation of the Lyapunov exponent, are of great importance for providing a measure of chaos in various dynamical systems. The range of applications includes, among various other dynamical systems already mentioned, chemical reaction dynamics. The Belousov–Zhabotinsky (BZ) reaction, for example, shows various features of a dynamical system that can go chaotic. The Lyopunov exponent provides a useful measure of chaos in this case also [7]. The determination of the maximal (i.e., 'principal') Lyapunov exponent has emerged as an extremely important tool to make long term predictions of any dynamical system, for example, from weather forecasting to predicting financial stock-market behavior; from projecting internet traffic, to foretelling the beating of the heart, from controlling manufacturing systems, to analysis of electrical networks. The study of chaos and fractals has become an integral part of all domains of physical, chemical and life sciences [2–6], all branches of engineering [8,9], financial market analysis [10], and even health care [11].

Problems with Solutions

P9.1:

Show that the golden ratio ϕ is irrational.

***Solution*:**

From Eq. 9.2C we have $\phi^2 - \phi - 1 = 0$. We use the method of 'reductio ad absurdum'.

Suppose ϕ is rational, then we can write $\phi = \frac{p}{q}, (q \neq 0)$, where p and q are co-prime integers.

$$\phi^2 - \phi - 1 = 0; \text{ hence, } \frac{p^2}{q^2} - \frac{p}{q} - 1 = 0, \text{ i.e., } p^2 - pq - q^2 = 0. \text{ Hence } p(p-q) = q^2,$$

or, $\frac{q^2}{p} = (p - q)$. This should be an integer and non-zero, since $p \neq q$.

However, if $p = 1$ then $q = \frac{1}{\phi}$ would be a non-integer. This is a contradiction, since p, q are co-prime integers. Hence our assumption is incorrect and ϕ is irrational.

P9.2:

Write a computer program to plot the ratio of successive terms of the following sequence and find its limiting value, after a large number of iterations.

$g_0 = 3, g_1 = 0, g_2 = 2,$

$g_n = g_{n-2} + g_{n-3}$, for $n \geq 3$.

This sequence is called the Perrin's sequence. Find the limiting value, $\psi = \lim_{n\to\infty} \frac{g_{n+1}}{g_n}$, referred to as the Plastic constant.

Solution:

You may write the program in any programming language, such as Python. Be careful about handling division by zero. The plot you will get will look like that shown in the figure below.

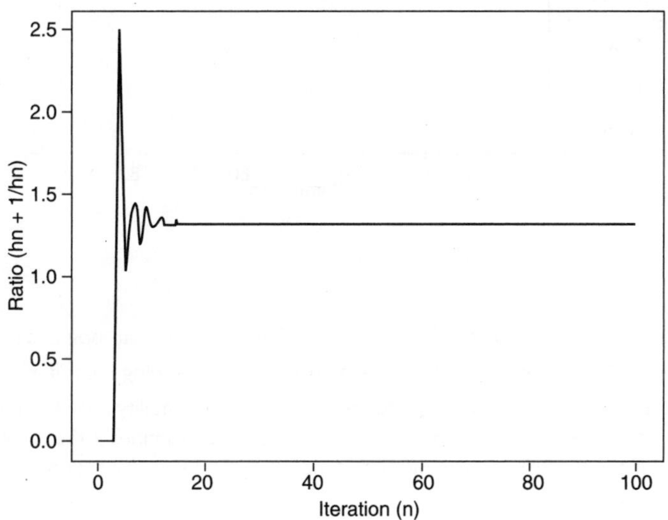

The limiting value of the ratio will turn out to be:

$$\psi = \lim_{\substack{n\to large \\ number}} \frac{g_{n+1}}{g_n} = 1.324717....$$

P9.3:

Find the limit of the ratio of successive terms of the following sequence:

$h_o = 0, h_1 = 0, h_2 = 1,$

$h_n = h_{n-1} + h_{n-2} + h_{n-3}$, for $n \geq 3$. This sequence is called the Tribonacci sequence.

Solution:

The limiting value of the ratio will turn out to be:

$$\chi = \lim_{n\to large} \frac{h_{n+1}}{h_n} = 1.83928....$$

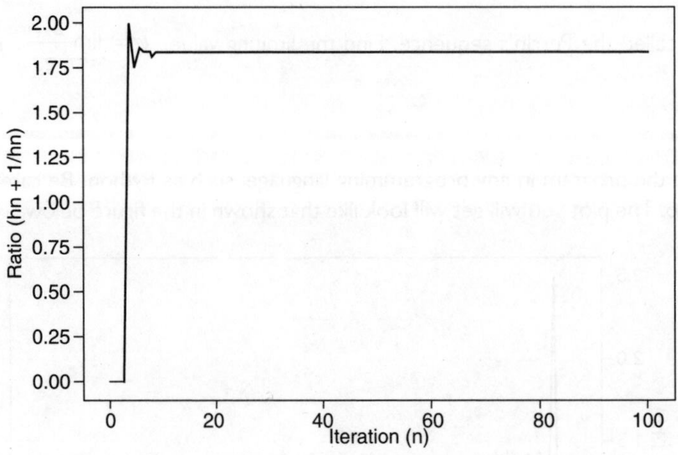

P9.4:

One can determine the 'fixed point' of an orbit by finding if at that point, say $x = x^*$, we have $\dot{x} = 0$. The method of linear stability analysis makes use of the following idea: If x^* is a fixed point, i.e., ($\dot{x} = 0$ at $x = x^*$) such that $\ddot{x}$ at $x = x^*$ is non-zero, we can use linear stability analysis. This method is based on the response of the system to a small perturbation to $x(t)$, and examines if the result decays with time, or grows with time. If the perturbation grows with time, then the fixed point is unstable, if it decays, then it is stable.

Using linear stability analysis, find the stable and unstable fixed points of the differential equation $\dot{N} = rN\left(1 - \frac{N}{K}\right) = f(N)$

***Solution*:**

(a) $\dot{N} = 0 \Rightarrow N^* = 0 \text{ or } N^* = K$ are fixed points.

$$f'(N) = r - \frac{2rN}{K}$$

At $N = 0$ $r > 0$

$\Rightarrow$ $N^* = 0$ is an unstable fixed point.

At $N = K$ $-r < 0$

$\Rightarrow$ $N^* = K$ is a stable fixed point.

P9.5:

Use linear stability analysis (explained in P9.4) to find the fixed points of the following differential equations and determine if they are stable: (a) $\dot{x} = \tan x$ (b) $\dot{x} = \ln x$

Solution:

(a) $f(x) = \tan x = 0 \Rightarrow x = \arctan(0) \Rightarrow x^* = n\pi \ n \in Z$

$$f'(x^*) = \sec^2 x \ \text{ at } \ x = n\pi$$

$$= \frac{1}{\cos^2(n\pi)}.$$

$x^* = n\pi$, is a stable fixed point if n is odd, and it is an unstable fixed point if n is even.

(b) $\dot{x} = \ln x = f(x) = 0$

$\Rightarrow x^* = 1$. $f'(x) = \frac{1}{x}$ at $x^* = 1$. Hence $x^* = 1$ is an unstable fixed point.

P9.6:

The behavior of many dynamical systems is modeled using Ordinary Differential Equations (ODE). Suppose a dynamical system is described by the differential equation: $\dot{x} = \sin x$

(a) Sketch the phase diagram $(\dot{x}, x)$ in a Cartesian framework.

(b) Find the fixed points of the given system, and find if they are stable.

(c) Solve $\dot{x} = \sin x$, by integration, and obtain an expression for t in terms of x.

Solution:

(a) $(\dot{x}, x)$ plot

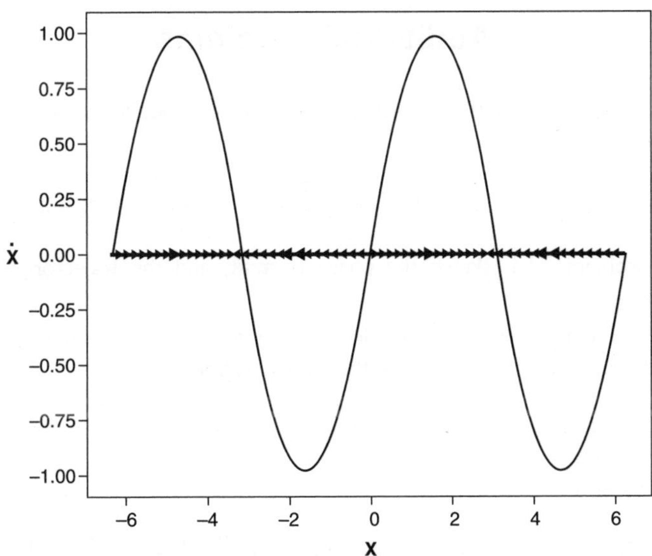

(a) Observe that the arrows in the above plot point away from x = 0, on both sides, while they point towards $x = \pm\pi$ on both sides of $\pm\pi$, thereby indicating the nature of the stability at these points. The method of linear stability analysis, explained in the problem P9.4, can be used further to verify this result.

(b) The points where $\dot{x} = 0$ are called fixed points. One can visualize the differential equation $\dot{x} = \sin x$ as a vector field representing the velocity of a particle at each x.

$$\dot{x} = 0 \Rightarrow \sin x = 0 \quad \therefore\ x = n\pi,\ n \in Z(n = ...-2,-1,0,1,2....).$$

From the above figure in (a), if the flow of the function moves towards a certain fixed point from both directions, it is a stable fixed point, and if the flow moved away from a certain fixed point in both directions, it is an unstable fixed point. If the flow moves toward from one direction and moves away in the other it is a half-stable fixed point, like a 'saddle' point.

$x_0 = 2n\pi,\ n \in Z$ i.e., $x_0 = 0, \pm 2\pi, \pm 4\pi$ ..... are unstable fixed points.

$x_0 = (2n+1)\pi,\ n \in Z$ i.e., $x_0 = 0, \pm \pi, \pm 3\pi$ ..... are stable fixed points.

(c) $\dot{x} = \sin x$

$$\frac{1}{\sin x}\frac{dx}{dt} = 1$$

$$\int_a^x \operatorname{cosec} x \frac{dx}{dt} dt = \int_0^t dt$$

Let $x(t = 0) = a$

$$\Rightarrow \quad t - 0 = \left[-\ln\left|\operatorname{cosec} x + \cot x\right|\right]_a^x; \quad t = \ln\left|\frac{\operatorname{cosec} a + \cot a}{\operatorname{cosec} x + \cot x}\right|.$$

Additional Problems

P9.7 For the previous problem (P9.6), write a small computer code to plot $x(t)$ vs. t for the following initial conditions:

(i) $x(t=0) = \pm\frac{\pi}{4}$, (ii) $x(t=0) = \pm\frac{\pi}{2}$ and (iii) $x(t=0) = \pi$ (iv) $x(t=0) = 0$.

P9.8 For the problem P9.4, sketch the phase portrait, and the trajectories for various initial conditions.

P9.9 Write a computer program to determine the limiting values of the series on the right hand side of the following expressions to check the equations below:

(a) $\pi = \lim_{n\to\infty} \dfrac{2^{4n}}{n\binom{2n}{n}^2}$ (b) $e = \lim_{n\to\infty}\left(1 + \dfrac{1}{n}\right)^n$

(c) $\pi = 4 \lim_{n\to\infty} \sum_{k=1}^{n} \dfrac{(-1)^{k+1}}{2k-1}$ (d) $e = \lim_{n\to\infty}\left[\dfrac{(n+1)^{n+1}}{n^n} - \dfrac{n^n}{(n-1)^{n-1}}\right]$

P9.10 Let $\dot{x} = x^c$, where c is a real constant and $x \geq 0$. Find all values of c for which $x = 0$ is a stable fixed point.

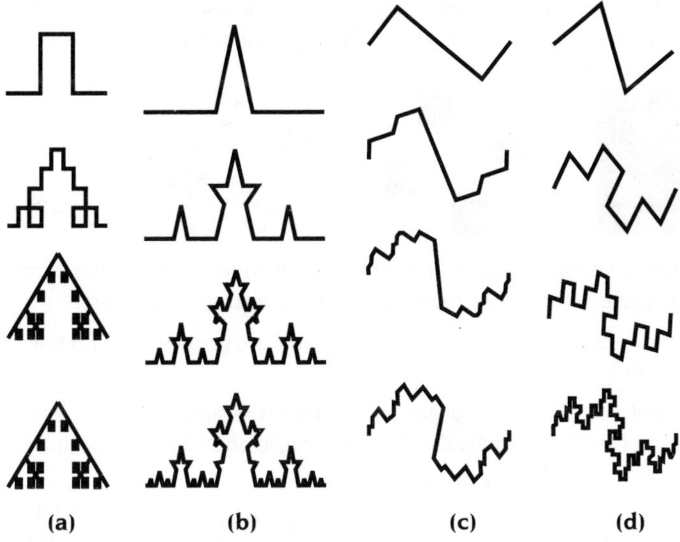

P9.11 What are the fractal dimensions of the figures: (a), (b), (c), and (d)?

P9.12 Suppose the dynamics of a certain artificial neural network is given by $x' = x - x^3$. The initial condition is $x(t = 0) = X$. Determine $Y = x(t \to \text{very large})$ when (i) $X < 0$ and (ii) $X > 0$. The output $Y = x(t \to \text{very large})$ is the attractor of the system $x' = x - x^3$. This is a very simple example of how dynamical neural networks perform information processing.

P9.13 Using the procedure described in the text, determine the Lyapunov exponent for the logistic difference equation. Also, plot the bifurcation diagram for the logistic map. Comment on the results.

P9.14 Plot the bifurcation diagram for $x_{n+1} = r \sin(\pi x_n)$. Numerically calculate the first Feigenbaum constant for (a) this map, and (b) for the logistic difference map. Compare the two results.

P9.15 For the map $x_{n+1} = r \sin(\pi x_n)$ calculate numerically the Lyapunov exponent and plot it.

P9.16 If the side of an initial cube is 2, find the volume and the surface area of the Menger sponge obtained from it after nth iteration. Find the same if the side is 'a'.

P9.17 Determine the volume and the area of the Menger sponge in the limit that the number of iterations goes to infinity.

P9.18 Our lung is a fractal. Assume that the surface area of the lungs is 100 m^2 and the volume is 5 litres. Find the surface area per unit volume of our lungs. Find the radius of a sphere with the same surface area per volume ratio.

P9.19 (a) Construct the Varahamihira's (Pascal's) triangle on a blank piece of paper. Choose a large sheet, and generate at least 8 rows of the Varahamihira's (Pascal's) triangle. Shade the odd-numbers in the triangle. You will notice that the Sierpinski triangle would emerge. (b) Write a code to generate the Varahamihira's (Pascal's) triangle with the odd-numbers in a different color.

P9.20 Write a code to determine if various complex numbers belong to a Mandelbrot set. Select two complex numbers one of which belongs to the Mandelbrot set, and the other outside it. Find the iterations z_n up to $n = 100$.

P9.21 If the perimeter and the area of the initial Sierpinski triangle is P_0 and A_0, respectively, find the perimeter and the area of the triangle after the nth iteration.

P9.22 Observe the first four iterations shown in the figure below. At each iteration, the squares are divided into sixteen smaller squares, and twelve of these are removed. What is the fractal dimension of this object?

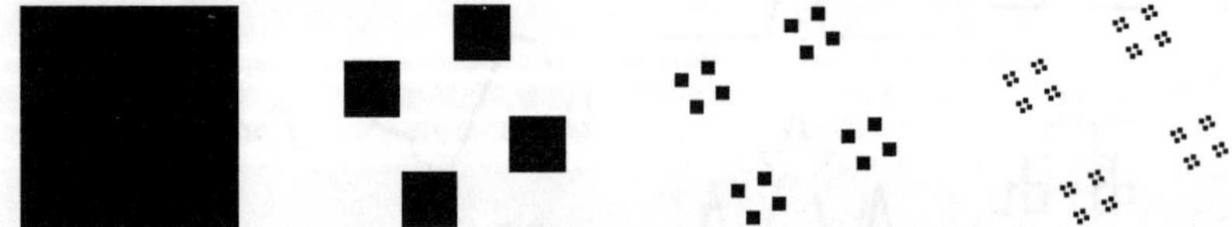

P9.23 Generate the Sierpinski gasket up to the $k = 5$ iteration. What is the fractal dimension of the Sierpinski gasket after infinite repetitions of the construction process?

P9.24 In a Sierpinski carpet, how many smallest squares are present after (a) the 4th iteration (b) the 50th iteration?

P9.25 In Varahamihira's (Pascal's) triangle, each entry is the sum of the two entries above it. Find in which row of the Varahamihira's (Pascal's) triangle does three consecutive entries occur such that their ratio is 3:4:5?

P9.26 Suppose the dynamics of a neural network is given by $x' = x - x^3$. With initial data given by $x(t = 0) = X$. If the output is $Y = x(t \to \infty)$, find Y when (a) $X < 0$ and (b) $X > 0$. Note that the output $(x \to \infty)$ is the attractor of the system $x' = x - x^3$. This is a simple example of how dynamical neural networks perform information processing. For a given input X, the output Y classifies X according to a specific property of X; i.e., the specific property is decided by the attractor of the system.

References

[1] 'Who was Fibonacci?' http://www.maths.surrey.ac.uk/hosted-sites/R.Knott/Fibonacci/fibBio.html. Accessed on 2 July 2018.

[2] Gleick, James. 1987. *Chaos: Making a New Science*. New York: Random House.

[3] Devaney, Robert L. 1992. *A First Course in Chaotic Dynamical Systems*. Boston: Addison-Wesley.

[4] Dewey, David 1996. *Introduction to the Mandelbrot Set – A Guide for People with Little Math Experience*. http://www.ddewey.net/mandelbrot/

[5] Peitgen, H.-O., and P. H. Richter. 1986. *The Beauty of Fractals*. Berlin: Springer-Verlag.

[6] Earnshaw, J. C., and D. Haughey. 1993. 'Lyapunov Exponents for Pedestrians'. *American Journal of Physics*. 61. 401. 61(5): 401-407.

[7] Rajans, Abiya, P. C. Deshmukh, and Neelima M. Gupte. *Lyapunov Exponents to Study Deterministic Chaotic Behavior in the Belousov–Zhabotinsky Reaction* (2019).

[8] Lévy-Véthel, Jacques, and Evelyne Lutton, eds. 2005. *Fractals in Engineering*. London: Springer-Verlag.

[9] McCaulay, Joseph. 2013. *Chaos, Dynamics, and Fractals*. Cambridge: Cambridge Univ. Press.

[10] Espirit, Soul. 1994. *Fractal Trading: Analyzing Financial Markets using Fractal Geometry and the Golden Ratio*. New York: McGraw Hill.

[11] Don, S., D. Chung, D. Min, and E. Choi. 2013. 'Analysis of Electrocardiogram Signals of Arrhythmia and Ischemia Using Fractal and Statistical Features.' *Journal of Mechanics in Medicine and Biology*. 13(01). 1350008. Downloaded from www.worldscientific.com on 06/22/14.

CHAPTER 10

Gradient Operator, Methods of Fluid Mechanics, and Electrodynamics

There is no greater burden than an unfulfilled potential.

—*Charles M. Schulz*

10.1 THE SCALAR FIELD, DIRECTIONAL DERIVATIVE, AND GRADIENT

In the discussion on Fig. 2.4 (Chapter 2), we learned that it was neither innocuous to define a vector merely as a quantity that has both direction and magnitude, nor to define a scalar simply as a quantity that has magnitude alone. It is not that the properties referred here of a scalar and a vector are invalid. Rather, it is to be understood that these *properties* do not provide an unambiguous *definition*. Only a *signature criterion* of a physical quantity can be used to define it. We therefore introduced, in Chapter 2, comprehensive definitions of the scalar as a tensor of rank 0, and of the vector as a tensor of rank 1. In this chapter we shall acquaint ourselves with the mathematical framework in which the laws of fluid mechanics and electrodynamics are formulated using vector algebra and vector calculus. In fact, the techniques are used not merely in these two important branches of classical mechanics, but also in very various other subdivisions of physics. The background material seems at times to be intensely mathematical, but that is only because the laws of nature engage a mathematical formulation very intimately, as we encounter repeatedly in the analysis of physical phenomena. There are many excellent books in college libraries from which one can master the mathematical methods. These topics are extremely enjoyable to learn; they help us develop rigorous insights in the laws of nature. The literature on these topics is vast. A couple of illustrative books [1, 2] are suggested for further reading.

We consider the example of a particular scalar function, namely the temperature distribution in a room. The temperature is a physical property at a particular point in space, such as the point P in Fig. 10.1. Depending on the distribution of the sources of heat in the region that surrounds the point P, temperature may be different from point to point in space, and also possibly from time to time. The reason the temperature at a point is a scalar, is that its value at that point is independent of where the observer's frame of reference is located, and also independent of how it is oriented. Thus, the temperature at the point P is essentially the same, regardless of the frame referenced being F, or F' (Fig. 10.1). The temperature is a *point function*, i.e., it is a function of the *particular point in space*. Being a scalar, it is called a *scalar point function*. When we describe it for *all* the points in a neighborhood, we have a *scalar field* defined in that region of space. Of further interest now is the question how the temperature changes from the point P to some other point in the scalar field? Clearly, when different sources (and sinks) of heat are scattered in the region, the variation in temperature in different directions from point P could be quite different from each other. In Fig. 10.1, the variation in temperature from P toward the fire would naturally be different from what it is toward the ice-cream which the child is having.

The *rate of change* of temperature with distance, *at the point P* in Fig. 1, is given by

$$\left[\frac{dT}{ds}\right]_P = \left[\lim_{\delta s \to 0} \frac{\delta T}{\delta s}\right]_P . \tag{10.1}$$

This rate depends on whether we consider the distance δs in the denominator to be in the direction from the point P *toward* the source of fire, or in some other direction. The *derivative* of the temperature with respect to distance is therefore a scalar, but its value depends on the *direction* in which the distance δs is referenced. The scalar, $\left[\frac{dT}{ds}\right]_P$, is therefore called the *directional derivative*. Nonetheless, the specific information regarding '*which specific direction*' is missing from Eq. 10.1. As its very name suggests, despite being a scalar, the directional derivative has a *directional* attribute. A direction that may be represented by a unit vector $\hat{e}$ must therefore be implicitly involved in some way.

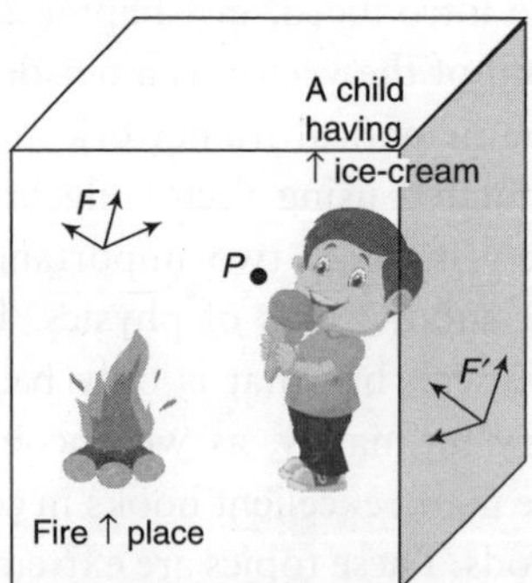

Fig. 10.1 If there is a fireplace in a room, and someone is having ice-cream around it, the temperature from the point P would change at a different rate towards the fireplace as toward the child having ice-cream. The value of temperature at any point does not depend on what the observer's frame of reference is, whether frame F, or another, such as frame F' that is arbitrarily displaced, and also turned, with respect to the previous one.

The magnitude of the directional derivative need not be the same in all directions; it may even be zero. For example, along the isotherm (i.e., a curve on which the temperature is constant), the directional derivative will be, of course, essentially zero. In general, when a scalar point function changes from point to point in a region, its rate of change with distance is *maximum* in a *specific direction* of a *particular* vector, called the *gradient* of the scalar field.

The scalar field that we shall now consider may be completely arbitrary. It could be the temperature field as we have considered above, or the height of a mountain, as was considered in Fig. 2.4 (Chapter 2). We shall denote this arbitrary scalar field by $\psi(\vec{r})$. The scalar field of considerable interest in this chapter is the 'potential field', but we shall introduce it after persisting for now with an arbitrary scalar field. The argument $(\vec{r})$ of $\psi(\vec{r})$ reminds us that the scalar field is a 'point' function; it may have different value at different points in space. In fact, it may change also with time. To indicate the dependence of the scalar field on time t, we may denote the scalar function by $\psi(\vec{r}, t)$. However, for the time being, we shall not worry about the time dependence. The position vector of the point P under consideration appears as an explicit argument, but only because it pins down the point P. The coordinate frame of reference in which the position vector is described is not important; only the point P is.

We know now that the directional derivative of $\psi(\vec{r})$ at a point P is denoted by $\left[\dfrac{d\psi}{ds}\right]_P$. The gradient of the scalar field, is written as $\vec{\nabla}\psi$. It is such a vector that its *component* along a unit vector $\hat{e}$ gives the directional derivative in that direction:

$$\left[\frac{d\psi}{ds}\right]_P = \hat{e}\cdot\vec{\nabla}\psi. \tag{10.2}$$

The scalar product in Eq. 10.2 has the cosine of the angle between $\vec{\nabla}\psi$ and $\hat{e}$. Its largest value therefore occurs when the directional derivative is taken in the direction of the gradient itself. The symbol $\vec{\nabla}$ is read as the 'gradient operator', or 'del operator', and sometimes also as the 'nabla' operator. For completeness, one should include a subscript P on $\vec{\nabla}\psi$, and write it as $[\vec{\nabla}\psi]_P$. For brevity, however, such a subscript is usually omitted. Higher derivatives with respect to coordinates may also exist, as the gradient may change from one point to another. You must remember that the gradient operator, $\vec{\nabla}$, is a vector *operator*. It is not a vector itself. When it operates on a scalar field ψ, it returns a vector result, namely the $\vec{\nabla}\psi$.

The gradient of a scalar field is especially useful when a scalar point function, varies at a different rate in different directions. Fig. 10.2 shows an obvious example. It shows a horse's saddle. Manifestly, from the point P, the slope of the saddle changes at a different rate with displacement in different directions. The slope is different along the length of the horse (i.e., along SPT) compared to that along the width of the horse (along UPW). The point P is an equilibrium point. However, it is an unstable equilibrium with regard to displacement along UPW, and it is a point of stable equilibrium with regard to a displacement along SPT. The value of the directional derivative of the height of a point on the saddle from the ground in different directions from the point P then depends on the direction $\hat{e}$ in which the directional derivative is sought. Now, the unit vector in any direction can be easily obtained by considering a tiny displacement vector $\overrightarrow{\delta r}$ in a chosen direction, dividing it by the magnitude $\delta s = |\overrightarrow{\delta r}|$ of that displacement vector, and then taking the limit $\delta s \to 0$:

$$\hat{e} = \lim_{\delta s \to 0} \frac{\overrightarrow{\delta r}}{\delta s}. \tag{10.3}$$

Hence, $$\left[\frac{d\psi}{ds}\right]_P = \hat{e} \cdot \vec{\nabla}\psi = \lim_{\delta s \to 0} \frac{\overrightarrow{\delta r}}{\delta s} \cdot \vec{\nabla}\psi, \tag{10.4a}$$

i.e., $$\lim_{\delta s \to 0}\left[\frac{\delta\psi}{\delta s}\right]_P = \left[\frac{d\psi}{ds}\right]_P = \lim_{\delta s \to 0} \frac{\overrightarrow{\delta r} \cdot \vec{\nabla}\psi}{\delta s}. \tag{10.4b}$$

Accordingly, we may write

$$[\delta\psi]_P = \left[\overrightarrow{\delta r} \cdot \vec{\nabla}\psi\right]_P, \tag{10.5a}$$

or, $$[d\psi]_P = \left[\overrightarrow{dr} \cdot \vec{\nabla}\psi\right]_P. \tag{10.5b}$$

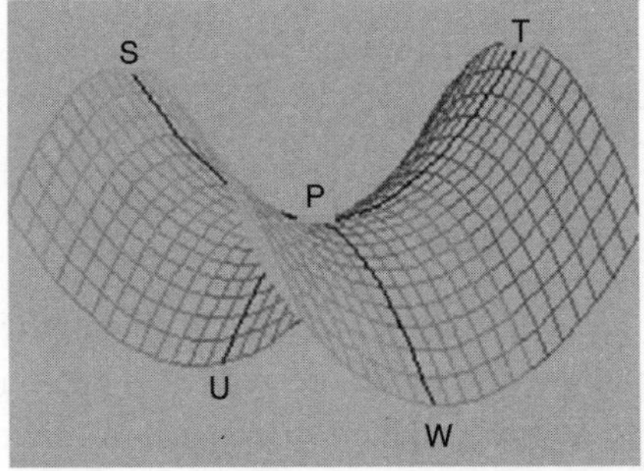

Fig. 10.2 The point P on a horse's saddle is a point of stable equilibrium with respect to displacements along SPT, and unstable equilibrium with respect to displacements along UPW. Such a point, which is both a stable and an unstable equilibrium, is called z 'saddle point'. The slope of the saddle is different in different directions from the saddle point. The value of the directional derivative at the point P is therefore different along the length of the horse, from that along the horse's width, and also from all intermediate directions.

Equation 10.5 represents an infinitesimal *difference* in the value of the scalar function at the point P, and its value at a neighboring point displaced through $\overrightarrow{\delta r}$. Now, the difference $\delta\psi$ is of course independent of the coordinate system in which the position vector is expressed. We may have described the position vector of P in the frame F (Fig. 10.1), or the frame F', which is arbitrarily displaced, and even turned, with respect to the frame F. Moreover, the scalar product on the right hand side of Eq. 10.5 is independent of whether the coordinate system employed is the Cartesian coordinate system, or cylindrical polar, or spherical polar coordinate system, or any other. We have discussed these coordinate systems in Chapter 2.

From Chapter 2 we know the exact expressions for the infinitesimal displacement vector $\vec{\delta r}$ in the more commonly used Cartesian, cylindrical polar, and the spherical polar coordinate systems. These are compiled in Table 10.1. Using these expressions, we can easily figure out how to express the gradient of the scalar field, $\vec{\nabla}\psi$, in the different coordinate systems. In the last column of Table 10.1, the following question is raised: what must be the expression for the gradient operator, $\vec{\nabla}$, in each coordinate system? To answer this question, all we now have to do is to use the fact that the difference $\delta\psi = (\vec{\delta r}) \cdot (\vec{\nabla}\psi)$ in any coordinate system which employs the unit base vectors $\{\hat{e}_1, \hat{e}_2, \hat{e}_3\}$, will have the form $\sum_{i=1}^{3} a_i b_i = \vec{a} \cdot \vec{b}$, with $\vec{a} = \sum_{i=1}^{3} \hat{e}_i a_i$ and $\vec{b} = \sum_{i=1}^{3} \hat{e}_i b_i$. This form holds good in *every* orthonormal basis $\{\hat{e}_1, \hat{e}_2, \hat{e}_3\}$, whether the Cartesian, the cylindrical polar, or the spherical polar coordinate systems, or any other.

Table 10.1 Expressions for an infinitesimal displacement vector in the commonly used coordinate systems.

Coordinate system	Infinitesimal displacement vector $\vec{\delta r}$	What must be the expression for the gradient operator, $\vec{\Delta}$ in each coordinate system, since $\delta\psi\,(r_{\rightarrow})$ must be as given below ↓ ?
Cartesian	$\vec{\delta r} = \hat{e}_x \delta x + \hat{e}_y \delta y + \hat{e}_z \delta z$	$\delta\psi(x, y, z) = \left(\frac{\partial\psi}{\partial x}\right)\delta x + \left(\frac{\partial\psi}{\partial y}\right)\delta y + \left(\frac{\partial\psi}{\partial z}\right)\delta z$
Cylindrical Polar	$\vec{\delta r} = \hat{e}_\rho \delta\rho + \hat{e}_\varphi \rho\delta\varphi + \hat{e}_z \delta z$	$\delta\psi(\rho, \varphi, z) = \left(\frac{\partial\psi}{\partial\rho}\right)\delta\rho + \left(\frac{\partial\psi}{\partial\varphi}\right)\delta\varphi + \left(\frac{\partial\psi}{\partial z}\right)\delta z$
Spherical Polar	$\vec{\delta r} = \hat{e}_\rho \delta r + \hat{e}_\theta r\delta\theta + \hat{e}_\varphi r \sin\theta\delta\varphi$	$\delta\psi(r, \theta, \varphi) = \left(\frac{\partial\psi}{\partial r}\right)\delta r + \left(\frac{\partial\psi}{\partial\theta}\right)\delta\theta + \left(\frac{\partial\psi}{\partial\varphi}\right)\delta\varphi$

It is now easy to *deduce* the expressions for $\vec{\nabla}\psi$ in the three coordinate systems, since only a specific, unique, decomposition of $\vec{\nabla}\psi$ along the basis vectors $\{\hat{e}_1, \hat{e}_2, \hat{e}_3\}$ would be compatible with the decomposition of $\vec{\delta r}$ in the corresponding basis to generate the correct expression for $\delta\psi$ (given in the last column of Table 10.1). In Table 10.2 the result is presented, which you would have hopefully already obtained. Note that we have not *defined* the gradient using the *forms* in Table 10.2. The definition of the gradient in Eq. 10.4 is *independent* of any coordinate system, as it must be so. The choice of a coordinate system can be a matter of convenience and may vary from one situation to another. The definition of a physical quantity must, however, be independent of such choices. The different expressions for the gradient in different coordinate systems, given in Table 10.2, have only resulted on using Eq. 10.5 in various coordinate systems. In deducing the expressions of the gradient in Table 10.2, the three coordinate systems have been treated secularly, as we certainly should. One can of course use other coordinate systems, such as the parabolic or any other curvilinear coordinate system, discussed in Chapter 2. The expressions for the gradient in those systems can be obtained using the above simple procedure. It is not necessary to memorize the expressions for the gradient operator in different coordinate systems, since we know now how to deduce the same quickly enough using Table 10.1. Note that each term in the last column of Table 10.2 has the dimension of inverse-length. It is nice to keep this in mind, since it could help a quick way of detecting any inadvertent error. The gradient operator has the dimension of inverse-length.

Table 10.2 Expressions for the gradient operator in commonly used coordinate systems.

Coordinate system	Expression for $\vec{\nabla}\psi(\vec{r})$ as would give the correct $\delta\psi$ in the last column of Table 10.1	Consequent expression for the gradient operator, $\vec{\nabla}$
Cartesian	$\vec{\nabla}\psi(\vec{r}) = \left[\hat{e}_x \frac{\partial}{\partial x} + \hat{e}_y \frac{\partial}{\partial y} + \hat{e}_z \frac{\partial}{\partial z}\right]\psi(\vec{r})$	$\vec{\nabla} = \hat{e}_x \frac{\partial}{\partial x} + \hat{e}_y \frac{\partial}{\partial y} + \hat{e}_z \frac{\partial}{\partial z}$
Cylindrical Polar	$\vec{\nabla}\psi(\vec{r}) = \left[\hat{e}_\rho \frac{\partial}{\partial \rho} + \hat{e}_\varphi \frac{1}{\rho}\frac{\partial}{\partial \varphi} + \hat{e}_z \frac{\partial}{\partial z}\right]\psi(\vec{r})$	$\vec{\nabla} = \hat{e}_\rho \frac{\partial}{\partial \rho} + \hat{e}_\varphi \frac{1}{\rho}\frac{\partial}{\partial \varphi} + \hat{e}_z \frac{\partial}{\partial z}$
Spherical Polar	$\vec{\nabla}\psi(\vec{r}) = \left[\hat{e}_r \frac{\partial}{\partial r} + \hat{e}_\theta \frac{1}{r}\frac{\partial}{\partial \theta} + \hat{e}_\varphi \frac{1}{r\sin\theta}\frac{\partial}{\partial \varphi}\right]\psi(\vec{r})$	$\vec{\nabla} = \hat{e}_r \frac{\partial}{\partial r} + \hat{e}_\theta \frac{1}{r}\frac{\partial}{\partial \theta} + \hat{e}_\varphi \frac{1}{r\sin\theta}\frac{\partial}{\partial \varphi}$

We shall now discuss a specific scalar field, namely the potential field. The term potential energy is familiar to students from their high-school physics courses. It is the energy an object has by virtue of its position; it renders a capacity in the body to do work. In Fig. 4.1 (Chapter 4), a 1-dimensional potential is plotted as a function of position. We have already discussed various kinds of equilibrium, shown on this plot. At a point of equilibrium, an object's momentum does not change. Thus, $\frac{d\vec{p}}{dt} = \vec{0}$ at all points of equilibrium, whether stable, unstable, or neutral. At these points, no force acts on the body, and its momentum is completely determined by its inertia of motion, which of course includes both the state of rest, or of uniform momentum. It is therefore clear that a *change* in the *potential* with respect to the position of the object is necessary to change the object's momentum. From an analysis of Fig. 4.1, the following four points may be appreciated:

(i) The derivative of the potential with respect to the coordinate x must be *non-zero* for the force deducible from it to change the object's momentum.

(ii) At a point of stable equilibrium, the force would be *toward* the equilibrium, whichever way (i.e., toward the left, or the right the object is displaced).

(iii) At a point of unstable equilibrium, the force would be *away* from the equilibrium point, whichever way the object is displaced.

(iv) Greater the magnitude of the slope $\frac{dV}{dx}$, greater will be the acceleration produced in the object.

The above points are recognized in a single, simple, equation which identifies the force acting on the body as

$$F = -\frac{dV}{dx}. \tag{10.6}$$

The negative sign in Eq. 10.6 is important. It is needed to provide the appropriate direction of force, as required by the points (ii) and (iii) above. As is readily recognized from Table 10.2, it is the gradient of the potential that provides the 3-dimensional equivalent expression for the force, no matter which coordinate system is used:

$$\vec{F} = -\vec{\nabla}V. \tag{10.7}$$

The potential function $V = V(\vec{r})$ is a scalar point function; it is defined at every point in the region of space under consideration. Essentially, we have a scalar field in the space under consideration. Its negative gradient at each point tells us what force would act on a body placed at that point. The negative gradient of the potential, $-\vec{\nabla}V$, then gives the force, at the very point where the gradient is taken. The force (Eq. 10.7), also generates a point function (defined at each point in space), which is a *vector* physical quantity. It produces a vector field at each point in space. More generally, we can have a tensor field of appropriate rank, defined at each point in space. Equation 10.7 is well reconciled with Newton's second law, $\vec{F} = -\frac{d\vec{p}}{dt}$, for conservative fields. The potential field is conservative, when the work done by the force is independent of the path along which an object is displaced by the force. Equivalently, the work done by a conservative force over a closed path is zero:

$$\oint \vec{F} \cdot \overrightarrow{dl} = 0. \tag{10.8a}$$

The 'circle' on the integration symbol in Eq. 10.8a denotes the mandatory requirement that the integration is carried out over a *closed* path (not necessarily circular). No matter what path a to b is taken to displace an object by the force that is obtained from the gradient of the potential as per Eq. 10.7, the work done over the closed path is just the sum

$$\oint \vec{F} \cdot \overrightarrow{dl} = \int_a^b \vec{F} \cdot \overrightarrow{dl} + \int_b^a \vec{F} \cdot \overrightarrow{dl} = 0. \tag{10.8b}$$

Manifestly, the path independence of work done is guaranteed by the fact that

$$\int_a^b \vec{F} \cdot \overrightarrow{dl} = -\int_b^a \vec{F} \cdot \overrightarrow{dl}, \tag{10.8c}$$

irrespective of the path a to b. The path b to a need not necessarily retrace backward the exact path a to b; only that the complete paths must be entirely within the region of the conservative potential.

Now, the fundamental interactions in nature, like the gravitational and the electromagnetic, are conservative interactions. We consider the displacement of a mass m by the gravitational force $G\frac{Mm}{r^2}\hat{u}$. The gravitational force is an attractive force exerted on the mass m by another mass M. We may also consider the electromagnetic force $q[\vec{E} + (\vec{v} \times \vec{B})]$, usually called the Lorentz force, on a charge q exerted by an electromagnetic field. In this expression, the electric intensity at the position $\vec{r}$ of the charge q is $\vec{E}(\vec{r})$, the magnetic induction field at that position is $\vec{B}(\vec{r})$, and the instantaneous velocity of the charge is $\vec{v}$. Now, it will be nice to develop our formalism with respect to the gravitational interaction in such a way that it is applicable to describe the dynamics of *all masses m* no matter how large or small. Likewise, we should develop the analysis with respect to electromagnetic interaction in a manner which would account for the dynamics of any charge q, no matter what its value is. It is therefore best to develop a general methodology that describes the effect of the gravitational field generated by the mass M, on a *test mass* of 1 unit (whatever be the system of units in which we carry out the analysis). The effect on an arbitrary mass m can then be obtained simply by scaling the results for the unit test mass. A similar thing can be done for electromagnetic interaction, by describing the dynamics of a *unit test charge* in the electromagnetic field of intensity $(\vec{E}, \vec{B})$. The gravitational force on the *unit test mass* due to the field generated by the mass M is $\vec{e}_g = G\dfrac{M}{r^2}\hat{u}$. This force, per unit mass, is called the *intensity* of the gravitational force, and it is in fact nothing but the acceleration due to gravity $\vec{e}_g = \vec{g}$. It has the dimensions of acceleration LT^{-2}. Likewise, the intensity of the electromagnetic force is the force per unit charge, $\vec{E} + (\vec{v} \times \vec{B})$.

One often suspects that a velocity-dependent force is non-conservative. The electromagnetic force depends on the velocity of the charge, but it is a conservative force since the work done by the velocity-dependent magnetic term, $\delta w = (\vec{v} \times \vec{B}) \cdot \overrightarrow{\delta r}$ is identically zero, since $\vec{v} = \lim\limits_{\delta t \to 0} \dfrac{\overrightarrow{\delta r}}{\delta t}$. Independence with respect to velocity is of course *not* the criterion that defines a conservative force. The defining criterion is simply Eq. 10.8; it has to do with the path-independence of the work done. Both gravitational and electromagnetic forces perform path-independent work while displacing the object on which they act. Notwithstanding this, we often encounter energy dissipation, but only because the losses come from our excluding from the equation of motion one or more object with which the object of interest may exchange energy. As explained in Chapter 4, friction results essentially from *unspecified* degrees of freedom.

Since the gravitational interaction between any pair of masses decreases with distance, we expect the interaction to vanish at infinite separation between the masses. Distances need to be referred to a reference frame, so we choose a coordinate system centered at the mass M, which is the source of gravitational field in the present case. The zero of the energy scale is then chosen to be at infinite distance from M. The attractive gravitational force that M would exert on a unit test mass placed at infinity would displace the test mass, and bring it closer to itself, say to a finite distance, r, from M. In doing so, the gravitational field of M would perform work on the unit test mass. This work done would be stored in the test mass as its capacity to do work, i.e., its potential. As much work will have to be done by an external agency on the test mass to take it back to infinite distance against the attractive force of gravity. We shall write the gravitational potential as

$$V_g = -G\frac{M}{r}. \tag{10.9}$$

It is easy to see that the magnitude, and the sign, of the right hand side of Eq. 10.9 are correct. As can be seen by using Eq. 10.7, the negative gradient of the potential given in Eq. 10.9 is

$$-\vec{\nabla}V_g = -\hat{e}_r \frac{\partial}{\partial r}[V_g(\vec{r})] = -\hat{e}_r \frac{\partial}{\partial r}\left(-\frac{GM}{r}\right) = -\hat{e}_r \frac{GM}{r^2}. \tag{10.10}$$

We have used the expression for gradient in the spherical polar coordinates, given in Table 10.2. The above equation shows that the direction of the force would be along $-\hat{e}_r$, which is *toward* the mass M, remembering that the spherical polar radial unit vector has its plus sign *away* from the center (Chapter 2). When M represents the mass of the Earth, $-\vec{\nabla}V_g = \vec{g}$, the acceleration due to gravity. We know that Eq. 10.10 is correct; it is nothing but the Newton's law of gravity. An opposite sign in Eq. 10.9 would result in the force being directed *away* from the center, which would be absurd. An *attractive* potential is therefore written with a *negative* sign; a *repulsive* potential with a *positive* sign. These sign conventions are beautifully tied up with the sign conventions of unit vectors in the coordinate systems so that together they make perfect sense.

Physical results must of course be independent of the choice of a coordinate system. Using the expression for the gradient in any of the coordinate systems, it can be easily verified that the *negative* gradient of the scalar potential field (Eq. 10.9) gives the gravitational force field at every point. We shall defer our discussion on the potential representing the electromagnetic interaction until Chapter 12. The electromagnetic potentials and the electromagnetic field will be found to have some similarities with the gravitational case, but also some differences, which are both interesting and instructive. In the meanwhile, we may comfortably anticipate that the electric field $\vec{E}(\vec{r})$, and the magnetic induction field $\vec{B}(\vec{r})$, each constitutes a vector field, consisting of a vector point function in a region of space.

10.2 THE 'IDEAL' FLUID, AND THE CURRENT DENSITY VECTOR FIELD

Now that you have developed some acquaintance with scalar and vector fields, we shall study the *fluid velocity field*. A 'fluid' is that state of matter that *flows*. It includes both liquids and gases. Our prototype of fluid flow may be, very simply, water flowing in a river, or in a hose pipe. The mechanics and dynamics of fluids are referred to as 'fluid mechanics' and 'fluid dynamics', respectively. Earlier studies often referred to the field as 'hydrodynamics', since the Greek word for water is *hýdōr*, and this term has continued in parallel use. However, fluids have extremely complicated dynamics, and the dynamics of water does not represent the dynamics of all fluids. In fact, even the description of the flow of water is not easy. At the molecular level, neighboring molecules of water cluster together through hydrogen bonding, constituting dimers and trimers. These tumble and flow together, but also break up forming new dimers and trimers dynamically. The dynamics of a fluid is therefore really complicated.

At the macroscopic level, fluids have physical properties which change *continuously* from one point to another. At the microscopic level, however, fluids have internal molecular, and atomic, structure. When we talk about moving from one 'point' in the fluid to another, we can get into serious difficulties if we were to refer to displacements of the size of atomic dimensions. The challenging study of fluids is therefore developed by making two specific approximations which are central, and crucial:

(i) the *continuum* hypothesis.
(ii) the assumption that the fluid under consideration is an *ideal* fluid.

First, we discuss the continuum approximation. Let us consider the density of the fluid at a point ($\vec{r}$) in the medium. It is given by the limiting value of the ratio of the mass to volume, in the limit in which the volume tends to zero: $\rho(\vec{r}) = \lim_{\delta V \to 0} \frac{\delta m}{\delta V}$. In this expression, δm is the mass of the fluid in the tiny volume element δV. We already have a problem here, since the mathematical limit $\delta V \to 0$ is obviously smaller than atomic dimensions. In this mathematical limit, the idea of a homogenous fluid, of course, breaks down completely. Nonetheless, we are mostly interested in properties over macro-scale, and not over the micro-scale. On the macro-scale, physical properties of the fluid get averaged out, and the *average* properties can be considered to be varying smoothly in the region of the fluid. In the *continuum hypothesis*, we remain content to deal with these average properties so that the physical properties like density and pressure of the fluid can be taken to vary continuously from point to point. We can seek and obtain the derivatives of these physical quantities with respect to position, and also with respect to time. Essentially, in this model, the limit $\delta V \to 0$ would not even refer to the approach to the small molecular sizes, let alone to the mathematical limit which would tend to a 'geometrical point'. Rather, it shall refer to approaching a size that is extremely small, but larger than the molecular dimensions. It would be the smallest conceivable size at which we may talk about a 'fluid particle' contained in that elemental size. It cannot be defined in terms of the molecular size, since the molecules of the fluid often form dimers and trimers, etc. The continuum hypothesis circumvents details with regard to the fluid structure on the molecular scale. This approximation facilitates development of powerful models to describe fluid dynamics. The fluid velocity at a point is then dealt within the continuum hypothesis. As such, at a point in the region of the fluid, a large number of fluid particles would be criss-crossing each other. Their average, however, has a smooth variation, and it is this average that we shall deliberate on within the framework of the continuum model of the fluid. This would then permit us to analyze the rate at which components of the velocity vector change with both space coordinates, and with time.

Since even water has rather complex properties, the theoretical formalism for fluid dynamics is built for a hypothetical fluid, which shall be called the 'ideal' fluid. It is defined by a few characteristic properties. These properties are so chosen that many fluids, though not all, have properties that come rather close to those qualities which define the ideal fluid. The theoretical formalism developed for the ideal fluid is then applicable, with minimal corrections, to a large number of real liquids. The definition of the 'ideal' fluid is chosen to be characterized by the following properties:

The 'ideal' fluid is defined by a few characteristic properties. These properties are so chosen that many fluids, (specially water) have properties that come rather close to these qualities. The theoretical formalism developed for the ideal fluid is then applicable, with minimal improvisation, to a large number of real liquids. The 'ideal' fluid is characterized by the following properties:

(i) its density $\rho(\vec{r}, t)$, in its rest frame,

(ii) its isotropic pressure $p(\vec{r}, t)$,

(iii) its velocity $\vec{v}(\vec{r}, t)$.

In addition to the above physical properties, an ideal fluid is considered to have the following properties:

(iv) it has zero viscosity,

(v) the stress in an ideal fluid is one of 'compression'; there is no 'shear stress' in it, and also no 'tension stress'. The terms compression, shear and stress are shortly defined below.

The physical state of the ideal fluid is completely characterized by five parameters: density $\rho(\vec{r}, t)$, pressure $p(\vec{r}, t)$, and the three components of velocity $\vec{v}(\vec{r}, t)$. The density of the *ideal* fluid is considered to remain invariant. The ideal fluid is *incompressible.* That it has no viscosity implies that layers of the ideal liquid do not exchange heat through friction by rubbing against each other. To appreciate the ideal fluid's characteristic 'compression stress' mentioned above, we must carefully define (i) 'tension', (ii) 'shear', and (iii) 'compression'. Toward that end, we consider a 'point' P (in the sense of the continuum hypothesis) within a fluid, as shown in Fig. 10.3. We will be interested in the force applied by the fluid at the point P, over an infinitesimal elemental area that passes through that point. In paticular, we shall examine the component of the force in a direction normal to the elemental area. Studying the force *per unit area*, in the limit $\delta A \to 0$, will help us study the fluid pressure at that particular 'point' P. It is now superfluous to remind that this is not the dimensionless point that we speak of in Euclidean geometry. It is the tiny volume that we employ in the continuum hypothesis. The continuum hypothesis enables using tools of differential calculus in studying continuous functions which describe various properties of the fluid. The limiting value of an elemental area $\delta A \to 0$ is also to be interpreted within the framework of the continuum hypothesis. Similarly, we shall interpret the limit $\delta A \to 0$ of a tiny volume element withing the framework of the continuum hypothesis.

An infinite number of surfaces pass through the point P. Some such illustrative surfaces are shown in Fig. 10.3. To be able to pick an unambiguous elemental area, we shall represent the elemental tiny area by a vector whose magnitude is equal to the size of the area, and whose direction is normal to surface element. The direction of the vector area is defined by the right-hand screw convention, described below. This convention uniquely describes the orientation of the surface area element.

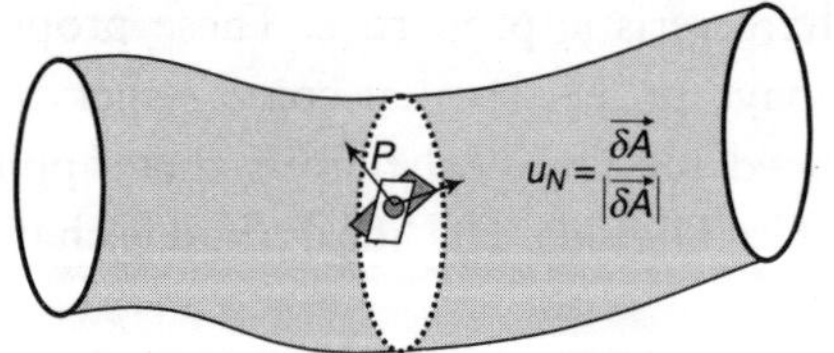

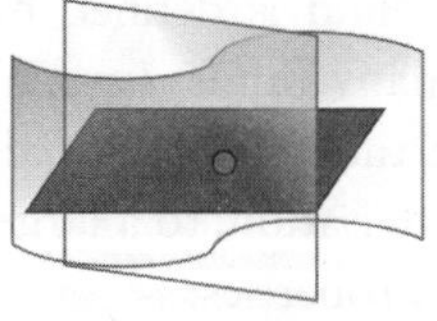

Fig. 10.3 At the point P inside the fluid, one may consider a tiny surface element passing through that point. An infinite number of surfaces of this kind can be considered.

The area of the tiny rectangular *flat* strip $ABCD$ (Fig. 10.4a) is represented by the vector $\overrightarrow{AB} \times \overrightarrow{BC}$. The direction of the surface element is given by the right hand screw rule for the cross product $\overrightarrow{AB} \times \overrightarrow{BC}$. This vector is in fact a pseudo- (i.e., axial-) vector, being the cross product of two polar vectors. The unit normal vector indicating the orientation of the strip is $\hat{u} = \frac{\overrightarrow{AB} \times \overrightarrow{BC}}{\left|\overrightarrow{AB} \times \overrightarrow{BC}\right|}$.

In fact, using this technique, the orientation of the surface at any particular point on it, can be defined unambiguously even if the surface is a large one, and/or it is arbitrarily ruffled, as long the surface is bounded by a well-defined, unambiguous edge as shown in Fig. 10.4b. As long as the boundary edge never crosses itself, an infinitesimal tiny loop *abcda* can be constructed *around* any point P on the surface (Fig. 10.4b).

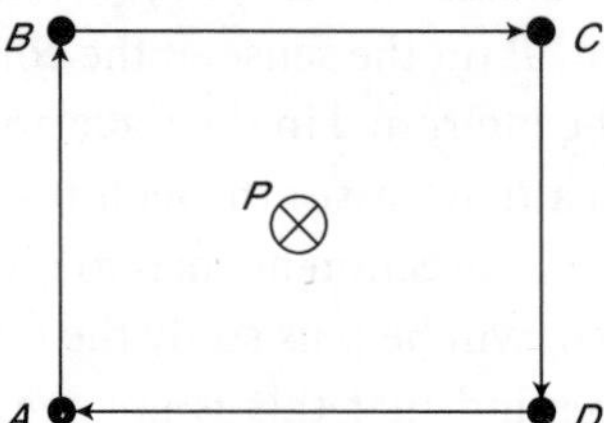

Fig. 10.4a The orientation of an infinitesimal tiny rectangle around a point P is given by the forward motion of a right hand screw as would traverse in the sense $ABCDA$. In the above figure, the orientation of the infinitesimal rectangle $ABCDA$ is into the plane of the paper at the point P, indicated by the tail of the arrow at P.

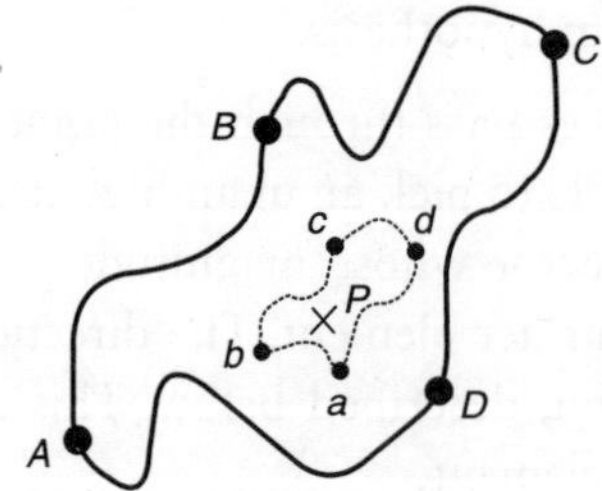

Fig. 10.4b The orientation of a surface at an arbitrary point P on it can be defined unambiguously even for a large ruffled surface, which may even have a zigzag boundary.

The path abcda can be *turned* and *expanded* (like an elastic strand), without getting *out* of the surface, in such a way that its edge (i.e., its perimeter) would coincide with the outer edge $ABCDA$. The right hand screw convention can then be applied to determine the orientation of the ruffled surface at the point A, by considering the rotation of the screw along the path $ABCDA$, or equivalently along abcda. The unit normal to the surface at the point P is then the direction $\hat{u}$ in which the right hand screw would advance as the screw is turned along abcda in the limit that the elemental area $\delta A \to 0$. A normal vector can therefore be drawn at any point on the surface, however, large or ruffled. Some examples of such surfaces are shown in Fig. 10.5. The only ambiguity in the direction of the orientation comes from how the edge is considered to be traversed by the right hand screw. If it goes along the path $A \to B \to C \to D \to A$, the orientation would be along $\hat{u}$. If it is opposite, i.e., along $A \to D \to C \to B \to A$, then the surface orientation is taken to be along $-\hat{u}$. The direction of the orientation of the surface is therefore uniquely fixed by picking the rotational sense in which the right hand screw is considered to turn. Note that the surfaces may be not merely curved, but also *ruffled up and down* in infinitely many different ways, as suggested in the various surfaces shown in Fig. 10.5. The only thing is that the path $ABCDA$ must not cross itself. If it did, the path would not enclose a properly orientable surface. We always consider the rim of a right hand screw to be turned along the edge (perimeter) of the orientable surface. The direction of the normal vector which represents the orientation of the surface is uniquely given by the *forward* motion of the screw.

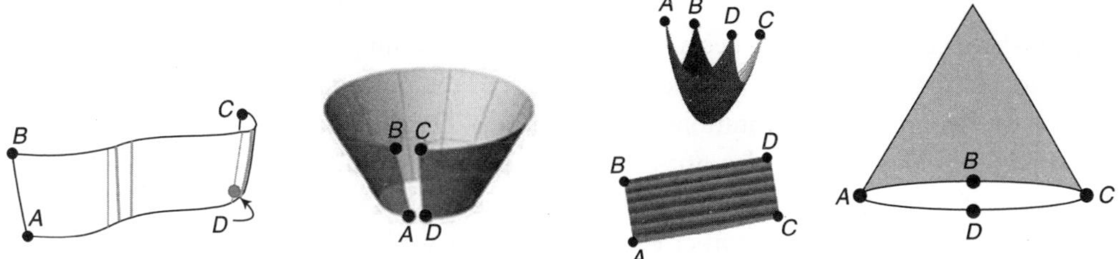

Fig. 10.5 All the surfaces shown in this figure are *orientable* according to the right hand screw convention irrespective of their curvatures and ruffles.

If the edge of the surface is traversed along the opposite way by the right hand screw, the orientation of the surface would change, becoming opposite as well. Even if the butterfly net shown in Fig. 10.6 is *pinched*, and only partially pulled 'out' of the rim, the orientation of the surface that constitutes the butterfly net can be defined without ambiguity at any point P on the net. The direction of the orientation of the net's surface remains fixed, irrespective of how the net is pinched, pulled and ruffled. This is because the right hand screw convention makes a reference only to the sense in which the edge (perimeter) of the surface is thought to be traversed. It uniquely fixes the orientation of the surface that makes up the net. Surfaces that can be oriented in this manner are called 'well-behaved', 'orientable' surfaces. A unique direction $\hat{u}$ which is normal to the surface can be assigned at each point on such surfaces. Identifying this direction is very important to recognize orientation of any surface.

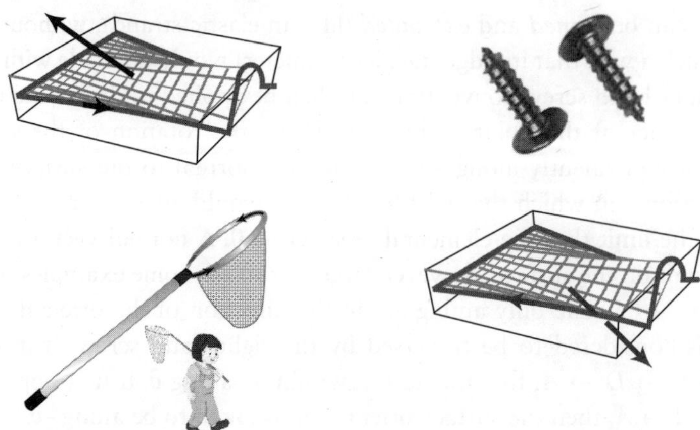

Fig. 10.6 The orientation of a surface is recognized as the *forward* motion of a *right hand screw*, when it is turned in a sense that is *along* the edge of the surface.

As you will soon discover, the identification of the orientation of a surface element plays an important role in the methods of vector calculus that we shall soon exploit to study fluid mechanics. We will study these methods in the rest of this chapter, and also in the next chapter. In fact, these methods are of great importance also for the study of electrodynamics (Chapter 12).

We alert the reader that not all surfaces are orientatble. The surface of a cylinder *open* at both ends is a classic example, because it does not have a unique edge which circumscribes the cylinder's surface (Fig. 10.7). Choosing the top, or the bottom, end of such a cylinder can give rise to incompatible orientations of the surface. The tray having a hole, shown in the middle panel of Fig. 10.7, is also not orientable. In fact, for the purpose of determining its orientation, it is no different from the cylinder open at both ends. A cylinder *closed* at one end is, of course, orientable. Another example of a *not*-well-behaved surface, is the Möbius strip. It can be easily constructed from a rectangular strip of paper, such as the strip *ABCD* shown in Fig. 10.5 (top left panel). One can bring the edges *AB* and *CD* together, and join them in two alternate ways: join *A* to *D*, and *B* to *C*, or *flip* one of the two edges of the strip, and then join *A* to *C*, and *B* to *D*. In the latter case, the twisted and joined strip you will get is called the Möbius strip. It is shown in the third panel of Fig. 10.7.

If you run your finger along the edge of the Möbius strip, you will find that the distinction between the 'top' edge and the 'bottom' edge vanishes. Likewise, if you run your finger on its surface, you find that the difference between the 'front' side and the 'back' side of the strip vanishes. The surface of the Möbius strip is therefore not orientable. It is a fascinating object. In the earlier days of machines, the Möbius strip was used as a fan belt to lessen surface wear and tear. In modern machines, materials with novel properties against wear and tear are used; the use of the Möbius strip in fan belts is no longer popular. The Möbius strip is of great importance in understanding the topology of manifolds. An object called the 'Klein bottle' is its analogue in higher dimension. The vanishing difference between the top edge and the bottom edge of the Möbius strip, and the disappearance of the difference between its front and the back side, marks only the beginnings of some mind-boggling shapes and constructions. The 2-dimensional drawings and 3-dimensional sculptures [3] by the illustratious and imaginative artist, M. C. Escher (1898–1972) are awsome; they are, in many ways, architypes of mathematical shapes and models like the Möbius strip and the Klein bottle in higher manifolds.

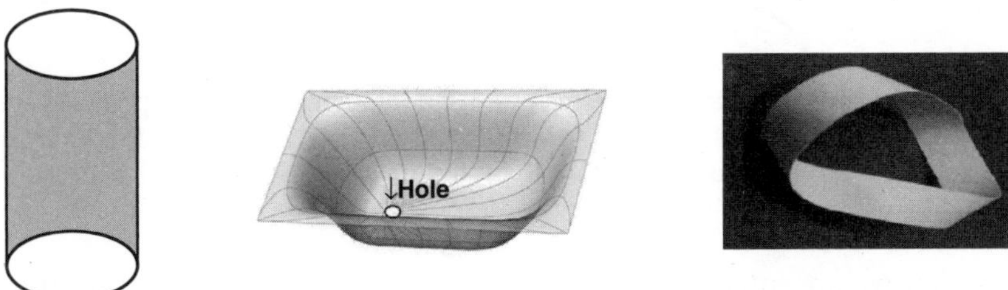

Fig. 10.7 The orientation of the surface cannot be defined unambiguously unless a unique edge bounds the surface. A cylinder open at both ends, or a surface that has one or more holes, are not orientable. A Möbius strip, is another example of a non-orientable surface. Such surfaces are said to be *not* well-behaved.

Having discussed the orientation of tiny surface elements, we are now ready to appreciate the last element to be discussed, which characterizes an 'ideal' fluid that in it the 'stress' is essentially 'compression'. We consider a well oriented tiny surface element through the point P as shown in Fig. 10.3. The fluid exerts a force $\vec{F}$ on that tiny elemental area $\overrightarrow{\delta A}$ of the strip, oriented along the unit normal vector $\hat{u}_N$ in accordance with the right hand screw rule. The force per unit area at the point P is called 'stress', denoted by $\vec{S}$. Three cases are of interest now:

(a) $\vec{S} \cdot \hat{u}_N = |\vec{S}|$, (10.11a)

(b) $\vec{S} \cdot \hat{u}_N = 0$, (10.11b)

(c) $\vec{S} \cdot \hat{u}_N = -|\vec{S}|$. (10.11c)

In the case (a), the stress is referred to as 'tension'. It is along the direction of the orientation of the surface element on which it acts. In the case (b), the stress is known as 'shear'. It is an indication of the stress being orthogonal to the orientation of the surface as would tend to shear the surface element. In the case 'c', the stress is opposite to the direction of the orientation of the surface. It is thus 'inward', and is called 'compression'. The distinguishing property of the 'ideal' fluid is that it applies a force per unit area surrounding any point in it such that the stress is necessarily one of 'compression'. Eq. 10.11c holds strictly for an ideal fluid.

We are now fully equipped to develop the formalism required to study fluid dynamics. In Fig. 10.8, we show a fluid passing through a well-defined region of space. For simplicity, we show the fluid getting into the region from the left, and getting out from the right. The fluid may, however, both enter and exit the region from various sides, not necessarily from just one side. Besides, the shape of the region may be quite irregular, not necessarily like the rectangular parallelepiped that is shown in this figure.

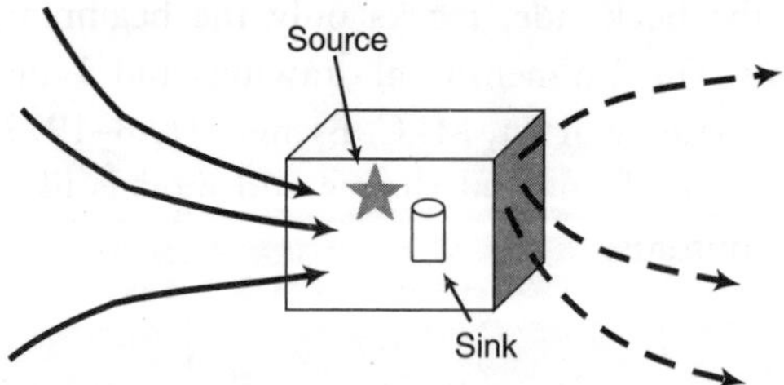

Fig. 10.8 This figure shows a fluid passing through a well-defined finite region of space. There may be in that region few sources from which additional fluid may be generated, and supplementary fluid springs out. There may also be some regions where the fluid becomes extinct, which we shall refer to as sinks.

In fluid dynamics, we are interested in knowing if there is net accumulation (positive, or negative) of the fluid in a tiny volume element surrounding a select point in the region, and what would happen in the limiting case when this volume element shrinks to zero, of course only within the framework of the continuum hypothesis. A certain mass of fluid can be considered to be crossing a tiny surface element δA in unit time. We shall study the fluid mass crossing unit area in unit time. The dimensions of this quantity will be $ML^{-2}T^{-1}$. The physical property of the fluid that represents this quantity is called the *current density vector*, or the *mass flux density vector*. It is defined as the product of the density of the fluid $\rho(\vec{r}, t)$ and its velocity $\vec{v}(\vec{r}, t)$. The dimensions of the density are ML^{-3} and that of the velocity are LT^{-1}, so their product has the dimensions $ML^{-2}T^{-1}$. Thus, the current density vector is given by

$$\vec{J}(\vec{r}, t) = \rho(\vec{r}, t)\, \vec{v}(\vec{r}, t). \tag{10.12}$$

The arguments $(\vec{r}, t)$ remind us that the physical quantities may, in principle, change from one point to another, and also from one instant of time to another. The techniques we are developing here are equally applicable to discuss a flow of electrically charged particles which constitute an electric current. In the electrical case, the current density vector will consist of the amount of electrical charge crossing per unit area per unit time. The dimensions of the electrical current density vector are $QL^{-2}T^{-1}$, where Q represents the dimension of the electrical charge. In the context of the electrical charge flow, $\vec{J}(\vec{r}, t)$ would represent the *charge flux density vector*. Our focal quantity of interest in this (and the next) chapter is the fluid mass flux (current) density vector $\vec{J}(\vec{r}, t)$, with dimensions $ML^{-2}T^{-1}$.

10.3 GAUSS' DIVERGENCE THEOREM, AND THE CONSERVATION OF FLUX

We consider an arbitrary vector field $\vec{K}(\vec{r}, t)$. It could be the fluid mass current density vector, or the electrical current density vector, or any other vector field. We first define the FLUX of the vector field $\vec{K}(\vec{r}, t)$ crossing an area $\delta S\hat{n}$ as the surface integral

$$\delta\phi = \iint_{\delta A} \vec{K}(\vec{r}) \cdot dS\hat{n}. \tag{10.13}$$

The surface integral will be over two degrees of freedom and is carried over the surface element under consideration, no matter curved or ruffled, as long as it is a well-behaved orientable surface. From our discussion on Fig. 10.4b, we also know that the size and shape of the surface also do not matter. We shall be interested in the *net* flux of the vector field $\vec{K}(\vec{r})$ through the *volume* element *ABCDHGEF* shown in Fig. 10.9. It is really important to remember that shape and size of the region are both irrelevant. In order to get net influx, and net outflow, of the vector field $\vec{K}(\vec{r})$ through the parallelepiped, we shall first estimate the same for the pair of faces which are orthogonal to the *Z*-axis; namely the faces *AGHD* and *BEFC*. We shall then do the same for the other two pairs of faces, orthogonal respectively to the *X*-axis, and the *Y*-axis.

While the unit normal to an 'open' surface is given by the right hand screw convention discussed above, the convention for the unit normal to a closed surface is that it is always the *outward* normal, i.e., coming *out* of the closed region. The unit normal to the top face *BEFC* is therefore along $\hat{e}_z$, and the unit normal to the bottom face *AGHD* is along $-\hat{e}_z$. Using Eq. 10.13, the *net* flux through the faces *AGHD* and *BEFC* is then given by the sum of the flux crossing these two faces:

$$\delta\phi_z = \iint_z \vec{K}(\vec{r}) \cdot dS\hat{n} = \left[K_z\left(x_0, y_0, z_0 + \frac{\delta z}{2} \right) - K_z\left(x_0, y_0, z_0 - \frac{\delta z}{2} \right) \right] \delta x\, \delta y. \quad (10.14a)$$

The first term in Eq. 10.14 has a plus sign. It corresponds to the flux crossing the top face *BEFC* whose unit normal vector is along $\hat{e}_z$, but the second term has a minus sign, since the unit normal to the bottom face *AGHD* is along $-\hat{e}_z$. Note also that the *z*-coordinate on the face *BEFC* is $z_0 + \frac{\delta z}{2}$, while that at the face *AGHD* is $= - \mathrm{l} - \frac{3}{2} + 5 = \frac{5}{2}$ In anticipation of the limit in which the volume δV of the parallelepiped would be sought to tend to zero, the *x*-coordinate and the *y*-coordinate are taken to be simply those of the point *P* itself. The area $\delta x\, \delta y$ is just the magnitude of the surface elements *BEFC* and *AGHD*, as shown in the figure. The subscript *z* on the elemental flux, on the left hand side of Eq. 10.14, denotes the fact that only the flux through faces orthogonal to the *Z*-axis are considered yet.

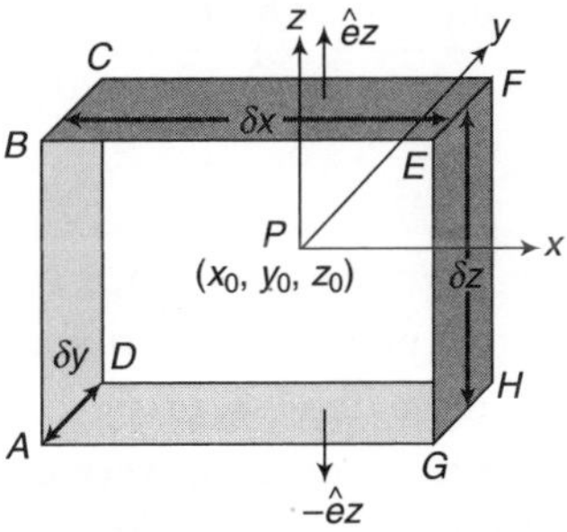

Fig. 10.9 The *volume* element *ABCDHGEF* is a closed region of space, fenced by the six faces of the parallelepiped. The six faces generate a closed space inside the surface of the six faces which close on themselves. The unit normal to each face is the outward normal.

We immediately see that the *net* flux through the faces orthogonal to the Z-axis is:

$$\delta\phi_z = \iint_z \vec{K}(\vec{r})\cdot dS\hat{n} = \left(\frac{\partial K_z}{\partial z}\delta z\right)\delta x\,\delta y = \left(\frac{\partial K_z}{\partial z}\right)\delta x\,\delta y\,\delta z = \left(\frac{\partial K_z}{\partial z}\right)\delta V, \quad (10.14a)$$

where $\delta V = \delta x\,\delta y\,\delta z$ is the volume of the elemental parallelepiped.

Similarly, the net flux through the other two pairs of faces which are respectively orthogonal to the X-axis and the Y-axis are:

$$\delta\phi_x = \iint_x \vec{K}(\vec{r})\cdot dS\hat{n} = \left(\frac{\partial K_x}{\partial x}\delta x\right)\delta y\,\delta z = \left(\frac{\partial K_x}{\partial x}\right)\delta y\,\delta z\,\delta x = \left(\frac{\partial K_x}{\partial x}\right)\delta V, \quad (10.14b)$$

and

$$\delta\phi_y = \iint_y \vec{K}(\vec{r})\cdot dS\hat{n} = \left(\frac{\partial K_y}{\partial y}\delta y\right)\delta z\,\delta x = \left(\frac{\partial K_y}{\partial y}\right)\delta z\,\delta x\,\delta y = \left(\frac{\partial K_y}{\partial y}\right)\delta V. \quad (10.14c)$$

The net flux of the vector field $\vec{K}(\vec{r})$, through the parallelepiped is then obtained by adding Eqs. 10.14a, 10.14b and 10.14c. Addition of the left hand sides of Eq. 10.14a, b, and c we get

$$\iint_z \vec{K}(\vec{r})\cdot dS\hat{n} + \iint_x \vec{K}(\vec{r})\cdot dS\hat{n} + \iint_y \vec{K}(\vec{r})\cdot dS\hat{n} = \oiint_{\substack{\text{surface}\\ \text{enclosing}\\ \text{the parallelopiped}}} \vec{K}(\vec{r})\cdot dS\hat{n}. \quad (10.15)$$

The circle on the surface integral notation on the right hand side above denotes the fact that the surface integral is over the *closed* surface (the closed surface of course does not have to be spherical).

Addingtherighthandsidesof Eqs.10.14a,b,andc,weget $\iiint\left(\frac{\partial K_x(\vec{r})}{\partial x}+\frac{\partial K_y(\vec{r})}{\partial y}+\frac{\partial K_z(\vec{r})}{\partial z}\right)dV$, wherein the triple integral is over the entire volume of the parallelepiped that is bound by the closed surface that appears on the right hand side of Eq. 10.15. This is the net flux through the parallelepiped. Along with Eq. 10.15, we therefore have an important result:

$$\oiint_{\substack{\text{surface}\\ \text{enclosing}\\ \text{a volume}\\ \text{element}}} \vec{K}(\vec{r})\cdot dS\hat{n} = \iiint\left(\frac{\partial K_x(\vec{r})}{\partial x}+\frac{\partial K_y(\vec{r})}{\partial y}+\frac{\partial K_z(\vec{r})}{\partial z}\right)dV. \quad (10.16)$$

As mentioned above, the unit normal to a surface at a point on a closed surface is the *outward normal* to the surface at that point. Its direction can therefore be different from point to point on a surface, depending on the flatness, or, the curvature of the surface. The physical quantity that appears in Eq. 10.16 has many, and very important, applications in physics. It represents the total flux of a vector field emanating from a closed region of space. Any region of arbitrary size and shape can be filled up with infinitesimal rectangular parallelepipeds as shown in Fig. 10.9. The results over each of the inner tiny elemental parallelepipeds can be added up (i.e., integrated) to get the final outcome for the full region. The final result of the volume integration is essentially the flux emanating from the total volume enclosed by the outermost closed surface, since unit normal to the inner adjacent cells are in opposite directions, resulting in cancelation of the flux. The left hand side of Eq. 10.16 represents the flux stemming out of the closed surface, no matter what the size or shape of the enclosed region is. It is obtained by adding $\vec{K}(\vec{r})\cdot\delta S\hat{n}$ at each point on the enfolding surface.

We can, however, enclose a point by a closed surface in an infinitesimal volume region and then seek the limit in which the volume element shrinks to zero. The physical quantity so obtained is a measure of the total flux originating, or diverging, out of that point. It gives us a scalar point function, called the *divergence* of the vector field $\vec{K}(\vec{r})$ at the point whose position vector is $(\vec{r})$. It is written as 'div $\vec{K}$', or as $\vec{\nabla}\cdot\vec{K}$. The latter notation is commonly used in vector calculus to denote the divergence of a vector field. The following expression therefore provides the formal *definition* of the divergence of a vector field:

$$\text{div }\vec{K} = \vec{\nabla}\cdot\vec{K} = \lim_{\delta V\to 0}\frac{\displaystyle\oiint_{\substack{\text{surface}\\ \text{enclosing}\\ \text{dV}}}\vec{K}(\vec{r})\cdot dS\hat{n}}{\delta V}. \tag{10.17}$$

Essentially, the divergence of a vector field at a point is the limiting value of the *net flux per unit volume* emanating from that point in the limit that the volume element surrounding the point shrinks to zero. This definition is clearly independent of any coordinate system, since only the left hand side of Eq. 10.16 has been referenced in its numerator. On using the right hand side of Eq. 10.16, we get the Cartesian *expression* for the divergence:

$$[\text{div }\vec{K}]_{\text{Cartesian}} = \frac{\partial K_x(\vec{r})}{\partial x}+\frac{\partial K_y(\vec{r})}{\partial y}+\frac{\partial K_z(\vec{r})}{\partial z} = [\vec{\nabla}\cdot\vec{K}]_{\text{Cartesian}}. \tag{10.18}$$

We emphasize the fact that Eq. 10.18 is the Cartesian *expression* of the divergence; it is not the *definition* of divergence. We emphasize that $\vec{\nabla}\cdot\vec{K}$ cannot be interpreted as a scalar product of two vectors, although the notation on the right hand side of Eq. 10.18 is perilously similar. We already know that $\vec{\nabla}$ is the gradient operator. It is a vector *operator*; it is not a vector itself. The gradient operator has been defined in Section 10.1 in the context of the directional derivative. The definition of the divergence is given by Eq. 10.17 in a form that is free from the choice of any coordinate system. The *definition* of a physical quantity cannot be given in the narrow framework of any particular coordinate system. The divergence of a vector field is an extremely important entity in fluid dynamics, electrodynamics, and in many other branches of Physics, including quantum mechanics.

The Cartesian *form* (Eq. 10.18) of the divergence in the Cartesian coordinate system has a weird similarity with the form of a scalar product of two vectors, $\vec{A}\cdot\vec{B} = A_xB_x + A_yB_y + A_zB_z$. This is recognized immediately on employing the Cartesian form of the gradient operator given in Table 10.2. The analogy of the Cartesian expression for the divergence with a scalar product, however, stops right there. If we invoke the commutation property of the scalar product of two vectors and try to find a meaning for the analogue of $\vec{B}\cdot\vec{A}$, we shall get $\left[K_x(\vec{r})\dfrac{\partial}{\partial x} + K_y(\vec{r})\dfrac{\partial}{\partial y} + K_z(\vec{r})\dfrac{\partial}{\partial z} \right]$, which is of course not a scalar point function. Instead, it is an *operator*, whose meaning would emerge only after we consider its application on an appropriate operand. At this point we observe that while the scalar product of two vectors ,$\vec{B}\cdot\vec{A}$, is exactly the same as that of $\vec{A}\cdot\vec{B}$, such is obviously not the case with $\vec{\nabla}\cdot\vec{K}$, i.e., $\vec{\nabla}\cdot\vec{K} \neq \vec{K}\cdot\vec{\nabla}$.

Equation 10.18 provides the expression for the divergence of a vector field in the Cartesian coordinate system. The corresponding expressions in the commonly used cylindrical polar and the spherical polar coordinates can be obtained easily, using the expressions for the gradient operator given in Table 10.2 of the above section. In the cylindrical polar coordinates, we shall have:

$$[\text{div}\,\vec{K}]_{\substack{\text{Cylindrical}\\ \text{Polar}}} = \vec{\nabla}\cdot\vec{K}(\vec{r}) = \left[\hat{e}_\rho \frac{\partial}{\partial\rho} + \hat{e}_\varphi \frac{1}{\rho}\frac{\partial}{\partial\varphi} + \hat{e}_z \frac{\partial}{\partial z} \right]$$

$$\cdot\,[\hat{e}_\rho K_\rho(\rho,\varphi,z) + \hat{e}_\varphi K_\varphi(\rho,\varphi,z) + \hat{e}_z K_z(\rho,\varphi,z)]. \quad (10.19a)$$

The expression for the divergence of a vector field involves a combination of (a) the vector algebra of taking a scalar product of two vectors, and (b) differential calculus, since partial derivative operators are also involved. This combination is what prevents the divergence from being interpreted merely as a scalar product of two vectors. The operators to obtain the partial derivatives would operate on every function placed at their right, including the unit vectors. In the Cartesian coordinate system, this does not matter as the Cartesian unit vectors are constants. It is not so in the cylindrical polar coordinate system. We therefore simplify the expression on the right hand side of Eq. 10.19, respecting this fact. The unit vectors of the cylindrical polar coordinates depend on the polar coordinates. The dependence of the cylindrical polar unit vectors $\{\hat{e}_\rho, \hat{e}_\varphi, \hat{e}_z\}$ on the coordinates has been discussed in Chapter 2. Accordingly, we get

$$\vec{\nabla}\cdot\vec{K}(\vec{r}) = \left[\hat{e}_\rho \frac{\partial}{\partial\rho} \right]\cdot[\hat{e}_\rho K_\rho(\rho,\varphi,z) + \hat{e}_\varphi K_\varphi(\rho,\varphi,z) + \hat{e}_z K_z(\rho,\varphi,z)]$$

$$+\left[\hat{e}_\varphi \frac{1}{\rho}\frac{\partial}{\partial\varphi} \right]\cdot[\hat{e}_\rho K_\rho(\rho,\varphi,z) + \hat{e}_\varphi K_\varphi(\rho,\varphi,z) + \hat{e}_z K_z(\rho,\varphi,z)] \quad (10.19b)$$

$$+\left[\hat{e}_z \frac{\partial}{\partial z} \right]\cdot[\hat{e}_\rho K_\rho(\rho,\varphi,z) + \hat{e}_\varphi K_\varphi(\rho,\varphi,z) + \hat{e}_z K_z(\rho,\varphi,z)],$$

i.e., $$\vec{\nabla}\cdot\vec{K}(\vec{r}) = \hat{e}_\rho\cdot\left\{\frac{\partial}{\partial\rho}\right\}[\hat{e}_\rho K_\rho(\rho,\varphi,z)+\hat{e}_\varphi K_\varphi(\rho,\varphi,z)+\hat{e}_z K_z(\rho,\varphi,z)]$$
$$+\hat{e}_\varphi\cdot\left\{\frac{1}{\rho}\frac{\partial}{\partial\varphi}\right\}[\hat{e}_\rho K_\rho(\rho,\varphi,z)+\hat{e}_\varphi K_\varphi(\rho,\varphi,z)+\hat{e}_z K_z(\rho,\varphi,z)] \tag{10.19c}$$
$$+\hat{e}_z\cdot\left\{\frac{\partial}{\partial z}\right\}[\hat{e}_\rho K_\rho(\rho,\varphi,z)+\hat{e}_\varphi K_\varphi(\rho,\varphi,z)+\hat{e}_z K_z(\rho,\varphi,z)].$$

Hence, using Eq. 2.10 (Chapter 2), we get

$$[\text{div}\,\vec{K}]_{\substack{\text{Cylindrical}\\ \text{Polar}}} = \vec{\nabla}\cdot\vec{K}(\vec{r}) = \frac{\partial}{\partial\rho}K_\rho(\rho,\varphi,z)$$
$$+\frac{1}{\rho}K_\rho(\rho,\varphi,z)+\frac{1}{\rho}\frac{\partial}{\partial\varphi}K_\varphi(\rho,\varphi,z)+\frac{\partial}{\partial z}K_z(\rho,\varphi,z). \tag{10.19d}$$

Similarly, to get the expression for the divergence of a vector in spherical polar coordinates, we use the expression for the gradient operator in Table 10.2:

$$[\text{div}\,\vec{K}]_{\substack{\text{Spherical}\\ \text{Polar}}} = \vec{\nabla}\cdot\vec{K}(\vec{r}) = \left\{\hat{e}_r\frac{\partial}{\partial r}+\hat{e}_\theta\frac{1}{r}\frac{\partial}{\partial\theta}+\hat{e}_\varphi\frac{1}{r\sin\theta}\frac{\partial}{\partial\varphi}\right\}$$
$$\cdot[\hat{e}_r K_r(r,\theta,\varphi)+\hat{e}_\theta K_\theta(r,\theta,\varphi)+\hat{e}_\varphi K_\varphi(r,\theta,\varphi)]. \tag{10.20a}$$

Finally, using Eq. 2.16 which gives the rate of change of the unit vectors of spherical polar coordinates with respect to the coordinates themselves, we get

$$[\text{div}\,\vec{K}]_{\substack{\text{Spherical}\\ \text{Polar}}} = \vec{\nabla}\cdot\vec{K}(\vec{r}) = \frac{1}{r^2}\frac{\partial}{\partial r}[r^2K_r(r,\theta,\varphi)]+\frac{1}{r\sin\theta}\frac{\partial}{\partial\theta}[K_\theta(r,\theta,\varphi)\sin\theta]$$
$$+\frac{1}{r\sin\theta}\frac{\partial}{\partial\varphi}K_\varphi(r,\theta,\varphi). \tag{10.20b}$$

Equations 10.18, 10.19 and 10.20 give the expressions of the divergence of a vector in the three commonly used coordinate systems, viz., the Cartesian, cylindrical polar, and the spherical polar coordinate systems. The *definition* of the divergence itself that given in Eq. 10.17, which is free from the choice of any coordinate system.

From the combination of Eq. 10.16 and Eq. 10.17, we get an extremely important theorem, known as the Gauss' divergence theorem. This theorem is named after the German mathematician–physicist, Johann Carl Friedrich Gauss (1777–1855). It states that the volume integral of divergence of a vector field over a volume-region of space is equal to the total outward flux of the vector field *across* the surface which wraps around that volume and enfolds it:

$$\iiint (\vec{\nabla} \cdot \vec{K}) dV = \underset{\substack{\text{closed surface} \\ \text{enclosing} \\ \text{a volume}}}{\oiint} \vec{K}(\vec{r}) \cdot dS\hat{n}. \tag{10.21}$$

The Gauss' divergence theorem applies to any vector field $\vec{K}(\vec{r})$ in a region of space in which it is well-defined. For the theorem to hold, components of the vector field must of course be continuous in the region of space under consideration, since their partial derivatives are involved in the integrand that appears under the volume integral. The volume integration must be carried out in the *entire* continuous region of space which is bound by the closed surface. Likewise, the surface integration must be carried out over the entire surface that encloses the region. We will find, in this (and the next two chapters) that the Gauss' theorem is a powerful tool in vector calculus. We shall now apply the divergence theorem to the mass flux (current) density vector $\vec{J}(\vec{r}, t)$, which we had introduced in Eq. 10.12:

$$\underset{\substack{\text{closed surface} \\ \text{enclosing} \\ \text{a volume}}}{\oiint} \vec{J}(\vec{r}, t) \cdot dS\hat{n} = \iiint \vec{\nabla} \cdot \vec{J}(\vec{r}, t)\, dV. \tag{10.22}$$

The current density vector is just the mass crossing per unit area per unit time. The left hand side is the integration of its flux, which gives the *total mass* M crossing from inside to outside the closed surface surrounding a volume region of space, *per unit time*. Thus, we get:

$$-\frac{dM}{dt} = \underset{\substack{\text{closed surface} \\ \text{enclosing} \\ \text{a volume}}}{\oiint} \vec{J}(\vec{r}, t) \cdot dS\hat{n} = \iiint \vec{\nabla} \cdot \vec{J}(\vec{r}, t)\, dV. \tag{10.23a}$$

If the physical quantity we considered was a flow of electrical charges, we would have $-\frac{dQ}{dt}$ instead of $-\frac{dM}{dt}$ on the left hand side of Eq. 10.23a:

$$-\frac{dQ}{dt} = \underset{\substack{\text{closed surface} \\ \text{enclosing} \\ \text{a volume}}}{\oiint} \vec{J}(\vec{r}, t) \cdot dS\hat{n} = \iiint \vec{\nabla} \cdot \vec{J}(\vec{r}, t))\, dV. \tag{10.23b}$$

The negative sign on the left hand sides of Eq. 10.23 is on account of the fact that the quantity that we have determined is a measure of the rate at which the mass (or charge) is being *depleted* from the region of space under study.

The total mass M flowing out of the region is of course the volume integral of the density:

$$M = \iiint \rho(\vec{r}, t) dV. \tag{10.24a}$$

In the case of the electrical charge flow, we shall have the total charge Q flowing out of the region:

$$Q = \iiint \rho(\vec{r}, t) dV. \tag{10.24b}$$

All along in our analysis of fluid properties like its density $\rho(\vec{r}, t)$, pressure $p(\vec{r}, t)$, and velocity of the fluid, $\vec{v}(\vec{r}, t)$, we considered the physical quantities at a particular instant of time, t, and at *a particular point P in space*, whose position vector in the laboratory frame of reference is $\vec{r}$. The position vector $\vec{r}$ always denoted the *same* fixed point in space. Different fluid particles merely would go past this point as they flow. Such a description of the fluid is called the Eulerian description, named after Leonhard Euler, who has been referred to even in the previous chapters. Thus, in the consideration of taking the derivative of a time-dependent function with respect to time, the operator $\frac{d}{dt} \equiv \frac{\partial}{\partial t}$, in so far as the operand has the form $f(\vec{r}, t)$. This is because $\vec{r}$ denotes a fixed point in space; it is not a function of time. The density in Eq. 10.24a is the mass density, where is in Eq. 10.24b it is the charge density. In both the cases, combining Eq. 10.23 and 10.24, and using the *Eulerian* equivalence $\frac{d}{dt} \equiv \frac{\partial}{\partial t}$, we get

$$-\frac{\partial}{\partial t} \iiint \rho(\vec{r}, t)\, dV = \underset{\substack{\text{closed surface}\\ \text{enclosing}\\ \text{a volume}}}{\oiint} \vec{J}(\vec{r}, t) \cdot dS\hat{n} = \iiint \vec{\nabla} \cdot \vec{J}(\vec{r}, t))\, dV, \tag{10.25a}$$

$$\text{i.e.,} \quad \iiint \left[-\frac{\partial \rho(\vec{r}, t)}{\partial t} \right] dV = \underset{\substack{\text{closed surface}\\ \text{enclosing}\\ \text{a volume}}}{\oiint} \vec{J}(\vec{r}, t) \cdot dS\hat{n} = \iiint \vec{\nabla} \cdot \vec{J}\,(\vec{r}; t))\, dV. \tag{10.25b}$$

We have switched the order of taking the derivative with respect to time with the integration over space variables on the left hand side. This is completely justified since differentiation with respect to time and integration with respect to space are completely independent mathematical processes. The two volume integrals in Eq. 10.25b (on the left hand side, and on the right hand side) are definite integrals over essentially the same region of space, and hence the corresponding integrands must be equal. We therefore get:

$$\vec{\nabla} \cdot \vec{J}\,(\vec{r}, t) + \frac{\partial \rho(\vec{r}, t)}{\partial t} = 0, \tag{10.26a}$$

$$\text{i.e.,} \quad \rho \vec{\nabla} \cdot \vec{v}\,(\vec{r}, t) + \vec{v}(\vec{r}, t) \cdot \vec{\nabla} \rho + \frac{\partial \rho(\vec{r}, t)}{\partial t} = 0. \tag{10.26b}$$

Equations 10.25 and 10.26 are completely equivalent; the former employs the integral form and involves whole-space properties, and the latter is the differential form in which the terms involved express properties at a particular point in space. They represent a *conservation principle*, viz., when there is no source and no sink in a region of space, the density of matter (or charge density) in a volume element can change *if and only if* the mass (or the charge) flows in, *or out*, of that region across the surface that encloses that volume region. The equivalent expressions Eqs. 10.25 and 10.26 are referred to as the *equation of continuity*. It essentially expresses a conservation principle. It is of great importance not only in fluid dynamics and electrodynamics, but also in many other branches of engineering and physics, including quantum mechanics.

The equation of continuity is a scalar equation which involves partial derivatives with respect to four independent variables, viz., three space coordinates and time. In a given situation, one may know the velocity of the fluid, from which the divergence of the velocity field can be determined. One can then solve the equation of continuity by integrating it with respect to time and determine the time-dependence of density of the fluid. It is challenging when it has to be used when the density of the fluid is known, and the velocity is to be determined. The velocity being a vector, it has three unknown scalar components. The problem posed by the equation of continuity does not meet the 'closure' requirement when the number of unknowns is larger than the number of equations available. We meet numerous such challenges in solving problems in fluid dynamics, and in modeling the dynamics of charge plasma in stellar atmosphere, laser plasma accelerators, tokomak environments, etc. In Chapter 11, we shall discuss how such complexities are addressed.

An alternative to the Eulerian viewpoint is the Lagrangian description of the fluid. This is named after Joseph Louis Lagrange, who has also been referred to in the previous chapters, especially in Chapter 6. In the Lagrangian description, we have the capability of tracking the dynamics of each fluid particle individually. In this description, $\vec{r}$ would not denote a fixed point in space, as in the Eulerian description. Instead, $\vec{r}(t)$ would denote the instantaneous position of a particle of the fluid moving in the flow. The Lagrangian description of the fluid would enable us to track the fluid particles from the headwaters of a river, downstream, all the way along the flow of the waters.*

The time dependence of the physical state of the system, characterized by its density, pressure and velocity, must be represented comprehensively in denoting these as functions of time. The physical quantities of interest depend on time explicitly, hence the time must appear as an argument on which they depend. This explicit time-dependence is denoted by writing the density as $\rho = \rho(t)$, the pressure as $p = p(t)$, and the velocity as $\vec{v} = \vec{v}(t)$. In addition to this explicit time-dependence, there is an additional time-dependence which the density, pressure and velocity have in the Lagrangian description, because they represent properties of fluid particles which are individually tracked. The distinct fluid particles are in motion, and depend

* On the sideline of this discussion, we observe that it is the Lagrangian description of fluid flow that can help us understand the very name of the Triveni Sangam (which literally means *Three-Braid-Confluence*). It is the name of the confluence of the three rivers Ganga, Yamuna and Saraswati, at Prayag. It is now determined by researchers [https://www.downtoearth.org.in/coverage/rivers/saraswati-underground-15455 (accessed on 23 November 2018)] that the path of the river Saraswati was from the Manna village near Badrinath, through the states of Haryana and Rajasthan, into the Arabian sea. The mainstream of Saraswati river never really passed through Prayag. How come, then, the Triveni Sangam at Prayag gets its name from the confluence of not just Ganga and Yamuna, but also Saraswati? This can be understood only from the *Lagrangian description* of fluid flow, not from the Eulerian viewpoint. The Lagrangian description enables tracking the individual waters of two rivers even if the mainstreams separate out even after only a short-lived confluence. Hundreds of kilometers before the waters of Yamuna reach the Triveni Sangam, the river Yamuna crosses the river Saraswati, even if only briefly. Yamuna then carries some of the waters of Saraswati from this *prior confluence* into the Triveni Sangam, thus accounting for the confluence of the three rivers, even if only two mainstreams rivers, namely Ganga and Yamuna, merge at Prayag.

on the position vector of the *moving* particles. The position vectors of the fluid particles change with time as the fluid moves, unlike the position vectors of points fixed in space in the Eulerian description. Thus, $\vec{r} = \vec{r}(t)$. We therefore indicate both the implicit and the explicit time-dependence by writing the density as $\rho = \rho(\vec{r}(t), t)$, the pressure as $p = p(\vec{r}(t), t)$ and the velocity as $\vec{v} = \vec{v}(\vec{r}(t), t)$. In writing the arguments as we have done, it is acknowledged that in the Lagrangian description the dynamics of each fluid particle is tracked. In the Eulerian description, on the other hand, we only observe physical properties of the fluid in a fixed region of space, consisting of fixed points, whose positions do not change.

A function of space and time (such as the pressure or the density of the fluid) in the Lagrangian description therefore can be represented as $\phi = \phi(\vec{r}(t), t)$. The dependence of ϕ on time described is then given by

$$\frac{d}{dt}[\phi(\vec{r}, t)] = \frac{d\phi(x, y, z, t)}{dt} = \left[\frac{\partial \phi}{\partial x}\frac{dx}{dt} + \frac{\partial \phi}{\partial y}\frac{dy}{dt} + \frac{\partial \phi}{\partial z}\frac{dz}{dt} + \frac{\partial \phi}{\partial t} \right],$$

$$= \frac{\partial \phi}{\partial t} + \vec{v} \cdot \vec{\nabla}\phi \equiv \left(\frac{\partial}{\partial t} + \vec{v} \cdot \vec{\nabla} \right) \phi. \tag{10.27}$$

The above relation holds for any (arbitrary) function ϕ, and hence we have the following operator equivalence:

$$\frac{d}{dt} \equiv \frac{\partial}{\partial t} + \vec{v} \cdot \vec{\nabla}. \tag{10.28}$$

To interpret the conservation principle embedded in the equation of continuity using the Lagrangian description, we note, from Eq. 10.26, that

$$0 = \frac{\partial \rho}{\partial t} + \vec{\nabla} \cdot (\rho \vec{v}) = \frac{\partial \rho}{\partial t} + \vec{v} \cdot \vec{\nabla}\rho + \rho \vec{\nabla} \cdot \vec{v} = \left(\frac{\partial}{\partial t} + \vec{v} \cdot \vec{\nabla} \right) \rho + \rho \vec{\nabla} \cdot \vec{v} = 0. \tag{10.29}$$

Hence, using Eq. 10.28,

$$\frac{d\rho}{dt} + \rho \vec{\nabla} \cdot \vec{v} = 0. \tag{10.30}$$

As opposed to the *partial* time derivative operator $\frac{\partial}{\partial t}$ which is appropriate for indicating the time-dependence of Eulerian variables, the Lagrangian time-derivative operator $\frac{d}{dt}$ (Eq. 10.28 and Eq. 10.30) is the *total* time derivative operator. It is also known as the 'convective derivative', or, the 'material time-derivative', or, as the 'substantive time-derivative' operator. The part $\frac{\partial}{\partial t}$ of this operator comes from the *explicit* dependence on time. The other part, $\vec{v} \cdot \vec{\nabla}$, comes from the *implicit* dependence on time. It is called the *advective* derivative operator. It involves the scalar product of the velocity (of the moving fluid-unit) with the spatial gradient of the physical quantity under consideration, since $\vec{r} = \vec{r}(t)$. Whereas Eq. 10.26 expresses the conservation principle (equation of continuity), in the Eulerian description, Eq. 10.30 expresses the same in the Lagrangian description.

10.4 VORTICITY, AND THE KELVIN–STOKES' THEOREM

You would have noticed while watching a river that leaves from nearby trees often fall in the river and they float and drift on the flowing water surface. Often, apart from drifting along the flow, the leaves also spin. In addition to the linear motion, they have a rotational motion. Isn't it interesting that when the flow of water is linear, unidirectional down the stream, the motion of the leaf is rotatory? This is because the linear velocity field must have, in one way or another, a concomitant rotational property. This may remind you of curling up primarily straight hair. The property of the velocity field that causes the spinning motion of the leaf floating on the fluid flow, is in fact called the 'curl of the velocity field'. It is also called the 'vorticity'.

The *curl of a vector field* $\vec{K}(\vec{r})$ is written as curl $\vec{K}(\vec{r})$. It is denoted as $\vec{\nabla} \times \vec{K}(\vec{r})$. It is a vector field, defined at each point (whose position vector is $\vec{r}$) in space. It can be *defined* uniquely by identifying the criteria that provide its components along a *complete* basis of unit vectors $\{\hat{u}_i(\vec{r}), i = 1, 2, 3\}$, irrespective of whether the unit vectors are fixed (as in Cartesian coordinate system), or position-dependent (as in cylindrical polar or spherical polar coordinate systems). We now provide the definition of the curl of a vector by defining its components along a complete set of orthonormal basis set of vectors.

The components of $\vec{\nabla} \times \vec{K}(\vec{r})$ along a complete set of orthonormal basis vectors, $\{\hat{u}_i(\vec{r}), i = 1, 2, 3\}$ are *defined* by

$$\hat{u}_i \cdot (\vec{\nabla} \times \vec{K}(\vec{r})) = \lim_{\delta S \to 0} \frac{\oint \vec{K}(\vec{r}) \cdot \overrightarrow{dl}}{\delta S}, \tag{10.31}$$

wherein the line integral on the right hand side is carried over a closed path that encircles an infinitesimal elemental vector area $(\delta S)\hat{u}_i$. The path integral in the numerator of the right hand side of Eq. 10.31 is called the *circulation* of the vector field $\vec{K}(\vec{r})$ along the path over which the line integral is determined. The term 'circulation' is appropriate, since the path encircles the point that is within that loop, at which the curl is defined. Essentially, each of the three components of the curl of a vector field is *defined* as the limiting value of the *circulation per unit area*. Very importantly, the direction of the unit vector $\hat{u}_i$ is *inseparably* linked to the sense in which the path of the line integral is traversed. The forward motion of a right hand screw that is turned *along* the integration path is exactly along the direction $\hat{u}_i$. The surface element enclosed by the path of integration is therefore orthogonal to the unit vector $\hat{u}_i(\vec{r})$. The path of the line-integration is closed (not necessarily circular), and it encloses an orientable surface whose normal at the specified point P is along $\hat{u}_i(\vec{r})$. This linkage of the orientation of the surface element uniquely connects the right hand side of Eq. 10.31 with the ith component of the curl, for i = 1,2,3. The basis set $\{\hat{u}_i(\vec{r}), i = 1, 2, 3\}$ is complete, it has orthonormal unit vectors. It could be a Cartesian basis, or a cylindrical polar coordinate basis, or a spherical polar coordinate basis, or any other. The definition of the curl of a vector field given in Eq. 10.31 is independent of any particular coordinate system, as it should be. The left hand side of Eq. 10.31 is a function of the specific point whose position vector is $\vec{r}$. It defines a scalar point function, which is the ith component of the curl $\vec{K}(\vec{r})$. The circulation $\oint \vec{K}(\vec{r}) \cdot \overrightarrow{dl}$ which appears on the right hand side of Eq. 10.31 is of course a scalar, but it requires for its determination the properties over the complete path around a point; it is therefore *not* a scalar *point* function. It does not therefore constitute a scalar *field*. Notwithstanding the fact that the right hand side is determined in the limit $\delta S \to 0$, the ratio $\dfrac{\oint \vec{K}(\vec{r}) \cdot \overrightarrow{dl}}{\delta S}$ does not blow up; it remains finite.

In particular, if the vector field $\vec{K}(\vec{r})$ happens to be a conservative force field $\vec{F}(\vec{r})$, then the circulation of the force field, $\oint \vec{F}(\vec{r}) \cdot \overrightarrow{dl}$, over a closed path is necessarily zero. By Eq. 10.31, all the components of the *curl* $\vec{F}(\vec{r})$ would be zero. A vector field whose curl is identically zero at all points in space is called an *irrotational* field. That the force field is irrotational, is then a completely equivalent criterion to determine if it is conservative. If, on the other hand, the curl of a vector field is not zero, then it is said to be *rotational*.

Let us now use the definition of the curl of a vector, Eq. 10.31 (for each component), to obtain its Cartesian expression. We do so by evaluating the right hand side of this equation, using the Cartesian coordinate system, using Fig. 10.10.

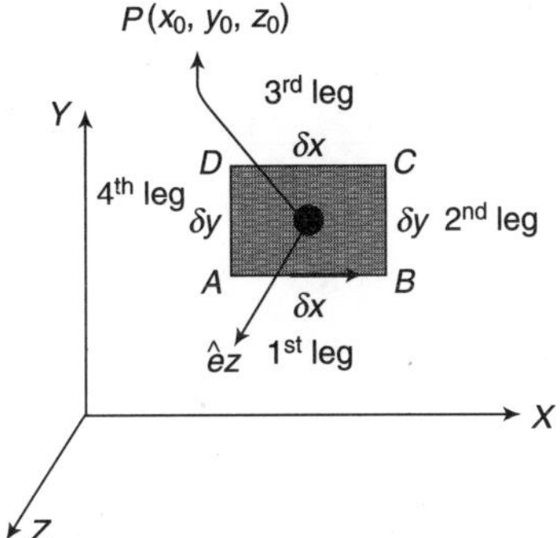

Fig. 10.10 This figure shows the closed path $A \to B \to C \to D \to A$ of the circulation as a sum over the four legs $A \to B$, $B \to C$, $C \to D$ and $D \to A$. Note that the counter-clockwise sense of traversing this path would advance a right hand screw along $\hat{e}_z$.

We now evaluate the circulation per unit area for the vector field at a point $P(x_0, y_0, z_0)$ along the path $A \to B \to C \to D \to A$ which *encircles* that point. The closed path is a sum over the four legs shown. The values of the components of the vector field $\vec{K}(\vec{r}) = \vec{K}(x, y, z)$ on the four legs are, however, different from the values of its components at the point $P(x_0, y_0, z_0)$. This is so because the leg BC is at a distance $\dfrac{\delta x}{2}$ to the right of the point $P(x_0, y_0, z_0)$, and likewise, the other three legs are also displaced away from $P(x_0, y_0, z_0)$. Since the path $A \to B \to C \to D \to A$ has a direction, we must keep track of the fact that the x-coordinate increases ($\delta x > 0$) along the leg 1, and decreases ($\delta x < 0$) along the leg 3. Likewise, the

y-coordinate increases ($\delta y > 0$) along the leg 2, and decreases ($\delta y > 0$) along the leg 4. Keeping track of these signs, the very definition (Eq. 10.31) of the components of curl of the vector field therefore gives us

$$\hat{e}_z \cdot (\vec{\nabla} \times \vec{K}(\vec{r})) = \frac{\begin{bmatrix} K_x\left(x_0, y_0 - \dfrac{\delta y}{2}, z_0\right) \\ -K_x\left(x_0, y_0 + \dfrac{\delta y}{2}, z_0\right) \end{bmatrix} \delta x + \begin{bmatrix} K_y\left(x_0 + \dfrac{\delta x}{2}, y_0, z_0\right) \\ -K_y\left(x_0 - \dfrac{\delta x}{2}, y_0, z_0\right) \end{bmatrix} \delta y}{\delta x \delta y}. \tag{10.32a}$$

Note that the coordinates (x, y, z) are displaced from (x_0, y_0, z) in accordance with their values on the respective legs of the integration segments. On the right hand side of Eq. 10.32a, the first term is the difference between the values of the x-component of the vector field on the path AB and CD, which are separated by a distance δy along the Y-axis. We consider vector fields which are sufficiently differentiable; i.e., at each point the vector has differentiable components which are continuous on the surface bound by a curve. Since the vector point function $\vec{K}(\vec{r})$ is continuous, the values of its components at the neighboring points are related through the first (partial) derivative *times* the distance between those points. Accordingly, we get

$$\hat{e}_z \cdot (\vec{\nabla} \times \vec{K}(\vec{r})) = \frac{\left[-\dfrac{\partial K_x}{\partial y}\delta y\right]\delta x + \left[\dfrac{\partial K_y}{\partial x}\delta x\right]\delta y}{\delta x\, \delta y}, \tag{10.32b}$$

$$\text{i.e.,} \quad (\vec{\nabla} \times \vec{K}(\vec{r}))_z = \hat{e}_z \cdot (\vec{\nabla} \times \vec{K}(\vec{r})) = \frac{\partial K_y}{\partial x} - \frac{\partial K_x}{\partial y}. \tag{10.32c}$$

In the last step, the elemental area $\delta x\, \delta y$ in the numerator has been canceled with the corresponding area in the denominator, relieving us from any issue about the value of the ratio arising from the denominator in the limit $(\delta x\, \delta y) = \delta S \to 0$.

The other two Cartesian components can be easily obtained by making cyclic changes $x \to y \to z \to x$, giving

$$\text{curl } \vec{K}(\vec{r}) = \vec{\nabla} \times \vec{K}(\vec{r}) = \left(\frac{\partial K_z}{\partial y} - \frac{\partial K_y}{\partial z}\right)\hat{e}_y + \left(\frac{\partial K_x}{\partial z} - \frac{\partial K_z}{\partial x}\right)\hat{e}_y + \left(\frac{\partial K_y}{\partial x} - \frac{\partial K_x}{\partial y}\right)\hat{e}_z. \tag{10.33}$$

This expression can be, albeit with some restraints (explained below), written as a determinant

$$\vec{\nabla} \times \vec{K} = \begin{vmatrix} \hat{e}_x & \hat{e}_y & \hat{e}_z \\ \dfrac{\partial}{\partial x} & \dfrac{\partial}{\partial y} & \dfrac{\partial}{\partial z} \\ K_x & K_y & K_z \end{vmatrix}. \tag{10.34}$$

The restraining factor is that interpreting the curl of a vector as a determinant is perilous, since we can surely interchange the last two rows of a determinant and get a determinant with a negative sign, but doing the same with the term on the right hand side of Eq. 10.34 would leave us with the partial derivative operators hanging at the end with nothing to operate on. So also, using the Cartesian expression in Eq. 10.33 to *define* the curl of a vector would lack rigor, since it is not at all a good idea to *define* a physical quantity in a manner that is specific to a coordinate frame of reference. One may, however, use any coordinate system to evaluate the curl (or divergence) of a vector, using the primary definitions (Eq. 10.17, for the divergence, and Eq. 10.31, for the curl) which are independent of the coordinate system.

The choice of a coordinate frame to solve a problem, as we have learned in Chapter 2, is a matter of convenience. It is therefore expeditious to obtain the detailed form of the curl of a vector field in the commonly used coordinate systems. Let us therefore obtain the expression for the curl of a vector in the spherical polar coordinate system. Using the expression for the gradient given in Table 10.2, we have

$$\vec{\nabla} \times \vec{K} = \left(\hat{e}_r \frac{\partial}{\partial r} + \hat{e}_\theta \frac{1}{r}\frac{\partial}{\partial \theta} + \hat{e}_\varphi \frac{1}{r\sin\theta}\frac{\partial}{\partial \varphi} \right) \times (\hat{e}_r K_r(r,\theta,\varphi) + \hat{e}_\theta K_\theta(r,\theta,\varphi) + \hat{e}_\varphi K_\varphi(r,\theta,\varphi)). \tag{10.35a}$$

Simplifying, we get

$$\begin{aligned}\vec{\nabla} \times \vec{K} &= \left(\hat{e}_r \frac{\partial}{\partial r} \right) \times (\hat{e}_r K_r(r,\theta,\varphi) + \hat{e}_\theta K_\theta(r,\theta,\varphi) + \hat{e}_\varphi K_\varphi(r,\theta,\varphi)) \\ &+ \left(\hat{e}_\theta \frac{1}{r}\frac{\partial}{\partial \theta} \right) \times (\hat{e}_r K_r(r,\theta,\varphi) + \hat{e}_\theta K_\theta(r,\theta,\varphi) + \hat{e}_\varphi K_\varphi(r,\theta,\varphi)) \\ &+ \left(\hat{e}_\varphi \frac{1}{r\sin\theta}\frac{\partial}{\partial \varphi} \right) \times (\hat{e}_r K_r(r,\theta,\varphi) + \hat{e}_\theta K_\theta(r,\theta,\varphi) + \hat{e}_\varphi K_\varphi(r,\theta,\varphi)).\end{aligned} \tag{10.35b}$$

i.e.,

$$\begin{aligned}\vec{\nabla} \times \vec{K} &= (\hat{e}_r) \times \left(\frac{\partial}{\partial r} \right) (\hat{e}_r K_r(r,\theta,\varphi) + \hat{e}_\theta K_\theta(r,\theta,\varphi) + \hat{e}_\varphi K_\varphi(r,\theta,\varphi)) \\ &+ (\hat{e}_\theta) \times \left(\frac{1}{r}\frac{\partial}{\partial \theta} \right) (\hat{e}_r K_r(r,\theta,\varphi) + \hat{e}_\theta K_\theta(r,\theta,\varphi) + \hat{e}_\varphi K_\varphi(r,\theta,\varphi)) \\ &+ (\hat{e}_\varphi) \times \left(\frac{1}{r\sin\theta}\frac{\partial}{\partial \varphi} \right) (\hat{e}_r K_r(r,\theta,\varphi) + \hat{e}_\theta K_\theta(r,\theta,\varphi) + \hat{e}_\varphi K_\varphi(r,\theta,\varphi)).\end{aligned} \tag{10.35c}$$

Remembering now that the unit vectors of spherical polar coordinate systems are not constant, and that their coordinate dependence is given in Chapter 2 (Eq. 2.16), we get

$$\vec{\nabla} \times \vec{K} = \hat{e}_r \frac{1}{r\sin\theta}\left\{\frac{\partial}{\partial\theta}(\sin\theta\, K_\varphi) - \frac{\partial K_\theta}{\partial\varphi}\right\} + \hat{e}_\theta \frac{1}{r}\left\{\frac{1}{\sin\theta}\frac{\partial K_r}{\partial\varphi} - \frac{\partial}{\partial r}(rK_\varphi)\right\} \quad (10.36)$$

$$+ \hat{e}_\varphi \frac{1}{r}\left\{\frac{\partial}{\partial r}(rK_\theta) - \frac{\partial K_r}{\partial\theta}\right\}.$$

The expression for the curl in cylindrical polar coordinates is similarly obtained, using the expression (from Table 10.2) for the gradient operator in the cylindrical polar coordinate system:

$$\begin{aligned}\vec{\nabla} \times \vec{K} = {} & [\hat{e}_\rho]\times\frac{\partial}{\partial\rho}[\hat{e}_\rho K_\rho(\rho,\varphi,z) + \hat{e}_\varphi K_\varphi(\rho,\varphi,z) + \hat{e}_z K_z(\rho,\varphi,z)] \\ & + [\hat{e}_\varphi]\times\left(\frac{1}{\rho}\frac{\partial}{\partial\varphi}\right)[\hat{e}_\rho K_\rho(\rho,\varphi,z) + \hat{e}_\varphi K_\varphi(\rho,\varphi,z) + \hat{e}_z K_z(\rho,\varphi,z)] \\ & + [\hat{e}_z]\times\frac{\partial}{\partial z}[\hat{e}_\rho K_\rho(\rho,\varphi,z) + \hat{e}_\varphi K_\varphi(\rho,\varphi,z) + \hat{e}_z K_z(\rho,\varphi,z)]. \end{aligned} \quad (10.37a)$$

Finally, on using Eq. 2.10 to see how the unit vectors of the cylindrical polar coordinate system change with the coordinates, we get the expression for the curl of a vector in the cylindrical polar coordinate system:

$$\vec{\nabla}\times\vec{K} = \hat{e}_\rho\left(\frac{1}{\rho}\frac{\partial K_z}{\partial\varphi} - \frac{\partial K_\varphi}{\partial z}\right) + \hat{e}_\varphi\left(\frac{\partial K_\rho}{\partial z} - \frac{\partial K_z}{\partial\rho}\right) + \hat{e}_z\frac{1}{\rho}\left[\frac{\partial(\rho K_\varphi)}{\partial\rho} - \frac{\partial K_\rho}{\partial\varphi}\right]. \quad 10.37b$$

The very definition of the curl, given in Eq. 10.31 for each of its components, provides the basis for establishing an important theorem of vector calculus, viz., the Stokes' theorem. For any well-behaved surface (discussed in Section 10.1), however, ruffled, we can slice it down into an infinite number of infinitesimal rectangular cells, as shown in Fig. 10.10.

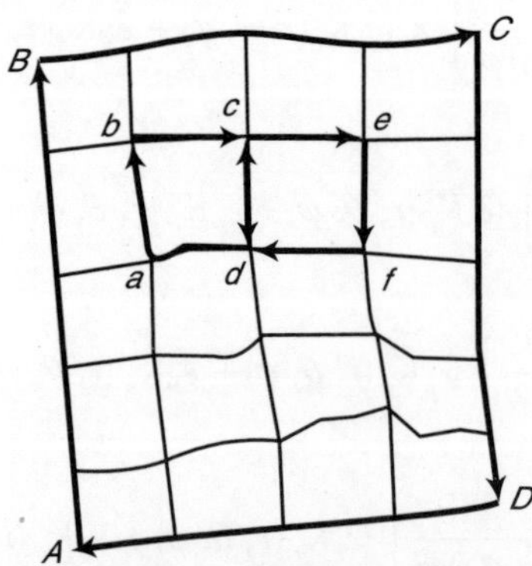

Fig. 10.11 A well-behaved surface, no matter whether flat, curved, or ruffled, can be sliced into tiny cells as shown in this figure.

Any finite area S can be split up into a large number of infinitesimal bits δS_i bound by tiny curves δC_i. Only well-behaved surfaces, as in Fig. 10.5 may be considered. The result will not be applicable for the kind of surfaces shown in Fig. 10.7. We can then apply the *definition* of the curl to each cell. Note that the contributions to the circulation from the paths of adjacent *inner* cells would cancel when the inner cells are oriented according to the right hand screw convention. From Fig. 10.11 we see that the contribution to the circulation from the path c to d of the cell on the left cancels the opposite contribution coming from the path d to c on the cell on its immediate right. On adding (i.e., integrating) the results for n number of cells which fill up the larger surface bound by the path $A \rightarrow B \rightarrow C \rightarrow D \rightarrow A$, we get

$$\oint_{ABCDA} \vec{K}(\vec{r}) \cdot \vec{dl} = \sum_{i=1}^{n} \oint_{\delta C_i} \vec{K}(\vec{r}) \cdot \vec{dl} = \sum_{i=1}^{n} \iint \text{curl}\vec{K}(\vec{r}) \cdot dS\hat{n}. \tag{10.38}$$

We have used the fact that the circulation over a path is just the component of the curl of the vector multiplied by the elemental surface area that is bound by the path, over which the circulation is determined. Adding (integrating) the results of Eq. 10.38 for all the n number of cells, $n \rightarrow \infty$, we get

$$\oint_{C} \vec{K}(\vec{r}) \cdot \vec{dl} = \iint curl\,\vec{K}(\vec{r}) \cdot dS\hat{n}. \tag{10.39}$$

The result stated in Eq. 10.39 is called the Kelvin–Stokes theorem, after George Gabriel Stokes (1819–1903) and Lord Kelvin (William Thomson, 1824–1907). This theorem was first stated in a letter by Lord Kelvin to Gabriel Stokes in July 1850. Stokes popularized this theorem, hence often it is known only as the Stokes theorem. It relates the line integral of a vector (about a closed curve), to the surface integral of its curl (over the area bound by that curve). Any surface bound by the closed curve will work; you can pinch the butterfly net and distort its shape in any which way, and it won't matter. You must remember that the orientation of the vectorial area element on the right hand side of Eq.10.39 is inseparably connected with the sense in which the line integral on the left hand side is determined, on account of the right hand screw convention that must be employed to relate the two.

The methods from vector calculus developed in this chapter are of great importance in analysing the dynamics of fluids, and also electrodynamics. In fact, these techniques are of great importance in many branches of physics and engineering. The next two chapters will reveal some of the important applications.

Problems with Solutions

P10.1:

Find the directional derivative of the function $f(x, y, z) = x^2yz$ in the direction $4\hat{e}_x - 3\hat{e}_z$ at point (1, –1, 1).

***Solution*:**

The unit vector in the direction of $4\hat{e}_x - 3\hat{e}_z$ is $\breve{e} = \dfrac{4\breve{e}_x - 3\breve{e}_z}{|4\breve{e}_x - 3\breve{e}_z|} = \dfrac{4\breve{e}_x - 3\breve{e}_z}{5}$.

Now, $\vec{\nabla} f = \frac{\partial f}{\partial x}\breve{e}_x + \frac{\partial f}{\partial y}\breve{e}_y + \frac{\partial f}{\partial z}\breve{e}_z = 2xyz\breve{e}_x + x^2z\breve{e}_y + x^2y\breve{e}_z$. Hence, the directional derivative is:

$$\left[\frac{df}{ds}\right]_{(x,y,z)} = \breve{e}.\vec{\nabla} f = \frac{1}{5}(4\breve{e}_x - 3\breve{e}_z)\cdot(2xyz\breve{e}_x + x^2z\breve{e}_y + x^2y\breve{e}_z) = \frac{1}{5}(8xyz - 3x^2y).$$

At the point $(1, -1, 1)$ the directional derivative in the direction of $4\hat{e}_x - 3\hat{e}_z$ is $\left[\frac{df}{ds}\right]_{(1,-1,1)} = -1$.

P10.2:

Find the potential function from which the following vector force field is derivable:

$$\vec{F} = (2x\cos y - 2z^3)\hat{i} + (3 + 2y\hat{e}_z - x^2\sin y)\hat{j} + (y^2\hat{e}_z - 6xz^2)\hat{k}.$$

***Solution*:**

$$\vec{F} = -\vec{\nabla}V = -\left(\frac{\partial V}{\partial x}\breve{e}_x + \frac{\partial V}{\partial y}\breve{e}_y + \frac{\partial V}{\partial z}\breve{e}_z\right)$$, where V is the scalar potential function.

For the given force field, we therefore have

$$-\frac{\partial V}{\partial x} = 2x\cos y - 2z^3; \quad -\frac{\partial V}{\partial y} = 3 + 2ye^z - x^2\sin y; \quad \text{and} \quad -\frac{\partial V}{\partial z} = y^2e^z - 6xz^2.$$

Integrating the first of the above relations, $V = 2xz^3 - x^2\cos y + f(x, z)$. Its partial derivative with respect to y gives: $\frac{\partial V}{\partial y} = x^2\sin y + \frac{\partial}{\partial y}[f(y, z)]$. Hence, $\frac{\partial f(y, z)}{\partial y} = -3 - 2ye^z$, and therefore, $f(y, z) = -3y - y^2e^z + g(z)$. Accordingly, $V = 2xz^3 - x^2\cos y - 3y - y^2e^z + g(z)$.

Its partial derivative with respect to z gives $\frac{\partial V}{\partial z} = 6xz^2 - y^2e^z + \frac{\partial g(z)}{\partial z}$.

Therefore, $\frac{\partial g(z)}{\partial z} = 0$; i.e., $g(z)$ = constant = C. Hence, $V = 2xz^3 - x^2\cos y - 3y - y^2e^z + C$.

P10.3:

Determine the path integral of the force field given by $\vec{F}(\vec{r}) = F(r)\,\hat{e}_\varphi$ over a circle of radius 'a' in the XY-plane, centered at the origin of the coordinate frame. Also, determine the surface integral of the force field on the surface of a hemisphere erected (with $z > 0$) on this circle.

***Solution*:**

$$\oint \vec{F}(\vec{r})\cdot\vec{dr} = \oint (F(r)\breve{e}_\varphi)\cdot(\rho d\varphi\breve{e}_\varphi) = F(r)a\oint(d\varphi) = 2\pi F(a)a\cdot$$

$$\iint \vec{\nabla}\times(F(r)\breve{e}_\varphi)\cdot\vec{dS} = \iint \vec{\nabla}\times(F(r)\breve{e}_\varphi)\cdot(r^2\sin\theta d\theta d\varphi\breve{e}_r) = \iint\left(\frac{F(r)\cos\theta}{r\sin\theta}\breve{e}_r\right)\cdot(r^2\sin\theta d\theta d\varphi\breve{e}_r)$$

Hence, $$\iint \vec{\nabla}\times(F(r)\breve{e}_\varphi)\cdot\vec{dS} = \iint F(r)\cos\theta r d\theta d\varphi = aF(a)\times\int_0^{\frac{\pi}{2}}\cos\theta d\theta\times\int_0^{2\pi}d\varphi = 2\pi aF(a).$$

P10.4:

Determine the total flux of a vector field $\vec{A} = x\hat{e}_x + y\hat{e}_y + z\hat{e}_z$ over the boundary of the surfaces of the seamless structure shown below.

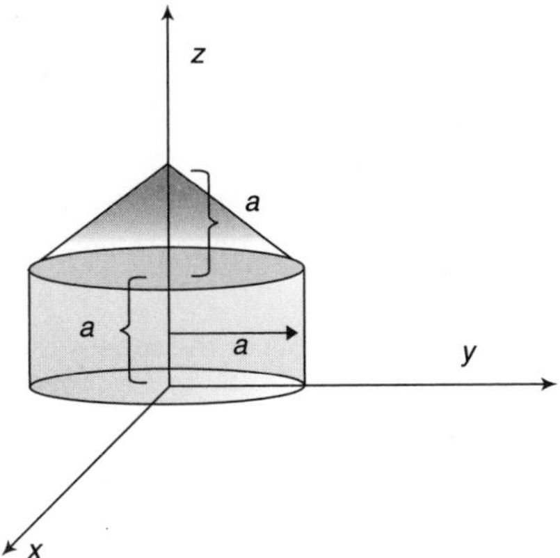

Solution:

Given $\vec{A} = x\breve{e}_x + y\breve{e}_y + z\breve{e}_z$. Hence: $\vec{\nabla}\cdot\vec{A} = 3$

For the given figure: $\phi = \iiint\limits_V (\vec{\nabla}\cdot\vec{A})dV = 3\iiint\limits_{\text{Cylinder}} dV + 3\iiint\limits_{\text{Cone}} dV$

$$\phi = 3\int_{\rho=0}^{a}\int_{\varphi=0}^{2\pi}\int_{z=0}^{a}\rho d\rho d\varphi dz + 3\int_{\varphi=0}^{2\pi}d\varphi\int_{z=a}^{2a}\left\{\int_{\rho=0}^{2a-z}\rho d\rho\right\}dz$$

($\because$ ρ at the surface of the cone varies with z as $\rho = 2a - z$ when we go from bottom to top of the cone)

$$\phi = 3\left[\left(\frac{\rho^2}{2}\Big|_0^a + \varphi\Big|_0^{2\pi} + z\Big|_0^a\right) + 2\pi\left(\int_{z=a}^{2a}\frac{\rho^2}{2}\Big|_0^{2a-z}dz\right)\right] = 3\left[\pi a^3 + \frac{\pi a^3}{3}\right] = 4\pi a^3$$

Let us calculate the flux of $\vec{A}$ through the surface of the structure using surface integral.

We have, $\vec{A} = x\breve{e}_x + y\breve{e}_y + z\breve{e}_z$

$= \rho\breve{e}_\rho + z\breve{e}_z$

Now, elemental surface on the curved surface (S_1) of the cylinder is

$$d\vec{s}_1 = (ad\varphi.dz)\breve{e}_\rho a$$

Total flux through the curved surface (S_1) of the cylinder

$$\int_{S_1}\vec{A}\cdot d\vec{s}_1 = a^2\int_{\varphi=0}^{2\pi}d\varphi\int_{z=0}^{a}dz = 2\pi a^3$$

Again, elemental surface on the bottom surface (S_1) of the cylinder is

$$d\vec{s}_2 = (\rho d\varphi . d\rho)\breve{e}_z$$

Total flux through the bottom surface (S_2) of the cylinder

$$\int_{S_2} \vec{A}\cdot d\vec{s}_2 = z\int_{\varphi=0}^{2\pi} d\varphi \int_{\rho=0}^{a} \rho d\rho = 0 \qquad (\because\ z = 0 \text{ for the bottom surface})$$

Now, for the flat surface on the top of the cylinder (which is the common surface between the cylinder and the cone), surface integral cancels due to the opposite unit normal due to the cylinder and the cone.

Now, elemental surface on the curved surface (S_3) of the cone is

$$d\vec{s}_3 = (\rho d\varphi . dl)\left(\frac{\breve{e}_\rho + \breve{e}_z}{\sqrt{2}}\right) \qquad (dl \text{ is the line element along the boundary of the curved surface})$$

Now, $$dl = \sqrt{(d\rho)^2 + (dz)^2} = dz\sqrt{1+\left(\frac{d\rho}{dz}\right)^2} = \sqrt{2}dz \qquad (\because\ \rho = 2a - z)$$

Hence, $$d\vec{s}_3 = (\rho d\varphi . dz)(\breve{e}_\rho + \breve{e}_z).$$

The flux through the curved surface (S_3) of the cone is

with, $$\vec{A}\cdot d\vec{s}_3 = (\rho^2 + \rho z)d\varphi dz = (4a^2 - 2az)d\varphi dz a,$$

$$\int_{S_3} \vec{A}\cdot d\vec{s}_3 = \int_{\varphi=0}^{2\pi} d\varphi \int_{z=a}^{2a} (4a^2 - 2az)dz = 2\pi a^3.$$

So, the total flux through the structure is,

$$\int_{S_1} \vec{A}\cdot d\vec{s}_1 + \int_{S_2} \vec{A}\cdot d\vec{s}_2 + \int_{S_3} \vec{A}\cdot d\vec{s}_3 = 2\pi a^3 + 0 + 2\pi a^3 = 4\pi a^3.$$

Thus, we can clearly see that, $\iiint_V (\vec{\nabla}\cdot\vec{A})dV = \oiint_S \vec{A}\cdot d\vec{s}$.

P10.5:

Consider a steady-state 2-dimensional incompressible fluid flow at velocity $\vec{v} = (x^2 + 3x - 4y)\hat{e}_x - (2xy + 3y)\hat{e}_y$. (a) Does the flow satisfy the equation of continuity?

(b) Determine the vorticity. (c) Locate all the stagnation points.

***Solution*:**

(a) In steady state, $\vec{\nabla}\cdot\vec{J}=0$. The fluid being incompressible, $\vec{\nabla}\cdot\vec{v}=\frac{\partial v_x}{\partial x}+\frac{\partial v_y}{\partial y}=0$. This equation is satisfied for $\vec{v}=(x^2+3x-4y)\hat{e}_x-(2xy+3y)\hat{e}_y$. Hence, the equation of continuity is satisfied.

(b) $\vec{\nabla}\times\vec{v}=\begin{vmatrix}\breve{e}_x & \breve{e}_y & \breve{e}_z\\ \frac{\partial}{\partial x} & \frac{\partial}{\partial y} & \frac{\partial}{\partial z}\\ x^2+3x-4y & 2xy+3y & 0\end{vmatrix}=\breve{e}_z(2y-4)\neq\vec{0}$. Hence, the flow is rotational.

(c) Stagnation points occur when both the components of the fluid flow are zero, hence when

$(x^2+3x-4y)=0$ and also $-(2xy+3y)=0$.

Now, when $(2xy+3y)=0$, we get $2y\left(x+\frac{3}{2}\right)=0$ and hence $y=0$ and/or $x=-\frac{3}{2}$.

Now, when, $(x^2+3x-4y)=0$, if, $x=-\frac{3}{2}$, then $y=-\frac{9}{16}$.

On the other hand, if $y=0$, then we have $x=0$ or $x=-3$.

Thus, there are three stagnation points: $\left(-\frac{3}{2},-\frac{9}{16}\right)$, (0, 0) and (–3, 0).

P10.6:

Determine the work done by the force field $\vec{F}=\cos\varphi\hat{e}_\rho-\rho\sin\varphi\hat{e}_\varphi-5\rho\hat{e}_z$ in moving an object along the paths C_1, C_2, C_3, C_4 (in that order), starting at the Cartesian coordinates (1, 0, 0). Obtain your answer by finding the work done by the force as a line integral over the closed path, and also as a surface integral, using the Kelvin–Stokes theorem.

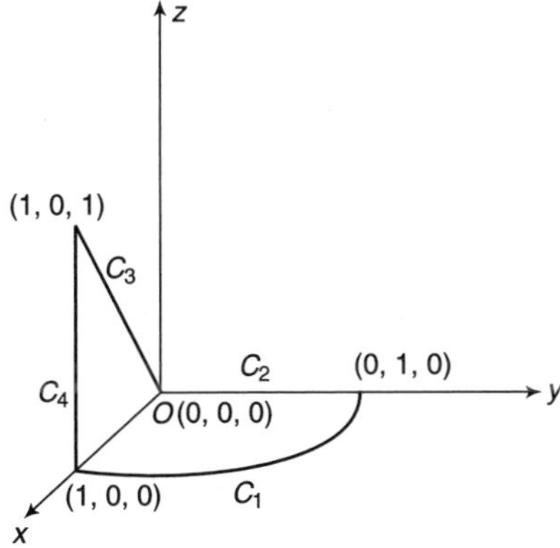

***Solution*:**

On C_1, $\rho=1$, $z=0$ and $0\leq\varphi\leq\frac{\pi}{2}$. Also: $d\vec{l}=\rho d\varphi\breve{e}_\varphi$.

Hence, $\vec{F}\cdot d\vec{l} = -\rho^2 \sin\varphi d\varphi = -\sin\varphi d\varphi$. Therefore, $\int_{C_1} \vec{F}\cdot d\vec{l} = \int_0^{\pi/2} -\sin\varphi d\varphi = -1$

On C_2, $x = 0$, $z = 0$ and y goes from 1 to 0. Now, on the y-axis: $\varphi = \frac{\pi}{2}$, $\sin\varphi = 1$.

Hence, $y = \rho\sin\varphi = \rho$. On C_2, ρ varies from 1 to 0. Thus, on C_2, $\vec{F}\cdot d\vec{l} = \cos\varphi d\rho = 0$.

Therefore, $\int_{C_2} \vec{F}\cdot d\vec{l} = 0$.

On C_3, $\int_{C_3} \vec{F}\cdot d\vec{l} = \int_0^1 (d\rho - 5\rho d\rho) = -\frac{3}{2}$

On C_4, $\int_{C_4} \vec{F}\cdot d\vec{l} = \int_1^0 -5d\rho = 5$. Hence, the total work done is $= -1 - \frac{3}{2} + 5 = \frac{5}{2}$

To use the Kelvin–Stokes theorem, we first determine the curl of the force.

$$\vec{\nabla}\times\vec{F} = 5\breve{e}_\rho - 2\sin\varphi\breve{e}_z + \frac{1}{\rho}\sin\varphi\breve{e}_z .$$

The surface enclosed by the closed path $(C_1 + C_2 + C_3 + C_4)$ is the sum of the two surfaces, S_1, which is the quarter-disc of unit radius in the first quadrant on the XY-plane) and S_2, which is the triangle in the ZX-plane with vertices (0, 0), (1, 1) and (1, 0).

Thus the total work done is $\iint_{S_1} (\vec{\nabla}\times\vec{F})\cdot d\vec{s} + \iint_{S_2} (\vec{\nabla}\times\vec{F})\cdot d\vec{s} = 0 + \frac{5}{2}$, same result as we get from the line integral.

Additional Problems

P10.7 Which of the following velocity fields satisfies the conservation of mass in an incompressible plane flow? The x-and the y- components of a 2-dimensional fluid flow are given in the form of the components (v_x, v_y) of the velocity $\vec{v}$:

(a) $(xy + y^2t, xy + x^4t)$, (b) $(4x^2y^3, -2xy^4)$, (c) $(u = 3xt, v = -3yt)$.

P10.8 Verify Stokes' theorem for $\vec{F} = (2x - y)\hat{e}_x - yz^2\hat{e}_y - y^2z\hat{e}_z$, where S is the upper half surface of the sphere $x^2 + y^2 = 1$ and C is its boundary.

P10.9 Determine the point(s) at which the function $f(x) = x^3 - 12xy + 8y^3$ has maximum, minimum and saddle point.

P10.10 For the vector field $\vec{A} = x^2\hat{e}_x + xy\hat{e}_y - yz\hat{e}_z$, verify the divergence theorem over a cube of unit side length. The cube is situated in the first octant $(x \geq 0, y \geq 0, z \geq 0)$ of a Cartesian coordinate system, with one corner at origin.

P10.11 A particle is taken along the path $A(0, 0, 0) \to B(d_1, 0, 0) \to C(d_1, d_2, 0), \to D(0, d_2, 0) \to A(0, 0, 0)$ against a force field given by $\vec{F}(x, y, z) = 4xy\hat{e}_x - 2x^2\hat{e}_y + 3z\hat{e}_z$, where the various constants have appropriate dimensions. Determine the work done by this force, and verify the Stokes theorem by finding the work done using both the line integral and the surface integral that appears in it.

P10.12 The Cartesian components of an incompressible, steady state velocity fluid flow at a point are given as $\vec{v} = (x^2 + y^2 + z^2, xy + yz + z, w = ?)$ in terms of the Cartesian coordinates of that point. Determine w such that the equation of continuity is satisfied.

P10.13(a) Express the equation of continuity in terms of the cylindrical polar coordinates. (b) Express the equation of continuity in terms of the spherical polar coordinates.

P10.14(a) Determine the divergence of a vector point function described by

$$\vec{A}(\vec{r}) = r\cos\theta\breve{e}_r + r\sin\theta\cos\varphi\breve{e}_\varphi + r\sin\theta\breve{e}_\theta.$$

(b) Find the flux of the above vector field over a closed surface that encloses a hemisphere of radius R resting on the XY-plane, with its center at origin. The hemisphere is located in the region $z \geq 0$.

P10.15 A vector field representing the velocity of a fluid in motion is given as $\vec{v}(\rho, \varphi, z) = K\vec{\nabla}\varphi$, where K is a positive constant of appropriate dimensions.

Sketch a few field lines for this velocity field in all Cartesian quadrants indicating direction of flow.

P10.16 A vector field is given by $\vec{A} = x^2\hat{e}_x + y^2\hat{e}_y + z^2\hat{e}_z$. Determine the surface integral $\oiint \vec{A} \cdot \vec{ds}$ over the closed surface of a cylinder $x^2 + y^2 = 16$ bounded by the planes $z = 0$, $z = 8$.

P10.17 Determine the directional derivative of the function $f(x, y, z) = \sqrt{(x^2 + y^2)}$ in the direction $\hat{e}_x + \hat{e}_y + \hat{e}_z$ at the point (2, 0, 1).

P10.18 Find the surface integral of the vector field $\vec{F} = 3x^2\hat{e}_x - 2y\hat{e}_y + z\hat{e}_z$ over the surface S that is the graph of $z = 2x - y$ over the rectangle $[0, 2] \times [0, 2]$.

P10.19 Determine if the force field $\vec{F} = x^2y\hat{e}_x + xyz\hat{e}_y - x^2y^2\hat{e}_z$ is conservative.

P10.20 A vector field is given by $\vec{F} = -y\hat{e}_x + z\hat{e}_y + x^2\hat{e}_z$. Determine its line integral over a circular path of radius R in the XY-plane, with its center at the origin. Determine if the Stokes' theorem holds good for this circular path, by considering the following surfaces:

(i) the plane of the circle, and (ii) surface of a cylinder of height h erected on the circle.

P10.21 Determine the curl of the following vectors fields: (a) $\rho\cos\varphi\hat{\varepsilon}_\rho - \rho\sin\varphi\hat{e}_\varphi$, (b) $\frac{1}{\rho}\breve{e}_\rho$.

P10.22 For the steady irrotational flow of incompressible, non-viscous liquids, if the flow occurs under gravity, find the value of pressure at an arbitrary height, z.

P10.23 The density field of a plane, steady state fluid flow given by $\rho(x_1, x_2) = kx_1x_2$ where k is constant. Determine the form of the velocity field if the fluid is incompressible.

P10.24 Determine the stationary points of the function $f(x, y) = 3x^2y + y^3 - 4x^2 - 4y^2 + 6$. Find its local minima, local maxima and saddle points.

References

[1] Arfken, G. B., H. J. Weber, and F. E. Harris. 2013. *Mathematical Methods for Physicists*. Amsterdam: Elsevier.

[2] Boas, Mary L. 2006. *Mathematical Methods in the Physical Sciences*. Hoboken, N.J.: John Wiley and Sons.

[3] https://www.mcescher.com/. Accessed on 20 November 2018.

CHAPTER 11

Rudiments of Fluid Mechanics

Surrender to the flow of the River of Life, yet do not float down the river like a leaf or a log. While neither attempting to resist life, nor to hurry it, become the rudder and use your energy to correct your course, to avoid the whirlpools and undertow.

—*Jonathan Lockwood Huie*

11.1 THE EULER'S EQUATION OF MOTION FOR FLUID FLOW

In the previous chapter, we studied the Kelvin–Stokes theorem. When restricted to just two dimensions, it is referred to as the Green's theorem which is therefore only a special case of the Kelvin–Stokes theorem. Using the Green's theorem, we are automatically led to another prodigious theorem in the analysis of complex functions, known as the *Cauchy's theorem*. The methods developed in Chapter 10 are of great importance in fluid dynamics, electrodynamics and also in quantum dynamics. Practical applications of these methods are abundantly found not merely in fluid dynamics but also in the study of plasma in stellar atmosphere, and the analysis of charge plasma produced by lasers. The techniques are therefore of great consequence in engineering and technology, apart of course, from basic sciences. In this chapter, we shall use these methods to develop the equation of motion for a classical fluid.

The central question in mechanics is how to describe the state of a system, and how the system evolves with time. We have learned in earlier chapters that the state of a material particle system is represented by its position and momentum in the phase space. The temporal evolution of the system is provided by its equation of motion, viz., the Newton's equation of motion, or equivalently by the Lagrange or the Hamilton's equation of motion, which are discussed in earlier chapters. Much of the study of classical mechanics is about setting up the equation of motion, and learning to solve the same. When the medium is the continuum fluid, the system is very complex. It cannot be described as particles, and its evolution cannot be simply described by the usual familiar form of the Newton's, Lagrange's, or Hamilton's equations, even if the basic tenets of classical mechanics remain applicable. A fluid consists of a large number of fluid 'particles', which is an idea that is not defined exactly the same way as that for a piece of sand. Hence the position and momentum which represent a point in phase space are not suitable to represent the state of a fluid. The state of a fluid is, instead, characterized by some other properties, like its density, pressure, velocity and temperature. Some thermodynamic considerations get therefore involved in describing its equation of motion. Since the velocity is a vector, we already have a set of six physical properties to determine to describe the state of the fluid. The position and momentum could be determined, for a particle with one degree of freedom, by the two first order Hamilton's differential equations of motion. In the case of the fluid, we obviously need more equations to determine its physical properties. The set of equations which must be solved to get the pressure, density, velocity and temperature of a fluid are usually referred to as the *governing equations* of fluid dynamics. The system of equations that represent the conservation of momentum, namely the Navier–Stokes equations, constitute a major part of this system, but often the entire set of governing equations are referred to as the Navier–Stokes equations. You can already see that the governing equations will be really hard to solve. First of all, there is a larger set of differential equations to be solved. The bigger challenge is that the differential equations involve the partial differential operators, non-linear terms, and are coupled. They pose a huge mathematical challenge.

In spite of the magnitude of the stiff mathematical conundrum, the fundamental principles which lead us to the family of the governing equations of fluid mechanics are deep rooted in the very foundations of classical mechanics to which you have already been exposed sufficiently in the previous chapters. In this chapter, we shall deal with some of the simplest forms of fluids, called the Newtonian neutral fluids. Being neutral, the machinery of dealing with the electromagnetic interactions, introduced later in Chapter 12, is not required to study this part of fluid dynamics. Besides, the fluid velocities of concern to us are low, typically well below the speed of sound—let alone the speed of light. Hence the relativistic mechanics, introduced in Chapter 13, is also not of any significant consequence to study the rudiments of fluid mechanics introduced in this chapter. The arduous task of solving a problem in fluid mechanics is brought down to a very approachable and enjoyable task. In this chapter, you will develop adequate comfort to address this task. To begin with, we shall make a number of approximations to describe the fluid, which are nonetheless good enough to describe a large number of real fluids, notably water.

In the previous chapter, as our first approximation we introduced the 'ideal' fluid. Compressibility of a fluid is an important consideration in its equation of motion. As such, all fluids (not just gases, but even liquids) are compressible to some extent or the other. However, to a significant extent, even air (and all gases) can be regarded as *incompressible*. When air is compressed, some of it tries to escape the region at the speed of sound. Hence for speeds of air-flow lower than the speed of sound, incompressibility is not a bad approximation, especially when the region holding the air is much larger than the part where air movement is induced. Experiments reveal that incompressibility approximation works best, when ratio of the *local speed of the fluid* to the *speed of sound* in that medium is less than 0.3. This ratio is called Mach number, after Ernst Mach (1838–1916). We shall consider incompressible fluids. It is, however, important to take the compressibility of a fluid into account for better accuracy, especially when one is analysing fluid dynamics in the context of high speed rocket engines, jet aircrafts, and also in some industrial applications. For details, the readers must refer to specialized books. There are several excellent books on fluid mechanics. The available literature is massive. The primary source and suggestion for supplementary reading for this chapter is the classic book by Landau and Lifshitz [1]. The book by White [2] is also excellent to learn many practical applications of the methods of fluid mechanics. The mathematical methods employed can be easily found in References [3,4]. The subject being vast, only rudimentary introduction is covered in this chapter. The methods developed in this chapter are introductory and simple, but have many applications.

We now proceed to set up the equation of motion for the ideal fluid, characterized by its density, pressure and velocity. We shall discuss the temperature of the fluid only in the passing, since it involves several inputs from thermodynamics which are outside our scope, and become important only when motion involves both mass and heat transfer. In the continuum hypothesis, a specific property about the nature of the stress in the 'ideal' fluid is that the stress at every point in the fluid is one of compression, and it is isotropic. The isotropic nature of stress in the ideal fluid was first recognized by Blaise Pascal (1623–1662). In his short life, Pascal did some amazing experiments and contributed significantly to our understanding of how fluids behave. Pascal's law states that *increase in pressure at an arbitrary point in a fluid placed in a container, no matter what shape the container has, is transmitted equally in all the directions to every other point in the container*. This property is incorporated in the defining description of an ideal fluid. The stress (force per unit area) at a point in it, i.e., compression, is purely isotropic. There is no 'shear', and no 'tension'. Being isotropic, the pressure, i.e., the compressional stress, has exactly the same value in all the directions. Being force per unit area, it is a vector, but since it is isotropic, direction is not important. It is therefore often *treated* as a scalar. A vector would need three components; but the pressure is specified merely by a single number (in appropriate units). The pressure gets transmitted evenly throughout the medium, but *not* the force itself, since the force is pressure multiplied by the cross-sectional area.

If we consider the forces at a point in a liquid, in a mathematical cell which may be of arbitrary shape and size, but chosen to be a rectangular parallelepiped only for illustrative purpose, the compressional inward pressure at that point is shown in Fig. 11.1. The unit normal to the surface that bounds a region of space is *outward*, and the stress is opposite to that, as shown in Eq. 10.11c.

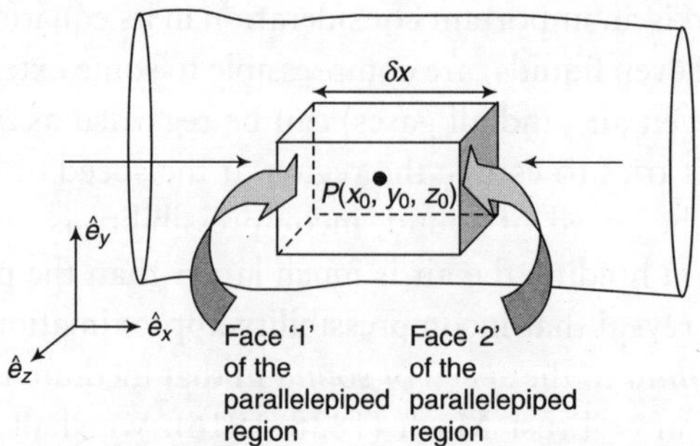

Fig. 11.1 In an ideal fluid, the stress is essentially one of compression. The forces by the neighboring regions at a point are all inward, toward the said point.

The force on the face '1' on the left hand side of the parallelepiped shown in Fig. 11.1, whose magnitude will be given by the product of the pressure on that face multiplied by the area of the face, will therefore be in the direction of the unit vector $\hat{e}_x$ whereas the force on the face '2' will be along $-\hat{e}_x$ in the Cartesian coordinate system shown in Fig. 11.1. The face '1' is, however, displaced by $\frac{\delta x}{2}$ to the left, and the face '2' is displaced by an equal amount to the right, of the point P. The force on the face '1' is therefore given by

$$\vec{F}(1) = \left\{ p(\vec{r}) - \left[\frac{\partial p}{\partial x} \right]_P \frac{\delta x}{2} \right\} (\delta y \delta z) \, (+\hat{e}_x), \tag{11.1a}$$

and that on face '2' is given by

$$\vec{F}(2) = \left\{ p(\vec{r}) + \left[\frac{\partial p}{\partial x} \right]_P \frac{\delta x}{2} \right\} (\delta y \delta z) \, (-\hat{e}_x). \tag{11.1b}$$

The sum of the forces at the point P through the pair of forces '1' and '2' is

$$\vec{F}(1) + \vec{F}(2) = -\left[\frac{\partial p}{\partial x} \right]_P \delta x (\delta y \, \delta z) \, (+\hat{e}_x) = -\left[\frac{\partial p}{\partial x} \right]_P \delta V \; \hat{e}_x. \tag{11.2a}$$

Summing over all the six faces, which are the three pairs of opposite faces, of the parallelepiped which surrounds the point P, we get

$$\sum_{i=1}^{6} \vec{F}(i) = -\left\{ \left[\frac{\partial p}{\partial x} \right]_P \hat{e}_x + \left[\frac{\partial p}{\partial y} \right]_P \hat{e}_y + \left[\frac{\partial p}{\partial z} \right]_P \hat{e}_z \right\} \delta V = -\vec{\nabla} p \; \delta V. \tag{11.2b}$$

The net force, from all directions, per unit volume that encloses the point P, in the limit that the volume element becomes infinitesimally small, therefore is

$$\text{Net force per unit volume} = \vec{\nabla} p. \tag{11.3}$$

This force originates from the compressional inward forces, applied by the fluid itself, that act at a point inside the medium. *It is due to the hydrostatic pressure.* As mentioned in Chapter 10, *hýdōr* in Greek means 'water'; but we shall continue to use this term, as is commonly done, for any 'ideal' fluid. As such, even water is not exactly an 'ideal' fluid, as mentioned earlier, but it does come pretty close to it. We shall denote the hydrostatic *force* per unit volume acting at a point P by $\vec{F}^H_{UV}$. The superscript 'H' is for 'hydrostatic', and the subscript 'UV' for 'unit volume'. Accordingly,

$$\vec{F}^H_{UV} = -\vec{\nabla} p. \tag{11.4}$$

Of course, an external force, such as gravity, would also act on the volume of the liquid in the parallelepiped. We shall denote such an external force, per unit volume, by $\vec{F}^E_{UV}$. The superscript 'E' here denotes the 'external' force. In the present case, it represents the gravitational pull by the Earth. It is given by

$$\vec{F}^E_{UV} = \lim_{\delta V \to 0} \frac{\vec{F}_{\text{external}}}{\delta V} = \lim_{\delta V \to 0} \left(\frac{\vec{F}_{\text{external}}}{\delta m} \right) \left(\frac{\delta m}{\delta V} \right) = \lim_{\delta V \to 0} \left(\frac{\vec{F}_{\text{external}}}{\delta m} \right) \rho(\vec{r}) = \vec{g} \rho(\vec{r}). \tag{11.5}$$

In the above equation, $\vec{g}$ is the acceleration due to gravity, and $\rho(\vec{r})$ is the fluid density.

The total hydrostatic *plus* external (gravity) force, per unit volume, $\vec{F}^{T}_{UV}$, (the superscript '*T*' denoting the 'total' force) acting on the parallelepiped, per unit volume, therefore is

$$\vec{F}^{T}_{UV} = -\vec{\nabla}p + \vec{g}\rho(\vec{r}). \tag{11.6}$$

As per Newton's second law, this force, of course, must be equal to the mass times acceleration, divided by the volume element, since we are considering forces per unit volume. The acceleration being referred to, would be that of the fluid 'particle' at the point $P(x_0, y_0, z_0)$ at an instant t. It would represent a real fluid 'particle' that is accelerating through the point P under the action of the combined hydrostatic and the external (gravitational) forces. This force therefore requires the Lagrangian description of the fluid, discussed in the previous chapter. This force, per unit volume, is $\lim\limits_{\delta V \to 0} \dfrac{\delta m}{\delta V}\dfrac{d\vec{v}}{dt}$, wherein $\dfrac{d\vec{v}}{dt}$ is the acceleration of the particle, in the Lagrangian description. This is just the acceleration that appears in the Newton's equation of motion, $\vec{F} = m\dfrac{d\vec{v}}{dt}$. Hence, we have

$$\lim_{\delta V \to 0} \frac{\delta m}{\delta V}\frac{d\vec{v}}{dt} = \vec{F}^{T}_{UV} = \rho(\vec{r})\frac{d\vec{v}}{dt} = -\vec{\nabla}p + \vec{g}\rho(\vec{r}). \tag{11.7}$$

Accordingly, dividing both sides by the density, we get

$$\frac{d\vec{v}(\vec{r},t)}{dt} = \frac{-\vec{\nabla}p}{\rho(\vec{r},t)} + \vec{g} = \frac{-\vec{\nabla}p}{\rho(\vec{r},t)} - \vec{\nabla}\Phi_g. \tag{11.8}$$

The above result is the Cauchy–Euler momentum equation. It gives the rate at which the momentum (per unit mass) changes. The differential operator that finds the derivative with respect to time appearing on the left hand side is the Lagrangian, material, total time-derivative operator. In Eq. 11.8, we have identified the acceleration due to gravity as the *force per unit mass*, derivable from the gravitational potential Φ_g, as we did earlier in the previous chapter (Eq. 10.7). The time derivative operator in Eq. 11.8 is not providing the rate at which the velocity is changing at some particular Eulerian point in space. Rather, $\dfrac{d\vec{v}(\vec{r},t)}{dt}$ gives the rate at which a particular fluid 'particle' is accelerated by a force acting on it, in the sense of Newton's principle of causality and determinism contained in the second law of mechanics. The time-derivative operator in Eq. 11.7 is therefore the same substantive, total time derivative operator, which we have discussed at length in the previous chapter (see the discussion between Eqs. 10.27 and 10.30). The difference between the material, time derivative operator, $\dfrac{d}{dt}$, and the 'local' time derivative operator, $\dfrac{\partial}{\partial t}$, has been discussed already in the previous chapter. The advective operator $(\vec{v}\cdot\vec{\nabla})$ must be added to the local time-derivative operator to take account of the material transport that is involved in the substantive time-derivative operator. Therefore, the equation of motion is given by

$$\left(\vec{v}\cdot\vec{\nabla} + \frac{\partial}{\partial t}\right)\vec{v}(\vec{r},t) = \frac{-\vec{\nabla}p}{\rho(\vec{r},t)} + \vec{g} = \frac{-\vec{\nabla}p}{\rho(\vec{r},t)} - \vec{\nabla}\Phi_g. \tag{11.9}$$

Equation 11.9 is known as the Euler Equation, after Leonhard Euler, who was a student of Johann Bernoulli, and a friend of Daniel Bernoulli. He was the first to arrive at the results in Eq. 11.9, in the year 1755. Euler worked at the St. Petersburg Academy of Sciences, Russia (1727–41, 1766–83), and at the Berlin Academy (1741–66). He made outstanding contributions to a vast range of fields in physics and mathematics including number theory, calculus, geometry, trigonometry, calculus of variation, analysis of rigid body motion, fluid dynamics, etc.

If we now focus our attention *only* on the forces due to the fluid (i.e., ignore the external force due to gravity), we get

$$\left(\vec{v}\cdot\vec{\nabla}+\frac{\partial}{\partial t}\right)\vec{v}(\vec{r},t)=\frac{-\vec{\nabla}p}{\rho(\vec{r},t)}. \tag{11.10}$$

Spatially *uniform* pressure in a fluid medium has a zero gradient, and would therefore not exert any net force on a fluid element. The Euler equations (Eqs. 11.9 and 11.10) provide the means to study the acceleration of the fluid, which is the primary quantity of interest in the consideration with regard to the *conservation of momentum*. In the spirit of the cause-effect relationship contained in Newton's second law, the momentum of a body changes only if a force acts on it, and at a rate that is equal to the force. It is this change in momentum that manifests as the acceleration term that is given by the Euler's equations for the *ideal* fluid. Energy dissipation effected by viscosity (friction) is ignored in this, but it provides an excellent starting point for solving real problems.

In the model of the fluid we are dealing with, heat exchanges between different parts of the fluid are not taken into account. Transfer of heat energy from different portions of the fluid rubbing against each other is therefore ignored. This amounts to assuming that the thermal conductivity of the ideal fluid is zero. With no heat exchange permitted, the motion of the ideal fluid is strictly adiabatic. A characteristic property of the *reversible* adiabatic process is that its entropy s remains constant:

$$\frac{ds}{dt}=0, \tag{11.11a}$$

i.e., $$\left(\vec{v}\cdot\vec{\nabla}+\frac{\partial}{\partial t}\right)s=0. \tag{11.11b}$$

The above equation describes the adiabatic motion of an ideal fluid. Multiplying Eq. 11.11b by the density ρ, we get

$$\rho\vec{v}\cdot\vec{\nabla}s+\rho\frac{\partial s}{\partial t}=0. \tag{11.12}$$

Multiplying now the equation of continuity (Eq. 10.26b) by the entropy s, we get

$$s\frac{\partial\rho}{\partial t}+s\rho\vec{\nabla}\cdot\vec{v}+s\vec{v}\cdot\vec{\nabla}\rho=0. \tag{11.13}$$

Adding Eq. 11.12 and Eq. 11.13, we get

$$\rho\frac{\partial s}{\partial t}+s\frac{\partial\rho}{\partial t}+s\rho\vec{\nabla}\cdot\vec{v}+s\vec{v}\cdot\vec{\nabla}\rho+\rho\vec{v}\cdot\vec{\nabla}s=0, \tag{11.14a}$$

i.e., $$\frac{\partial(s\rho)}{\partial t}+s\rho\vec{\nabla}\cdot\vec{v}+\vec{v}\cdot\vec{\nabla}(s\rho)=0, \tag{11.14b}$$

i.e., $$\frac{\partial(s\rho)}{\partial t}+\vec{\nabla}\cdot(s\rho\vec{v})=0. \tag{11.15}$$

The form of the result in Eq. 11.15 is similar to the equation of continuity (Eq. 10.26). The latter gave the rate of change of the density of the fluid in terms of the divergence of the current density vector ($\vec{J}=\rho\vec{v}$), i.e., the mass flux density vector. Equation 11.15 gives the rate of change of $s\rho\vec{v}$ in terms of the divergence of

$$\vec{\xi}_{sfd}=s\vec{J}=s\rho\vec{v}, \tag{11.16}$$

which is called the *entropy flux density vector*. Most often, the fluid's entropy is homogeneous throughout the medium, and it is also independent of the time t. The state of the fluid is then said to be *isentropic*. The equation of motion for the entropy flux density then takes a simpler form.

Now, the differential increment in w, the heat function per unit mass (enthalpy), is related to the change in entropy and that in pressure through the following relation:

$$dw=Tds+v_s dp=Tds+\frac{1}{\rho}dp. \tag{11.17a}$$

The quantity $v_s=\frac{1}{\rho}$ is the specific volume. For the isentropic process, therefore,

$$\frac{dp}{\rho}=dw \text{ i.e. } \frac{\vec{\nabla}p}{\rho}=\vec{\nabla}w. \tag{11.17b}$$

Using Eq. 11.17b in the Euler equation, Eq. 11.9, we get

$$\frac{\partial\vec{v}(\vec{r},t)}{\partial t}+(\vec{v}\cdot\vec{\nabla})(\vec{v}(\vec{r},t))=-\vec{\nabla}w-\vec{\nabla}\Phi_g=-\vec{\nabla}(w+\Phi_g). \tag{11.18}$$

The Euler's equation of motion (Eqs. 11.9 and 11.10) for the fluid involves the fluid's density, pressure, and velocity. It is possible to extract from the Euler's equation an essential form of the equation of motion in terms of *only* the velocity. Toward this end, we will now use the following vector identity:

$$\vec{\nabla}(\vec{A}\cdot\vec{B})=(\vec{A}\cdot\vec{\nabla})\vec{B}+(\vec{B}\cdot\vec{\nabla})\vec{A}+\vec{A}\times(\vec{\nabla}\times\vec{B})+\vec{B}\times(\vec{\nabla}\times\vec{A}). \tag{11.19a}$$

Using the above identity, replacing both the vectors $\vec{B}$ and $\vec{A}$ with the fluid velocity $\vec{v}$, we get

$$(\vec{v}\cdot\vec{\nabla})\,\vec{v} = \frac{1}{2}\vec{\nabla}\,(\vec{v}\cdot\vec{v}) - \vec{v}\times(\vec{\nabla}\times\vec{v}). \tag{11.19b}$$

Using this result in Eq. 11.18, we get

$$\frac{\partial\vec{v}(\vec{r},t)}{\partial t} + \frac{1}{2}\vec{\nabla}\left(\vec{v}\cdot\vec{v}\right) - \vec{v}\times(\vec{\nabla}\times\vec{v}) = -\vec{\nabla}(w+\Phi_g). \tag{11.20}$$

Taking now the curl of both sides of Eq. 11.20, we get

$$\frac{\partial}{\partial t}\{\vec{\nabla}\times\vec{v}(\vec{r},t)\} - \vec{\nabla}\times\{\vec{v}\times(\vec{\nabla}\times\vec{v})\} = \vec{0}. \tag{11.21}$$

We have used, on the right hand side of Eq. 11.21, the fact that the curl of the gradient of a scalar is identically zero, no matter what the scalar field is. Also, in the first term, we have swapped the positions of the time-derivative operator and the space-derivative operators. This is legitimate, since the two operations are completely independent of each other. The result presented in Eq. 11.21 is straight out of the Euler's equation of motion for the fluid flow. However, while the Euler equation (Eqs. 11.9 and 11.10) involved both the pressure and the density, the result in Eq. 11.21 is in terms of only the velocity.

The above result can be written in a more compact form as

$$\frac{\partial\vec{\Omega}}{\partial t} - \vec{\nabla}\times(\vec{v}\times\vec{\Omega}) = \vec{0}, \tag{11.22}$$

$$\text{where,}\quad \vec{\Omega} = \vec{\nabla}\times\vec{v}, \tag{11.23}$$

is the *vorticity* of the velocity field. As can be easily seen, the curl of the velocity field of a fluid is (twice) the angular velocity. It is this property that provides the spinning motion to a leaf that is floating on a river that may have only a unidirectional flow. It is for this reason that the curl of a vector is called its *vorticity*. The vorticity of the velocity field of the fluid plays an important role in the equation of motion for the fluid flow.

The equation of motion for the fluid can then be solved by applying boundary conditions. As stated earlier, the ideal fluid is described by five parameters, viz., its density, pressure and the three components of its velocity. These can be obtained by solving the three Euler equations (one for each component of the vector equation), the equation of continuity, and the equation that describes the adiabatic motion of the ideal fluid. When the problem includes an analysis of the temperature, an additional equation, which comes from the first law of thermodynamics for the conservation of energy, is also required to solve the governing equations.

11.2 FLUIDS AT REST, 'HYDROSTATICS'

Although the term hydrostatics seems to refer to water, it is generally used for the subject of all fluids that are at *rest*. Now, a fluid exerts pressure in all directions inside the medium, and also at the boundary of a surface which contains it; it exerts a pressure perpendicular to the confining surface. We shall examine properties of the fluid when the fluid velocity $\vec{v} = 0$. The Euler's equation (Eq. 11.9), for fluids at rest, becomes

$$\vec{0} = \frac{-\vec{\nabla} p}{\rho(\vec{r}, t)} + \vec{g}, \tag{11.24}$$

where $\vec{g} = -\vec{\nabla}\Phi_g$, as in Eq. 11.8, with $\Phi_g = -\frac{GM}{r}$ being the gravitational potential. The relation Eq. 11.24 expresses hydrostatic equilibrium under the gravitational potential $\Phi_g = -\frac{GM}{r}$. The isobaric surface (i.e., surface on which the pressure is the same) will therefore be perpendicular to the direction of the *local* gravity. It is for this reason that a fluid's *open* surface is horizontal. However, if the fluid is rotated, the centrifugal term (Chapter 3) would add to Earth's gravity. It will therefore change the direction of the local gravity. The fluid's free surface then becomes parabolic instead of horizontal.

Let us consider a fluid in a region of space in which the acceleration due to gravity is uniform. We may consider a Cartesian coordinate system in which the XY-plane is horizontal, and the acceleration due to the gravity is $\vec{g} = -g\hat{e}_z$. Then, by symmetry, the partial derivatives of the pressure with respect to x and y coordinates are immediately seen to be zero. Equation 11.24 then reduces to a scalar equation:

$$\frac{dp(\vec{r}, t)}{dz} = -g\rho(\vec{r}, t). \tag{11.25}$$

It was Laplace, who was the first one to obtain this relationship which describes the dependence of the pressure on height (or, depth below the open surface). Integration with respect to z, assuming that there is no significant variation in the density, gives the pressure as a linear function of z:

$$p(z) = -\rho g z + c, \tag{11.26a}$$

where c is the constant of integration.

If we presume that the pressure at the free surface of the fluid, at the height $z = z_0$ (see Fig. 11.2), is p_0, then,

$$p_0 = p_{(z=z_0)} = -\rho g z_0 + c,$$

i.e., $c = p_0 = \rho g z_0$.

Accordingly,

$$p(z) = -\rho g z + p_0 + \rho g z_0 = p_0 + \rho g(z_0 - z) = p_0 + \rho g h. \tag{11.26b}$$

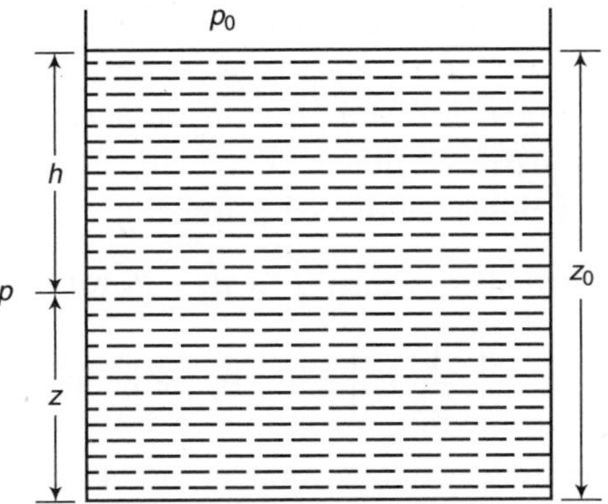

Fig. 11.2 The pressure at a point inside the fluid at rest depends only on the depth of that point below the free surface of the fluid.

The pressure at a point in a static fluid depends *only* on the depth of that point below the surface (Fig. 11.2). This property enables *leveraging* a force by transmitting it through fluid conduits (Fig. 11.3). This is easily achieved by adjusting the cross-sectional areas of the conduits. A narrow column of the fluid can be used to raise even a large weight, which seems paradoxical. A force can be effectively transmitted as desired by designing the constrictions of the container of the fluid. A classic application of hydraulic leveraging of a force is a car lift. One often needs to get under the car, for example, to change its exhaust muffler. It is much easier to operate if you raised the car above your head instead of crawling under it in the narrow clearance space under the car. Now, raising the huge weight of the car needs a huge upward force to be generated. This is achieved by distributing a force applied on a fluid in a very narrow pipe, through a well-designed hydraulic conduit system, to cylinders having a large cross-sectional area.

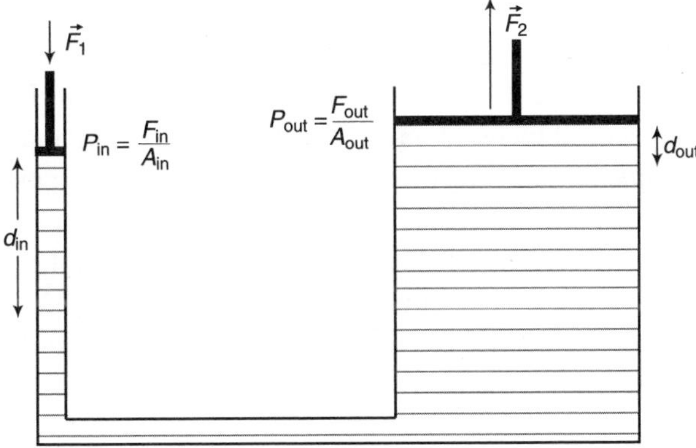

Fig. 11.3 The hydraulic press does wonders by exploiting the leveraging of the force applied by a fluid pressure. By suitably designing it, one can lift large weights. It works on the elementary principle of equalizing the fluid pressure, even if its advantages seem almost paradoxical.

A smaller force F_1 can be thus amplified to counter a large weight by producing a much bigger force that can oppose and overcome a large weight, exploiting hydraulic leveraging through the differential area, *pressure at a given depth being equal.* The energy expended in pushing the piston in the narrow cylinder containing an incompressible fluid results in the work done $F_1 \times d_1$, and this is conserved (assuming no dissipation). The work done on the piston in the larger cylinder is then $F_2 \times d_2 = P_2 A_2 d_2 = P_1 A_2 d_2 = \frac{F_1}{A_1} A_2 d_2$. We see immediately that the force $F_2 = \left(\frac{A_2}{A_1} F_1 \right) > F_1$. By designing the machine, a tremendous advantage in favour of generating a counter force F_2 can be achieved, enhanced by the *ratio* of the two areas. One can thus lift a huge weight using a simple hydraulic pedal. Using this principle, automobiles and dentist's chairs are raised, and ceramics and food products are compacted. Other applications include those in metal forming operations, carbon fiber molding etc. An actual hydraulic system has a somewhat more complicated design than is shown in the schematic diagram in Fig. 11.3. It employs a reservoir to admit and store the incompressible fluid, and an arrangement of ball valves, etc. The system must be well machined, since an air-bubble will spoil the transmission of pressure which is central to the functioning of this machine. The hydraulic brakes also work on the same principle. A small force on the brake pedal can be transmitted to the brake shoes through hydraulic lines to operate the brakes effectively over large areas. This results in effective stopping of the vehicle. The Bramah press, named after its inventor Joseph Bramah (1748–1814), also works on the Pascal's principle of uniform distribution of the pressure. It is the forerunner of a whole range of applications in hydraulic engineering.

Very often, we are interested in the *changes* in the pressure with reference to p_0, which often is the atmospheric pressure. It could be of course due to something else as well. It is these *changes* that one usually measures. Hence $\rho g h$ is called the gauge pressure, while $p(z)$ is the absolute pressure. However, when the density of the fluid is changing with depth (or height), one must carry out an appropriate integration of Eq. 11.25.

The deeper one goes in a liquid, the higher would be the pressure. The result of Eq. 11.26 shows that the pressure would increase linearly with the *depth* h in an incompressible liquid. Any object that is *completely* immersed in a liquid therefore experiences an upward force, since the pressure at its bottom is *higher* than at its top. This results in an upward force, called the buoyant force. It is essentially arising out of the differential pressure at the bottom, and the top, of the object immersed. Thus, on an object partially or completely immersed in a liquid, two forces act: its downward gravitational weight mg, and the upward buoyant upward force. The magnitude of the upward buoyant force is always equal to the weight of the fluid that is *displaced* by the immersed object, since it is exactly this displaced amount of fluid that maintains the hydrostatic balance in the *absence* of the immersed object. The upward buoyant force on the object is therefore given by

$$F_B = W_{df} = m_{df} g = V_{df} \rho_f g. \tag{11.27a}$$

In the above equation, $\vec{g}$ is the acceleration due to gravity, m_{df} is the mass of the displaced fluid, V_{df} is the volume of the displaced fluid, and ρ_f is the density of the fluid. Essentially, the buoyant force which an immersed object would experience is just the force that the fluid would experience prior to displacement, under equilibrium conditions. This force is therefore the 'pressure times area' integrated over the closed surface that bounds the volume of the fluid displaced by the immersed object.

Thus, $\vec{F}_B = -\oint\!\!\oint p(\vec{r})\, dS\hat{n}$. (11.27b)

The vector $\hat{n}$ is the unit normal vector on the surface, directed inward into the body immersed, which may be different for each point on the surface. We can easily determine the surface integral in the above equation by applying the Gauss' divergence theorem to a vector $p(\vec{r})\vec{c}$, where $p(\vec{r})$ is the pressure and $\vec{c}$ is an *arbitrary* constant vector:

$$\oint\!\!\oint (\vec{c}p)\cdot dS\hat{n} = \iiint (\vec{\nabla}\cdot(\vec{c}p))dV = \iiint p\vec{\nabla}\cdot\vec{c}dV + \iiint \vec{c}\cdot\vec{\nabla} p dV = \iiint \vec{c}\cdot\vec{\nabla} p dV = \vec{c}\cdot\iiint \vec{\nabla} p dV.$$

We have used in the above result the obvious fact that the divergence of a constant vector vanishes. Hence, $\vec{c}\cdot\oint\!\!\oint p(\vec{r})dS\hat{n} = \vec{c}\cdot\iiint \vec{\nabla} p dV$. Finally, since the direction of the constant vector $\vec{c}$ is arbitrary, we must have

$$\vec{F}_B = -\oint\!\!\oint p(\vec{r})dS\hat{n} = \iiint \vec{\nabla} p dV = -\rho_f \vec{g}\iiint dV = -\rho_f \vec{g} V_{df}. \quad (11.27c)$$

The minus sign on the right hand side of Eq. 11.27 is important. It provides the upward buoyant force that opposes the downward force due to gravity. The physical principle contained in Eq. 11.27 is named after Archimedes (287–212 BCE), who discovered it. The shape of the immersed object is not relevant, since the volume integration in Eq. 11.27 does not depend on that. It is only when the object is completely immersed in the fluid that the volume of the displaced fluid is equal to the total volume of the immersed object. By displacing a large amount of the liquid, as ships do, the upward buoyant force can be increased enough to counter the downward weight, enabling the object to float. The weight is, of course, independent of the part immersed in water; it is only the mass times the acceleration due to gravity. The magnitude of the buoyant force can, however, be augmented by suitably *shaping* the ship. A tiny coin made of the same material sinks, since it displaces only a small volume of the liquid. Adequate buoyant force does not counter its weight. When an object floats, its entire weight is balanced by the fluid's buoyant force, but the weight consists not merely of the material of the ship, but plenty of empty space in which there is only air, at a density that is much lower than the density of the water on which the ship floats.

In arriving at Eq. 11.26, we had ignored the variation of density with depth. This is not a bad approximation for liquids, which are mostly incompressible. The density of the sea water does increase somewhat with depth, but not a whole lot. Sea-water is nearly incompressible. Hence it is alright to ignore the variation of density with depth. We should remember, however, that when there is a huge mass of the fluid involved, the density of the fluid cannot be assumed to be constant. The deeper one goes into the fluid the force exerted by the mass of the fluid above it could affect the density of the fluid.

For gases, including Earth's atmosphere, it becomes important to consider the change in its density with altitude. This is because of the fact that the gases are significantly compressible. The density and pressure of Earth's atmosphere does change with height, and so does the temperature. The problem therefore becomes rather complicated. The thermodynamic properties of the atmosphere also change significantly in the troposphere, and in the ozone layer above it. At even higher altitudes, in the stratosphere, and the mesosphere, and then in the thermosphere at the very top, the thermodynamic properties are quite different. Unlike liquids, the density of the atmosphere then varies with the altitude. The atmosphere must therefore be regarded as a compressible fluid, As a first approximation, without considering the complicated thermodynamic properties, we shall consider the atmosphere to be described by the following relationship between its pressure p, density ρ, and the absolute temperature T:

$$p = \rho RT, \tag{11.28}$$

R being the individual gas constant for the particular gas, which is just the universal gas constant divided by the gas molecular weight.

If the Earth's atmosphere is in a state of rest, at mechanical equilibrium, then the atmospheric pressure at a point P would depend only on the altitude of that point. Any lateral differences in the pressure would cause the atmosphere to flow, and it would *not* be in the state of mechanical equilibrium. From Eq. 11.25 and Eq. 11.28, it then follows that any variation in the density would be given by

$$\frac{dp}{dz} = -g\rho = -g\frac{p}{RT}. \tag{11.29}$$

Hence, $$\int_a^b \frac{dp}{p} = -\frac{g}{R}\int_a^b \frac{dz}{T(z)}, \tag{11.30}$$

i.e., $$\ln\frac{dp}{p} = -\frac{g}{R}\int_a^b \frac{dz}{T(z)} = -\frac{g}{R}\int_a^b \frac{\left(1-\frac{Bz}{T_0}\right)^{-1} dz}{T_0}, \tag{11.31}$$

wherein a linear temperature drop is assumed, described by

$$T(z) = (T_0 - Bz). \tag{11.32}$$

Up to 11 km, the linear temperature drop is well described by Eq. 11.32, with T_0 = 288.16°K (i.e., ~15°C) as the absolute temperature at sea level. The coefficient B = 0.0065 K/m. The temperature, however, remains unchanged (at –56.5°C) between 11 km and ~20 km. Above it, the temperature has somewhat complicated dependence on the altitude. From Eq. 11.32, which is valid in the troposphere, we find that

$$p = p_a\left(1-\frac{Bz}{T_0}\right)^{\frac{g}{RB}}. \tag{11.33}$$

For air, the dimensionless number $\frac{g}{RB}$ is 5.26. We therefore see that the pressure reduces with altitudes. At altitudes above the troposphere, the pressure continues to drop, but a bit slower than indicated in the above relation. Since the pressure drops with altitude, it is often significantly lower at the top of the mountains than it is at the sea level. Mount Everest has a height over 8.8 km, and a mountaineer who scales it, is prone to bleed, since his blood would tend to flow from a higher pressure inside his body to a lower pressure outside. For the same reason, airplanes, when flying at high altitudes, have to artificially maintain the cabin pressure to what it is normally at the sea level.

Now, the differential increment in the thermodynamic potential per unit mass, often called the Gibbs free energy, is given by

$$d\Phi_{tp} = -sdT + v_s dp = -sdT + \frac{1}{\rho}dp. \tag{11.34}$$

The subscript '*tp*' on the left hand side represents the term for the thermodynamic potential. If the fluid is in a state of thermal equilibrium, it would have uniform temperature. From Eq. 11.23 and Eq. 11.34 it then follows that

$$\vec{\nabla}\Phi_{tp} = \frac{\vec{\nabla}p}{\rho} = \vec{g} = -\vec{\nabla}\Phi_g, \tag{11.35a}$$

where Φ_g is the gravitational potential. Hence,

$$\vec{\nabla}(\Phi_{tp} + \Phi_g) = \vec{0}. \tag{11.35b}$$

It immediately follows from the above result that

$$\Phi_{tp} + \Phi_g = c, \text{ a constant.} \tag{11.36}$$

Now, the gravitational potential at a distance r from the center of the Earth is

$$\Phi_g(\vec{r}) = -\int_{\infty}^{r}\left(-\frac{GM}{r'^2}\right)dr' = GM\int_{\infty}^{r}\left(\frac{1}{r'^2}\right)dr' = GM\left(\frac{-1}{r}\right) = -\frac{GM}{r}. \tag{11.37a}$$

The potential can, however, be referenced to an arbitrary zero of the potential scale. Thus, with reference to the potential at the Earth's surface, which is at a distance r_E, the Earth's radius, from its center, we have

$$\Phi_g(r_E + z) - \Phi_g(r_E) = -\frac{GM}{r_E + z} + \frac{GM}{r_E} = \frac{GM}{r_E}\left[1 - \left(1 + \frac{z}{r_E}\right)^{-1}\right], \tag{11.37b}$$

i.e., $$\Phi_g(r_E + z) - \Phi_g(r_E) \simeq \frac{GM}{r_E}\left[1 - \left(1 - \frac{z}{r_E}\right)\right] = gz, \tag{11.37c}$$

where $g = \frac{GM}{r_E^{\,2}}$ is the acceleration due to gravity, assumed to be uniform in the vicinity of the Earth. Referring therefore to the gravitational potential, Φ_g, at a height z *above* the Earth's surface, Eq. 11.36 therefore reads

$$\Phi_{tp} + gz = c, \text{ a constant.} \tag{11.38}$$

Taking the volume integral of the divergence of Eq. 11.35a, we get

$$\iiint \left(\vec{\nabla}\cdot\frac{\vec{\nabla}p}{\rho}\right) dV = \iiint (\vec{\nabla}\cdot\vec{g})\, dV = \oiint \vec{g}\cdot\hat{n}dS. \tag{11.39a}$$

In the last step, we have used the Gauss' divergence theorem.

Hence,

$$\iiint \left(\vec{\nabla}\cdot\frac{\vec{\nabla}p}{\rho}\right) dV = \oiint \frac{GM}{r_E^{\,2}}(-\hat{e}_r)\cdot(+\hat{e}_r)dS = -\oiint \frac{GM}{r_E^{\,2}} r^2 d\Omega \simeq -4\pi GM, \tag{11.39b}$$

since $r^2 \simeq r_E^{\,2}$.

$$\text{Hence, } \iiint \left(\vec{\nabla}\cdot\frac{\vec{\nabla}p}{\rho}\right) dV = -4\pi GM = \iiint (-4\pi G\rho)\, dV. \tag{11.39c}$$

Equating now the integrands of the above *definite* volume integrals, we get

$$\vec{\nabla}\cdot\left(\frac{\vec{\nabla}p}{\rho}\right) = \vec{\nabla}\cdot\vec{g} = -4\pi G\rho. \tag{11.40a}$$

Using Eq. 10.20b for the expression for the divergence of a vector field in spherical polar coordinates, we get

$$\frac{1}{r^2}\frac{d}{dr}\left[r^2\left(\frac{1}{\rho}\frac{dp}{dr}\right)\right] = -4\pi G\rho. \tag{11.40b}$$

Instead of the partial derivative operator $\frac{\partial}{\partial r}$, we have now used the operator $\frac{d}{dr}$, since all the *operands* now depend only on the radial distance from the origin.

We shall now try to determine the $\iiint (\vec{\nabla}\cdot\vec{g})dV$ as a volume integral. In Eq. 11.37b, we determined it as a surface integral using the Gauss' divergence theorem. To do that, we first determine the divergence of the gravitational force per unit mass, i.e., the divergence of the acceleration due to gravity. Oddly though, *blind use* of Eq. 10.20 to determine $\vec{\nabla}\cdot\vec{g} = \vec{\nabla}\cdot\left(-\frac{GM}{r^2}\hat{e}_r\right)$ turns out to be zero, instead of $-4\pi G\rho$, which we got in Eq. 11.40. The apparent discrepancy comes from the fact that the integration of the divergence of $\frac{\vec{r}}{r^3}$

involves a special trick, necessitated by the fact that $\frac{\vec{r}}{r^3}$ blows up at the point $r = 0$, which is included in the region of volume integration. This integration must be carried out using a special technique, introduced by Paul Adrien Maurice Dirac (1902–1984). Dirac interpreted the divergence of $\frac{\vec{r}}{r^3}$ as a *generalized distribution*:

$$-\vec{\nabla}\cdot\vec{\nabla}\left(\frac{1}{r}\right) = \vec{\nabla}\cdot\frac{\vec{r}}{r^3} = \vec{\nabla}\cdot\frac{\hat{e}_r}{r^2} = 4\pi\delta^3(\vec{r}). \tag{11.41}$$

In the above equation, $\delta^3(\vec{r}) = \delta(x)\delta(y)\delta(z)$. It is called the Dirac δ (delta) function; defined below. In one-dimension, the Dirac $\delta(x)$ function (rather, 'distribution') is defined by the following sifting property:

$$f(0) = \int_{-\infty}^{+\infty} f(x)\,\delta(x)\,dx, \tag{11.42a}$$

i.e., $$f(a) = \int_{-\infty}^{+\infty} f(x)\,\delta(x-a)\,dx. \tag{11.42b}$$

There are various functions which are consistent with the properties under integration expressed in Eq. 11.42a and Eq. 11.42b. All of these functions have a *spike* at $x = a$. Three representations of the Dirac δ function are shown in Fig. 11.4. In some sense, it is like the charge density, $\rho(\vec{r}) = \lim\limits_{\delta V \to 0} \frac{\delta q}{\delta V}$. One would think that the ratio $\frac{\delta q}{\delta V}$ will blows up to infinity in the limit that the volume element in the denominator goes to zero. Yet, when it is integrated over whole space, it gives the finite charge

$$q = \iiint \rho(\vec{r})\,d^3\vec{r}. \tag{11.43}$$

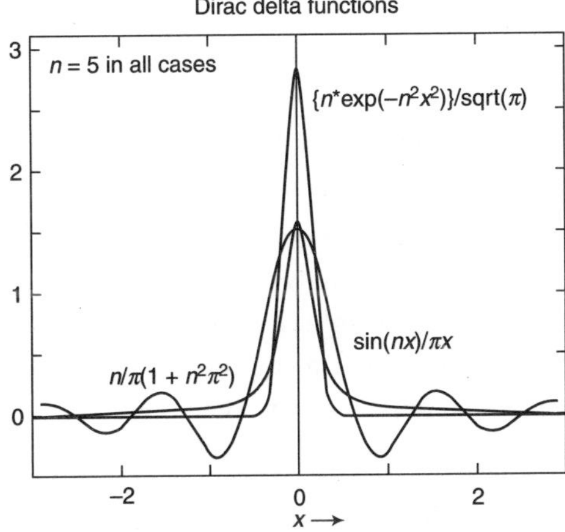

Fig. 11.4 The Dirac-δ 'function' is in fact better known as a distribution or a generalized function. It is narrow, and sharply peaked. There are various representations of the Dirac-δ which, under integration, are consistent with Eq. 11.42. Three of these representations are shown in this figure.

The acceleration due to the Earth's gravity must be seen as the integrated result of the acceleration of a unit mass at coordinate $\vec{r}$ resulting from the integration of an infinite number of forces acting on it, exerted by tiny infinitesimal elements of mass $\rho(\vec{r}')dV'$(density multiplied by the elemental volume element) spread throughout the Earth. Here, $\rho(\vec{r}')$ is the mass density at the point $\vec{r}'$, inside the Earth. The point whose position vector is $\vec{r}'$, is therefore called the 'source' point, and the point $\vec{r}$ is called the 'field' point. Each of these tiny elements at $\vec{r}'$ will result in an acceleration of the unit mass at the field point, directed toward the source point $\vec{r}'$. It will therefore be given by $-G[\rho(\vec{r}')dV']\dfrac{(\vec{r}-\vec{r}')}{|\vec{r}-\vec{r}'|^3}$.

Integrating over *all* the infinitesimal source elements inside the Earth, we get

$$\vec{\nabla}\cdot\vec{g}(\vec{r}) = -\vec{\nabla}\cdot\iiint G\rho(\vec{r}')\frac{(\vec{r}-\vec{r}')}{|\vec{r}-\vec{r}'|^3}dV'. \tag{11.44}$$

Now, the divergence operator involves differentials with respect to the coordinates of the field point. Mathematically, this process is completely *independent* of the integration with respects to the source coordinates. Also, the gradient with respect to the field coordinates would leave the source density alone; the latter depends only on the source coordinates. Hence, the divergence of the term $\dfrac{(\vec{r}-\vec{r}')}{|\vec{r}-\vec{r}'|^3}$ in the integrand can be carried out first. Accordingly,

$$\vec{\nabla}\cdot\vec{g}(\vec{r}) = -G\iiint \rho(\vec{r}')\left[\vec{\nabla}\cdot\frac{(\vec{r}-\vec{r}')}{|\vec{r}-\vec{r}'|^3}\right]dV' = -G\iiint \rho(\vec{r}')\,4\pi\delta^3(\vec{r}-\vec{r}')\,dV'. \tag{11.45}$$

Hence, $\vec{\nabla}\cdot\vec{g}(\vec{r}) = -4\pi G\rho(\vec{r})$, (11.46a)

i.e., using Eq.11.35a, $(\vec{\nabla}\cdot\vec{\nabla})\Phi_g = \vec{\nabla}^2\Phi_g = 4\pi G\rho(\vec{r})$, (11.46b)

which is the same result that we had got in Eq. 11.40.

A star is often modeled as a gaseous sphere in hydrostatic equilibrium in which the gravitational force is balanced by pressure gradients. The results which we have obtained are therefore of great importance in analyzing gaseous self-gravitating fluidic matter in astrophysics. Of course, stellar atmospheric fluid has many other properties, inclusive of compressibility, charge, etc., for which a number of corrections must be made. Along with the laws of the electromagnetic theory (Chapter 12, 13), one then has to use magneto-hydrodynamics.

11.3 STREAMLINED FLOW, BERNOULLI'S EQUATION, AND CONSERVATION OF ENERGY

We shall discuss the principle of conservation of energy in this section. To begin with, we consider the steady flow. By 'steady state', what we mean is that the velocity of the fluid at a particular point in space does not change with time; i.e., $\frac{\partial \vec{v}(\vec{r},t)}{\partial t} = 0$.

Equation 11.20 then reduces to

$$\frac{1}{2}\vec{\nabla}(\vec{v}\cdot\vec{v}) - \vec{v}\times(\vec{\nabla}\times\vec{v}) = -\vec{\nabla}\,(w+\Phi_g)\,. \tag{11.47}$$

Fluid flow very often is streamlined. A streamline is a mathematical curve in the fluid space such that the fluid velocity $\vec{v}(\vec{r}, t)$ at every point on this curve is along tangent to the streamline. Now, the displacement vector $\overrightarrow{ds} = (\hat{e}_x dx + \hat{e}_y dy + \hat{e}_z dz)$ of the fluid particles is essentially along the fluid velocity vector $\vec{v}(\vec{r}) = \hat{e}_x v_x(\vec{r}) + \hat{e}_y v_y(\vec{r}) + \hat{e}_z v_z(\vec{r})$. Therefore, the cross product of these two vectors vanishes:

$$\vec{0} = \overrightarrow{ds}\times\vec{v} = (\hat{e}_x dx + \hat{e}_y dy + \hat{e}_z dz)\times(\hat{e}_x v_x + \hat{e}_y v_y + \hat{e}_z v_z), \tag{11.48a}$$

i.e., $$\vec{0} = \hat{e}_x(v_z dy + v_y dz) + \hat{e}_y(v_x dz + v_z dx) + \hat{e}_z(v_y dx + v_x dy). \tag{11.48b}$$

Recognizing that each component of the above vector equation is independently zero, we get

$$v_z dy = v_y dz \Rightarrow \frac{dy}{v_y} = \frac{dz}{v_z}, \tag{11.49a}$$

$$v_x dz = v_z dx \Rightarrow \frac{dz}{v_z} = \frac{dx}{v_x}, \tag{11.49b}$$

and $$v_y dx = v_x dy \Rightarrow \frac{dx}{v_x} = \frac{dy}{v_y}. \tag{11.49c}$$

Therefore, combining the above results,

$$\frac{dx}{v_x(\vec{r})} = \frac{dy}{v_y(\vec{r})} = \frac{dz}{v_z(\vec{r})}. \tag{11.50}$$

Now, the tangents to the paths of the fluid particles are along the pathlines, and the tangent to the streamlines are along the directions of the velocities. When the fluid flow is steady, the pathlines of the fluid particles are along the streamlines, which do not vary with time. The conservation of energy relation is exhibited on taking the component of Eq. 11.47 along the direction of the velocity. Representing the direction of the velocity by the unit vector $\hat{v}$ along it, this component is

$$\hat{v}\cdot\left[\frac{1}{2}\vec{\nabla}(\vec{v}\cdot\vec{v})-\vec{v}\times(\vec{\nabla}\times\vec{v})\right]=\hat{v}\cdot[-\vec{\nabla}(w+\Phi_g)]. \tag{11.51a}$$

$$\text{Hence, } \hat{v}\cdot\left[\frac{1}{2}\vec{\nabla}(\vec{v}\cdot\vec{v})+\vec{\nabla}(w+\Phi_g)\right]=0. \tag{11.51b}$$

Essentially therefore, $\hat{v}$, the direction of the fluid velocity, is orthogonal to the vector for the gradient, $\vec{\nabla}\left[\frac{1}{2}(\vec{v}\cdot\vec{v})+(w+\Phi_g)\right]$. This is the gradient of the energy per unit mass, which is $E=\left[\frac{1}{2}(\vec{v}\cdot\vec{v})+(w+\Phi_g)\right]$. Any change in $\left[\frac{1}{2}(\vec{v}\cdot\vec{v})+(w+\Phi_g)\right]$ is therefore orthogonal to the streamline. Along the streamline, we therefore have

$$\left[\frac{1}{2}(\vec{v}\cdot\vec{v})+(w+\Phi_g)\right]=\kappa_s,\ \text{a constant for every streamline.} \tag{11.52}$$

Furthermore, if the fluid flow is steady, and also irrotational, then from Eq. 11.42 it immediately follows that

$$\left[\frac{1}{2}(\vec{v}\cdot\vec{v})+(w+\Phi_g)\right]=\kappa,\ \text{a constant,} \tag{11.53a}$$

$$\text{i.e.,}\quad \left[\frac{1}{2}(\vec{v}\cdot\vec{v})+(w+gz)\right]=\kappa. \tag{11.53b}$$

We had used a subscript 's' on the constant used in Eq. 11.52 since it was obtained by showing that any change in $\left[\frac{1}{2}(\vec{v}\cdot\vec{v})+(w+\Phi_g)\right]$ is orthogonal to the streamline; but no such *special orientation* of the gradient of $\left[\frac{1}{2}(\vec{v}\cdot\vec{v})+(w+\Phi_g)\right]$ is involved when the fluid flow is in steady state, and is *also* irrotational. The constancy in Eq. 11.48 therefore extends to the entire steady state and irrotational velocity field; it is not limited to a streamline as in the case of Eq. 11.52. Equation 11.53 is called Bernoulli's equation, named after Daniel Bernoulli (1700–1782), son of Johann Bernoulli (1667–1748), who had posed the brachistochrone problem (Chapter 6).

Even though the subject of fluid dynamics is an extremely complicated one, the fundamental properties of the dynamics of a fluid are well described by the equations that describe the conservation of (i) mass, (ii) energy, and (iii) momentum. The principle of the conservation of mass constitutes the equation of continuity (Eq. 10.26). In order to develop the equation for the conservation of energy, we must determine the dependence of the energy content on a volume element of the fluid on time. We must determine the equation of motion for the *energy flux density vector*, $\vec{\xi}_{efd}$, defined in a manner similar to Eq. 11.16 in which we introduced the entropy flux density vector:

$$\vec{\xi}_{efd} = \left[\frac{1}{2}(\vec{v}\cdot\vec{v}) + \varepsilon\right]\vec{J} = \left[\frac{1}{2}(\vec{v}\cdot\vec{v}) + \varepsilon\right]\rho\vec{v} = \left[\frac{1}{2}\rho(\vec{v}\cdot\vec{v}) + \rho\varepsilon\right]\vec{v} = e_{fd}\vec{v}. \tag{11.54}$$

In the above relation, ε is the internal energy per unit mass. It is related to the heat function per unit mass, w (enthalpy), through the relation,

$$w = \varepsilon + \frac{p}{\rho}. \tag{11.55}$$

The energy flux would represent the amount of energy crossing unit area orthogonal to the fluid velocity in unit time. The energy density (i.e., energy of unit volume) of the fluid, originating from the *kinetic* and the *internal* energy parts, which we shall denote by e_{kid} is

$$e_{\text{kid}} = \frac{1}{2}\rho v^2 + \rho\varepsilon. \tag{11.56}$$

$$\frac{\partial e_{\text{kid}}}{\partial t} = \frac{\partial}{\partial t}\left(\frac{1}{2}\rho v^2\right) + \frac{\partial}{\partial t}(\rho\varepsilon). \tag{11.57}$$

$$\frac{\partial}{\partial t}\left(\frac{1}{2}\rho v^2\right) = \frac{\partial}{\partial t}\left[\frac{1}{2}\rho(\vec{v}\cdot\vec{v})\right] = \frac{v^2}{2}\frac{\partial\rho}{\partial t} + \rho\left(\vec{v}\cdot\frac{\partial\vec{v}}{\partial t}\right),$$

$$\frac{\partial}{\partial t}\left[\frac{1}{2}\rho(\vec{v}\cdot\vec{v})\right] = -\frac{1}{2}v^2\vec{\nabla}\cdot(\rho\vec{v}) + \rho\left(\vec{v}\cdot\frac{\partial\vec{v}}{\partial t}\right).$$

Using now the Euler's equation, Eq. 11.10, for $\frac{\partial\vec{v}(\vec{r},t)}{\partial t}$, we get

$$\frac{\partial}{\partial t}\left(\frac{1}{2}\rho v^2\right) = -\frac{1}{2}v^2\vec{\nabla}\cdot(\rho\vec{v}) - \vec{v}\cdot\vec{\nabla}p - \rho\vec{v}\cdot(\vec{v}\cdot\vec{\nabla})\,\vec{v}(\vec{r},t).$$

From Eq. 11.17a: $\vec{\nabla}p = \rho\vec{\nabla}w - \rho T\vec{\nabla}s$.

Hence, $\frac{\partial}{\partial t}\left(\frac{1}{2}\rho v^2\right) = -\frac{1}{2}v^2\vec{\nabla}\cdot(\rho\vec{v}) - \rho\vec{v}\cdot\vec{\nabla}w + \rho T(\vec{v}\cdot\vec{\nabla}s) - \rho\vec{v}\cdot(\vec{v}\cdot\vec{\nabla})\vec{v}(\vec{r},t)$.

Using now Eq. 11.19b (for irrotational flow), we get

$$\vec{v}\cdot(\vec{v}\cdot\vec{\nabla})\vec{v} = \frac{1}{2}\vec{v}\cdot\vec{\nabla}(v^2).$$

$$\frac{\partial}{\partial t}\left(\frac{1}{2}\rho v^2\right) = -\frac{1}{2}v^2\vec{\nabla}\cdot(\rho\vec{v}) - \rho\vec{v}\cdot\vec{\nabla}w - \frac{1}{2}\rho\vec{v}\cdot\vec{\nabla}(v^2) + \rho T(\vec{v}\cdot\vec{\nabla}s).$$

$$\frac{\partial}{\partial t}\left(\frac{1}{2}\rho v^2\right) = -\frac{1}{2}v^2\vec{\nabla}\cdot(\rho\vec{v}) - \rho\vec{v}\cdot\vec{\nabla}\left(w + \frac{v^2}{2}\right) + \rho T(\vec{v}\cdot\vec{\nabla}s). \tag{11.58}$$

From Eq. 11.55, we have

$$d\varepsilon = d\left(w - \frac{p}{\rho}\right) = dw - d\left(\frac{p}{\rho}\right) = Tds + \frac{1}{\rho}dp - \frac{1}{\rho}dp + p\frac{1}{\rho^2}d\rho$$

Hence,

$$d\varepsilon = Tds + \frac{p}{\rho^2}d\rho \text{ and } \rho d\varepsilon = \rho Tds + \frac{p}{\rho}d\rho$$

$$d(\rho\varepsilon) = \rho d\varepsilon + \varepsilon d\rho = \rho Tds + \frac{p}{\rho}d\rho + \varepsilon d\rho = \rho Tds + wd\rho$$

$$\frac{\partial}{\partial t}(\rho\varepsilon) = \rho T\frac{\partial s}{\partial t} + w\frac{\partial \rho}{\partial t}$$

Using the equation for the adiabatic motion of an ideal fluid (Eq. 11.11b), and the equation for continuity (Eq. 10.26), we therefore get the following result for the rate at which the product ($\rho\varepsilon$) varies with time:

$$\frac{\partial}{\partial t}(\rho\varepsilon) = \rho T(-\vec{v}\cdot\vec{\nabla}s) + w(-\vec{\nabla}\cdot\vec{j}) = -[w\vec{\nabla}\cdot\vec{j} + \rho T\vec{v}\cdot\vec{\nabla}s]. \tag{11.59}$$

Combining Eqs. 11.53 and 11.54,

$$\frac{\partial e_{\text{kid}}}{\partial t} = -\frac{1}{2}v^2\vec{\nabla}\cdot(\rho\vec{v}) - \rho\vec{v}\cdot\vec{\nabla}\left(w + \frac{v^2}{2}\right) + \cancel{\rho T(\vec{v}\cdot\vec{\nabla}s)} - w\vec{\nabla}\cdot\vec{j} - \cancel{\rho T\vec{v}\cdot\vec{\nabla}s} \tag{11.60a}$$

i.e.,
$$\frac{\partial e_{\text{kid}}}{\partial t} = -\left(w + \frac{v^2}{2}\right)\vec{\nabla}\cdot\vec{j} - \vec{j}\cdot\vec{\nabla}\left(w + \frac{v^2}{2}\right). \tag{11.60b}$$

Hence,
$$\frac{\partial e_{\text{kid}}}{\partial t} = -\vec{\nabla}\cdot\left\{\left(w + \frac{v^2}{2}\right)\vec{j}\right\}. \tag{11.60c}$$

Integrating over the whole volume:

$$\iiint \frac{\partial e_{\text{kid}}}{\partial t}dV = -\iiint \vec{\nabla}\cdot\left\{\left(w + \frac{v^2}{2}\right)\vec{j}\right\}dV = -\oint\!\!\oint\left\{\left(w + \frac{v^2}{2}\right)\vec{j}\right\}\cdot\hat{n}dA \tag{11.61a}$$

i.e.,
$$\iiint \frac{\partial e_{\text{kid}}}{\partial t}dV = -\oint\!\!\oint \vec{e}_{fdv}\cdot\hat{n}dA. \tag{11.61b}$$

In Eq. 11.61b, we have introduced the vector $\vec{e}_{fdv} = \left(w + \frac{v^2}{2}\right)\vec{j},$ (11.62)

called the *energy flux density vector*.

We must remember that w (enthalpy) is the heat function per unit mass, which is related to the internal energy per unit mass, ε, by Eq. 11.55, whereas the sum of the kinetic plus the internal energy is given by Eq. 11.56. Using the latter in Eq. 11.61, we get

$$\frac{\partial}{\partial,t}\iiint e_{\text{kid}}dV = -\oiint\left\{\left(\varepsilon+\frac{v^2}{2}\right)\vec{J}\right\}\cdot\hat{n}dA - \oiint\left\{\frac{p}{\rho}\vec{J}\right\}\cdot\hat{n}dA, \tag{11.63a}$$

i.e., $$\frac{\partial}{\partial t}\iiint e_{\text{kid}}dV = -\oiint\left\{\left(\varepsilon+\frac{v^2}{2}\right)\vec{J}\right\}\cdot\hat{n}dA - \oiint p\vec{v}\cdot\hat{n}dA. \tag{11.63b}$$

The left hand side is the time-derivative of the volume integral of the energy density. It is the rate at which the total energy is changing. The right hand side is a sum of (i) the *net* (kinetic plus internal) energy transported by the fluid across the surface and (ii) the work done by the fluid's pressure within the surface. We will conclude this chapter by modifying the Euler's equation to include the viscosity effects and develop the Navier–Stokes equation. That would complete the foundation of laying down the governing equations for fluid motion.

11.4 NAVIER–STOKES EQUATION(S): GOVERNING EQUATIONS OF FLUID MOTION

The Euler equations (Eqs. 11.9 and 11.10) express how the forces acting on a fluid result in the change in momentum (per unit mass) of the flowing fluid. Essentially, it is an expression of the fact that the momentum of a system is conserved in the absence of forces acting on it. The equilibrium is, however, disturbed when one or more forces act on the fluid, no matter what the sources of the forces are. This is because forces are essentially additive. Only the 'ideal' fluid was, however, considered in formulating the Euler's equation. A real fluid has a non-zero viscosity. As a result of this, the system is dissipative. Different layers of fluid exchange energy as they rub against each other. Before we include the viscosity effects, we shall first *rewrite* the Euler equation (11.10) using *indices* for the components of the vector fields:

$$\left(\sum_{i=1}^{3} v_i \frac{\partial}{\partial x_i}\right)\sum_{k=1}^{3}\hat{e}_k v_k + \frac{\partial}{\partial t}\sum_{k=1}^{3}\hat{e}_k v_k = -\frac{1}{\rho}\sum_{k=1}^{3}\hat{e}_k\frac{\partial p}{\partial x_k}, \tag{11.64}$$

If the external field due to gravity is concerned, then we must use the form in Eq. 11.9, and we shall have to add $\sum_{k=1}^{3}\hat{e}_k g_k$ on the right hand side.

Thus, for the kth component, the Euler equation (without the gravity term) becomes

$$\left(\sum_{i=1}^{3} v_i \frac{\partial v_k}{\partial x_i}\right) + \frac{\partial v_k}{\partial t} = -\frac{1}{\rho}\frac{\partial p}{\partial x_k}, \tag{11.65a}$$

We simplify the notation by following the 'double index' summation convention. If an index appears twice, then a sum over that index must be carried out.

The Euler equation for the kth component then becomes

$$v_i \frac{\partial v_k}{\partial x_i} + \frac{\partial v_k}{\partial t} = -\frac{1}{\rho}\frac{\partial p}{\partial x_k}, \tag{11.65b}$$

$$\text{i.e.,} \quad \rho \frac{\partial v_k}{\partial t} = -\frac{\partial p}{\partial x_k} - \rho v_i \frac{\partial v_k}{\partial x_i}, \tag{11.65c}$$

Now,

$$\frac{\partial}{\partial t}(\rho v_k) = \rho \frac{\partial v_k}{\partial t} + v_k \frac{\partial \rho}{\partial t}. \tag{11.66a}$$

We can use the Euler's equation (Eq. 11.65c) to get the first term in Eq. 11.66, and we can use the equation to continuity (Eq. 10.26) to express the time-derivative of the density in the second term in Eq. 11.61. Accordingly, we get

$$\frac{\partial}{\partial t}(\rho v_k) = -\frac{\partial p}{\partial x_k} - \rho v_i \frac{\partial v_k}{\partial x_i} - v_k \frac{\partial(\rho v_i)}{\partial x_i}, \tag{11.66b}$$

$$\text{i.e.,} \quad \frac{\partial}{\partial t}(\rho v_k) = -\frac{\partial p}{\partial x_k} - \frac{\partial(\rho v_k v_i)}{\partial x_i}. \tag{11.66c}$$

The relationships expressed in Eq. 11.66 are essentially the same as in the Euler's result for the ideal fluid. The simplest way to appreciate the corrections due to viscous diffusion terms that must be accounted for a non-ideal and real fluid is to first rewrite the Euler results using a tensor notation. This will enable us extend the Euler's equations to include the effects of viscosity, thus leading us to what are known as the Navier–Stokes equations. The second term in Eq. 11.66c already has two indices, k and i, of which the index i gets summed over. The first term in Eq. 11.66c is therefore also written as a summation over a dummy index i by exploiting the property of the double-indexed Kronecker δ_{ki}. It is equal to unity when the two indices are the same, and it is zero when the indices of the Kronecker δ_{ki} are different from each other. Hence,

$$\frac{\partial p}{\partial x_i} = \delta_{ki}\frac{\partial p}{\partial x_i}. \tag{11.67}$$

The index i is summed over, and Eq. 11.67 naturally produces, on this summation, the first term in Eq. 11.66c.

$$\text{Therefore,} \quad \frac{\partial}{\partial t}(\rho v_k) = -\delta_{ki}\frac{\partial p}{\partial x_i} - \frac{\partial(\rho v_k v_i)}{\partial x_i} = -\left[\frac{\partial(p\delta_{ki})}{\partial x_i} + \frac{\partial(\rho v_k v_i)}{\partial x_i}\right], \tag{11.68a}$$

$$\text{i.e.,} \quad \frac{\partial}{\partial t}(\rho v_k) = -\frac{\partial}{\partial x_i}(p\delta_{ki} + \rho v_k v_i). \tag{11.68b}$$

We now *define* the *momentum flux density tensor for the ideal fluid* as

$$\Pi_{ki} = p\delta_{ki} + \rho v_k v_i \tag{11.69}$$

Equation 11.68b then takes the compact form

$$\frac{\partial}{\partial t}(\rho v_k) = -\frac{\partial \Pi_{ki}}{\partial x_i}. \tag{11.70}$$

The above expressions are valid only for the 'ideal' fluid in which the momentum transfer is considered to be totally reversible. It will not hold good for a viscous fluid. When the viscosity of the fluid medium must be taken into account, the *momentum flux density tensor* must be *redefined*. Instead of Eq. 11.69, it is therefore defined for a viscous fluid by *adding* a viscosity-dependent term:

$$\Pi_{ki} = p\delta_{ki} + \rho v_k v_i - \sigma'_{ki}. \tag{11.71}$$

The term that is added, σ'_{ki}, is called the *viscosity stress tensor*. For simplicity, we also define the *stress tensor*, as

$$\sigma_{ki} = p\delta_{ki} + \sigma_{ki}. \tag{11.72}$$

The momentum flux density tensor for a viscous medium then becomes

$$\Pi_{ki} = \rho v_k v_i - \sigma_{ki}. \tag{11.73}$$

Internal friction would result only when there is relative motion between different parts of the fluid. This would result from various fluid particles moving at velocities different from each other's. The stress tensor then takes the following form, known as the *constitutive relation* for Newtonian fluid:

$$\sigma_{ki} = -p\delta_{ki} + \mu\left(\frac{\partial v_k}{\partial x_i} + \frac{\partial v_i}{\partial x_k}\right). \tag{11.74}$$

In the above equation μ is the coefficient of viscosity of the fluid. The above constitutive relation is applicable for incompressible fluids. The Euler equation is now *modified*, and we must solve the adjusted equation for the viscous fluid, called the Navier–Stokes equation. It is named after George Gabriel Stokes (1816–1903) and Louis Marie Henri Navier (1785–1836), who independently derived them:

$$\rho(\vec{r}, t)\left[(\vec{v}\cdot\vec{\nabla}) + \frac{\partial}{\partial t}\right]\vec{v} = -\vec{\nabla}p + \mu(\vec{\nabla}\cdot\vec{\nabla})\vec{v}. \tag{11.75a}$$

For compressible fluids, the constitutive relations take a more complicated form. These relations are necessary to supplement the various other relationships (like the equation of continuity, the Euler–Navier–Stokes equations) to solve the equations of motion for the fluid. The constitutive relations are specific to each fluid. If the external field due to gravity is also considered, then the Navier–Stokes equation becomes

$$\rho(\vec{r}, t)\left[(\vec{v}\cdot\vec{\nabla}) + \frac{\partial}{\partial t}\right]\vec{v} = -\vec{\nabla}p + \mu(\vec{\nabla}\cdot\vec{\nabla})\vec{v} + \rho\vec{g}. \tag{11.75b}$$

We emphasize here that the science of *matter that flows* is very complex. As such, we have restricted our analysis to what are called 'Newtonian fluids'. These are characterized by the fact that the shear stress in these fluids is linearly proportional to the shear strain. Water, air, oil, gasoline, most other common fluids belong to the category of Newtonian fluids. There of course are materials, including some that we often classify as solids, which *flow* in a manner that is unlike the Newtonian fluids. Non-Newtonian fluids are then studied under the general canopy of the science of rheology, rather than fluid dynamics. Typical non-Newtonian fluids include materials like paints, toothpaste, quicksand, cornflour starch in water, molten polymers, etc.

In general, the Navier–Stokes equation is very difficult to solve. The above form is valid only for incompressible fluids. If the fluid is compressible, the components of the shear stress take a complicated form. The Cartesian form of Eq. 11.75 provides three equations:

$$\rho(\vec{r},t)\left[v_x\frac{\partial v_x}{\partial x}+v_y\frac{\partial v_x}{\partial y}+v_z\frac{\partial v_x}{\partial y}+\frac{\partial v_x}{\partial t}\right]=-\frac{\partial p}{\partial x}+\mu\left(\frac{\partial^2}{\partial x^2}+\frac{\partial^2}{\partial y^2}+\frac{\partial^2}{\partial z^2}\right)v_x+\rho g_x, \quad (11.76a)$$

$$\rho(\vec{r},t)\left[v_x\frac{\partial v_y}{\partial x}+v_y\frac{\partial v_y}{\partial y}+v_z\frac{\partial v_y}{\partial y}+\frac{\partial v_y}{\partial t}\right]=-\frac{\partial p}{\partial y}+\mu\left(\frac{\partial^2}{\partial x^2}+\frac{\partial^2}{\partial y^2}+\frac{\partial^2}{\partial z^2}\right)v_y+\rho g_y, \quad (11.76b)$$

and

$$\rho(\vec{r},t)\left[v_x\frac{\partial v_z}{\partial x}+v_y\frac{\partial v_z}{\partial y}+v_z\frac{\partial v_z}{\partial y}+\frac{\partial v_z}{\partial t}\right]=-\frac{\partial p}{\partial z}+\mu\left(\frac{\partial^2}{\partial x^2}+\frac{\partial^2}{\partial y^2}+\frac{\partial^2}{\partial z^2}\right)v_z+\rho g_z. \quad (11.76c)$$

The *system of equations*, known as the Navier–Stokes equations, provides the governing equations for the fluid flow. They include, as per common convention, the equation of continuity that represents the conservation of mass, and the three scalar equations representing the conservation of momentum. The governing equations also include an extension of the Bernoulli's equation that provides the time-dependent conservation of energy, based on thermodynamic considerations such as the first law of thermodynamics. The governing equations have the four independent variables, namely the time, and the three space coordinates. The fluid density, pressure, temperature and the three components of velocity constitute the six dependent variables to be determined from the six coupled partial differential equations. The modern field of computational fluid dynamics (CFD) often refers to all of these six governing equations as the *family of Navier–Stokes equations*, though earlier literature referenced only the extension of the Euler's momentum equation, to include the viscosity, as the Navier–Stokes equation.

Setting up and solving the Navier–Stokes system of equations is amongst the most widely formulated problems in science and engineering. These are coupled partial differential equations. One of the biggest problems in mathematics, for which a millennium prize is available, is to uniquely solve the Navier–Stokes equations. The solutions have to rely on numerical methods. Even if these solutions are not exact, and rely on intensive computational applications, they are of great value in designing automobiles and airplanes in which fluid motion is the central phenomenology to analyze. In a large number of situations which involves heat and mass transfer, the continuum hypothesis is applicable, and one must take recourse to solving the Navier–Stokes equations. Analysis of ocean water currents, wind-throw in forests and in mountains, particle movement in the atmosphere, etc., all require solving the Navier–Stokes equations.

The Navier–Stokes equations are manifestly intimidating, even if in the above form we have already employed various approximations. We have assumed that the fluid is incompressible and has uniform viscosity. The Navier–Stokes equations help understand the consequences of body-forces, and surface forces, acting on fluid elements in control volume spaces. Any system of differential equations requires the boundary conditions and the initial conditions to be spelled out. These involve the values of the fluid velocity and the pressure at the initial time, and the values and their derivatives to be provided at the boundary of the region. The boundary conditions are typically referred to as the Dirichlet or the von Neumann boundary conditions. You may refer more specialized literature (such as Reference [1, 2]) to learn further about the methods to solve Navier–Stokes equations with appropriate boundary conditions. The utility of such a complex system of equations comes from the fact that in many real situations, some viable approximations are possible, and the solutions not only provide insight into the physical situation, but actually help solve engineering problems for a variety of applications. The Navier–Stokes equations are used to model diverse situations in fluid dynamics, including the flow of blood in arteries and veins, airflow around automobiles and airplanes, water flow around river and oceanic vessels, fluid flow in tubes, pipes, etc. With appropriate boundary conditions they are used to understand weather patterns, and along with the laws of electrodynamics, they have important applications in magneto-hydrodynamics, plasma oscillations, etc. Along with the thermodynamic equation for the conservation of energy, the set of the coupled, non-linear partial differential equations cannot be solved analytically. The challenges are computationally intensive. The formalism we have presented in this chapter develops confidence that physicists and mathematicians have developed innovative methods to obtain solutions to very difficult problems. This study is instructive and inspiring, but extremely challenging.

We have seen the important role played by the divergence and the curl of a fluid velocity field in determining the fluid flow. Both the divergence and the curl of the velocity fields are crucial to determine the fluid velocity field. These properties are of far reaching consequence not just in fluid dynamics, but also in the study of electrodynamics, quantum theory, and in several other subjects. The importance of *divergence* and *curl* is revealed by a theorem in vector calculus, namely the *Helmholtz theorem*, which states that given these properties of a vector field, and appropriate boundary conditions, the vector field is completely specified. Of course, appropriate boundary conditions are of crucial importance. In the next chapter, we shall employ the divergence and the curl of a vector field to study a very different phenomenology, namely the electromagnetic interaction. The essence of the laws of electromagnetic interactions is contained in the Maxwell's equations, which provide the curl and the divergence of the electromagnetic field. In the next chapter, we shall study this very important and exciting subject, even if only briefly.

Problems with Solutions

P11.1:

Determine if the 2-dimensional fluid flow whose Cartesian velocity components (v_x, v_y) are given below, constitutes a steady state flow of an incompressible fluid.

(a) $(3x^2 + 2x, -y(6x + 2))$ (b) $\left(\dfrac{x^2}{2y^2}, -\dfrac{x}{y}\right)$ (c) $(r^3 + r^2 \sin\theta, -4r^3\theta + 3r^2 \cos\theta)$

Solution:

For an incompressible steady state flow of an incompressible fluid, the divergence of the velocity must vanish.

(a) $\dfrac{d}{dx}(3x^2 + 2x) = 6x + 2$ and $\dfrac{d}{dy}(-y(6x+2) = -6x - 2$. Hence $\text{div}(\vec{v}) = 0$; the fluid flow is a steady state flow.

(b) $\dfrac{d}{dx}\left(\dfrac{x^2}{2y^2}\right) = \dfrac{2x}{y^2}$ and $\dfrac{d}{dy}\left(-\dfrac{x}{y}\right) = \dfrac{x}{y^2}$. Hence $\text{div}(\vec{v}) \neq 0$; the fluid flow is not in steady state.

(c) $\text{div}(\vec{v}) = \dfrac{\partial}{\partial r}(r^3 + r^2\sin\theta) + \dfrac{r^3 + r^2\sin\theta}{r} + \dfrac{1}{r}\dfrac{\partial}{\partial\theta}(-4r^3\theta + 3r^2\cos\theta)$;

i.e., $\text{div}(\vec{v}) = (3r^2 + 2r\sin\theta) + (r^2 + r\sin\theta) - 4r^2 - 3r\sin\theta = 0$. Hence the fluid flow is a steady state flow.

P11.2:

Find the value of the electric current if a charge $q = (2x^2t^2 + y \cos t)$ coulombs flows such that the flow velocity of the probe with which we are measuring the charge is $(3\breve{e}_x + 4\breve{e}_y)\text{ms}^{-1}$ at ($x = 6$, $y = 8$) at the time $t = 3$ sec.

Solution:

The current is $\dfrac{Dq}{Dt} = \dfrac{\partial q}{\partial t} + (\vec{v}\cdot\vec{\nabla})q = (4x^2t - y\sin t) + (3\breve{e}_x + 4\breve{e}_y)\cdot\vec{\nabla}(2x^2t^2 + y\cos t)$,

i.e., $\dfrac{Dq}{Dt} = (4x^2t - y\sin t) + (3\breve{e}_x + 4\breve{e}_y)\cdot(4xt^2\breve{e}_x + \cos t\,\breve{e}_y)$,

i.e., $\dfrac{Dq}{Dt} = (4x^2t - y\sin t) + (12xt^2 + 4\cos t) = 1084.412$ amperes.

P11.3:

The velocity field of a fluid flow is given by $\vec{v} = (2x^3y)\breve{e}_x + (-3x^2y^2 + x^3)\breve{e}_y$. Determine if this fluid flow represents a fluid flow in steady state, assuming that the fluid is incompressible, neglecting gravitational and viscous effects. Furthermore, determine the pressure gradient.

Solution:

$\vec{\nabla}\cdot\vec{v} = \vec{\nabla}\cdot[(2x^3y)\breve{e}_x + (-3x^2y^2 + x^3)\breve{e}_y] = 6x^2y - 6x^2y = 0$. Hence the fluid flow is in steady state, since the fluid is given to be incompressible.

$$\frac{d\vec{v}(\vec{r},t)}{dt} = \frac{\partial\vec{v}}{\partial t} + (\vec{v}\cdot\vec{\nabla})\vec{v} = \frac{-\vec{\nabla}p}{\rho(\vec{r},t)} = \frac{-\vec{\nabla}p}{\rho(\vec{r},t)}.$$

Hence, along the x direction,

$$-\frac{1}{\rho}\frac{\partial P}{\partial x} = (2x^3y)\frac{\partial}{\partial x}(2x^3y) + v\frac{\partial}{\partial y}(2x^3y) + \frac{\partial}{\partial t}(2x^3y) = (2x^3y)(6x^2y) + (-3x^2y^2 + x^3)(2x^3) + 0$$

Hence, $\frac{\partial P}{\partial x} = -\rho(6x^5y^2 + 2x^5)$.

Along y direction, $-\frac{1}{\rho}\frac{\partial P}{\partial y} = v_x\frac{\partial}{\partial x}(-3x^2y^2 + x^3) + v_y\frac{\partial}{\partial y}(-3x^2y^2 + x^3) + \frac{\partial}{\partial t}(-3x^2y^2 + x^3)$,

i.e., $-\frac{1}{\rho}\frac{\partial P}{\partial x} = (2x^3y)(-6xy^2 + 3x^2) + (-3x^2y^2 + x^2)(-6x^2y) + 0$.

Hence, $\frac{\partial P}{\partial y} = -\rho(6x^5y + 6x^4y^3 - 6x^4y)$.

Therefore, the pressure gradient is: $\vec{\nabla}P = \breve{e}_x[-\rho(6x^5y^2 + 2x^5)] + \breve{e}_y[-\rho(6x^5y + 6x^4y^3 - 6x^4y)]$.

P11.4:

For an incompressible fluid, find the locus of all the points having an acceleration of magnitude 7 if the Cartesian components of the velocity $\vec{v}$ are $v_x = y^2$ and $v_y = x^2$.

Solution:

$a_x = \frac{dv_x}{dt} = 2y\frac{dy}{dt}$ and $a_y = \frac{dv_y}{dt} = 2x\frac{dx}{dt}$. Hence, the magnitude of the acceleration is given by: $|a| = \sqrt{a_x^2 + a_y^2} = \sqrt{4y^2{v_y}^2 + 4x^2{v_x}^2} = 7$. The required locus is $2yx\sqrt{x^2 + y^2} = 7$.

P11.5:

The velocity of a fluid in a steady state incompressible 2-dimensional flow field is: $\vec{v} = v_x\breve{e}_x + v_y\breve{e}_x$ with $v_x = 2y$, $v_y = 4x$.

(a) Determine the equation of the corresponding streamlines.

(b) Sketch several streamlines. Indicate the direction of flow along the streamlines.

Solution:

From the equation of continuity, $\vec{\nabla}\cdot(\rho\vec{v})=\frac{\partial\rho}{\partial t}$, we get, for an incompressible fluid in steady state flow: $\vec{\nabla}\cdot\vec{v}=0$. Hence: $\frac{\partial v_x}{\partial x}+\frac{\partial v_y}{\partial y}=0$. Now, for the streamlines, we have $\left[\frac{dx}{v_x(\vec{r})}=\frac{dy}{v_y(\vec{r})}\right]=$ $\psi=$ constant.

Suppose now that we can find a function $\psi=\psi(x, y)$, in such a way that the following relations hold for a 2-dimensional flow: $v_x=\frac{\partial\psi}{\partial y}$, $v_y=-\frac{\partial\psi}{\partial x}$. Since, we always have $\frac{\partial\psi}{\partial x}\frac{\partial\psi}{\partial y}-\frac{\partial\psi}{\partial y}\frac{\partial\psi}{\partial x}=0$ (for any well-behaved function), we see that the continuity equation $\left(\frac{\partial v_x}{\partial x}+\frac{\partial v_y}{\partial y}=0\right)$ is automatically satisfied, and therefore it ***need not be solved***. Finding the ***stream function*** is therefore useful. It was introduced by Lagrange. We have used it here for a 2-dimensional flow, but it can be introduced also for a 3-dimensional flow; nonetheless it has a rather complicated form for the 3-dimensional case. The stream function $\psi=\psi(x, y)$ has the beautiful property that along the line for which $d\psi=\frac{\partial\psi}{\partial x}dx+\frac{\partial\psi}{\partial y}dy=v_x dx-v_y dy=0$, we have $\frac{dx}{v_x}=\frac{dy}{v_y}$, which is the equation of the streamline.

Thus, ψ: constant defines the streamlines; it is for this reason that the function $\psi=\psi(x, y)$ is called the *stream function*.

In the given problem, $v_x=2y$, i.e., $v_x=\frac{\partial\psi}{\partial y}=2y$, hence, $y=y^2+f(x)$.

Also, $v_y=-\frac{\partial\psi}{\partial x}=4x$, hence: $y=-2x^2+g(y)$.

Therefore, the streamlines are given by $\psi=y^2-2x^2$, a constant.

(b) The streamline is given by ψ: constant. Hence, $\psi=y^2-2x^2$; i.e., $\frac{y^2}{\psi}-\frac{x^2}{\psi/2}=1$.

We get an equation that describes a hyperbola. The streamlines are sketched in the figure below, $\psi(x=0, y=0)=0$.

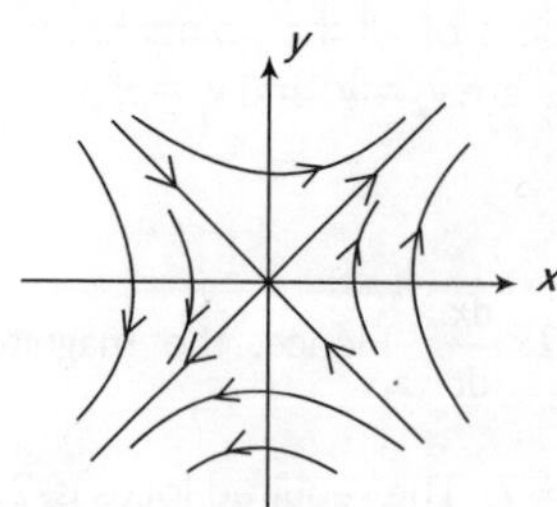

P11.6:

Consider a steady-state 2-dimensional incompressible fluid flow at velocity $\vec{v}=(x^2+3x-4y)\,\breve{e}_x-(2xy+3y)\breve{e}_y$. Determine the ***stream function*** for this fluid flow.

Solution:

We have $v_x = \dfrac{\partial \psi}{\partial y}$; i.e., $\partial \psi = v_x \partial y$ Integrating, we get

$$\psi = \int v_x \partial y = \int (x^2 + 3x - 4y)\partial y = x^2 y + 3xy - 2y^2 + C(x).$$

Further, since $v_y = -\dfrac{\partial \psi}{\partial x}$, we get: $v_y = -2xy - 3y - C'(x)$. Since it is given that $v_y = -2xy + 3y)$, we conclude that $C'(x) = 0$ and hence C: constant. Thus, $\psi = x^2y + 3xy - 2y^2 + C$. Likewise, we have $v_y = -\dfrac{\partial \psi}{\partial x}$; i.e., $\partial \psi = -v_y \partial x$.

Integrating, we get: $\psi = -\int v_y \partial x = \int (2xy + 3y)\partial x = x^2 y + 3xy + D(y)$. From this, we can get $v_x = \dfrac{\partial \psi}{\partial y} = x^2 + 3x + D'(y)$ which means that $\dfrac{dD}{dy} = -4y$. Hence, $D = -2y^2 + k(y)$.

and the stream function is: $\psi = x^2y + 3xy - 2y^2$.

Additional Problems

P11.7 A fluid is shown in the adjacent figure. Note that the cross-section of the fluid's conduit changes; it is different at the position 'I' from what it is at position 'II'. At the position 'I', the velocity of the fluid is 2 m s^{-1} and the area of the cross section is 3 m^2. At the position 'II', the area of the cross section is 8 m^2. Determine the difference in pressure (1) P_2-P_1 and (2) P_3-P_1, where the points '1', '2' and '3' are as shown in the figure. The density of the fluid is given to be 0.6 g cm^{-3}.

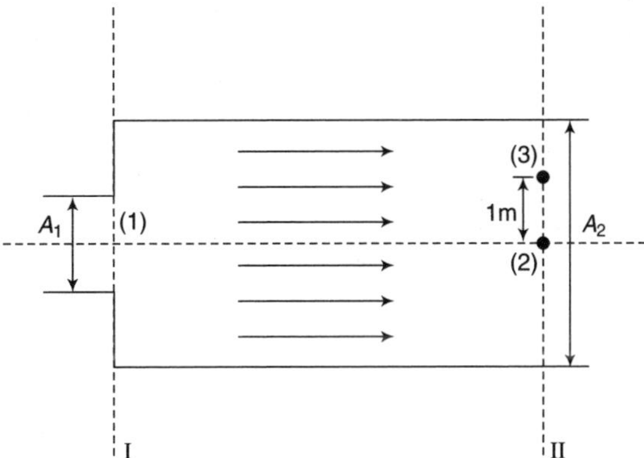

P11.8 A water fountain is built inside a park. The boundaries of the fountain are at a radius of 2 m. Find the velocity at the tip such that, the water is just able to touch the foot of the boundary, find the maximum height that the water can reach.

P11.9 A man is holding a venturi meter inside a moving car. The venturi meter is placed in such a way that the velocity of the air at the throat is equal to the velocity of the car. Find the speed of the car if the liquid rises up to 7 cm. It is given that the density of the fluid is 0.7 g cm^{-3}.

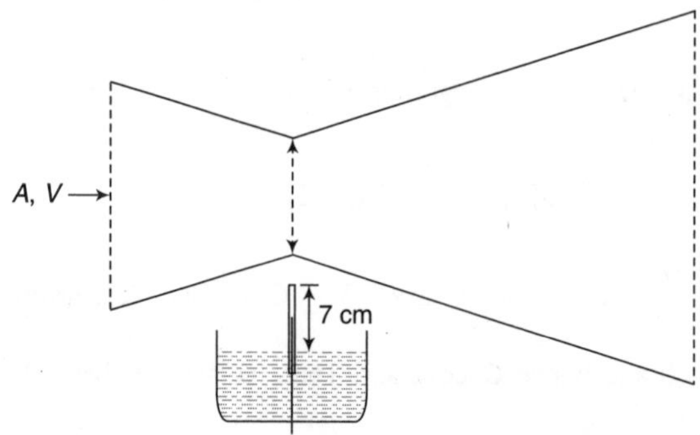

P11.10 The radial and the polar components of the velocity of a fluid flow are given as: $v_r = \frac{ur^2}{\sin\theta}$, $v_\theta = -\frac{4ur^2\theta}{\sin\theta}$; $v_\theta = -\frac{4ur^2\theta}{\sin\theta}$. Determine σ'_{ki}, the viscous stress tensor in the Navier–Stokes Equation for a Newtonian incompressible fluid.

Hint: Use Eq. 11.72 and Eq. 11.74.

P11.11 A cone is pointing towards a water hose. The velocity of the water is V_ω at the end of the hose, the diameter is d and the half angle of the cone is α. Find the force needed to keep the cone in place.

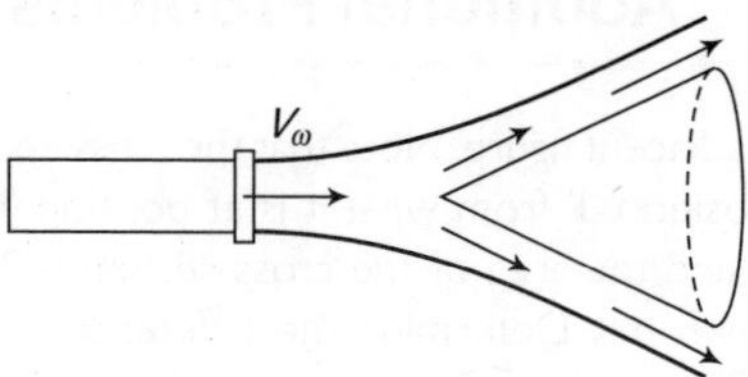

P11.12 In the figure there are two fluids of densities ρ_1 and ρ_1 respectively. The first fluid enters from the left face and exits from the right face. The second fluid is present in the U-tube and cannot exit the setup. The diameters d_1= 4 cm, d_2 =15 cm, of the cross sections 1 and 2 are given, and ρ_1= 3.4 g cm^{-3}. Find the velocity of the fluid if the fluid in the U-tube has a height difference of 10 cm between its two surfaces.

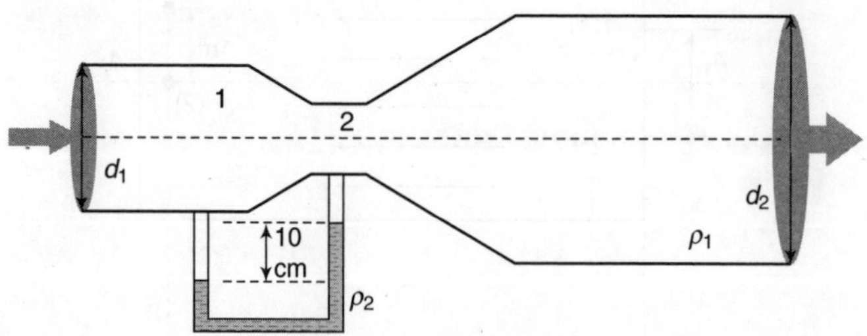

P11.13 A balloon is filled with hot air of 60°C. Its diameter is 10 m. The environmental temperature is 0°C. The pressure outside and inside the balloon is 10^5 Pa. The weight of the balloon material can be neglected. Determine the buoyant force.

P11.14 The figure given below is that of a half cone submerged under water. Find the forces acting on the cone, as well as their directions.

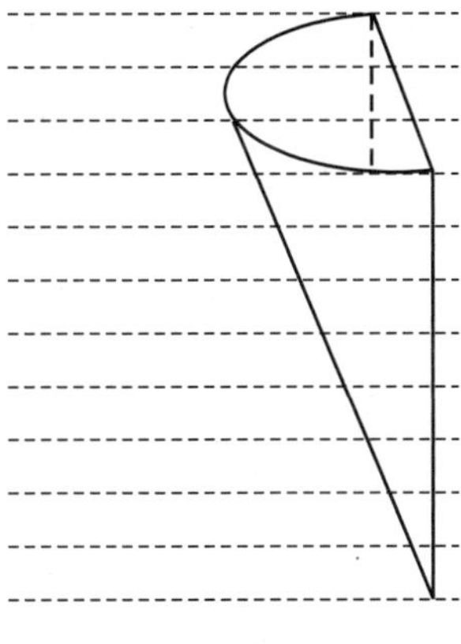

P11.15 Find the stream function for the fluid velocity field $\vec{v} \equiv (v_x, v_y, v_z)$ with $v_x = a(2x^2y - y^2)$, $v_y = -2axy^2$, and $v_z = 0$. Sketch the result.

P11.16 A water pump is used to pump water to a height of 4 m (see adjacent figure). Find the power required to drive the pump if the mass flow rate is 0.7 kg s^{-1}. Also find the velocity in the two pipes shown in the figure, if d_1 = 100 mm and d_2 = 400 mm.

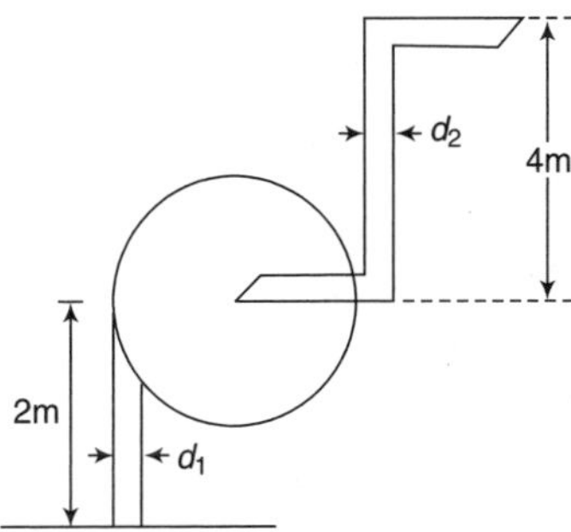

P11.17 The density field of a plane, steady state, fluid flow given by $\rho(x_1, x_2) = kx_1x_2$ where k is constant.

(a) Determine the form of the velocity field if the fluid is incompressible.

(b) Find the equation of the path line.

P11.18 Find the total acceleration of the particle of a fluid described by its Eulerian fluid velocity vector field given by $\vec{v} = 3t\breve{e}_x + x^2z\breve{e}_y + ty^2\breve{e}_z$.

P11.19 What power is needed to drive the shaft of a glide bearing with 2880 l/min, when the shaft is 60 mm wide, 100 mm long and the gap between bearing and shaft is 0.2 mm? (μ_{oil} = 0.01 kg/ms) How is it possible to decrease this power?

P11.20 Oil flow rate of qv = $2 \cdot 10^{-4}$ m^3/s has to be transported through a 10 m long straight pipe (ρ = 800 kg/m^3, $\nu = 10^{-4}$ m^2/s). The available pressure difference is not more than $2 \cdot 10^5$ Pa. Determine the diameter D [mm] of the pipe.

P11.21 D'Alembert's paradox: the drag force on any 3-dimensional body moving uniformly in a potential flow is zero. However, this is actually not the case; since the flow past 3-dimensional bodies are not 'potential flows'. Determine the pressure force exerted by a steady fluid flow on a solid sphere. [Hint: Find the velocity potential in spherical polar coordinates. Find the non-zero components of the fluid velocity. Use Bernoulli's theorem to express the pressure force on the solid sphere in terms of the fluid velocity. Get the pressure distribution on the sphere.]

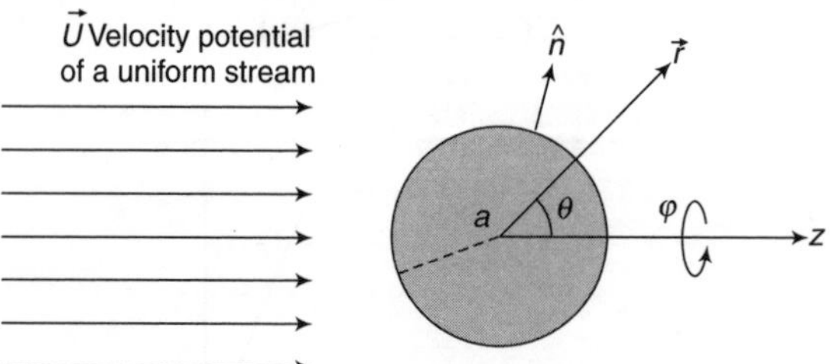

P11.22 Bubble oscillations: The sound of a "babbling brook" is due to the oscillations (compression/expansion) of air bubbles entrained in a stream. The pitch or tone of the sound depends on the size of the bubbles. Find the frequency of the small oscillations the bubble undergoes.

[Hint: Take a bubble of radius *b(t)*. At the surface of the bubble the velocity of the fluid is $u_r = \dfrac{db}{dt} = \dot{b}$. Now model the oscillation using a potential flow due to a point source/sink of fluid at the center of the bubble:

$\phi(r,t) = -\dfrac{k(t)}{r}$. Use the boundary condition at the surface $r = a$. Apply Bernoulli's theorem as $r \to \infty$; also get all body forces at the surface of the bubble and combine them. Ignore gravity. If the gas inside the bubble of mass m goes through adiabatic changes, the equation of state can be given as: $p_g = K\rho_g^{\gamma}$ where, $\rho_g = \dfrac{3m}{4\pi b^3}$, and K is a constant. The adiabatic index γ depends on the gas. Now use pressure continuity at the bubble surface].

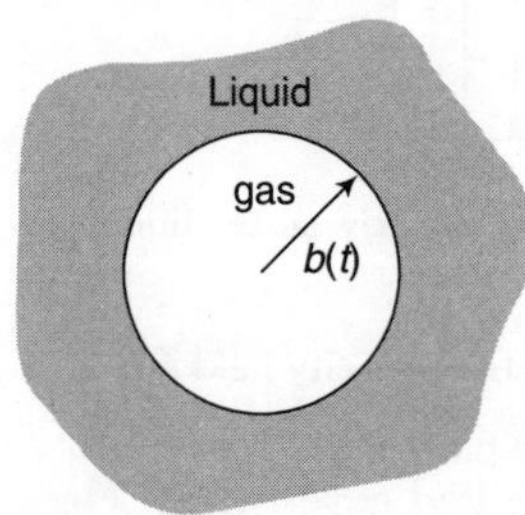

References

[1] Landau, L. D., and E. M. Lifshitz. 2008. *Fluid Mechanics*. Amsterdam: Elsevier, India.

[2] White, Frank M. 2017. *Fluid Mechanics*. 8th Edition. New York: McGraw Hill.

[3] Arfken, G. B., H. J. Weber, and F. E. Harris. 2013. *Mathematical Methods for Physicists*. Amsterdam: Elsevier.

[4] Boas, Mary L. 2006. *Mathematical Methods in the Physical Sciences*. Hoboken, N.J.: John Wiley and Sons.

CHAPTER 12

Basic Principles of Electrodynamics

The special theory of relativity owes its origins to Maxwell's equations of the electromagnetic field.... Since Maxwell's time, physical reality has been thought of as represented by continuous field, and not capable of any mechanical interpretation. This change in the conception of reality is the most profound and the most fruitful that physics has experienced since the time of Newton.

—*Albert Einstein*

12.1 THE UNIQUENESS AND THE HELMHOLTZ THEOREMS: BACKDROP OF MAXWELL'S EQUATIONS

The development of the idea of a 'field' in physics has a troubled, exciting, and long history. Today it is at the very center of all physical theories. Modern physics freely employs the notion of the field such as the gravitational field, the electromagnetic field, the spinor field, the tensor field, the quantum field, the Higgs field. Both Kepler and Newton struggled to account for the mechanism involved in the Sun's influence over vast distances that determined the motion of planets. In their works the seeds of a fields theory were already sown. It was perhaps in the work of Euler on fluid mechanics that the velocity field was rigorously analyzed. Both its divergence and the curl wẻre studied extensively. However, the fluid velocity consisted of flowing matter, whereas the presence of matter at a point where a field resides is redundant. The works of Kepler (1571–1630), Newton (1643–1727), Euler (1707–1783), and of Coulomb (1736–1806), Ampere (1775–1836), Oersted (1777–1851), and Faraday (1791–1867) made crucial contributions to the development of the notion of the 'field'. It was formally organized later by James Clerk Maxwell (1831–1879). Said Maxwell: "*The theory I propose may therefore be called a theory of the Electromagnetic Field because it has to do with the space in the neighborhood of the electric or magnetic bodies.*" The electromagnetic field theory developed by Maxwell provides both the divergence and the curl of the two vector fields $\vec{E}(\vec{r}, t)$ and $\vec{B}(\vec{r}, t)$, where $\vec{E}(\vec{r}, t)$ is the electric field intensity and $\vec{B}(\vec{r}, t)$ is the magnetic flux density field. These four equations (together with the boundary conditions and the initial conditions of the fields) provide all the information we need to solve any problem in classical electrodynamics. The reason these field equations are important is because from their solutions we can obtain the forces that act on a charged particle and on electric currents. These forces are required to solve the classical equation of motion, such as the Newton's law.

Now, a scalar field associates a *scalar* with each point in space; a vector field has a *vector* associated with each point in a region of space. These may change from point to point in that region, and also from one instant of time to another. We are most often concerned with fields which are continuous with respect to spatial displacements and temporal changes, and possess continuous derivatives. A scalar field is represented by a number (with dimensions that are appropriate to the physical quantity that it represents, in suitable units). A vector field is specified by the three components of the physical quantity it stands for, and how they transform when viewed from various frames of reference which are rotated with respect to each other. We now pose the following question: if we have the knowledge about the divergence of a vector field, and also about its curl, can we *uniquely* know the vector field? The divergence of a vector has in it the partial derivatives of the field component which give the rate at which the components change *along the coordinates*, and the curl has terms which provide the rate of change of the components *side-ways*. If we therefore obtain the mathematical equations which provide both the divergence and the curl of a vector field, we can *expect* that the answer to the above question is in the affirmative. Of course, we must expect that it will also be necessary to know the values of the fields at the boundaries of the region, and also know the initial conditions. It is therefore important to ask if the equations for the divergence and the curl of the fields $\vec{E}(\vec{r},t)$ and $\vec{B}(\vec{r},t)$ would give us sufficient information to determine the electromagnetic fields *uniquely*. At the end, as mentioned above, it is these *fields* which will be employed in the classical equation of motion, to track the motion of a charged particle under the action of forces exerted on it by the electromagnetic field.

Two theorems in vector calculus [1, 2], namely the '*uniqueness theorem*' and the '*Helmholtz theorem*' would reveal to us the far-reaching importance of the divergence and the curl of a vector field. These theorems provide the logical backdrop of the succinct relationships provided by the Maxwell's equations. Before we get to Maxwell's equations, we shall therefore first establish these two important theorems.

In this chapter, Maxwell's equations are presented in the framework of the aforementioned two theorems, instead of the historical context of the empirical laws of electricity and magnetism developed *separately* prior to Maxwell's work. The empirical laws of importance were formulated by Coulomb, Biot, Savart, Oersted, Ampere, and by Faraday and Lenz. We shall show, *after* introducing the four equations of Maxwell, that all the earlier empirical laws automatically fall under the canopy of the Maxwell's equations.

Theorem 1 (Uniqueness Theorem)

Given (i) the divergence and (ii) the curl of a vector field in a simply connected region, i.e., a domain in which one can shrink every closed path continuously into a point without plunging out of that domain, and (iii) if the vector field's normal components at the boundaries of this region are known, *then* the vector field is *uniquely* specified (within that domain). This theorem is referred to as the *uniqueness theorem*.

Let us therefore consider a vector field $\vec{K}(\vec{r})$ whose divergence and the curl are given:

$$\vec{\nabla}\cdot\vec{K}(\vec{r}) = \vec{s}(\vec{r}), \tag{12.1a}$$

$$\text{and} \quad \vec{\nabla}\times\vec{K}(\vec{r}) = \vec{c}(\vec{r}). \tag{12.1b}$$

We further assume that the normal components $K_{\perp}(\vec{r}_b)$ of the vector field $\vec{K}(\vec{r})$ on the boundary, i.e., for all the points on the boundary, are known. The theorem requires us to show that if there is any other vector field $\vec{K}'(\vec{r})$ for which Eqs.12.1a and b hold, and for which $K'_{\perp}(\vec{r}_b) = K_{\perp}(\vec{r}_b)$ at all the points on the boundary, then the difference $\vec{W}(\vec{r}) = \vec{K}(\vec{r}) - \vec{K}'(\vec{r})$ must vanish. In other words, the vector field is uniquely specified by the conditions (i), (ii) and (iii) as stated in this theorem. This would establish the fact that the three conditions, (i), (ii) and (iii) *uniquely* specify the vector field $\vec{K}(\vec{r})$.

Now, by our assumption, since both $\vec{K}'(\vec{r})$ and $\vec{K}(\vec{r})$ satisfy Eqs. 12.1a and b, we must have

$$\vec{\nabla}\cdot\vec{K}(\vec{r}) = \vec{s}(\vec{r}) = \vec{\nabla}\cdot\vec{K}'(\vec{r}), \tag{12.2a}$$

$$\text{hence,} \quad \vec{\nabla}\cdot\{\vec{K}(\vec{r}) - \vec{K}'(\vec{r})\} = \vec{\nabla}\cdot\vec{W}(\vec{r}) = 0. \tag{12.2b}$$

$$\text{Likewise,} \quad \vec{\nabla}\times\vec{K}(\vec{r}) = \vec{c}(\vec{r}) = \vec{\nabla}\times\vec{K}'(\vec{r}), \tag{12.2c}$$

$$\text{hence,} \quad \vec{\nabla}\times\{\vec{K}(\vec{r}) - \vec{K}'(\vec{r})\} = \vec{\nabla}\times\vec{W}(\vec{r}) = \vec{0}. \tag{12.2d}$$

Now a vector whose divergence is zero is called solenoidal, and one whose curl is zero, is called irrotational. The *difference* vector $\vec{W}(\vec{r})$ is therefore both solenoidal and irrotational. Since an irrotational vector can always be written as the gradient of a scalar function, we can write

$$\vec{W}(\vec{r}) = \vec{\nabla}\psi. \tag{12.3}$$

On the surface of the region, the normal component of $\vec{W}(\vec{r})$ is given by

$$W_{\perp} = \vec{W}(\vec{r})\cdot\hat{n} = [\vec{K}(\vec{r}) - \vec{K}'(\vec{r})]\cdot\hat{n} = \vec{K}(\vec{r})\cdot\hat{n} - \vec{K}'(\vec{r})\cdot\hat{n} = K_{\perp} - K'_{\perp} = 0. \tag{12.4}$$

Now, for two *arbitrary* scalar fields $f_1(\vec{r})$ and $f_2(\vec{r})$, we have

$$\vec{\nabla}\cdot[f_1(\vec{r})\vec{\nabla}f_2(\vec{r})] = f_1(\vec{r})\vec{\nabla}\cdot\vec{\nabla}f_2(\vec{r}) + \vec{\nabla}f_1(\vec{r})\cdot\vec{\nabla}f_2(\vec{r}). \tag{12.5}$$

Hence, equating the volume integrals of the two sides, we get

$$\iiint \vec{\nabla}\cdot[f_1(\vec{r})\vec{\nabla}f_2(\vec{r})]dV = \iiint f_1(\vec{r})\vec{\nabla}\cdot\vec{\nabla}f_2(\vec{r})dV + \iiint \vec{\nabla}f_1(\vec{r})\cdot\vec{\nabla}f_2(\vec{r})dV.$$

Using now the Gauss' divergence theorem for the left hand side:

$$\oiint [f_1(\vec{r})\vec{\nabla}f_2(\vec{r})]\cdot\hat{n}dS = \iiint f_1(\vec{r})\vec{\nabla}\cdot\vec{\nabla}f_2(\vec{r})dV + \iiint \vec{\nabla}f_1(\vec{r})\cdot\vec{\nabla}f_2(\vec{r})dV.$$

Choosing now $f_1(\vec{r}) = f_2(\vec{r}) = \psi(\vec{r})$,

$$\oiint [\psi(\vec{r})\vec{\nabla}\psi(\vec{r})]\cdot \hat{n}dS = \iiint \psi(\vec{r})\vec{\nabla}\cdot\vec{\nabla}\psi(\vec{r})dV + \iiint \vec{\nabla}\psi(\vec{r})\cdot\vec{\nabla}\psi(\vec{r})dV,$$

$$\oiint [\psi(\vec{r})W_{\perp}]dS = \iiint \psi(\vec{r})\vec{\nabla}\cdot\vec{W}(\vec{r})dV + \iiint \vec{W}(\vec{r})\cdot\vec{W}(\vec{r})dV.$$

The left hand side is zero on count of the fact that $W_{\perp} = 0$ (Eq. 12.4). The first term on the right hand side is zero on count of Eq. 12.2b.

Hence, $$\iiint \vec{W}(\vec{r})\cdot\vec{W}(\vec{r})dV = 0. \qquad (12.6)$$

The integrand being *non-negative*, the only way the above equation will hold is that the function $\vec{W}(\vec{r}) = \vec{0}$. This would guarantee that $\vec{K}(\vec{r}) = \vec{K}'(\vec{r})$. The *uniqueness* $\vec{K}'(\vec{r}) = \vec{K}(\vec{r})$ of the vector function satisfying the conditions (i), (ii) and (iii) is thus established.

Theorem 2 (Helmholtz Theorem):

A vector field $\vec{K}(\vec{r})$, whose divergence and curl are both known (say, given by Eq. 12.1), and which both asymptotically (i.e., as $r \to \infty$) vanish, can always be written as the sum of a solenoidal vector (i.e., a vector whose divergence is zero), and an irrotational vector (i.e., a vector whose curl is zero).

Since, the curl of a gradient of a scalar field is identically zero, irrespective of the scalar field is, the irrotational vector mentioned above can be chosen to be $-\vec{\nabla}\psi(\vec{r})$, where $\psi(\vec{r})$ is a scalar field. Likewise, since the divergence of the curl of a vector is always zero, the solenoidal vector mentioned above can be chosen to be $\vec{\nabla} \times \vec{G}(\vec{r})$, where $\vec{G}(\vec{r})$ is a vector field. The Helmholtz theorem therefore states that a vector field $\vec{K}(\vec{r})$ whose divergence and curl are both known (given by Eq. 12.1), and which both asymptotically (i.e., as $r \to \infty$) vanish, can be expressed as

$$\vec{K}(\vec{r}) = [-\vec{\nabla}\psi(\vec{r})]_I + [\vec{\nabla} \times \vec{G}(\vec{r})]_S. \qquad (12.7)$$

The subscripts '*I*' and '*S*', above, stand respectively for the fact that $[-\vec{\nabla}\psi(\vec{r})]$ is irrotational for an arbitrary scalar field, and that $[\vec{\nabla} \times \vec{G}(\vec{r})]$ is solenoidal for every vector field $\vec{G}(\vec{r})$. Now, since the divergence and the curl of $\vec{K}(\vec{r})$ are known to be given by Eq. 12.1, we can use them to *choose* $\psi(\vec{r})$ and $\vec{G}(\vec{r})$, both being hitherto arbitrary:

$$\psi(\vec{r}) = \frac{1}{4\pi}\iiint \frac{s(\vec{r}\,')}{|\vec{r} - \vec{r}\,'|}d^3V', \qquad (12.8a)$$

and $$\vec{G}(\vec{r}) = \frac{1}{4\pi}\iiint \frac{\vec{c}(\vec{r}\,')}{|\vec{r} - \vec{r}\,'|}d^3V'. \qquad (12.8b)$$

We shall refer to the point $(\vec{r})$ as the field point, and the point $(\vec{r}\,')$as the source point, separated by $\vec{R}' = \vec{r} - \vec{r}\,'$, shown in Fig. 12.1. The volume (triple) integration is over the source region, represented by the *primed* variable $(\vec{r}\,')$ in our notation. The integration volume element (dV') for the integration is therefore also labeled with a prime in Eq. 12.8.

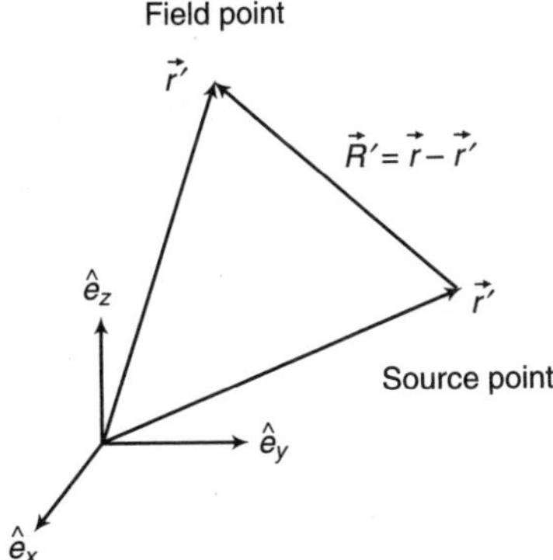

Fig. 12.1 A source at the point $\vec{r}'$ generates an influence at the field point $\vec{r}$. The physical influence diminishes as an inverse function of the distance between the source point and the field point.

Thus, if only we can ensure that $\vec{K}(\vec{r})$ can indeed be written as Eq. 12.7 (as is yet to be proved), then we shall have, since the divergence of a curl of any vector field is always zero,

$$\vec{\nabla}\cdot\vec{K}(\vec{r}) = -\vec{\nabla}\cdot\vec{\nabla}\psi(\vec{r}), \tag{12.8c}$$

and also,

since the curl of the gradient of any vector field is always zero,

$$\vec{\nabla}\times\vec{K}(\vec{r}) = \vec{\nabla}\times(\vec{\nabla}\times\vec{G}(\vec{r})). \tag{12.8d}$$

Hence, if only we can now show that

$$-\vec{\nabla}\cdot\vec{\nabla}\psi(\vec{r}) = \vec{s}(\vec{r}), \tag{12.9a}$$

and that $\vec{\nabla}\times[\vec{\nabla}\times\vec{G}(\vec{r})] = \vec{c}(\vec{r}),$ (12.9b)

then we would have established that $\vec{K}(\vec{r})$ can be convincingly written as Eq. 12.7, and the enunciation stated in the Helmholtz theorem will be confirmed and thus established.

Now,

$$-\vec{\nabla}\cdot\vec{\nabla}\psi(\vec{r}) = -\vec{\nabla}\cdot\vec{\nabla}\left\{\frac{1}{4\pi}\iiint\frac{s(\vec{r}')}{|\vec{r}-\vec{r}'|}d^3V'\right\} = -\left\{\frac{1}{4\pi}\iiint s(\vec{r}')(\vec{\nabla}\cdot\vec{\nabla})\left(\frac{1}{|\vec{r}-\vec{r}'|}\right)d^3V'\right\}, \tag{12.10a}$$

since the gradient is with respect to the (unprimed) field point, and the integration is with respect to the (primed) source points. The operator $\vec{\nabla}\cdot\vec{\nabla}$ involves operators for partial derivatives with respect to the coordinates of the *field* point, and hence cannot affect the physical properties at the *source* locations. Using now Eq. 11.41, we immediately see that

$$-\vec{\nabla}\cdot\vec{\nabla}\psi(\vec{r}) = -\left\{\frac{1}{4\pi}\iiint s(\vec{r}')(4\pi\delta(\vec{r}-\vec{r}')d^3V'\right\} = s(\vec{r}). \tag{12.10b}$$

Hence, Eq. 12.9a holds.

We now proceed to check if Eq. 12.9b also holds.

Now, $\vec{\nabla}\times[\vec{\nabla}\times\vec{G}(\vec{r})]=\vec{\nabla}\{\vec{\nabla}\cdot\vec{G}(\vec{r})\}-(\vec{\nabla}\cdot\vec{\nabla})\vec{G}(\vec{r}).$ (12.10c)

We shall determine the two terms in Eq. 12.10c separately, and accordingly we write

i.e., $\vec{\nabla}\times[\vec{\nabla}\times\vec{G}(\vec{r})]=\vec{T}_1-\vec{T}_2,$ (12.10d)

where $\vec{T}_1=\vec{\nabla}\{\vec{\nabla}\cdot\vec{G}(\vec{r})\}=\text{grad }\{\text{div }\vec{G}(\vec{r})\},$ (12.11a)

and $\vec{T}_2=(\vec{\nabla}\cdot\vec{\nabla})\vec{G}(\vec{r}).$ (12.11b)

First, we shall determine div $\vec{G}(\vec{r})=\vec{\nabla}\cdot\vec{G}(\vec{r})$. Subsequently, we shall determine its gradient.

$$\vec{\nabla}\cdot\vec{G}(\vec{r})=\vec{\nabla}\cdot\left\{\frac{1}{4\pi}\iiint\frac{\vec{c}(\vec{r}\,')}{|\vec{r}-\vec{r}\,'|}d^3V'\right\}=\frac{1}{4\pi}\iiint\vec{\nabla}\cdot\left\{\frac{\vec{c}(\vec{r}\,')}{|\vec{r}-\vec{r}\,'|}\right\}d^3V', \tag{12.12a}$$

i.e.,
$$\vec{\nabla}\cdot\vec{G}(\vec{r})=\frac{1}{4\pi}\iiint\vec{c}(\vec{r}\,')\cdot\vec{\nabla}\left\{\frac{1}{|\vec{r}-\vec{r}\,'|}\right\}d^3V+\frac{1}{4\pi}\iiint\left\{\frac{1}{|\vec{r}-\vec{r}\,'|}\right\}\{\vec{\nabla}\cdot\vec{c}(\vec{r}\,')\}d^3V'. \tag{12.12b}$$

Hence,
$$\vec{\nabla}\cdot\vec{G}(\vec{r})=\frac{1}{4\pi}\iiint\vec{c}(\vec{r}\,')\cdot\vec{\nabla}\left\{\frac{1}{|\vec{r}-\vec{r}\,'|}\right\}d^3V'=-\frac{1}{4\pi}\iiint\vec{c}(\vec{r}\,')\cdot\vec{\nabla}'\left\{\frac{1}{|\vec{r}-\vec{r}\,'|}\right\}d^3V', \tag{12.12c}$$

since $\vec{\nabla}\cdot\vec{c}(\vec{r})=0,$ (12.13a)

and
$$\vec{\nabla}'\left\{\frac{1}{|\vec{r}-\vec{r}\,'|}\right\}=-\vec{\nabla}\left\{\frac{1}{|\vec{r}-\vec{r}\,'|}\right\}. \tag{12.13b}$$

Now, the integrand in Eq. 12.12c is

$$\vec{c}(\vec{r}\,')\cdot\vec{\nabla}'\left\{\frac{1}{|\vec{r}-\vec{r}\,'|}\right\}=\vec{\nabla}'\cdot\left\{\vec{c}(\vec{r}\,')\frac{1}{|\vec{r}-\vec{r}\,'|}\right\}-\frac{1}{|\vec{r}-\vec{r}\,'|}\vec{\nabla}'\cdot\vec{c}(\vec{r}\,'),$$

and therefore

$$\vec{\nabla}\cdot\vec{G}(\vec{r})=-\frac{1}{4\pi}\iiint\left[\vec{\nabla}'\cdot\left\{\vec{c}(\vec{r}\,')\frac{1}{|\vec{r}-\vec{r}\,'|}\right\}-\frac{1}{|\vec{r}-\vec{r}\,'|}\vec{\nabla}'\cdot\vec{c}(\vec{r}\,')\right]d^3V'. \tag{12.14a}$$

The right hand side can be written as a sum of two volume integrations:

$$\vec{\nabla}\cdot\vec{G}(\vec{r})=-\left[\frac{1}{4\pi}\iiint\vec{\nabla}'\cdot\left\{\vec{c}(\vec{r}\,')\frac{1}{|\vec{r}-\vec{r}\,'|}\right\}d^3V'\right]+\left[\frac{1}{4\pi}\iiint\frac{1}{|\vec{r}-\vec{r}\,'|}\vec{\nabla}'\cdot\vec{c}(\vec{r}\,')d^3V'\right]. \tag{12.14b}$$

The second term is zero, since the vector field $\vec{c}$, is the curl of the vector field $\vec{K}$ (Eq. 12.1b), and the divergence of a curl is always zero. Using the Gauss' divergence theorem on the first term,

$$\vec{\nabla}\cdot\vec{G}(\vec{r})=-\frac{1}{4\pi}\oiint\left\{\vec{c}(\vec{r}\,')\frac{1}{|\vec{r}-\vec{r}\,'|}\right\}\cdot\hat{n}dS'. \tag{12.14c}$$

The above integral has to be evaluated on the asymptotic surface at a distance $r' \to \infty$. Now, any physical field diminishes asymptotically, whereas the elemental integration area $dS' = r'^2\, d\Omega$, increases as the square of the distance. Hence, if the vector field $\vec{c}$ diminishes at least as fast as the square of the inverse-distance, then the other factor $\frac{1}{|\vec{r} - \vec{r}'|}$ in the integrand would guarantee that the integral goes to zero. Thus, $\vec{\nabla} \cdot \vec{G}(\vec{r}) = 0$, and the term $\vec{T}_1$ in Eq. 12.10d goes to zero. We therefore get

$$\vec{\nabla} \times [\vec{\nabla} \times \vec{G}(\vec{r})] = -\vec{T}_2 = -(\vec{\nabla} \cdot \vec{\nabla})\vec{G}(\vec{r}) = -(\vec{\nabla} \cdot \vec{\nabla}) \left[\frac{1}{4\pi} \iiint \frac{\vec{c}(\vec{r}')}{|\vec{r} - \vec{r}'|} d^3V' \right], \tag{12.15a}$$

i.e.,
$$\vec{\nabla} \times [\vec{\nabla} \times \vec{G}(\vec{r})] = -\left[\frac{1}{4\pi} \iiint \vec{c}(\vec{r}') \left\{ (\vec{\nabla} \cdot \vec{\nabla}) \frac{1}{|\vec{r} - \vec{r}'|} \right\} d^3V' \right], \tag{12.15b}$$

and hence, using Eq. 11.41, we get

$$\vec{\nabla} \times [\vec{\nabla} \times \vec{G}(\vec{r})] = \left[\frac{1}{4\pi} \iiint \vec{c}(\vec{r}') \{4\pi\delta(\vec{r} - \vec{r}')\} d^3V' \right] = \vec{c}(\vec{r}). \tag{12.15c}$$

Equation 12.10b and Eq. 12.15c demonstrate that both the requirements sought in Eqs. 12.9a and 12.9b are satisfied. Hence, an arbitrary vector field $\vec{K}(\vec{r})$ can indeed be written as $\vec{K}(\vec{r}) = \vec{\nabla}\psi(\vec{r}) + \vec{\nabla} \times \vec{G}(\vec{r})$, thus establishing the Helmholtz theorem (Eq. 12.7).

As mentioned earlier, the consequences of the uniqueness theorem, and the Helmholtz theorem, in the theory of electrodynamics are far reaching. Applications of these theorems are embodied in the very fundamental equations of the electrodynamics theory, namely the Maxwell's equations, introduced in the next section.

12.2 MAXWELL'S EQUATIONS

The fundamental equations of electrodynamics, which comprehensively combine the laws of electricity and magnetism in a seamless framework and provide the inspiration for the theory of relativity, are named after James Clerk Maxwell (1831–1879), an outstanding polymath. Albert Einstein had placed photographs of three scientists in his study—those of Isaac Newton, Michael Faraday and James Clerk Maxwell. One of the biggest challenges in modern physics today is the search for a Grand Unified Theory, GUT, which seeks to unify the electro-weak and the strong interactions, and the Theory of Everything, TOE, which would integrate the GUT with gravity. The TOE would sew in the know-how of all physical fields, inclusive of all matter (and anti matter), and energy (also dark matter and dark energy), in a single framework. If Maxwell's work is regarded at an outstanding high pedestal, it is because in his work we find the very first comprehensive understanding of physical fields, and also fully exhaustive integration of the electrical and the magnetic phenomena. Maxwell thus laid out the path on which physicists are still marching today, in search of what may turn out to be everything that we can know about the physical universe, and/or the multiverse?.

The discovery of the electrical and the magnetic interactions has a very old history. The attraction and repulsion between magnets, and orientation of small magnets in the Earth's magnetic field, was observed perhaps over a thousand years B. Observation of the triboelectric phenomena (effects of static electricity that result from friction between some materials) also dates back to ~600 BCE, in the experiments carried out by the Greek philosopher, Thales of Miletus (624–546 BCE). Our understanding of the electromagnetic interaction was slow and incremental, through the ancient times and the middle ages, and the first major leaps came only in the sixteenth and seventeenth centuries, through the work of William Gilbert (1544–1603), Robert Boyle (1627–1691), Otto von Guericke (1602–1686), and Benjamin Franklin (1785–1788).

A comprehensive record of the historical developments is not presented here, not even symbolically. However, the studies of some scientists whose works mark some of the major milestones are mentioned in this section, only to get an overview of the significant time periods when some of the significant advances were made. Physical models of the electromagnetic phenomena which encapsulate *all* earlier works were expressed in *semi-empirical* laws developed by Charles Coulomb (1736–1806), Jean-Baptiste Biot (1774–1862), Félix Savart (1791–1841), André-Marie Ampère (1775–1836), Emil Lenz (1804–1865), and Michael Faraday (1791–1867). The theory developed by James Clerk Maxwell (1831–1879) provides a *backward integration* of the works of Coulomb, Oersted, Ampere, Faraday, and Lenz. Maxwell's theory, in fact, does much more than that. It not only consolidates all electromagnetic phenomena in a unified theory which accounts for the generation of magnetic fields from changing electric fields, and vice versa, but it also foreshadowed Einstein's special theory of relativity. Furthermore, Maxwell's theory projects the role of symmetry as an underlying consideration of the laws of nature, which has now become one of the major cornerstones of modern physics. As mentioned earlier, Maxwell's theory is the forerunner of the search for the GUT (and the eventual TOE), of which some of the best minds among the sharpest physicists today are in hot pursuit.

Totally bypassing the details of the path-breaking work of Coulomb, Oersted, Ampere, Faraday, and Lenz, we state the Maxwell's equations for the electromagnetic field $(\vec{E}(\vec{r}, t), \vec{B}(\vec{r}, t))$. These equations give the electric field intensity $\vec{E}(\vec{r}, t)$ and the magnetic flux density field $\vec{B}(\vec{r}, t)$ at the spatial coordinate $(\vec{r})$ at the instant of time t. As one would expect from the uniqueness theorem, and the Helmholtz theorem, the electromagnetic field will be described by the divergence and the curl of the electric field intensity $\vec{E}(\vec{r}, t)$, and the divergence and the curl of the magnetic flux density field $\vec{B}(\vec{r}, t)$. Thus there are four equations that we need, which tell us how the electromagnetic field $(\vec{E}(\vec{r}, t), \vec{B}(\vec{r}, t))$ varies from point to point in space, and from one instant of time to another. These are therefore partial *coupled* differential equations that one may solve, with appropriate boundary values, and initial conditions, to get the unique solution for $(\vec{E}(\vec{r}, t), \vec{B}(\vec{r}, t))$. The above uniqueness theorem ensures that this can indeed be achieved, from the divergence and the curl of each of the two fields, $\vec{E}(\vec{r}, t)$ and $\vec{B}(\vec{r}, t)$, with suitable boundary conditions and initial values. The Helmholtz theorem further guarantees that the electromagnetic fields are expressible as a sum of a solenoidal vector field and an irrotational vector field. The vector fields $(\vec{E}(\vec{r}, t), \vec{B}(\vec{r}, t))$ are then derivable from appropriate scalar and vector fields, called the scalar and vector *potentials*, $(\phi(\vec{r}, t), \vec{A}(\vec{r}, t))$.

We shall now introduce the four fundamental equations of electrodynamics, the Maxwell's equations. These are, as stated above, partial differential coupled equations, in which the dynamics of the electric field intensity $\vec{E}(\vec{r}, t)$ and that of the magnetic flux density field $\vec{B}(\vec{r}, t)$ is coupled, making them inseparable entities of a larger physical entity, referred to as the electromagnetic field tensor of rank 2, which we shall introduce in the next chapter.

Now, the divergence of a vector field is a *scalar point function*, and the curl of a vector field is a *vector point function*. The left hand sides of the Maxwell's equations stand for the divergence and the curl of the electromagnetic fields $(\vec{E}(\vec{r}, t), \vec{B}(\vec{r}, t))$, and the right hand sides provide us information about the physical quantities which represent the corresponding scalar point function (in the case of the divergence of the fields), and the vector point function (in the case of the curl of the fields). The equations for the divergence of the electric field intensity $\vec{E}(\vec{r}, t)$ and the magnetic flux density field $\vec{B}(\vec{r}, t)$ are

$$\vec{\nabla} \cdot \vec{E}(\vec{r}, t) = \frac{\rho(\vec{r}, t)}{\varepsilon_0}, \tag{12.16a}$$

and $$\vec{\nabla} \cdot \vec{B}(\vec{r}, t) = 0. \tag{12.16b}$$

The discussion below will associate the first equation of Maxwell, Eq. 12.16a, with the historic Coulomb–Gauss law. This equation states that divergence of the electric field at a point in space is equal to the $\frac{1}{\varepsilon_0}$ times the electric charge density at that point. Maxwell's second equation, Eq. 12.16b, declares that in contrast with the occurrence of positive and negative electrical charges in nature, there is no analogous magnetic monopole. As per Eq. 12.16b, the divergence of the magnetic flux density field is zero.

The equations for the curl of the electric field intensity $\vec{E}(\vec{r}, t)$, and of the magnetic flux density field $\vec{B}(\vec{r}, t)$, are

$$\vec{\nabla} \times \vec{E}(\vec{r}, t) = -\frac{\partial \vec{B}(\vec{r}, t)}{\partial t}, \tag{12.16c}$$

and $$\vec{\nabla} \times \vec{B}(\vec{r}, t) = \mu_0 \left(\vec{J}(\vec{r}, t) + \varepsilon_0 \frac{\partial \vec{E}(\vec{r}, t)}{\partial t} \right) = \mu_0 \vec{J}(\vec{r}, t) + \mu_0 \varepsilon_0 \frac{\partial \vec{E}(\vec{r}, t)}{\partial t}. \tag{12.16d}$$

The third equation of Maxwell, Eq. 12.16c, embodies all the consequences of the semi-empirical Faraday–Lenz laws, as demonstrated below. Maxwell's third equation states that the curl of the electric field at a point in space is equal to the negative time-derivative of the magnetic flux density field at that point. Finally, Maxwell's fourth equation (Eq. 12.16d) states that the curl of the magnetic flux density field is equal to the sum of the μ_0 times the electric current density vector at that point, and $\mu_0\varepsilon_0$ times the time-derivative of the electric field at that point. It will be seen below that this law not only contains within it the semi-empirical law of Oersted–Ampere, but goes beyond it to include time-dependent effects. Maxwell's equations are manifestly coupled partial differential equations. In the above equations, μ_0 and ε_0 are universal constants which represent properties of vacuum. $\varepsilon_0 = \frac{1}{36\pi} \times 10^{-9}$ Farad/meter is the electrical permittivity of free space, and $\mu_0 = 4\pi \times 10^{-7}$ Henry/meter is the magnetic permeability of free space.

Using the Gauss' divergence theorem (Chapter 10, Eq. 10.21), the volume integral of Eq. 12.16a gives

$$\left.\begin{aligned} &\iiint \vec{\nabla} \cdot \vec{E}(\vec{r}, t) dV = \frac{1}{\varepsilon_0} \iiint \rho(\vec{r}, t) dV, \\ \text{i.e.,} \quad &\oiint \vec{E}(\vec{r}, t) \cdot \hat{n} dS = \frac{q_{enc}(\vec{r}, t)}{\varepsilon_0}. \end{aligned}\right\} \tag{12.17a}$$

In the above result, the volume integral of the charge density gives the electric charge. It is the net charge enclosed by the surface that bounds the region over which the volume integral is carried out. Similarly, the volume integration of Eq. 12.16b gives:

$$\left.\begin{aligned} &\iiint \vec{\nabla} \cdot \vec{B}(\vec{r}, t) dV = 0, \\ \text{i.e.,} \quad &\oiint \vec{B}(\vec{r}, t) \cdot \hat{n} dS = 0. \end{aligned}\right\} \tag{12.17b}$$

Since these relations hold good for arbitrary regions of space, including the tiniest ones, Eq. 12.17b rules out the possibility of a magnetic analogue of an electrical charge. It dismisses the possibility of a magnetic monopole.

Similarly, using the Kelvin–Stokes theorem (Chapter 10, Eq. 10.39) the integral over an arbitrary 'open' surface of Eq. 12.16c gives

$$\left.\begin{aligned} &\iint \vec{\nabla} \times \vec{E}(\vec{r}, t) \cdot \hat{n} dS = -\frac{\partial}{\partial t} \iint \vec{B}(\vec{r}, t) \cdot \hat{n} dS, \\ \text{i.e.,} \quad &\oint \vec{E}(\vec{r}, t) \cdot \overrightarrow{d\ell} = -\frac{\partial \Phi_m}{\partial t}. \end{aligned}\right\} \tag{12.17c}$$

The surface integral of Eq. 12.16d gives

$$\left.\begin{aligned} &\iint \vec{\nabla} \times \vec{B}(\vec{r}, t) \cdot \hat{n} dS = \mu_0 \iint \vec{J}(\vec{r}, t) \cdot \hat{n} dS + \mu_0 \varepsilon_0 \frac{\partial}{\partial t} \iint \vec{E}(\vec{r}, t) \cdot \hat{n} dS, \\ \text{i.e.,} \quad &\oint \vec{B}(\vec{r}, t) \cdot \overrightarrow{d\ell} = \mu_0 I + \mu_0 \varepsilon_0 \frac{\partial \Phi_e}{\partial t}. \end{aligned}\right\} \tag{12.17d}$$

Again, we have made use of the Kelvin–Stokes theorem. The surface integral of vector field through a surface is just the flux of that field, which we have defined and discussed in Chapter 10, Eq. 10.13. Accordingly, in Eq. 12.17c,

$$\Phi_m = \iint \vec{B}(\vec{r}, t) \cdot \hat{n} dS, \tag{12.18a}$$

is just the *magnetic induction flux* through a surface area,

and likewise, in Eq. 12.17d, we have,

$$\Phi_e = \iint \vec{E}(\vec{r}, t) \cdot \hat{n} dS, \tag{12.18b}$$

which is the *electric* flux through the surface area over which the integration is carried out.

'*I*', in Eq. 12.17d, is just the total current, since it is the surface integral of the current density vector:

$$I = \iint \vec{J}(\vec{r}, t) \cdot \hat{n} dS. \tag{12.18c}$$

Equations 12.17a, b, c, and d are essentially integral forms of the four differential equations of Maxwell. The differential forms (Eqs. 12.16a, b, c, and d) provide information at a *particular point* in space, whereas the integral forms (Eqs. 12.17a, b, c, and d) provide properties over *extended space* over which the global integration is carried out.

To appreciate Maxwell's contribution to *extending* the earlier empirical laws, we first deliberate on the equations for the curl of the electric field, and then on the curl of the magnetic flux density field. After that, we shall discuss the equations for the divergence of the electric field, and that of the magnetic flux density field. Fittingly, we shall briefly retreat to some of the most fascinating experiments which provide the framework for the all-important Ohm's law, and also the Kirchhoff's laws in D.C. electric circuits. Next, very quickly in small incremental steps, we shall consider time-dependent experiments. Specifically, we shall introduce the Faraday–Lenz experiments which paved the way for the development of Maxwell's equations. *These are some of the most amazing experiments in physics*. You will soon find, from the discussion below, why Richard Feynman [2] has referred to the outcome of these experiments as absolutely unique. To appreciate Feynman's amazement, pointed out below, we shall first briefly revisit properties of electric circuits in which a steady current flows, driven by a D.C. source, such as a battery (for example). The Ohm's law states that the electrical current *I* that flows in a resistor *R* is proportional to the voltage source *V* across it. The relationship between I, V, and R is given by

$$V = RI = \left(\rho \frac{\ell}{A}\right)(JA) = \frac{1}{\sigma}\ell J, \tag{12.19a}$$

wherein the resistance *R* is proportional to the length ℓ of the resistor, and inversely proportional to its cross-sectional area *A*. The proportionality is given by the resistivity $\rho = \frac{1}{\sigma}\Sigma$, being the conductivity of the material of the resistor. In Eq. 12.19, we have identified, following the discussion in Chapter 10 (Eq. 10.12), that the current is the product of the current density and the cross-sectional area: $I\hat{i} = \vec{J}A$. In this relation, the direction of the current is given by the unit vector $\hat{i}$.

Hence, $$J = \sigma \frac{V}{\ell} = \sigma \frac{V}{\ell} = \sigma \left(\frac{1}{\ell}\right)\left(\frac{F\ell}{q}\right) = \sigma E, \tag{12.19b}$$

i.e., $$\vec{J} = \sigma \vec{E}. \tag{12.19c}$$

In writing Eq. 12.19b, we have used the fact that the voltage difference is just the work done by the electric force to carry a *unit* charge from one point to the other.

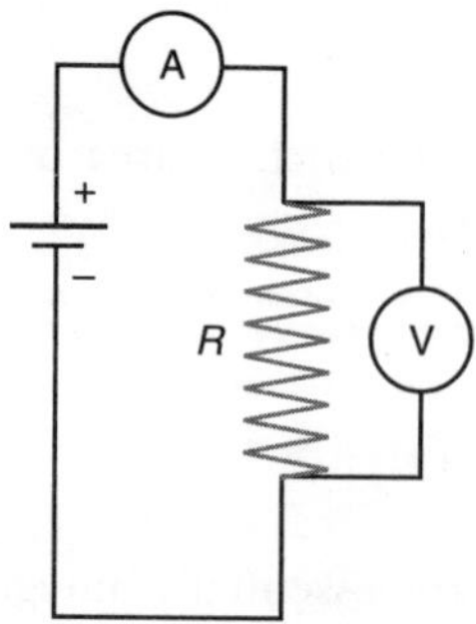

Fig. 12.2a The Ohm's law states that in the above D.C. electrical circuit, $V = RI$ where V is the voltage across the resistor R, and I is the current in the circuit. The resistance offered by the ammeter A is negligible, and the current passing through the voltmeter is also negligible.

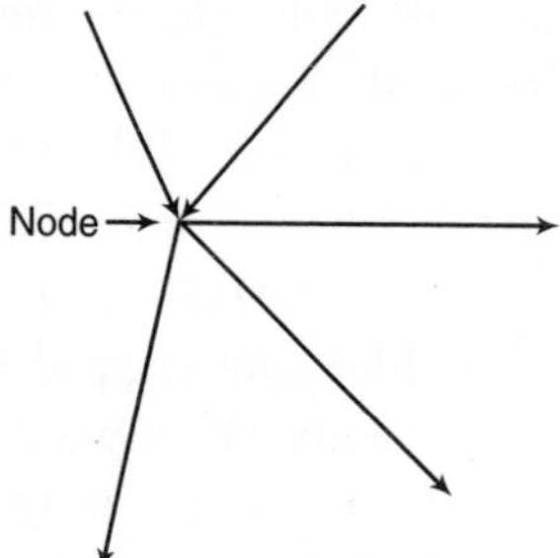

Fig. 12.2b Kirchhoff's first law, also called the current law, states that at a node (also known as the branch point) in a D.C. circuit, the sum of incoming current is equal to the sum of the outgoing current. The inward/outward directions of the current flow at the node are indicated by the arrows in the above diagram.

i.e., $\sum_j I_j^{\text{inward}} + \sum_k I_k^{\text{outward}} = 0.$

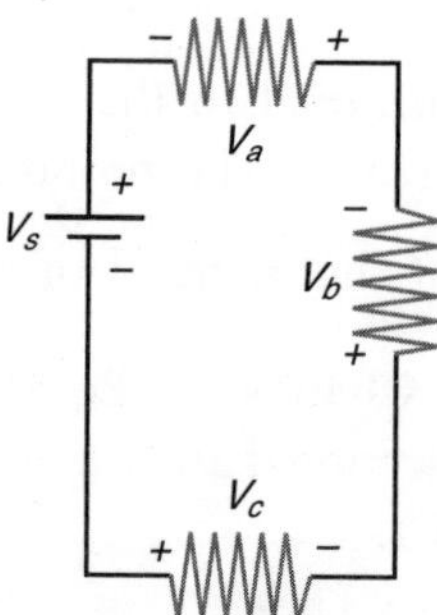

Fig. 12.2c Kirchhoff's second law, also known as the voltage law, states that the arithmetic sum of the voltage differences across all the elements, including the source, in a D.C. circuit is always zero.

i.e., $\sum_k V_k = 0.$

Equation 12.19b is an alternative statement of the Ohm's law, completely equivalent to Eq. 12.19a. Essentially, it is the voltage *difference* between two points on a conductor that is responsible for the force that drives a charge along the direction of the intensity of the electric field $\vec{E}$ which comes into being because of the voltage difference. This force is given by

$$\vec{F}_e = q\vec{E}. \tag{12.20}$$

The subscript 'e' on the force stands for the fact that this force is due to the electric field. In the absence of a voltage difference generated by a source such as a battery, one would therefore not expect a current to flow in a circuit. The electric current that flows in the Faraday–Lenz experiments, described below, is best appreciated in this context. Maxwell's equation will include the phenomenology of Faraday–Lenz's historic discovery naturally. In doing so, Maxwell's work goes well past the physics of the D.C. circuits. It would go not merely beyond the Ohm's law, but also beyond the Kirchhoff's laws. In particular, in a D.C. circuit, there being no accumulation of charges at any node (Fig. 12.2b), the first law of Kirchhoff states that the sum of the electrical currents at the node (branch point) in an electrical circuit is zero:

$$\sum_j I_j^{\text{inward}} + \sum_k I_k^{\text{outward}} = 0. \tag{12.21a}$$

This statement is completely equivalent to the recognition that the current density vector $\vec{J}$ is solenoidal at the node:

$$\vec{\nabla}\cdot\vec{J} = 0. \tag{12.21b}$$

The second law of Kirchhoff (Fig. 12.2c) is

$$\sum_k V_k = 0. \tag{12.21c}$$

This law tells us that if the D.C. voltage source V_S in Fig. 12.2c is zero, then the only way Eq. 12.21c would be satisfied is that the voltage drop across *each* resistor in that figure would be zero, and no current can then flow in the circuit. Maxwell's theory, as you will soon find below, would go beyond, even as it *continues to include*, the results summarized in Eq. 12.19 and Eq. 12.21.

Fig. 12.3a corresponds to the situation that is relevant to Eq. 12.21c. This experiment has an electrical circuit, just a loop of a metallic conducting wire, fixed in a *field-free* space. There is no material source of voltage in the circuit apparatus. As expected, no current flows in the circuit and the ammeter shows no deflection. So far so good, but we must now get ready for the surprise that is now just about to come. In the experiment shown schematically in Fig. 12.3b, we have the apparatus placed in a constant uniform magnetic flux density field. The direction of this field is *into* the plane of the figure, represented by $\vec{B}\otimes$. The cross inside the circle in this notation represents an arrow pointing into the plane of the figure. The connector PQ in the circuit (Fig. 12.3b) can be made to slide toward the left, or toward the right. The sliding segment PQ connects to the top and the bottom segments to complete the circuit as shown in Fig. 12.3b. When the connector PQ slides, a current flows in the circuit, and the ammeter shows a deflection. This current *cannot* result from a voltage drop in a battery, there being no battery in the circuit. Instead, it results from the physical movement of the charge q in the segment PQ at a velocity $\vec{v}$, in the presence of a magnetic flux density field. In the present experiment, the charge q in the segment PQ is imparted a motion by *dragging* the circuit-element PQ laterally at a velocity $\vec{v}$. The resulting force is given by

$$\vec{F}_m = q(\vec{v} \times \vec{B}). \tag{12.22}$$

The subscript 'm' on the force in the above equation serves a *dual* purpose: it stands for the fact that it corresponds to the charge in *m*otion, in a *magnetic flux density field* $\vec{B}$. That the force $\vec{F}_m$ would set up a current in the circuit of Fig. 12.3b is now understood by recognizing that a unit positive charge would pick up a horizontal velocity at which the circuit element PQ is dragged toward the right, or left, in the diagram. The charge would then be influenced by the force $\vec{F}_m$. This force would accelerate the charge in the direction of $q(\vec{v} \times \vec{B})$. When the magnetic field is into the plane of the figure, and the segment PQ is dragged to the right, the direction of the force $q(\vec{v} \times \vec{B})$ on a positive charge would be in the direction from the point Q to the point P. The resulting acceleration of charge disturbs the charge balance in the conductor, causing a clockwise current to flow in the circuit. Of course, it is the negatively charged electrons which actually flow in just the opposite direction, but we are referring here to the conventional positive direction of the electric current. This current is detected in the ammeter in Fig. 12.3b. If the segment PQ is dragged to the left, an anticlockwise current would flow, as $q(\vec{v} \times \vec{B})$ will be in the direction from the point P to the point Q. The flow of the electric current in the experiment described in Fig. 12.2a was due to the electric field generated because of the voltage drop across the battery in Fig. 12.2a. The current in the experiment in Fig. 12.3b is due to the motion of the circuit element PQ. The currents in these two situations (Figs. 12.2a and 12.3b) are then a result of the combination of the forces described in Eq. 12.20, and Eq. 12.22, called the *Lorentz force*:

$$\vec{F} = q(\vec{E} + \vec{v} \times \vec{B}). \tag{12.23}$$

The $q(\vec{v} \times \vec{B})$ force is commonly called as the *motional EMF* ('EMF' stands for 'electromotive *force*'). It causes a current to flow in the circuit, and we must therefore acknowledge a source of the electric field to be present in the circuit that would drive the current. It is this source that is called the motional emf. It plays a role similar to a source of a voltage difference in a circuit that would drive a current.

At this point, with regard to the experiment sketched in Fig. 12.3c, we now enter a most astonishing phenomenology in the electromagnetic interaction. In this experiment, all the circuit elements are left alone, completely unmoved. The magnetic flux density *field* $\vec{B}\otimes$ itself is, however, dragged sideways. This can be done, for example, by dragging to the left or right the horse-shoe magnet itself in which the apparatus is placed. Again, on dragging the magnetic flux density field $\vec{B}$ to the left (or right), a current flows in the circuit. This is really an amazing result, considering the fact that there is neither a (*material*) voltage source in the circuit, nor is any circuit element (like the segment PQ in Fig. 12.3b) imparted a physical velocity. There must therefore be a hitherto unreferenced, unseen, source of voltage that drives the current.

This is really startling, since (a) there is no (*material*) voltage source in the circuit (like a battery), and (b) the all the circuit elements are held totally unmoved. *Neither* the voltage drop that may come from the Ohm's and the Kirchhoff's laws, *nor* the Lorentz force (Eq. 12.23) for a charge in motion, therefore can be invoked to account for the current that is detected in the ammeter. There can be no 'motional emf', as there is no motion of any circuit element, yet current flows in the circuit.

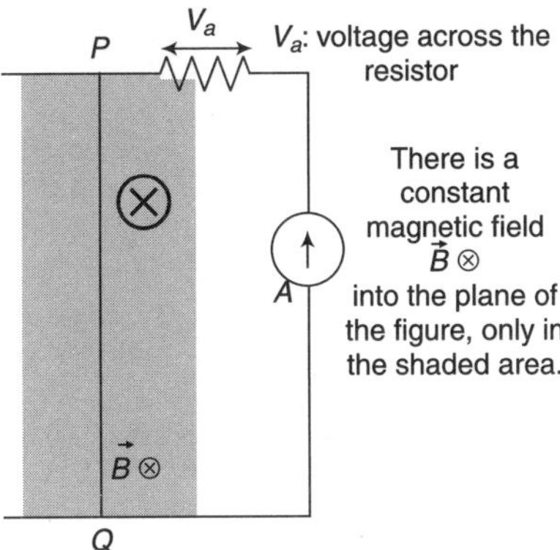

Fig. 12.3a There is no D.C. source in the circuit. By the Kirchhoff's second law, we have $\sum_k V_k = 0$.
As one would expect, no current therefore flows in the circuit and the ammeter shows no deflection.

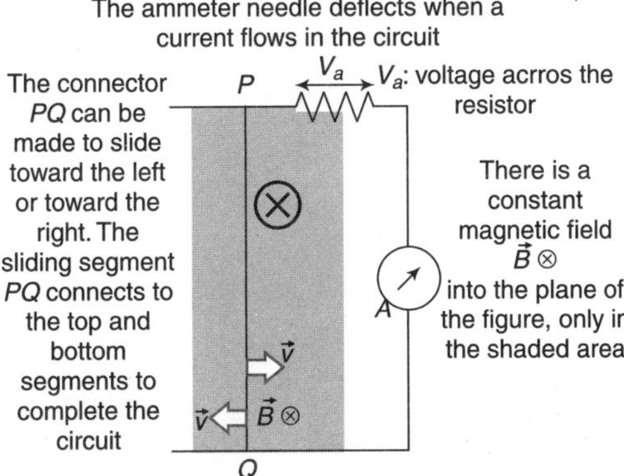

Fig. 12.3b In this experiment, the apparatus is placed in a region of space consisting of a constant magnetic flux density field into the plane of the figure, shown by $\vec{B}\otimes$. The symbol $\vec{B}\otimes$ represents the magnetic field going into the plane of the figure. The segment *PQ* of the circuit can be physically dragged to the right, or to the left. A current flows in the circuit, even if there is no D.C. voltage source in the circuit, when the segment *PQ* is made to slide. When it slides to the right, a clockwise current flows, as $\vec{v} \times \vec{B}$ points along *QP*, from *Q* to *P*. When it slides to the left, anticlockwise current flows, as $\vec{v} \times \vec{B}$ now points from *P* to *Q*.

Finally, in the experiment described in Fig. 12.3d, *neither* is any part of the apparatus moved, *nor* is the magnetic flux density field displaced laterally, and of course, there is absolutely no source like a battery in the circuit. Instead, it is only the strength (magnitude) of the magnetic flux density field that is, however, varied with time. The mechanism of obtaining the time-dependent magnetic field is not important here, but you may think of different ways of achieving this. We are only interested in the *consequences* of the time-varying magnetic field, regardless of how it is generated. There is absolutely no physical motion of any kind between any of the circuit elements and the magnetic flux density field, yet a current flows and is detected in the ammeter.

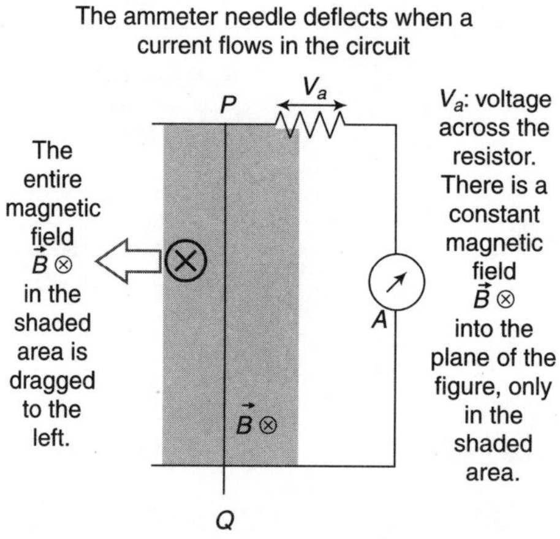

The ammeter needle deflects when a current flows in the circuit
P
V_a
The entire magnetic field $\vec{B}\otimes$ in the shaded area is dragged to the left.
V_a: voltage across the resistor. There is a constant magnetic field $\vec{B}\otimes$ into the plane of the figure, only in the shaded area.
A
$\vec{B}\otimes$
Q

Fig. 12.3c In the experiment shown here, the apparatus is left alone, unmoved, but the magnetic flux density field itself is dragged to the left. This can be done, for example, dragging the magnet which generated the magnetic flux density field in which the apparatus is placed.

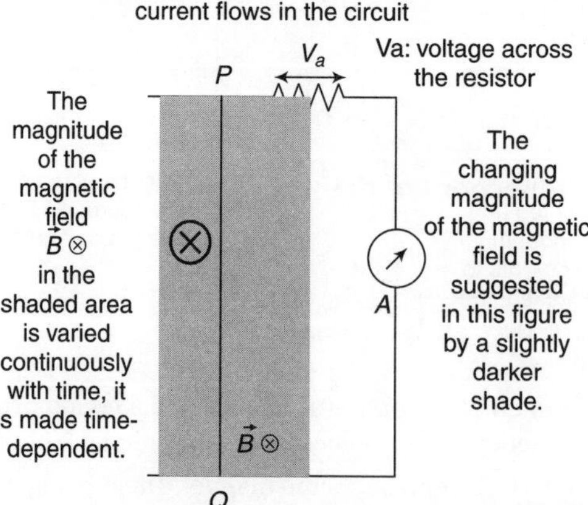

Fig.12.3d In the experiment shown here, there is no movement of the apparatus of the circuit, nor the field. However, the magnetic flux density field is changed as a function of time, suggested by a darker shade in the region of the magnetic field. In this case also, the ammeter shows a deflection, notwithstanding the fact there is no physical movement of any part of the apparatus, or of the magnetic flux density field .

The experiments schematically shown in Figs. 12.3a, b, c, and d contain the quintessence of the trail-blazing experiments done by Michael Faraday (1791–1867) and Emil Khristianovich Lenz (1804–1865). The deflection in the ammeter mentioned in the experiments described in Fig. 12.3c and Fig. 12.3d demands that there has to be a totally *new* driving mechanism which empowers the current to flow. Historically, the driving mechanism which is responsible for the current detected in these experiments is called the *induced EMF* (electromotive force). This is the source of the voltage that would drive the current in these experiments. The induced EMF is *neither* coming from any material device, like a voltage source battery, *nor* from the initial physical *movement* of an electrical charge. It is an effect induced by the time-variation of the magnetic field in the experiment. The time variation of the flux of the magnetic induction field that crosses the circuit is involved in the experiments described in Fig. 12.3c and Fig. 12.3d. The current seen in the experiment described in Fig. 12.3b is accounted for by the Lorentz force law (Eq. 12. 23) in terms of the 'motional emf'. On the other hand, the current in the experiments described in Fig. 12.3c and Fig. 12.3d, is accounted for by a different mechanism, enunciated as empirical laws, prompted by careful observations, and named after Faraday and Lenz who carried out the experiments painstakingly.

Faraday's First Law:

Whenever a conductor is placed in a varying magnetic flux density field, an EMF gets induced across the conductor. If the conductor is in a closed circuit, then an induced current flows through it. This is called Faraday's first law of electromagnetic induction. It is the formal statement that linked a magnetic phenomenon with an electrical phenomenon.

Faraday's Second Law:

Faraday's second law of electromagnetic induction states that, the magnitude of induced EMF is equal to the rate of change of the magnetic induction flux through the circuit.

Lenz's Law:

Lenz's law of electromagnetic induction states that, when an EMF is induced according to Faraday's law, the polarity (direction) of that induced EMF is such that it opposes the cause of its production.

We therefore have an extra-ordinary situation. We use the Lorentz force law (Eq. 12.23) to account for the flow of current in the experiments in Fig. 12.2a, and Fig. 12.3b. However, to account for the experiments in Fig. 12.3c and Fig. 12.3d, we need to invoke the semi-empirical Faraday–Lenz laws (for induced emf) which have absolutely nothing to do with the Lorentz force law (for the motional emf). Richard Feynman [2] points out that there is "... *no other place in physics where such a simple and accurate general principle requires for its real understanding an analysis in terms of two different phenomena*." Maxwell made a giant leap in formulating the laws of electrodynamics by making a bold modification of the historical empirical laws of electricity and magnetism to include the time-dependent phenomena which would *naturally* account for the Faraday–Lenz semi-empirical laws. This is immediately seen in Eq. 12.16c. Taking the surface integral of Eq. 12.16c, we get

$$\iint \vec{\nabla} \times \vec{E}(\vec{r}, t) \cdot \hat{n} dS = -\frac{\partial}{\partial t} \iint \vec{B}(\vec{r}, t) \cdot \hat{n} dS, \tag{12.24a}$$

i.e., using the Kelvin–Stokes theorem (Chapter 10, Eq. 10.39),

$$\oint \vec{E}(\vec{r}, t) \cdot \overrightarrow{d\ell} = -\frac{\partial \Phi_m}{\partial t}, \tag{12.24b}$$

where Φ_m is the magnetic induction *flux* (defined earlier in Chapter 10, Eq. 10.13). Equation 12.24 represents the work done by the *time-dependent* magnetic flux density field to carry a unit positive test charge round the circuit loops, in the experiments depicted in Fig. 12.3c and Fig. 12.3d. The energy for this is drawn from the mechanism that produces the time-dependent magnetic flux density field. This mechanism is therefore capable of setting up a current in the electrical circuit loop even if (i) there is no material device in the apparatus that produces a voltage difference and (ii) the Lorentz force is *not* applicable to account for the induced current. The minus sign on the right hand side of Eq. 12.16c (and hence in Eq. 12.24a and Eq. 12.24b) is most appropriate; it is completely in agreement with the direction of the induced EMF which determines the direction of the induced current in accordance with the Lenz's law mentioned above. The first of the 'curl' equations (Eq. 12.16c) among the equations of Maxwell therefore subsumes the semi-empirical Faraday–Lenz laws. We can regard Eq. 12.16c as Maxwell's incredibly glorious *ansatz* which accounts for the Faraday–Lenz effect. We must, however, get ready to get impressed even more, since Maxwell's equations do lot more than this.

The integral form of the second 'curl' equation of Maxwell gives us the circulation of the magnetic field along a closed path. We consider the application of this equation to a special case in which (i) we have a *steady-state* (i.e., *time-independent*), and (ii) consider a very long (infinitely long), straight conducting wire carrying a constant current 'I'. By symmetry, the circulation of the magnetic flux density field can be easily evaluated along a circular path in a plane perpendicular to the straight current, with the current carrying conductor at the center of this circle. Fig. 12.4a shows the infinite straight conductor, coming *out* of the plane of the figure, directed toward us. In the cylindrical polar coordinates, the current is seen to be coming toward us, from behind the figure. The axis of the cylindrical polar coordinate system is then chosen to be along the direction of the current. The result of the second 'curl' equation (that for the magnetic induction field) among Maxwell's four equations, on using the Stokes-Kelvin theorem (Eq. 12.17d), is

$$B(2\pi\rho) = \mu_0 I, \tag{12.25a}$$

i.e., $$B = \frac{\mu_0}{(2\pi\rho)} I. \tag{12.25b}$$

This is just what we would obtain from the application of the Biot–Savart–Oersted–Ampere law, which was formulated semi-empirically earlier. The Biot–Savart law states that a steady current ('steady' means 'time-independent') I in an infinitesimal current element of segment $d\ell'$ located at a point whose position vector is $\vec{r}'$ (where lies therefore the *source point*) produces a magnetic flux density field at a point P whose position vector is $\vec{r}$ (where lies the *field point*). The net magnetic flux density field at P is therefore obtained from the *integration* of the effects generated by the tiny current source elements in segments $d\ell'$ over the *entire* closed current loop (Fig. 12.4b):

$$\vec{B}(\vec{r}) = \frac{\mu_0}{4\pi} I \oint \frac{\hat{u}(\vec{r}') \times (\vec{r} - \vec{r}')}{\left|\vec{r} - \vec{r}'\right|^3} d\ell' . \tag{12.26}$$

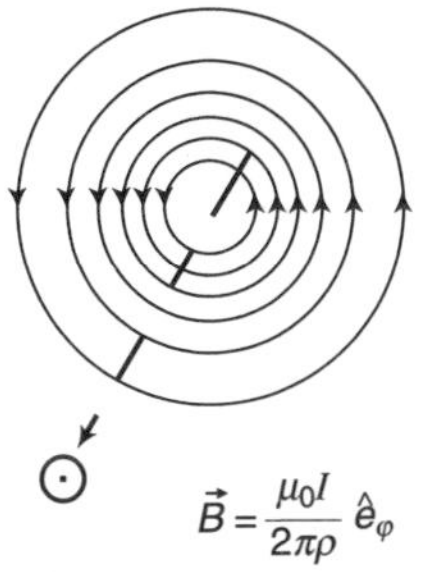

$$\vec{B} = \frac{\mu_0 I}{2\pi\rho}\hat{e}_\varphi$$

Fig. 12.4a An infinitely long straight wire carrying a steady current is perpendicular to the plane of the figure. The direction of the current is toward us. It generates a magnetic induction field around it as shown. (Ampere–Maxwell law).

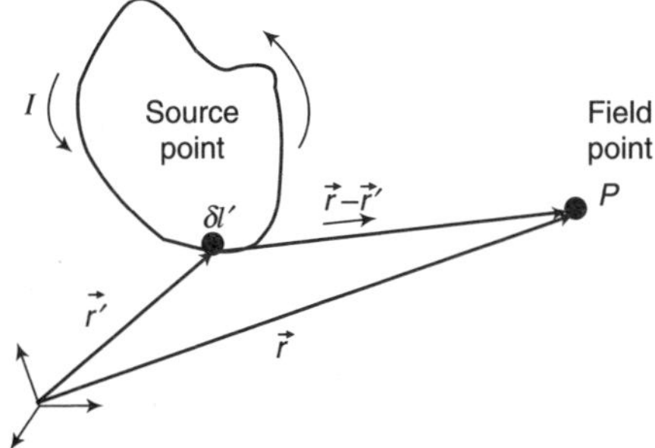

Fig. 12.4b Every tiny element of a current carrying closed loop generates a magnetic induction field at the field point which is, on using the principle of superposition, merely the vector addition of the field generated by all the elements of the integrated loop. (Biot–Savart–Maxwell).

The integration is over the entire current loop. It is important to note that $\hat{u}(\vec{r}')$ is the *direction of the current* at the source element whose position vector is $\vec{r}'$. This would change, in general, from point to point in space depending on the shape of the wire in which the current flows. The result in Eq. 12.25, commonly known as the *Ampere's law*, is the outcome of the Biot–Savart law applied to an infinite straight conductor. Of course, we obtained it from the Maxwell's equations (in particular Eq. 12.17d), which therefore incorporates the semi-empirical Biot–Savart–Oersted–Ampere laws, just as the Maxwell's equations inherently assimilated the Faraday–Lenz laws.

In this chapter, we have not followed the path of historical development of the subject. The Biot–Savart law was formulated in 1820, and the Ampere's law in 1823. Maxwell's equations, in their final form, were published in 1873. Maxwell's equations provide backward integration of the Biot–Savart–Oersted–Ampere laws, and also the Faraday–Lenz laws. The four equations of Maxwell provide the most general statement of classical electrodynamics, making it redundant to learn the Faraday–Lenz or the Biot–Savart–Oersted–Ampere laws separately. We have seen above how the '*curl*' equations subsume the semi-empirical laws that were known earlier. Similarly, Maxwell's equation for the divergence of the electric field automatically incorporates the earlier knowledge of the semi-empirical Coulomb's law, which was formulated much earlier, in 1785. This is easily seen from the Maxwell's Eq. 12.17a, which is nothing but the integral form of Eq. 12.16a. Applying this to the special case in which a point charge q at the center of a sphere, we get, exploiting spherical symmetry of the electric intensity field that would be generated by the point charge,

$$\oiint \vec{E}(\vec{r}, t)\cdot \hat{n} dS = E 4\pi r^2 = \frac{q}{\varepsilon_0}, \tag{12.27a}$$

which means that

$$E = \frac{q}{4\pi\varepsilon_0 r^2}. \tag{12.27b}$$

The above result is essentially the statement of the Coulomb's inverse square law. The intensity is just the force per unit charge, hence the force on a charge Q would be

$$F = \frac{Qq}{4\pi\varepsilon_0 r^2}. \tag{12.27c}$$

Equation 12.27c is perhaps the most familiar, and popular, form in which the Coulomb's law is seen.

The Maxwell's equations contain within them the sum and the substance of *all* the semi-empirical laws of electricity and magnetism, namely the (i) Coulomb's law, (ii) the Biot–Savart–Oersted–Ampere's laws, and the (iii) Faraday–Lenz laws, as discussed above. The *other* 'divergence' equation of Maxwell, viz., Eq. 12.16b (and its integral form in Eq. 12.17b) dismisses the existence of a magnetic monopole. It essentially means that there is no magnetic analogue of the electrical charge, which appears on the right hand side of Eq. 12.16a. By including this in the scheme of his four equations, Maxwell effected the subliminal inclusion of the momentous role of *symmetry* in the laws of nature, apart from providing the *required* divergence and the curl of both the fields, $\vec{E}(\vec{r}, t)$ and $\vec{B}(\vec{r}, t)$. Maxwell's equations integrate the electromagnetic phenomena, and thus constitute the first unified field theory, which is the forerunner of subsequent unification formalisms. Maxwell's equations also laid the seeds of the special theory of relativity, as we shall see in the next chapter, developed by Albert Einstein about four decades after Maxwell's work.

12.3 THE ELECTROMAGNETIC POTENTIALS

Solutions to the Maxwell's equations provide the electromagnetic fields. These may then be used in the Lorentz force law (Eq. 12.23) for applications in solving the equation of motion for a charged particle in an electromagnetic field. From the two theorems we discussed in Section 1, we know that the electromagnetic vector fields can be determined, using appropriate boundary conditions, from the Maxwell's equations which provide the curl and the divergence of both the vector fields. Furthermore, from the Helmholtz theorem, we can write each field as a sum of a solenoidal field, and an irrotational field.

We know that the divergence of the magnetic flux density field $\vec{B}$ itself is always zero. It can therefore be written as the curl of a vector field, as the divergence of the curl of a vector is necessarily solenoidal. Thus,

$$\vec{B}(\vec{r},t) = \vec{\nabla} \times \vec{A}(\vec{r},t). \tag{12.28}$$

The magnetic vector induction field is $\vec{B}(\vec{r},t)$ therefore *derivable* from the vector field $\vec{A}(\vec{r},t)$. The vector field is called the *magnetic vector potential field*. Using Eq. 12.28 in the Maxwell's equation (Eq. 12.16c) for the curl of the electric field intensity, we get:

$$\vec{\nabla} \times \vec{E}(\vec{r},t) = -\frac{\partial \vec{B}(\vec{r},t)}{\partial t} = -\frac{\partial [\vec{\nabla} \times \vec{A}(\vec{r},t)]}{\partial t} = -\vec{\nabla} \times \frac{\partial \vec{A}(\vec{r},t)}{\partial t}, \tag{12.29a}$$

i.e.,
$$\vec{\nabla} \times \left[\vec{E}(\vec{r},t) + \frac{\partial \vec{A}(\vec{r},t)}{\partial t} \right] = \vec{0}. \tag{12.29b}$$

In other words, $\left\{ \vec{E}(\vec{r},t) + \frac{\partial \vec{A}(\vec{r},t)}{\partial t} \right\}$ constitutes an irrotational vector field. Since the curl of the gradient of a vector field is identically zero, we may write

$$\vec{E}(\vec{r},t) + \frac{\partial \vec{A}(\vec{r},t)}{\partial t} = -\vec{\nabla}\phi(\vec{r},t), \tag{12.30a}$$

i.e.,
$$\vec{E}(\vec{r},t) = -\left[\vec{\nabla}\phi(\vec{r},t) + \frac{\partial \vec{A}(\vec{r},t)}{\partial t} \right]. \tag{12.30b}$$

We find that the electric field intensity is *derivable* from a *combination* of the (negative of) $\frac{\partial \vec{A}(\vec{r},t)}{\partial t}$, the time-derivative of the magnetic vector potential, and from the (negative of) $\vec{\nabla}\phi$, the (negative) gradient of the electric scalar potential $\phi(\vec{r},t)$. In the case of *steady-state* electrodynamics, i.e., when the electromagnetic phenomena are *independent of time*, the electric field intensity $\vec{E}(\vec{r},t)$ is derivable from the electric scalar function alone, as its the (negative) gradient, and it is irrotational. In *time-dependent* phenomena, it is rather the *sum* $\left[\vec{E}(\vec{r},t) + \frac{\partial \vec{A}(\vec{r},t)}{\partial t} \right]$ that is irrotational.

We have now seen that the electromagnetic field is derivable from the electromagnetic (scalar and the vector) potentials $\{\phi(\vec{r}, t), \vec{A}(\vec{r}, t)\}$. One can determine the fields from the potentials, and inversely, with appropriate boundary conditions, one can determine the potentials from the fields, by exercising the inverse mathematical operation of integration. The importance of the potentials, however, emerges not merely from the convenience in carrying out some mathematical operations, but go beyond that. The potentials acquire tangible credibility as one proceeds to apply them in quantum mechanical processes. Much more than the fields, it is the potentials which become physical quantities of great merit. This is dramatically revealed in the quantum Aharonov–Bohm effect [2,4,5]. The Aharonov–Bohm effect is well beyond the scope of this book. We therefore continue to discuss some other rudimentary aspects of the electromagnetic potentials.

We have mentioned above that while the electromagnetic fields appear in the Lorentz force equation (Eq. 12.23) which can be used in the equation of motion (such as Newton's). Nonetheless, a given electromagnetic field is derivable from alternative sets of potentials. The alternatives come from the fact that the magnetic induction field $\vec{B}(\vec{r}, t)$ in Eq. 12.28 remains invariant if we changed the vector potential as

$$\vec{A}(\vec{r}, t) \to \vec{A}(\vec{r}, t) - \vec{\nabla}\psi, \tag{12.31a}$$

where ψ is an arbitrary scalar field. However, this would affect the electric intensity field, which becomes, using Eq. 12.30b:

$$\vec{E}(\vec{r}, t) = -\left[\vec{\nabla}\phi(\vec{r}, t) + \frac{\partial\{\vec{A}(\vec{r}, t) - \vec{\nabla}\psi\}}{\partial t}\right] = -\left[\vec{\nabla}\left\{\phi(\vec{r}, t) - \frac{\partial\psi}{\partial t}\right\} + \frac{\partial\{\vec{A}(\vec{r}, t)\}}{\partial t}\right].$$

Hence, a transformation of the scalar field, given by

$$\phi(\vec{r}, t) \to \phi(\vec{r}, t) - \frac{\partial\psi(\vec{r}, t)}{\partial t}, \tag{12.31b}$$

would leave the electric intensity field invariant.

In other words, the pair of the potentials, $\{\phi(\vec{r}, t), \vec{A}(\vec{r}, t)\}$, from which the field $\{\vec{E}(\vec{r}, t), \vec{B}(\vec{r}, t)\}$ are derivable (using Eq. 12.28 and Eq. 12.30) is not unique. The transformations of the electromagnetic potentials given in Eq. 12.31a and 12.31b leave the fields unchanged. Hence the natural question we are prompted to ask is how should we choose the potentials? Now the vector field $\vec{A}(\vec{r}, t)$ is determined by its curl and its divergence, of which the $\vec{\nabla} \times \vec{A}(\vec{r}, t)$ must be chosen such that it gives the magnetic induction field $\vec{B}(\vec{r}, t)$, as per Eq. 12.28. We must now decide how to choose the $\vec{\nabla} \cdot \vec{A}(\vec{r}, t)$. Our primary search is for appropriate potentials $\{\phi(\vec{r}, t), \vec{A}(\vec{r}, t)\}$ that are consistent with the Maxwell's equations. By now, we have only fixed one property of these potentials, namely that the $\vec{\nabla} \times \vec{A}(\vec{r}, t) = \vec{B}(\vec{r}, t)$. To determine the electromagnetic potentials which are consistent with the Maxwell's equations, we must find the differential equations that are satisfied by $\{\phi(\vec{r}, t), \vec{A}(\vec{r}, t)\}$. The primary challenge for this comes from the fact that Maxwell's *field* equations are *coupled* partial differential equations. The partial differential equations for the electromagnetic *potentials also* are coupled. In order to solve them, one must therefore employ some special tactics.

Taking the negative divergence (i.e., with a minus sign) of the electric field intensity, expressed as Eq. 12.30b, and using Maxwell's first equation (Eq. 12.16a), we have

$$(\vec{\nabla}\cdot\vec{\nabla})\phi(\vec{r},t)-\frac{\partial}{\partial t}\{\vec{\nabla}\cdot\vec{A}(\vec{r},t)\}=-\frac{\rho}{\varepsilon_0}. \tag{12.32}$$

In the above differential equation for the electric scalar potential $\phi(\vec{r},t)$, we have a term in the time-derivative of magnetic vector potential. Let us see what kind of differential equation we shall get for the magnetic vector potential. From Eq. 12.28, we have

$$\vec{\nabla}\times\vec{B}=\vec{\nabla}\times(\vec{\nabla}\times\vec{A})=\vec{\nabla}(\vec{\nabla}\cdot\vec{A})-(\vec{\nabla}\cdot\vec{\nabla})\vec{A}. \tag{12.33a}$$

The above relation must of course agree with Maxwell's fourth equation (Eq. 12.16d):

$$\vec{\nabla}\times\vec{B}(\vec{r},t)=\mu_0\left(\vec{J}(\vec{r},t)+\varepsilon_0\frac{\partial\vec{E}(\vec{r},t)}{\partial t}\right)=\mu_0\vec{J}(\vec{r},t)-\mu_0\varepsilon_0\frac{\partial}{\partial t}\left\{\vec{\nabla}\phi(\vec{r},t)+\frac{\partial\vec{A}(\vec{r},t)}{\partial t}\right\}. \quad 12.33b$$

Hence, equating the right hand sides of Eq. 12.32a/33a and Eq. 12.32b/33b,

$$\vec{\nabla}(\vec{\nabla}\cdot\vec{A})-(\vec{\nabla}\cdot\vec{\nabla})\vec{A}=\mu_0\vec{J}(\vec{r},t)-\mu_0\varepsilon_0\frac{\partial}{\partial t}\left\{\vec{\nabla}\phi(\vec{r},t)+\frac{\partial\vec{A}(\vec{r},t)}{\partial t}\right\}. \tag{12.34}$$

This time around, we got a differential equation for the vector potential, but it has a term in the scalar potential as well. The second order partial differential equations (Eq. 12.32 and 12.34) for the electromagnetic potentials are thus coupled, and not easy to solve. In order to develop procedures to decouple them, we first merely re-write Eq. 12.34 as follows:

$$\left[(\vec{\nabla}\cdot\vec{\nabla})\vec{A}-\mu_0\varepsilon_0\frac{\partial^2\vec{A}(\vec{r},t)}{\partial t^2}\right]-\vec{\nabla}\left[(\vec{\nabla}\cdot\vec{A})+\mu_0\varepsilon_0\frac{\partial\phi(\vec{r},t)}{\partial t}\right]=-\mu_0\vec{J}(\vec{r},t). \tag{12.35}$$

The vector field $\vec{A}(\vec{r},t)$ requires both its curl and the divergence to be specified (Helmholtz theorems), of which its curl must give us the magnetic flux density field $\vec{B}$. We have already made that choice; there is no escape from it now. Two distinctive preferences for the divergence of the magnetic vector potential can, however, be fruitfully exercised. These options provide alternative *measures* of the divergence, $\vec{\nabla}\cdot\vec{A}$. Since 'to measure' is 'to gauge', these are referred to as alternative gauges. The two gauges are

(i) Lorenz gauge: $(\vec{\nabla}\cdot\vec{A})=-\mu_0\varepsilon_0\frac{\partial\phi(\vec{r},t)}{\partial t}$ i.e., $(\vec{\nabla}\cdot\vec{A})+\mu_0\varepsilon_0\frac{\partial\phi(\vec{r},t)}{\partial t}=0,$ (12.36)

(ii) Coulomb gauge: $\vec{\nabla}\cdot\vec{A}=0.$ (12.37)

In the Coulomb gauge, the term in the magnetic vector potential in the differential equation for the electric scalar potential (Eq. 12.32) drops out, and thus readily provides a decoupled equation for the scalar potential:

$$(\vec{\nabla}\cdot\vec{\nabla})\,\phi(\vec{r},t)=-\frac{\rho(\vec{r},t)}{\varepsilon_0}. \tag{12.38a}$$

This is just what one would get for the steady-state phenomena, and corresponds to the *instantaneous* Coulomb's law. It is for this reason that the choice represented by Eq. 12.37 is called the Coulomb gauge.

However, the potential at the field point is affected by changes in the charge density at the source point instantly, at infinite speed. This would seem to violate the relativistic limits (discussed in the next chapter), but the dilemma seen here is only an apparent one; the physical influence is perceptible not from the potential, but from the electric field intensity (Eq. 12.30b). The latter has a time-dependence, which prevents any violation of the relativistic limit.

The equation for the magnetic vector potential, Eq. 12.35, in the Coulomb gauge becomes

$$\Box^2 \vec{A}(\vec{r},t) = -\mu_0 \vec{J}(\vec{r},t) + \mu_0 \varepsilon_0 \vec{\nabla}\left\{\frac{\partial \phi(\vec{r},t)}{\partial t}\right\}, \tag{12.38b}$$

where the 'box' operator is

$$\vec{\nabla}\cdot\vec{\nabla} - \mu_0 \varepsilon_0 \frac{\partial^2}{\partial t^2} \equiv \Box^2. \tag{12.39}$$

The differential equation for the magnetic vector potential thus continues to have in it the (time-derivative of) electric scalar potential. It therefore remains difficult to solve when time-dependence of the potentials is of importance. Nonetheless, in time-independent phenomena, the second term on the right hand side of Eq. 12.38b vanishes, and this brings in some comfort.

In the Lorenz gauge, named after Ludvig Valentin Lorenz (1829-1891), using Eq. 12.36 in Eq. 12.32, we get

$$\left[(\vec{\nabla}\cdot\vec{\nabla}) - \mu_0 \varepsilon_0 \frac{\partial^2}{\partial t^2}\right] \phi(\vec{r},t) = -\frac{\rho}{\varepsilon_0}, \tag{12.40a}$$

and Eq. 12.35 becomes

$$\left[(\vec{\nabla}\cdot\vec{\nabla}) - \mu_0 \varepsilon_0 \frac{\partial^2}{\partial t^2}\right] \vec{A}(\vec{r},t) = -\mu_0 \vec{J}(\vec{r},t). \tag{12.40b}$$

The Lorenz gauge is sometimes called, perhaps partially inappropriately, as the Lorentz gauge. The partial rationale comes from the fact is that the Lorenz gauge is Lorentz invariant. We shall meet the Lorentz transformations in the next chapter.

Both the differential equations (12.40a and 12.40b) for the electric scalar potential and the magnetic vector potential have become decoupled. We can write both of the above equations in a compact form using the 'box' operator:

$$\Box^2 \phi(\vec{r},t) = -\frac{\rho(\vec{r},t)}{\varepsilon_0}, \tag{12.41a}$$

$$\Box^2 \vec{A}(\vec{r},t) = -\mu_0 \vec{J}(\vec{r},t). \tag{12.41b}$$

We shall now restrict ourselves to steady-state (i.e., time-independent) phenomena. For time-varying fields, one may use the generalized Helmholtz theorem [3]. For steady-state electromagnetic processes, Maxwell's equations for the vorticity of the electric field intensity, and of the magnetic flux density field, reduce to the following:

$$\vec{\nabla} \times \vec{E} = \vec{0}, \tag{12.42a}$$

and $\vec{\nabla} \times \vec{B} = \mu_0 \vec{J}.$ (12.42b)

The differential equations for the electric scalar potential and the magnetic vector potential are then given by

$$(\vec{\nabla} \cdot \vec{\nabla})\phi(\vec{r}) = -\frac{\rho(\vec{r})}{\varepsilon_0}, \tag{12.43a}$$

and $(\vec{\nabla} \cdot \vec{\nabla})\, \vec{A}(\vec{r}) = \mu_0 \vec{J}(\vec{r}).$ (12.43b)

The above equations are known as Poisson's equations, after Simeon Denis Poisson (1781–1840), who would say that "*Life is good for only two things, discovering mathematics, and teaching mathematics.*" When the right hand sides of the Poisson's equations are zero, as would happen in the absence of the charge and current density, they reduce to what are known as Laplace's equation:

$$(\vec{\nabla} \cdot \vec{\nabla})\phi = 0, \tag{12.44a}$$

$$(\vec{\nabla} \cdot \vec{\nabla})\, \vec{A} = \vec{0}, \tag{12.44b}$$

We assert now, with the guarantee that we shall *verify* this assertion very soon (below), that the solution to the Poisson equation for the electric scalar potential (Eq. 12.44a) is given by

$$\phi(\vec{r}) = \frac{1}{4\pi\varepsilon_0} \iiint \frac{\rho(\vec{r}\,')}{|\vec{r} - \vec{r}\,'|} dV' = \frac{1}{4\pi\varepsilon_0} \iiint \frac{\rho(\vec{r}\,')}{R'} dV', \tag{12.45}$$

wherein $R' = |\vec{r} - \vec{r}\,'| = [(\vec{r} - \vec{r}\,')\cdot(\vec{r} - \vec{r}\,')]^{\frac{1}{2}},$ (12.46)

is the distance between the source point and the field point (Fig. 12.1).

Likewise, the solution to the Poisson equation for the magnetic vector potential (Eq. 12.44b) is

$$\vec{A}(\vec{r}) = \frac{\mu_0}{4\pi} \iiint \frac{\vec{J}(\vec{r}\,')}{|\vec{r} - \vec{r}\,'|} dV' = \frac{\mu_0}{4\pi} \iiint \frac{\vec{J}(\vec{r}\,')}{R'} dV'. \tag{12.47}$$

A steady state volume current density vector $\vec{J}(\vec{r}')$ is presumed in the above solution. In the case of a line-current, the corresponding solution would be

$$\vec{A}(\vec{r}) = \frac{\mu_0 I}{4\pi}\oint \frac{dl'(\vec{r}')[\hat{i}(\vec{r}')]}{|\vec{r} - \vec{r}'|}. \tag{12.48}$$

In the above equation, $\hat{i}(\vec{r}')$ is a unit vector in the direction of the current at the source point $(\vec{r}')$ located in the line-current. In writing these expressions, it may be tempting to express the current as a *vector*, since it of course has a magnitude, and also a direction. In this book, we however choose *not* to do that, since we cannot add two currents at a node (such as in the Kirchhoff's circuit, Fig. 12.2b) using the laws of addition of vectors. We therefore indicate the direction of the current flow by the unit vector $\hat{i}(\vec{r}')$ and not as $\frac{\vec{I}(\vec{r}')}{|\vec{I}(\vec{r}')|}$. The latter cannot really be defined. Finally, when the current flow is along a surface, we have

$$\vec{A}(\vec{r}) = \frac{\mu_0}{4\pi}\iint \frac{K(r')\hat{i}\,dS'}{|\vec{r} - \vec{r}'|}. \tag{12.49}$$

As promised, we now show that Eq. 12.45 indeed provides the solution to the Poisson's equation. Operating on both sides of Eq. 12.45 by the Laplacian operator $\vec{\nabla}\cdot\vec{\nabla} = \vec{\nabla}^2$, we get

$$(\vec{\nabla}\cdot\vec{\nabla})\phi(\vec{r}) = \frac{1}{4\pi\varepsilon_0}\iiint (\vec{\nabla}\cdot\vec{\nabla})\frac{\rho(\vec{r}')}{|\vec{r} - \vec{r}'|}dV' = \frac{1}{4\pi\varepsilon_0}\iiint \rho(\vec{r}')(\vec{\nabla}\cdot\vec{\nabla})\left\{\frac{1}{|\vec{r} - \vec{r}'|}\right\}dV', \tag{12.50a}$$

Using now the Dirac-δ function (Chapter 11, Eq. 11.41), we can immediately see that Eq. 12.45 is the solution to the Poisson equation, since,

$$(\vec{\nabla}\cdot\vec{\nabla})\phi(\vec{r}) = \frac{1}{4\pi\varepsilon_0}\iiint \rho(\vec{r}')[-4\pi\delta^3(\vec{r} - \vec{r}')]dV' = -\frac{\rho(\vec{r})}{\varepsilon_0}, \tag{12.50b}$$

which is *exactly* that we needed to ascertain.

That the expressions given in Eqs. 12.47, 12.48 and 12.49 are indeed the solutions to the Poisson equation for the corresponding magnetic vector potentials can also be seen similarly, using the Dirac-δ integration. However, in general the integrals in the solutions to the Poisson's equation are not easy to solve. What comes in handy is an approximation scheme. It works extremely efficiently in most situations, especially when our interest lies in obtaining the potentials and fields at points which are sufficiently far from their source. To develop the approximation scheme, we note the fact that the solutions to the Poisson's equation, both for the electric scalar potential and the magnetic vector potential, have the inverse of the distance R' (Eq. 12.46) between the source point and the field point. Subsequent integration over the entire source region is of course required. The efficacy of the solution comes from the fact that the in all of these solutions, the inverse distance can be written as a series in increasing powers of $\frac{r'}{r}$, which is the ratio of the distance of the source point from the origin of the coordinate frame of reference, to the distance of the field point from the origin. At field points that are sufficiently far away, as seen below, the powers series converges, often rapidly, making it relatively easy to determine the potentials, even if only approximately.

Let us consider a source of the potential that is distributed over a region, such as shown in Fig. 12.5.

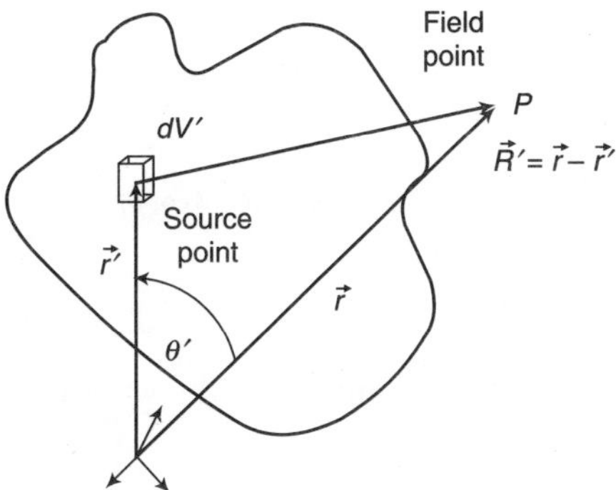

Fig. 12.5 When the field point is far away from the source region, its inverse distance from any point in the source region is given by the power series expansion in Eq. 12.54. Often, the first few terms suffice.

The distance $R' = |\vec{r} - \vec{r}'|$ between the source point and the field point is given by

$$R'^2 = (\vec{r} - \vec{r}')^2 = r^2\left[1 + \frac{r'^2}{r^2} - 2\frac{r'}{r}\cos\theta'\right] = r^2\left[1 + \frac{r'}{r}\left\{\frac{r'}{r} - 2\cos\theta'\right\}\right] = r^2(1+\xi), \quad (12.51)$$

where $$\xi = \frac{r'}{r}\left(\frac{r'}{r} - 2\cos\theta'\right). \quad (12.52)$$

Hence, $$(R')^{-1} = r^{-1}(1+\xi)^{-\frac{1}{2}} = r^{-1}\left(1 - \frac{1}{2}\xi + \frac{3}{8}\xi^2 - \frac{5}{16}\xi^3 + \cdots\right). \quad (12.53)$$

We see that the inverse-distance between the source point and the field point is a power series in the ratio $\frac{r'}{r}$. Coefficients in this expansion include various functions of $\cos\theta'$, where θ' is the angle between the position vector, $\vec{r}$, of the field point, and that of the source point, $\vec{r}'$ (Fig. 12.5). The functions of $\cos\theta$ which are involved are called the Legendre polynomials [1,4]. We shall not discuss the several interesting and important properties of the Legendre polynomials, denoted as $P_\ell(\cos\theta')$ for $\ell = 0, 1, 2, 3, \cdots$. We refer the reader to some other sources on mathematical physics, such as the Reference [1]. However, we do list, in Table 12.1, the first few Legendre polynomials that are of immediate interest. For $r \gg (a = r'_{\max})$, $\xi \ll 1$, we can therefore write the inverse distance as

$$(R')^{-1} = \frac{1}{|\vec{r} - \vec{r}'|} = \frac{1}{r}\sum_{\ell=0}^{\infty}\left(\frac{r'}{r}\right)^{\ell} P_\ell(\cos\theta'). \quad (12.54)$$

Table 12.1 Legendre Polynomials.

ℓ	$P_\ell(\cos\theta)$
0	1
1	$\cos\theta$
2	$\frac{1}{3}(3\cos^2\theta - 1)$
3	$\frac{1}{2}(5\cos^3\theta - 3\cos\theta)$
...	...

Substituting from Eq. 12.53 in Eq. 12.44, and using the values of the Legendre polynomials [1] in Table 12.1, we get

$$\phi(\vec{r}) = \frac{1}{4\pi\varepsilon_0}\begin{bmatrix} \frac{1}{r}\iiint \rho(\vec{r}\,')(r')^0 P_0(\cos\theta')dV' + \frac{1}{r^2}\iiint \rho(\vec{r}\,')(r')^1 P_1(\cos\theta')dV' + \\ \frac{1}{r^3}\iiint \rho((\vec{r}\,'))(r')^2 P_2(\cos\theta')dV' + \\ \sum_{\ell=3}^{\infty}\frac{1}{r^{\ell+1}}\iiint \rho(\vec{r}\,')(r')^\ell P_\ell(\cos\theta')dV' \end{bmatrix}, \quad (12.55a)$$

$$\text{i.e.,}\quad \phi(\vec{r}) = \frac{1}{4\pi\varepsilon_0}\frac{1}{r}\begin{bmatrix} Q + \iiint \rho(\vec{r}\,')\left(\frac{r'}{r}\right)^1 P_1(\cos\theta')dV' + \\ \iiint \rho(\vec{r}\,')\left(\frac{r'}{r}\right)^2 P_2(\cos\theta')dV + \\ \sum_{\ell=3}^{\infty}\iiint \rho(\vec{r}\,')\left(\frac{r'}{r}\right)^\ell P_\ell(\cos\theta')dV' \end{bmatrix}. \quad (12.55b)$$

The first term, $\frac{1}{4\pi\varepsilon_0}\frac{Q}{r}$, provides the monopole term, which gives the potential at the field point as would be generated by the total integrated charge of the charge distribution located at the origin of the coordinate system. The second term, likewise, corresponds to the potential at the field point as would be generated by a point-sized electric dipole, placed at the origin of the coordinate frame of reference (Fig. 12.6). It is given by

$$V_{\text{dipole}} = \frac{1}{4\pi\varepsilon_0}\frac{1}{r^2}\iiint \rho(\vec{r}\,')(r')^1 P_1(\cos\theta')dV' = \frac{1}{4\pi\varepsilon_0}\frac{1}{r^2}\iiint \rho(\vec{r}\,')r'\cos\theta'\,dV', \quad (12.56a)$$

$$\text{i.e.,}\quad V_{\text{dipole}} = \frac{1}{4\pi\varepsilon_0}\frac{1}{r^2}\iiint \rho(\vec{r}\,')(\hat{e}_r\cdot\vec{r}\,')dV' = \frac{1}{4\pi\varepsilon_0}\frac{\hat{e}_r\cdot\iiint \vec{r}\,'\rho(\vec{r}\,')dV'}{r^2}. \quad (12.56b)$$

Hence, $$V_{\text{dipole}}(\vec{r}) = \frac{1}{4\pi\varepsilon_0}\frac{\hat{e}_r \cdot \vec{p}_d}{r^2}, \tag{12.56c}$$

where $\vec{p}_d = \iiint \vec{r}' \rho(\vec{r}')dV'$ is the dipole moment of the charge distribution in the volume over which the integration is carried out. Subsequent terms in the infinite series in Eq. 12.55 have similar interpretations. These terms become progressively less important, since for large distances of the field point, $\left(\frac{r'}{r}\right)^{\ell}$ diminishes and becomes progressively weaker. As many terms can be retained as may be required for the accuracy that is desired.

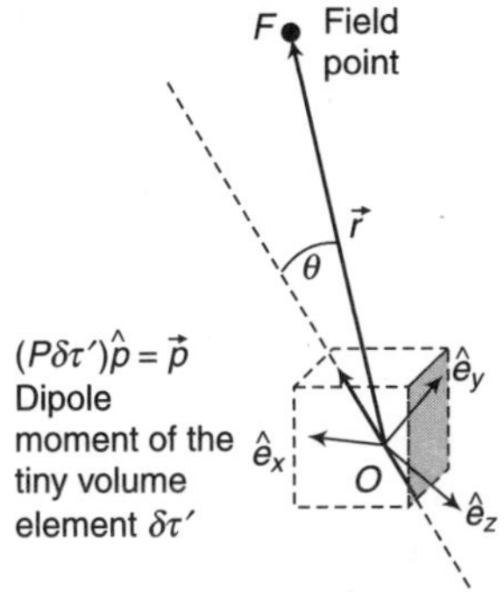

Fig. 12.6 The potential at the field point F whose position vector is due to a point sized dipole moment placed at the origin of a coordinate system $\frac{1}{4\pi\varepsilon_0}\frac{\hat{e}_r \cdot \vec{p}_d}{r^2}$.

We can of course apply the expansion of the one-over-distance term (Eq. 12.54) in the expressions for the magnetic vector potentials also, given in Eqs. 12.47, 48 and 49. These three forms correspond respectively to the three cases when we have a volume, line, or a surface current.

For simplicity, we shall illustrate the approximation scheme only for the case of the magnetic vector potential generated by the *line* current. Using Eq. 12.54 in Eq. 12.48, we get

$$\vec{A}(\vec{r}) = \frac{\mu_0 I}{4\pi}\frac{1}{r}\oint dl'(\vec{r}')[\hat{i}(\vec{r}')]\sum_{\ell=0}^{\infty}\left(\frac{r'}{r}\right)^{\ell} P_\ell(\cos\theta'), \tag{12.57a}$$

i.e., $$\vec{A}(\vec{r}) = \frac{\mu_0 I}{4\pi}\left[\begin{array}{l}\frac{1}{r}\oint (r')^0 P_0(\cos\theta')\overrightarrow{dl'} + \frac{1}{r^2}\oint (r')^1 P_1(\cos\theta')\overrightarrow{dl'} + \\ \sum_{\ell=2}^{\infty}\frac{1}{r^{\ell+1}}\oint (r')^{\ell} P_\ell(\cos\theta')\overrightarrow{dl'}\end{array}\right]. \tag{12.57b}$$

We now look closely at the first few of the above terms, for different values of ℓ.

For $\boxed{\ell = 0}$ $$\frac{\mu_0 I}{4\pi}\frac{1}{r}\oint (r')^0 P_0(\cos\theta')\overrightarrow{dl'} = \vec{0}. \tag{12.58}$$

This corresponds to the Maxwell's equation for the divergence of the magnetic flux density field, which essentially vanishes; there being no magnetic monopole analogue of the electric charge.

$$\text{For } \boxed{\ell = 1} \quad \frac{\mu_0 I}{4\pi}\frac{1}{r^2}\oint (r')^1 P_1(\cos\theta)\overrightarrow{dl'} = \frac{\mu_0 I}{4\pi}\frac{1}{r^2}\oint r' \cos\theta\, \overrightarrow{dl'} = \frac{\mu_0 I}{4\pi}\frac{1}{r^2}\oint (\hat{r}\cdot\vec{r}')\overrightarrow{dl'}. \tag{12.59}$$

In order to interpret this term and get an insight into it, we use a vector identity:

$$\vec{\nabla}'(\vec{A}\cdot\vec{B}) = \vec{A}\times(\vec{\nabla}'\times\vec{B}) + \vec{B}\times(\vec{\nabla}'\times\vec{A}) + (\vec{A}\cdot\vec{\nabla}')\vec{B} + (\vec{B}\cdot\vec{\nabla}')\vec{A}. \tag{12.60a}$$

The prime on the gradient operator denotes the fact that the partial differential operators in the gradient would operate only on the primed coordinates, which are the coordinates of the source point (Fig. 12.5). Using now this identity for the integrand in Eq. 12.59,

$$\vec{\nabla}'(\hat{r}\cdot\vec{r}') = \hat{r}\times(\vec{\nabla}'\times\vec{r}') + \vec{r}'\times(\vec{\nabla}'\times\hat{r}) + (\hat{r}\cdot\vec{\nabla}')\vec{r}' + (\vec{r}'\cdot\vec{\nabla}')\hat{r} = (\hat{r}\cdot\vec{\nabla}')\vec{r}'. \tag{12.60b}$$

$$\text{Hence,}\quad \vec{\nabla}'(\hat{r}.\vec{r}') = (\hat{r}\cdot\vec{\nabla}')\vec{r}' = \left(\hat{r}_x\frac{\partial}{\partial x'} + \hat{r}_y\frac{\partial}{\partial y'} + \hat{r}_z\frac{\partial}{\partial z'}\right)(x'\hat{e}_x + y'\hat{e}_y + z'\hat{e}_z) = \hat{r}. \tag{12.60c}$$

Note that in the first bracket, the scalar components of the unit vector that points in the direction of the field point appear. This unit vector, then, appears on the right side of Eq.12.60c. The ancillary result that we have obtained above plays an important role in understanding the term for $\ell = 1$. To appreciate this point, we now consider a vector $\vec{\Gamma} = (\hat{r}\cdot\vec{r})\vec{\kappa}$, where $\vec{\kappa}$ is an arbitrary constant vector. From the Kelvin–Stokes theorem (Chapter 10, Eq. 10.39), we have

$$\iint \vec{\nabla}\times\{(\hat{r}.\vec{r}')\vec{\kappa}\}\cdot\hat{n}dS = \oint\{(\hat{r}.\vec{r}')\vec{\kappa}\}\cdot\overrightarrow{d\ell}, \tag{12.61a}$$

$$\text{i.e.,}\quad \iint (\hat{r}.\vec{r}')\vec{\nabla}\times\vec{\kappa})\cdot\hat{n}dS - \iint \vec{\kappa}\times\vec{\nabla}(\hat{r}.\vec{r}')\cdot\hat{n}dS = \oint\{(\hat{r}.\vec{r}')\vec{\kappa}\}\cdot\overrightarrow{d\ell}. \tag{12.61b}$$

Since the curl of a constant vector is always zero, the first term on the left hand side vanishes. We can interchange the positions of the dot-product, and the cross-product, in the scalar triple product that appears in the integrand of the second term. Further, recognizing the fact that $\vec{\kappa}$ is a constant vector, the dot product with it can be pulled *out* of the integration process, giving

$$-\vec{\kappa}\cdot\iint\vec{\nabla}(\hat{r}.\vec{r}')\times\hat{n}dS = \vec{\kappa}\cdot\oint(\hat{r}.\vec{r}')\overrightarrow{d\ell}. \tag{12.61c}$$

Again, as $\vec{\kappa}$ is not merely a constant, but also *arbitrary*, it has a random orientation. Therefore, we must have

$$-\oint(\hat{r}.\vec{r}')\overrightarrow{d\ell'} = \iint\vec{\nabla}'\,\hat{r}.\vec{r}')\times\hat{n}dS' = \iint\hat{r}\times\hat{n}dS' = \hat{r}\times\iint\hat{n}dS'. \tag{12.61d}$$

We can now use Eq. 12.61d in Eq. 12.59, and get

$$\boxed{\ell = 1}\quad \frac{\mu_0 I}{4\pi}\frac{1}{r^2}\oint(\vec{r}')^1 P_1(\cos\theta)\overrightarrow{dl'} = \frac{\mu_0 I}{4\pi}\frac{1}{r^2}\oint(\hat{r}.\vec{r}')\overrightarrow{dl'} = \frac{\mu_0}{4\pi r^2}[(I\iint\hat{n}dS'(\vec{r}'))\times\hat{r}]. \tag{12.62}$$

Now, $$I\iint \hat{n}dS'(\vec{r}') = I\vec{\alpha} = \vec{m} \tag{12.63}$$

is nothing but the 'magnetic dipole moment' of a tiny current loop which encloses an area $\vec{\alpha}$ (Fig. 12.7). The 'dipole' term, which corresponds to $\ell = 1$, of the magnetic vector potential therefore is given by

$$[\vec{A}_{\ell=1}(\vec{r})]_{\text{dipole}} = -\frac{\mu_0}{4\pi}\frac{1}{r^2}\hat{r} \times I\iint \hat{n}dS'(\vec{r}') = \frac{\mu_0}{4\pi}\frac{\vec{m} \times \hat{r}}{r^2}. \tag{12.64}$$

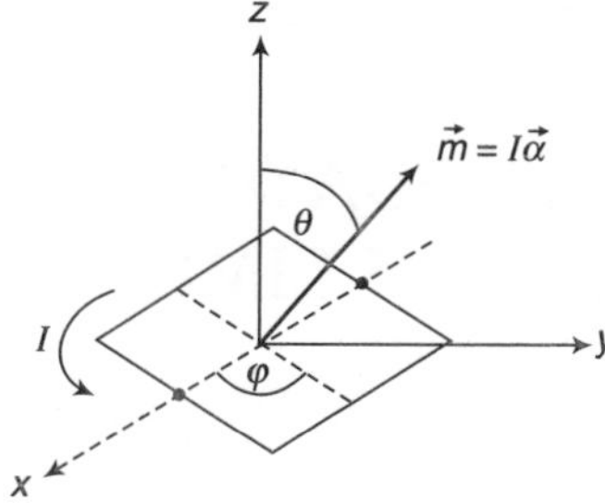

Fig. 12.7 A tiny closed circuit loop carrying a current I generates a magnetic field that is identical, to that generated by a point-sized bar magnet whose dipole moment would be $\vec{m} = I\vec{\alpha}$. Here, $\vec{\alpha}$ is the area enclosed by the loop.

In effect, the above expression represents the magnetic vector potential at the field point due to a tiny point-sized current loop in which the current flowing is I, and the loop encloses an area $\vec{\alpha}$. The magnetic dipole moment of the current loop is $\vec{m} = I\vec{\alpha}$. The elemental area $\vec{\alpha} = \iint \hat{n}dS'(\vec{r}')$ is a pseudo-vector (defined in Chapter 2), i.e., an axial vector, whose direction is given by the right-hand screw convention. When the screw is turned along the direction in which the current flows, the forward movement of the screw provides the direction of the axial-vector-area, and hence also of the dipole magnetic moment. Obviously, the magnetic dipole moment is also not a polar vector; it is an axial vector, as is the magnetic flux density field $\vec{B}$. In this analysis, a charge in motion, i.e., an electric current, is recognized as the *source* of the magnetic induction field. In the next section, we shall admit magnetic materials in our discussion, and these would also generate a magnetic field, just as a flowing current.

The series of terms in Eq. 12.55, and in Eq. 12.57, is called the multipole expansion, respectively of the electric scalar and the magnetic vector, potential. It saves us from the trouble of evaluating the difficult, often impossible, integral involved in the solution of the Poisson's equation, yet get the solution to a desired degree of accuracy by evaluating the much simpler leading terms.

Finally, we observe that in the Coulomb gauge (Eq. 12.37), the electric field intensity (Eq. 12.30b) has a particularly straight forward interpretation. It is given by a sum of an irrotational part, $-\vec{\nabla}\phi(\vec{r}, t)$, and a solenoidal part, $-\frac{\partial \vec{A}(\vec{r}, t)}{\partial t}$. This is just what one would expect from the Helmholtz theorem. However, in this gauge, when we consider time-varying phenomena, the solution to the Poisson equation (Eq. 12.38) in the form given by Eq. 12.55

apparently hoodwinks the relativistic limit. This is because changes at the source (represented by the *primed* coordinates in Eq. 12.55) would instantly influence the potential at the field point (represented by the *unprimed* coordinate in Eq. 12.55), at infinite speed. This is, however, only an apparent problem, since tangible effects are generated by the *field* which includes the time-derivative of the vector potential, and this would remove the ostensible contradiction. The Lorenz gauge is more adept in these situations. We conclude this section by pointing out that solutions to Eq. 12.41 can be systematically obtained using time-dependent 'Green's function', which yield the 'retarded potentials'. You will learn about these from specialized books, and in advance courses on electrodynamics.

12.4 MAXWELL'S EQUATIONS FOR MATTER IN THE CONTINUUM LIMIT

Maxwell's equations, Eq. 12.16a, b, c, and d (and their equivalent integral forms, Eq. 12.17a, b, c, and d) are universal. For immediate reference we compile them in Table 12.2. These equations are applicable to every situation, which is what is meant by stating that they are *universal*. The simpler relations for the steady state (i.e., time-independent) phenomena are implicitly contained in them already.

Table 12.2 Maxwell's equations of electrodynamics.

Equation Number	Divergence and the Curl of $\{E(\vec{r}, \vec{t}), B(\vec{r}, \vec{t})\}$	Popular name of the equation
1 (Eq. 12.16a)	$\vec{\nabla} \cdot \vec{E}(\vec{r}, t) = \dfrac{\rho(\vec{r}, t)}{\varepsilon_0}$	Coulomb–Gauss–Maxwell law
2 (Eq. 12.16b)	$\vec{\nabla} \cdot \vec{B}(\vec{r}, t) = 0$	Absence of magnetic monopole
3 (Eq. 12.16c)	$\vec{\nabla} \times \vec{E}(\vec{r}, t) = -\dfrac{\partial \vec{B}(\vec{r}, t)}{\partial t}$	Faraday–Lenz–Maxwell law
4 (Eq. 12.16d)	$\vec{\nabla} \times \vec{B}(\vec{r}, t) = \mu_0 \left\{ \vec{J}(\vec{r}, t) + \varepsilon_0 \dfrac{\partial \vec{E}(\vec{r}, t)}{\partial t} \right\}$	Oersted–Ampere–Maxwell law

Source of the electric field $\vec{E}(\vec{r}, t)$: electric charge, and/or time-varying magnetic field.
Source of the magnetic field $\vec{B}(\vec{r}, t)$: electric current, and/or time-varying electric field.

Maxwell's equations being universal, all electromagnetic dynamical problems can be solved using these equations, which of course need appropriate boundary conditions to be spelled out. These equations already go beyond the semi-empirical laws of Coulomb, Faraday–Lenz and Oersted–Ampere, which are all subsumed in the physics of the Maxwell's equations. Whereas recognizing a time-varying magnetic field as an alternative source of an electric field was known to be necessitated by the semi-empirical Faraday–Lenz law, the inclusion of the term $\varepsilon_0 \frac{\partial \vec{E}}{\partial t}$ in addition to the Amperian current $\vec{J}$ in the fourth equation was an outstanding ingenious creation of Maxwell, prompted by making the relationships for the curl of $\vec{E}(\vec{r}, t)$ and the curl of a $\vec{B}(\vec{r}, t)$ appear to be symmetric. Symmetry plays an exceptionally important role in understanding the laws of nature. The beginnings of this path-breaking contrivance can be traced to the symmetry in Maxwell's equations, recognized by Einstein, as discussed in the next chapter.

Let us highlight now the following two upshots of the Maxwell's equations:

(i) The electric field $\vec{E}(\vec{r}, t)$ generated by a time-dependent magnetic field, is *no different* from that produced by electric charges. The electric charges considered here may be at rest, but they are also *free* to flow, under a voltage difference, and thereby constitute an electric current.

(ii) The magnetic induction field $\vec{B}(\vec{r}, t)$ produced by a time-dependent electric flux, is no different from that produced by flowing electric charges, i.e., by an electric current.

It turns out that macroscopic matter at rest is capable of generating an electromagnetic field due to *alternate* physical mechanisms, which are *different* from those considered above in (i) and (ii). Some materials produce an electric field $\vec{E}(\vec{r}, t)$, which is absolutely no different from that considered in (i), even if this field is *not* due to either of the two processes described above (i). Also, some materials are capable of producing a magnetic field $\vec{B}(\vec{r}, t)$ due to their intrinsic properties, involving phenomenology that is totally distinct from what has been deliberated above, in (ii). We shall now discuss these alternative mechanisms that generate an electromagnetic field, and incorporate the same in Maxwell's equations.

Matter is often neutral not because there are no charges in it, but because it consists of equal amounts of positive charge (in its protons, inside the atomic nuclei), and negative charge (in the atomic electrons, outside the nuclei). In polarized dielectric materials, the integrated charge remains zero, but the *centers* of the positive and the negative charges are slightly displaced from each other. Polarized dielectric materials therefore consist of tiny electric dipoles which produce an electric field $\vec{E}(\vec{r}, t)$ identical to that considered in (i). This electric field is due to *displaced* electric charges which constitute the tiny dipoles. These charges are *not* free to flow, and hence cannot constitute an electric current; they merely result from tiny displacements of the positive and negative charge centers of neutral matter. Unlike the electric charges which are *free* to flow and constitute a current, the charges resulting from displacements of centers of positive and negative charges remain localized in a tiny elemental space. They remain essentially *bound* to that region. The electric dipole moment resulting from these displaced bound charges, however, constitutes an additional *source* of the electric field, derivable from a potential that has a form exactly identical to $V_{\text{dipole}}(\vec{r}) = \frac{1}{4\pi\varepsilon_0}\frac{\hat{e}_r \cdot \vec{p}_d}{r^2}$ (Eq. 12.56c). Its source is, however, not the electric charge that can flow and constitute a current. Instead, it is due to displaced charges in a dielectric material that remain essentially bound in a tiny region of the dielectric material. This extra source, when a polarized dielectric material is present, must be included in the phenomenology of the Maxwell's equations. After all, there is no difference in the electric field produced by these bound charges from that produced by (i) the electric charges considered hitherto which are *free* to flow and constitute an electric current and (ii) a Faraday–Maxwell magnetic flux.

Furthermore, even if it can be *without the need* of any detailed knowledge of the quantum theory, we urge the readers to seamlessly admit in their analytical skills the fact that atomic electrons have intrinsic properties called the quantum orbital angular momentum, and also the quantum spin angular momentum. These are intrinsic, innate natural properties of electrons, but they can be fully understood only by invoking the quantum theory. Most of the young readers of this book would have only a cursory knowledge of the quantum theory, if at all any. A detailed knowledge of the quantum theory is, however, not necessary to describe the phenomena we are about to present. Suffice it is to admit in our scheme of Maxwell's equations an additional source of the magnetic field $\vec{B}(\vec{r}, t)$ which arises from the innate quantum angular momentum of the electrons, associated with which is a magnetic moment. This quantum magnetic moment has exactly the same effect as an electric current in a tiny loop (Fig. 12.7), except that this magnetic moment is entirely due to intrinsic quantum magnetic moment; it is *not* due to a flowing current, *nor* due to a time-varying electric flux.

Some classical models, such as electric currents constituted by an electron circling the nucleus, and spinning about itself, have been used in the literature to account for the quantum orbital angular momentum. We will, however, refrain from using such artificial models; they are really misleading. The quantum angular momentum (both 'orbital' and 'spin') simply *cannot* be modeled correctly using *any* classical picture. It is best to wait for a formal course on quantum mechanics to learn about this exciting phenomenon. All we need at this stage is to incorporate in the Maxwell's equations an additional term which will account for these tiny innate magnetic moments in some materials due to their quantum angular momenta. However, only some materials exhibit a macroscopic effect due to these quantum magnetic moments. In other materials, their cumulative vector addition to zero disables any macroscopic manifestation of the magnetic moments. The materials which exhibit macroscopic magnetism are called, therefore, *magnetic materials*. Their further classification as ferromagnets, ferrimagnets, paramagnets, etc., is not of importance to our focus here on the Maxwell's equations.

All we need to admit in the discussion on Maxwell's equations is that magnetic materials consist of quantum sources of magnetic moments, which produce magnetic fields, derivable from a magnetic vector potential $\vec{A}(\vec{r})_{\text{dipole}} = \dfrac{\mu_0}{4\pi}\dfrac{\vec{m}_d \times \hat{r}}{r^2}$, in accordance with Eq. 12.64. The magnetic dipole moment $\vec{m}_d$ we are now referring to does *not* owe its origin to an electric current loop, as in Fig. 12.7. Instead, it originates from the integrated bulk, innate, quantum properties exhibited by the magnetic material.

Notwithstanding the intricacies of the quantum processes involved at the *microscopic* level, The integrated bulk effects of (i) the electric dipole moments $\vec{p}_d$ due to the bound charges in the dielectric materials, and of (ii) the magnetic dipole moments $\vec{m}_d$ associated with the quantum angular momenta of the electrons in magnetic materials, can be taken easily into account using methods which we are already well familiar with. The methods we shall apply are somewhat similar to the considerations in the continuum hypothesis in fluid dynamics (Chapters 10 and 11). We shall consider these properties in a tiny region of space, δV in the limit $\delta V \to 0$. This limit will, however, be understood in the continuum hypothesis; not in the strict mathematical sense in which the limit $\delta V \to 0$. The mathematical limit will make the volume element smaller than atomic sizes. The continuum limit would allow that limiting volume to retain bulk properties resulting from the electric dipole moment of the dielectric material in the electrical case, and from the magnetic dipole moments of the magnetic materials in the magnetic case. The bulk properties we have mentioned above are formally defined, in the continuum hypothesis, in terms of the polarization $\vec{P}(\vec{r}')$ property of the dielectric materials, and the magnetization $\vec{M}(\vec{r}')$ property of the magnetic materials; both of these are, defined below.

Polarization $\vec{P}(\vec{r}')$ is a property of the dielectric materials defined as the electric dipole moment per unit volume, due to displaced bound charges in the dielectric material. It is given by

$$\vec{P}(\vec{r}') = \lim_{\delta\tau' \to 0} \frac{\overrightarrow{\delta p_d}}{\delta\tau'}, \tag{12.65}$$

where $\vec{r}'$ is the position vector of a point in the dielectric material. $\delta\tau'$ is a tiny volume element surrounding the source point at the position vector $\vec{r}'$.

Likewise, the magnetization $\vec{M}(\vec{r}')$ is a property of the magnetic materials defined as the magnetic dipole moment per unit volume due to the quantum angular momenta in the magnetic material. It is given by

$$\vec{M}(\vec{r}') = \lim_{\delta\tau' \to 0} \frac{\overrightarrow{\delta m_d}}{\delta\tau'}, \tag{12.66}$$

where $\vec{r}'$ is the position vector of a point in the magnetic material.

We shall now proceed to *rephrase* the Maxwell's equations (Table 12.2), accounting for the macroscopic sources (Eqs. 12.65 and 12.66) of electric and the magnetic fields. The rephrasing necessitates the explicit manifestation in the Maxwell's equations of the sources indicated in Eqs. 12.65 and 12.66. The underlying content of Maxwell's equations, however, remains essentially the same.

The net electric dipole moment in the tiny volume element $\delta\tau'$ surrounding a source point S (Fig. 12.8) in the dielectric material is $\vec{P}(\vec{r}')\delta\tau'$. This dipole would generate an electric field at the field point F whose position vector is $\vec{r}$. The net dipole potential is given by an expression like Eq. 12.56. However, Eq. 12.56 stood for the dipole potential of a point-sized dipole placed at the origin of the coordinate system. In the present case (Fig. 12.8), the dipole moment $\vec{P}(\vec{r}')\delta\tau'$ is spread all over the bulk polarized material. We must therefore integrate over the entire region of the polarized material. The integrands will have essentially the same form, as in Eq. 12.56c. Hence, the net effective dipolar potential at the field point due to the polarized dielectric medium is

$$V(\vec{r}) = \frac{1}{4\pi\varepsilon_0} \iiint\limits_{\substack{\text{volume}\\\text{polarized}\\\text{medium}}} \frac{\hat{R}\cdot\oplus\vec{P}(\vec{r}')}{\left|\vec{R}\right|^2} d\tau' = \frac{1}{4\pi\varepsilon_0} \iiint\limits_{\substack{\text{volume}\\\text{polarized}\\\text{medium}}} \frac{(\vec{r}-\vec{r}')\cdot\vec{P}(\vec{r}')}{\left|\vec{r}-\vec{r}'\right|^3} d\tau \tag{12.67a}$$

$$\text{i.e.,}\quad V(\vec{r}) = \frac{1}{4\pi\varepsilon_0} \iiint\limits_{\substack{\text{volume}\\\text{polarized}\\\text{medium}}} \vec{P}(\vec{r}')\cdot\vec{\nabla}'\left(\frac{1}{R}\right) d\tau', \tag{12.67b}$$

where $R = \left|\vec{R}\right| = \left|\vec{r}-\vec{r}'\right|$ and the unit vector $\hat{R} = \dfrac{\vec{r}-\vec{r}'}{\left|r-r'\right|}$. I am sure you would have noted that in Eq. 12.67, both the differentiation and the integration, are with respect to the primed coordinates, i.e., the coordinates in the *source* region. We can of course write the result in Eq. 12.67b as a difference of two terms, by considering the divergence $\vec{\nabla}'\cdot\left(\dfrac{\vec{P}(\vec{r}')}{R}\right)$ and get

$$V(\vec{r}) = \frac{1}{4\pi\varepsilon_0} \iiint\limits_{\substack{\text{volume}\\\text{polarized}\\\text{medium}}} \vec{\nabla}'\cdot\left(\frac{\vec{P}(\vec{r}')}{R}\right) d\tau' - \frac{1}{4\pi\varepsilon_0} \iiint\limits_{\substack{\text{volume}\\\text{polarized}\\\text{medium}}} \left(\frac{1}{R}\right)\vec{\nabla}'\cdot\vec{P}(\vec{r}')d\tau'. \tag{12.67c}$$

Using now the Gauss' divergence theorem to express the first term as a surface integral,

$$V(\vec{r}) = \frac{1}{4\pi\varepsilon_0} \oiint \left(\frac{\vec{P}(\vec{r}')\cdot\vec{n}'}{R}\right) dS' + \frac{1}{4\pi\varepsilon_0} \iiint\limits_{\substack{\text{volume}\\\text{polarized}\\\text{medium}}} \frac{(-\vec{\nabla}'\cdot\vec{P}(\vec{r}'))}{R} d\tau'. \tag{12.68a}$$

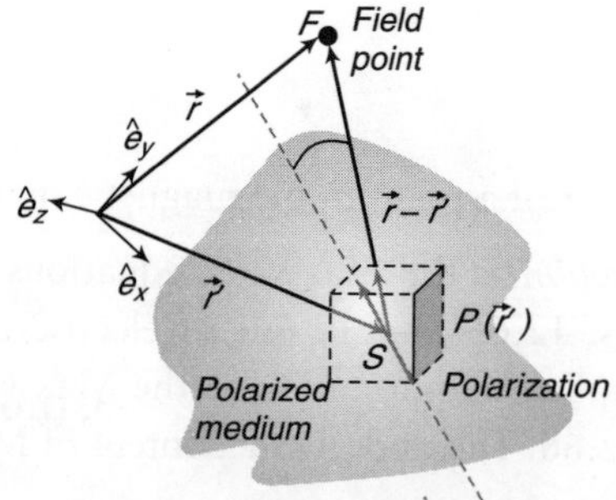

Fig. 12.8 The effective electric intensity field generated at a field point F due to a polarized dielectric (electrically neutral) medium is the net integrated effect of tiny dipoles in infinitesimal volume elements throughout the dielectric source medium. These electric dipoles are due to displaced centers of localized bound positive and negative charges which add up to a net zero charge of the total integrated medium.

which we write as

$$V(\vec{r}) = \frac{1}{4\pi\varepsilon_0} \oiint \left(\frac{\sigma_b(\vec{r}')}{R}\right) dS' + \frac{1}{4\pi\varepsilon_0} \iiint\limits_{\substack{\text{volume}\\\text{polarized}\\\text{medium}}} \frac{(\rho_b(\vec{r}'))}{R} d\tau', \tag{12.68b}$$

with $\sigma_b(\vec{r}') = \vec{P}(\vec{r}') \cdot \hat{n}'$, (12.69a)

and $\rho_b(\vec{r}') = -\vec{\nabla}' \cdot \vec{P}(\vec{r}')$. (12.69b)

Since these *bound* charges are only arising out of displacement of electric and negative charges in the overall dielectric material which is neutral on the whole, we have

$$\underset{\substack{\text{closed}\\\text{surface}\\\text{that encloses}\\\text{the volume}}}{\oiint} \sigma_b(\vec{r}')dS' = - \underset{\substack{\text{volume}\\\text{polarized}\\\text{medium}}}{\iiint} \rho_b(\vec{r}')d\tau'. \tag{12.70}$$

Inside the 'source' region where the dielectric material is present, we have used the primed coordinate. At such a point, the first of the Maxwell's equation (Eq. 12.16a) can be now re-written to include both the *free* charge, which can be made to flow and constitute a current, and the *bound* charge, which remains confined to a region, and is a result of the polarization of the dielectric material. Thus,

$$\vec{\nabla}' \cdot \vec{E}(\vec{r}') = \frac{\rho(\vec{r}')}{\varepsilon_0} = \frac{\rho_f(\vec{r}') + \rho_b(\vec{r}')}{\varepsilon_0} = \frac{\rho_f(\vec{r}')}{\varepsilon_0} + \frac{\rho_b(\vec{r}')}{\varepsilon_0}. \tag{12.71}$$

The subscript 'f' denotes the free charge density, and the subscript 'b' denotes the bound charge density. We had introduced the primed-coordinate to represent points in the source region to distinguish them from the coordinates of the field point. Such a distinction is no longer necessary in Eq. 12.71 which stands for properties at a point in the dielectric medium. There is no need therefore to use the prime. Therefore,

$$\vec{\nabla} \cdot (\varepsilon_0 \vec{E}) = \rho_f + \rho_b = \rho_f - \vec{\nabla} \cdot \vec{P}. \tag{12.72a}$$

Equivalently, the above result can be written as

$$\vec{\nabla} \cdot \vec{D} = \rho_f, \tag{12.72b}$$

wherein we have introduced the vector field

$$\vec{D} = \varepsilon_0 \vec{E} + \vec{P}, \tag{12.73}$$

which is called the *electric displacement field.*

It is clear that its volume integral over the source region, (from Eq. 12.72b) gives

$$\underset{\substack{\text{volume}\\\text{polarized}\\\text{medium}}}{\iiint} \vec{\nabla} \cdot \vec{D} d\tau' = \underset{\substack{\text{volume}\\\text{polarized}\\\text{medium}}}{\iiint} \rho_f(\vec{r}')d\tau' = Q_{f,\text{enclosed}}, \tag{12.74}$$

where $Q_{f,\text{ enclosed}}$ is the total charge in the dielectric medium that can be dislodged, put in a flow, constituting a current.

For materials in which the polarization is proportional to the electric intensity field, we have

$$\vec{P} = \varepsilon_0 \chi \vec{E}. \tag{12.75}$$

Such materials are called *linear dielectrics*, and χ is called the *dielectric susceptibility.*

Combining Eq. 12.73 and Eq. 12.75, we get

$$\vec{D} = \varepsilon_0\vec{E} + \vec{P} = \varepsilon_0(1 + \chi)\vec{E} = \varepsilon\vec{E}, \tag{12.76}$$

with $\varepsilon = \varepsilon_0(1 + \chi)$,

which is called the *permittivity* of the dielectric material. (12.77a)

Its ratio with the permittivity of free space is

$$\frac{\varepsilon}{\varepsilon_0} = (1 + \chi) = \varepsilon_r, \tag{12.77b}$$

which is also called the *relative permittivity* or as the *dielectric constant* of the material. The relation expressed in Eq. 12.76 is specific to the material, representing the material's response to the electric field. It is therefore a constitutive relation.

We now consider the cumulative effect of the magnetization of the material due to the tiny effective dipole moments, discussed in Eq. 12.66. These magnetic moments are then spread out in the entire magnetic material, as shown in Fig. 12.9.

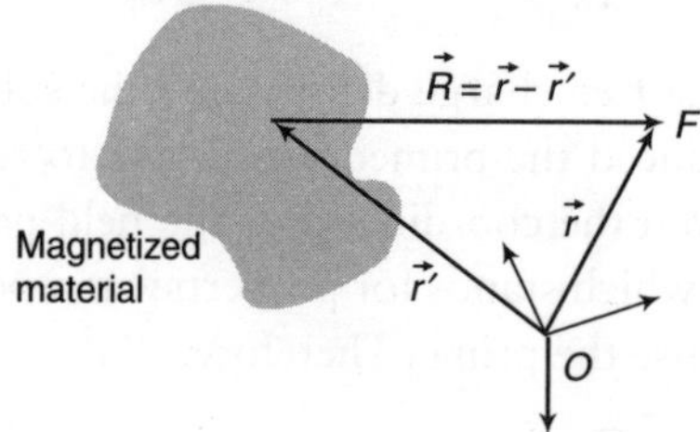

Fig. 12.9 A magnetic material generates a magnetic induction field at the field point due to the magnetic moment arising out of the quantum angular momentum of its electrons. The net field is, however, expressible as the integrated effect of tiny magnetic dipoles located in infinitesimal volume elements in the entire material.

A tiny volume element $\delta\tau'$ in the magnetized material then has a magnetic moment given by $\vec{M}(\vec{r}')\delta\tau'$, where the magnetization $\vec{M}(\vec{r}')$ is given by Eq. 12.66. The vector potential at a field point due to a magnetized material is then the cumulative integrated effect of the dipole moments in each tiny volume elements in the magnetized material.

Thus,

$$\vec{A}(\vec{r})\Big]_{\substack{\text{due to}\\ \text{magnetized}\\ \text{bulk}\\ \text{material}}} = \frac{\mu_0}{4\pi}\iiint \frac{\vec{M}(\vec{r}') \times \hat{R}}{R^2} d\tau'. \tag{12.78}$$

Since the differentiation with respect to the coordinates of the source point is obviously independent from that with respect to the coordinates of the field points, and since the distance between source point and the field point is given by $R = \left|\vec{R}\right| = \left|\vec{r} - \vec{r}'\right|$, we have

$$\vec{\nabla}\left(\frac{1}{R}\right) = -\frac{\hat{R}}{R^2}, \text{ and } \vec{\nabla}'\left(\frac{1}{R}\right) = \frac{\hat{R}}{R^2}. \tag{12.79}$$

Eq. 12.78 can therefore be written equivalently as a difference between two terms:

$$\vec{A}(\vec{r})\Big]_{\substack{\text{due to}\\\text{magnetized}\\\text{bulk}\\\text{material}}} = \frac{\mu_0}{4\pi}\iiint\left[\frac{1}{R}\vec{\nabla}'\times\vec{M}(\vec{r}')-\vec{\nabla}'\times\left\{\frac{1}{R}\vec{M}(\vec{r}')\right\}\right]d\tau', \tag{12.80a}$$

i.e.,

$$\vec{A}(\vec{r})\Big]_{\substack{\text{due to}\\\text{magnetized}\\\text{bulk}\\\text{material}}} = \frac{\mu_0}{4\pi}\iiint\frac{\vec{\nabla}'\times\vec{M}(\vec{r}')}{R}d\tau' - \frac{\mu_0}{4\pi}\iiint\vec{\nabla}'\times\frac{\vec{M}(\vec{r}')}{R}d\tau'. \tag{12.80b}$$

A tiny digression in which we introduce two auxiliary vectors enable us interpret the magnetic vector potential at the field point generated by the magnetized material in a particularly fruitful manner. For simplicity, we tentatively introduce two auxiliary vectors,

$$\vec{Q}(\vec{r}) = \frac{\vec{M}(\vec{r})}{R}, \tag{12.81a}$$

and

$$\vec{G}(\vec{r}) = \vec{Q}(\vec{r})\times\vec{K} = \frac{\vec{M}(\vec{r})}{R}\times\vec{K}, \tag{12.81b}$$

where $\vec{K}$ is an *arbitrary constant vector*. We now use the Gauss' divergence theorem for the vector field $\vec{G}(\vec{r})$, to get

$$\iiint\limits_{\substack{\text{volume}\\\text{region}}}\vec{\nabla}\cdot\{\vec{Q}(\vec{r})\times\vec{K}\}dV = \oiint\limits_{\substack{\text{enclosing}\\\text{surface}}}\{\vec{Q}(\vec{r})\times\vec{K}\}\cdot dS\hat{n}. \tag{12.82}$$

Using now the vector identity,

$$\vec{\nabla}\cdot(\vec{A}\times\vec{B}) = \vec{B}\cdot\vec{\nabla}\times\vec{A} - \vec{A}\cdot\vec{\nabla}\times\vec{B}, \tag{12.83}$$

for $\vec{\nabla}\cdot\{\vec{Q}(\vec{r})\times\vec{K}\}$, we get

$$\iiint\limits_{\substack{\text{volume}\\\text{region}}}\vec{K}\cdot(\vec{\nabla}\times\vec{Q}(\vec{r}))dV - \iiint\limits_{\substack{\text{volume}\\\text{region}}}\vec{Q}(\vec{r})\cdot(\vec{\nabla}\times\vec{K})dV = \oiint\limits_{\substack{\text{enclosing}\\\text{surface}}}\{\vec{Q}(\vec{r})\times\vec{K}\}\cdot dS\hat{n}, \tag{12.84a}$$

i.e.,
$$\iiint\limits_{\substack{\text{volume}\\\text{region}}}\vec{K}\cdot\vec{\nabla}\times\vec{Q}(\vec{r})dV = -\oiint\limits_{\substack{\text{enclosing}\\\text{surface}}}\vec{K}\cdot\vec{Q}(\vec{r})\times dS\hat{n}, \tag{12.84b}$$

since the curl of a constant vector is zero. You would have noticed, of course, that on the right hand side, we have a scaler triple product of three vectors in which we have interchanged the positions of the dot product and the cross product, and also reversed the order of the cross product which is responsible for the negative sign that is seen on the right hand side of Eq. 12.84b.

Now, since $\vec{K}$ is a constant vector, we get

$$\vec{K}\cdot\iiint\limits_{\substack{\text{volume}\\\text{region}}}\vec{\nabla}\times\vec{Q}(\vec{r})dV = -\vec{K}\cdot\oiint\limits_{\substack{\text{enclosing}\\\text{surface}}}\vec{Q}(\vec{r})\times dS\hat{n}, \tag{12.84d}$$

and hence, $$\underset{\substack{\text{volume}\\\text{region}}}{\iiint} \vec{\nabla} \times \vec{Q}(\vec{r})dV = -\underset{\substack{\text{enclosing}\\\text{surface}}}{\oiint} \vec{Q}(\vec{r}) \times dS\hat{n}. \tag{12.85a}$$

Since we have used the primed coordinates for the source region, we rewrite the above result as

$$\underset{\substack{\text{volume}\\\text{region}}}{\iiint} \vec{\nabla}' \times \vec{Q}(\vec{r}')dV' = -\underset{\substack{\text{enclosing}\\\text{surface}}}{\oiint} \vec{Q}(\vec{r}') \times dS'\hat{n}. \tag{12.85b}$$

Using now the original magnetization vector instead of the auxiliary vector that we had tentatively introduced, we get

$$\underset{\substack{\text{volume}\\\text{region}}}{\iiint} \vec{\nabla}' \times \frac{\vec{M}(\vec{r}')}{R}dV' = -\underset{\substack{\text{enclosing}\\\text{surface}}}{\oiint} \cdot \frac{\vec{M}(\vec{r}')}{R} \times dS'\hat{n}. \tag{12.86}$$

Using now Eq. 12.80b and Eq. 12.86, we get

$$\vec{A}(\vec{r})\Big]_{\substack{\text{due to}\\\text{magnetized}\\\text{bulk}\\\text{material}}} = \frac{\mu_0}{4\pi}\iiint \frac{\vec{\nabla}' \times \vec{M}(\vec{r}')}{R}dV' + \frac{\mu_0}{4\pi}\underset{\substack{\text{enclosing}\\\text{surface}}}{\oiint} \frac{\vec{M}(\vec{r}') \times \hat{n}(\vec{r}')}{R}dS'. \tag{12.87}$$

We now rewrite the integrands by defining

$$\vec{J}_b(\vec{r}') = \vec{\nabla}' \times \vec{M}(\vec{r}'), \quad \text{i.e.,}\ \vec{J}_b = \vec{\nabla} \times \vec{M}, \tag{12.88a}$$

called the *bound* volume current density, and

$$\vec{K}_b(\vec{r}') = \vec{M}(\vec{r}') \times \hat{n}(\vec{r}'), \quad \text{i.e.,}\ \vec{K}_b = \vec{M} \times \hat{n}, \tag{12.88b}$$

called the *bound* surface current density.

It is easy to see, using dimensional analysis, that $\vec{\nabla} \times \vec{M}$ has the dimensions of the volume current density and that $\vec{M} \times \hat{n}$ has the dimensions of the surface current density. The currents we are now referring to, do not involve any physical flow of charges. They are simply the result of the bulk magnetic moment of the material which has its origins in the quantum effects, but effectively represented by equivalent dipole moments. No flow of charges constituting an electric current is involved here; hence $\vec{J}_b = \vec{\nabla} \times \vec{M}$ is called *bound* volume current density and $\vec{K}_b = \vec{M} \times \hat{n}$ is called the *bound* surface current density. In terms of the bound current densities, the effective magnetic vector potential at a field point is therefore

$$\vec{A}(\vec{r})\Big]_{\substack{\text{due to}\\\text{magnetized}\\\text{bulk}\\\text{material}}} = \frac{\mu_0}{4\pi}\iiint \frac{\vec{J}_b(\vec{r}')}{R}dV' + \frac{\mu_0}{4\pi}\underset{\substack{\text{enclosing}\\\text{surface}}}{\oiint} \frac{\vec{K}_b(\vec{r}')}{R}dS'. \tag{12.89}$$

The Oersted–Ampere's law was already modified by Maxwell to include the effect of the time-dependent electric field (Eq. 12.16d, presented in the fourth row of Table 12.2). The additional effects due to bulk magnetization originating from quantum properties of the magnetic material must now be added, by including its effect represented by the bound volume current density, $\vec{J}_b = \vec{\nabla} \times \vec{M}$, in Maxwell's equation for the curl of the magnetic induction field. It must now be rewritten as

$$\vec{\nabla} \times \vec{B} = \mu_0 \vec{J}_{\text{free}} + \mu_0 \vec{J}_{\text{displacement}} + \mu_0 \vec{J}_{\text{bound}}, \tag{12.90a}$$

i.e., $$\vec{\nabla} \times \vec{B} = \mu_0 \vec{J}_{\text{free}} + \mu_0 \varepsilon_0 \frac{\partial \vec{E}}{\partial t} + \mu_0 (\vec{\nabla} \times \vec{M}). \tag{12.90b}$$

In effect, we have

$$\vec{\nabla} \times (\vec{B} - \mu_0 \vec{M}) - \mu_0 \varepsilon_0 \frac{\partial \vec{E}}{\partial t} = \mu_0 \vec{J}_{\text{free}}, \tag{12.90c}$$

i.e., $$\vec{\nabla} \times \left(\frac{\vec{B}}{\mu_0} - \vec{M} \right) - \varepsilon_0 \frac{\partial \vec{E}}{\partial t} = \vec{J}_{\text{free}}. \tag{12.90d}$$

The vector $\left(\frac{\vec{B}}{\mu_0} - \vec{M} \right)$ contains the essential magnetic bulk property of the medium in the continuum hypothesis, and is called the *magnetic intensity field*, defined as

$$\vec{H} = \frac{\vec{B}}{\mu_0} - \vec{M}. \tag{12.91}$$

Magnetic properties of materials are usually compiled in terms of the relationship between the magnetic field intensity $\vec{H}$ and the magnetization $\vec{M}$. Typically, this relation has the following form:

$$\vec{M}(\vec{r}) = \chi_m \vec{H}(\vec{r}), \tag{12.92a}$$

so that we shall have

$$\vec{B} = \mu_0 \vec{H} + \vec{M} = (\mu_0 + \mu_0 \chi_m)\vec{H} = \mu_0(1 + \chi_m)\vec{H} = \mu \vec{H}. \tag{12.92a}$$

χ_m is called the *magnetic susceptibility* of the material, and $\mu = \mu_0(1 + \chi_m)$ is called the magnetic permeability of the material. From Eq. 12.90d, it follows that

$$\vec{\nabla} \times \vec{H} - \varepsilon_0 \frac{\partial \vec{E}}{\partial t} = \vec{J}_{\text{free}}. \tag{12.93}$$

We can now update Table 12.2 for Maxwell's equations to include the consequences of bulk matter in the continuum hypothesis. As we do so, the essential content of the Maxwell's equations will remain the same. The expressions for the charge and the current however get modified, so that bulk properties find an explicit manifestation in terms of the equivalent bound charges and the bound currents introduced above. The Maxwell's equations for matter in the continuum limit are consolidated and tabulated in Table 12.3.

The method which we have employed above has made it possible to represent extremely intricate properties of matter, which would in fact require a quantum mechanical treatment, to be described in terms of *average* properties that *bulk* matter manifests. Toward this end, the continuum hypothesis that is employed in defining the polarization $\vec{P}$ and the magnetization $\vec{M}$ has of course played a paramount role.

Table 12.3 Maxwell's equations of electrodynamics—*matter matters!*

Equation Number	Divergence and the Curl of $\{E(\vec{r},\vec{t}), B(\vec{r},\vec{t})\}$	Popular name of the equation
1 (Eq. 12.73)	$\vec{\nabla}\cdot\vec{D}=\rho_f$ $\vec{D}=\varepsilon_0\vec{E}+\vec{P}=\varepsilon_0(1+\chi)\vec{E}=\varepsilon\vec{E}$	Coulomb–Gauss–Maxwell law
2 (Eq. 12.16b)	$\vec{\nabla}\cdot\vec{B}(\vec{r},t)=0$	Absence of magnetic monopole
3 (Eq. 12.16c)	$\vec{\nabla}\times\vec{E}(\vec{r},t)=-\dfrac{\partial\vec{B}(\vec{r},t)}{\partial t}$	Faraday–Lenz–Maxwell law
4 (Eqs. 12.90 and 12.93)	$\vec{\nabla}\times\vec{B}(\vec{r},t)-\mu_0\vec{\nabla}\times\vec{M}(\vec{r},t)-\mu_0\varepsilon_0\dfrac{\partial\vec{E}(\vec{r},t)}{\partial t}=\mu_0\vec{J}_f(\vec{r},t)$ $\vec{\nabla}\times\vec{H}(\vec{r},t)-\varepsilon_0\dfrac{\partial\vec{E}(\vec{r},t)}{\partial t}=\vec{J}_f(\vec{r},t)$ $\vec{H}=\dfrac{\vec{B}}{\mu_0}-\vec{M}$	Oersted–Ampere–Maxwell law

Source of the electric field $\vec{E}(\vec{r},t)$: electric charge, polarized dielectric material, and/or time-varying magnetic field.
Source of the magnetic field $\vec{B}(\vec{r},t)$: electric current, magnetic material, and/or time-varying electric field.

While solving the Maxwell's equations (Table 12.3), one certainly requires the boundary and the interface conditions for the region in which the electromagnetic field of interest is of interest. Various types of the boundary conditions are used to solve the differential equations. Essentially, the boundary refers to the surface beyond which the electromagnetic field does not exist. Depending on the nature of the field, whether it is a scalar field or a vector field, the values of the scalar field or the tangential component of a vector field are prescribed on the boundary. Such boundary conditions are called the Dirichlet boundary conditions. Alternatively, the normal component of the gradient of the scalar field, and the tangential component if the curl of the vector field are provided as the boundary conditions. These are called the Neumann boundary conditions.

Also of importance are the interface conditions when the electromagnetic field exists in regions whose intrinsic material properties, such as the electrical permittivity and the magnetic permeability are different across a surface of separation between two media. The differential forms of the Maxwell's equations are of course not useful to determine the interface conditions. It is the integral forms that are used, along with the Gauss' divergence theorem and the Kelvin–Stokes theorem that are useful. The tiny volume element over which the volume integral in the Gauss' divergence theorem is carried out is usually called the

Gaussian pill box. The Kelvin–Stokes theorem is also useful in determining the boundary conditions, since it can be used over a closed loop that straddles the surface of separation between the two media. The closed path over which the path integral is carried out is often called the Amperian loop. Gaussian pill box and the Amperian loop are both shown in Fig. 12.10.

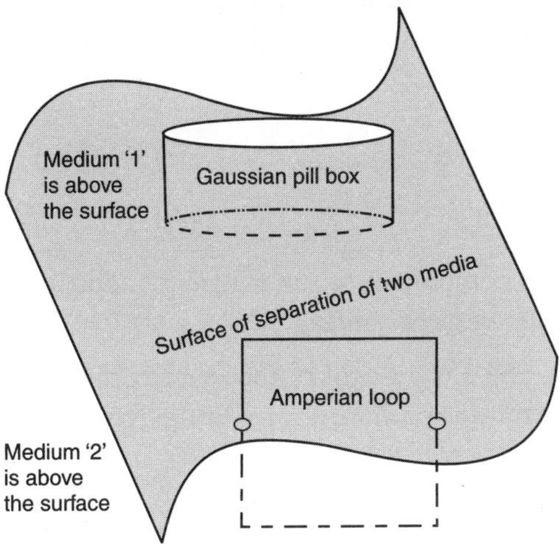

Fig. 12.10 The integral forms of the Maxwell's equations for $\vec{E}$, $\vec{D}$ and for $\vec{B}$, $\vec{H}$ can be employed along with the Gauss' divergence theorem to determine the expressions for the integrals over the Gaussian 'pill box' which straddles the surface of separation of the two media, shown in this figure. Likewise, the equations can be used along with the Kelvin–Stokes theorem over an Amperian loop which straddles the surface of separation.

To obtain the interface conditions, we shall now use the integral form of Eq. 1 in Table 12.3 (i.e., Eq. 12.73). Accordingly, we get

$$\iiint \vec{\nabla} \cdot \vec{D} dV = \iiint \rho_f dV = Q_f = \sigma_f a. \tag{12.94a}$$

The integral is determined in the limit that the volume element shrinks to zero. The net charge enclosed in the pill box is merely the surface charge density σ_f multiplied by the cross sectional area 'a' that is straddled by the pill box. The integration is carried out in the region of space of the Gaussian pill box (Fig. 12.10). Q_f, in the above equation, is the integrated *free* charge (as opposed to the bound charge) that is enclosed by the surface which bounds the Gaussian pill box. 'a' is the cross-sectional area of the disk end of the Gaussian pill box which is sandwiched between the two disks, which are in the two media across the surface of separation as shown in Fig. 12.10. The volume integral of the divergence of the displacement vector field, by the Gauss' divergence theorem, along with Eq. 12.94a gives

$$\oiint \vec{D} \cdot \hat{n} dS = \sigma_f a. \tag{12.94b}$$

Determining the above surface integral, we get

$$D_1^{\perp} a - D_2^{\perp} a = \sigma_f a, \tag{12.94c}$$

i.e., $$D_1^{\perp} - D_2^{\perp} = \sigma_f. \tag{12.95}$$

Essentially, the above result means that the normal components of the displacement vector field across a surface of separation between two media differ by the surface charge density on the surface of separation.

Next, we shall use the second equation of Table 12.3 (Eq. 12.16b). This equation declares the absence of magnetic monopole. Hence, applying Gauss' theorem for the divergence of the magnetic induction field in the region enclosed by the Gaussian pill box, we get

$$\oiint \vec{B} \cdot \hat{n} dS = 0, \tag{12.96a}$$

$$\text{i.e.,} \quad B_1^{\perp} a - B_2^{\perp} a = 0, \tag{12.96b}$$

$$\text{and hence,} \quad B_1^{\perp} - B_2^{\perp} = 0. \tag{12.97}$$

We find that the normal component of the magnetic induction field is continuous across the surface of separation between two media.

Furthermore, applying the Kelvin–Stokes theorem to the third equation in Table 12.3 (i.e., Eq. 12.16c), carrying out the path integral over the Amperian loop shown in Fig. 12.10, we get

$$\oint \vec{E}(\vec{r},t) \cdot \overrightarrow{d\ell} = \iint \vec{\nabla} \times \vec{E}(\vec{r},t) \cdot \overrightarrow{dS} = \iint \left\{ -\frac{\partial \vec{B}(\vec{r},t)}{\partial t} \right\} \cdot \overrightarrow{dS} = -\frac{\partial}{\partial t} \iint \vec{B}(\vec{r},t) \cdot \overrightarrow{dS} = 0. \tag{12.98a}$$

Now, in the limit that the side of the loop that straddles the two media becomes zero, the magnetic flux crossing the Amperian loop would vanish, and thus $\iint \vec{B}(\vec{r},t) \cdot \overrightarrow{dS} = 0$. Hence, from Eq. 12.98a, we get

$$E_1^{\parallel} \ell - E_2^{\parallel} \ell = 0, \tag{12.98b}$$

where ℓ is the length of one side of the Amperian loop that is completely on one (or the other) medium. Hence,

$$E_1^{\parallel} - E_2^{\parallel} = 0. \tag{12.99}$$

In other words, the tangential components of the electric intensity field across the surface of separation between two media are equal.

Finally, to determine the interface conditions on the intensity of the magnetic field $\vec{H}$, we use the Kelvin–Stokes theorem for the fourth equation in Table 12.3 over an Amperian loop (Fig. 12.11).

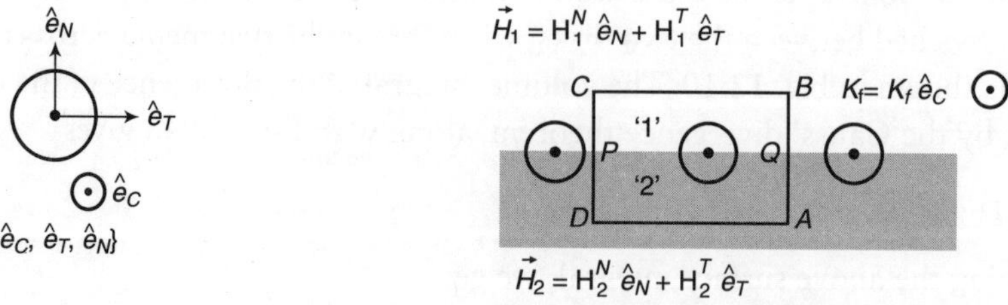

Fig. 12.11 The surface current flows in the direction of $\hat{e}_C$. $\{\hat{e}_C, \hat{e}_T, \hat{e}_N\}$ forms a right handed basis set of unit vectors. The surface current density is $\vec{K}_f = K_f \hat{e}_C$. We consider an Amperian loop along AQBCPDA in the plane of $\hat{e}_T, \hat{e}_N$.

Carrying out the path integration over the Amperian loop shown in Fig. 12.11 over the closed path shown in this figure, A→Q→B→C→P→D→A, we get, using Eq. #4 in Table 12.3 (i.e., Eq. 12.93), and the Kelvin–Stokes theorem:

$$\oint_{AQBCPDA} \vec{H}\cdot\vec{dl} = \iint \vec{\nabla}\times\vec{H}\cdot dS\hat{e}_C = \iint \vec{J}_f\cdot dS\hat{e}_C = I_f^{\text{enclosed}} = (K_f s), \tag{12.100a}$$

where $s = BC = DA$; $h = AB = CD$,

and $\vec{K}_f = K_f\hat{e}_C$ is the surface current density; i.e., surface current per *transverse* length.

Accordingly,

$$\oint_{AQBCPDA} \vec{H}\cdot\vec{dl} = \oint_{AQ} \vec{H}\cdot\vec{dl} + \oint_{QB} \vec{H}\cdot\vec{dl} + \oint_{BC} \vec{H}\cdot\vec{dl} + \oint_{CP} \vec{H}\cdot\vec{dl} + \oint_{PD} \vec{H}\cdot\vec{dl} + \oint_{DA} \vec{H}\cdot\vec{dl} = (K_f s). \tag{12.100b}$$

Hence,

$$\oint_{AQBCPDA} \vec{H}\cdot\vec{dl} = \cancel{H_2^N\frac{h}{2}} + \cancel{H_1^N\frac{h}{2}} - H_1^T s - \cancel{H_1^N\frac{h}{2}} - \cancel{H_2^N\frac{h}{2}} + H_2^T s = (K_f s), \tag{12.100c}$$

i.e., $$\oint_{AQBCPDA} \vec{H}\cdot\vec{dl} = (H_2^T - H_1^T)s = (K_f s). \tag{12.100d}$$

This result is best written in the vector notation as

$$\hat{n}\times(\vec{H}_2 - \vec{H}_1) = \vec{K}_f. \tag{12.100e}$$

The interface boundary conditions are compiled and tabulated in Table 12.4.

Table 12.4 Interface boundary conditions for $\vec{E}$, $\vec{D}$, $\vec{B}$, $\vec{H}$.

Equation Number	Interface boundary conditions		
	Scalar form	Vector form	Comments
Eq. 12.95	$D_1^\perp - D_2^\perp = 0$	$\hat{n}\cdot(\vec{D}_1 - \vec{D}_2) = 0$	If a surface charge resides on the boundary of an interface between two media, then the normal component of the electric displacement fields is discontinuous at the interface boundary. The difference in its value across the interface is equal to the surface charge density.
Eq. 12.97	$B_1^\perp - B_2^\perp = 0$	$\hat{n}\cdot(\vec{B}_1 - \vec{B}_2) = 0$	The normal component (i.e., orthogonal to the surface of separation) of the magnetic induction is continuous across the interface.
Eq. 12.99	$E_1^\parallel - E_2^\parallel = 0$	$\hat{n}\times(\vec{E}_2 - \vec{E}_1) = 0$	The tangential component of the electric field (i.e., the component along the interface surface) is continuous across the interface between two media.
Eq. 12.100	$(H_2^T - H_1^T) = K_f$	$\hat{n}\times(\vec{H}_2 - \vec{H}_1) = \vec{K}_f$	When a surface current flows at the surface of separation, for example, between a dielectric material and a conducting material, then the tangential component of the magnetic intensity field changes discontinuously across the surface. The difference is equal to the transverse surface current density.

In the first section of the next chapter, we discuss the electromagnetic waves and the Poynting vector along which the electromagnetic energy flows. Along with the interface boundary conditions above, the electromagnetic wave phenomena provide a simple basis for the laws of reflection and refraction. The first section in the next chapter also sets up the framework for the special theory of relativity, which is presented in the rest of Chapter 13.

Problems with Solutions

P12.1:

Using Coulomb's law, show that $\oiint \vec{E}(\vec{r},t)\cdot \check{n}dS = \frac{q_{enc}(\vec{r},t)}{\varepsilon_0}$ (Eq. 12.17a) wherein the integration is over an *arbitrary* closed surface and $q_{enc}(\vec{r},t)$ is the total electric charge fenced *inside* the closed surface.

Solution:

Consider am infinitesimal tiny patch on an arbitrary closed surface which encloses a charge q, as shown in Fig. P12.1(a). The ***outward*** normal to this tiny patch is along $\hat{n}$, which, in general, makes an angle ξ with the unit vector $\hat{u}$ which is along the line from the charge to the patch. The (pseudo-)vector area of the tiny patch is $\overrightarrow{dS} = dS\hat{n}$. The solid angle this patch subtends at the charge q is $d\Omega = \frac{\check{u}\cdot\overrightarrow{dS}}{\left|\vec{r}-\vec{r}'\right|^2} = \frac{dS\cos\xi}{\left|\vec{r}-\vec{r}'\right|^2}$, as seen from Fig. P12.1(b).

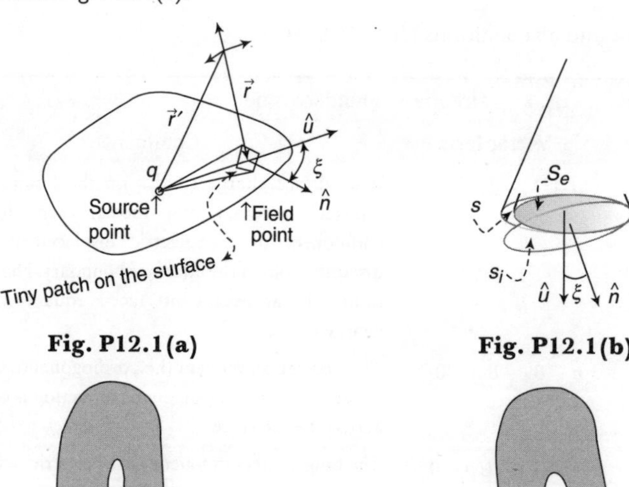

Fig. P12.1(a)

Fig. P12.1(b)

Fig. P12.1(c)

Fig. P12.1(d)

Hence, $\oiint \vec{E}(\vec{r},t)\cdot \breve{n} dS = \oiint \left(\frac{q}{4\pi\varepsilon_0} \frac{\breve{u}}{|\vec{r}-\vec{r}'|^2} \right) \cdot \overrightarrow{dS} = \frac{q}{4\pi\varepsilon_0} \oiint \frac{dS\cos\xi}{|\vec{r}-\vec{r}'|^2} = \frac{q}{4\pi\varepsilon_0} \oiint d\Omega = \frac{q}{\varepsilon_0}$. The result we are looking for follows from the fact that the above result is independent of the actual position of the charge within the closed surface, including as in Fig. P12.1(c). The right hand side becomes, on using the principle of superposition of the electric intensity, $\frac{q_{enc}}{\varepsilon_0}$; $q_{enc}(\vec{r}, t)$ being the total electric charge ***enclosed***. If the charge is outside the closed surface, as in Fig. P12.1(d), we can see that the integral is zero.

P12.2:

Show that the confidence level in our contention that the force between two charges goes as inverse-square of the distance between the charges is related to the confidence with which we know the rest mass of the photon.

Solution:

The confidence in the inverse-square-law comes from our belief that the Coulomb potential goes as ***one-over-distance***. We therefore ask what would happen if this potential is even slightly modified by scaling the numerator: $V(r) \rightarrow \frac{e^{-\left(\frac{r}{\lambda}\right)}}{r} = \frac{e^{-\left(\frac{r}{\frac{h}{\mu c}}\right)}}{r} = \frac{e^{-\left(\frac{r\mu c}{h}\right)}}{r}$. We have scaled the potential by $e^{-\left(\frac{r}{\lambda}\right)}$, where λ must have the dimension of length. A reasonable candidate for this is the de Broglie wavelength, $\lambda = \frac{h}{p} = \frac{h}{\mu c}$, where μ would be the rest-mass of the (virtual) photon which mediates the interaction between the charges. We see that $\mu \rightarrow 0$ therefore corresponds to the exact one-over-distance potential, i.e., the one-over-distance-square force. The rate at which the potential between two charges diminishes with distance is thus related to how accurately we know the rest-mass of the photon.

> *"Because classical Maxwellian electromagnetism has been one of the cornerstones of physics during the past century, experimental tests of its foundations are always of considerable interest. Within that context, one of the most important efforts of this type has historically been the search for a rest mass of the photon....."*
>
> *—The Mass of the Photon by Liang-Cheng Tu, et al., 2005. Rep. Prog. Phys. 68. 77–130.*

P12.3:

Sketch the electric intensity, and the electric potential, as a function of distance from the center of the charge of:

(a) a ***hollow*** uniformly charged spherical surface, of ***surface*** charge density σ, and

(b) ***a solid*** uniformly charged sphere, of ***volume*** charge density distance ρ.

The radius of the sphere (a) and (b) both is R, and the total charge in both the cases is Q.

Solution:

We use Eq. 12.17a, $\oiint \vec{E}(\vec{r},t)\cdot \breve{n}dS = \frac{q}{\varepsilon_0}$, to determine the electric intensity at an arbitrary distance from the center of the sphere, and then integrate it to get the potential at a point.

Case (a): $E(r>R) = \frac{1}{4\pi\varepsilon_0}\frac{Q}{r^2} = \frac{1}{4\pi\varepsilon_0}\frac{\sigma 4\pi R^2}{r^2} = \frac{1}{\varepsilon_0}\frac{\sigma R^2}{r^2}\Big]_{r>R}$, and $E(r<R) = 0]_{r<R}$

Inside the spherical surface charge, there is no charge, hence: $E(r<R) = 0]_{r<R}$. *Just outside* the spherical surface charge: $E(r=R^+) = \frac{1}{4\pi\varepsilon_0}\frac{Q}{r^2} = \frac{1}{4\pi\varepsilon_0}\left[\frac{\sigma 4\pi R^2}{r^2}\right]_{r=R^+} = \frac{\sigma}{\varepsilon_0}\Big]_{r=R^+}$ At larger distances, the electric intensity drops as per the inverse square of the distance from the center of the sphere.

Case (b): $\boxed{s>R}:[E(s)]_{s>R} = \frac{1}{4\pi\varepsilon_0}\frac{Q}{s^2} = \frac{1}{4\pi\varepsilon_0}\frac{\rho\left(\frac{4}{3}\pi R^3\right)}{s^2} = \frac{1}{3\varepsilon_0}\frac{\rho R^3}{r^2}.$

$$\boxed{s<R}: \underset{\substack{\text{surface at}\\ \text{distance } s}}{\oiint} \vec{E}(\vec{r})\cdot\vec{dS} = \underset{\substack{\text{volume}\\ \text{enclosed}}}{\iiint} \vec{\nabla}\cdot\vec{E}(\vec{r})d^3\vec{r} = \frac{1}{\varepsilon_0}\underset{\substack{\text{volume}\\ \text{enclosed}}}{\iiint}\rho d^3\vec{r} = \frac{4}{3\varepsilon_0}\pi s^3\rho = \frac{4}{3\varepsilon_0}\pi s^3\frac{Q}{\frac{4}{3}\pi R^3} = \frac{Q}{\varepsilon_0}\frac{s^3}{R^3}$$

$E(4\pi s^2) = \frac{Q}{\varepsilon_0}\frac{s^3}{R^3}$, i.e., $E = \frac{Q}{4\pi\varepsilon_0}\frac{s}{R^3}$. The electric intensity increases linearly with s till s = R, and then drops as per the inverse square of the distance from the center of the sphere.

$$\phi(r<R) = -\int_{-\infty}^{r} E(r)dr = -\int_{-\infty}^{R}\frac{1}{4\pi\varepsilon_0}\left(\frac{Q}{r^2}\right)dr - \int_{R}^{r}\frac{1}{4\pi\varepsilon_0}\left(\frac{Qs}{R^3}\right)ds = \frac{1}{4\pi\varepsilon_0}\frac{Q}{R} - \frac{1}{4\pi\varepsilon_0}\frac{Q}{R^3}\left(\left[\frac{s^2}{2}\right]_R^r\right)$$

$$\phi(r<R) = \frac{1}{4\pi\varepsilon_0}\frac{Q}{R} - \frac{1}{4\pi\varepsilon_0}\frac{Q}{2R^3}(r^2-R^2) = \frac{1}{4\pi\varepsilon_0}\frac{Q}{R} - \frac{1}{4\pi\varepsilon_0}\frac{Q}{2R^3}r^2 + \frac{1}{4\pi\varepsilon_0}\frac{QR^2}{2R^3}$$

$$\phi(r<R) = \frac{1}{4\pi\varepsilon_0}\frac{3}{2}\frac{Q}{R} - \frac{1}{4\pi\varepsilon_0}\frac{Q}{2R^3}r^2 = \frac{1}{4\pi\varepsilon_0}\frac{Q}{2R}\left(3 - \frac{r^2}{R^2}\right).$$

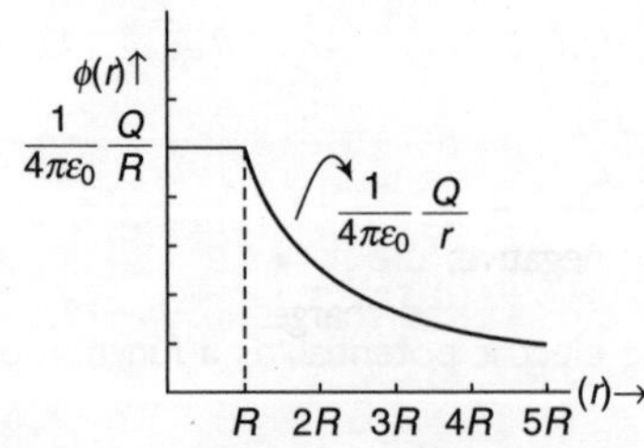

(a): The potential remains invariant inside the hollow shell. The electric intensity is zero inside the shell. Outside the shell, the intensity drops as per the inverse distance-square outside the sphere. The potential drops as one over distance outside the sphere.

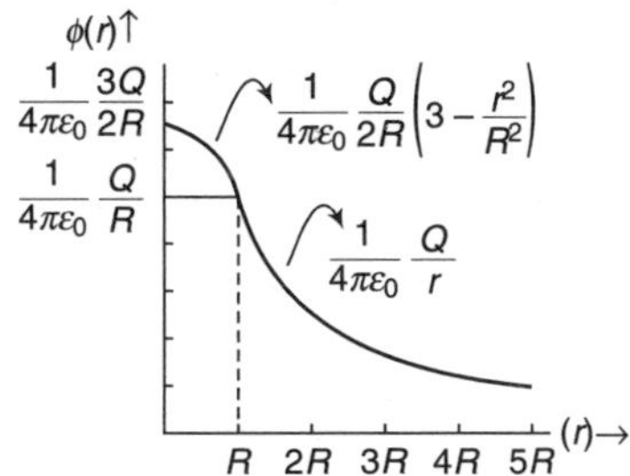

(b): The potential varies with distance ***both*** inside and outside a uniformly charged sphere. The electric intensity first increases linearly from the center to the edge of the sphere, and then drops as per the inverse distance-square outside the sphere.

P12.4:

(a) Prove that the intensity of an electric field in a conductor placed in a time-independent electric field is zero.

(b) Prove that electric charge on a conductor resides only on the surface of a conductor, no matter what shape, and the intensity of the electric field is normal to the surface at each point.

(c) If a charge $+q$ is placed inside a cavity in a conductor, prove that it produces an effect as if the charge $+q$ is distributed on the surface of the conductor.

Solution:

(a) Positive and negative charges that belong to the neutral matter inside the conductor reorganize quickly when the conductor is placed in an external electric field (See the figure given below, Fig. P12.4(a)). The reorganization of the charges generates an electric field in the opposite direction as shown till the induced field cancels the external field, and equilibrium is achieved quickly. The electric field inside the conductor is therefore zero.

(b) Since the electric field inside the conductor is zero, its divergence at all points inside is also zero. Hence, by Eq. 12.16a, the charge at all points inside is zero. Any net charge on the conductor must therefore be smeared only on its surface. Under equilibrium condition, the charge would just sit on the surface, smeared all over, so the surface of the conductor, no matter what shape, is an equipotential surface. Any change in the potential must therefore be orthogonal to the surface, and the electric intensity is then normal to the surface at each point (see the figure below, Fig. P12.4 (b)).

(c) A positive charge in the cavity causes the conducting electrons in the conductor to move closer toward it inducing a net negative charge on the inner surface of the cavity as shown in Fig. P12.4(c). The net charge inside any Gaussian surface (such as in Fig. P12.4(c)) is zero. The over-all charge neutrality of matter in the conducting material spreads a net positive charge on the outer surface, exactly equal to the value of the charge in the cavity, by the conservation of the electrical charge. If the charge in the cavity is negative, the conducting electrons would move away to the outer surface, reversing the sign of the induced charges in Fig. P12.4(c). The electric field will be normal to the surface, just as in the case (b).

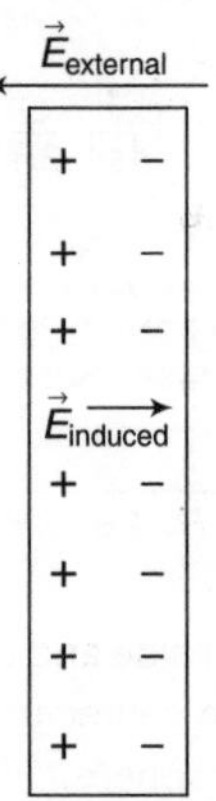

Fig. P12.4(a) A conductor in a time-independent external electric field.

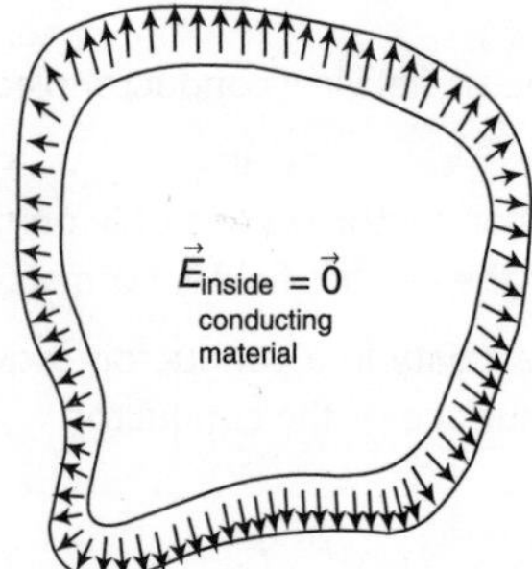

Fig. P12.4(b) A charge on conductor spreads to the outer surface. The field inside is zero, and that outside is normal to the surface of the conductor.

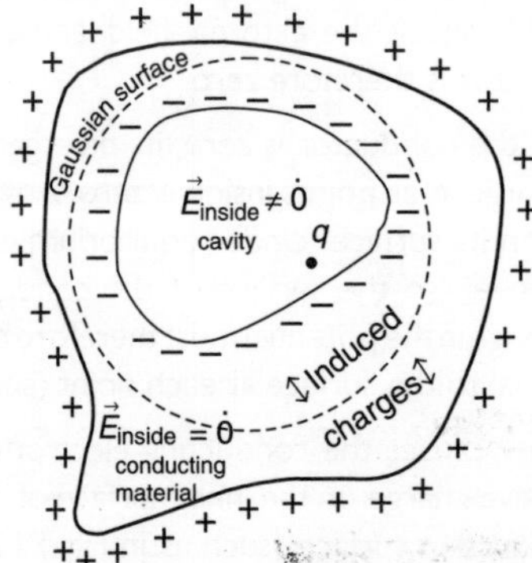

Fig. P12.4(c) A charge inside a cavity in a conductor produces an electric field inside the cavity, but there is no field in the conducting material itself. Outside the conductor, the field is similar to the case (b).

P12.5:

A free (i.e., mobile) charge is implanted in a sphere made of dielectric material (dielectric constant ε_r (Eq. 12.77b) so that this charge is distributed throughout the sphere at a uniform volume charge density ρ_f. Determine the energy of this system.

Solution:

Every increment $\delta\rho_f$ of the implanted free charge requires work δW to be done, which would add up to the energy of the system.

$$\delta W = (\delta\rho_f)\phi(\vec{r}) = (\vec{\nabla}\cdot\delta\vec{D})\phi(\vec{r}),\ \text{since } \vec{\nabla}\cdot\vec{D} = \rho_f \quad \text{(Eq. 12.73)}$$

Now, $\vec{\nabla}\cdot(\phi\delta\vec{D}) = \phi\vec{\nabla}\cdot(\delta\vec{D}) + \vec{\nabla}\phi\cdot\delta\vec{D}$. Therefore, $\delta W = \vec{\nabla}\cdot(\phi\delta\vec{D}) - \vec{\nabla}\phi\cdot\delta\vec{D}$

Hence, the required energy of the system is given by the integral over whole space $\iiint \delta W dV$.

$W = \iiint \vec{\nabla}\cdot(\phi\delta\vec{D})dV + \iiint \vec{E}\cdot\delta\vec{D}dV = \oint\!\!\oint \phi\delta\vec{D}\cdot\vec{dS} + \iiint \vec{E}\cdot\delta\vec{D}dV = \iiint \vec{E}\cdot\delta\vec{D}dV$, since the surface integral goes to zero when integration is taken over all space.

Again, $\vec{E}\cdot\delta\vec{D} = \varepsilon\vec{E}\cdot\delta\vec{E} = \vec{D}\cdot\delta\vec{E} = \frac{1}{2}\vec{D}\cdot\vec{E} = \frac{1}{2}\varepsilon E^2$. HENCE: $\delta W = \frac{1}{2}\iiint \vec{D}(\vec{r})\cdot\vec{E}(\vec{r})dV$.

To get $\vec{D}(\vec{r})$ in the dielectric sphere in which the mobile/free charge is implanted, we use Eq. 12.74:

$$\boxed{\text{For } s<R}: \oint\!\!\oint \vec{D}\cdot\vec{dS} = \iiint \vec{\nabla}\cdot\vec{D}d\tau' = \iiint \rho_f(\vec{r}')d\tau' = \rho_f\iiint d\tau' = \rho_f\frac{4}{3}\pi s^3$$

i.e., $\boxed{\text{For } s<R}: D4\pi s^2 = \rho_f\frac{4}{3}\pi s^3 \Rightarrow \vec{D}(\vec{r}) = \frac{\rho_f}{3}r\breve{e}_r$. Hence, $\boxed{\text{For } r<R}: \vec{E}(\vec{r}) = \frac{\rho_f}{3\varepsilon}r\breve{e}_r = \frac{\rho_f}{3\varepsilon_0\varepsilon_r}r\breve{e}_r$.

$$\boxed{\text{For } s>R}: \oint\!\!\oint \vec{D}\cdot\vec{dS} = \varepsilon_0\oint\!\!\oint \vec{E}\cdot\vec{dS} = \rho_f \underset{\substack{\text{charge}\\ \text{enclosed}}}{\iiint} d\tau' = \rho_f\frac{4}{3}\pi R^3$$

i.e., $\boxed{\text{For } r>R}: D4\pi r^2 = \rho_f\frac{4}{3}\pi R^3 \Rightarrow \vec{D}(\vec{r}) = \frac{\rho_f}{3}\frac{R^3}{r^2}\breve{e}_r$. Hence, $\boxed{\text{For } r>R}: \vec{E}(\vec{r}) = \frac{\rho_f}{3\varepsilon_0}\frac{R^3}{r^2}\breve{e}_r$

$$W = \frac{1}{2}\iiint \vec{D}(\vec{r})\cdot\vec{E}(\vec{r})dV = \left[\frac{4\pi}{2}\int_{r=0}^{R}\left(\frac{\rho_f}{3}r\breve{e}_r\right)\cdot\left(\frac{\rho_f}{3\varepsilon_0\varepsilon_r}r\breve{e}_r\right)r^2dr\right] + \left[\frac{4\pi}{2}\int_{r=R}^{\infty}\left(\frac{\rho_f}{3}\frac{R^3}{r^2}\breve{e}_r\right)\cdot\left(\frac{\rho_f}{3\varepsilon_0}\frac{R^3}{r^2}\breve{e}_r\right)r^2dr\right]$$

$$W = \frac{4\pi}{2}\frac{\rho_f^2}{9\varepsilon_0\varepsilon_r}\int_{r=0}^{R} r^4dr + \frac{4\pi}{2}\frac{\rho_f^2}{9\varepsilon_0}R^6\int_{r=R}^{\infty}\frac{1}{r^2}dr = \frac{2\pi}{9}\frac{\rho_f^2}{\varepsilon_0\varepsilon_r}\frac{R^5}{5} + \frac{4\pi}{2}\frac{\rho_f^2}{9\varepsilon_0}R^6\left[\frac{-1}{r}\right]_R^{\infty} = \frac{2\pi}{9\varepsilon_0}\rho_f^2R^5\left(\frac{1}{5\varepsilon_r}+1\right).$$

P12.6:

A spherical charge distribution is described by a charge density that has azimuthal symmetry but depends on (r, θ) of the spherical polar coordinate system. It is given by:

$\rho(r,\theta) = k\frac{R}{r^2}(R-2r)\sin\theta$, where k is a constant. Determine the leading term in the multipole expansion of the potential on the Z-axis at a distance $z \gg R$.

Solution:

The multipole expansion is given by Eq. 12.55. The monopole term for $\ell = 0$ and the dipole term for $\ell = 1$ both integrate to zero. The first non-zero term, which is therefore the leading term, is the quadrupole term, for $\ell = 2$:

$$\boxed{\ell=2\text{ term}}:\ \phi_{\ell=2}=\frac{1}{4\pi\varepsilon_0}\frac{1}{r^3}\iiint\rho(\vec{r}\,')(r')^2P_2(\cos\theta')dV'=\frac{1}{4\pi\varepsilon_0}\frac{1}{r^3}\iiint\rho(\vec{r}\,')(r')^2\left\{\frac{1}{2}(3\cos^2\theta'-1)\right\}dV'$$

$$\phi_{\ell=2}(\vec{r})=\frac{1}{4\pi\varepsilon_0}\frac{1}{r^3}\iiint\left\{k\frac{R}{r'^2}(R-2r')\sin\theta'\right\}(r')^2\left\{\frac{1}{2}(3\cos^2\theta'-1)\right\}r'^2dr'\sin\theta' d\theta' d\phi$$

$$\phi_{\ell=2}(\vec{r})=\frac{1}{4\pi\varepsilon_0}\frac{1}{r^3}\int_0^{2\pi}d\phi\int_{r'=0}^{\infty}\int_{\theta'=0}^{\pi}\left\{k\frac{R}{r'^2}(R-2r')\sin\theta'\right\}(r')^2\left\{\frac{1}{2}(3\cos^2\theta'-1)\right\}r'^2dr'\sin\theta\, d\theta$$

$$\phi_{\ell=2}(\vec{r})=\frac{2\pi}{4\pi\varepsilon_0}\frac{1}{2}\frac{kR}{r^3}\int_{r'=0}^{R}(R-2r')r'^2dr'\int_{\theta'=0}^{\pi}(3\cos^2\theta'-1)\sin^2\theta' d\theta'$$

$$\phi_{\ell=2}(\vec{r})=\frac{2\pi}{4\pi\varepsilon_0}\frac{1}{2}\frac{kR}{r^3}\int_{r'=0}^{R}(Rr'^2-2r'^3)dr'\int_{\theta'=0}^{\pi}(3\cos^2\theta'-1)\sin^2\theta' d\theta'$$

$$\phi_{\ell=2}(z;z\gg R)=\frac{1}{4\varepsilon_0}\frac{kR}{z^3}\left(\frac{R^4}{3}-\frac{2R^4}{4}\right)\left(-\frac{\pi}{8}\right)=\frac{1}{4\varepsilon_0}\frac{kR}{z^3}\frac{(4-6)R^4}{12}\left(-\frac{\pi}{8}\right)=\frac{\pi}{4\varepsilon_0}\frac{kR^5}{48z^3}$$

Additional Problems

P12.7 Prove that if a function $(\vec{\nabla}\cdot\vec{\nabla})\psi(\vec{r})=0$ satisfies the Laplace's equation *and* satisfies a complete set of boundary conditions, then it is uniquely specified.

P12.8 Wires are connected to the centers of a parallel circular plates capacitor. The radius of the circular plates is R. (of radius a), which form a capacitor. The separation δ between the plates is much less than R. When a current flows out of the wires, the surface charge on the capacitor plates remains uniform at each instant of time, but of course it increases at the plates get charged and decreases as the plates get discharged. (a) Find the electric field between the plates as a function of time. (b) Find the displacement current density vector $\vec{i}_d=\mu_0\varepsilon_0\frac{\partial\vec{E}}{\partial t}$ through a circle of radius $s < R$ half-way between the circular plates, with the plane of the circle being parallel to the plates. (c) With reference to the plane of this circle, determine the magnetic induction field at a distance s from the line joining the centers of the two plates.

P12.9(a) Find the vector potential $\vec{A}$ whose curl would give you the following magnetic induction field: $\vec{B}=B\hat{e}_z$ along the z-axis for $\rho<\rho_0$ and $\vec{B}=\vec{0}$ for $\rho>\rho_0$ (cylindrical polar coordinates $\{\rho,\varphi,z\}$ used). The infinitely long solenoid has such a field. (b) Can a solenoid having a finite length have such a field? [Comment: Part (a) of this example will lead you to a situation where the vector potential is non-zero in a region where the magnetic induction field is zero. Classical experiments cannot detect such a vector potential, even if it is non-zero. The vector potential, however, manifests itself in a quantum experiment known as the Aharonov–Bohm effect. This topic is clearly beyond the scope of this book, but you will presumably revisit this problem when you learn quantum mechanics.]

P12.10(a) Find the vector potential at a point (i) $\vec{r}$ with $r < R$, and (ii) $\vec{r}$ with $r > R$, of a spherical shell of radius R which is spinning at an angular velocity $\vec{\omega} = \omega\hat{e}_z$ if the shell carries a surface charge of uniform density σ. (b) Is the potential continuous at $r = R$? (c) Find the magnetic induction field $\vec{B}$ in both the regions, (i) $r < R$, and (ii) $r > R$,

P12.11 Consider an infinite rectangular sheet having a finite thickness Δ. The permittivity of the material of this block is ε, and it carries a uniform volume charge density ρ. Use the Gauss–Maxwell equation to determine the displacement field $\vec{D}$ and the electric field $\vec{E}$ (i) inside the block and (ii) outside the block.

P12.12 Show that the magnetic intensity field $\vec{H}$ inside a conducting material satisfies the diffusion equation (Eq. 4.38, Chapter 4).

Hint: Use Eq. 12.93, ignoring the time-dependent term.

P12.13 Two spherical cavities, of radii a and b, are carved out from the interior of an electrically neutral conducting sphere of radius R as shown in figure. At the center of each cavity, point charges q_a and q_b are placed.

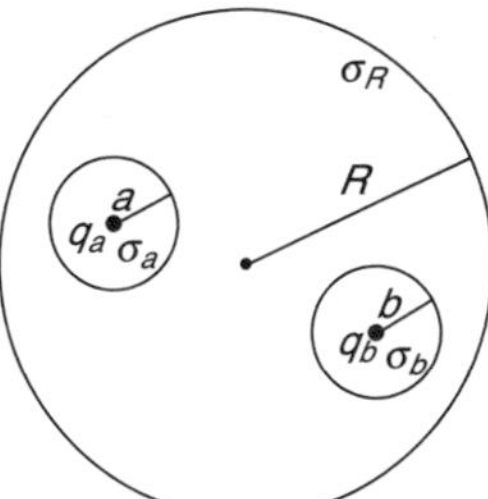

(a) Find the surface charge densities σ_a, σ_b, *and* σ_R induced by the charges q_a, q_b.

(b) Determine the intensity of the electric field outside the conductor.

(c) What is the field inside each cavity?

(d) What is the force on the charges q_a and q_b?

(e) Answers to which of the above questions would change if a third charge, q_c, were brought near the conductor?

P12.14 A static electric charge is distributed in a spherical shell of inner radius R_1 and outer radius R_2. The electric charge density is given by $\rho = a + br$, where r is the distance from the center of the shell. (a) Find the electric field everywhere. (b) Find the electric potential and the energy density for $r < R_1$. Consider the zero of the potential to be at infinity.

P12.15(a) Show that when the current in an isolated circuit element changes, the magnetic induction flux through that circuit element generated by that very current also changes. As a result of this, a 'back' emf is generated. Show, using the Biot–Savart law to determine that the rate at which the flux changes with current is a constant (called 'self-inductance', represented by the letter L), that the induced emf is $\varepsilon = -L\dfrac{dI}{dt}$.

(b) Consider now two circuit coils placed next to each other. Let the number of turns in the first coil be n_1 and that in the second coil be n_2, and let the currents through them be, respectively, I_1 and I_2. We shall denote by φ_{ij} the flux through one turn of the coil i due to the current Ij, with $i = 1, 2$. A changing current in coil 1 induces an emf in the second: $\mathcal{E}_{21} = -M_{21}\dfrac{dI_1}{dt}$ where M_{21} is called the 'mutual inductance'. Prove that $M_{21} = M_{12}$, which can be denoted by M.

(c) Prove that $M = k\sqrt{L_1L_2}$ where $0 \leq k \leq 1$ and that this is a constant that depends only on the geometrical elements of the two circuits.

(d) A magnetic field is given by $\vec{B}(z,t) = \beta\left(\frac{t}{t_0}\right)\left(\frac{z}{z_0}\right)\breve{e}_x$. Determine the emf it induces in a loop of wire loop in a square shape placed in the YZ-plane with one of its corners at the origin.

P12.16(a) A charge distribution in a sphere of radius R has a constant charge density ρ inside the sphere and zero outside. Use the Poisson equation to determine the potential it generates at field points.

(b) A sphere radius a has a volume charge density, given by $\rho_c = \begin{Bmatrix} \kappa\frac{r}{a}; (0 \leq r \leq a), \\ 0; (r > a) \end{Bmatrix}$, where κ is a constant of appropriate dimensions. Find the electric potential using the Poisson equation in the entire region. The zero of the potential may be considered to be at infinity. Sketch the potential as a function of r, distance from the center of the sphere.

P12.17 A hollow conical surface carries a uniform surface charge σ. The height of the cone is h, and the radius of the rim at the top is R. Find the potential difference between points "a" (the vertex) and "b" (the center of the top).

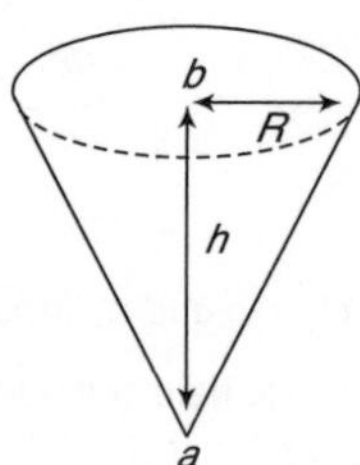

P12.18 A large parallel-plate capacitor has uniform surface charge density σ on the upper plate, and $-\sigma$ on the lower plate. The assembly moves at a constant speed v, as shown in figure.

(a) Find the magnetic field between the plates, also above, and below them.

(b) Find the magnetic force per unit area on the upper plate, including its direction.

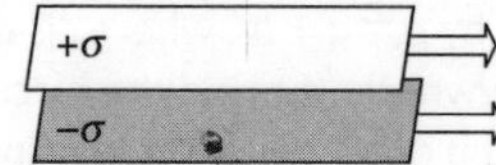

P12.19 A slab of thickness $2a$, extending from $z = -a$ to $z = +a$, carries a uniform volume current density $\vec{J} = J\hat{e}_x$. Determine the magnetic field as a function of z, both inside and outside the slab.

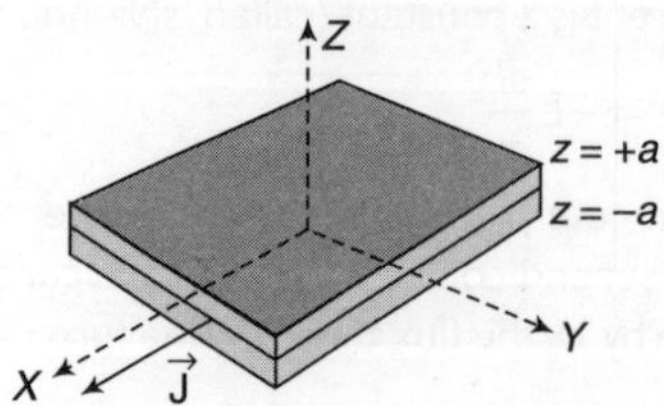

P12.20 A cube of side L is placed with its three sides parallel to the Cartesian coordinate axes and one of its corners at the origin. The cube is in the quadrant where $x \geq 0$, $y \geq 0$, $z \geq 0$. A flat thin square sheet of size L^2 is placed inside the cube, parallel to the XY-plane, at $z = \frac{L}{2}$. This sheet carries a charge density $\sigma = -xy$. Determine the electric flux through the six faces of the cube.

P12.21 Two long coaxial solenoids of different radii each carries a current I. in directions opposite to each other. opposite directions, as shown in figure. The inner solenoid has a radius ρ_1 and has N_1 turns per unit length. The outer solenoid has a radius ρ_2 and has N_2 turns per unit length. Find the magnetic field in the three regions: (i) inside the inner solenoid, (ii) between the two solenoids and (iii) outside the outer solenoid.

P12.22 Consider an excellent, perfect, conducting fluid, such as a fluid consisting of charged particles. Its resistance may be assumed to be zero. Prove that the magnetic flux crossing a closed loop that is moving along with the fluid cannot change with time.

P12.23 The magnetic field intensity in a medium '1' is $\vec{H}_1 = 3\breve{e}_x + 4\breve{e}_y + 2\breve{e}_z \; A/m$. The relative permeability of this medium is $\mu_{r,1} = \frac{\mu_1}{\mu_0} = 5000$. The medium '1' resides in the region $z \leq 2$ at which the medium changes and the medium '2' resides above it. There is no current at the surface. The medium '2' has a relative permeability $\mu_{2,1} = \frac{\mu_2}{\mu_0} = 2000$. Determine the magnetic field intensity in the medium 2.

P12.24 A surface charge resides on the surface of a dielectric material placed in a region where there is nothing else, except for this sphere. The electric permittivity of the dielectric material of the sphere is ε. The surface charge density is expressed in a spherical polar coordinate system, centerd at the center of the sphere, and given by $\sigma(\vec{r}) = f_r(r)f_\theta(\theta)f_\varphi(\varphi)\; C/M^2$. The electric displacement vector in the dielectric material is $\vec{D}(\vec{r}) = D_r\hat{e}_r + D_\theta\hat{e}_\theta + D_\varphi\hat{e}_\varphi$. Find the electric intensity vector and the displacement vector just outside the surface the sphere at the point whose position vector is $(r + \delta)\hat{e}_r$, where δ is an infinitesimal positive distance.

References

[1] Arfken, G. B., H. J. Weber, and F. E. Harris. 2013. *Mathematical Methods for Physicists.* Amsterdam: Elsevier.

[2] Feynman, R. P. 2011. *The Feynman Lectures on Physics* Vol. 2, New Millennium edition. New York: Basic Books.

[3] Davis, M. 2006. 'A Generalized Helmholtz Theorem for Time-varying Vector Fields.' *Am. J. Phys.* 74–76. DOI: 10.1119/1.2121756.

[4] Griffiths, David J. 2015. *Introduction to Electrodynamics.* London: Pearson.

CHAPTER 13

Introduction to the Special Theory of Relativity

This velocity is so nearly that of light that it seems we have strong reason to conclude that light itself (including radiant heat and other radiations) is an electromagnetic disturbance in the form of waves propagated through the electromagnetic field according to electromagnetic laws.

—*James Clerk Maxwell*

13.1 ENERGY AND MOMENTUM IN ELECTROMAGNETIC WAVES

Maxwell's equations from the previous chapter (Eqs. 12.16a, b, c, and d) can be written in another form. This alternate form is known as the wave form, for reasons that will become clear shortly. In that form, they would become the cornerstone of a revolution in physics. We now embark upon an exotic journey toward this extra-ordinary development that took place over a hundred years ago, and our first step begins with the Maxwell's equations:

$$\vec{\nabla}\cdot\vec{E}(\vec{r},t)=\frac{\rho(\vec{r},t)}{\varepsilon_0}, \tag{13.1a}$$

$$\vec{\nabla}\cdot\vec{B}(\vec{r},t)=0, \tag{13.1b}$$

$$\vec{\nabla}\times\vec{E}(\vec{r},t)=-\frac{\partial\vec{B}(\vec{r},t)}{\partial t}, \tag{13.1c}$$

and
$$\vec{\nabla}\times\vec{B}(\vec{r},t)=\mu_0\vec{J}(\vec{r},t)+\mu_0\varepsilon_0\frac{\partial\vec{E}(\vec{r},t)}{\partial t}. \tag{13.1d}$$

We now take the *curl* of the equation for the *curl of the electric field*:

$$\vec{\nabla}\times(\vec{\nabla}\times\vec{E}(\vec{r},t))=-\frac{\partial}{\partial t}(\vec{\nabla}\times\vec{B}(\vec{r},t))=-\mu_0\frac{\partial\vec{J}(\vec{r},t)}{\partial t}-\frac{\partial}{\partial t}\left\{\mu_0\varepsilon_0\frac{\partial\vec{E}(\vec{r},t)}{\partial t}\right\}, \tag{13.2a}$$

i.e.,
$$\vec{\nabla}\times(\vec{\nabla}\times\vec{E}(\vec{r},t))=-\mu_0\frac{\partial\vec{J}(\vec{r},t)}{\partial t}-\mu_0\varepsilon_0\frac{\partial^2\vec{E}(\vec{r},t)}{\partial t^2}. \tag{13.2b}$$

We have interchanged, in the operation for the *curl*, the order in which we take the space derivative and the derivative with respect to time. Similarly, taking the *curl* of the equation for the *curl of the magnetic induction field*, we get,

$$\vec{\nabla} \times (\vec{\nabla} \times \vec{B}(\vec{r}, t)) = \mu_0 \vec{\nabla} \times \vec{J}(\vec{r}, t) - \mu_0 \varepsilon_0 \frac{\partial^2 \vec{B}(\vec{r}, t)}{\partial t^2}, \tag{13.3a}$$

i.e.,
$$\vec{\nabla}(\vec{\nabla} \cdot \vec{B}(\vec{r}, t)) - (\vec{\nabla} \cdot \vec{\nabla})\vec{B}(\vec{r}, t) = \mu_0 \vec{\nabla} \times \vec{J}(\vec{r}, t) - \mu_0 \varepsilon_0 \frac{\partial^2 \vec{B}(\vec{r}, t)}{\partial t^2}. \tag{13.3b}$$

We shall now consider these equations in a source-free region, i.e., in vacuum. Both the electric charge density ρ, and the current density $\vec{J}$, are then zero. Accordingly, we get, from Eq. 13.2b and Eq. 13.3b,

$$(\vec{\nabla} \cdot \vec{\nabla})\vec{E}(\vec{r}, t) = \mu_0 \varepsilon_0 \frac{\partial^2 \vec{E}(\vec{r}, t)}{\partial t^2}, \tag{13.4a}$$

and
$$(\vec{\nabla} \cdot \vec{\nabla})\vec{B}(\vec{r}, t) = \mu_0 \varepsilon_0 \frac{\partial^2 \vec{B}(\vec{r}, t)}{\partial t^2}. \tag{13.4b}$$

If the symmetry in Maxwell's equations (Eqs. 13.1a, b, c, and d) appealed to you, then Eqs. 13.4a and b would impress even more. We get identical equations for the electric field, and for the magnetic field. This symmetry, as will be seen in this chapter, would play a huge role in bringing about an extra-ordinary breakthrough in physics. Some of you would have recognized already that Eqs. 13.4a, b have exactly the same form as Eq. 4.30b of Chapter 4. Yes, indeed, what Eqs. 13.4a and 13.4b are telling us is that the electric field and the magnetic induction field satisfy a wave equation, just like Eq. 4.30b. Comparison between Eq. 4.30b and Eq. 13.4a and b tells us, merely at a glance, that the electric vector field, and the magnetic induction vector field, are waves traveling at the speed

$$c = \frac{1}{\sqrt{\mu_0 \varepsilon_0}}. \tag{13.5}$$

One can easily calculate this speed, since we know the values for the electric permittivity of free space, and we also know its magnetic permeability. These are, from Reference [1],

$$\varepsilon_0 = 8.854187817 \times 10^{-12} \text{ Farad per meter,} \tag{13.6a}$$

and $\mu_0 = 4\pi \times 10^{-7}$ Henry per meter (or Newton per square-Amepere). (13.6b)

The speed of the electromagnetic waves is then given by

$$c = \frac{1}{\sqrt{\mu_0 \varepsilon_0}} = 2.99792458 \times 10^8 \text{ meters per second.} \tag{13.7}$$

It is this result which prompted Maxwell to exclaim that "*This velocity is so nearly that of light that it seems we have strong reason to conclude that light itself (including radiant heat and other radiations) is an electromagnetic disturbance in the form of waves propagated through the electromagnetic field according to electromagnetic laws.*"

The numerical values of the permittivity and the permeability of free space used in Eqs. 13.6a and b, and in Eq. 13.7 are the updated values currently accepted, after continuous revisions prompted by advances in experimental and theoretical research. The journey of carrying out accurate measurements, and obtaining their theoretical estimates, is a continuing one, and remains at the very heart of pushing the frontiers of our understanding of the fundamental laws of nature.

We are aware, from Chapter 4, that a wave transports *energy* through space. In the context of the physical disturbance that travels as a wave through a medium, such as a string, or air, or water, the energy is transported merely by the *nudge* of the series of neighboring oscillators. This mechanism requires the wave to be therefore described by a function of *both* space and time, which is just what we have in Eq. 13.4a and b. However, the wave propagation of the electromagnetic field is very much in *free space*, in *vacuum*. We dismiss, therefore, the presence of any material medium for the propagation of the electromagnetic waves. In one of the most beautiful experiments ever performed, Albert A. Michelson and Edward Morley, in 1887, found that the electromagnetic waves did not require any medium. This description of the vacuum is almost accurate, but there are important amendments to be made to it as our understanding of physics would advance from the classical to the quantum laws, as mentioned later in this section.

The wave equations (Eqs. 13.4a and b) are second order differential equations in which the second derivative with respect to space coordinates are proportional to the second derivative with respect to time. All of you are familiar with the trigonometric sine and cosine functions, which have the property that their second derivatives are proportional to their original form. Hence, a superposition of these linearly independent functions offers itself as a suitable candidate as a solution to the wave equation. It is most convenient to write these solutions in the exponential form, because in this form, differential operations turn out to be equivalent to appropriate scalar or vector multiplications. In particular, when the functions of space and time are expressed in their exponential forms, we have the following operator equivalence:

$$\vec{\nabla} \equiv i\vec{k}, \tag{13.8a}$$

and $$\frac{\partial}{\partial t} \equiv -i\omega. \tag{13.8b}$$

For example, we shall have $\vec{\nabla}\cdot\vec{G}\,(\vec{r},t) = i\vec{k}\cdot\vec{G}\,(\vec{r},t)$ and $\vec{\nabla}\times\vec{G}\,(\vec{r},t) = i\vec{k}\times\vec{G}\,(\vec{r},t)$ when solutions in the exponential form are employed.

We therefore write the solution to the wave equation (Eq. 13.4a) for the electric intensity field as

$$\vec{E}(\vec{r},t) = \hat{e}_E E_0 e^{i(\vec{k}\cdot\vec{r}-\omega t+\delta_E)} = \hat{e}_E\{E_0 e^{i\delta_E}\}e^{i(\vec{k}\cdot\vec{r}-\omega t)} = \hat{e}_E\tilde{E}e^{i(\vec{k}\cdot\vec{r}-\omega t)}, \tag{13.9a}$$

i.e., $$\vec{E}(\vec{r},t) = \hat{e}_E\tilde{E}e^{i(\vec{k}\cdot\vec{r}-\omega t)}, \tag{13.9b}$$

where $$\tilde{E} = E_0 e^{i\delta_E}, \tag{13.10}$$

may be regarded as the complex amplitude of the wave which includes the phase δ_E, which hitherto is considered to be an arbitrary angle. The vector $\vec{k} = k\hat{k} = \frac{2\pi}{\lambda}\hat{k}$ is the wave vector, λ being the wavelength of the wave, and $\hat{k}$ the direction of propagation of the wave. The frequency of oscillation of the wave is $\nu = \frac{\omega}{2\pi}$.

Similarly, the solution to the wave equation (Eq. 13.4b) for the magnetic induction field is:

$$\vec{B}(\vec{r}, t) = \hat{e}_B B_0 e^{i(\vec{k}\cdot\vec{r} - \omega t + \delta_B)} = \hat{e}_B \{B_0 e^{i\delta_B}\} e^{i(\vec{k}\cdot\vec{r} - \omega t)} = \hat{e}_B \tilde{B} e^{i(\vec{k}\cdot\vec{r} - \omega t)}, \tag{13.11a}$$

i.e., $$\vec{B}(\vec{r}, t) = \hat{e}_B \tilde{B} e^{i(\vec{k}\cdot\vec{r} - \omega t)}, \tag{13.11b}$$

where $$\tilde{B} = B_0 e^{i\delta_B}, \tag{13.12}$$

with δ_B being a hitherto arbitrary angle, similar to that in the solution to the wave equation for the electric field.

We shall now analyze the nature of the solutions to Maxwell's wave equations, Eq. 13.9 and Eq. 13.11. In particular, the following questions are of importance:

(i) What is the shape of the wave-front?

(ii) In which direction does the wave travel?

(iii) What are the phases δ_E and δ_B?

(iv) What is the relation, if any, between (a) the polarization direction $\hat{e}_E$ of the electric field, (b) the polarization direction $\hat{e}_B$ of the magnetic induction field, and (c) $\hat{k}$, which is the direction of propagation of the wave-front?

To answer the first question above, we recognize from Eq. 13.9 and Eq. 13.11 that the surface of constant phase is given by

$$(\vec{k}\cdot\vec{r} - \omega t + \delta) = \kappa, \tag{13.13a}$$

i.e., $$\vec{k}\cdot\vec{r} = \kappa + \omega t - \delta, \tag{13.13b}$$

wherein κ is a constant phase angle, and the phase δ is equal to δ_E for the electric vector intensity field, and δ_B for the magnetic vector induction field. At this point of discussion in this chapter, we are yet to be aware of the relation, if any, between δ_E and δ_B. The right hand side of Eq. 13.13b is independent of the position vector. Hence, if we take an arbitrary point on the surface of constant phase, we must have

$$\hat{k}\cdot\vec{r} = \frac{1}{k}(\kappa + \omega t - \delta_{E,B}). \tag{13.13c}$$

The surface of constant phase, as we can see from Eq. 13.13c, progresses linearly with time t.

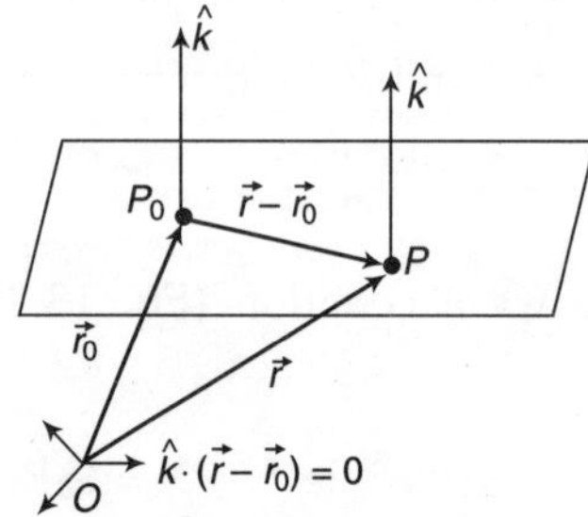

Fig. 13.1 The wave-front of the electromagnetic wave propagating in space is a flat plane that advances as time progresses in the direction $\hat{k}$, which is essentially normal to this plane. A straight line between any two points on this plane is orthogonal to the direction of propagation of this plane.

We therefore conclude that for two points on the surface of constant phase, whose position vectors are respectively $\vec{r}$ and $\vec{r}_0$, as shown in Fig. 13.1, we must have

$$\hat{k}\cdot(\vec{r}-\vec{r}_0)=0, \tag{13.14a}$$

$$\text{i.e.,}\quad \hat{k}\cdot\vec{r}=\hat{k}\cdot\vec{r}_0. \tag{13.14b}$$

Therefore, from Eq. 13.14, we see that the wave fronts which represent solutions to the wave equation for the electric vector field intensity, and also the magnetic vector induction field, are flat planes, which advance as time flows, since the location of the wave-front advances linearly with time. Both the electric field and the magnetic induction field wave-fronts advance linearly with time. However, we do not know yet whether the electric plane and the magnetic plane will be the same, since we are yet to determine the relationship between δ_E and δ_B.

Now, using Eqs. 13.8a and b, and on using Maxwell's equations in free space, we get the following revealing relations:

$$\vec{\nabla}\cdot\vec{E}(\vec{r},t)=0 \Rightarrow i\vec{k}\cdot\vec{E}(\vec{r},t)=0 \Rightarrow \sphericalangle(\vec{k},\vec{E})=\frac{\pi}{2}, \tag{13.15a}$$

$$\text{and}\quad \vec{\nabla}\cdot\vec{B}(\vec{r},t)=0 \Rightarrow i\vec{k}\cdot\vec{B}(\vec{r},t)=0 \Rightarrow \sphericalangle(\vec{k},\vec{B})=\frac{\pi}{2}. \tag{13.15b}$$

Essentially, we have established above that the electric vector intensity field and the magnetic induction vector field are both orthogonal to the direction $\hat{k}$ in which the electric field and the magnetic field waves travel.

Using now Eqs. 13.8a and b, along with the Maxwell's equations for the curl of the electric field and the magnetic induction field, we get

$$\vec{\nabla} \times \vec{E}(\vec{r}, t) = -\frac{\partial \vec{B}(\vec{r}, t)}{\partial t} \Rightarrow i\vec{k} \times \vec{E}(\vec{r}, t) = i\omega\vec{B}(\vec{r}, t),$$

$$\Rightarrow \vec{k} \times \vec{E}(\vec{r}, t) = \omega\vec{B}(\vec{r}, t), \tag{13.16a}$$

and $$\vec{\nabla} \times \vec{B}(\vec{r}, t) = \mu_0\varepsilon_0 \frac{\partial \vec{E}(\vec{r}, t)}{\partial t} \Rightarrow i\vec{k} \times \vec{B}(\vec{r}, t) = -i\omega\mu_0\varepsilon_0\vec{E}(\vec{r}, t),$$

$$\Rightarrow \vec{k} \times \vec{B}(\vec{r}, t) = -\omega\mu_0\varepsilon_0\vec{E}(\vec{r}, t). \tag{13.16b}$$

From Eqs. 13.15a and b and 13.16a and b, we conclude that the set of unit vectors $\{\hat{e}_E, \hat{e}_B, \hat{k}\}$ form a Cartesian-type right handed orthogonal system.

Again, from Eq. 13.16a, we get

$$\vec{k} \times \vec{E}(\vec{r}, t) = \omega\vec{B}(\vec{r}, t) \Rightarrow \vec{k} \times \hat{e}_E\tilde{E}e^{i(\vec{k}\cdot\vec{r} - \omega t)} = \omega\hat{e}_B\tilde{B}e^{i(\vec{k}\cdot\vec{r} - \omega t)}$$

$$\Rightarrow \vec{k} \times \hat{e}_E\tilde{E} = \omega\hat{e}_B\tilde{B}.$$

Hence, $$\frac{\tilde{E}}{\tilde{B}} = \frac{\omega}{k}, \tag{13.17a}$$

which is the ratio of the complex amplitudes of the electric field and the magnetic induction field, as per Eqs. 13.10 and 13.12.

Similarly, from Eq. 13.16b, we get

$$\frac{\tilde{E}}{\tilde{B}} = \frac{k}{\omega\mu_0\varepsilon_0}. \tag{13.17b}$$

From Eq. 13.17a and Eq. 13.17b, it therefore follows that

$$\frac{\tilde{E}}{\tilde{B}} = \frac{k}{\omega\mu_0\varepsilon_0} = \frac{\omega}{k} \Rightarrow \frac{\omega^2}{k^2} = \frac{1}{\mu_0\varepsilon_0}, \tag{13.18a}$$

which essentially guarantees that

$$\frac{\omega}{k} = \frac{2\pi\nu}{(2\pi / \lambda)} = \nu\lambda = \frac{1}{\sqrt{\mu_0\varepsilon_0}} = c, \tag{13.18b}$$

which is the speed (Eq. 13.7) at which the waves propagate.

Now, using Eqs. 13.16a and 13.16b, we can now write the expressions for the magnetic induction field,

$$\vec{B} = \frac{1}{\omega}\vec{k} \times \vec{E} = \frac{k}{\omega}\hat{k} \times \vec{E} = \frac{1}{c}\hat{k} \times \vec{E}, \tag{13.19a}$$

and the electric field $\vec{E} = \frac{k}{\omega\mu_0\varepsilon_0}\vec{B}\times\hat{k} = c\vec{B}\times\hat{k}$. (13.19b)

Finally, since $c = \frac{1}{\sqrt{\mu_0\varepsilon_0}} = \frac{\tilde{E}}{\tilde{B}} = \frac{E_0 e^{i\delta_E}}{B_0 e^{i\delta_B}} = \frac{E_0}{B_0}e^{i(\delta_E - \delta_B)}$, (13.20a)

on noting now that c is real, we conclude that $\delta_E = \delta_B$. (13.20b)

The electromagnetic wave is therefore a transverse wave, with the electric vector and the magnetic vector oscillating *in phase* in a plane, orthogonal to the direction of propagation. As discussed earlier, the unit vectors $\{\hat{e}_E, \hat{e}_B, \hat{k}\}$ form a right-handed Cartesian-type basis set of vectors. The propagation of the electromagnetic wave takes place at a speed given by Eq. 13.7. This wave propagation is depicted in Fig. 13.2.

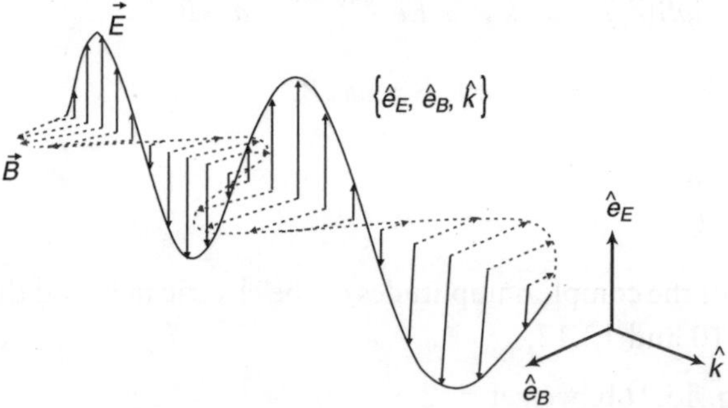

Fig. 13.2 The direction of the magnetic induction field vector and the electric field vector are perpendicular to each other, and to the direction of propagation of the electromagnetic field. In particular, $\vec{E} = c\vec{B}\times\hat{k}$, and $\vec{B} = \frac{1}{c}\hat{k}\times\vec{E}$. The unit vectors $\{\hat{e}_E, \hat{e}_B, \hat{k}\}$ form a right handed coordinate system.

We shall now proceed to deliberate on the mechanism involved in the transfer of energy in the electromagnetic wave. In the early days of the theory of electrodynammics, this seemed very unlikely, since there is no material (at least in the sense in which classical physics is understood) particle that undergoes oscillations as electromagnetic energy is transported by these waves, similar to the case of the waves on strings, or in a medium such as a fluid. We therefore need to understand the properties of the medium in some details, and therefore turn to the characteristic properties of the medium through which the electromagnetic waves pass. These properties are contained in the *specific* properties of the medium, and hence in the constitutive relations. Essentially, a constitutive relation shows the response of the medium to the applied field, similar to the relationship that describes the Hooke's law (Chapter 4, Eq. 4.5a) for elastic materials, or the stress tensor relationship (Chapter 11, Eq. 11.74) for a Newtonian fluid. In the present context, we therefore re-examine the constitutive relations given in the previous chapter (Eqs. 12.76 and 12.92a). These relations are:

$$\vec{D} = \varepsilon_0 \vec{E} + \vec{P} = \varepsilon_0 (1 + \chi) \vec{E} = \varepsilon \vec{E}, \tag{13.21a}$$

and $$\vec{B} = \mu_0 \vec{H} + \vec{M} = (\mu_0 + \mu_0 \chi_m) \vec{H} = \mu_0 (1 = \chi_m) \vec{H} = \mu \vec{H}. \tag{13.21b}$$

As we see, the physical property ε of the medium is a measure of the degree to which it *permits* the electric field to pass through it, and is hence called its permittivity. Likewise, μ is a measure of the degree to which the medium is *permeable* by a magnetic field, and hence called its permeability. Permittivity of dielectric materials comes from its response to an electric field which permits it to get polarized. As discussed in the previous chapter, this happens because the centers of the positive electric charge and the negative electric charge in the medium get displaced, and constitute effectively a multipole charge configuration. Similarly, net magnetization results when the intrinsic magnetic moments of the material do not cancel out and add up to a net magnetic moment, derivable from a multipole vector potential. Unlike matter, vacuum cannot be polarized by an electric field, nor magnetized by a magnetic field, and hence these relations become

$$\vec{D} = \varepsilon_0 \vec{E}, \tag{13.22a}$$

and $$\vec{B} = \mu_0 \vec{H}, \tag{13.22b}$$

with ε_0 as the electric permittivity, and μ_0 as the magnetic permeability of the free space. It is nice, of course, that vacuum is *permitive* to the passage of the electric field, and is *permeable* by the magnetic field. Notwithstanding the detailed mechanism that imparts the properties of electric permitivity and the magnetic permeability of the vaccum, it is essentially these properties that permit the electromagnetic field to advance through vaccum. Without these properties, light from distant stars would not travel through empty space. We would not get the sunlight carrying electromagnetic energy, and we just wouldn't be. Of course, one must remember that between the two properties ε_0 and μ_0, only one is independent, since the two are related by Eq. 13.5, 13.7 the speed of light being a fundamental constant of nature.

Having argued that vacuum would have to have both permittivity and permeability, we must admit that vacuum is an extremely intriguing entity. In common language, we recognize it as free space, devoid of all matter. In more accurate theories of the laws of nature, in particular the quantum theory, this is however strictly not quite correct. Dirac's formulation of the relativistic quantum mechanics, provides for particles and antiparticles to be created and destroyed over extremely short periods, in and out of the vacuum. The quantum uncertainty principle bails out the conservation of energy and matter, in case you are worried about the creation and destruction of these particle-antiparticle pairs. These are advanced considerations; they go well beyond the scope of this book. Advanced readers among you may wish to read about a recent work, by Urban et al. [5], in which the authors argue that the permittivity and the permeability of vacuum has its basis in polarization and magnetization of the particle-antiparticle pairs that are created out of the vacuum.

We will now use $(\vec{E}, \vec{H})$ to describe the electromagnetic field, instead of $(\vec{E}, \vec{B})$. We know, of course, that for vacuum, $\vec{B} = \mu_0 \vec{H}$. The preference for the pair $(\vec{E}, \vec{H})$ is because of the fact that the characteristic impedence of vacuum is given by the ratio E/H, and that the momentum carried by the electromagnetic field is given by $(\vec{E} \times \vec{H})$. The former result is seen readily by using the fact that the characteristic impedance of a transmission line is given by

$$Z_{TL} = \sqrt{\frac{R_{TL} + i\omega L_{TL}}{G_{TL} + i\omega C_{TL}}}, \tag{13.23a}$$

where R_{TL}, L_{TL}, G_{TL} and C_{TL} are respectively the resistance, inductance, conductance, and the capacitance, *per unit length*. To avoid digressing too much, we refer the reader to some other source, such as References [2] through [6], for a detailed discussion leading to Eq. 13.23a. Suffice it does for the present purpose to see that the characteristic impedance of free space is given by a relation similar to Eq. 13.23a, only simpler. The impedance of free space is given, by the square root of the ratio of its *inductance per unit length* (permeability) to its *capacitance per unit length* (permittivity):

$$Z_0 = \sqrt{\frac{L_0}{C_0}} = \sqrt{\frac{\mu_0}{\varepsilon_0}} = \mu_0 c = \mu_0 \frac{E}{B} = \frac{E}{H} \simeq 377 \text{ Ohms}, \tag{13.23b}$$

since $\varepsilon_0 = 8.85 \times 10^{-12}$ Farad per meter,

and $\mu_0 = 4\pi \times 10^{-7}$ Henry per meter.

We now consider the energy in the electromagnetic field. It is obtained as the volume integral of the electromagnetic energy density. It should not surprise us that this is written as a volume integral over the whole space. One must therefore define energy density as a scalar point function, and then integrate it over whole space.

In any wave propagation, the energy is proportional to the square of the amplitude of the wave, so we must expect the electric energy density to be proportional to E^2, and the magnetic energy density to be proportional to B^2. Let us therefore write the electric energy density, and the magnetic energy density, to be respectively given by

$$u_{eed} = \kappa_{e,\varepsilon_0} E^2, \tag{13.24a}$$

and $$u_{med} = \kappa_{m,\mu_0} B^2. \tag{13.24b}$$

κ_{e,ε_0} is the constant of proportionality between E^2 and the electric energy density which we have denoted by u_{eed}. We expect this constant of proportionality to be determined by the electric permittivity of free space, which is therefore indicated as a subscript in Eq. 13.24. Similarly, κ_{m,μ_0} (in Eq. 13.24b) is the constant of proportionality between B^2 and the magnetic energy density u_{med}.

Guided by the *symmetry* in the Maxwell's equations for $\vec{E}(\vec{r}, t)$ and $\vec{B}(\vec{r}, t)$, we expect the electric energy density to be equal to the magnetic energy density, and hence

$$\kappa_{e,\varepsilon_0} E^2 = \kappa_{m,\mu_0} B^2. \tag{13.25}$$

Therefore, we must have

$$\frac{E^2}{B^2} = \frac{\kappa_{m,\mu_0}}{\kappa_{e,\varepsilon_0}} = c^2 = \frac{1}{\mu_0 \varepsilon_0}, \tag{13.26}$$

where we have used Eq. 13.20. We can now easily identify the proportionality constants κ_{m,μ_0} and κ_{e,ε_0}, and write

$$\kappa_{m,\mu_0} = \frac{1}{2\mu_0}, \tag{13.27a}$$

and $$\kappa_{e,\varepsilon_0} = \frac{\varepsilon_0}{2}. \tag{13.27b}$$

Except for the factor $\frac{1}{2}$ in the above equations, which we have not explained, we have comfortably identified the proportionality constants κ_{m,μ_0} and κ_{e,ε_0}. The factor $\frac{1}{2}$ can also be justified easily, not merely because it gives a form similar to the familiar forms for the kinetic energy and the potential energy densities, but also because it is easy to show that the energy density in an electrical circuit having a capacitor is given by $u_{eed} = \frac{\varepsilon_0}{2}E^2$, and that in a circuit containing an inductor, the magnetic energy density is $u_{med} = \frac{1}{2\mu_0}B^2$. Specialized books on electrodynamics justify these expressions more rigorously. The energy in the electromagnetic field is therefore given by the volume integral of the electromagnetic energy density, $u_{\text{emed}} = u_{\text{eed}} + u_{\text{med}}$. Thus,

$$U_{\text{Electromagnetic}} = \iiint u_{\text{emed}}\, dV = \iiint \left(\frac{\varepsilon_0 E^2}{2} + \frac{B^2}{2\mu_0} \right) dV. \tag{13.28}$$

In Section 4.3, Chapter 4, we have pointed out that the transport of energy through a region of space is an essential property that waves have. The electromagnetic waves achieve this even through vacuum, carrying energy. We will now proceed to show that the electromagnetic waves transport energy, and also momentum, in the direction of $\vec{E} \times \vec{H}$.

Since the electromagnetic wave carries energy, we must expect it to be able to do mechanical work on an electric charge. We now estimate the mechanical work δW_M done by the EM field on a charge element δq, in a volume element $\delta\tau$, over a time-interval $\overrightarrow{\delta t}$, in displacing the charge through $\overrightarrow{\delta\ell}$. The work done will be given by

$$\delta W_M = \vec{F}\cdot\overrightarrow{\delta\ell} = \delta q(\vec{E} + \vec{v}\times\vec{B})\cdot\frac{\overrightarrow{\delta l}}{\delta t}\delta t = \delta q\vec{E}\cdot\vec{v}\delta t = (\rho\delta\tau)\vec{E}\cdot\left(\frac{\vec{J}}{\rho}\right)\delta t = (\delta\tau)\vec{E}\cdot\vec{J}\delta t. \tag{13.29}$$

In the above relation, we have identified the total charge as $\delta q = \rho\delta\tau$; i.e., charge density times the volume element. Also, the current density vector $\vec{J} = \rho\vec{v}$ has been used in the above equation. The term $\vec{E}\cdot\vec{J}$ is readily interpreted from the above expression as the work done per unit time, per unit volume; i.e., as power delivered to the charge per unit volume. Thus, the *rate* at which work is done on *all the charges* in the integrated region is given by

$$\frac{\partial W_M}{\partial t}=\iiint_{\substack{\text{whole}\\\text{space}}} d\tau\vec{E}(\vec{r})\cdot\vec{J}(\vec{r}). \tag{13.30}$$

We will now use the fourth equation in Table 12.2 (Eq. 12.16d), to get the current density vector:

$$\vec{J}(\vec{r},t)=\frac{1}{\mu_0}\vec{\nabla}\times\vec{B}(\vec{r},t)-\varepsilon_0\frac{\partial\vec{E}(\vec{r},t)}{\partial t}. \tag{13.31}$$

Hence, $$\frac{\partial W_M}{\partial t}=\frac{1}{\mu_0}\iiint_{\substack{\text{whole}\\\text{space}}} d\tau\vec{E}(\vec{r})\cdot\vec{\nabla}\times\vec{B}(\vec{r},t)-\varepsilon_0\iiint_{\substack{\text{whole}\\\text{space}}} d\tau\vec{E}(\vec{r})\cdot\frac{\partial\vec{E}(\vec{r},t)}{\partial t}. \tag{13.32}$$

Now, $$\vec{E}(\vec{r})\cdot\vec{\nabla}\times\vec{B}(\vec{r},t)=\vec{B}(\vec{r},t)\cdot[\vec{\nabla}\times\vec{E}(\vec{r})]-\vec{\nabla}\cdot[\vec{E}(\vec{r})\times\vec{B}(\vec{r},t)], \tag{13.33}$$

therefore,

$$\frac{\partial W_M}{\partial t}=\frac{1}{\mu_0}\iiint_{\substack{\text{whole}\\\text{space}}} d\tau\begin{Bmatrix}\vec{B}(\vec{r},t)\cdot[\vec{\nabla}\times\vec{E}(\vec{r})]-\\\vec{\nabla}\cdot[\vec{E}(\vec{r})\times\vec{B}(\vec{r},t)]\end{Bmatrix}-\varepsilon_0\iiint_{\substack{\text{whole}\\\text{space}}} d\tau\vec{E}(\vec{r})\cdot\frac{\partial\vec{E}(\vec{r},t)}{\partial t}. \tag{13.34}$$

Again, using Maxwell's equation for the curl of the electric intensity field, we get

$$\frac{\partial W_M}{\partial t}=-\frac{1}{\mu_0}\iiint_{\substack{\text{whole}\\\text{space}}} d\tau\begin{Bmatrix}\vec{B}(\vec{r},t)\cdot\dfrac{\partial\vec{B}(\vec{r},t)}{\partial t}+\\\vec{\nabla}\cdot[\vec{E}(\vec{r})\times\vec{B}(\vec{r},t)]\end{Bmatrix}-\varepsilon_0\iiint_{\substack{\text{whole}\\\text{space}}} d\tau\vec{E}(\vec{r})\cdot\frac{\partial\vec{E}(\vec{r},t)}{\partial t}, \tag{13.35a}$$

i.e., $$\frac{\partial W_M}{\partial t}=-\iiint_{\substack{\text{whole}\\\text{space}}} d\tau\begin{Bmatrix}\dfrac{\vec{B}(\vec{r},t)\cdot\vec{B}(\vec{r},t)}{2\mu_0}+\\\varepsilon_0\dfrac{\vec{E}(\vec{r},t)\cdot\vec{E}(\vec{r},t)}{2}\end{Bmatrix}-\iiint_{\substack{\text{whole}\\\text{space}}} d\tau\vec{\nabla}\cdot[\vec{E}(\vec{r})\times\vec{H}(\vec{r},t)]\}. \tag{13.35b}$$

Hence, $$\frac{\partial W_M}{\partial t}=-\frac{\partial}{\partial t}\iiint_{\substack{\text{whole}\\\text{space}}} d\tau u_{\text{emed}}-\iiint_{\substack{\text{whole}\\\text{space}}} d\tau\,\vec{\nabla}\cdot[\vec{E}(\vec{r})\times\vec{H}(\vec{r},t)]\}, \tag{13.35c}$$

i.e., $$\frac{\partial W_M}{\partial t}=-\frac{\partial}{\partial t}\iiint_{\substack{\text{whole}\\\text{space}}} d\tau u_{emed}-\iiint_{\substack{\text{whole}\\\text{space}}} d\tau\vec{\nabla}\cdot\vec{S}(\vec{r},t)\}, \tag{13.36}$$

where $\vec{S}(\vec{r},t)=\vec{E}(\vec{r},t)\times\vec{H}(\vec{r},t)$. $\quad$ (13.37)

Using now the Gauss' divergence theorem,

$$\frac{\partial W_M}{\partial t} = -\frac{\partial U_{em}}{\partial t} - \oiint \vec{S}(\vec{r}, t)\cdot \hat{n} dA. \tag{13.38}$$

The result in the above Eq. 13.38 is in fact a statement of the conservation of energy. It states that the rate at which mechanical work is done on the charges by the electromagnetic field, is equal to the rate at which the total electromagnetic energy in the field diminishes, *less* the rate at which the electromagnetic energy escapes out of the surface enclosing the region of space. This conservation principle is known as the *Poynting theorem*, named after John Henry Poynting (1852–1914). The vector $\vec{S}(\vec{r}, t) = \vec{E}(\vec{r}, t) \times \vec{H}(\vec{r}, t)$ is called the Poynting vector.

Expressing now the net mechanical work done as a volume integral of the density of the mechanical work done,

$$W_M = \iiint w_{\text{med}}(\vec{r})\, d\tau, \tag{13.39}$$

where w_{med} is the mechanical work density,

$$\text{we get, } \iiint \left\{ \frac{\partial}{\partial t}(w_{\text{med}} + u_{\text{emed}}) \right\} d\tau = -\iiint \vec{\nabla}\cdot\vec{S}(\vec{r})\, d\tau. \tag{13.40}$$

The volume integrals are definite integrals over whole space, and therefore their integrands are equal to each other:

$$\frac{\partial}{\partial t}(w_{\text{med}} + u_{\text{emed}}) + \vec{\nabla}\cdot\vec{S}(\vec{r}, t) = 0. \tag{13.41a}$$

The above relation is identical to the equation of continuity,

$$\frac{\partial \rho}{\partial t} + \vec{\nabla}\cdot\vec{J}(\vec{r}) = 0. \tag{13.41b}$$

We therefore conclude that the electromagnetic energy flows very much the same way as the equation of continuity (discussed in Chapter 10) describes the flow of charge/mass.

We can see that the Poynting vector is given by

$$\vec{S}(\vec{r},t) = \vec{E}(\vec{r},t) \times \frac{\vec{B}(\vec{r}, t)}{\mu_0} = \hat{k}(|\vec{E}(\vec{r},t)|)\left(\frac{|\vec{E}(\vec{r}, t)|}{c\mu_0}\right) = \hat{k}\frac{1}{c\mu_0}\left|\vec{E}(\vec{r}, t)\right|^2 = \hat{k}c\varepsilon_0\left|\vec{E}(\vec{r}, t)\right|^2, \tag{13.42a}$$

$$\text{i.e., } \vec{S}(\vec{r}, t) = \hat{k}c\varepsilon_0 E_0^2 \cos^2(\vec{k}\cdot\vec{r} - \omega t + \delta). \tag{13.42b}$$

Hence, taking the averag , we get

$$\langle\vec{S}\rangle = \frac{c\varepsilon_0 |E_0|^2}{2}\hat{k}. \tag{13.43}$$

The quantity $\frac{c\varepsilon_0 |E_0|^2}{2}$ is the avserage power per unit area delivered by the electromagnetic wave. We have stated above that between the two constants ε_0 and μ_0, only one is fundamental, since the two are related by the speed of light (Eq. 13.5). Usually, it is ε_0 that is regarded as this fundamental quantity, since it is clear from Eq. 13.43 that the power carried by the electromagnetic radiation would be zero if ε_0 was to be zero. It should now be clear that the electric permittivity of free space (and hence the magnetic permeability of the free space) are essential properties for the passage of electromagnetic power, and also energy, through vacuum. It is also not surprising that only one of the two properties, ε_0 and μ_0, is fundamental; it is only a further celebration of the unity of the electrical and the magnetic phenomena.

The dimensions of the Poynting vector are those of energy *flux*; i.e., energy transfer per unit area, per unit time, given by $[\vec{S}(\vec{r})] = \frac{[\text{power}]}{[\text{area}]}$. Carrying on with the dimension analysis further, as is very fruitful in many problems in physics, especially in electrodynamics, we see that the dimensions of $\frac{\vec{S}}{c^2}$ are given by

$$\left[\frac{\vec{S}}{c^2}\right] = [\mu_0\varepsilon_0\vec{S}] = \frac{ML^2T^{-2}}{T} \times \frac{1}{L^2} \times \frac{T^2}{L^2} = ML^{-2}T^{-1}, \tag{13.44}$$

which are essentially the dimensions $MLT^{-1} \times L^{-3} = ML^{-2}T^{-1}$, of the momentum density.

The transverse electromagnetic waves therefore carry both energy and momentum along the direction of the Poynting vector. The momentum of the electromagnetic Poynting vector would of course impact an object and push it, just as a caroms striker would transfer its momentum to the coin it collides with. This is exactly what happens when sunlight falls on the loosened dust particles of a comet, which therefore appear as the comet's *tail*, usually called the comet's *coma*, which the sunlight *pushes* as it illuminates it. The coma therefore always points *away* from the direction of the sunrays (Fig. 13.3), under the radiation pressure. The momentum of the electromagnetic radiation from the Sun impacts the comet's dust, and pushes it in the direction of the incident momentum of the sunlight. The same principle is taken advantage of in the laser-cooling of atoms, but this application is clearly beyond our scope.

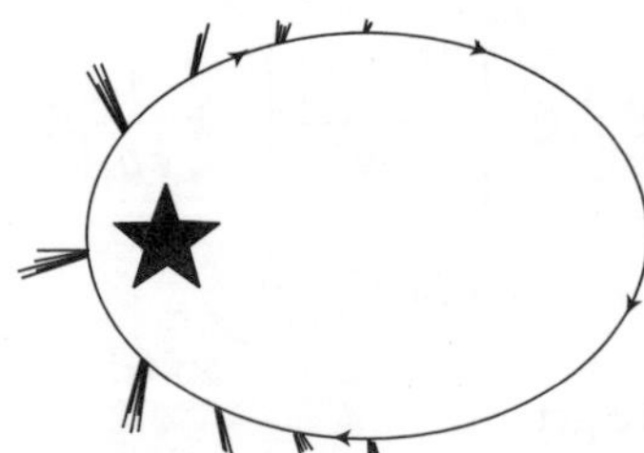

Fig. 13.3 A comet consists of somewhat loosened dust particles which are blown away by the pressure of the electromagnetic radiation from the Sun. Scattered light from this part of the comet therefore appears as a tail, always pointing *away* from the Sun.

13.2 LORENTZ TRANSFORMATIONS: COVARIANT FORM OF THE MAXWELL'S EQUATIONS

The most astounding result we have got above is that electromagnetic waves travel at the speed $c = \frac{1}{\sqrt{\mu_0 \varepsilon_0}} = 2.99792458 \times 10^8$ meters per second. This is in complete contradiction with Galilean–Newtonian scheme of mechanics, in which the speed of a moving object must be referenced to the observer's frame of reference. Galilean–Newtonian mechanics would require this speed to diminish, or increase, depending on whether the observer is moving in the direction of the electromagnetic radiation, or away from it. Equation 13.5 leaves no room to accommodate this, making Maxwell's equations incompatible with Galilean–Newtonian mechanics. It prompted the search for new, non-Galilean, laws of transformation between two frames of reference moving at a constant velocity with respect to each other, as would leave the Maxwell's equations invariant. This must have been a disturbing demand, since the Galilean transformations were at the very heart of Newtonian mechanics. They would ensure that the laws of physics, i.e., the principle of causality in Newton's second law, would hold in every inertial frame of reference (Eq. 3.4, Chapter 3).

Physics was in for a revolution. It demanded transformation laws to connect variables in different coordinate frames of reference moving at a constant velocity with each other that would leave the speed of the electromagnetic waves invariant. Now, to get the same speed of light in the two inertial frames moving at a constant velocity with respect to each other, *speed* being *distance over time*, it becomes imperative that under the new transformation laws, distance between two points in space must change, and so should time-intervals. This would require transformation of both space and time coordinates. The new transformation laws that would be required can be summarized as follows:

(a) Transform both space and time coordinates.

(b) Transformation equations must agree with Galilean transformations in appropriate limits (which will turn out to be $v \ll c$), since the success of the Galilean transformations could not be wished away in Newtonian mechanics.

(c) Ensure that speed of light is the same in all inertial frames of reference.

We therefore consider two Cartesian frames of reference, S and $\bar{S}$, such that their respective coordinate axes are right on top of each other at the *initial* time at which the frame $\bar{S}$ shoots off at a constant velocity $\vec{v} = v\hat{e}_x$ with respect to the frame S, along their common X-axis (Fig. 13.4).

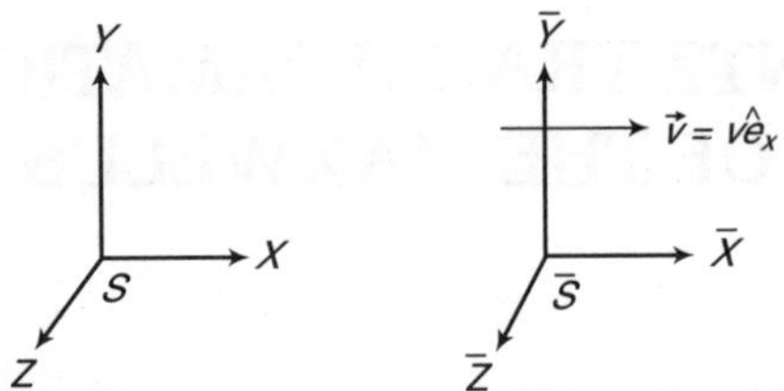

Fig. 13.4 The $\overline{S}$ frame of reference, with coordinate axes $\{\overline{X}, \overline{Y}, \overline{Z}\}$, moves at a constant velocity $\vec{v} = v\hat{e}_x$ with respect to the frame S, whose coordinate axes are $\{X, Y, Z\}$. The respective axes are parallel to each other.

The transformations we are looking for are the Lorentz transformations, named after Hendrik Antoon Lorentz (1853–1928). They transform both space and time coordinates (x, y, z, t) in the frames S and $(\overline{x}, \overline{y}, \overline{z}, t)$ in the frame $\overline{S}$ as per the following laws:

$$\overline{x} = \gamma(x - vt) \quad ; \quad x = \gamma(\overline{x} + v\overline{t}), \tag{13.45a}$$

$$\overline{y} = y \quad ; \quad y = \overline{y}, \tag{13.45b}$$

$$\overline{z} = y \quad ; \quad z = \overline{z}, \tag{13.45c}$$

$$\text{and } \overline{t} = \gamma\left(t - \frac{vx}{c^2}\right) \quad ; \quad t = \gamma\left(\overline{t} + \frac{v\overline{x}}{c^2}\right). \tag{13.45d}$$

In the above transformations,

$$\gamma = \frac{1}{\sqrt{1 - \frac{v^2}{c^2}}} = \frac{1}{\sqrt{1 - \beta^2}}, \tag{13.46a}$$

$$\text{with } \beta = \frac{v}{c}. \tag{13.46b}$$

The Lorentz transformations manifestly address the requirement (a), mentioned above. That the requirement (b) is also satisfied is immediately found on observing that when $v \ll c$, $\beta \to 0$ and $\gamma \to 1$. Accordingly, $\overline{x} \to (x - vt)$ and $\overline{t} \to t$; i.e., we recover the Galilean transformations. It now remains for us to show that the speed of Maxwell's wave is c in *both* the frames S and $\overline{S}$. It will be convenient now to use the tensor notation introduced in Section 2 of Chapter 2. Since space and time both are transformed in the Lorentz transformations, we shall now write them with numbered indices, denoting a 4-vector (i.e., a vector with 4 components in a 4-dimensional space) as (x^0, x^1, x^2, x^3) which would respectively represent $(\tau = ct, x, y, z)$. Note that $x^0 = ct$ has the dimension of length, same as that of the other coordinates.

The transformations represented in Eq. 13.45a, b, c, and d can be written in a compact form in the following matrix equation:

$$\begin{bmatrix} \bar{x}^0 \\ \bar{x}^1 \\ \bar{x}^2 \\ \bar{x}^3 \end{bmatrix} = \begin{bmatrix} \gamma & -\gamma\beta & 0 & 0 \\ -\gamma\beta & \gamma & 0 & 0 \\ 0 & 0 & 1 & 0 \\ 0 & 0 & 0 & 1 \end{bmatrix} \begin{bmatrix} x^0 \\ x^1 \\ x^2 \\ x^3 \end{bmatrix}. \tag{13.47}$$

We have used the notation $(\bar{x}^0, \bar{x}^1, \bar{x}^2, \bar{x}^3)$ for the coordinates in the moving frame of reference $\bar{S}$.

It is important to examine the geometry of the 4-dimensional space we are now confronted with. More appropriately, we should now be talking about spacetime rather than just space, since the Lorentz transformations scramble them. In Section 2.3 of Chapter 2, we learned that the geometry of space (now upgraded to *spacetime*) will be characterized by the procedure to determine the distance between two points in the spacetime. You will expect, therefore, that it will be the 'metric', discussed in Section 2.3 of Chapter 2, which would characterize the structure of the spacetime. In the 3-dimensional Euclidean space, the metric **g**, defined by Eq. 2.24, provided the measure of the distance between 2 points. It was given, when the Cartesian coordinate system is used, by:

$$\mathbf{g} = g_{ij} = \begin{bmatrix} 1 & 0 & 0 \\ 0 & 1 & 0 \\ 0 & 0 & 1 \end{bmatrix}, \tag{13.48a}$$

which produces the measure of the distance between two points as the square root of the 3-dimensional scalar product:

$$\overrightarrow{ds} \cdot \overrightarrow{ds} = ds^2 = \sum_{i=1}^{3} \sum_{j=1}^{3} dx_i g_{ij} dx_j = dx_1^2 + dx_2^2 + dx_3^2, \tag{13.48b}$$

where $g_{ij} = \delta_{ij}$, δ_{ij} being the Kronecker-δ. (13.49)

Extension of Eq. 13.48 to a 4-dimensional Euclidean spacetime necessitates the upper limit of the double sum to be 4, instead of 3. However, let us not jump ahead of ourselves to conclude that this will serve our purpose. In fact, it will not, for reasons described below. Let us analyze the g-metric with $g_{ij} = \delta_{ij} \; \forall \; i, j = 1, 2, 3, 4$, which would correspond to this naïve extension of the 3-dimensional Euclidean space to 4-dimensional spacetime. Just as the measure of distance between two points in the usual 3-dimensional Euclidean space is given by the Pythagoras theorem, wherein the distance-square between two points is given by the 3-dimensional scalar product (Eq. 13.48b) which gives the sum of the squares of the orthogonal projections of $\overrightarrow{ds}$, the measure of the distance between two events in the spacetime S is given by scalar product in the 4-dimensional space, which would give the squared distance of an infinitesimal step in the 4-dimensional Euclidean space:

$$\overrightarrow{d\xi}\cdot\overrightarrow{d\xi} = d\xi^2 = (dw)^2 + (dx)^2 + (dy)^2 + (dz)^2, \tag{13.50a}$$

wherein dw would be interpreted as the projection of the 4-dimensional displacement along the fourth orthogonal axis in the 4-dimensional Euclidean space.

It is obvious that the g-metric for this geometry would be a 4x4 unit matrix:

$$[g_{\mu\nu}]_{\text{Euclid } 4d} = \begin{bmatrix} 1 & 0 & 0 & 0 \\ 0 & 1 & 0 & 0 \\ 0 & 0 & 1 & 0 \\ 0 & 0 & 0 & 1 \end{bmatrix}. \tag{13.50b}$$

The 4-dimensional scalar product in Eq. 13.50a for which the g-metric is given by Eq. 13.50b is, however, *not* appropriate to develop the special theory of relativity. Instead, the following metric is required, for reasons that would soon become clear:

$$[g_{\mu\nu}] = \begin{bmatrix} 1 & 0 & 0 & 0 \\ 0 & -1 & 0 & 0 \\ 0 & 0 & -1 & 0 \\ 0 & 0 & 0 & -1 \end{bmatrix} = [g^{\mu\nu}]. \tag{13.51}$$

In fact, alternatively, the following g-metric also works:

$$[g_{\mu\nu}] = \begin{bmatrix} -1 & 0 & 0 & 0 \\ 0 & 1 & 0 & 0 \\ 0 & 0 & 1 & 0 \\ 0 & 0 & 0 & 1 \end{bmatrix} = [g^{\mu\nu}]. \tag{13.52}$$

In the literature, you will find one or the other forms, Eq. 13.51 or Eq. 13.52, of the g-metric being used. In this book, we shall use the form of the g-metric as given in Eq. 13.51. It is represented by stacking its diagonal elements in a row, (1 –1 –1 –1), called appropriately as the *signature* of the g-metric.

We shall employ the following equivalence between the 4-dimensional spacetime coordinates and the traditional notation for the coordinates in the Cartesian coordinate system:

$$x^0 = ct,\ x^1 = x,\ x^2 = y \text{ and } x^3 = z. \tag{13.53}$$

Having chosen the *signature* of the g-metric, we see that the distance between two '*points*' (called hereafter as '*events*', rather than 'points') in the 4-dimensional spacetime S will be given, instead of Eq. 13.50a, by the 4-dimensional scalar product,

$$\overrightarrow{ds}\cdot\overrightarrow{ds} = ds^2 = dx^\mu dx_\mu = dx^\mu g_{\mu\upsilon} dx^\upsilon, \tag{13.54a}$$

where the g-metric is given by Eq. 13.51. In the above expression for the distance between two events, we have used the spacetime coordinates introduced in Eq. 13.53. Accordingly,

$$ds^2 = (dx^\mu)\, g_{\mu\upsilon}\, (dx^\upsilon) = [dx^0 \quad dx^1 \quad dx^2 \quad dx^3] \begin{bmatrix} 1 & 0 & 0 & 0 \\ 0 & -1 & 0 & 0 \\ 0 & 0 & -1 & 0 \\ 0 & 0 & 0 & -1 \end{bmatrix} \begin{bmatrix} dx^0 \\ dx^1 \\ dx^2 \\ dx^3 \end{bmatrix}, \tag{13.54b}$$

$$\text{i.e.,} \quad ds^2 = [dx^0 \quad dx^1 \quad dx^2 \quad dx^3] \begin{bmatrix} dx^0 \\ -dx^1 \\ -dx^2 \\ -dx^3 \end{bmatrix}, \tag{13.54c}$$

$$\text{i.e.,} \quad ds^2 = (dx^0)^2 - (dx^1)^2 - (dx^2)^2 - (dx^3)^2. \tag{13.54d}$$

The advantages of (i) introducing the contravariant and covariant vectors, and (ii) defining the g-metric to provide the signature of the spacetime, are now expressly manifest. Used together, they enable the extension of the Euclidean notion about distance between two points to *non-Euclidean* 4-dimensional spacetime. Sandwiching $g_{\mu\upsilon}$ from Eq. 13.51 in Eq. 13.54b has enabled us attain the benefits (i) and (ii). The Einstein summation convention of summing over the index that appears twice has been used. Notwithstanding the similarities with the Pythagoras description of this 'distance', the differences are also of great significance: (a) we are now dealing with the 4-dimensional spacetime continuum, and (b) the signature of the g-metric has a mix of plus and minus signs, which makes the spacetime non-Euclidean. In the frame $\overline{S}$, the corresponding measure of the distance would be given by

$$\overline{ds}^2 = (\overline{d\tau})^2 - (\overline{dx})^2 - (\overline{dy})^2 - (\overline{dz})^2. \tag{13.55a}$$

The bar-coordinates from the frame $\overline{S}$ used above are related to the unbar-coordinates in the frame S (Fig. 13.4), by the Lorentz transformations (Eq. 13.45).

Hence, we get

$$\overline{ds}^2 = \left\{ c\gamma\left(dt - \frac{v dx}{c^2} \right) \right\}^2 - \{d\gamma(dx - vdt)\}^2 - (dy)^2 - (dz)^2. \tag{13.55b}$$

After a little bit of simplification of the terms, we get

$$\overline{ds}^2 = d\tau^2 - dx^2 - dy^2 - dz^2 = ds^2. \tag{13.55c}$$

Equation 13.55c has a very beautiful form; it ensures that the measure for the distance between two events in the 4-dimensional spacetime that we chose is appropriate under the Lorentz transformations. A difference in the sign between the time-like coordinate and the space-like coordinate is essential to ensure that the speed of the electromagnetic waves will turn out to be the same in the frame S and $\overline{S}$. This property makes the 4-dimensional spacetime we require to be *different* from the simple *extension* of a 3-dimensional Euclidean space to a

4-dimensional Euclidean space. It is for this reason that the 4-dimensional spacetime we are employing is said to be non-Euclidean.

Both the positive and the negative signs appear in Eq. 13.53 (also therefore in Eqs. 13.54 and 13.55). All the three possibilities are of interest, as per the sum of the four terms in these equations being

(a) positive, when the spacetime interval is called *time like*,

(b) or zero, when the spacetime interval is called *light like*,

(c) negative, when the spacetime interval is called *space like*.

We have now found a prescription to interpret *distance* between two *events* in the 4-dimensional spacetime, called the spacetime continuum. It is satisfactorily defined by Eq. 13.54, or equivalently by Eq. 13.55. It is essentially the same in the frame S and $\bar{S}$. The interpretation of speed, normally understood as distance-over-time, however, needs to be done carefully, since now space-like and time-like coordinates *both* transform in the 4-dimensional spacetime continuum. This situation necessitates a *re-definition* of speed. We shall return to this fascinating question in Section 13.3, but, for now we first examine if the Maxwell's equations are invariant under the Lorentz transformations, since they are not invariant under the Galilean transformations.

In order to study the response of the Maxwell's equations to the Lorentz transformations, we must inspect how the components of the electric intensity field $\vec{E}$, and the components of the magnetic induction field $\vec{B}$, transform. Hitherto we have treated them as 3-vectors, (i.e., vectors having 3 components). In ordinary 3-dimensional Euclidean space, we would have considered their transformation properties as per the transformation laws for the components of a 3-vector, taking care, however, that the electric intensity field $\vec{E}$ is a polar vector field, but the magnetic induction field $\vec{B}$ is an axial vector field. This scheme cannot be expected to work in the 4-dimensional non-Euclidean space–time continuum. Under the Lorentz transformations, it turns out that the Maxwell's equations remain invariant in all inertial frames of reference, but the electromagnetic field components transform as elements of a second *rank tensor*, and not as components of two separate 3-vectors. This is mathematically elegant, physically enlightening, since it treats the electric and the magnetic fields completely in an *equivalent* manner. This is fittingly in the spirit of Maxwell's theory which effortlessly unifies electric and the magnetic phenomena.

We have already seen how the *distance* between two *events* in the space–time 4-dimensional space–time continuum transforms under the Lorentz transformation. We shall quickly spell out how an arbitrary 4-vector in the spacetime continuum transforms, and then proceed to discuss how a second rank tensor transforms, under the Lorentz transformations. We shall then be ready to appreciate the expression of the electromagnetic field as a second rank tensor. The g-metric must be expected to play a crucial role in understanding and interpreting the Lorentz transformations. The g-metric, which was introduced in Section 2.4 of Chapter 2, is now seen to play a valuable role in contracting the summations over the double-index while preserving and highlighting the structure of the spacetime continuum.

Now, the relationship between the contravariant components and the covariant components of a four vector is provided by the g-metric

$$x^{\mu} = g^{\mu\nu} x_{\nu}. \tag{13.56a}$$

The g-metric has *raised* an index in the above equation.

Likewise, $x_{\mu} = g_{\mu\nu} x^{\nu}$, (13.56b)

and the g-metric has now *lowered* an index.

The above relations are only a compact form of a matrix equation that is readily recognized on using the Einstein convention of summing over the repeated index:

$$\begin{bmatrix} x^0 \\ x^1 \\ x^2 \\ x^3 \end{bmatrix} = \begin{bmatrix} 1 & 0 & 0 & 0 \\ 0 & -1 & 0 & 0 \\ 0 & 0 & -1 & 0 \\ 0 & 0 & 0 & -1 \end{bmatrix} \begin{bmatrix} x_0 \\ x_1 \\ x_2 \\ x_3 \end{bmatrix}, \tag{13.56c}$$

$$\begin{bmatrix} x_0 \\ x_1 \\ x_2 \\ x_3 \end{bmatrix} = \begin{bmatrix} 1 & 0 & 0 & 0 \\ 0 & -1 & 0 & 0 \\ 0 & 0 & -1 & 0 \\ 0 & 0 & 0 & -1 \end{bmatrix} \begin{bmatrix} x^0 \\ x^1 \\ x^2 \\ x^3 \end{bmatrix}. \tag{13.56d}$$

Now, an arbitrary contravariant 4-vector would transform from the frame S to the frame $\bar{S}$ by a transformation law, similar to Eq. 13.37:

$$\begin{bmatrix} \bar{a}^0 \\ \bar{a}^1 \\ \bar{a}^2 \\ \bar{a}^3 \end{bmatrix} = \begin{bmatrix} \gamma & -\gamma\beta & 0 & 0 \\ -\gamma\beta & \gamma & 0 & 0 \\ 0 & 0 & 1 & 0 \\ 0 & 0 & 0 & 1 \end{bmatrix} \begin{bmatrix} a^0 \\ a^1 \\ a^2 \\ a^3 \end{bmatrix}, \tag{13.57a}$$

i.e., $\forall \mu$, i.e., for $\mu = 0, 1, 2, 3$, we have

$$\bar{a}^{\mu} = \Lambda^{\mu}{}_{\nu} a^{\nu}, \tag{13.57b}$$

where $\gamma = \dfrac{1}{\sqrt{1 - \dfrac{v^2}{c^2}}}$ and $\beta = \dfrac{v}{c}$, introduced already in Eqs. 13.46a and b,

$$\text{and} \quad \Lambda^{\mu}_{\nu} = \begin{bmatrix} \gamma & -\gamma\beta & 0 & 0 \\ -\gamma\beta & \gamma & 0 & 0 \\ 0 & 0 & 1 & 0 \\ 0 & 0 & 0 & 1 \end{bmatrix}. \tag{13.58}$$

As you can see, crucial information about the nature of the space–time transformation of the four vector from the frame S to the moving inertial frame $\bar{S}$ is contained in the matrix Λ^{μ}_{ν}. In the 4-vector notation as is appropriate for the spacetime continuum, the EM potentials (Section 3, Chapter 12) are given by

$$A^{\mu}; (\mu = 0, 1, 2, 3) = \left(\frac{\phi(\vec{r}, t)}{c}, \vec{A}(\vec{r}, t) \right). \tag{13.59}$$

The electromagnetic fields are derivable from these potentials, and the derivatives are now written in the tensor notation as

$$F^{\mu\nu} = \frac{\partial A^{\nu}}{\partial x_{\mu}} - \frac{\partial A^{\mu}}{\partial x_{\nu}} = \partial^{\mu} A^{\nu} - \partial^{\nu} A^{\mu}, \tag{13.60}$$

wherein we have used a further compact notation,

$$\partial^{\mu} \equiv \frac{\partial}{\partial x_{\mu}} \text{ and } \partial_{\mu} \equiv \frac{\partial}{\partial x^{\mu}}. \tag{13.61}$$

The indices μ, v each takes four values, which are 0, 1, 2, and 3. The electromagnetic field (Eq. 13.61) is then expressed as a second rank tensor, which is written as a 4×4 matrix, whose elements are $F^{\mu\nu}$. We can see from Eq. 13.61 that the elements of the second rank tensor become zero when $\mu = v$, and that they change sign on interchanging μ and v. The second rank electromagnetic field tensor is therefore antisymmetric.

The charge density $\rho(\vec{r}, t)$ and the current density $\vec{J}(\vec{r}, t)$ which appear in the Maxwell's equations can be expressed as 4-vector:

$$J^{\mu} = \begin{bmatrix} J^0 = \rho(\vec{r}, t)c \\ J^1 = J_x(\vec{r}, t) \\ J^2 = J_y(\vec{r}, t) \\ J^3 = J_z(\vec{r}, t) \end{bmatrix} = \begin{bmatrix} \rho(\vec{r}, t)c \\ \vec{J}(\vec{r}, t) \end{bmatrix}. \tag{13.62}$$

The continuity equation (Eq. 10.26),

$$\vec{\nabla} \cdot \vec{J}(\vec{r}, t) + \frac{\partial \rho(\vec{r}, t)}{\partial t} = 0, \tag{13.63a}$$

is as valid for the fluid mass current density vector $\vec{J}(\vec{r}, t) = \vec{v}(\vec{r}, t)\, \rho(\vec{r}, t)$ for the fluid mass density $\rho(\vec{r}, t)$, as for the electric charge density $\rho(\vec{r}, t)$, when $\vec{J}(\vec{r}, t) = \vec{v}(\vec{r}, t)\, \rho(\vec{r}, t)$ would be the electric current density vector. Using the 4-vector notation, we have

$$0 = \frac{\partial \rho}{\partial t} + \vec{\nabla} \cdot \vec{J} = \frac{\partial (\rho c)}{\partial (ct)} + \vec{\nabla} \cdot \vec{J} = \frac{\partial J^0}{\partial x^0} + \frac{\partial J^1}{\partial x^1} + \frac{\partial J^2}{\partial x^2} + \frac{\partial J^3}{\partial x^3} \equiv \frac{\partial J^{\mu}}{\partial x^{\mu}} \equiv \partial_{\mu} J^{\mu}. \tag{13.63b}$$

The electromagnetic field tensor $F^{\mu\nu}$, from Eq. 13.60, can now be fully constructed:

$$F^{\mu\nu} = \begin{bmatrix} F^{00} = 0 & F^{01} = \frac{E_x}{c} & F^{02} = \frac{E_y}{c} & F^{03} = \frac{E_z}{c} \\ F^{10} = -F^{01} & F^{11} = 0 & F^{12} = B_z & F^{13} = -B_y \\ F^{20} = -F^{02} & F^{21} = -F^{12} & F^{22} = 0 & F^{23} = B_x \\ F^{30} = -F^{03} & F^{31} = -F^{13} & F^{32} = -F^{23} & F^{33} = 0 \end{bmatrix}, \tag{13.64a}$$

$$\text{i.e.,} \quad F^{\mu\nu} = \begin{bmatrix} 0 & \frac{E_x}{c} & \frac{E_y}{c} & \frac{E_z}{c} \\ -\frac{E_x}{c} & 0 & B_z & -B_y \\ -\frac{E_y}{c} & -B_z & 0 & B_x \\ -\frac{E_z}{c} & B_y & -B_x & 0 \end{bmatrix}. \tag{13.64b}$$

As we know, the Maxwell's equations involve relationships between the partial derivatives of the components of the electric and the magnetic fields with respect to space and time. They also involve the charge density and the vector current density. We shall first demonstrate that two of the Maxwell's equations, those which involve the charge density and the vector current density, are obtained readily from the following equation for the electromagnetic field tensor $F^{\mu\nu}$:

$$\frac{\partial F^{\mu\nu}}{\partial x^{\nu}} = \mu_0 J^{\mu}, \tag{13.65}$$

$$\text{where} \quad J^{\mu} = (c\rho, J_x, J_y, J_z). \tag{13.66}$$

Let us expand Eq. 13.65 for $\mu = 0$:

$$\text{Thus, for } \mu = 0\text{:} \quad \frac{\partial F^{0\nu}}{\partial x^{\nu}} = \sum_{\nu=0}^{3} \frac{\partial F^{0\nu}}{\partial x^{\nu}} = \frac{\partial F^{00}}{\partial x^0} + \frac{\partial F^{01}}{\partial x^1} + \frac{\partial F^{02}}{\partial x^2} + \frac{\partial F^{03}}{\partial x^3} = \mu_0 J^0, \tag{13.67a}$$

$$\text{i.e.,} \quad \frac{1}{c}\left[\frac{\partial E_x}{\partial x} + \frac{\partial E_y}{\partial y} + \frac{\partial E_z}{\partial z}\right] = \mu_0 c\rho, \tag{13.67b}$$

$$\text{i.e.,} \quad \frac{\partial E_x}{\partial x} + \frac{\partial E_y}{\partial y} + \frac{\partial E_z}{\partial z} = \mu_0 c^2 \rho = \frac{\rho}{\varepsilon_0}. \tag{13.67c}$$

In other words, we have recovered

$$\vec{\nabla}\cdot\vec{E}=\frac{\rho}{\varepsilon_0}, \tag{13.68}$$

which is essentially the Coulomb–Gauss–Maxwell law.

Now, expanding Eq. 13.65 for $\mu = 1$ we get

$$\text{For } \mu=1: \frac{\partial F^{1\nu}}{\partial x^\nu}=\sum_{\nu=0}^{3}\frac{\partial F^{1\nu}}{\partial x^\nu}=\frac{\partial F^{10}}{\partial x^0}+\frac{\partial F^{11}}{\partial x^1}+\frac{\partial F^{12}}{\partial x^2}+\frac{\partial F^{13}}{\partial x^3}, \tag{13.69a}$$

$$\text{i.e., } \mu=1: \frac{\partial F^{1\nu}}{\partial x^\nu}=-\frac{1}{c^2}\frac{\partial E_x}{\partial t}+\frac{\partial B_z}{\partial y}-\frac{\partial B_y}{\partial z}=\left[-\frac{1}{c^2}\frac{\partial \vec{E}}{\partial t}+\vec{\nabla}\times\vec{B}\right]_{x\text{-component}}=\mu_0 J^1. \tag{13.69b}$$

$$\text{For } \mu=2: \frac{\partial F^{2\nu}}{\partial x^\nu}=-\frac{1}{c^2}\frac{\partial E_y}{\partial t}+\frac{\partial B_x}{\partial z}-\frac{\partial B_z}{\partial x}=\left[-\frac{1}{c^2}\frac{\partial \vec{E}}{\partial t}+\vec{\nabla}\times\vec{B}\right]_{y\text{-component}}=\mu_0 J^2. \tag{13.69c}$$

$$\text{For } \mu=3: \frac{\partial F^{2\nu}}{\partial x^\nu}=-\frac{1}{c^2}\frac{\partial E_z}{\partial t}+\frac{\partial B_y}{\partial x}-\frac{\partial B_x}{\partial y}=\left[-\frac{1}{c^2}\frac{\partial \vec{E}}{\partial t}+\vec{\nabla}\times\vec{B}\right]_{z\text{-component}}=\mu_0 J^3. \tag{13.69d}$$

Combining the results for $\mu = 1, 2, 3$, we recover the Ampere–Maxwell law:

$$\vec{\nabla}\times\vec{B}=\mu_0\vec{J}+\frac{1}{c^2}\frac{\partial\vec{E}}{\partial t}=\mu_0\vec{J}+\mu_0\varepsilon_0\frac{\partial\vec{E}}{\partial t}. \tag{13.70}$$

In Eq. 13.68 and Eq. 13.70, we have recovered two of the four Maxwell's equations from the tensor form Eq. 13.65. From the $\mu = 0$ index, we got one scalar equation, corresponding to the Coulomb–Gauss–Maxwell law, and from the $\mu = 1, 2, 3$, indices, we got the vector equation corresponding to the Ampere–Maxwell law.

We now have to get the other two Maxwell's equation, of which one is a scalar equation, which dismisses magnetic monopoles by declaring the divergence of the magnetic induction field to be zero. The other, and the last of the Maxwell's equations to be recovered is the vector equation for the Faraday–Lenz–Maxwell law. Between these two equations, we have 4 scalar equations, which we can obtain from a *dual* anti symmetric electromagnetic field tensor $G^{\mu\nu}$. To obtain $G^{\mu\nu}$, we recognize the fact that the electromagnetic 4-vector potential can be written, instead of Eq. 13.59, as

$$\tilde{A}^\mu;(\mu=0,1,2,3)=\left(\vec{A}(\vec{r},t),\frac{\phi(\vec{r},t)}{c}\right). \tag{13.71a}$$

The charge density $\rho(\vec{r},t)$ and the current density $\vec{J}(\vec{r},t)$ which appear in the Maxwell's equations can then be expressed, instead of Eq. 13.62, as the following 4-vector:

$$J^{\mu} = \begin{bmatrix} J^0 = J_x(\vec{r}, t) \\ J^1 = J_y(\vec{r}, t) \\ J^2 = J_z(\vec{r}, t) \\ J^3 = \rho(\vec{r}, t)c \end{bmatrix} = \begin{bmatrix} \vec{J}(\vec{r}, t) \\ \rho(\vec{r}, t)c \end{bmatrix}. \tag{13.71b}$$

This alternative representation is just as valid as that in Eq. 13.62 which was used earlier. This representation also recovers the equation of continuity for us:

$$0 = \vec{\nabla} \cdot \vec{J} + \frac{\partial \rho}{\partial t} = \vec{\nabla} \cdot \vec{J} + \frac{\partial (\rho c)}{\partial (ct)} = \frac{\partial J^0}{\partial x^0} + \frac{\partial J^1}{\partial x^1} + \frac{\partial J^2}{\partial x^2} + \frac{\partial J^3}{\partial x^3} \equiv \frac{\partial J^{\mu}}{\partial x^{\mu}} \equiv \partial_{\mu} J^{\mu}, \tag{13.71c}$$

just like Eq. 13.63, however, only with the indices 0, 1, 2, 3 rotated to 1, 2, 3, 0.

The *dual*, equivalent and alternative, second rank anti symmetric field tensor $G^{\mu\nu}$ can now be readily constructed, as before from the potential $\tilde{A}^{\mu}$, using

$$G^{\mu\nu} = \frac{\partial \tilde{A}^{\nu}}{\partial x_{\mu}} - \frac{\partial \tilde{A}^{\mu}}{\partial x_{\nu}} = \partial^{\mu} \tilde{A}^{\nu} - \partial^{\nu} \tilde{A}^{\mu}, \tag{13.72}$$

$$\text{i.e.,} \quad G^{\mu\nu} = \begin{bmatrix} G^{00} = 0 & G^{01} = B_x & G^{02} = B_y & G^{03} = B_z \\ G^{10} = -G^{01} & G^{11} = 0 & G^{12} = \frac{-E_z}{c} & G^{13} = -\frac{-E_y}{c} \\ G^{20} = -G^{02} & G^{21} = -G^{12} & G^{22} = 0 & G^{23} = \frac{-E_x}{c} \\ G^{30} = -G^{03} & G^{31} = -G^{13} & G^{32} = -G^{23} & G^{33} = 0 \end{bmatrix}. \tag{13.73a}$$

We could have constructed this alternate second rank tensor more readily by interchanging $\frac{\vec{E}}{c} \to \vec{B}$ and $\vec{B} \to -\frac{\vec{E}}{c}$.

Essentially, we can see that the $G^{\mu\nu}$ tensor is

$$G^{\mu\nu} = \begin{bmatrix} 0 & B_x & B_y & B_z \\ -B_x & 0 & -\frac{E_z}{c} & -\frac{E_y}{c} \\ -B_y & \frac{E_z}{c} & 0 & -\frac{E_x}{c} \\ -B_z & \frac{E_y}{c} & \frac{E_x}{c} & 0 \end{bmatrix}. \tag{13.73b}$$

We can now obtain the remaining two of the Maxwell's equations from the partial derivative operations on the elements of the second rank antisymmetric dual electromagnetic field tensor $G^{\mu\nu}$, giving

$$\frac{\partial G^{\mu\nu}}{\partial x^{\nu}} = 0. \tag{13.74}$$

For $\quad \mu = 0: \dfrac{\partial G^{0\nu}}{\partial x^{\nu}} = \sum_{\nu=0}^{3} \dfrac{\partial G^{0\nu}}{\partial x^{\nu}} = \dfrac{\partial G^{00}}{\partial x^{0}} + \dfrac{\partial G^{01}}{\partial x^{1}} + \dfrac{\partial G^{02}}{\partial x^{2}} + \dfrac{\partial G^{03}}{\partial x^{3}} = \vec{\nabla}\cdot\vec{B} = 0. \quad (13.75)$

For $\quad \mu = 1: \dfrac{\partial G^{1\nu}}{\partial x^{\nu}} = \sum_{\nu=0}^{3} \dfrac{\partial G^{1\nu}}{\partial x^{\nu}} = \dfrac{\partial G^{10}}{\partial x^{0}} + \dfrac{\partial G^{11}}{\partial x^{1}} + \dfrac{\partial G^{12}}{\partial x^{2}} + \dfrac{\partial G^{13}}{\partial x^{3}}, \quad (13.76a)$

i.e., for $\quad \mu = 1: \dfrac{\partial G^{1\nu}}{\partial x^{\nu}} = -\dfrac{1}{c}\dfrac{\partial B_x}{\partial t} - \dfrac{1}{c}\dfrac{\partial E_z}{\partial y} + \dfrac{1}{c}\dfrac{\partial E_y}{\partial z}. \quad (13.76b)$

Hence, for $\mu = 1: \dfrac{\partial G^{1\nu}}{\partial x^{\nu}} = -\dfrac{1}{c}\left[\dfrac{\partial \vec{B}}{\partial t} + \vec{\nabla}\times\vec{E}\right]_{x\text{-component}} = 0, \quad (13.76c)$

for $\quad \mu = 2: \dfrac{\partial G^{2\nu}}{\partial x^{\nu}} = -\dfrac{1}{c}\left[\dfrac{\partial \vec{B}}{\partial t} + \vec{\nabla}\times\vec{E}\right]_{y\text{-component}} = 0, \quad (13.76d)$

for $\quad \mu = 3: \dfrac{\partial G^{3\nu}}{\partial x^{\nu}} = -\dfrac{1}{c}\left[\dfrac{\partial \vec{B}}{\partial t} + \vec{\nabla}\times\vec{E}\right]_{z\text{-component}} = 0. \quad (13.76e)$

Combining the results for $\mu = 1, 2, 3$, we now recover the Faraday–Lenz–Maxwell law:

$$\frac{\partial \vec{B}}{\partial t} + \vec{\nabla}\times\vec{E} = \vec{0}. \tag{13.77}$$

From the second rank antisymmetric dual electromagnetic field tensor $G^{\mu\nu}$, in Eqs. 13.75 and 13.77, we have recovered the remaining two of the Maxwell's four laws. *Thus, all the four laws of Maxwell are embodied in the two tensor equations, Eq. 13.65 and Eq. 13.74.*

We must now check if the Maxwell's equations are invariant under the Lorentz transformations. For that, we examine the transformation properties of a general anti symmetric second rank tensor from the frame S to the frame $\overline{S}$ (Fig. 13.4). Let us write an arbitrary anti symmetric second rank tensor as

$$T^{\mu\nu} = \begin{bmatrix} T^{00} = 0 & T^{01} & T^{02} & T^{03} \\ T^{10} = -T^{01} & T^{11} = 0 & T^{12} & T^{13} \\ T^{20} = -T^{02} & T^{21} = -T^{12} & T^{22} = 0 & T^{23} \\ T^{30} = -T^{03} & T^{31} = -T^{13} & T^{32} = -T^{23} & T^{33} = 0 \end{bmatrix}. \tag{13.78}$$

Then, this tensor would transform to $\bar{T}^{\mu\nu}$ in the frame $\bar{S}$ according to the following law:

$$\bar{T}^{\mu\nu} = \Lambda^{\mu}{}_{\lambda}\Lambda^{\nu}{}_{\sigma}T^{\lambda\mu}. \tag{13.79}$$

The transformation given in Eq. 13.58 has to be applied twice, since we now have two indices in the second rank tensor.

The electromagnetic field tensor $F^{\lambda\sigma}$ therefore transforms from the frame S to that in the frame $\bar{S}$ according to the following law:

$$\bar{F}^{\mu\nu} = \Lambda^{\mu}{}_{\lambda}\Lambda^{\nu}{}_{\sigma}F^{\lambda\sigma} = \Lambda\mathbb{F}\Lambda^{T}, \tag{13.80a}$$

$$\text{i.e.,}\quad \bar{F}^{\mu\nu} = \begin{bmatrix} \gamma & -\gamma\beta & 0 & 0 \\ -\gamma\beta & \gamma & 0 & 0 \\ 0 & 0 & 1 & 0 \\ 0 & 0 & 0 & 1 \end{bmatrix} \begin{bmatrix} 0 & \frac{E_x}{c} & \frac{E_y}{c} & \frac{E_z}{c} \\ -\frac{E_x}{c} & 0 & B_z & -B_y \\ -\frac{E_y}{c} & -B_z & 0 & B_x \\ -\frac{E_z}{c} & B_y & -B_x & 0 \end{bmatrix} \begin{bmatrix} \gamma & -\gamma\beta & 0 & 0 \\ -\gamma\beta & \gamma & 0 & 0 \\ 0 & 0 & 1 & 0 \\ 0 & 0 & 0 & 1 \end{bmatrix}. \tag{13.80b}$$

$$\text{Therefore,}\quad \bar{F}^{\mu\nu} = \begin{bmatrix} 0 & \frac{E_x}{c} & \frac{\gamma E_y}{c} - \gamma\beta B_z & \frac{\gamma E_z}{c} + \gamma\beta B_y \\ -\frac{E_x}{c} & 0 & -\frac{\gamma\beta E_y}{c} + \gamma B_z & -\frac{\gamma\beta E_z}{c} - \gamma B_y \\ -\frac{\gamma E_y}{c} + \gamma\beta B_z & \frac{\gamma\beta E_y}{c} - \gamma B_z & 0 & B_x \\ -\frac{\gamma E_z}{c} - \gamma\beta B_y & \frac{\gamma\beta E_z}{c} + \gamma B_y & -B_x & 0 \end{bmatrix}. \tag{13.80c}$$

By comparing the corresponding elements of $F^{\lambda\mu}$ from Eq. 13.64b with those of $\bar{F}^{\mu\nu}$ in Eq. 13.80c, we see that the transformation of the electromagnetic field tensor from the frame S to $\bar{S}$ is consistent with the components of the electric intensity and the magnetic induction vector field, themselves transforming as follows:

$$\bar{E}_1 = E_1 \quad ; \quad \bar{B}_1 = B_1, \tag{13.81a}$$

$$\bar{E}_2 = \gamma(E_2 - \beta c B_3) \quad ; \quad \bar{B}_2 = \gamma\left(B_2 + \frac{\beta}{c}E_3\right), \tag{13.81b}$$

$$\text{and}\quad \bar{E}_3 = \gamma(E_3 + \beta c B_2) \quad ; \quad \bar{B}_3 = \gamma\left(B_3 - \frac{\beta}{c}E_2\right). \tag{13.81c}$$

Similarly, the dual field tensor $G^{\lambda\sigma}$ transforms from the frame S to the frame $\bar{S}$ according to the following law:

$$\bar{G}^{\mu\nu} = \Lambda^{\mu}{}_{\lambda}\Lambda^{\nu}{}_{\sigma}G^{\lambda\sigma} = \Lambda\mathbb{G}\Lambda^{T}, \tag{13.82a}$$

$$\text{i.e., } \bar{G}^{\mu\nu} = \begin{bmatrix} \gamma & -\gamma\beta & 0 & 0 \\ -\gamma\beta & \gamma & 0 & 0 \\ 0 & 0 & 1 & 0 \\ 0 & 0 & 0 & 1 \end{bmatrix} \begin{bmatrix} 0 & B_x & B_y & B_z \\ -B_x & 0 & -\frac{E_z}{c} & \frac{E_y}{c} \\ -B_y & \frac{E_z}{c} & 0 & -\frac{E_x}{c} \\ -B_x & -\frac{E_y}{c} & \frac{E_x}{c} & 0 \end{bmatrix} \begin{bmatrix} \gamma & -\gamma\beta & 0 & 0 \\ -\gamma\beta & \gamma & 0 & 0 \\ 0 & 0 & 1 & 0 \\ 0 & 0 & 0 & 1 \end{bmatrix}. \tag{13.82b}$$

The result of this transformation also confirms the transformations of the components of the electromagnetic field tensor as per Eq. 13.81a, b, and c.

Equations 13.81a, b, and c are exceptionally beautiful. The Maxwell's equations in their 3-vector notation expressed the components of the electric intensity field vector in terms of components of the *functions* of the magnetic induction field vector, and vice-versa. The results of Eq. 13.81a, b, and c are even more dramatic. We find in these equations that the components of the electric intensity field and the magnetic induction field are actually *scrambled*. Need there be any further elaboration of the *unity* of the electric and the magnetic phenomena? We have actually witnessed, in Eq. 13.81a, b, and c the superposition, the admixing, of the components of the electric field and the magnetic induction field. The break-up of the electromagnetic unified field into an electric part and a magnetic part in one inertial frame of reference is effectively a *different* break-up in another inertial frame. Thus, what is an electric field for one observer is a mix of the electric and the magnetic field for another observer, even if he is in frame of reference that is moving at a *constant* velocity with respect to the first observer. Both the observers must be expected to see the same physics, although the break-up of the effect of the Lorentz force $\vec{F} = q(\vec{E} + \vec{v} \times \vec{B})$ into the electric part and the magnetic part would of course be different, since the velocity of the charge would be different in the two frames in relative motion with respect to each other. A static charge in one frame is an electric current in a frame moving with respect to the first. Since both the frames of reference are essentially inertial frames, no pseudo-force (as discussed in Chapter 3) is involved; only the description of the physical force in terms of the electric and the magnetic components is different in the two inertial frames.

In the Reference [8], dynamics of various charged particles moving at different velocities in an electromagnetic field is analyzed. It is shown in the simulation described in this work that the dynamics seen by an observer in one frame of reference is essentially the same as that seen by an observer in a different inertial frame of reference. Trajectories of charged particles in different inertial frames of reference are determined. The trajectories in the two frames can be compared by transforming the trajectory in one inertial frame to the other using the Lorentz transformations given in Eq. 13.56. The trajectories are also obtained, in the Reference [8], independently in the second frame by solving the equation of motion with the *transformed electromagnetic field* using Eqs. 13.81a, b, and c. The *congruence* of the trajectories obtained in these two *alternative* methods is in complete harmony with the invariance under the Lorentz transformation of the Maxwell's equations for the electromagnetic field tensor. By all means these results are stunning. The underlying phenomenology for these results is the fact that laws of physics are the same for observers in inertial frames, moving at constant velocity with respect to each other. If it was this alone, physicists have known it earlier, and therefore expected, since the times of Galileo–Newton. However, in the Galileo–Newton scheme, the relative velocity of the moving frames needed to be compensated (added or subtracted) to compare an object's velocity in the different frames. We have discussed this earlier in Chapter 3. In particular, you may recollect the discussion on Fig. 3.1 in this regard. The new astounding element that is now required to be incorporated in the scheme is the *constancy of the speed of light* in all inertial frames, *irrespective of their relative motions.* This was a totally shocking situation. It took a 26-years old clerk at the Swiss patent office in Bern to take on this shocking but exquisite result head on, and interpret it boldly. His theory changed physics as never before. It continues to influence developments on the frontiers of physics, a hundred years later. It is the dominant factor to reckon with for the understanding of the origin, and the evolution, of the universe.

13.3 SIMULTANEITY, TIME DILATION, AND LENGTH (LORENTZ) CONTRACTION

Maxwell's equations are valid in all inertial frames of reference which speed away from each other at constant velocities. The only consideration to be kept in mind is, that, what is an electric field in one frame, is a mix of the electric and the magnetic field in another that is moving away at a constant velocity with respect to the previous. The components of the electromagnetic field which are parallel to the relative motion of the two inertial frames remain unchanged, as per Eq. 13.81a. The components orthogonal to the relative motion of the two frames transform as per Eqs. 13.81b and c. Once these transformations are made, the Maxwell's equations given by Eq. 13.65 and Eq. 13.74 remain invariant. In particular, the solution to the Maxwell's wave equation in free space continues to be a wave-front traveling at the same speed, $\frac{1}{\sqrt{\mu_0\varepsilon_0}} = c$, determined only by properties of the vacuum; i.e., by nothing.

Now, if this speed remains the same in inertial frames moving with respect to each other, speed being distance-over-time, Albert Einstein concluded that space-intervals (distance) and time-intervals must both change, together in tandem, so that the speed of light remains the same. However, this also necessitates a re-interpretation of speed. The obscure results become easily comprehensible on recognizing that events that are simultaneous for one observer, are not so for another moving with respect to the previous. Thus we have the outstanding formalism of the Special Theory of Relativity (STR), developed in the year 1905, which is commonly referred as the *annus mirabilis* (wonderful year), in which Einstein not only formulated the STR, but also laid the foundations of the quantum theory, and made other important contributions. Excellent records about the *annus mirabilis* are readily available for further reading.

To appreciate the STR, we shall first examine what is meant by simultaneity. In particular, we ask if it means the same for two observers moving with respect to each other. We shall therefore compare the observations made by two observers, Achal, who is a bystander on the ground, and Pravasi, who is a traveler on a train passing at a constant velocity with respect to the ground, as shown in Fig. 13.5. The train would cross Achal in front of him, from right to left as shown. in Fig. 13.5. Achal and Pravasi are both looking toward the far side of the train, behind it, and see two *cloud-to-earth* lightning bolts strike the ground, behind the train. In Sanskrit, the word 'ACHAL' means 'unmoved', or, stationary, and 'PRAVASI' means a traveler. Achal sees the two bolts hit the ground *simultaneously*. He finds absolutely no time-gap between the two lightning bolts strike the ground. However, Pravasi is on the moving train, and she registers the lightning *LF* a tad bit *before* the lightning *LR* (Fig. 13.5). Pravasi's inference that there is that tiny little bit delay in the lightning *LR* relative to *LF* is natural, since she is moving in the direction *toward LF*, and *away* from *LR*. The conclusion of Achal is also natural in his ground-based frame of reference. We therefore conclude that two events that are simultaneous for one observer are *not* so for another observer who is in motion relative to the previous observer. What is at the heart of this conclusion is the fact that light travels at *exactly* the same finite speed, irrespective of the observer's frame, whether it is the ground-based frame of reference, or the moving-train-based frame. No new assumption is made; we have only used what we have already learned from Maxwell's equations in the previous two sections of this chapter, that the speed of light is the same in all inertial frames.

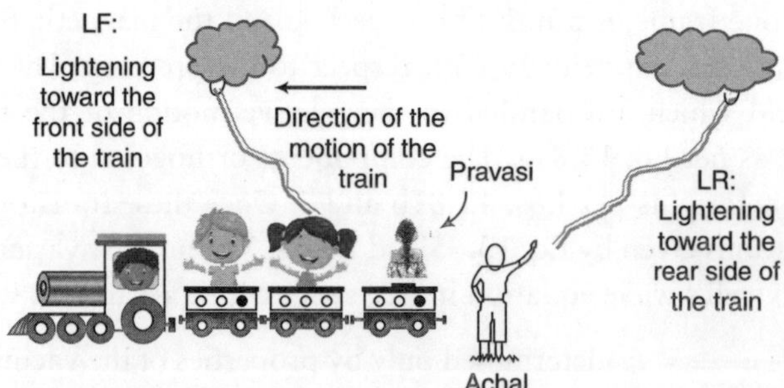

Fig. 13.5 The observer Achal is on our side of the train, unmoved, in a stationary frame of reference. He waves at passengers on a train that goes from right to left at a constant velocity with respect to the ground. Pravasi is a passenger on the train, looking away, in the same direction as Achal. Both Achal and Pravasi see two *cloud-to-earth* lightning bolts, *LF* and *LR*, hit the Earth.

Constancy of the speed of light in all inertial frames challenges the notion of simultaneity, and the concomitant notions of space-intervals, and time-intervals. We therefore take a close look at how clocks measure time-intervals.

We recollect from Chapter 3 that Galileo discovered the periodicity of the oscillations of a pendulum by counting the oscillations against his pulse rate. Any periodic event serves as a clock and measures time-interval. Since we are concerned directly with the speed of light, we shall consider a clock in which time-intervals are measured between two events, one in which a light burst is emitted from a source S, and the other when this light burst is detected at a detector D. Between these events, the light burst travels over a fixed path, and then this act repeats. Fig. 13.6 shows such a clock. This clock is viewed in two different frames of reference, R and M. The frame R is stationary, ground-based, and the frame M moves at a constant velocity with respect to R as shown in the figure. In the clock-frame (M), the light burst is emitted by the light source S, travels straight up, gets reflected by the mirror, and is then received at the detector D. Both the source S and the detector D can be considered to be point-sized. As soon as the light burst is detected at D, the source S emits the next burst, and the act repeats periodically, turning the device into a clock, which we shall refer to as a light-ray clock. Time intervals can then be measured in terms of the number of these periodic events. Any clock would serve the purpose, as all clocks can be calibrated with reference to some reference clock. We shall continue to work with the light-ray clock. A ground-based observer in the reference rest frame R, however, sees the ray of light travel *not* straight up, but along an oblique path from the source S to the mirror, and then again along the oblique path from the mirror to the detector D. Remember that the positions of S and D are the same; these are point-sized devices located exactly at the same place in the frame M. This is a thought experiment, and we do not worry about how to manufacture such a device. What is important, after all, is the understanding of the space and time intervals, and not the fabrication process that would give us this clock.

In the moving clock frame M, between the emission and detection of the light burst, the light travels a distance of 2h and would take a time-interval $(\Delta t)_M = \frac{2h}{c}$. The moving frame M is the one in which the light-ray clock is placed, so we shall call this time-interval as the *proper time*. The time interval $(\Delta t)_R$ between the emission and detection of the light burst in the ground-based rest frame R, however, is different. We first determine the time the light-ray would take to reach the mirror at the top after it is emitted by the source. As the light ray hits the mirror at the top, as seen in the ground-based rest frame R the clock has moved to the right through a distance $v\frac{(\Delta t)_R}{2}$. The time interval $(\Delta t)_R$ is therefore obtained by solving the Pythagoras relation,

$$\left(c\frac{(\Delta t)_R}{2} \right)^2 = h^2 + \left(v\frac{(\Delta t)_R}{2} \right)^2. \tag{13.83}$$

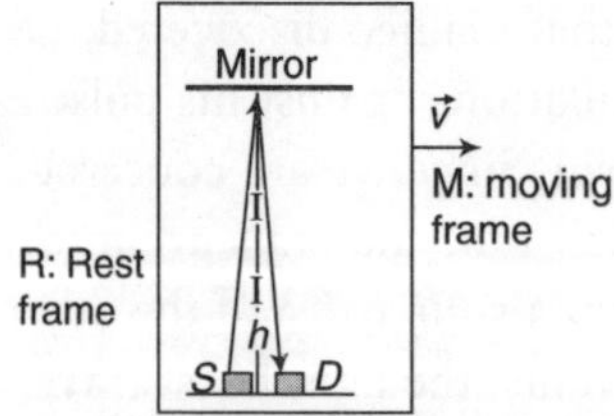

Fig. 13.6 The light clock is placed in a frame of reference M that moves at a constant velocity $\vec{v} = v\,\hat{e}_x$ with respect to the rest frame R. The time-interval measured by the clock in the frame M is called the *proper time*, since the clock is at rest in this frame.

Hence $$\left(c\frac{(\Delta t)_R}{2}\right)^2 = h^2 + \left(v\frac{(\Delta t)_R}{2}\right)^2, \tag{13.84a}$$

$$(\Delta t)_R = \frac{2h/c}{\sqrt{1-\frac{v^2}{c^2}}} = \frac{(\Delta t)_M}{\sqrt{1-\beta^2}} = \gamma(\Delta t)_M, \tag{13.84b}$$

where $$\beta = \frac{v}{c}, \text{ and } \gamma = \frac{1}{\sqrt{1-\frac{v^2}{c^2}}} = \frac{1}{\sqrt{1-\beta^2}}, \tag{13.85}$$

Note that $v > c$ must be ruled out, as it would render Eq. 13.84b senseless because of the scale factor in the two measures of the time interval. We therefore dismiss the possibility of any object having a speed greater than the speed of light.

Since $$\beta < 1,\ \gamma > 1, \tag{13.86}$$

$$(\Delta t)_R > (\Delta t)_M. \tag{13.87}$$

Thus, corresponding to N_R number of heart-beats of a healthy person in the rest frame R, the heart of her twin in the moving frame M would have beaten only $N_M = \dfrac{N_R}{\gamma}$ times. The twin in the moving frame M would therefore age less. The result in Eq. 13.87, no matter how disturbing, is a simple consequence of the constancy of the speed of light in the frames R and M. No new idea has been added to this. It leads us naturally to the inescapable conclusion that moving clocks go slow, compared to stationary clocks. Time-interval between two *ticks* of the clock is longer in the moving frame M, compared to that in a stationary clock such as the one in the rest frame R. This phenomenology is called the *relativistic time-dilation*. This condition is completely dictated by the non-Galilean constancy of the speed of light in all inertial frames of reference. This would mean, amongst various fascinating things, that if one of the twins travels, then the home-bound sibling would age *quicker* than the traveling one. We shall come back to this issue shortly. For the moment, we need to reconcile with yet another mindboggling effect.

We consider measuring distance between two stars, a large star L, and a small star s, shown in Figs. 13.7a and b. The distance is however measured by two observers, one is home-bound, in the frame R, and another is in a rocket frame M, which is moving with respect to the rest frame R at a constant velocity $\vec{v}$.

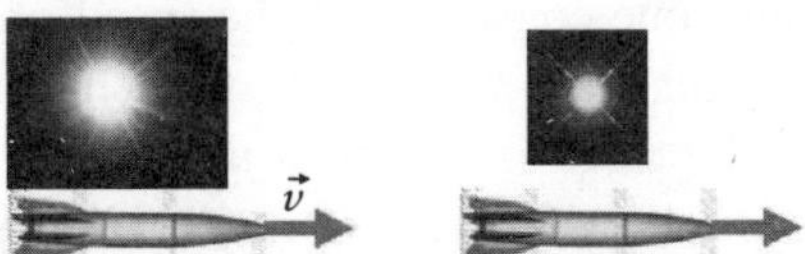

Fig. 13.7a In the rest frame R, the stars are at rest and the rocket moves from left to right at the velocity $\vec{v}$.

Fig. 13.7b In the moving frame M of the rocket, the stars move from right to left at the velocity $-\vec{v}$.

In the rocket frame of reference, which is the moving frame M, the stars are fixed; it is the observer who is moving. An observer in the rocket-frame would note the time when he crossed the large star L, and then again notes the time in his clock when he crosses the small star s. For this observer of course, it would seem that the stars are moving from right to left. Let us say that the time-interval between crossing the large star and the small star, measured in his clock, is $(\Delta t)_M$. The distance between the large star and the small star, measured by the observer in the rocket frame, would therefore be

$$(\delta\ell)_M = v(\Delta t)_M, \tag{13.88}$$

where $(\Delta t)_M$ is the *proper time*, since this is the frame in which the clock is placed. It is the clock's own frame, and hence this time interval is the proper time. The observer on the Earth, i.e., in the rest frame R, would use the rocket as a marker. He would record the time in his earth-based clock when he sees the rocket cross the large star L, and then again record the time in his clock when he sees the rocket cross the small star s. The observer in the rest-frame R therefore would conclude that the distance between the two stars is

$$(\delta\ell)_R = v(\Delta t)_R. \tag{13.89}$$

Hence,

$$v = \frac{(\delta\ell)_R}{(\Delta t)_R} = \frac{(\delta\ell)_M}{(\Delta t)_M}. \tag{13.90}$$

Now, using Eq. 13.84b, i.e., $(\Delta t)_R = \gamma(\Delta t)_M$, we get

$$\frac{(\delta\ell)_R}{\gamma(\Delta t)_M} = \frac{(\delta\ell)_M}{(\Delta t)_M}, \tag{13.91}$$

and hence $(\delta\ell)_M = \dfrac{(\delta\ell)_R}{\gamma}$. (13.92a)

Since $\gamma > 1$, $(\delta\ell)_M < (\delta\ell)_R$. (13.92b)

The contraction of the space interval in the moving frame of reference is called the *Lorentz contraction,* or as the *length contraction*. This is again a completely non-Galilean outcome, and its origin, goes back to the same constancy of the speed of light in all inertial frames which was also responsible for the *time-dilation*. Just as the time-interval $(\Delta t)_M$ is called the *proper time interval* (since the clock is at rest in the moving frame), the space interval $(\delta\ell)_R$ is called the *proper space interval*, since the stars are fixed, at rest, in this frame.

13.4 THE TWIN PARADOX, MUON LIFETIME, AND THE MASS–ENERGY EQUIVALENCE

Seetha and Geetha are identical twins. Geetha stays at home, and Seetha travels in the direction of the unit vector $\hat{e}_x$ of a Cartesian coordinate system, in a rocket at a speed $\frac{4}{5}c$ for 3 years measured in her own clock, placed in the rocket-clock. Seetha's clock is in the moving frame M. The clock is at rest in this frame; hence the time interval it measures is the *proper time*. Geetha's home-based clock is in the rest-frame R. It measures the corresponding time interval as $(\Delta t)_R = \gamma(\Delta t)_M = \frac{(\Delta t)_M}{\sqrt{1-\beta^2}}$, with $\beta = \frac{v}{c} = 0.8$. This value of β gives $\gamma = \left(\sqrt{1-\frac{v^2}{c^2}}\right)^{-1} = \left(\sqrt{1-\frac{(0.8c)^2}{c^2}}\right)^{-1}$, i.e., $\gamma = \frac{5}{3}$. Geetha's home-bound clock would therefore record a time interval $(\Delta t)_R = \frac{5}{3}(\Delta t)_M$ and hence $(\Delta t)_R = \frac{5}{3}(3 \text{ years}) = 5 \text{ years}$. In essence, during the time-interval through which Seetha traveled, and aged by 3 years, her home-bound twin Geetha aged by 5 years. As discussed above, this is because the home-bound sibling ages *quicker* than the traveling one. We know that moving clocks tick slower. This is *therefore* not even paradoxical, even if it sounds somewhat funny, given the fact that we already have met the reality of the constancy of the speed of light in all inertial frames of reference. It is precisely this veracity that had led us naturally to Eq. 13.84b, viz., $(\Delta t)_R = \gamma(\Delta t)_M$, with $\gamma = \left(\sqrt{1-\frac{v^2}{c^2}}\right)^{-1}$, and $\gamma > 1$, because $v < c$. Mindless of the oddity that home-bound Geetha has aged through 5 years while her sibling has aged through only 3 years, we do not, therefore, have any paradox. *Well, not yet.*

Let us examine what happens next. Seetha now turns around, and returns at the same speed, thus taking another 3 years measured, of course, in her clock in the rocket frame M. During this period, homebound Geetha's clock advances by another 5 years. Thus, during Seetha's round trip home-bound Geetha would age by 10 years, and traveling Seetha by only 6 years. Even this is not really a paradox, of course, as it is only the same time-dilation that is responsible for this kinky situation. Since time-dilation is a natural, logical, consequence of the ground reality that the speed of light is the same in all inertial frames of reference, we still do not have a paradox. *Well, not even yet.*

Nonetheless, the caption of this section is the *twin-paradox*, and it is surely not intended to mislead the readers. We do have, in fact, a paradox; or at least one that seems paradoxical, considering the symmetry, equivalence, principle: from Seetha's perspective, it is Geetha who appears to be the traveling sibling, and therefore it would be *Geetha who would be younger* than Seetha. Now, we do have a paradox, both Seetha and Geetha cannot be right when their inferences contradict each other. The paradox now before us is posed by this equivalence principle, and not by time-dilation. The incompatibility of Seetha's conclusions compared to Geetha's conclusion is the paradox that we must now resolve.

Let us therefore analyze the events described above from Seetha's perspective. She sees Geetha travel along $-\hat{e}_x$ at 0.8c. Seetha now clocks 3 years in her wait, measuring this time-interval in her own clock, and then takes off in the same direction to catch up with Geetha, who continues in her travel at the same earlier speed. We must first determine the speed at which Seetha must travel to catch up with Geetha in 3 *additional* years, as per Seetha's clock.

Let us consider a happy evening when you and your Dad plan to go for a dinner at a restaurant that is 5 km away. Your table is booked for 9pm, say. Your Dad starts out at 7 pm, and walks at the rate of 3 km/hour for one hour, and then, at the rate of 2 km/hour to reach the restaurant at 9 pm. You start an hour after your Dad, at 8 pm, but you must meet your Dad at the restaurant at 9 pm. You must therefore go to the restaurant at the speed of (3 + 2 = 5) km/hour. You had to go at the sum of the speeds of your Dad. If we now use the same logic to determine the speed at which Seetha must travel to catch up with her sister Geetha, then she would need to travel at $\frac{4}{5}c + \frac{4}{5}c = 1.6c$. Now, this would be impossible, as Seetha, of course, *cannot* travel faster than the speed of light, as already pointed out above (Eqs. 13.86, 13.87).

In our hurry, however, we have made a mistake above in applying the law of addition of velocities. We added the velocities using the Galilean law of addition of velocities. We must, however, use the Lorentz law to add the relative velocities. Let us therefore consider two frames of reference, S and S' whose respective origins O and O′ are coincident at $t = 0, t' = 0$. Their respective coordinate axes X, Y, Z and X′, Y′, Z′ of S and S' are aligned (Fig. 13.8). At $t' = 0$, the frame S' takes off along their common X, X′ axes, at a constant velocity $u_x\hat{e}_x$, as shown in Fig. 13.8.

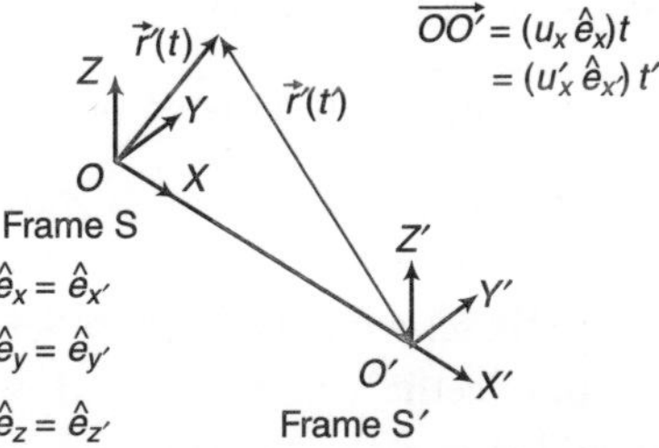

Fig. 13.8 Speed is distance over time. However, the Lorentz transformations scramble space and the time coordinates. The Lorentz transformations must therefore be accounted for while determining the relative speeds while dividing a 'distance-interval' by the 'time-interval'.

The velocity of any object whose instantaneous coordinates are (x, y, z) in the frame S is

$$\vec{v} = \frac{d\vec{r}}{dt} = \frac{dx}{dt}\hat{e}_x + \frac{dy}{dt}\hat{e}_y + \frac{dz}{dt}\hat{e}_z. \tag{13.93a}$$

The corresponding velocity in the frame S' is

$$\vec{v}' = \frac{d\vec{r}'}{dt'} = \frac{dx'}{dt'}\hat{e}_{x'} + \frac{dy'}{dt'}\hat{e}_{y'} + \frac{dz'}{dt'}\hat{e}_{z'} = \frac{dx'}{dt'}\hat{e}_x + \frac{dy'}{dt'}\hat{e}_y + \frac{dz'}{dt'}\hat{e}_z. \tag{13.93b}$$

Now, using the Lorentz transformations,

$$\frac{dx'}{dt'} = \frac{d(\gamma(x - u_x t))}{d(\gamma(t - \frac{u_x x}{c^2}))} = \frac{d(x - u_x t)}{d(t - \frac{u_x x}{c^2})}, \tag{13.94a}$$

hence, $$\frac{dx'}{dt'} = \frac{v_x - u_x}{1 - \frac{u_x v_x}{c^2}}. \tag{13.95}$$

Now, from Seetha's perspective, Geetha would travel along $-\hat{e}_x$, and hence, instead of Eq. 13.95, noting the 'sign' corresponding to the direction of the travel, we shall have

$$\frac{dx'}{dt'} = \frac{v_x + u_x}{1 + \frac{u_x v_x}{c^2}}. \tag{13.96}$$

We shall also have, $V'_y = V_y$ and $V'_z = V_z$.

Hence, the relative velocity according to Lorentz transformations is

$$v_{\text{relative}} = \frac{v_1 + v_2}{1 + \frac{v_1 v_2}{c^2}} = \frac{\left(\frac{4}{5}\right)c + \left(\frac{4}{5}\right)c}{1 + \frac{\left(\frac{4}{5}\right)c\left(\frac{4}{5}\right)c}{c^2}} = \frac{\left(\frac{8}{5}\right)c}{1 + \frac{16}{25}} = \frac{40}{41}c. \tag{13.97}$$

This speed is huge, but unlike the speed 1.6c that we got from Galilean relativity (which is higher than the speed of light), the relative speed (Eq. 13.97) that we now get from Lorentz transformations is less than the speed of light. Seetha now clocks 3 years in her own clock, and shoot off toward Geetha (i.e. along $-\hat{e}_x$) at a speed of $\frac{40}{41}c$. Subsequently, after 3 more years (according to her own, i.e., Seetha's, clock), she is to catch up with Geetha. We already know that while Seetha clock's 6 years in her clock, Geetha would clock 10 years in her own clock. To distinguish between these, we may refer to Seetha's clock as the *S-clock* and Geetha's as the *G-clock*. Now, from Seetha's perspective, it is Geetha who is traveling, and hence it is the G-clock which would record the *proper time*, $\Delta\tau = 10$ years. Seetha must then expect to travel for $\Delta\tau = \frac{\Delta\tau}{\sqrt{1-\beta^2}}$ with $\beta = \frac{v}{c} = \frac{4}{5} = 0.8$.

Thus, $\Delta t = \dfrac{\Delta\tau}{\sqrt{1-\beta^2}} = \dfrac{10}{\sqrt{1-\left(\dfrac{4}{5}\right)^2}} = \dfrac{10}{\sqrt{0.36}} = \dfrac{10}{0.6} \simeq 16.6667$ years (in S-clock). (13.98)

Let us now determine the distance that Geetha would have traveled during this period. Since distance is speed multiplied by time, a naïve calculation would suggest that the distance Seetha would need to catch up with Geetha is

$$\tilde{d} = (0.8c)\times 16.66667 = 0.8\times\left(c \text{ in } \frac{ly}{\text{year}}\right)\times 16.66667\,\text{years} \simeq 13.333336\ ly. \quad (13.99)$$

We have used the unit ly for 'light years' for distance, and 'year' for time.

However, to Seetha, this distance, would appear to be Lorentz-contracted, and therefore given by

$$d = \tilde{d}\sqrt{1-\beta^2} = 13.333337\times\sqrt{1-\left(\frac{40}{41}\right)^2} = 13.333336\times\sqrt{\frac{1681-1600}{1681}}ly, \quad (13.100a)$$

i.e., $d = 2.926829\ ly.$ (13.100b)

This is excellent, since Seetha would take, at the speed $\left(\dfrac{40}{41}\right)c$, which is slightly *less* than the speed of light, just a little bit *more* than 2.926829 years, according to her own S-clock. The exact time interval turns out to be

$$\text{time} = \frac{\text{distance}}{\text{speed}} = \frac{2.926829\ ly}{\left(\dfrac{40}{41}\times 1\right)\dfrac{ly}{yr}} = 2.926829\times\frac{41}{40} = 3\ years. \quad (13.101)$$

From Seetha's perspective also, then, it is Geetha who must age by 10 years in the G-clock while Seetha aged by 6 years in the S-clock. The symmetry-equivalence principle is salvaged neatly, and the paradox is completely removed effectively, very much within the framework of the STR (Special Theory of Relativity). Please note that we have *not* invoked any breakdown of the inertial nature of the two frames of reference due to change in momentum; neither in the first consideration, when Seetha turns around to return to meet her sister, nor in the second consideration, when Seetha waits for 3 years as per her own S-clock and then start out to catch up with her twin. An intermediate acceleration of the frame of reference has been interpreted by some analysts as going beyond the considerations of the inertial frames of reference employed between the Lorentz transformations. In the next chapter, we introduce elements of the GTR (General Theory of Relativity) in which we shall learn the correspondence between acceleration and a source of gravity. One might then be tempted to suspect if the resolution of the twin paradox requires GTR, and that STR is not enough. Nonetheless, in the above analysis, we make no use of the GTR. The twin paradox is completely resolved within the framework of the STR. In fact, the *turn-around* (in the first consideration) and the *delayed-start* (in the second consideration) can be completely done away with in our analysis, thus dismissing any allusion of an intermediate acceleration of the frame of reference. Seetha's acceleration, in either of the two considerations, can be easily completely done away with by considering in our thought-experiment a third observer, Lalitha, the missing sibling from the triplets their mom had delivered. Lalitha would then fly past both Seetha and Geetha and communicate the time-intervals measured in their respective clocks. Lalitha, the super-girl, would fly past both of them and share with each sibling what time-intervals their respective clocks had recorded. The resolution of the apparent breakdown of the principle of equivalence (symmetry) that appeared like a paradox is thus complete well within the framework of the STR. In doing so, both time-dilation and the Lorentz contraction has been made use of. The only paradox, if any, is that we never needed the two observers to be the twins Seetha and Geetha. They could have been twin brothers, Ram and Shyam, or they could be any two observers, not necessarily twins, nor even siblings. The two observers could be just about any two persons, in two inertial frames of reference moving with respect to each other. They could be siblings, cousins, friends, or even enemies, but it is fun to think of them as twins and refresh memories of hilarious movies about the confusion created by twins that you may have seen.

Actually, the Lorentz contraction and the time dilation appears for any considerations of time and space intervals, which form a continuum intermingled by the Lorentz transformations (Eq. 13.45). It is a very real phenomenon, and it actually takes place in spacetime. It applies to Seetha–Geetha's ageing, as well as the life-time of excited states of some particles. Excited states of particles are *resonances* which couple with their environments and may decay into fragments. This process is quantum mechanical, but for our immediate purpose we need not get into the quantum dynamics that governs this process, which is fully governed by the quantum uncertainty principle. Skipping the quantum theoretical considerations, let us ask how on earth we can see the tracks of the muons, which have a mass of $\sim 207m_e$ [5], have a mean lifetime of $\sim 2.2\mu s$ in their rest-frame, and decay into electrons and neutrinos, converting mass into energy. They travel at $\sim 0.994c$, and at this speed they could travel about $\sim 660m$ in their lifetime. Since they are produced in the cosmic rays reactions in the upper atmosphere, about 10 to 15 km above the Earth's surface, there would be very little chance to detect them on Earth. The little chance comes from the fact that the mean-life is only a statistical property, and some muons may live longer and make it to the Earth. These surviving muons would take $\sim 33.53\ \mu s$ to cover the 10 km distance and reach the Earth. A small fraction, $\sim 2.39 \times 10^{-7}$, of the muons could, however, make it to the Earth's surface this way. You can easily set up this experiment in your institute's physics laboratory using a cloud chamber. A simple classroom tutorial experiment is described in Reference [9] which describes how this experiment can be easily set up. Such an experiment would show, in your very own laboratory, that *very many* muon tracks are detected in this experiment, on the Earth's surface. The observations far exceed the fraction $\sim 2.39 \times 10^{-7}$ estimated above. However, when you account for the *augmented life-time* of the muons due to time-dilation, you will find that the fraction of the muons reaching the Earth becomes ~ 0.188, and this is quite consistent with the easy detection of the muon tracks on the Earth's surface. The exact same *increased fraction* results in the muon's moving frame of reference, calculated using the *Lorentz contraction* of the distance to the

Earth. Very many sophisticated high-technology experiments have been done now, and in every such experiment, the consequences of time-dilation and length contraction are borne out exactly.

In the next chapter, we shall introduce the GTR, the General Theory of Relativity. The formalism we have discussed in the present chapter is known as the STR, the Special Theory of Relativity. The reason for this nomenclature will become clear in the next chapter. We shall find that STR is indeed a special case of the GTR. Before we wrap up this chapter, we shall secure for us yet another important consequence of the STR which Albert Einstein obtained for us. It is the equivalence between mass and energy, which had been previously regarded as totally separate entities, for which *separate* conservation principle would apply.

Length contraction and time-dilation, which are concurrent upshots of the STR constancy of the speed of light, necessitated the recognition of the *proper time interval* and the *proper space interval*. We now need to introduce the *proper velocity*. It is defined as

$$\vec{\eta} = \text{proper velocity} = \frac{\text{proper length}}{\text{proper time}}. \tag{13.102}$$

Thus, the proper velocity, is

$$\vec{\eta} = \frac{d\vec{r}}{d\left(\frac{t}{\gamma}\right)} = \gamma\frac{d\vec{r}}{dt} \quad ; \quad \gamma = \frac{1}{\sqrt{1-\frac{v^2}{c^2}}} \quad ; \quad \gamma \geq 1. \tag{13.103}$$

We shall denote the 3-component vector for the proper velocity in Eq. 13.103 as *part* of a 4-component velocity 4-vector,

$$\eta^{\mu}, (\text{with } \mu = 0, 1, 2, 3) = (\eta^0, \eta^1, \eta^2, \eta^3) = (\gamma mc, \gamma m\vec{v}) = \gamma m(c, \vec{v}). \tag{13.104}$$

Thus, $$\eta^0 = \frac{dx^0}{d(t/\gamma)} = \gamma\frac{dx^0}{dt} = \gamma\frac{d(ct)}{dt} = \gamma c, \tag{13.105a}$$

$$\eta^1 = \frac{dx^1}{d(t/\gamma)} = \gamma\frac{dx^1}{dt} = \gamma\frac{d(x)}{dt} = \gamma v_x, \tag{13.105b}$$

$$\eta^2 = \frac{dx^2}{d(t/\gamma)} = \gamma v_y, \tag{13.105c}$$

and $$\eta^3 = \frac{dx^3}{d(t/y)} = \gamma v_z. \tag{13.105d}$$

We can clearly see that this definition of the proper velocity is appropriate, since,

$$\eta^{\mu}\eta_{\mu} = \gamma^2 c^2 - \gamma^2 v^2 = \gamma^2(c^2 - v^2) = \frac{c^2}{(c^2 - v^2)}(c^2 - v^2) = c^2, \tag{13.106}$$

which is a manifestly invariant quantity in all inertial frames. The introduction of the *proper velocity* now enables us to define the *proper momentum* which would be appropriate for the STR. It is given, naturally, by

$$p^{\mu},\ (\mu = 0, 1, 2, 3) = m\eta^{\mu} = (\gamma mc, m\vec{\eta}) = (\gamma mc, \gamma\vec{v}) = \gamma m(c, \vec{v}). \tag{13.107}$$

Thus, $p^0 = m\eta^0 = m\gamma c,$ (13.108a)

$p^1 = m\eta^1 = m\gamma v_x,$ (13.108b)

$p^2 = m\eta^2 = m\gamma v_y,$ (13.108c)

and $p^3 = m\eta^3 = m\gamma v_z.$ (13.108d)

Note that the mass m is essentially the rest mass of the particle under consideration. We readily see that this definition of the proper momentum is also very natural and appropriate for the STR, since

$$p^{\mu}p_{\mu} = m^2\gamma^2c^2 - m^2\gamma^2v^2 = m^2\gamma^2(c^2 - v^2) = \frac{m^2c^2}{(c^2 - v^2)}(c^2 - v^2) = m^2c^2. \tag{13.109a}$$

Again, $p^{\mu}p_{\mu} = m^2c^2$ is manifestly invariant in all the inertial frames. The length of the momentum 4-vector is conserved in all inertial frames of reference.

We now recognize that

$$m^2c^2 = p^{\mu}p_{\mu} = m^2\gamma^2c^2 - m^2\gamma^2v^2 = \frac{E^2}{c^2} - \vec{p}.\vec{p}. \tag{13.109b}$$

We therefore have $E = \gamma mc^2 = mc^2\left(1 - \frac{v^2}{c^2}\right)^{-\frac{1}{2}}.$ (13.110a)

If $V = 0$, we have $\gamma = 1$, and we immediately see that Eq. 13.110 is a special case of Eq. 13.109b.

Thus, in general, $E = \gamma mc^2 = mc^2\left(1 + \frac{1}{2}\frac{v^2}{c^2} + \frac{\frac{1}{2}\left(\frac{1}{2}+1\right)}{2!}\left(\frac{v^2}{c^2}\right)^2 + \cdots\right),$ (13.110b)

i.e., $E = \gamma mc^2 = mc^2 + \frac{1}{2}mv^2 + \frac{3}{8}m\frac{v^4}{c^2} + \frac{5}{16}m\frac{v^6}{c^4} + \cdots.$ (13.110c)

The first term mc^2 is the rest mass energy. Subsequent terms come from the motion, of which the leading term $\frac{1}{2}mv^2$ is the Galilean kinetic energy. All subsequent terms become progressively smaller, since $v << c$. We are mostly interested in *changes* in the kinetic energy, since the non-relativistic equation of motion is invariant under the gauge transformation of the potential. The Galileo–Newton approximation $\frac{1}{2}mv^2$ for the kinetic energy therefore works pretty well. The Galileo–Newton formalism is hence not absurd; it is an excellent approximation to the fundamental laws of nature.

If we now apply this relation for a photon, we get

$$E = \gamma mc^2 = \left[\frac{1}{\sqrt{1-\frac{v^2}{c^2}}} mc^2 \right]_{m=0}^{v=c} = \frac{0}{0}, \tag{13.111}$$

since the rest mass of a photon is zero, and its speed is of course the speed of light. The result of Eq. 13.111 is *thankfully indeterminate*; it prompts us to realize that for a photon, we should *not* be using this relation to determine its energy. Nonetheless, for the photon too, Eq. 13.109 holds, and its rest mass being zero, we have

$$\frac{E^2}{c^2} = \left|\vec{p}\right|^2. \tag{13.112}$$

Using now the de Broglie relation from wave–mechanics for the massless photons

$$\lambda = \frac{h}{p}, \tag{13.113}$$

where λ is the wavelength of the photon, p its momentum and h the Planck's constant, we get

$$E = pc = \frac{h}{\lambda}c = h\frac{c}{\lambda} = h\nu. \tag{13.114}$$

This result is not only much better than that from Eq. 13.111, it is actually as it should be. Note that we were led to it by $E = \gamma mc^2$, which we found to be indeterminate. If we had interpreted the energy equivalence as $E = mc^2$, we would have got all photons to have zero energy. It is obviously important to write one of the most famous physics equations unambiguously, fully correctly. We must interpret the mass-energy equivalence as $E = \gamma mc^2$, and *not* as $E = mc^2$. Can one get around this by defining a relativistic mass as $m_{\text{Relativistic}} = \gamma m$ (with m as the rest mass)? We could then write $E = m_{\text{Relativistic}}c^2$ instead of $E = \gamma mc^2$. However, we would then end up introducing the relativistic mass, and also the relativistic energy. That seems completely redundant, since mass and energy are equivalent. We need to comprehend only one of the two, either the energy, or the mass. If the relativistic energy is introduced in our formalism, mass is automatically included, being equivalent; and also, of course, vice versa. There need then be no separate principles of the conservation of energy and matter. It is therefore appropriate to introduce only *one* unambiguous mass, m, which is *essentially the rest mass*. The energy-mass equivalence is then expressed as $E = \gamma mc^2$, and *not* as $E = mc^2$ (which gives zero energy for photons), *nor* as $E = m_{\text{Relativistic}}c^2$. The seeds of the mass-energy equivalence, time-dilation and Lorentz contraction are all found in the constancy of the speed of light of Maxwell's electromagnetic waves, and the fact that laws of physics are the same in all inertial frames. More than anything else, *including* the Michelson-Morley experiment, it would be the Maxwell's equations that led to the formulation of the STR [10].

It is the symmetry in Maxwell's equations that epitomizes the secrets of the mind-boggling laws of relativity. The importance of symmetry thus begins with Maxwell–Einstein. It is a dominant principle in the laws of nature; profoundly stated in Noether's theorem, which we have referred to earlier (in particular in Chapters 1, 6, 7, and 8).

The STR, however much beautiful it is, has some limitations, which we shall meet in the next chapter. It is because one needs a larger theoretical model, of which the STR is only a *special* case. The larger scheme is the General Theory of Relativity, which Albert Einstein presented ten years after the STR. In the next chapter, we provide glimpses of the GTR, and discuss some of its triumphs.

Problems with Solutions

P13.1:

Using the boundary conditions for the electromagnetic field at the surface of separation of two dielectric media, obtain the optical laws of reflection and refraction. Assume that there is no surface charge density or any surface current density at the interface.

Solution:

When an electromagnetic wave is incident on a surface of separation between two dielectric media, a part of the incident electromagnetic energy is reflected and a part is transmitted. The transmitted wave often is not collinear with the incident direction; it is refracted. Let us represent the incident, reflected and transmitted electromagnetic waves as follows:

$$\text{Incident wave:} \quad \vec{E}_i = \vec{\tilde{E}}_i e^{i(\vec{k}_i\cdot\vec{r}-\omega t)} \text{ and } \vec{B}_i = \vec{\tilde{B}}_i e^{i(\vec{k}_i\cdot\vec{r}-\omega t)}, \tag{P13.1.1}$$

$$\text{Reflected wave:} \quad \vec{E}_r = \vec{\tilde{E}}_r e^{i(\vec{k}_r\cdot\vec{r}-\omega t)} \text{ and } \vec{B}_r = \vec{\tilde{B}}_r e^{i(\vec{k}_r\cdot\vec{r}-\omega t)}, \tag{P13.1.2}$$

$$\text{Transmitted wave:} \quad \vec{E}_t = \vec{\tilde{E}}_t e^{i(\vec{k}_t\cdot\vec{r}-\omega t)} \text{ and } \vec{B}_t = \vec{\tilde{B}}_t e^{i(\vec{k}_t\cdot\vec{r}-\omega t)}, \tag{P13.1.3}$$

where $\vec{\tilde{E}}_i = \breve{e}_i\tilde{E}_i = \breve{e}_i E_i e^{i\delta_E}$, $\vec{\tilde{B}}_i = \breve{b}_i\tilde{B}_i = \breve{b}_i B_i e^{i\delta_B}$ etc., and $\omega = \omega_i = \omega_r = \omega_t$. The frequencies are the radiation frequencies which are determined at the very source of the electromagnetic radiation; they remain unchanged. The basis $\{\hat{e}_i, \hat{b}_i\hat{k}\}$ constitutes a right-handed set of unit vectors. $\hat{e}_i$, $\hat{b}_i$ denote the directions along which the electric and the magnetic vectors of the electromagnetic wave are polarized. We shall consider the electromagnetic incident wave to be plane polarized, although the treatment below can be generalized for other polarization states, and also for unpolarized waves. In Fig. P13.1A, the electric vector is considered to be in the plane of incidence, hence ***parallel*** to it. This is called '*P*' polarization. The magnetic vectors are then perpendicular to the plane of incidence, so this orientation is also called the TM polarization. We can also consider the case when it is the magnetic vector that is in the plane of incidence, and it is the electric vector that is in the transverse plane. This orientation is therefore called the TE polarization, or the '**S**' polarization. The letter '*S*' stands for the German word 'senkrecht' which means '***perpendicular***'. The reflection and transmission of TE polarized electromagnetic waves is shown in Fig. P13.1B.

Let us consider the incident wave to be '*P*' (TM) polarized (Fig. P13.1A). The wave numbers of the incident, reflected and the transmitted waves are given by

$$\left|\vec{k}_i\right| = \frac{2\pi}{\lambda_1} = \frac{2\pi\nu}{c_1} = \frac{\omega n_1}{c_{\text{vacuum}}} = \left|\vec{k}_r\right|;\ \left|\vec{k}_t\right| = \frac{2\pi}{\lambda_t} = \frac{2\pi\nu}{c_2} = \frac{\omega n_2}{c_{\text{vacuum}}};\ \text{hence: } k_i = k_r = \frac{c_2}{c_1}k_t = \frac{n_1}{n_2}k_t,$$

since the refractive index is given by: $n = \dfrac{c_{\text{vaccum}}}{c_{\text{medium}}} = \sqrt{\dfrac{\mu\varepsilon}{\mu_0\varepsilon_0}}$.

Now, at the interface of the medium '1' and '2', at $y = 0$, we have

$\vec{E}_1 = \vec{E}_i + \vec{E}_r,\ \vec{E}_2 = \vec{E}_t$, and similarly: $\vec{B}_1 = \vec{B}_i + \vec{B}_r,\ \vec{B}_2 = \vec{B}_t$.

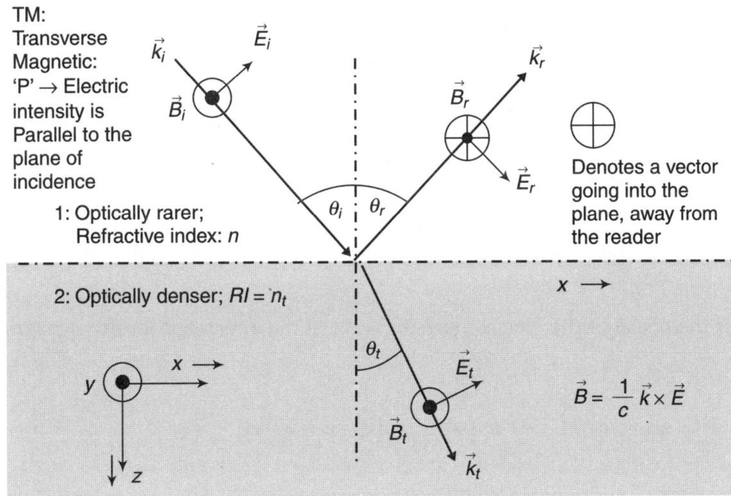

Fig. P13.1A In the case of TM polarization (*P* polarization), the incident magnetic induction field vector is perpendicular to the plane of incidence. In this figure, it points towards the reader. The reflected magnetic induction field vector points away from the reader, into the plane of incidence. The electric field intensity vectors of the incident, reflected, and transmitted wave are all in the plane of incidence.

Since there is no surface charge density, nor any current density, at the interface, the ***first*** of the interface conditions, summarized in Table 12.4 of Chapter 12, applies, and reduces to the following:

$\hat{n} \cdot (\varepsilon_1\vec{E}_1 - \varepsilon_2\vec{E}_2) = 0$. Hence, $\hat{n} \cdot (\varepsilon_1(\vec{E}_i + \vec{E}_r) - \varepsilon_2\vec{E}_t) = 0$, at $y = 0$. The electromagnetic waves only get reflected or transmitted at the surface of separation; there is no interaction at the surface that can ***change*** the plane of the polarization vectors. We must therefore have

$$\varepsilon_1(-\tilde{E}_i \sin\theta_i e^{i(\vec{k}_i\cdot\vec{r}-\omega t)} + \tilde{E}_r \sin\theta_r e^{i(\vec{k}_i\cdot\vec{r}-\omega t)}) - \varepsilon_2(-\tilde{E}_t \sin\theta_t e^{i(\vec{k}_i\cdot\vec{r}-\omega t)}) = 0\,. \qquad \text{(P13.1.4)}$$

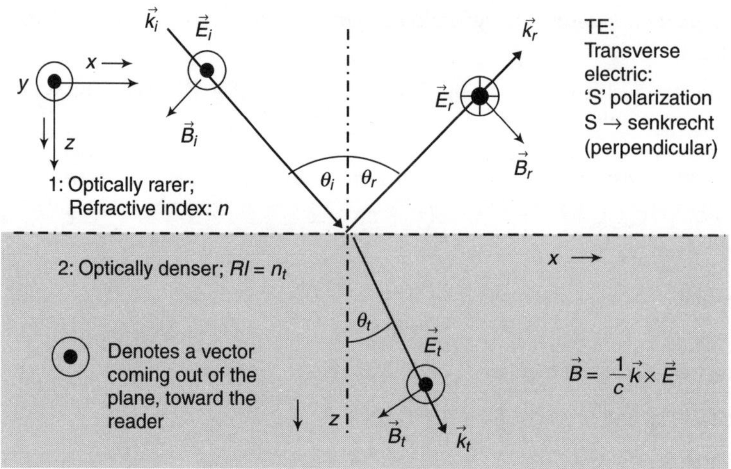

Fig. P13.1B In the case of TE polarization (S polarization), the electric field vector is perpendicular to the plane of incidence for each of the incident, reflected and the transmitted wave. The electromagnetic wave field is transverse, for all the waves, for which Eq. 13.19a, b hold, but the speed at which it travels is different in the two media.

The above relation holds good for all values of time, and also for all values of (x, y) on the surface of separation, at $z = 0$. This requires the phase factors to be equal, since the entire dependence on the space and time is only in the phases.

Continuity of the phase of the electromagnetic wave at the interface therefore guarantees that

$$[\vec{k}_i \cdot \vec{r}]_{z=0} = [\vec{k}_r \cdot \vec{r}]_{z=0} = [\vec{k}_t \cdot \vec{r}]_{z=0}$$

i.e., $[k_{i,x}x + k_{i,y}y]_{z=0} = [k_{r,x}x + k_{r,y}y]_{z=0} = [k_{t,x}x + k_{t,y}y]_{z=0}$

Now, x, y are arbitrary coordinates, so the relation must remain valid for all values of these coordinates, including $x = 0$ and $y = 0$. Hence: $k_{i,x} = k_{r,x} = k_{t,x}$, and $k_{i,y} = k_{r,y} = k_{t,y}$. This is an important conclusion. Essentially, the result we got proves that ***the incident ray, the reflected ray and the transmitted ray all lie in the same plane***, which is the first law of reflection, refraction and transmission.

Again, $k_{i,x} = k_{r,x} = k_{t,x} \Rightarrow k_i \sin\theta_i = k_r \sin\theta_r = k_t \sin\theta_t$, (P13.1.5)

i.e., $\omega\dfrac{n_1}{c_{\text{vacuum}}}\sin\theta_i = \omega\dfrac{n_1}{c_{\text{vacuum}}}\sin\theta_r = \omega\dfrac{n_2}{c_{\text{vacuum}}}\sin\theta_t$.

Therefore, $n_1 \sin\theta_i = n_1 \sin\theta_r = n_2 \sin\theta_t$

Hence, $\sin\theta_i = \sin\theta_r \Rightarrow \boxed{\theta_i = \theta_r}$ (law of reflection). (P13.1.6)

Finally, we have $\boxed{\dfrac{\sin\theta_{\text{inc}}}{\sin\theta_{\text{rfr}}} = \dfrac{n_2}{n_1} = \dfrac{c_{\text{med. '1'}}}{c_{\text{med. '2'}}} = \sqrt{\dfrac{\mu_2\varepsilon_2}{\mu_1\varepsilon_1}}}$ (Snell's law of refraction). (P13.1.7)

These results are valid for both 'P' (TM) and 'S' (TE) modes, though only 'P' polarization was considered above.

P13.2:

(a) If the incident electromagnetic wave is plane polarized with the *electric vector in the plane of the incidence* ('*P*' polarization, or TM polarization), then prove that transmitted amplitude and the reflected amplitude depend on the angle of incidence.

(b) Prove that the transmitted wave is always in phase with the incident wave and that the reflected wave is either in-phase or out of phase.

Solution:

(a) From the third interface condition, $E_1^{\|} - E_2^{\|} = 0$, in Table 12.4 of Chapter 12, applied to the reflection and transmission geometry of the '*P*' polarized electromagnetic waves (Fig. P13.2A), gives us:

$$(\tilde{E}_i \cos\theta_i + \tilde{E}_r \cos\theta_r) - (\tilde{E}_t \cos\theta_t) = 0, \quad \text{(P13.1.8)}$$

i.e., $$\cos\theta_i (\tilde{E}_i + \tilde{E}_r) - \cos\theta_t(\tilde{E}_t) = 0, \text{ since } \theta_i = \theta_r;$$

or, $$\tilde{E}_i + \tilde{E}_r = \frac{\cos\theta_t}{\cos\theta_i}\tilde{E}_t = \alpha\tilde{E}_t\,, \quad \text{(P13.1.7)}$$

where $$\alpha = \frac{\cos\theta_t}{\cos\theta_i}.$$

From the fourth interface condition, $H_1^T - H_2^T = 0$, in Table 12.4, we get

$$0 = H_1^{\|} - H_2^{\|} = \frac{B_1^{\|}}{\mu_1} - \frac{H_2^{\|}}{\mu_2} = \frac{E_1^{\|}}{c_1\mu_1} - \frac{E_2^{\|}}{c_2\mu_2} = \frac{\tilde{E}_i - \tilde{E}_r}{c_1\mu_1} - \frac{\tilde{E}_t}{c_2\mu_2}\,,$$

i.e., $$\tilde{E}_i - \tilde{E}_r = \frac{c_1}{c_2}\frac{\mu_1}{\mu_2}\tilde{E}_t = \frac{n_2}{n_1}\frac{\mu_1}{\mu_2}\tilde{E}_t = \beta\tilde{E}_t\,, \quad \text{(P13.1.8)}$$

where we have defined $\beta = \frac{n_2}{n_1}\frac{\mu_1}{\mu_2}$.

Using Eq. P13.1.7 and Eq. P13.1.8, we have

$$\tilde{E}_t = \frac{2}{\alpha + \beta}\tilde{E}_i\,, \quad \text{(P13.1.9a)}$$

and $$\tilde{E}_r = \frac{\alpha - \beta}{\alpha + \beta}\tilde{E}_i\,. \quad \text{(P13.1.9b)}$$

The equations Eq. P13.1.9a and Eq. P13.1.9b give the amplitudes, respectively, of the transmitted and the reflected waves in terms of the incident amplitude. They depend on the angle of incidence since α depends on the angle of incidence (and the refractive index), which determines the angle at which the transmitted wave is bent toward the normal. The above two equations (see also equivalent relations in the Problem 13.7, below) for the TM, '*P*', polarized waves, and two additional relations, which come from a similar analysis of the '*S*' (TE) polarized (Problem 13.8, below) incident electromagnetic wave (Fig. B, above), constitute a set of 'four' relationships, known as the Fresnel equations. They were first obtained by Augustin Jean Fresnel (1788 to 1827).

Tragically, Fresnel died very young, 4 years before Maxwell was even born. He made outstanding contributions to the wave theory, especially to optics and the theory of diffraction. He was a civil engineer, and designed lighthouse lenses, among his very many brilliant contributions to science and technology.

(b) From Eq. 13.1.9a, it follows that the transmitted wave is always in phase with the incident wave. Likewise, from Eq. 13.1.9b, it follows that the reflected wave is either in-phase, or out of phase, with the incident wave, depending on $\alpha > \beta$ or $\alpha < \beta$.

P13.3:

(a) Prove that at the grazing angle of incidence $\theta_i = \frac{\pi}{2}$ there is zero transmission. The entire electromagnetic energy that is incident on the interface is reflected back. (This phenomenon is called the 'total internal reflection').

(b) If the incident electromagnetic wave is '*P*' polarized, then determine the angle of incidence at which the reflected intensity is zero.

Solution:

(a) At $\theta_i = \frac{\pi}{2}$ (grazing angle), $\alpha = \frac{\cos\theta_t}{\cos\theta_i} \to \infty$. Essentially, this makes $\alpha \gg \beta$, and hence, $\tilde{E}_r = \frac{\alpha-\beta}{\alpha+\beta}\tilde{E}_i \simeq \frac{\alpha}{\alpha}\tilde{E}_i = \tilde{E}_i$. The entire incident amplitude is reflected. This is the phenomenon of total internal reflection, as occurs in a mirage.

(b) At $\alpha = \beta$, $\tilde{E}_r = \frac{\alpha-\beta}{\alpha+\beta}\tilde{E}_i = 0$. The reflected intensity becomes zero. There being no reflection, the entire incident energy is totally transmitted into the second medium. This angle is called the BREWSTER angle. If the incident intensity consists of unpolarized light, it has both the TE and the TM components. The reflected intensity of the TM '*P*' component becomes zero at the Brewster angle. The reflected light is therefore TE, '*S*', polarized.

P13.4:

Obtain the expression for the frequency perceived by an observer if the observer is moving at a relativistic speed *u* (i) toward the source of an electromagnetic radiation (ii) away from the source of the radiation.

Solution:

The phenomenology under consideration is the Doppler effect. First we obtain the non-relativistic expression for the frequency perceived by a moving observer. We illustrate the case (i) in the middle panel of the figure shown below, and the case (ii) in the lowest panel. We consider the movement of the observer through a distance ut_0 during one time period t_0 of the oscillation of the source.

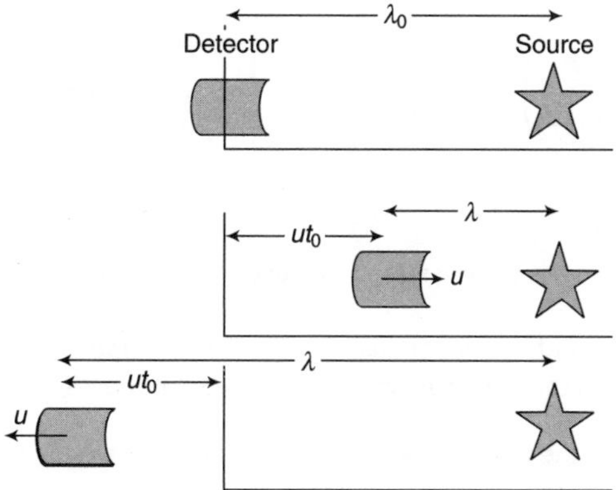

The distance between two consecutive points representing the same phase of the oscillation in the observer's frame will be:

$\lambda = \lambda_0 \mp ut_0$. Hence, $\frac{v}{\nu} = \lambda_0 \mp ut_0 = \frac{v}{\nu_0} \mp \frac{u}{\nu_0} = \frac{v \mp u}{\nu_0}$,

where v is the speed at which the wave travels.

The frequency of oscillation perceived by the observer is therefore, $\boxed{\nu = \frac{v}{v \mp u}\nu_0}$.

$\nu > \nu_0$: when the detector/observer is moving ***toward*** the source (called "BLUE SHIFT").

$\nu < \nu_0$: when the detector/observer is moving ***away*** from the source (called "RED SHIFT").

Now we consider the Doppler effect for the electromagnetic radiation detected by an observer moving at a relativistic speed u. The light source emits electromagnetic radiation of wavelength $\lambda_0 = \frac{c}{\nu_0}$. If the source moves at a constant speed u ***toward*** an observer, the observer detects the wave-amplitude maxima separated by a shorter wavelength (higher frequency: blue shift). If the source moves at a constant speed u ***away from the observer***, the observer detects the wave-amplitude maxima separated by a longer wavelength (lower frequency: red shift). The wavelength perceived by the observer therefore is

$\lambda = \lambda_0 \mp ut_0 = \frac{c}{\nu_0} \mp ut_0 = ct_0 \mp ut_0 = (c \mp u)t_0$, where the – sign corresponds to the case when the source moves toward the observer, and the + plus sign corresponds to the case when the source moves away from the observer. Without any loss of generality, we can write this as $\lambda = \lambda_0 - ut_0 = (c - u)t_0$, with u taken as positive when the source is moving toward the detector, and negative when it moves away from it.

Hence, $\frac{c}{\nu} = (c-u)\frac{1}{\nu_0}$. Therefore, $\nu = \frac{c}{c-u}\nu_0 = \frac{c}{c-u}\frac{1}{t_0}$.

We now account for the relativistic time dilation, through the factor $\gamma = \frac{1}{\sqrt{1-\frac{u^2}{c^2}}}$.

Accordingly, we get

$$\nu = \frac{c}{c-u}\frac{1}{\gamma T_0} = \left(\frac{c}{c-u}\right)\nu_0\sqrt{1-\frac{u^2}{c^2}} = \left(\frac{1}{1-\frac{u}{c}}\right)\nu_0\sqrt{\left(1+\frac{u}{c}\right)\left(1-\frac{u}{c}\right)} = \nu_0\sqrt{\frac{\left(1+\frac{u}{c}\right)}{\left(1-\frac{u}{c}\right)}}.$$

Hence: $\nu = \nu_0\sqrt{\frac{(c+u)}{(c-u)}}$. This is the relativistic expression for the frequency perceived by the observer.

The sign in the numerator and the denominator would be reversed if u is taken as positive when the source is moving away from the detector, and negative when it moves toward it. We can easily see that this result approximates to the non-relativistic Doppler shift by carrying out an expansion of $\sqrt{1-\frac{u^2}{c^2}}$:

$$\nu = \nu_0\left(\frac{1}{1-\frac{u}{c}}\right)\sqrt{1-\frac{u^2}{c^2}} \simeq \nu_0\frac{1-\frac{1}{2}\frac{u^2}{c^2}}{1-\frac{u}{c}} \approx \nu_0\frac{1}{1-\frac{u}{c}} = \frac{c}{c-u}\nu_0.$$

P13.5:

A light source that is in motion at a velocity $\vec{u}$, emits electromagnetic radiation which is received by a detector. In general, the line along which the detector is placed, which is the line along which the electromagnetic wave travels may or may not be parallel or antiparallel to the velocity of propagation of the source. (a) Determine the Doppler shift if the line of detection is orthogonal to the direction of motion of the source. (b) Determine the angle at which there is (practically) no Doppler shift, that is, the frequency of the radiation remains (almost) unchanged. (c) Show that if the geometry set for the critical angle that results from part (b), the speed of the source can be varied resulting in a shift from Doppler red-shift to blue-shift and vice versa.

Solution:

(a) We have seen above that: $\nu = \frac{c}{c-u}\frac{1}{\gamma T_0} = \frac{1}{1-\frac{u}{c}}\frac{\nu_0}{\gamma}$.

Hence, the circular frequency of this radiation is $\omega = \frac{c}{c-u}\frac{\omega_0}{\gamma}$.

$$\nu = \nu_0\left(\frac{1}{1-\frac{u}{c}}\right)\sqrt{1-\frac{u^2}{c^2}} = \nu_0\left(\frac{1}{1-\frac{u}{c}}\right)\frac{1}{\gamma} \text{ since } \gamma = \frac{1}{\sqrt{1-\frac{u^2}{c^2}}}.$$

Therefore, $\omega = \omega_0\left(1-\frac{u}{c}\right)^{-1}\frac{1}{\gamma} \simeq \omega_0\left(1+\frac{u}{c}\right)\frac{1}{\gamma} = \frac{1}{\gamma}\left(\omega_0+\frac{\omega_0 u}{c}\right) = \frac{1}{\gamma}(\omega_0+k_0u)$,

since $\omega_0 = 2\pi\nu_0 = 2\pi\frac{c}{\lambda_0} = k_0c$.

More generally, when the direction of motion of the source is not collinear with the direction in which the electromagnetic radiation is emitted, the term k_0u in the above expression is replaced by $\vec{k}_0 \cdot \vec{u}$.

Depending on the angle, $\measuredangle(\vec{k}_0, \vec{u})$, $\vec{k}_0 \cdot \vec{u}$ can be positive or negative. The Doppler shift can accordingly be a blue-shift (toward higher frequency) or a red-shift (toward lower frequency). We have resolved the momentum vector of the electromagnetic radiation along the line of observation in which the detector is placed (called the parallel component), and an orthogonal component (called the transverse component). Thus, when $\vec{u} = \vec{u}_\perp$, $\omega = \dfrac{\omega_0}{\gamma}$ and we get a red shift in the case of the transverse Doppler effect, with $\lambda = \lambda_0\gamma$.

(b) $\omega = \omega_0 \dfrac{\left(1 + \dfrac{\vec{k}_0 \cdot \vec{u}}{\omega_0}\right)}{\gamma} = D\omega_0$, wherein we have introduced the Doppler factor,

$$D = \frac{\left(1 + \dfrac{\vec{k}_0 \cdot \vec{u}}{\omega_0}\right)}{\gamma} = \left(1 + \frac{k_0 u}{\omega_0}\cos\xi\right)\left(1 - \frac{u^2}{c^2}\right)^{-\frac{1}{2}} \simeq \left(1 + \frac{u}{c}\cos\xi\right)\left(1 + \frac{1}{2}\frac{u^2}{c^2}\right),$$

$$D = \left(1 + \frac{u}{c}\cos\xi\right)\left(1 + \frac{1}{2}\frac{u^2}{c^2}\right) = 1 + \frac{1}{2}\frac{u^2}{c^2} + \frac{u}{c}\cos\xi + \frac{1}{2}\left(\frac{u}{c}\right)^3\cos\xi \simeq 1 + \frac{u}{c}\left(\frac{1}{2}\frac{u}{c} + \cos\xi\right).$$

We can see that at $\xi = -\cos^{-1}\left(\dfrac{u}{2c}\right)$, $D \cong 1$ and there is practically no Doppler shift.

(c) At the critical angle, $\dfrac{1}{2}\dfrac{u}{c} + \cos\xi = 0$. At a slightly lower speed, $\dfrac{1}{2}\dfrac{u}{c} + \cos\xi < 0$ and at a slightly higher speed, $\dfrac{1}{2}\dfrac{u}{c} + \cos\xi > 0$ which changes the sign of the Doppler factor, resulting in a red-shift to blue-shift transition, without changing the geometrical set up of the experiment.

P13.6:

(a) Determine the kinetic energy of an electron moving at 0.99c. Determine the ratio of the relativistic kinetic energy to the non-relativistic kinetic energy.

(b) In the core of the Sun, protons fuse to form the helium atom releasing some light particles like positrons and the neutrinos in a multi-step nuclear reaction. When four protons participate in the fusion of helium in this manner, the net mass of the product is *less* than the combined mass of the four protons by 4.57×10^{-29} kg. How much energy is obtained from the lost mass?

Solution:

(a) $E = \gamma mc^2 = mc^2 \left(1 + \dfrac{1}{2}\dfrac{v^2}{c^2} + \dfrac{\frac{1}{2}\left(\frac{1}{2}+1\right)}{2!}\left(\dfrac{v^2}{c^2}\right)^2 + \ldots\right)$, i.e. $E = (RE)_{\substack{\text{Rest}\\ \text{Mass}\\ \text{Energy}}} + (KE)_{\substack{\text{Relativistic}\\ \text{Kinetic}\\ \text{Energy}}} = mc^2 + (KE)_{\substack{\text{Relativistic}\\ \text{Kinetic}\\ \text{Energy}}}$.

Since $\gamma = \dfrac{1}{\sqrt{1 - \dfrac{0.99c^2}{c^2}}} = 7.0888$, $(KE)_{\substack{\text{Relativistic}\\ \text{Kinetic}\\ \text{Energy}}} = (\gamma - 1)mc^2 = (7.0888 - 1)mc^2 = 6.0888\, mc^2$,

where mc^2 = rest mass energy of the electron.

$m = 9.11 \times 10^{-31}$ kg and $c = 3 \times 10^8$ ms^{-1}. Therefore $(KE)_{\substack{\text{Relativistic}\\ \text{Kinetic}\\ \text{Energy}}} \simeq 499.22 \times 10^{-15}$ J

The non-relativistic kinetic energy is: $(KE)_{\substack{\text{NON-Relativistic}\\ \text{Kinetic}\\ \text{Energy}}} = \dfrac{1}{2} m(0.99c)^2 = 39.69 \times 10^{-15}$ J

$$\frac{(KE)_{\substack{\text{Relativistic}\\ \text{Kinetic}\\ \text{Energy}}}}{(KE)_{\substack{\text{Relativistic}\\ \text{Kinetic}\\ \text{Energy}}}} \simeq 12.6.$$

(b) The fusion of the protons to yield helium and release energy is a complicated multi-step process. The important steps are:

Two protons fuse together to produce deuterium, ^2H. A positron, a neutrino and some energy is released.

The deuterium fuses with another proton. ^3He. Some energy is released.

Two ^{3}He fuse together to produce ^4He. Two protons and energy is released.

The total mass of the final products is less than that of the ingredients that go into the nuclear multi-step fusion process by 4.57×10^{-29} kg. This mass is converted to energy according to the mass-energy equivalence, which 4.57×10^{-29} kg $\times 9 \times 10^{16} = 41.13 \times 10^{-13}$ J.

Additional Problems

P13.7 For the TM, '*P*', polarized electromagnetic waves (Fig. A of Problem P13.1), show that the reflection coefficient $R^P_{TM} = \dfrac{E_r}{E_i} = \dfrac{n_i \cos\theta_t - n_t \cos\theta_i}{n_i \cos\theta_t + n_t \cos\theta_i}$ and the transmission coefficient $T^P_{TM} = \dfrac{E_t}{E_i} = \dfrac{2n_i \cos\theta_t}{n_i \cos\theta_t + n_t \cos\theta_i}$. We have used $n_i = n_r = n_1$,, the refractive index of the medium '1' from which the electromagnetic wave impinges on the interface between the dielectric medium '1' and '2', and $n_t = n_2$ is the refractive index of the medium '2'.

P13.8 For the TE, '*S*', polarized electromagnetic waves (Fig. P13.1B of Problem P13.1), show that the reflection coefficient $R^P_{TM} = \dfrac{E_r}{E_i} = \dfrac{n_i \cos\theta_r - n_t \cos\theta_t}{n_i \cos\theta_r + n_t \cos\theta_t} = \dfrac{n_i \cos\theta_i - n_t \cos\theta_t}{n_i \cos\theta_i + n_t \cos\theta_t}$ and the transmission coefficient $T^P_{TM} = \dfrac{E_t}{E_i} = \dfrac{2n_i \cos\theta_r}{n_i \cos\theta_r + n_t \cos\theta_t} = \dfrac{2n_i \cos\theta_i}{n_i \cos\theta_i + n_t \cos\theta_t}$.

P13.9 Determine the reflected intensity $I_R = |R|^2$, and the transmitted intensity, $I_T = |T|^2$ for (i) TM, '*P*', case and (ii) TE, '*S*'. Determine also the sum $I_R + I_T$.

P13.10 Show, for the case of TE, '*S*', polarized electromagnetic waves (Fig. P13.1B of Problem P13.1), that $\frac{n_i}{n_t} = \frac{\sin\theta_t}{\sin\theta_i}$.

P13.11 According to measurements made by NASA satellites, the solar irradiance is 1,360 watts per square meter. It is the average intensity of the electromagnetic energy arriving at the top of the Earth's atmosphere on the side that faces the Sun. How much pressure (force per unit area) would the solar radiation exert?

P13.12 Show that the total angular momentum of the electromagnetic field is given by: $\vec{J} = \varepsilon_0 \iiint \sum_{j=1,2,3} E_j(\vec{r}\times\vec{\nabla})A_j dV + \varepsilon_0 \iiint (\vec{E}\times\vec{A})dV$. [The first term is the 'orbital' angular momentum of the electromagnetic field and the second term as the 'spin' angular momentum].

P13.13 Show that the orbital angular momentum of the electromagnetic field is independent of the polarization.

P13.14 Show that the spin angular momentum of the electromagnetic field depends on its polarization.

P13.15 Sketch the relativistic momentum of an electron as a function of its speed as it changes from 0 to 0.99c.

P13.16 Sketch the relativistic energy of an electron as a function of its speed as it changes from 0 to 0.99c.

P13.17 A given 12-volts car battery is able to supply two ,amps for 20 hours. It is thus rated at 40 ampere-hours. Does the mass of the battery change when it is fully charged from zero-charge? By what fraction of its mass is this increase of the mass of the battery is 10 kg when there is no charge?

P13.18 An atomic clock in a jet plane measures a time interval of 3000 s when the plane moves at a speed of 1000 km/hr. What would be the corresponding time interval recorded by an identical clock on the Earth's surface?

P13.19 A meter rod moves at 0.95c along the X-axis. The rod is oriented to make an angle of 300 with the X-axis. (a) Determine the length of the rod as measured by a stationary observer. (b) What angle does an observer in a stationary frame thinks the rod makes with the X-axis.

P13.20(a) At what speed, and in which direction, would a galaxy *A* be moving with respect to a galaxy *B* if a spectral line measured at 500 nm in '*B*' is blue-shifted, and measures 400 nm?

(b) At what speed, and in which direction, would a galaxy *C* be moving with respect to the galaxy *B* if a spectral line measured at 500 nm in '*B*' is red-shifted and measures 600 nm?

References

[1] Mohr, Peter J., David B. Newell, and Barry N. Taylor. 2016. 'CODATA Recommended Values of the Fundamental Physical Constants: 2014.' *Reviews of Modern Physics* 88: 20. DOI: 10.1103/RevModPhys.88.035009.

[2] Griffiths, David J. 2013. *Introduction to Electrodynamics* 4th edition. New York: Pearson New International Edition.

[3] Reitz, J. R., F. J. Milford, and R. W. Christy. 1979. *Foundations of Electromagnetic Theory.* Boston: Addison-Wesley.

[4] Feynman, R. P. 1964. *The Feynman Lectures on Physics* Vol II. Boston: Addison- Wesley.

[5] Lorrain, Paul, and Dale Corson. 1970. *Electromagnetic Fields and Waves* 2nd edition. New York: W. H. Freeman.

[6] Purcell, E. M. 1981. *Electricity and Magnetism Course, Berkeley Physics Course,* Vol 2. New York: McGraw-Hill Inc.

[7] Urban, M., F. Couchot, X. Sarazin, and A. Djannati-Atai. 2013. 'The Quantum Vacuum as the Origin of the Speed of Light.' *EPJ manuscript.* https://arxiv.org/pdf/1302.6165.pdf. Accessed on 2 January 2019.

[8] Das, P. Chaitanya, G. Srinivasa Murty, K. Satish Kumar, T. A. Venkatesh, and P. C. Deshmukh. 2004. 'Motion of Charged Particles in Electromagnetic Fields and Special Theory of Relativity.' *Resonance.* 9(7): 77–85.

[9] Kumar, Voma Uday, Gnaneswari Chitikela, Niharika Balasa, and P. C. Deshmukh. 2019. 'Revisiting Table-Top Demonstration of Relativistic Time-Delay and Length-Contraction'. *Bulletin of the Indian Association of Physics Teachers (IAPT)* 6(2): 172–179.

[10] John Stachel. 'How did Einstein Discover Relativity?' *AIP Center for History of Physics.* http://www.aip.org/history/einstein/emc1.htm. Accessed on 22 May 2016.

CHAPTER 14

A Glimpse of the General Theory of Relativity

If at first the idea is not absurd, then there is no hope for it.

—*Albert Einstein*

14.1 GEOMETRY OF THE SPACE–TIME CONTINUUM

The gravitational interaction is the earliest physical interaction that humans have registered. The earliest speculations about just what is the nature of gravity were not merely wrong, but absurdly far-fetched. Ancient philosophers even conjectured that the earth is the natural abode of things, and objects fall down when they are dropped just as horses return to their stables. Various theories of gravity were proposed, and the one that lasted much is that developed by Isaac Newton in the seventeenth century. Newton's work on gravity integrated the dynamics of astronomical objects with that of falling apples or coconuts, determined by one common principle (Chapter 8). We celebrate this principle as Newton's *one-over-distance-square* law of gravity.

An amazing consequence of the constancy of the speed of light in all inertial frames of reference that we studied in the previous chapter is the *time-dilation* and *Lorentz contraction* (also called the *length contraction*). The phenomenon that is responsible for the traveling twin to age slower than the home-bound twin holds for *any* and *every* object in motion. We have already noted that this happens to decaying muons. Essentially, the faster you move through space, the slower you move through time, in the spacetime continuum.

We all enjoy raising our speed, covering more distance in lesser, and lesser, time. Let us therefore ask, to what extent can we speed up an object? We ask if there is a natural limit for this. If you look back into the relations for time-dilation and the length contraction in the previous chapter, you will recognize that if $v = c$, the effect of time-dilation would be such that the traveling twin will simply stop ageing. Time would stop for her; time freezes. The effect of Lorentz contraction would also be total; she would think that the rest of the universe has spatially contracted to a point. She is therefore already *everywhere* (along the line of motion). All of these dramatic aftermaths are because of a simple fundamental property that the speed of the headlight of a car coming *toward* you at a velocity v is no different from that of the tail light of another that is *receding* away from you. Thus, the time intervals in a frame of reference moving at $v = c$ would simply stop. The query about an object *moving* through space in such a situation, as time *changes*, therefore becomes moot. The upshot of this physical reality, tested repeatedly in every experiment, is the fact that nothing can move faster than the speed of light. The conclusion would be no different from the point of view of the observer moving at $v = c$. In her frame, the length of the universe along which the frame M moves, would have Lorentz-shrunk to a point. She being already *everywhere*, no further motion is of any relevance. If this seems mind-boggling, it is only because we have allowed our intuition to be built on the assumption that the speed of light is *infinite*. Reality is bound to appear to be *counter-intuitive*, when intuition itself is misguided.

Having seen that time freezes at $v = c$, we have to concede that nothing can go faster than light; just nothing. No physical object, nor any physical influence, can go faster than this speed. The only physical objects that can go at the speed $v = c$ are those whose *rest mass is zero* (Eq. 13.110). Any particle whose rest mass is zero travels at $v = c$. Electromagnetic waves travel at this speed essentially because the photon has a zero rest mass. Of course, the speed is related to the electric permittivity of the medium, and hence in a different medium, light propagates at a speed different from that in vacuum. This phenomenology is responsible for the refraction of waves when light goes across a surface of separation between two media.

If no physical influence travels faster than $v = c$, it is then pertinent to ask what would happen to the Earth's orbit if the Sun's gravitational pull on it is suddenly switched off? After all, the most dominant effect of the Sun's gravity is that the Earth goes around it in Kepler–Newton orbits. However, since light takes a little over 8 minutes to reach the Earth from the Sun, we do not expect the Earth to experience any physical effect, for at least this duration, if the Sun's gravitational pull on the Earth is switched off. Yet, we expect that if a child is revolving a tiny mass wound to a string, as in Fig. 14.1, then the mass would be whirled off at a tangent as soon as the string snaps. The different response of the Earth to switching off the Sun's gravity is on account of the limitations in our understanding of gravity, as governed by Newton's law which makes the interaction between two masses *instantaneous*. This is incompatible with the finite limit on the speed of a physical influence that we now have to reconcile with. Essentially, this calls for going beyond Newton's law of gravity. This was no meek challenge, and *even* for Albert Einstein, to find a way out of this impasse, it took a full decade after he formulated the Special Theory of Relativity (STR) in his miracle year 1905. After this long struggle, Einstein published an article *Die Grundlage der allgemeinen Relativitätstheorie* (The Formal Foundation of the General Theory of Relativity) in the *Annalen der Physik*. This triggered a revolution in man's view of the universe. In the STR itself, we had already gone beyond the Newton–Galileo principle of relativity; now one must go even beyond that.

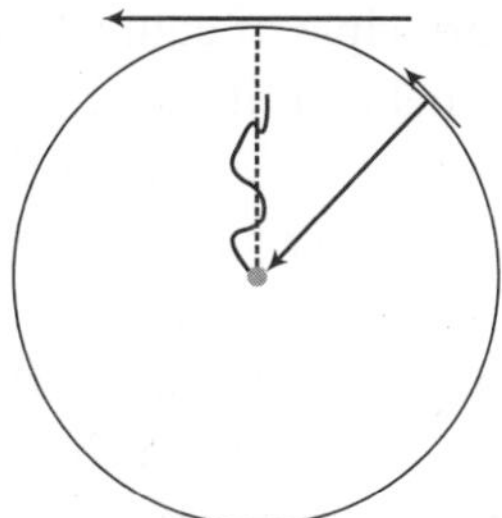

Fig. 14.1 If you are whirling a ball tied to a string in a circular trajectory, and suddenly the string snaps, the ball would dart off *instantly* along a tangent to the circle from the point at which the string snaps.

The recognition of c as a fundamental constant of nature giving rise to the non-Euclidean spacetime continuum, time-dilation, length-contraction, and the mass–energy equivalence, had already shaken our understanding of the physical world. Ten years later, the General Theory of Relativity (GTR) brought about a further revolution, enabling a deeper understanding of the geometry of the spacetime continuum. It turned out that the geometry of the STR spacetime is only a special case of the geometry of the GTR, hence the names STR, and the GTR.

The Newtonian notion of gravity as the force of attraction between two masses in accordance with the inverse square law and the product of the masses guided us in all the applications we have considered in the earlier chapters in this book; most notably in Chapter 8. Once we get used to this idea, any departure from it, even if correct, seems shocking. It needed a radical idea, not just shocking but even '*absurd*', to improve our understanding of gravity beyond Newton. Just like the incredible philosophy that went into the STR, and the revolutionary interpretation of the photoelectric effect that consolidated the foundation of quantum mechanics, the ground-breaking new theory of gravity also came from Einstein, only about a decade later.

GTR, in some sense, also has its origin in understanding the importance of symmetry in the physical laws. We have seen that symmetry appeared as an important element in our understanding of the law of conservation of momentum. In Chapter 1, we actually obtained Newton's third law from symmetry principles. In Chapter 6, we learned that the momentum canonically conjugate to a cyclic coordinate is conserved. We also acquainted ourselves with the dynamical symmetry of the Kepler–Newton law of gravity. Emmy Noether put all this together in the famed theorem, known after her [1, 2]. However, in some sense, it all started with Einstein. It was him who saw the symmetry in Maxwell's equations which led to the recognition of the constancy of the speed of light, *manifest in Maxwell's wave equation*, as a fundamental principle. This led to the STR. In the GTR, Einstein came up with yet another symmetry principle: *In a free fall in the gravitational field, a body is unable to feel its own weight. Thus, if a body is accelerating in free space at the value of g meter per second per second, then its motion must be described in exactly the same way as that of an object on the surface of the Earth, accelerating toward the Earth at g meter per second per second. This is an 'equivalence principle', guided yet again by the importance of symmetry.*

An amazing consequence of the principle of equivalence is how it impacts the trajectory of a beam of light in a gravitational field. To illustrate this, we refer to Fig. 14.2 [3]. There is of course plenty of excellent literature available on this subject. Readers can find these sources easily. We shall follow the elementary pedagogical discussion from the Reference [3]. Fig. 14.2 shows that if a beam of light is traveling from left to right directly *across* a rocket accelerating 'upward' in free space, at g ms^{-2}, its trajectory appears to go along a curve as the rocket accelerates upward. Of course, there is no such thing like 'up' and 'down' in the free space. The direction in which the rocket is accelerating is considered 'upward'.

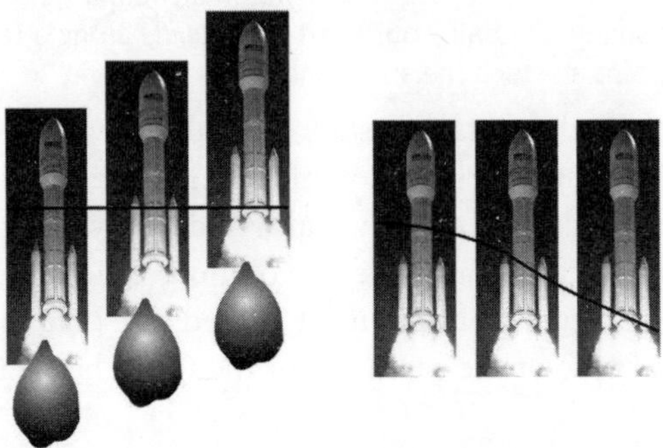

Fig. 14.2 Anything accelerating in free space at the value of g meter per second per second acts in exactly the same way as an object accelerating toward the Earth at g meter per second per second. The figure on the left side shows light moving across a rocket which is accelerating at g meter per second per second. The figure on the right shows light moving across a rocket standing at rest on the Earth. All objects on Earth accelerate toward it, at g meter per second per second, including light that is traveling across the rockets at rest on Earth. This figure is only schematic, and of course not to scale.

The principle of equivalence then suggests that if a non-accelerating rocket is placed *motionless* on the surface of the Earth, then light must respond to Earth's gravity in exactly the same way, i.e., it must get bent along a curve as it travels, even if the bend would be unnoticeable. Now, who might ever expect a pencil ray of light to bend under gravity, just like a ball thrown across is accelerated downward, and hence falls on the Earth? The principle of equivalence however necessitated a modification of the spacetime continuum that we used in the previous chapter on the special theory of relativity. When an object falls under gravity on Earth, one might consider this process to be totally equivalent to the Earth accelerating upward toward the object. The object accelerating down toward the Earth is the same as the Earth accelerating up toward the object. This makes perfect sense if the Earth were *flat*. However, since the Earth is round, we run into an awkward situation that antipodes (Fig. 14.3) would need to be accelerating in opposite directions even as the distance between their feet remained invariant.

Fig. 14.3 The equivalence principle suggests that an apple falling at the North Pole is equivalent to the Earth accelerating *upward* (in this diagram) toward the apple, but an apple falling at the South Pole would likewise require the Earth to be accelerating *downward*. How could both of these happen if the distance between the antipodes remains invariant, equal to the diameter of the Earth?

In 1912, Einstein figured that the impasse posed by the difficulty presented in Fig. 14.3 could be resolved if the geometry of the spacetime is not required to be *flat*, but *curved* [4]. One would then need a new theory of gravity; which Einstein would formulate. The new theory would have to be such that it would accommodate Newton's law of gravity, just as the Lorentz transformations (Chapter 13) would reduce to the Galilean in the limit $c \rightarrow \infty$. Einstein then came up with such an equation, now recognized as the Einstein Field Equation (EFE). The EFE accommodates Newton's equation in the limiting case. It requires the spacetime to be curved, and it predicts that a beam of light would *bend* under gravity. The prediction that a beam of light would actually bend under gravity, that light also *falls* under gravity just like apples and coconuts from the trees, based on the principle of equivalence, seemed too farfetched. Would the strength of logic in Einstein's thought experiment be enough to accept this? An experimental test seemed impossible, since the Earth's gravitational pull is not strong enough to measurably bend a pencil of light. In 1919, Dyson, Eddington and Davidson [5] provided the first experimental test for Einstein's *principle of equivalence*. The experiment was brilliantly conceived; it is easily one of the most beautiful experiments ever carried out. Dyson, Eddington and Davidson realized that only a much higher value of the acceleration due to gravity than what the Earth generates could bend light measurably. Only the Sun, the most massive object in our neighborhood, could produce a noticeable bending of light. The light to be bent of course would have to come from *behind* the Sun. Edington's experiment was performed almost exactly a century ago, on 29 May 1919.

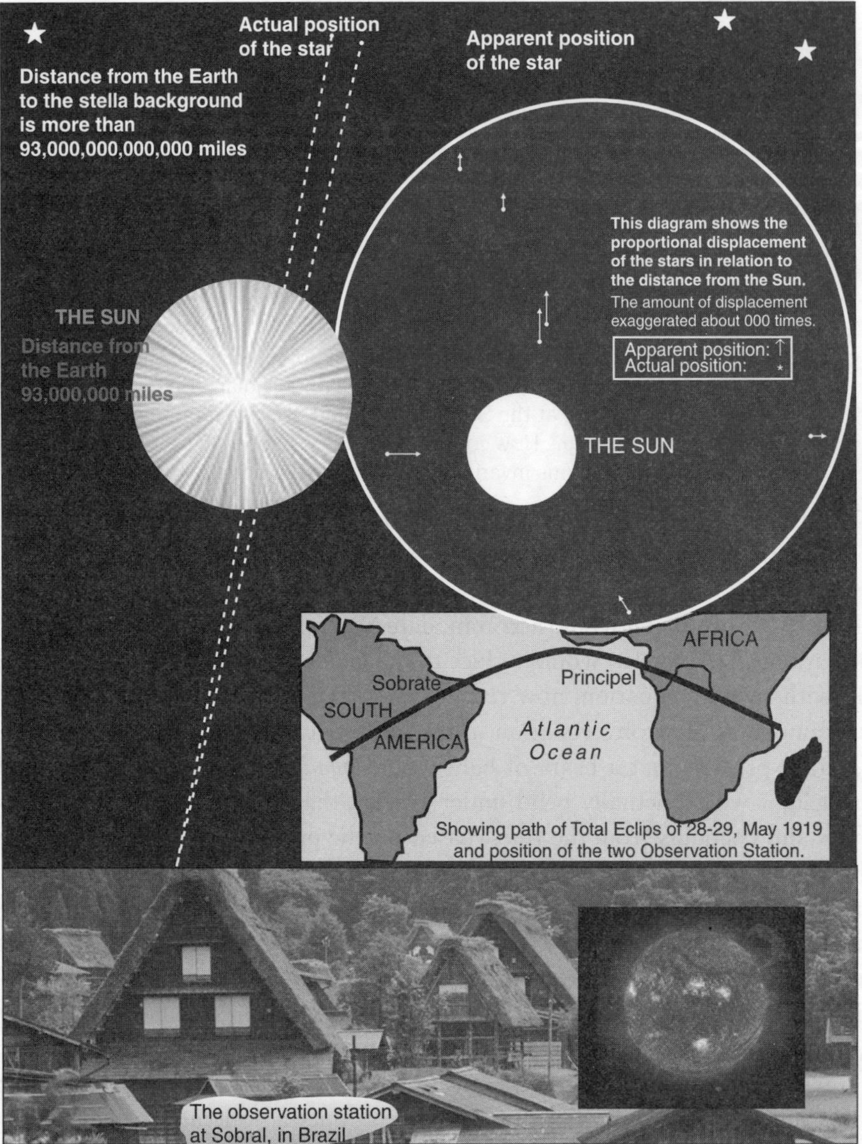

Fig. 14.4 Schematic diagram representing Eddington's experiment [5] of 29 May 1919. This experiment was the very first one that was in accordance with Einstein's principle of equivalence.

However, how could one ever detect this? The Sun's own light is so powerful that it completely dominates the sky. Light coming from any star, or from a cluster of stars, behind the Sun would get completely subjugated by the sunlight. There would be no way of knowing whatever happened to the light that came from *behind* the Sun. Dyson, Eddington, and Davidson [5] then prepared to execute a brilliant idea. They decided to look for light that came from *behind* the Sun, during the solar eclipse of 29 May 1919, so that sunlight itself would not dwarf it.

Two scientific expeditions were carried out to perform similar experiments during the eclipse. The total eclipse could be seen only from the Earth's equatorial belt, so one expedition was led by Davidson at Sobral (Brazil). The other expedition was led by Eddington at the Island of Sao Tamo and Principe (Gulf of Guinea). The eclipse lasted slightly under 7 minutes, during which very many photographs of the stars around the Sun's corona were taken. Dyson analyzed the data from Sobral, and Eddington carried out the analysis of the data from the Island of Príncipe. It took several months to complete the analysis. The data from both the sets of observations led to the *same* incredible conclusion: the Sun's gravity had, in fact, bent light, exactly in accordance with Einstein's principle of equivalence. Gravity and acceleration were thus experimentally found to be completely equivalent. Several romantic accounts of these historic experiments are available in the literature.

The conclusion from these experiments was unambiguous: light does *fall* under gravity, just as the Earth falls toward the Sun, the Moon toward the Earth, and apples and coconuts fall on the Earth. How at all can we reconcile this with the fact that a photon has *zero* rest mass? The only way to address this would be to conclude that the *fall* of light and apples is *not* governed by the Newton's law of gravity which requires the product of the masses in the numerator of the force of gravitational attraction between two masses. The *fall*, i.e., the bending of light, therefore had to have a totally *different* origin. Einstein, in his GTR, proposed that it is the curved *geometry* of the spacetime continuum which determines the trajectories of objects.

Now, a fundamental consideration in understanding the *geometry* of space, rather, the spacetime continuum, is the metric g. It tells us how the distance between two points in the space (i.e., the spacetime) is measured. The Pythagoras theorem gives us a measure of this distance in the 3-dimensional Euclidean space (see Eqs. 2.22, 2.29, and 2.30 from Chapter 2). STR required a modification of the notion of this distance to suit the non-Euclidean 4-dimensional spacetime continuum. The 'interval' between two spacetime events is given in terms of the scalar-product of the 4-vectors, given by Eq. 13.54, 13.55 (Chapter 13), characterized by the signature (+1 −1 −1 −1) of the metric as

$$ds^2 = dx^\mu\, g_{\mu\upsilon}\, dx^\upsilon = \begin{bmatrix} dx^0 & dx^1 & dx^2 & dx^3 \end{bmatrix} \begin{bmatrix} 1 & 0 & 0 & 0 \\ 0 & -1 & 0 & 0 \\ 0 & 0 & -1 & 0 \\ 0 & 0 & 0 & -1 \end{bmatrix} \begin{bmatrix} dx^0 \\ dx^1 \\ dx^2 \\ dx^3 \end{bmatrix}. \tag{14.1}$$

If we use the spherical polar coordinates (see Chapter 2), the signature of the g-metric (from Eq. 2.29) must correspond now to the 4-dimensional non-Euclidean signature employed in Eq. 14.1. The interval between two non-Euclidean spacetime events therefore becomes

$$ds^2 = dx^\mu g_{\mu\nu} dx^\nu = \begin{bmatrix} dx^0 & dx^1 & dx^2 & dx^3 \end{bmatrix} \begin{bmatrix} 1 & 0 & 0 & 0 \\ 0 & -1 & 0 & 0 \\ 0 & 0 & -r^2 & 0 \\ 0 & 0 & 0 & -r^2 \sin^2\theta \end{bmatrix} \begin{bmatrix} dx^0 \\ dx^1 \\ dx^2 \\ dx^3 \end{bmatrix}. \tag{14.2}$$

The g-metric used in Eq. 14.1, 14.2 was employed in the STR. It is appropriate for a 'flat' spacetime, but not for the 'curved' spacetime. The spacetime interval given by the following expression turns out to be appropriate to describe the *curved* spacetime continuum [6]:

$$ds^2 = \sum_{\mu,\nu} g_{\mu\nu}\,(dx^\mu)(dx^\nu) = U\,dt^2 - V\,dr^2 - r^2 d\theta^2 - r^2 \sin^2\theta d\phi^2. \tag{14.3}$$

The metric tensor that corresponds to the above spacetime interval is given by

$$g_{\mu\nu}(x) = \begin{bmatrix} U & 0 & 0 & 0 \\ 0 & -V & 0 & 0 \\ 0 & 0 & -r^2 & 0 \\ 0 & 0 & 0 & -r^2\sin^2\theta \end{bmatrix}. \tag{14.4}$$

The structure of the metric tensor is what determines the flatness, or the curvature, of the spacetime continuum. According to the GTR, a massive body 'curves' the spacetime continuum. The inertial trajectories of objects in gravitational fields generated by masses do *not* result from the Newton's inverse-square-law of gravity. These trajectories are *geodesics* that are determined by the *geometry* of the spacetime continuum.

14.2 EINSTEIN FIELD EQUATIONS

In order to introduce the Einstein Field Equations (EFE) [6, 7, 8], we first introduce the Christoffel symbols. These are of two kinds:

First kind - $$\Gamma_{\alpha\rho\sigma} = \frac{1}{2}(\partial_\sigma g_{\alpha\rho} + \partial_\rho g_{\alpha\sigma} - \partial_\alpha g_{\rho\sigma}), \tag{14.5a}$$

Second kind - $$\Gamma^{\mu}{}_{\rho\sigma} = g^{\mu\alpha}\Gamma_{\alpha\rho\sigma} = \frac{1}{2}g^{\mu\alpha}(\partial_\sigma g_{\alpha\rho} + \partial_\rho g_{\alpha\sigma} - \partial_\alpha g_{\rho\sigma}). \tag{14.5b}$$

See NOTES on Christoffel symbols, and on covariant derivative of tensors, at the bottom of this chapter, just before the references.

Also, we introduce the Riemann tensor,

$$R^{\rho}{}_{\beta\alpha\gamma} = \partial_\alpha\Gamma^{\rho}{}_{\gamma\beta} - \partial_\gamma\Gamma^{\rho}{}_{\alpha\beta} + \Gamma^{\lambda}{}_{\gamma\beta}\Gamma^{\rho}{}_{\alpha\lambda} - \Gamma^{\lambda}{}_{\alpha\beta}\Gamma^{\rho}{}_{\gamma\lambda}. \tag{14.6}$$

One may also determine the fully covariant Riemann tensor,

$$\left.\begin{aligned} R_{\mu\beta\alpha\gamma} &= g_{\rho\mu}R^{\rho}{}_{\beta\alpha\gamma} \\ &= \frac{1}{2}(\partial_\alpha\partial_\beta g_{\mu\gamma} + \partial_\gamma\partial_\mu g_{\alpha\beta} - \partial_\alpha\partial_\mu g_{\gamma\beta} - \partial_\gamma\partial_\beta g_{\mu\alpha}) + g_{\rho\lambda}(\Gamma^{\lambda}{}_{\mu\gamma}\Gamma^{\rho}{}_{\alpha\beta} - \Gamma^{\lambda}{}_{\mu\alpha}\Gamma^{\rho}{}_{\gamma\beta}) \end{aligned}\right\}. \tag{14.7}$$

It has the following symmetry properties:

$$R_{\mu\beta\alpha\gamma} = -R_{\beta\mu\alpha\gamma}, \tag{14.8a}$$

$$R_{\mu\beta\alpha\gamma} = -R_{\mu\beta\gamma\alpha}, \tag{14.8b}$$

$$R_{\mu\beta\alpha\gamma} = R_{\alpha\gamma\mu\beta}.$$

Thus, $$R_{\mu\alpha\beta\gamma} + R_{\mu\gamma\alpha\beta} + R_{\mu\beta\gamma\alpha} = 0. \tag{14.9}$$

A second rank tensor, called the Ricci tensor, symmetric in its two indices, can be built from the Riemann tensor:

$$R_{\beta\gamma} = g^{\mu\alpha} R_{\mu\beta\alpha\gamma}, \tag{14.10a}$$

$$R_{\beta\gamma} = R_{\gamma\beta}. \tag{14.10b}$$

The two indices of the Ricci tensor can be contracted, to obtain the Ricci scalar R,

$$R = g^{\beta\gamma} R_{\beta\gamma}. \tag{14.10c}$$

We can now introduce the Einstein Field Equation (EFE), normally written as [6, 7, 8, 9]:

$$G_{\mu\nu} = R_{\mu\nu} - \frac{1}{2} R g_{\mu\nu} = \frac{8\pi G}{c^4} T_{\mu\nu}. \tag{14.11a}$$

In the above relation, $G_{\mu\nu}$ is called the 'Einstein Tensor', $T_{\mu\nu}$ is called the 'Stress-Energy Tensor', and $R_{\mu\nu}$ is the 'Ricci Tensor'. The EFE (Eq. 14.11) has 16 components. It represents a family of 16 equations. However, the metric and the stress tensors are symmetric about the diagonal, hence only 10 of these equations are independent. The Einstein tensor is therefore symmetric:

$$G_{\mu\nu} = G_{\nu\mu}. \tag{14.11b}$$

Thus, there are only 10 unique components of the equation of motion. These 10 components are a system of 10 highly non-linear, highly coupled, partial differential equations which are difficult to solve. It seems best to regard the EFE as an *ansatz*. It provides an acceptable picture of the dynamics in the physical universe. The EFE, as mentioned above, not only predicts the bending of light by gravity, but also comfortably allows the Newtonian limit of dynamics [10] to be obtained from Einstein's GTR. The commentary by John Wheeler on Eq. 14.11 provides an excellent description of the EFE. Wheeler observed that the left-hand-side of the EFE has properties dealing with the curvature of spacetime, while the right-hand-side dealt with matter-energy; thus the spacetime (left-hand-side) determines how matter would move, while matter (the right-hand-side) determines how the spacetime curves. Accordingly, the EFE interprets the motion of matter under gravity to be along what are known as inertial *geodesics* which are determined by the curvature of spacetime. The Newtonian notion of 'force' *proportional to the product of the masses and inversely proportional to the square of the distance between the masses* is abandoned. Instead, gravity results from the curvature of the spacetime caused by the larger mass. The Riemann curvature tensor provides a description of the curvature of spacetime which accounts for how the tangent vectors slide around the manifold. Other sources, such as Ref. [10], may be consulted for further details about the Einstein tensor, the stress-energy tensor and the Ricci tensor.

14.3 GTR COMPONENT OF PRECESSION OF PLANETS

We shall restrict ourselves to discuss the Schwarzschild solution to the EFE. Karl Schwarzschild (1873–1916) gave the first exact solution to Einstein's field equation. It is very unfortunate that he died very young, soon after this work. He employed a coordinate system, akin to the spherical polar coordinate system, that is now named after him. Besides, other important concepts like the Schwarzschild metric, Schwarzschild radius, Schwarzschild black holes and also Schwarzschild wormholes are named after him. Most importantly, the Schwarzschild solution accounts for the precession of the perihelion of planetary orbits in our solar system in terms of the GTR.

The Schwarzschild solution [11, 12, 13, 14] to the Einstein Field Equation deals with the *exterior spacetime* of a spherically symmetric body. To appreciate this, we consider the two-body interaction between the Sun and the planet Mercury. Mercury's mass, being so tiny compared to the larger mass of the Sun, is considered to have no effect on the latter. The right hand side of the EFE (Eq. 14.11) is then set to zero:

$$R_{\mu\nu} - \frac{1}{2} R g_{\mu\nu} = 0. \tag{14.12}$$

The spacetime interval (Eq. 14.3) is considered in the Schwarzschild solution to be given by

$$ds^2 = \left(1 - \frac{2Gm}{c^2 r}\right) c^2 dt^2 - \frac{1}{1 - \dfrac{2Gm}{c^2 r}} dr^2 - r^2 d\theta^2 - r^2 \sin^2 \theta d\varphi^2. \tag{14.13}$$

In the above equation, m is the solar mass. Since a force derivable from a potential is now abandoned with, the Lagrangian (Chapter 6) for the solar field is *potential free*:

$$L = \frac{1}{2} m v^2 = \frac{1}{2} m g_{\alpha\beta} \dot{x}^\alpha \dot{x}^\beta, \tag{14.14a}$$

where $\dot{x}^\alpha = \dfrac{dx^\alpha}{d\tau}$ and $\dot{x}^\beta = \dfrac{dx^\beta}{d\tau}$, τ being the 'proper' time. In the polar coordinate system employed by Schwarzschild, we have $x^0 = t, x^1 = r, x^2 = \theta, x^3 = \varphi$. The Lagrangian is therefore given by

$$L = \frac{1}{2} m \left[\left(1 - \frac{2m}{r}\right) \dot{t}^2 - \frac{1}{1 - \dfrac{2m}{r}} \dot{r}^2 - r^2 (\dot{\theta}^2 + \sin^2 \theta \dot{\varphi}^2) \right]. \tag{14.14b}$$

For each of the four coordinates $a = 0, 1, 2, 3$, we can solve the Euler–Lagrange equations (Section 6.1, Chapter 6):

$$\frac{\partial L}{\partial x^a} - \frac{d}{d\tau}\left(\frac{\partial L}{\partial \dot{x}^a}\right) = 0. \tag{14.15}$$

As discussed in details in Chapter 6, the Lagrangian formulation departs strongly from the linear stimulus-response principle in Newtonian dynamics. Instead of the cause-effect formulation of Newtonian mechanics, the trajectories of objects in motion are accounted for by the variational principle, by requiring that the *action* $S = \int_{t_1}^{t_2} Ldt$, is an extremum. The major difference now, from the principle illustrated in Chapter 6, is that the Lagrangian is *completely free* from the potential. The trajectory of an object is now described by *inertial geodesics* in a curved spacetime, in accordance with the EFE, 14.11. This approach accurately accounts for the precession of planetary orbits that Newtonian dynamics could not fully account for. Manipulation of the Euler–Lagrange equations, for $a = 0$, 2 and 3, as detailed in Reference [14] (specifically equations 18 through 28 therein), results in the following differential equation:

$$\frac{d^2u}{d\varphi^2} + u = \frac{mG}{h^2} + \frac{3mG}{c^2}u^2. \tag{14.16a}$$

In the above equation, $u = \frac{1}{r}$, $r = |\vec{r}|$, and $\vec{r}$ denotes the position vector of a planet with respect to the Sun, m is the mass of the Sun, $\vec{h} = \vec{r} \times \vec{v}$ is the 'specific' angular momentum, and φ is the angle made by $\vec{r}$ with the Keplerian major axis. You will immediately recognize that Eq. 14.16a is a minor departure from Eq. 8.20b from Chapter 8. We have used a slightly different notation in this chapter compared to that in Chapter 8, but you will see the correspondence immediately. The main difference is the additional quadratic term in Eq. 14.16a. It is this quadratic term that gives solutions that are slightly *different* from the orbits having various eccentricities that we discussed in Chapter 8. It is precisely this difference that accounts for the precession of planetary orbits. In particular, mercury's orbit about the Sun is not *exactly* elliptic, as would be expected from Newton's law of gravity. There is a departure from the strict ellipse. First of all, the Laplace–Runge–Lenz vector (Eq. 8.14b, Chapter 8) is conserved only for a strict one-over-distance potential. The 'fixed ellipse' is a consequence of the dynamical symmetry [Section 8.2, Chapter 8] of the classical Newtonian inverse square law force, in the two-body interaction. There are other planets in the solar system, so we do not have an exact dynamical symmetry. This results in a precession of mercury's orbit. The rotation of the major axis of Mercury's orbit around the Sun was observed to be ~5599.7 seconds of an arc, per earth-century. Various corrections [15] that must be made to the two-body Kepler–Newton problem account for much of the precession of mercury's orbit. These corrections, summarized in Table 14.1, account for *most* of the precession, but not all of it. The precession itself is schematically shown in Fig. 14.5 [from References 16a, b].

Table 14.1 The precession of the major axis of Mercury's elliptic orbit is partially accounted for by various factors listed below. However, it leaves a residual correction factor, which can only be accounted for using Einstein's field equations of the GTR. Reference [15].

Angle in arc-second per earth-century accounted for	Cause
5025.6	Coordinate (precession of the equinoxes)
0531.4	Gravitational tugs due to other planets
0000.0254	Oblateness of the Sun
Sum of above three: 5557.0254	*Combined effect of the above three causes*

The difference between the observed rotation and what is accounted for in Table 14.1 is ~43 seconds of an arc per earth-century.

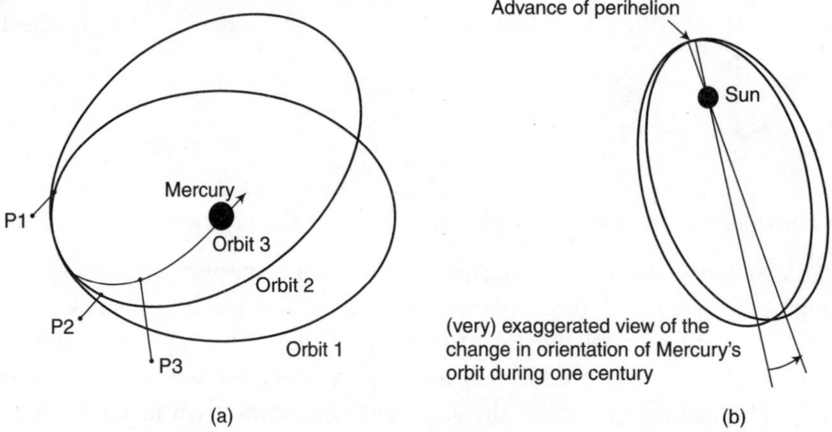

Fig. 14.5 Schematic representation of the excess angle $\theta_{\text{precession}}$ of Eqs. 14.16b and c. Fig. 14.5a is from Reference [16a], and Fig. 14.5b is from Reference [16b]). Of the annual (i.e., over Mercury-year) advance of the planet Mercury about its trajectory around the Sun resulting in 'precession' of the Keplerian elliptic orbit.

Attempts to account for this residual precession included speculations on additional phantom planets, including an inner one between Mercury and the Sun, but all such attempts failed. The angle of precession is so very small that most physicists might ignore it completely; but not Einstein, of course. It happily turns out that the solution to Eq. 14.16a, based on the Einstein Field Equation, finally accounts for the residual precession, as shown below. Equation 14.16a can be solved using perturbative methods. The ratio of the two terms on the right hand side, each of which has the dimension L^{-1}, is $\dfrac{3mGu^2/c^2}{mG/h^2}=\dfrac{3h^2u^2}{c^2}=\dfrac{3h^2}{c^2r^2}$. Using the Kepler–Newton data for the approximately–elliptic orbit of Mercury around the Sun, we shall first determine the ratio $\dfrac{3h^2}{c^2r^2}$. Using the semi-major axis of the Mercury's approximately elliptic orbit, $r = b$, the speed of light given by $c = 299792458\ ms^{-1}$, and the specific angular momentum of Mercury about the Sun to be given by $h = 2.756740382 \times 10^{15}\ m^2 s^{-1}$, we get $\frac{3h^2}{c^2 b_{\|}^{\,2}} \simeq 0.756 \times 10^{-7}$. This is a rather small number, and hence it can be dealt with as a tiny perturbative correction. To use perturbation theory, we shall use a dimensionless perturbation parameter, $\lambda = \frac{3m^2G^2}{h^2c^2}$. Equation 14.12a thus becomes

$$\frac{d^2u}{d\varphi^2} + u = \frac{mG}{h^2} + \lambda \frac{h^2u^2}{mG}. \tag{14.16b}$$

A solution to the above differential equation can now be sought in the form

$$u = u_o + \lambda u_1 + O(\lambda^2), \tag{14.17}$$

where in $O(\lambda^2)$ represents terms of the order of λ^2.

Substituting Eq. 14.17 in Eq. 14.16b, we get

$$\frac{d^2u_0}{d\varphi^2} + u_0 = \frac{mG}{h^2} - \lambda\left(\frac{d^2u_1}{d\varphi^2} + u_1 - \frac{h^2u_0^{\,2}}{mG}\right) - O(\lambda^2). \tag{14.18}$$

If only the perturbation parameter λ were zero, Eq. 14.18 would be exactly the same as Eq. 8.20b of Chapter 8. The solution to Eq. 14.18, to first order in the perturbation theory, is then given by

$$u \approx \frac{mG}{h^2}[1 + \varepsilon \cos\{\varphi(1 - \lambda)\}]. \tag{14.19}$$

It is obvious that $\lambda = 0$ would result in a fixed ellipse. The non-zero value of λ, however tiny, has essentially come from the manipulation of the Euler–Lagrange equation and has its origin strictly in the Einstein Field Equation of the General Theory of Relativity. The trajectory repeats, but *not* on tracing an angle 2π, as expected for a closed elliptic Keplerian orbit. Instead, the periodicity occurs for a slightly larger angle, on tracing an angle given by

$$\Theta \simeq 2\pi(1 + \lambda). \tag{14.20a}$$

It is this larger angle that is responsible for the (residual) precession of the planet Mercury. The *excess* angle, represents the *advance* of the planet that is *over* and above what is accounted for in Table 14.1, in the SI units, is given by

$$\theta_{\text{precession}} = 2\pi\lambda = \frac{6\pi G^2m^2}{c^2h^2}. \tag{14.20b}$$

The subscript 'precession' on the angle given in the above equation represents the precession of the planet over and above what is given in Table 14.1. In 'geometrized' units, $c = 1$, $G = 1$, and the excess angle is,

$$\theta_{\text{precession}} = 2\pi\lambda = \frac{6\pi m^2}{h^2}. \tag{14.20c}$$

The actual calculation for the angle $\theta_{\text{precesion}}$ can now be carried out. This is easily done using the primary Kepler–Newton data on Mercury's orbit, even if we are actually interested in determining the correction to it. From Kepler's second law, we have

$$T_{\text{Mercury}} = \frac{2\pi b_{\parallel} b_{\perp}}{h}, \tag{14.21a}$$

where $b_{\parallel}$ and $b_{\perp}$ are respectively the semi-major and the semi-minor axis of the Kepler orbit. T_{Mercury} is the time taken by the planet Mercury to complete one Kepler-orbit round the Sun; it is what we must call as the *Mercury-year*. Thus,

$$T^2_{\text{Mercury}} = \frac{4\pi^2 b_{\parallel}^{\,4}(1-\varepsilon^2)}{h^2}, \tag{14.21b}$$

where the eccentricity of the Kepler orbit is denoted by ε. Again, from Kepler's third law, we have

$$T^2_{\text{Mercury}} = \frac{4\pi^2 b_{\parallel}^{\,3}}{G(m_s + m_m)}. \tag{14.21c}$$

Using now $m_m << m_s$, we get

$$G^2 m_s^{\,2} = \frac{4\pi^2 b_{\parallel}^{\,2} h^2}{T^2_{\text{Mercury}}(1-\varepsilon^2)}. \tag{14.21d}$$

Now using Eq. 14.21d in Eq. 14.20b, we get

$$\theta_{\text{precession}} = \frac{6\pi}{c^2 h^2}\left(\frac{4\pi^2 b_{\parallel}^{\,2} h^2}{T^2_{\text{Mercury}}(1-\varepsilon^2)}\right) = \frac{24\pi^3 b_{\parallel}^{\,2}}{c^2 T^2_{\text{mercury}}(1-\varepsilon^2)}. \tag{14.22a}$$

We can now employ the data from astronomy to calculate $\theta_{\text{precesion}}$. The semi-major axis of Mercury's elliptic orbit around the Sun is $b_{\parallel} = 57.91 \times 10^9\ m$. The speed of light is $c = 299792458\ ms^{-1}$, and $T_{\text{Mercury}} = 87.9691$ *earth-days* $= 7600530.24\ s$. The orbit's eccentricity ε is 0.20563593. Using these values in Eq. 14.18a, we get

$$\theta_{\text{precession}} = \frac{24 \times (3.14159265359)^3 \times (57.91 \times 10^9\,\text{m})^2}{(299792458\ \text{ms}^{-1})^2\ (7600530.24\text{s})^2\ (1-(0.20563593)^2}\ ,$$

i.e., $\theta_{\text{precession}} = 5.018836253 \times 10^{-7}$ Radians per Mercury-year. (14.22b)

In terms of time-intervals measured in units of the Earth-year, or Earth-century, this becomes

$\theta_{\text{precession}} = 2.083836249 \times 10^{-6}$ Radians per Earth-year,

i.e. $\theta_{\text{precession}} = 2.083836249 \times 10^{-4}$ Radians per Earth-century,

or, $\theta_{\text{precession}} = 42.98224839$ seconds of an arc per Earth-century, (14.22c) since 1 *Radian* = 206265 seconds of an arc. The result in Eq. 14.22c is pretty much the very same *residual* difference that was not accounted for in Table 14.1. That GTR accounts for this difference exactly is a major triumph of Einstein's theory, just like Eddington's experiment described above.

The above formalism is applicable for other planets as well, of course, not just to Mercury. The angles of precession 14.22 of the other planets, is independent of the planet's mass; it depends only on the orbit parameters and the 'specific' angular momentum of the particular planet's orbit. This result is akin to Galileo's classic experiment mentioned in Section I, performed in 1589; stones of different masses dropped from the tower of Pisa accelerate equally under free fall. Notwithstanding the approximations employed, the calculations presented above agree very well with those from more rigorous treatment, such as in Straumann's book [6]. The corresponding result from Straumann is

$$\theta^S_{\text{precession}} = 3\pi \frac{R_s}{b_{\|}(1-\varepsilon^2)}. \tag{14.23}$$

In the left hand side of Eq. 14.23, the superscript S represents reference to Straumann, and on the right hand side, R_s is the Schwarzschild radius given by

$$\left.\begin{aligned} R_s &= \frac{2Gm}{c^2} = \frac{2\times 6.6740831\times 10^{-11}\,\text{m}^3\,\text{kg}^{-1}\text{s}^{-2}\times 1988500\times 10^{24}\,\text{kg}}{(299792458\ \text{ms}^{-1})^2} \\ &= \frac{2.654282849\times 10^{20}\ \text{m}^3\text{s}^{-2}}{8.987551787\times 10^{16}\,\text{m}^2\text{s}^{-2}} \end{aligned}\right\},$$

$$\text{i.e.,}\quad R_s = 2953.287961\ m. \tag{14.24}$$

Using the result from Eq. 14.24 in Eq. 14.23, we get

$$\theta^S_{\text{precession}} = 42.9807196 \text{ seconds of an arc per earth-century.} \tag{14.25}$$

The result in Eq. 14.22 is thus in good agreement with Straumann's result.

The GTR angle of precession for the planet Mercury, and also for Earth and for Venus, calculated (Reference [3]) using the same method, is given in Table 14.2.

Table 14.2 The residual unaccounted precession of the major axis of Mercury's orbit is accounted for by Einstein's field equations of the GTR.

Planet	GTR Angle in arc-second per earth-century [3]	Literature value [16, 17, 18]	Percentage difference
Mercury	43.3	42.9	0.9
Earth	3.7	3.8	2.6
Venus	8.5	8.6	1.2

In Fig. 14.6, we depict the GTR advance of the perihelion of Mercury, from Reference [3].

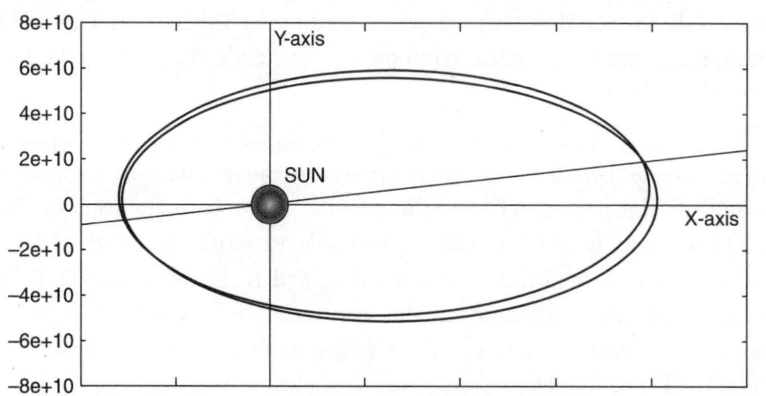

Fig. 14.6 Plot obtained from our simulation code depicting the precession of the perihelion of a planet orbiting the Sun over two mercury-years. The advance of the perihelion of Mercury over hundred earth-years is only 43 arc-seconds; so the effect is magnified in the above figure to dramatize it by scaling up $\theta_{\text{precession}}$ by a factor 5×10^5, thus exaggerating the turn of the major axis.

14.4 GRAVITY, BLACK HOLES, AND GRAVITATIONAL WAVES

Notwithstanding its limitations, one must applaud Newtonian mechanics which accounts for 5557.0254 *(out of the observed 5599.7)* arc-seconds precession of Mercury's major axis per earth-century (Table 14.1, above). Newton's law of gravity is, however, challenged by two major actualities: [i] it predicts instantaneous gravitational influence, and [ii] it fails to account for observations, no matter how weak, such as the (residual) precession of the planetary orbits. With regard to these two concerns, Newton's law fails completely. That the laws of Newton account well-enough for a vast majority of events in our daily lives is because the Newton's law of gravity can be seen as a *not-too-bad* (in fact, rather good) approximation [10] to the EFE of GTR. The GTR furnishes an accurate account of gravity. Most importantly, GTR interprets gravity in terms of the curvature of spacetime, and not as a force/interaction between two masses.

The bending of light around the Sun (Fig. 14.2) and the correct account of the residual correction of ~43 arc-seconds per earth-century to mercury's precession were amongst the first evidences in support of Einstein's GTR. In the citation for the Nobel Prize that was awarded to Einstein, an explicit reference only to his explanation of the photoelectric effect was made. The reference to Einstein's theory of relativity, if any, was only indirect. Specifically, the citation in the Nobel Prize award to Einstein's work read: "... *for his services to theoretical physics, and especially for his discovery of the law of the photoelectric effect.*" Einstein's ideas on relativity were totally out of the blue, completely counter-intuitive as seemed then. His thinking was so revolutionary that it took a lot of time for the scientific community to finally accept Einstein's theory.

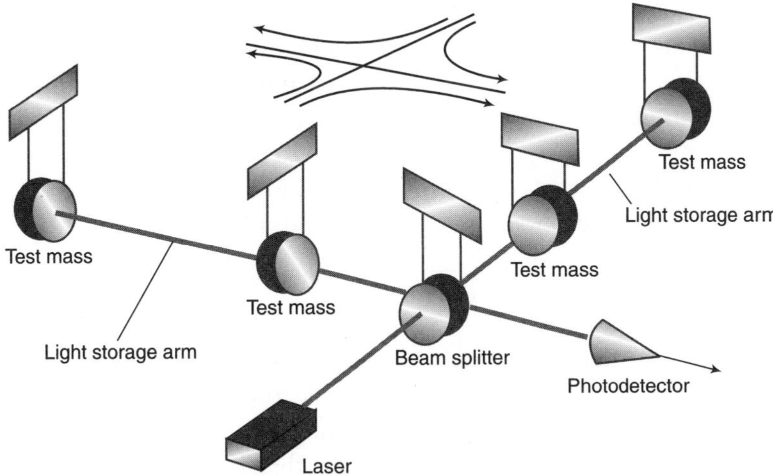

Fig. 14.7 The hanging mirrors of the equal arm Michelson laser interferometer serve as gravitational test masses. An incident gravitational wave shown by the stress pattern coming down from above, stretches one arm of the interferometer and compresses the other in accordance with the GTR. This results in a difference in the time taken by light to traverse the different arms which manifests as a measurable interference pattern. The high precision experiment measure a phase shift to a few billions of an interference fringe.

Einstein's theory of Relativity has been validated repeatedly over the past hundred years. EFE predictions that are borne out by sophisticated experiments include the detection of black holes, and the gravitational waves. Extremely high precision experiments carried out at the Laser Interferometer Gravitational-Wave Observatory [19–22] (LIGO) made international headlines only recently. The successful detection of gravitational waves by the LIGO (Fig. 14.7) is the result of a major collaborative experimental program between the California Institute of Technology and the Massachusetts Institute of Technology. Two ultra-high precision experiments are carried out at a distance 3,030 km apart, one at Hanford in the state of Washington, and the other at Livingston in the state of Louisiana, in the LIGO experiment, and the results are correlated to test the compatibility of their predictions. The first detection of the gravitational waves came on 14th September 2015. The gravitational wave signal that was detected was referred to as GW150914. Essentially, Einstein's theory of the GTR predicts that gravitational waves would compress space in one direction, and stretch the same in an orthogonal direction. This would result in a change in lengths of the two orthogonal arms of the LIGO interferometer. The change is very tiny, only about the ten-thousandths fraction of the size of a proton, but the high precision LIGO interferometer is capable of detecting it. The gravitational wave GW150914 was actually produced by the merger of two black holes that took place over a billion light years away. One of two black holes was 36 times, and the other 29 times, the mass of our Sun.

Einstein's Field Equations of the General Theory of Relativity go well beyond the explanation of the bending of light by matter, or of the precession of planetary orbits. In fact, GTR goes even beyond observing cosmic events like the merger of black holes through the detection of the gravitational waves. The EFE seeks to unravel the deepest cosmic mysteries about the origins, and the evolution, of the Universe. Einstein, however, discovered that his equation did not predict a static (stationary) universe. At that time, almost everybody believed that the universe was stationary. EFE predicted a universe that would either expand, or contract, To obtain a static condition, Einstein therefore needed to counter the gravitational attraction between astronomical masses. Completely on an ad-hoc basis, then, Einstein inserted a repulsive term. With the inclusion of this extra term, Eq. 14.11 then gets modified to:

$$R_{\mu\nu} - \frac{1}{2} R g_{\mu\nu} + \Lambda g_{\mu\nu} = \frac{8\pi G}{c^4} T_{\mu\nu}. \tag{14.26}$$

The constant Λ, in the third term above, is called the Cosmological constant. Its objective was merely to add a repulsive term that would make astronomical bodies would move apart from each other. The repulsive term would then compete with the gravitational attraction, and thereby maintain a static balance, which Einstein then believed, was required.

In the 1920s, Edwin Hubble (1889–1953) carried out groundbreaking observations, which, however, established that the universe existed well beyond the Milky Way, and that there were very many galaxies out there. In March 1929 Hubble went on to publish his findings, based on his observations of the red-shift of light coming to us from distant galaxies, that not only were there many galaxies, but that the farther a galaxy was from us, the faster it moved away from us. In fact, Hubble predicted that the recessional speed of a galaxy was directly proportional to its distance from us:

$$v = H_0 d. \tag{14.27}$$

The above expression is known as the *Hubble* (or *Hubble–Lemaître*) law. In it, v is the recessional velocity of a galaxy, and d its distance from our galaxy. H_0 is called the Hubble constant; its value is estimated to be between 50 and 100 km/sec/Mpc, where 1 parsec is 1.262 light-year. Since Einstein's original equation (one without the Cosmological constant Λ) allowed for an expanding universe, he felt that injecting the cosmological constant was not necessary, and hence doing so was a blunder. Einstein therefore removed the cosmological constant. In retrospect, however, it now seems that not inserting, but removing the cosmological constant was a blunder, since it is now found that the universe not only expands, but the distant galaxies are actually *accelerating* away from each other. The discovery of the accelerating universe is so important, that the 2011 Nobel Prize was awarded to Saul Perlmutter, Brian Schmidt and Adam Reiss for discovering this. One does not know what causes this *anti gravity* and what energizes the acceleration of the expanding universe. The source of this acceleration has come to be known as the *dark energy*, only because we do not really know what it is. Search to account for the dark energy and the dark matter (Chapter 8) are amongst the most important problems that modern physics must face. Fully relativistic numerical methods are being employed to address these challenges. The multi-institute LIGO experimental collaboration has now been expanded to include several other countries. For example, the LIGO-India [21, 22] collaboration (IndIGO consortium) will set up an advanced interferometric gravitational wave detector in India. This will provide improved accuracy observations for Gravitational Wave (GW) astronomy. Future challenges include the reconciliation between the quantum physics and the theory of relativity. Even though much remains to be known, and understood, we seem close to a major breakthrough that would result in a giant leap in our understanding of the universe, or the multiverse, if there is one.

Finally, one must celebrate the fact that the theory of relativity (both the STR and the GTR), just like the other astonishing thought and theory, viz., the quantum physics, developed in the early part of the twentieth century, have made an incredible impact on our lives. Our life-style and behavior is largely affected by how the theories of relativity and quantum physics have changed the world we live in. It is not that relativity is important only if we are concerned with objects that move at a significantly large fraction of the speed of light. Even a particle at rest has a quantum property called its intrinsic *spin* angular momentum, which requires for its description a relativistic formulation of the quantum theory. Properties of matter, which constitute us and things around us, would be totally different without this *spin*. All the gadgets we use in our daily life operate on the principles of these two extra-ordinary theories, developed by Einstein and others. Even the Global Positioning System (GPS) requires an accuracy of the atomic clocks that must be calculated by accounting for not merely the STR effects, but also the GTR. Gravitational blue-shift occurs when a clock moves to a *lower* altitude, and thus your head ages faster [23] than your feet. If a satellite clock ignores this, the error caused would build up and result in a navigational error of more than 11 km [24] over just one day. For further studies of the GTR, and to get into the quantum theory, the readers must now be referred to other texts.

Problems with Solutions

P14.1:

Explain why the Einstein Field Equations are identically zero in a Minkowski spacetime.

Solution:

A Minkowski spacetime is given by the metric whose elements are constants, $\eta_{\mu\nu} = diag(1, -1, -1, -1)$. If the gravitational field is described by the Einstein Field Equations, $G_{\mu\nu} = R_{\mu\nu} - \frac{1}{2} R g_{\mu\nu}$, then each term is built from the Riemann tensor $R^{\rho}{}_{\beta\alpha\gamma} = \partial_{\alpha}\Gamma^{\rho}_{\gamma\beta} - \partial_{\gamma}\Gamma^{\rho}_{\alpha\beta} + \Gamma^{\lambda}_{\gamma\beta}\Gamma^{\rho}_{\alpha\lambda} - \Gamma^{\lambda}_{\alpha\beta}\Gamma^{\rho}_{\gamma\lambda}$, which is constructed in terms of the Christoffel symbols $\Gamma_{\alpha\rho\sigma} = \frac{1}{2}(\partial_{\sigma} g_{\alpha\rho} + \partial_{\rho} g_{\alpha\sigma} - \partial_{\alpha} g_{\rho\sigma})$, which in turn depends on this metric. In the Minkowski spacetime, we have ordinary derivatives of constants, which give zero: $\Gamma_{\alpha\rho\sigma} = \frac{1}{2}(\partial_{\sigma}\eta_{\alpha\rho} + \partial_{\rho}\eta_{\alpha\sigma} - \partial_{\alpha}\eta_{\rho\sigma}) = 0$. The Ricci tensor $R_{\mu\nu}$ and the Ricci scalar R are therefore both zero, and the Einstein Field equations go identically to zero: $G_{\mu\nu} = 0$.

P.14.2:

Einstein sought to recover Newton's gravitation law in an appropriate limit from the field equations. This is usually achieved by expressing the field equations in an alternative form: $R_{\mu\nu} = \frac{8\pi G}{c^4}\left(T_{\mu\nu} - \frac{1}{2} g_{\mu\nu} T \right)$. Using this form, obtain the Einstein Field Equations in the conventional form.

Solution:

Consider the alternative form given in the question and contract it with $g^{\mu\nu}$:

$$g^{\mu\nu} R_{\mu\nu} = \frac{8\pi G}{c^4}\left(g^{\mu\nu} T_{\mu\nu} - \frac{1}{2} g^{\mu\nu} g_{\mu\nu} T \right).$$

Recall that repeated indices are summed over, following Einstein's summation convention.

Now, $g^{\mu\nu} g_{\mu\nu} = 4$, which implies: $R = -T\frac{8\pi G}{c^4}$. Substituting for T in the given equation, we get:

$R_{\mu\nu} - \frac{1}{2} g_{\mu\nu} R = \frac{8\pi G}{c^4} T_{\mu\nu}$, which is the usual form of the Einstein's field equation.

P.14.3:

In four dimensions, the Riemann curvature tensor $R_{\mu\nu\alpha\beta}$ has $4^4 = 256$ elements from the 4 indices. Use the symmetry properties of this tensor to determine how many of these elements are both non-zero and unique (i.e. independent).

Solution:

From the antisymmetry property, $R_{\mu\beta\alpha\gamma} = -R_{\beta\mu\alpha\gamma}$ and $R_{\mu\beta\alpha\gamma} = -R_{\mu\beta\gamma\alpha}$ we see that when two indices in the antisymmetry pair are the same, $R_{\mu\nu\alpha\beta}$ would be zero. For example, $R_{11\alpha\beta} = 0$. Hence, only permutations of '1234' are non-zero, as well as permutations that have numbers repeated across the first pair and the second pair. In each pair, we have $n = 4$ numbers with $r = 2$ chosen, while the number cannot repeat, which leaves 12 possible permutations. Again, since $R_{12\alpha\beta} = -R_{21\alpha\beta}$, these components are not unique, and can be dispensed with. Based on the anti symmetry properties, in each anti symmetric pair, we are thus left with 6 unique permutations. We thus have $6 \times 6 = 36$ possible unique permutations. Furthermore, the pairs are symmetric under the interchange $R_{\mu\beta\alpha\gamma} = R_{\alpha\gamma\mu\beta}$. This leads to a further reduction (from 36) of the number of unique pairs. For example, $R_{1234} = R_{3412}$, and again uniqueness is lost. We have 6 pairs (12, 13, 14, 23, 24, 34), choosing $r = 2$, resulting in 15 repeated permutations. This reduces the number of unique indices to $36 - 15 = 21$.

Now we consider the identity $R_{\mu\alpha\beta\gamma} + R_{\mu\gamma\alpha\beta} + R_{\mu\beta\gamma\alpha} = 0$. Of the 21 components we have from the previous step, we have 3 which follow this identity, with indices 1234, 1423, 1342. For example, $R_{1234} + R_{1423} + R_{1342} = 0$. Any one of these three can be therefore replaced by a combination of the other two. This leaves us with $21 - 1 = 0$ unique components. Therefore, the Riemann tensor has a total of 20 unique, non-zero elements. Essentially, this number can be deduced also from permutation laws which would give $N = \frac{d^2(d^2-1)}{12} = \frac{4^2(4^2-1)}{12} = 20$ for the Riemann tensor of dimension 4.

P.14.4:

Given the equation of motion of a free particle in inertial co-ordinate system as $m\frac{d^2x^\mu}{d\tau^2}=0$ $[d\tau^2=-\frac{d\tau^2}{c}$ is the proper time of the particle], show that under general co-ordinate transformations $x^\mu \rightarrow y^\alpha$, the equation of motion becomes the geodesic equation.

Solution:

The given equation is: $\frac{d^2x^\mu}{d\tau^2}=0$. Under the co-ordinate transformation, the velocity vector is given as: $\frac{dx^\mu}{d\tau}=\frac{\partial x^\mu}{\partial y^\alpha}\frac{dy^\alpha}{d\tau}$. Using chain rule, differentiation again we get

$$\frac{d^2x^\mu}{d\tau^2}=\frac{\partial x^\mu}{\partial y^\alpha}\left[\frac{d^2y^\alpha}{d\tau^2}+\Gamma^\alpha_{\beta\gamma}\frac{dy^\beta}{d\tau}\frac{dy^\gamma}{d\tau}\right],$$

where we define $\Gamma^\alpha_{\beta\gamma}=\frac{\partial y^\alpha}{\partial x^\mu}\frac{\partial^2 x^\mu}{\partial y^\beta \partial y^\gamma}$ [Refer Eq 14.5.]. Thus under a general co-ordinate transformation, the equation of motion has now become

$$\frac{d^2y^\alpha}{d\tau^2}+\Gamma^\alpha_{\beta\gamma}\frac{dy^\beta}{d\tau}\frac{dy^\gamma}{d\tau}=0.$$

P.14.5:

Recall the definition of covariant derivative from the text:

a. For a contravariant vector: $\nabla_\alpha u^\beta=\partial_\alpha u^\beta+\Gamma^\beta_{\alpha\gamma}u^\gamma$.

b. For a covariant vector: $\nabla_\alpha u_\beta=\partial_\alpha u_\beta-\Gamma^\gamma_{\alpha\beta}u_\gamma$.

c. For a rank 2 contravariant tensor: $\nabla_\alpha T^{\beta\gamma}=\partial_\alpha T^{\beta\gamma}+\Gamma^\beta_{\alpha\delta}T^{\delta\gamma}+\Gamma^\gamma_{\alpha\delta}T^{\beta\delta}$.

d. For a rank 2 covariant tensor: $\nabla_\alpha T_{\beta\gamma}=\partial_\alpha T_{\beta\gamma}-\Gamma^\delta_{\alpha\beta}T_{\delta\gamma}-\Gamma^\delta_{\alpha\gamma}T_{\beta\delta}$.

e. For a scalar, this becomes: $\nabla_\alpha f=\partial_\alpha f$.

Using this, and the fact that covariant derivatives satisfy Leibniz rule, check that the following properties hold:

a. $\nabla_\alpha \delta^\gamma_\beta=0$, where δ^γ_β is the Kronecker delta.

b. $[\nabla_\alpha,\nabla_\beta]\phi=0$, where ϕ is a scalar field.

Solution:

a. $\nabla_\alpha\delta^\gamma_\beta=\partial_\alpha\delta^\gamma_\beta+\Gamma^\gamma_{\alpha\eta}\delta^\eta_\beta-\Gamma^\eta_{\alpha\beta}\delta^\gamma_\eta=0+\Gamma^\gamma_{\alpha\beta}-\Gamma^\gamma_{\alpha\beta}=0$

b. $[\nabla_\alpha, \nabla_\beta]\phi = \nabla_\alpha \nabla_\beta \phi - \nabla_\beta \nabla_\alpha \phi = \nabla_\alpha \partial_\beta \phi - \nabla_\beta \partial_\alpha \phi$

i.e. $[\nabla_\alpha, \nabla_\beta]\phi = \partial_\alpha \partial_\beta \phi - \Gamma^{\gamma}_{\alpha\beta} \partial_\gamma \phi - \partial_\beta \partial_\alpha \phi + \Gamma^{\gamma}_{\alpha\beta} \partial_\gamma \phi = 0,$

as partial derivatives commute, and the Christoffel symbols are symmetric in the lower indices.

P.14.6:

Show that $\nabla_\alpha g_{\beta\gamma} = 0$.

Solution:

The Christoffel symbol $\Gamma^{\gamma}_{\alpha\beta}$ is chosen in such a way that the covariant derivative of the metric tensor becomes zero, in which case the symbol is metric compatible.

In principle, if the Christoffel symbol transforms in a manner such that the covariant derivative given in the question is non-zero, then the symbol is metric incompatible.

To have a unique connection on the manifold, we need two additional conditions:

i. It should be metric compatible.
ii. It should be symmetric in lower two indices.

That is how we get equation 14.5 of the book, which is used in the present problem.

$$\nabla_\alpha g_{\beta\gamma} = \partial_\alpha g_{\beta\gamma} - \Gamma^{\delta}_{\alpha\beta} g_{\delta\gamma} - \Gamma^{\delta}_{\alpha\gamma} g_{\beta\delta} \tag{i}$$

$\Gamma^{\alpha}_{\beta\gamma}$ is defined in Eq 14.5.

$$\Gamma^{\alpha}_{\beta\gamma} = \frac{1}{2} g^{\alpha\delta} (\partial_\beta g_{\gamma\delta} + \partial_\gamma g_{\beta\delta} - \partial_\delta g_{\beta\gamma}).$$

Substitute this in equation (i) to get,

$$\nabla_\alpha g_{\beta\gamma} = \partial_\alpha g_{\beta\gamma} - \frac{1}{2} g^{\delta\varepsilon} g_{\delta\gamma} (\partial_\alpha g_{\beta\varepsilon} + \partial_\beta g_{\alpha\varepsilon} - \partial_\varepsilon g_{\alpha\beta}) - \frac{1}{2} g^{\delta\varepsilon} g_{\beta\delta} (\partial_\alpha g_{\gamma\varepsilon} + \partial_\gamma g_{\alpha\varepsilon} - \partial_\varepsilon g_{\alpha\gamma}). \tag{ii}$$

Note that $g^{\delta\varepsilon} g_{\delta\gamma} = \delta^{\varepsilon}_{\gamma}$ and $g^{\delta\varepsilon} g_{\beta\delta} = \delta^{\varepsilon}_{\beta}$. Using this identity in Eq. (ii) we get

$$\nabla_\alpha g_{\beta\gamma} = \partial_\alpha g_{\beta\gamma} - \frac{1}{2} \delta^{\varepsilon}_{\gamma} (\partial_\alpha g_{\beta\varepsilon} + \partial_\beta g_{\alpha\varepsilon} - \partial_\varepsilon g_{\alpha\beta}) - \frac{1}{2} \delta^{\varepsilon}_{\beta} (\partial_\alpha g_{\gamma\varepsilon} + \partial_\gamma g_{\alpha\varepsilon} - \partial_\varepsilon g_{\alpha\gamma}) = 0.$$

P.14.7:

Consider the contravariant metric tensor $g^{\mu\nu}$ and the covariant Riemann tensor $R_{\mu\nu\alpha\beta}$. Using the symmetry properties of these tensors, show that $g^{\mu\nu} R_{\mu\nu\alpha\beta} = 0$.

Solution:

We shall use the symmetry property of the metric tensor $g^{\mu\nu} = g^{\nu\mu}$ and also the antisymmetry property of the Riemann tensor $R_{\mu\nu\alpha\beta} = -R_{\nu\mu\alpha\beta}$.

$$g^{\mu\nu}R_{\mu\nu\alpha\beta} = \frac{1}{2}g^{\mu\nu}R_{\mu\nu\alpha\beta} - \frac{1}{2}g^{\mu\nu}R_{\nu\mu\alpha\beta} = \frac{1}{2}g^{\mu\nu}R_{\mu\nu\alpha\beta} - \frac{1}{2}g^{\nu\mu}R_{\nu\mu\alpha\beta}$$
$$= \frac{1}{2}g^{\mu\nu}R_{\mu\nu\alpha\beta} - \frac{1}{2}g^{\mu\nu}R_{\mu\nu\alpha\beta} = 0.$$

The contraction between two symmetric and two antisymmetric tensor indices is always zero.

Additional Problems

P14.8 Determine the transformation law for the Christoffel symbols of the first kind $\Gamma_{\alpha\rho\sigma}$ and the second kind $\Gamma^{\mu}_{\rho\sigma}$. Do either of these symbols transform like a tensor? The required transformations are given in the notes on the Christoffel symbols given at the end of Chapter 14.

P.14.9 In P.14.3 we have shown that $\partial_\mu\phi(x)$ is a rank-1 covariant tensor. $\phi(x)$ is a scalar, which means it is a rank-0 tensor. Hence operating ∂_μ on $\phi(x)$ increases its rank from 0 to 1. If we operate ∂_μ on a covariant rank-1 tensor V_α, will we get a new covariant tensor of rank-2?

P.14.10 Consider an arbitrary general co-ordinate transformation $x^\mu \to y^\alpha$. We will denote the metric in x^μ co-ordinates as $g_{\mu\nu}$, and the metric in y^α co-ordinates as $g'_{\alpha\beta}$. We have the condition $g_{\mu\nu}dx^\mu dx^\nu = g'_{\alpha\beta}dy^\alpha dy^\beta$. Then show that the definitions of $\Gamma^{\alpha}_{\beta\gamma} = \frac{\partial y^\alpha}{\partial x^\mu}\frac{\partial^2 x^\mu}{\partial y^\beta \partial y^\gamma}$ as given in P.14.4 is equivalent to the definition given in Eq. 14.5.

P.14.11 Determine the Ricci curvature of the space described by the metric $ds^2 = dr^2 + r^2d\theta^2$, i.e., $g_{\mu\nu} \equiv \begin{pmatrix} 1 & 0 \\ 0 & r^2 \end{pmatrix}$.

P.14.12 In one of its strongest forms, the Equivalence Principle asserts that there is no experiment which can distinguish between uniform acceleration and a uniform gravitational field. Using this statement, qualitatively justify the following statements:

(a) A gravitational field deflects light.

(b) Clocks run slower in a gravitational field than in the absence of gravity.

P.14.13 Find the geodesics in the plane in polar co-ordinates.

P.14.14 The Riemann curvature tensor is defined in terms of the non-commutation of two covariant derivatives as follows:

$$(\nabla_\gamma\nabla_\alpha - \nabla_\alpha\nabla_\gamma)u_\beta = R^{\rho}{}_{\beta\alpha\gamma}u_\rho.$$

Using the definition of the covariant derivative, obtain the Riemann tensor.

P.14.15 Compute the geodesic equations for the Schwarzschild metric [Refer Eq. 14.13].

P.14.16 Geodesic Equation from the Action Principle

Consider the action

$$S[x] = \int \frac{1}{2}g_{\mu\nu}\frac{dx^\mu}{d\lambda}\frac{dx^\nu}{d\lambda}d\lambda$$

Apply the variation $x^\mu \to x^\mu + \delta x^\mu$, and obtain the equation of motion. See that it is the geodesic equation.

NOTES on Christoffel symbols and on covariant derivative of tensors:

(A) The Christoffel symbols do not transform as tensors. Rather, they transform as follows:

$$\text{- first kind - } \bar{\Gamma}_{\alpha\rho\sigma} = \frac{\partial x^\gamma}{\partial \bar{x}^\alpha}\frac{\partial x^\beta}{\partial \bar{x}^\rho}\frac{\partial x^\lambda}{\partial \bar{x}^\sigma}\Gamma_{\lambda\beta\lambda} + g_{\gamma\beta}\frac{\partial x^\gamma}{\partial \bar{x}^\alpha}\bar{\partial}_\sigma\frac{\partial x^\beta}{\partial \bar{x}^\rho},$$

$$\text{- second kind - } \bar{\Gamma}^{\mu}_{\ \rho\sigma} = \frac{\partial \bar{x}^\mu}{\partial x^\xi}\frac{\partial x^\beta}{\partial \bar{x}^\rho}\frac{\partial x^\lambda}{\partial \bar{x}^\sigma}\Gamma^{\xi}_{\ \beta\lambda} + \frac{\partial \bar{x}^\mu}{\partial x^\xi}\bar{\partial}_\sigma\frac{\partial x^\xi}{\partial \bar{x}^\rho}.$$

(B) Relation between partial derivative of the metric and the Christoffel symbols:

$$\partial_\sigma g_{\alpha\rho} = \Gamma_{\alpha\rho\sigma} + \Gamma_{\rho\alpha\sigma} = g_{\alpha\lambda}\Gamma^{\lambda}_{\ \rho\sigma} + g_{\rho\lambda}\Gamma^{\lambda}_{\ \alpha\sigma}$$

(C) Covariant derivative: $\nabla_\mu u^\alpha = \partial_\mu u^\alpha + \Gamma^{\alpha}_{\ \nu\mu}u^\nu$; $\nabla_\mu u^\alpha = \dfrac{\partial \bar{x}^\rho}{\partial x^\mu}\dfrac{\partial x^\alpha}{\partial \bar{x}^\nu}\bar{\nabla}_\rho \bar{u}^\nu$

$$\nabla_\mu u_\alpha = \partial_\mu u_\alpha - \Gamma^{\lambda}_{\ \mu\alpha}u_\lambda.$$

This transforms like a mixed tensor.

(D) Covariant derivative of a second rank tensors:

$$\nabla_\gamma T^{\alpha\beta} = \partial_\gamma T^{\alpha\beta} + \Gamma^{\alpha}_{\ \gamma\lambda}T^{\lambda\beta} + \Gamma^{\beta}_{\ \gamma\lambda}T^{\alpha\lambda};\ \ \nabla_\gamma T_{\alpha\beta} = \partial_\gamma T_{\alpha\beta} - \Gamma^{\lambda}_{\ \gamma\alpha}T_{\lambda\beta} + \Gamma^{\lambda}_{\ \gamma\beta}T_{\alpha\lambda}$$

References

[1] Deshmukh, P. C., and J. Libby. 2010. 'Symmetry Principles and Conservation Laws in Atomic and Subatomic Physics–1.' *Resonance* 15(9): 832–842.

[2] Deshmukh, P. C., and J. Libby. 2010. 'Symmetry Principles and Conservation Laws in Atomic and Subatomic Physics–2.' *Resonance* 15(10): 926–940.

[3] Deshmukh, P. C., K. J. Pillay, T. S. Raju, S. Dutta, and T. Banerjee. 2017. 'GTR Component of Planetary Precession.' *Resonance* 22(6): 577–596. https://doi.org/10.1007/s12045-017-0499-5

[4] Hawking, Stephen. 2001. *The Universe in a Nutshell*. New York: Bantam Books.

[5] Dyson, F. W., A. S. Eddington, and C. Davidson. 1920. 'A Determination of the Deflection of Light by the Sun's Gravitational Field, from Observations Made at the Total Eclipse of May 29, 1919.' *Philosophical Transactions of the Royal Society of London, Series A*. 220(571–581): 291–333.

[6] Straumann, N. 2004. *General Relativity with Applications to Astrophysics*. Berlin: Springer-Verlag.

[7] Landau, L. D., and E. M. Lifshitz. 1980. *Fluid Mechanics*. Boston: Addison-Wesley.

[8] Bredberg, Irene, Cynthia Keeler, Vyacheslav Lysov, and Andrew E. Strominger. 2012. 'From Navier–Stokes to Einstein.' *Journal of High Energy Physics* 2012(7): 146.

[9] Romatschke, Paul. 2010. 'New Developments in Relativistic Viscous Hydrodynamics.' *International Journal of Modern Physics E* 19(01): 1-53. https://doi.org/10.1142/S0218301310014613

[10] Padmanabhan, T. 2008. 'Schwarzschild Metric at a Discounted Price.' *Resonance.* 13(4): 312–318.

[11] Vojinovic, Marco. 2010. *Schwarzschild Solution in General Relativity.* http://gfm.cii.fc.ul.pt/events/lecture_series/general_relativity/gfm-general_relativity-lecture4.pdf Accessed on 28 May 2016

[12] Tolish, Alexander. *General Relativity and the Newtonian Limit.* http://www.math.uchicago.edu/~may/VIGRE/VIGRE2010/REUPapers/Tolish.pdf. Accessed on 28 May, 2016.

[13] Rydin, Roger A. 2011. 'The Theory of Mercury's Anomalous Precession.' *Proceedings of the NPA*: 8: 501–506.

[14] Pollock, Chris. *Mercury's Perihelion.* http://www.math.toronto.edu/~colliand/426_03/Papers03/C_Pollock.pdf

[15] Chaturvedi, S., R. Simon, and N. Mukunda. 2006. 'Space, Time and Relativity.' *Resonance* 11(7): 14–29.

[16a] 'The Advance of the Perihelion of Mercury.' http://www.astro.cornell.edu/academics/courses/astro201/merc_adv.htm. Accessed on 4 June 2016.

[16b] Christian Magnan. 'Advance of the Perihelion of Mercury in General Relativity.' http://www.lacosmo.com/PrecessionOfMercury/index.html. Accessed on 4 June 2016.

[17] Rydin, Roger A. 2011. 'The Theory of Mercury's Anomalous Precession.' *Proceedings of the NPA*: 501–506.

[18] Biswas, Abhijit, and R. S. Mani Krishnan. 2008. 'Relativistic Perihelion Precession of Orbits of Venus and the Earth.' *Central European Journal of Physics* 6(3): 754–758.

[19] 'Observation of Gravitational Waves from a Binary Black Hole Merger.' https://www.ligo.caltech.edu/system/media_files/binaries/301/original/detection-science-summary.pdf. Accessed on 5 June, 2016.

[20] Barish, B. C., and R. Weiss. 1999. 'LIGO and the Detection of Gravitational Waves.' *Physics Today* October: 44.

[21] 'Indian Initiative in Gravitational Wave Observations.' http://www.gw-indigo.org . Accessed on 23 July 2016.

[22] 'LIGO Scientific Collaboration.' http://www.ligo.org/. Accessed on 22 May 2016.

[23] Kaku, Michio. 'Why Your Head is Older than Your Feet.' https://www.youtube.com/watch?v=bqlUNGb_aQ4/. Accessed on 31 January 2019.

[24] Ashby, Neil. 2002. 'Relativity and the Global Positioning System.' *Physics Today* 55(5): 41–47.

Index